Threats to the Arctic

Threats to the Arctic

SCOTT ELIAS
Institute of Arctic and Alpine Research
University of Colorado
Boulder, Colorado, United States

ELSEVIER

Elsevier
Radarweg 29, PO Box 211, 1000 AE Amsterdam, Netherlands
The Boulevard, Langford Lane, Kidlington, Oxford OX5 1GB, United Kingdom
50 Hampshire Street, 5th Floor, Cambridge, MA 02139, United States

Notices
Knowledge and best practice in this field are constantly changing. As new research and experience broaden our
understanding, changes in research methods, professional practices, or medical treatment may become
necessary.

Practitioners and researchers must always rely on their own experience and knowledge in evaluating and using
any information, methods, compounds, or experiments described herein. In using such information or methods
they should be mindful of their own safety and the safety of others, including parties for whom they have a
professional responsibility.

To the fullest extent of the law, neither the Publisher nor the authors, contributors, or editors, assume any
liability for any injury and/or damage to persons or property as a matter of products liability, negligence or
otherwise, or from any use or operation of any methods, products, instructions, or ideas contained in the
material herein.

Library of Congress Cataloging-in-Publication Data
A catalog record for this book is available from the Library of Congress

British Library Cataloguing-in-Publication Data
A catalogue record for this book is available from the British Library

ISBN: 978-0-12-821555-5

For information on all Elsevier publications visit our website at
https://www.elsevier.com/books-and-journals

Publisher: Oliver Walter
Acquisitions Editor: Marisa LaFleur
Editorial Project Manager: Devlin Person
Production Project Manager: Kiruthika Govindaraju
Cover Designer: Alan Studholme

Typeset by TNQ Technologies

Introduction

We need to save the Arctic not because of the polar bears, and not because it is the most beautiful place in the world, but because our very survival depends upon it.

LEWIS GORDON PUGH

LAND, SEA, AND ICE

The Arctic Circle rings the globe at 66° 33′ 44″ N latitude. The Arctic region comprises about 30 million km^2 of land and sea north of the Arctic Circle (Fig. 1). The northern polar latitudes are quite different from the southern polar latitudes, where the continent of Antarctica dominates the polar region. The Arctic regions, on the other hand, are almost an equal mix of land and sea. The Arctic Ocean covers 15.6 million km^2—a little over half of the Arctic. The combined lands of the Arctic cover approximately 14.5 million km^2—a little less than half of the Arctic. The terrestrial regions of the Arctic are divided into the northern sectors of the Eurasian and North American continents, the island of Greenland, and numerous High-Arctic island groups.

The greatest number of Arctic islands is found in the Canadian Arctic Archipelago, which consists of 36,563 islands above the Arctic Circle in Canada. These islands encompass 1.4 million km^2 of land. Almost two-thirds of this land is found on the six largest of these islands, including Baffin Island (507,451 km^2), Ellesmere Island (196,236 km^2), Devon Island (55,247 km^2), Axel Heiberg Island (43,178 km^2), Melville Island (42,149 km^2), and Prince of Wales Island (33,339 km^2).

Greenland is the world's largest island, at 2.166 million km^2. The island spans climate zones from sub-Arctic in the south to High Arctic in the north (59°−83°N latitude). The Greenland ice sheet covers 81% of the land, containing about 2.8 million km^3 of ice. The thickness of the ice sheet ranges from about 2 to 3 km, with the ice flowing from the center toward the coasts.

The islands of the Russian Arctic are all situated inside the Arctic Circle, scattered through the Barents, Kara, Laptev, East Siberian, Chukchi, and Bering Seas. There are 56 major islands and hundreds of small ones. They cover about 215,500 km^2 in total.

Svalbard is the northernmost archipelago of Europe, an unincorporated area of Norway. It consists of 61,022 km^2 of land spread over nine major islands.

Arctic Ocean Circulation

The Arctic Ocean is the smallest of the world's oceans. Situated entirely within the Arctic Circle, it contains deep basins and extensive shelf areas. For centuries, the hallmark of the Arctic Ocean has been its perennial (multiyear) sea ice.

The Arctic Ocean is getting fresher. In other words, more freshwater is entering the Arctic Basin than had entered it previously. The sources of this freshwater are twofold. One is simply increased precipitation over the Arctic Ocean, both in the form of snow in winter and as rain in the summer. These changes in precipitation are mainly due to global warming, which is causing more storminess in the high latitudes. The other source of additional freshwater entering the Arctic basin is runoff from streams flowing North into the Arctic basin from the adjacent continents. While the Arctic Ocean represents only about 2% of the global ocean in terms of volume and surface area, it receives a disproportionate amount of freshwater from rivers. In fact, the Arctic Ocean collects over 11% of global river discharge. Some of this additional runoff is coming from increased precipitation over the land, and some of it is due to the melting of permafrost in these regions, which is releasing moisture into the rivers that flow into the Arctic Ocean and adjacent seas.

The net transport of water due to coupling between wind and surface waters is called Ekman transport. This water mass transport system, taken across the width of an ocean, is considerable, matching the volume of water transported by major ocean currents, such as the Gulf Stream. This wind-driven exchange of surficial and deep ocean waters ultimately drives the interior circulation of ocean waters.

Thermohaline Circulation

The northward transport of heat extends throughout the whole Atlantic Ocean so that South Atlantic

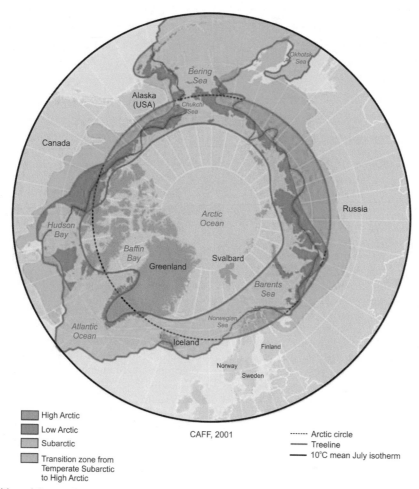

High Arctic
Low Arctic
Subarctic
Transition zone from
Temperate Subarctic
to High Arctic

CAFF, 2001

-------- Arctic circle
——— Treeline
——— 10°C mean July isotherm

FIG. 1 Map of the northern high latitudes, showing the High-Arctic, Low-Arctic, and Sub-Arctic regions, as well as the location of the Arctic Circle, the position of the northern tree line, and the position of the 10°C mean July isotherm, which is associated with the physiological limit of trees at the tree line. (From Conservation of Arctic Flora and Fauna (CAFF)., c. 2001. Definitions of the Arctic Region. https://www.arcticcentre.org/EN/arcticregion/Maps/definitions.)

Ocean heat is transported toward the equator. This meridional heat transport is associated with the Atlantic Meridional Overturning Circulation (AMOC), an overturning cell in which northward transport of warm upper ocean water is balanced by a southward flow of cooler deep water. The continual influx of warm water into the North Atlantic polar ocean keeps the regions around Iceland and southern Greenland mostly free of sea ice year-round. A key process in the maintenance of the AMOC is deep convection in the subpolar North Atlantic, especially in the Labrador and Nordic Seas. Here, winter cooling of relatively salty surface waters leads to a downward flow that drives the newly cooled waters away from the surface, where they eventually form the cold and salty North Atlantic Deep Water (NADW). As part of the global thermohaline conveyor belt, the AMOC has been a relatively stable component of the greater oceanic circulation system. This oceanic conveyor belt, called "thermohaline circulation" (*thermo* for temperature; *haline* for salinity), will presumably keep functioning as long as Arctic waters stay cold. But as Arctic waters start to warm, many oceanographers are concerned that the Atlantic thermohaline conveyor belt may either weaken in intensity or shut down altogether. Either way, the climates on the entire Atlantic region and adjacent landmasses would change dramatically.

Pacific Circulation

The entrance of Pacific waters (PW) into the Arctic is through Bering Strait. Unlike the inflow of Atlantic waters, PW are the dominant source of nutrient inputs into the Arctic Ocean.

Once the PW enters the Arctic Basin, it flows east, eventually entering the North Atlantic. Observations have revealed persistent flows of relatively fresh PW to the North Atlantic through Fram Strait (east of Greenland) and the passages through the Canadian Arctic Archipelago. This is a very slow process. Based on various numerical models, the transit time for PW to cross the Arctic Ocean and enter the North Atlantic is 10–15 years, and the time for the PW outflow to reach a quasi-equilibrium state is about 20 years.

The entrance of Atlantic currents into the Arctic brings warmer, saltier waters through Fram Strait and the Barents Sea. These routes are both wider and deeper than Bering Strait, and they transport about ten times more water than the inflow from the Pacific. Eurasian rivers contribute roughly two-thirds of the freshwater entering the Arctic. Arctic Ocean waters flow out only into the Atlantic, via the western side of Fram Strait, or through the channels of the Canadian Arctic Archipelago.

Sea Ice: Going, Going, Gone

Sea ice forms in the Arctic Ocean and adjacent seas from autumn of 1 year till the early summer of the following year. Prior to the 21st century, sea ice covered half of the Arctic Ocean during the summer months. Since 2007, the minimum sea-ice cover has decreased by around 40%. Average sea-ice thickness thinned by 65%, from 3.59 to 1.25 m between 1975 and 2012. Prior to the 1980s–90s, annual layers of sea ice would build up for several years, thickening the ice pack considerably. However, the extent and thickness of sea ice have been decreasing rapidly in recent years, and multiyear ice has almost disappeared from Arctic waters.

The presence of Arctic sea ice greatly affects the stratification of the water column, creating a fresher surface layer when it melts and causing the mixing of surface waters as brine is excluded from ice formation during the freezing process. Arctic Ocean waters are more-or-less uniformly cold (less than 2°C), so the density of Arctic Ocean waters is primarily determined by salinity.

CLIMATE

Other than the famously cold temperatures, the most defining feature of the Arctic climate is the strong seasonality in sunlight that reaches the surface. It is this strong seasonality in insolation (incoming solar radiation) that drives the large seasonal cycle in surface temperatures in the Arctic. In the winter, the surface receives very little or no insolation and the surface loses energy in the form of infrared radiation that is released back to space. Much of the summer sees round-the-clock sun, bringing rapid warming in May and June, and inaugurating the greening of the tundra.

Meteorological data are scarce in the High Arctic, because of the lack of weather stations. The Canadian Arctic Archipelago has two permanent high latitude stations: Alert and Resolute. The Alert station is on the northern end of Ellesmere Island—the furthest north location in Canada at 82.5°N latitude. The Resolute station is on Cornwallis Island at 74.7°N. The two localities have very similar climates. Above-freezing temperatures occur only in June to August, and mean daily temperatures do not exceed 5°C. The average annual temperature is −17.6°C at Alert and −16.2°C at Resolute. Precipitation is incredibly low at both sites, as it is throughout much of the Canadian Arctic Archipelago.

The northernmost weather station in Greenland is called Station Nord, situated near the north coast of the island at 81.7°N. The climate profile of Station Nord is similar to that of the Alert station, across the Nares Strait on Ellesmere. Average summer temperatures do not exceed 4°C, the summer season is short, and the average annual temperature is −16.2°C. The average daily temperature in March is about −28°C. Station Nord receives 135 mm of precipitation per year—another polar desert region.

Svalbard, although situated in the High Arctic, receives some warmth from the West Spitsbergen Current (WSC), a branch of the Gulf Stream. The North Atlantic Current flows along the coast of Norway and continues north, where it is called the WSC. Almost 60% of the water entering the Arctic Ocean comes by way of the WSC. By the time the WSC reaches Svalbard, its surface temperature is about 2.75°C. Even this moderate level of warmth is sufficient to alter the climate of this archipelago. For instance, at the Barentsburg weather station on Spitsbergen (78°N), above-freezing temperatures extend a few weeks into September. The mean annual temperature is −5.9°C, and the station receives 424 mm of precipitation per year—more than triple the amount that falls on the northern stations in Canada and Greenland.

Further east, at the Golomyanniy Meteorological Station in the Severnaya Zemlya archipelago of Russia (79.5°N, 90.6°E), the temperature regime is quite similar to those in High-Arctic Canada and Greenland: mean annual temperature of −15°C, mean July temperature just above freezing, and an average February temperature of $-27°C$. However, this archipelago receives a similar amount of precipitation to Svalbard: 420 mm per year. This moisture comes from storms that drift northeast from the North Atlantic.

Continuing east across the International Dateline, the village of Utqiagvik (formerly Barrow) Alaska lies at the northernmost point of the United States ($71.3°N$). Because of its relatively lower latitude, this locality has above-freezing average temperatures from June through mid-September. The temperature and precipitation profiles are quite similar to those of the Alert and Resolute stations in Canada.

ARCTIC SOILS

Most Arctic soils are in permafrost. This is ground that remains frozen for at least two consecutive years. Permafrost occurs both on land and beneath the shallow waters of the Arctic continental shelves. Its thickness ranges from less than 1 m to greater than 1000 m, the latter occurring in the coldest parts of Siberia. During the summer, the uppermost layer of the ground thaws, forming an "active layer." The thickness of the active layer on Arctic landscapes varies from about 0.3 to 4 m. Permafrost regions occupy approximately 22.8 million km^2, mostly between latitudes of $60°N$ and $68°N$. Permafrost declines sharply in regions north of $67°N$ because land gives way to the Arctic Ocean.

Because the active layer thaws and refreezes both on an annual basis and even within the summer season, the instability of the surface and the growth of ice crystals have strong impacts on Arctic vegetation. Nearly all the Arctic receives so little precipitation that is properly considered a desert. Yet despite this low precipitation, tundra environments may be rather wet at the surface and have many thaw lakes. Permafrost stops the drainage of moisture below the active layer, so almost all of it remains at the surface. Also, very low temperatures keep surface moisture from evaporating back to the atmosphere.

TERRESTRIAL VEGETATION

Because of the harsh climates and brief growing season, the biological productivity of the vegetation cover of Arctic landscapes is quite low. The best measure of this is net primary productivity (NPP), the rate of carbon flux from the atmosphere into plants, minus the carbon they use in respiration. In Low-Arctic tundra environments, NPP averages about $600-1000 \text{ g/m}^2$. In the High Arctic, most regions have less than 50% plant cover, with NPP values of 100 g/m^2 or less. For comparison, the average NPP of boreal forest regions is 3500 g/m^2; temperate grasslands average 2000 g/m^2 of NPP.

As one regional example, let us consider the High- and Low-Arctic vegetation of Canada. The High-Arctic vegetation of Canada is dominated by four plant associations. These include (1) crustose lichens and mosses; (2) cushion plants and forbs (an herbaceous flowering plant other than grass); (3) dry grasses, mosses, and cushion plants; and (4) prostrate dwarf shrubs and cryptogams (ferns, mosses, liverworts, lichens, algae, and fungi). All of these plants grow very close to the ground, thus avoiding the desiccation and tissue damage caused by wind-blown snow crystals. In contrast to the sparsity of vegetation types in the High Arctic, the tundra vegetation of the Canadian Low Arctic comprises seven different plant associations, ranging from wet grasses and mosses in wetlands to low and tall shrubs (willow, alder, and others) and tussock tundra grasses.

MAMMALS

As might be expected, mammal biodiversity decreases with latitude. This phenomenon is more pronounced with land mammals than marine mammals because food resources for land mammals decrease to almost nothing in the northernmost regions of the High Arctic. Also, because salt water freezes at $-1.8°C$, ocean temperatures never fall below that level. This means that the marine mammals never face water temperatures less than $-1.8°C$, even in winter. In contrast, land mammals face air temperatures down to $-50°C$ or lower in some regions of the Arctic.

Marine Mammals

A total of 35 species of marine mammals are found in Arctic waters. These include two species of sea lions, the walrus, eight species of seals, three species of right whales, five species of rorquals, the gray whale, the beluga and narwhal, the sperm whale, and four species of beaked whale. Three species of dolphins and two species of porpoise swim in Arctic waters. Polar bears are rightly considered marine mammals because they spend much of the year at sea, hunting from pack ice. More than half of the species of the Arctic marine mammal fauna swim in High-Arctic waters, though none are restricted to these polar latitudes.

Terrestrial Mammals

There are 63 species of terrestrial mammals on Arctic landscapes. These include a surprising diversity of shrews. There are 4 species of Arctic hares, 3 species of pikas, 3 squirrels, the beaver, and 18 species of voles and lemmings. The large grazers include the Eurasian elk, moose, and reindeer (caribou), as well as two species of sheep and the muskox. Predators include the gray wolf, coyote, and two species of fox. There are two species of bears and two species of lynx. Rounding out the list of predators are five species of mustelids, ranging from the least weasel, weighing in at 25 g, to the wolverine, weighing an average of 136 kg.

Only about one-quarter of the Arctic terrestrial mammals are found in the High Arctic. Of these, only the Wrangel Island collared lemming is restricted to the High Arctic, as it is endemic to this High-Arctic island off the northeast coast of Siberia. All the other High-Arctic species also inhabit the Low-Arctic regions.

A BRIEF HISTORY OF EXPLORATION

The outside world was quite slow to gain significant knowledge of Arctic geography. That knowledge was hard-won through recent centuries, as the Arctic was extremely remote, the land was frozen much of the year, and the seas were covered with pack ice (Fig. 2). The earliest European maps of the northern high latitudes were more fiction than fact. For instance, a 1680 map by British cartographer Moses Pitt displays a complete lack of information for the central and western regions of Arctic Canada (Fig. 3). A 1772 map of northwestern North America by the de Vaugondy family depicts a fictitious Northwest Passage (Fig. 4).

Scottish cartographer John Thomson removed the possibility of a Northwest Passage from his 1814 map of the Northern Hemisphere (Fig. 5). Stanford's 1887 map of the Arctic is perhaps the first to offer a more-or-less accurate depiction of Arctic geography (Fig. 6).

For the European explorers of the 17th, 18th, and 19th centuries, the Arctic was never a destination. It was simply a means to an end. European monarchs funded voyages of exploration into arctic waters to find alternative trading routes to China, either a Northwest Passage along the coast of North America or a Northeast Passage along the coast of Siberia. If a European country could (1) find such a passage, and (2) declare ownership of the passage, then that country would become very rich.

Seeking a Northwest Passage

Arctic sea explorers were brave souls, dedicated to their missions. Of the captains discussed here (Fig. 7) all but two perished during expeditions. One of the earliest of these efforts was led by Dutch explorer Willem Barents, after whom the Barents Sea is named, who made three voyages in search of a Northeast Passage. He discovered Spitsbergen in 1596 and continued east to the Kara Sea. But on his 1596 voyage, his ship became trapped in the ice pack, and he and his crew were forced to overwinter on Novaya Zemlya. They were the first Western Europeans to survive a High-Arctic winter. Weakened by scurvy, they set out for home in June 1597, but Barents died on the way, although most of his crew survived.

FIG. 2 The sun setting over the Arctic sea ice pack, as observed during the Beaufort Gyre Exploration Project in October 2014. (From https://www.innovations-report.com/earth-sciences/wintertime-arctic-sea-ice-growth-slows-long-term-decline-nasa/.)

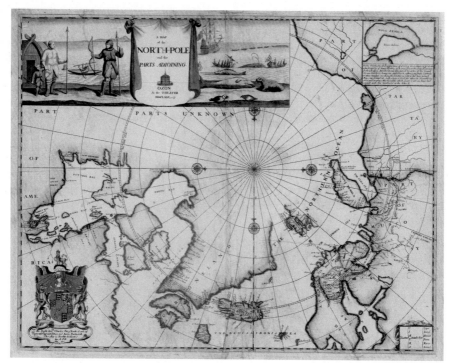

FIG. 3 1680 map by British cartographer Moses Pitt displays a complete lack of information for the central and western regions of Arctic Canada. (From Pitt, M., c. 2020. 1680 Map of the Arctic. https://www.123rf.com/photo_14986600_north-pole-and-adjoining-lands-old-map-created-by-moses-pitt-published-in-oxford-1680.html.)

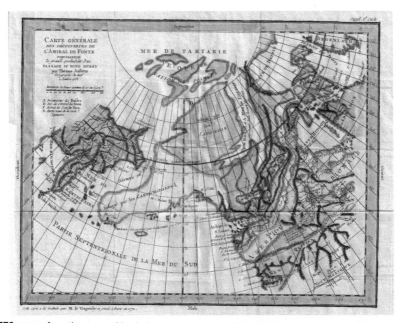

FIG. 4 1772 map of northwestern North America by the de Vaugondy family depicts a fictitious Northwest Passage. (From de Vaugondy., c. 2020. 1772 Map of Northwestern North America. https://commons.wikimedia.org/wiki/File:1772_Vaugondy_-_Diderot_Map_of_North_America_%5E_the_Northwest_Passage_-_Geographicus_-_NordetOuestAmerique-vaugondy-1772.jpg.)

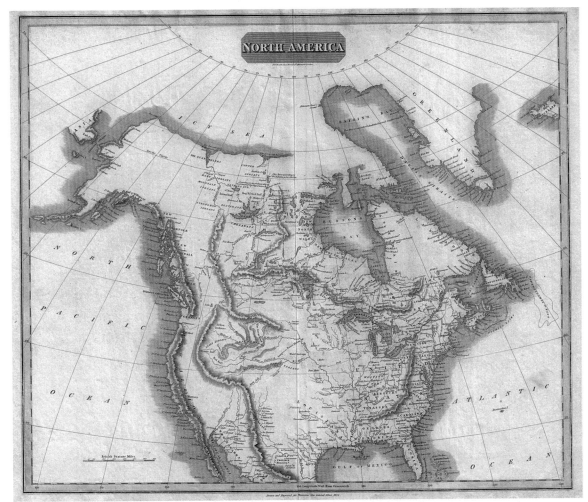

FIG. 5 Scottish cartographer John Thomson removed the Northwest Passage from his 1814 map of the Northern Hemisphere. (From Thomson, J., c. 2020. 1814 Map of the Northern Hemisphere. https://commons.wikimedia.org/wiki/File:1814_Thomson_Map_of_North_America_-_Geographicus_-_NorthAmerica-thomson-1814.jpg.)

A decade later (1607–10) English navigator Henry Hudson made three westward voyages across the North Atlantic in search of the Northwest Passage to Asia through the Arctic Ocean. On his final expedition on the ship *Discovery*, he entered Hudson Bay and mapped the shoreline. Like Barents, Hudson's ship became trapped in the ice, and the crew was forced to overwinter on the shore of James Bay. When the ice cleared in spring, Hudson wanted to continue exploring, but his crew wanted to return home. They mutinied and set Hudson, his son, and some crewmen adrift in a small boat. They were never heard from again.

Seeking a Northeast Passage

In the 18th century, the Russians attempted to chart a Northeast Passage from the Siberian coast to the Atlantic. Starting in 1732, the Russian Admiralty organized the Great Northern Expeditions to find the Northeast Passage. Although a passage was not found, the expeditions, led by Danish explorer Vitus Bering, mapped much of the Siberian coastline. In June 1741, Bering led a voyage of exploration from Kamchatka, sailing southeast. They sailed into the Gulf of Alaska on August 20. He surveyed the southwestern coast of Alaska, the Alaska Peninsula, and the Aleutian Islands.

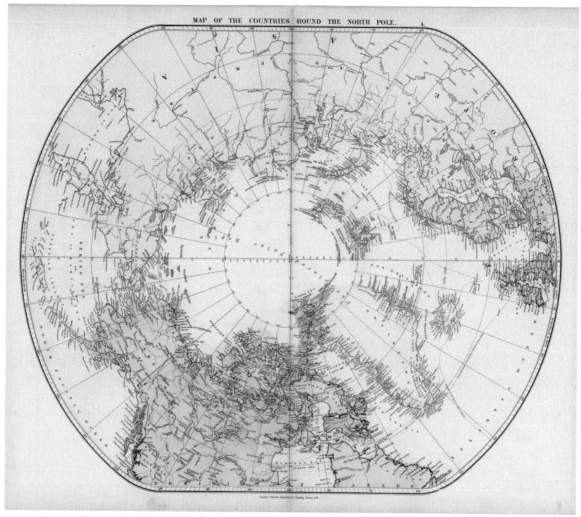

FIG. 6 Stanford's 1887 map of the Arctic—the first more-or-less accurate depiction of Arctic geography. (From American Geographical Society Library., c. 1887. Stanford's 1887 Map of the Arctic. https://agslibraryblog.wordpress.com/page/2/.)

Bering succumbed to scurvy on the return voyage, and in November, his ship ran aground on Bering Island, near Kamchatka. The survivors of the voyage reported back to the Russian authorities on an excellent opportunity for fur-trading along the Pacific coast of Alaska.

The Little Ice Age
Unbeknownst to the European explorers, their attempts to find a Northwest Passage in the first half of the 19th century were doomed to fail because of extreme sea-ice conditions associated with a climatic episode known as the Little Ice Age. Volcanic eruptions and reduced solar radiation caused global cooling between the 13th and

the 15th centuries. The resulting accelerated formation of sea ice in the Northern Seas triggered a positive feedback loop that fostered the Little Ice Age. Arctic sea-ice cover cooled regional climates, causing its persistence for much of every year until after 1850. Explorers looking for the Northwest Passage could not have chosen a worse time to enter arctic waters.

Last of the 18th-Century Attempts
English naval captain James Cook led part of his final voyage of exploration (1776—79) along the west coast of North America, as far north as Bering Strait (70° 41′N) in search of the Northwest Passage. There he

Willem Barents
1550-1597

Henry Hudson
ca 1570-1611

Vitus Bering
1681-1741

James Cook
1728-1779

John Franklin
1786-1847

James Clark Ross
1800-1862

FIG. 7 Sea captains who led early voyages into Arctic waters. Top left: Willem Barents. Top right: Henry Hudson. Middle left: Vitus Bering. Middle right: James Cook. Lower left: John Franklin. Lower right: James Clark Ross. (All images are in the public domain.)

ran into impenetrable pack ice and had to return south. His observations helped demonstrate the separation between the Asian and American continents. Shortly thereafter (February 1779), Cook was killed by natives of the Hawaiian Islands. A few decades later, the British Navy began another series of disastrous attempts to chart a Northwest Passage.

Into the 19th Century

In 1819, British naval officer William Parry undertook his first of these voyages. His was the first expedition to enter the Canadian Arctic Archipelago. The ship, the HMS *Hecla*, reached 110°W before ice prevented further exploration. The crew had to overwinter for 10 months on Melville Island. Also, in 1819, English captain John Franklin was chosen to lead the Coppermine expedition overland from Hudson Bay to chart the north coast of Canada eastward from the mouth of the Coppermine River. He planned to meet up with Parry who was coming by sea. Unfortunately, the expedition ended disastrously with 11 of its 20 members of the expedition losing their lives, most of them dying from starvation. The nine survivors, including Franklin, were forced to eat lichen and even attempted to eat their own leather boots.

Parry began his second voyage in search of the Northwest Passage in 1821 with the British naval ships *Hecla* and *Fury*. The expedition passed through Hudson Strait and explored the islands west of Baffin Island. Caught in pack ice that autumn, they were forced to overwinter on Winter Island (66° 16′N 83° 04′W). The party met a group of native Inuit who told them of a strait that led to the sea in the west. Once set free from the ice, Parry made his way to what is now known as the Fury and Hecla Strait. He found the strait choked with ice, although explorations on foot revealed a body of water to the west. The expedition remained for a second winter in the Arctic, hoping that the strait would clear of ice. In this, they were disappointed, and the expedition was forced to return to England in 1823. A subsequent attempt to pass through the Canadian High Arctic islands (1824—25) likewise ended in failure, as Parry's ship *Fury* ran aground and had to be abandoned. Parry took all the crew on board *Hecla* and they returned to England.

The British Quest for the Northwest Passage

The British quest for the Northwest Passage resumed in 1831 when James Clark Ross, who had sailed with Parry on the previous expeditions, resumed the search for the Northwest Passage. It was during this trip that a small party led by Ross located the position of the North Magnetic Pole on the Boothia Peninsula, the northernmost point of land of mainland Canada. However, the expedition was trapped in pack ice for several winters, and Ross eventually had to abandon his ship. The expedition finally returned home in 1835.

The Doomed Franklin Expedition

The most tragic of the European expeditions to find the Northwest Passage was led by Captain Sir John Franklin. The expedition, comprising 129 men, departed from England in 1845 aboard the HMS *Erebus* and HMS *Terror*. That autumn, the two ships became icebound in Victoria Strait near King William Island in the Canadian Arctic. The entire expedition, comprising 129 men, was lost (Fig. 8). Since their disappearance, multiple rescue expeditions and discoveries by other explorers, scientists, and interviews from native Inuit peoples pieced together what likely happened to Franklin and his crew. They spent the first winter on Beechey Island (74° 43′N 091° 51′W), where three crew members died and were buried. After traveling down Peel Sound through the summer of 1846, the two ships froze into the pack ice off King William Island in September.

FIG. 8 Top: The Franklin expedition was lost after setting sail in 1845 to find the Northwest Passage. Sketch artist, Lt. S. Gurney Creswell, 1854. Below: Photo of Franklin Camp on Beechey Island, Nunavut Canada. Three graves (L—R) commemorate John Torrington, William Braine, and John Hartnell. A fourth headstone marks the grave of a sailor named Thomas Morgan who came later in a Franklin search expedition and died at the camp. Photo by Gordon Leggett. (From Creswell, S., c. 1856. One of Franklin's Ships, Trapped in the Ice. https://www.historicmysteries.com/the-doomed-franklin-expedition.)

The crew apparently spent the next 2 years there, waiting for the pack ice to release their ships. According to a note dated April 25, 1848, and left on the island by two crew members, Franklin had died on June 11, 1847; the crew had wintered off King William Island in 1846–47 and 1847–48, and the remaining crew had planned to begin walking out of the Arctic on April 26, 1848, heading toward the Back River on the Canadian mainland. In 2014, a Parks Canada expedition located the wreck of the *Erebus* in shallow waters west of O'Reilly Island (68.04°N, 98.97°W), south of King William Island in Queen Maud Gulf. Two years later, the Arctic Research Foundation found the wreck of the *Terror* south of King William Island. None of the remains of the sailors who tried to walk out of the Arctic have ever been found.

Twentieth Century Voyages

The final chapter in the European quest for the Northwest Passage took place at the turn of the 20th century. In 1903, Norwegian explorer Roald Amundsen led a small crew aboard the *Gjøa* on a voyage that lasted to 1906. Ice-bound during 1904, the Gjøa finally set sail once more in August 1905, passing through Simpson Strait south of King William Island, clearing the Canadian Arctic Archipelago a few days later. The *Gjøa* passed through Bering Strait and eventually landed at Nome, Alaska. However, the discovery of a passage for commercial shipping, the original motive for finding the North-West Passage, remained elusive. Parts of Amundsen's journey were in waters only 1 m deep. While the very shallow draft of the *Gjøa* allowed its passage in such shallow waters, it would take the effect of global warming to open up the possibility of deeper sea routes for large ships, starting in 2007.

Importance of Local Knowledge

Of course, there were groups of people who knew about the ice-clogged conditions blocking a Northwest Passage during the Little Ice Age, though the European navigators very rarely sought their expertise. The Inuit peoples of northern Greenland and the Canadian High Arctic knew that the Arctic seas were too choked with pack ice in the 18th and 19th centuries to allow ships to make a Northwest Passage to the Pacific. In fact, there were Inuit peoples who witnessed the sinking of Franklin's ships *Erebus* and *Terror* in the 1840s. Oral traditions passed down to the modern generation helped Parks Canada find the shipwrecks in recent years.

For the most part, European governments, navies, and explorers considered Native peoples of the Arctic (and elsewhere) to be ignorant savages, still living in the stone age. They did not stop to consider that these same Native peoples had found ways to survive in the extreme environments that were causing their "modern," "civilized" expeditions to fail disastrously.

RESOURCE EXTRACTION AND NUCLEAR WASTE

Much of this book documents the climatic connections between the Arctic and the rest of the world. It is ironic that these remote, frozen regions should play one of the central roles in the global climate. As mentioned earlier, peoples of European descent have scarcely considered the Arctic as a *destination*. Instead, the Arctic has always been a region from which to extract valuable resources, starting with the Canadian fur trade from 1608, commercial whaling from 1611 to 1914, precious metals from 1870 onward, and oil and gas since World War II.

Because the Arctic is so physically remote from population centers, the Russian government used parts of their Arctic territory (notably Novaya Zemlya) for nuclear weapons testing, both below and above ground, from 1955 onward. This region and the adjacent Kara Sea have been used for decades as dumping grounds for Russian nuclear waste, including 19 surface ships, submarines, and barges loaded with radioactive waste that were dumped here, 14 nuclear reactors, and 17,000 containers filled with radioactive waste (Fig. 9). Novaya Zemlya has served as a dumping ground for radioactive waste from 16 Soviet-era nuclear reactors. Salbu et al. (1997) studied sediment and seawater samples from these nuclear waste sites around Novaya Zemlya and found that the waste containers were leaking radioactive materials. The highest concentrations of ^{137}Cs, ^{60}Co, ^{90}Sr, and ^{239,240}Pu have been observed in sediments collected close to dumped containers in Abrosimov and Stepovogo fjords.

WHY THE ARCTIC IS IMPORTANT TO CLIMATE CHANGE

To briefly summarize here, the Arctic is an important element of the global climate because of its *teleconnections* to regions further south. These long-distance connections are formed through the flow of air streams and ocean currents. Positive feedback loops act as amplifiers of Arctic climate change. As discussed by Miller et al. (2010), the dominant Arctic feedbacks display differences in their seasonal and spatial expressions, and their timescales vary greatly. Seasonal snow and sea ice cover are relatively rapid feedbacks with seasonal response times. Vegetation and permafrost feedbacks operate on timescales of decades to centuries. The

FIG. 9 Above: K-219 was a nuclear ballistic missile submarine of the Soviet Navy, equipped with either 32 or 48 nuclear warheads. In 1986 while on patrol off the coast of Bermuda, the vessel suffered an explosion and fire in a missile tube. Three days later the submarine sank with all nuclear weapons still aboard. Photo courtesy of the US Navy, in the public domain. Below: Hull of derelict Soviet nuclear submarine, K-159. The submarine sank while being towed to Murmansk. When it sank, the K-159 drowned nine Russian sailors who were aboard to plug leaks and the boat carried 800 kg of spent uranium fuel still in its two pressurized water reactors to a depth of 246 m. (From Hull of Derelict Soviet Submarine K-159., c. 2020. https://en.wikipedia.org/wiki/Soviet_submarine_K-159.)

slowest feedbacks operate on millennial timescales. These are associated with the growth and decay of continental ice sheets. The slowest feedbacks operate on millennial timescales. These are associated with the growth and decay of continental ice sheets.

Albedo is defined as the reflectivity of solar radiation from a surface. Dark surfaces absorb most of the sunlight that falls on them (low albedo), while white surfaces reflect much of this light back out to space (high albedo). Because fresh snow and sea ice have very high albedos, large changes in their seasonal and areal extent will have strong influences on the planetary energy balance. In a global warming scenario, increased surface air temperatures lead to a reduction in Arctic snow and sea-ice cover, causing a reduction in the planetary albedo, stronger absorption of solar radiation, and hence, a further rise in global mean temperature. This a positive feedback loop—it is self-reinforcing. Conversely, a lowering of global temperatures leads to a buildup of Arctic snow and sea ice. The increased albedo thus causes a further fall in temperature. These feedback loops stemming from changes in Arctic albedo cause global temperature changes to be amplified in the Arctic. Given that the Arctic is characterized by its seasonal snow cover and sea ice, it follows that the albedo feedback will be strongly expressed in this region. Another factor that contributes to the amplification of Arctic temperature changes is the low-level temperature inversion that characterizes the Arctic region for much of the year (Fig. 10). A layer of cold air sits atop a layer of warmer air, trapping it at the surface and strongly limiting vertical mixing in the atmosphere. This, in turn, helps focus the effects of heating near the surface.

THE FRAGILITY OF ARCTIC TERRESTRIAL ENVIRONMENTS
Permafrost Conditions

Under the current global warming conditions, Arctic permafrost is not as stable as it was in the past. In fact, Arctic permafrost melting is predicted to increase substantially in the next few decades (Overland et al., 2019). Model projections show a 20% decrease in Northern Hemisphere near-surface permafrost area, from the current 15 to 12 million km^2 by 2040, regardless of which IPCC RCP scenario is invoked. The IPCC is the Intergovernmental Panel on Climate Change, and

FIG. 10 Winter temperature inversion, Tanana Valley, Alaska. (Photo by the author.)

RCP, representative concentration pathway, is a global greenhouse gas atmospheric concentration trajectory adopted by the IPCC for its recent assessment reports.

The Fragility of Tundra Biota

Tundra plants are highly adapted to survive the very short growing season, low summer temperatures, extremely cold winters, and lack of moisture. Many tundra plants are perennials. This life strategy allows them time to slowly accumulate the metabolic energy needed to set seeds, a process that may take several summers. In one sense, this makes tundra plants "tough." They are survivors, living in a hostile environment. But examined from a different perspective, tundra plants are "tender." They are adapted to a narrow set of environmental conditions and are thus highly vulnerable to environmental changes that exceed their adaptations. Some of this boils down to low-temperature physiology. If a tundra plant species has a metabolism adapted to operating most effectively at 5°C (a typical summer temperature in the High Arctic), then when temperatures climb to 10°C, this cold-adapted plant goes into thermal stress as its enzymes (organic catalysts) either slow or stop functioning altogether.

Similarly, Arctic insects have evolved enzymes that facilitate chemical reactions at low temperatures (Georlette et al., 2004). The package of enzymes in cold-adapted insects has high catalytic efficiency. However, this efficiency comes at a price: their effectiveness breaks down at higher temperatures. One might be tempted to think that cold-adapted insects would thrive in the kind of warm temperatures experienced in the temperate zone, but this is not the case. Because cold-adapted metabolism is facilitated by enzymes that work *only* at low temperatures, Arctic insects exposed to temperatures above about 15–20°C often die because of metabolic failure, as their enzymes become ineffective and even break down (Georlette et al., 2004).

Vulnerability to Pollution and Disturbance

Tundra vegetation and soils are particularly vulnerable to chemical pollution. Toxic chemicals stay in the shallow active layer of soils and are very slow to disperse or break down. The permafrost boundary prevents chemicals from moving down to deeper layers. Near-freezing soil temperatures in the active layer slow the breakdown of pollutants, keeping soils toxic for decades or centuries after initial contamination. Atmospheric circulation brings air pollution from lower latitudes, and the pollutants tend to concentrate in the Arctic.

Oil pollution

Oil production in Arctic Eurasia focuses on western Siberia, where 68% of Russian gas and oil extraction take place. The largest production centers are in the Ob River basin, where 300 oil fields have pumped 65 billion barrels of oil since 1965. The Russian government conservatively estimates that at least 5% of the oil recovered from western Siberian wells has been spilled on the tundra surface since the 1960s. This amounts to 3 billion barrels of oil spilled in the Ob River and adjacent basins. Studies by regional ecologists have determined that there is no fish life in the Ob, Nadym, Pur, and Sob rivers because of pollution from oil production. The pollution is not limited just to the oil itself. Salts and heavy metals are in the brine that comes to the surface with oil from wells. These by-products are dumped on the land. The Russian oil industry tacitly acknowledged the level of water pollution in oil field regions when it installed water purification plants on-site, as the local water is unsafe to drink (Salmina, 2010).

Tundra vegetation is also particularly vulnerable to physical disturbances. Arctic plants recover with difficulty after disturbances, because they grow very slowly, and may only set seed every few years. Heavy vehicles driven over the tundra degrade the permafrost (Fig. 11), leading to the melting of frozen ground (thermokarst). Deep scars from this melting persist for many decades. Roads, buildings, and other structures must be insulated from the permafrost, or the heat they generate causes the ground beneath to melt.

THE FRAGILITY OF ARCTIC MARINE ENVIRONMENTS

Substantial loss of summer and multiyear sea ice has increased the dominance of thinner first-year ice. Also, later ice formation and earlier ice melt have been documented around the Arctic. These are some of the regional consequences of global warming, to be discussed more thoroughly in several subsequent chapters. However, all of these sea-ice changes have biological consequences for the organisms of the Arctic ocean.

Because sea ice has dominated the surface of the Arctic Ocean for many thousands of years, a great number of organisms rely on the presence of sea ice to complete their life cycles. These organisms range from microscopic plankton to walrus and polar bears. The future survival of this sea-ice biota will depend on whether they will be able to follow the receding ice edges and stay with the ice (e.g., ringed seals). Not all

FIG. 11 Aerial images of a seismic trail made in the winter of 1985 in the 1002 area of the Arctic National Wildlife Refuge, near Simpson Cove. The image on the left was taken in July 1985. The image on the right was taken in July 1999 — 15 years after the disturbance. (Photos courtesy of US Fish and Wildlife Service, in the public domain.)

species will survive the permanent diminution of sea-ice cover. Walrus, which are bottom feeders, rely on sea-ice cover relatively close to shore. This forms a platform from which they can dive for food. They cannot dive to the depths in the middle of Arctic Ocean to get to their benthic food sources. Not all sea life in Arctic waters relies on sea-ice cover. Pelagic or benthic life may be able to survive without sea ice.

Pollution of Arctic Seas

In recent decades, the Arctic Ocean has become polluted, mainly from sources outside of the Arctic. This pollution takes many forms. One of these is ocean acidification which is happening throughout the world but just happens to be more damaging to life in the Arctic Ocean. According to an in-depth assessment of Arctic Ocean acidification (AMAP, 2013), ocean acidification throughout the world is the direct result of inputs of CO_2 from the atmosphere. Atmospheric CO_2 levels are the highest they have been in the last 3 million years, and ocean surface CO_2 increases have followed the atmospheric increases. Consequently, the oceans are becoming more acidic (i.e., their pH is going down). The Arctic is intrinsically susceptible to ocean acidification because it has a low capacity to buffer changes in pH so that it will show greater changes in ocean acidification as CO_2 increases. Also, rapid

warming and ice melt are accelerating ocean acidification over most of the Arctic.

Mercury contamination

Mercury is another insidious type of pollution in the Arctic. Monomethyl mercury is formed in aquatic environments after the initial deposition of inorganic mercury. This compound is highly toxic, and it has the unfortunate property of accumulating in the tissues of marine wildlife. As mercury moves up the food chain, its concentration increases in animal tissues. This process of biomagnification of the most toxic form of mercury has been observed in Arctic marine wildlife. For example, over the past 25 years, mercury levels have increased by a factor of four in Arctic ringed seals and Beluga whales (Andersson et al., 2008). Dissolved gaseous mercury (DGM) concentrations measured near the North Pole were 10 times higher than the concentrations measured in the North Atlantic Ocean, pointing to the accumulation of DGM in the Arctic Ocean. Some of this mercury is entering Arctic seas through river discharge. For instance, elevated DGM concentrations were measured near the mouth of the Mackenzie River (Andersson et al., 2008). In addition, elevated DGM concentrations were found in ice-covered areas of the Arctic Ocean. This may indicate that the sea ice acts as a barrier for mercury exchange between the Arctic Ocean and the atmosphere.

Persistent organic pollutants

The term "persistent organic pollutants" (POPs) refers to organic compounds that are toxic, accumulate in animal tissues, and are resistant to degradation. They may persist in the environment for decades or centuries. Some of these compounds have been released to the environment as pollutants (polycyclic aromatic hydrocarbon [PAHs], dioxins) or have been produced intentionally and used in different applications (organochlorine pesticides, polychlorinated biphenyls [PCBs]).

PAHs are compounds formed during the combustion of fossil fuels. These pollutants can be found throughout the world, in air, water and soils. In spite of recent global declines in PAH concentrations, two decades of PAH measurements at multiple Arctic sites failed to show a significant decrease (Yu et al., 2019). Climate model simulations indicate that climate change may enhance the volatilization of lighter PAHs in the Arctic, thus delaying the expected decline. Furthermore, Arctic PAH concentrations are likely to increase from regional emissions due to human activities in the North as a result of warming, such as increased shipping, tourism, and resource extraction.

PCBs were produced from 1929 to the 1980s for use as insulating materials in electric equipment (capacitors and transformers), components of lubricants, paints, and plasticizers. In many countries, the use of PCBs was banned in the 1970s. Studies on PCB concentration trends show peak concentrations mostly in the 1940s–70s. The maximum half-life of PCBs in the environment is estimated at 48 years.

Dioxins (polychlorinated dibenzodioxins) are organic compounds that are formed during the combustion of organic materials containing chlorine atoms. Dioxins are very resistant to biodegradation and their half-life in the environment may exceed 100 years.

CONCLUSION

Hopefully, this brief introduction will give the reader a sense that the Arctic is a complex physical and biological system. The organisms that live here do not just suffer through life in the Arctic, they thrive here. Their physiologies, adaptations, and life cycles are geared to low temperatures, months of daylight followed by months of darkness, snow cover on land and ice cover on the sea. Native peoples of the North have likewise adapted themselves and their cultures to thrive in the Arctic. Thus, Arctic ecosystems and peoples functioned well for many thousands of years until Europeans arrived for the purpose of extracting wealth from the region. The attempts by the British to find a Northwest Passage all ended in frustration or tragedy (Fig. 12). Because the Arctic is so remote, European impacts on Arctic environments were fairly minimal until the middle of the 20th century. Following World War II, nuclear weapons testing, oil and gas extraction, and mineral mining entered the Arctic. The human footprint in the Arctic has been growing ever since. Not only that, but contamination of the Arctic with anthropogenic pollutants is now coming from the rest of the world through water and air.

Belatedly we are discovering that the Arctic is more important than we once thought. Not only is global warming heating the Arctic at approximately twice the rate it is doing in the rest of the world, but the rapidly

FIG. 12 Man proposes, god disposes, an 1864 painting by Edwin Landseer. The work was inspired by the search for Franklin's lost expedition which disappeared in the Arctic after 1845. The painting is in the collection of Royal Holloway, University of London.

warming Arctic is starting to affect the rest of the world as well. It turns out that the climate system is truly global in nature and that what happens in the Arctic affects the climate system in much of the world. The rest of this book describes how humans have altered the Arctic to this point and what the warmer Arctic is doing and will continue to do to the rest of the world.

REFERENCES

Arctic Monitoring and Assessment Programme (AMAP), 2013. AMAP Assessment 2013: Arctic Ocean Acidification. Oslo, Norway. viii + 99 pp. ISBN:978-82-7971-082-0.

American Geographical Society Library, 1887. Stanford's 1887 Map of the Arctic. https://agslibraryblog.wordpress.com/page/2/.

Andersson, M.E., Sommar, J., Gårdfeldt, K., Lindqvist, O., 2008. Enhanced concentrations of dissolved gaseous mercury in the surface waters of the Arctic Ocean. Marine Chemistry 110 (3–4), 190–194. https://doi.org/10.1016/j.marchem.2008.04.002.

Conservation of Arctic Flora and Fauna (CAFF), 2001. Definitions of the Arctic Region. https://www.arcticcentre.org/EN/arcticregion/Maps/definitions.

Creswell, S.G., 1856. One of Franklin's Ships, Trapped in the Ice. https://www.historicmysteries.com/the-doomed-franklin-expedition.

de Vaugondy, 2020. 1772 Map of Northwestern North America. https://commons.wikimedia.org/wiki/File:1772_Vaugondy_-_Diderot_Map_of_North_America_%5E_the_Northwest_Passage_-_Geographicus_-_NordetOuestAmerique-vaugondy-1772.jpg.

Georlette, D., Blaise, V., Collins, T., D'Amico, S., Gratia, E., Hoyoux, A., Marx, J.C., Sonan, G., Feller, G., Gerday, C., 2004. Some like it cold: biocatalysis at low temperatures. FEMS Microbiology Reviews 28 (1), 25–42. https://doi.org/10.1016/j.femsre.2003.07.003.

Hull of Derelict Soviet Submarine K-159, 2020. https://en.wikipedia.org/wiki/Soviet_submarine_K-159.

Miller, G.H., Alley, R.B., Brigham-Grette, J., Fitzpatrick, J.J., Polyak, L., Serreze, M.C., White, J.W.C., 2010. Arctic amplification: can the past constrain the future? Quaternary Science Reviews 29 (15–16), 1779–1790. https://doi.org/10.1016/j.quascirev.2010.02.008.

Overland, J., Dunlea, E., Box, J.E., Corell, R., Forsius, M., Kattsov, V., Olsen, M.S., Pawlak, J., Reiersen, L.O., Wang, M., 2019. The urgency of Arctic change. Polar Science 21, 6–13. https://doi.org/10.1016/j.polar.2018.11.008.

Petty, A., 2018. Sun Setting Over the Arctic Sea Ice Pack. NASA. https://www.innovations-report.com/earth-sciences/wintertime-arctic-sea-ice-growth-slows-long-term-decline-nasa/.

Pitt, M., 2020. 1680 Map of the Arctic. https://www.123rf.com/photo_14986600_north-pole-and-adjoining-lands-old-map-created-by-moses-pitt-published-in-oxford-1680.html.

Salbu, B., Nikitin, A.I., Strand, P., Christensen, G.C., Chumichev, V.B., Lind, B., Fjelldal, H., Bergan, T.D.S., Rudjord, A.L., Sickel, M., Valetova, N.K., Føyn, L., 1997. Radioactive contamination from dumped nuclear waste in the Kara Sea - results from the joint Russian-Norwegian expeditions in 1992-1994. Science of the Total Environment 202 (1–3), 185–198. https://doi.org/10.1016/S0048-9697(97)00115-0.

Salmina, Y., 2010. River Pollution in Oil Production Areas in Siberia, Vol. 4. Novosibirsk Region Social Committee for Water Protection, Novosibirsk.

Thomson, J., 2020. 1814 Map of the Northern Hemisphere. https://commons.wikimedia.org/wiki/File:1814_Thomson_Map_of_North_America_-_Geographicus_-_NorthAmerica-thomson-1814.jpg.

Yu, Y., Katsoyiannis, A., Bohlin-Nizzetto, P., Brorström-Lundén, E., Ma, J., Zhao, Y., Wu, Z., Tych, W., Mindham, D., Sverko, E., Barresi, E., Dryfhout-Clark, H., Fellin, P., Hung, H., 2019. Polycyclic aromatic hydrocarbons not declining in Arctic air despite global emission reduction. Environmental Science and Technology 53 (5), 2375–2382. https://doi.org/10.1021/acs.est.8b05353.

Contents

Arctic Seas

The next eight chapters will deal with various aspects of the Arctic Ocean and adjacent northern seas. The first of these chapters will deal the history of sea ice. As discussed in the introduction, sea ice cover in the Arctic was mainly considered a hindrance to navigation by those searching for a northwest passage to Asia. Before the invention of satellites, it was difficult to map the extent of sea ice with any precision. Satellite imagery came about the 1960s and 1970s, just as Arctic sea ice conditions were starting to change. The sea ice chapter traces the history of this phenomenon through recent centuries and examines predictions being made for the future sea ice cover in the Arctic.

The second of the Arctic Seas chapters deals with rising sea surface temperatures (SSTs) in the northern high latitudes. This chapter deals with causes and amplification of Arctic SSTs, the warming of the water column and Arctic seas, the regional ramifications of rising Arctic Ocean temperatures, and interactions between SSTs and sea ice cover. As with sea ice mapping, the systematic, accurate measurement of Arctic Ocean see temperatures has only been possible in the past few decades.

The third Arctic Seas chapter concerns changes in ocean circulation patterns that may come about because of the rapid warming of the Arctic Ocean. The pace and amplitude of such changes remain unknown. We are essentially conducting and ongoing experiment with the world's oceans. Does point, the only thing we can do is to create computer model simulations of what might happen and then watch is the day to come in over the next few decades. Therefore, this chapter examines these ocean circulation models. Specific aspects of future predictions discussed here include the effects of cold, fresh Arctic meltwater (from melting glaciers and ice sheets) on North Atlantic thermohaline circulation, predicted changes to North Pacific Ocean circulation, and how changes in the Beaufort Gyre of the Arctic Ocean may affect northwest Europe.

The fourth chapter in the section deals with sea level change and what will come about if the Greenland ice sheet (GIS) melts. This chapter will examine the connection between global sea level and the GIS, focusing in on predicted sea level changes under different model scenarios of GIS melting. Finally, I will examine the effects of global ocean warming on sea levels.

The fifth chapter in this section deals with the impacts of ocean acidification on Arctic marine ecosystems. As discussed in the introduction, Arctic waters are becoming acidified along with the rest of the world's oceans. Unfortunately, the problem of ocean acidification is made worse in Arctic waters because of decreased carbonate saturation. This will lead to the dissolution of calcium carbonate shells of marine plankton and mollusks, as well as physiological stress on these organisms. The chapter will also examine the stresses of ocean water acidity on Arctic fish and marine mammals.

The sixth chapter in this section examines the impacts of chemical pollution on Arctic marine ecosystems, beginning with the history of pollution of Arctic Seas, which began in earnest in the latter half of the 20th century. The chapter will trace the encroachment of toxic chemicals into the Arctic and then examine the effects of polychlorinated biphenyls, lead, and other toxins on marine invertebrates, marine fish, and mammals.

The seventh chapter in this section assesses the impacts of overfishing in Arctic and sub-Arctic waters since the 1950s. The chapter begins with a history of the Arctic fishery, focusing on the damage done by the international fleet of large trawler fishing ships, which led to the unprecedented collapse of North Atlantic cod stocks in recent decades. An example of enlightened resource conservation is provided in a discussion of the Canadian government's response to the cod stock collapse. Finally, the ecological consequences of overfishing are examined in a discussion on predatory fish depletion and the resulting changes in marine ecosystems.

The final chapter in the Arctic Seas section deals with the impacts of global shipping on Arctic Ocean ecosystems. Various aspects of this multifaceted problem are examined. These include the impacts of ocean pollution (waste disposal, fuel leaks, plastic) on marine life, the effects of noise pollution effects on marine mammals, the emission of black carbon from marine diesel engines, and the disturbances of shipping on fish and marine mammal migrations, feeding, and reproduction.

CHAPTER 1

Loss of Sea Ice

IMPORTANCE OF SEA ICE COVER

The permanent cover of pack ice is a unique feature of the central Arctic Ocean. This thin layer of ice radically modifies the characteristics of the Arctic Ocean surface by influencing the absorption and reflection of sunlight and modifying the exchanges of heat and moisture between the ocean and atmosphere. Thus, Arctic Sea ice, though a regional phenomenon, exerts influences on the global heat balance (Melling, 2012). The ice is subject to extreme oscillations between wintertime darkness and summertime sunlight. Likewise, because the Arctic Ocean basin is more-or-less surrounded by land, the sea ice has limited opportunities to drift into lower latitudes and melt. In winter, ice also forms in the seas adjacent to the Arctic Ocean, including the Sea of Okhotsk, the Bering Sea, Baffin Bay, Hudson Bay, the Greenland Sea, and the Labrador Sea. Arctic sea ice generally reaches its maximum extent each March and its minimum extent each September. Arctic Ocean sea surface temperature (SST) generally reaches a maximum in August. Of course, warmer SSTs contribute to the melting of sea ice. August mean sea surface temperatures in 2020 were ~1-3°C warmer than the 1982-2010 August mean over most of the Arctic Ocean (Timmermans and Ladd, 2018).

The annual cycle of sea ice waxing and waning drives salt downward in the water column, lowering the salinity of Arctic surface waters. The resulting low salinity of Arctic outflows has an impact on the **thermohaline circulation** of the global ocean. The presence and character of sea ice also shapes the ice-adapted oceanic ecosystems of northern seas.

When sea ice forms in the polar regions, the surrounding seawater increases in salinity due to salt rejection during ice formation, and it decreases in temperature. The increase in the density of cold saline (i.e., thermohaline) water directly beneath the ice triggers the sinking of the water mass down the continental slope and the spreading of the water masses to other parts of the ocean.

Links Between Land and Sea

The interconnectedness of Arctic sea ice and the global ocean works both ways. Accordingly, the areal extent and seasonal duration of ice cover are strongly affected by the net balance and fluxes of energy through the ocean's surface. Changes in the global climate system, such as atmospheric temperatures, are already modifying the thermal environment of the Arctic Ocean, forcing a reduced sea ice regime.

Albedo

In the warming Arctic, snow and ice cover on land and sea ice cover over the Arctic Ocean and adjacent seas are currently diminishing. Snow and ice cover are essential to the maintenance of cold conditions in the Arctic. These white surfaces reflect solar energy back into space. This reflectivity is called **albedo**. Snow has an even higher albedo than sea ice. Thick sea ice supports a cover of snow and reflects as much as 90% of the incoming solar radiation (Fig. 1.1).

Without snow cover on the sea ice, the ice reflects 50%–70% of the incoming energy. If the snow cover on sea ice melts, shallow melt ponds form. These ponds have an albedo of approximately 0.2–0.4, reducing the overall surface albedo to about 0.75. As melt ponds grow and deepen, the surface albedo drops to 0.15. The open ocean reflects only 6% of the incoming solar radiation and absorbs the rest. Because this system forms a positive feedback loop (Fig. 1.2), both summer and winter sea ice cover are predicted to disappear from the Arctic Ocean in the coming decades.

As sea ice cover declines, the exposed sea waters and adjacent atmosphere warm, since the dark surface of the water absorbs more solar energy. This oceanic heating spreads onto the adjacent Arctic lands, causing the ground surfaces to warm, contributing to the melting of permafrost (Fig. 1.3). When permanently frozen Arctic soils melt, the organic content of the soils decomposes, releasing large quantities of greenhouse gases (GHGs) into the atmosphere.

Threats to the Arctic. https://doi.org/10.1016/B978-0-12-821555-5.00005-X

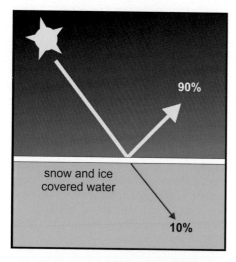

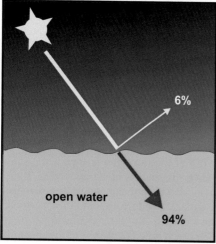

FIG. 1.1 Diagrams showing the effects of albedo on Arctic Ocean surfaces with and without sea ice cover. (Data from National Snow and Ice Data Center, 2020a. Quick Facts on Icebergs. https://nsidc.org/cryosphere/quickfacts/icebergs.html; National Snow and Ice Data Center, 2020b. Arctic Sea Ice News and Analysis, September, 2020. https://nsidc.org/arcticseaicenews/, in the public domain; From National Snow and Ice Data Center, 2019a. Arctic Science News/Thermodynamics: Albedo. https://nsidc.org/cryosphere/seaice/processes/albedo.html.)

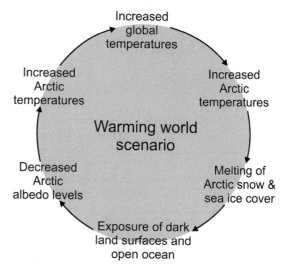

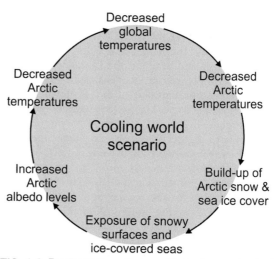

FIG. 1.2 Feedback loops that amplify Arctic warming (top) and Arctic cooling (below).

The carbon now held in the northern circumpolar region accounts for approximately 50% of all the below-ground global organic soil carbon. Arctic and subarctic soils are the repository of a greater amount of carbon than that held in the atmosphere or elsewhere in the biosphere (Tarnocai et al., 2009). The GHG warming caused by the release of soil carbon from melting permafrost will contribute to increased warming of the entire Arctic, thus completing a positive feedback loop as additional sea ice cover melts (Vaks et al., 2020).

Arctic climate system study

In the early 1990s prior to the Arctic Climate System Study (ACSYS), scientific interest focused on the stability of Arctic sea ice. It became apparent at that time that credible predictions of future Arctic ice conditions would require increased knowledge of pack ice and its interactions with ocean and atmosphere. At that time, the observations needed to generate useful computer models, even for short timescales (seasons, years, decades), did not exist.

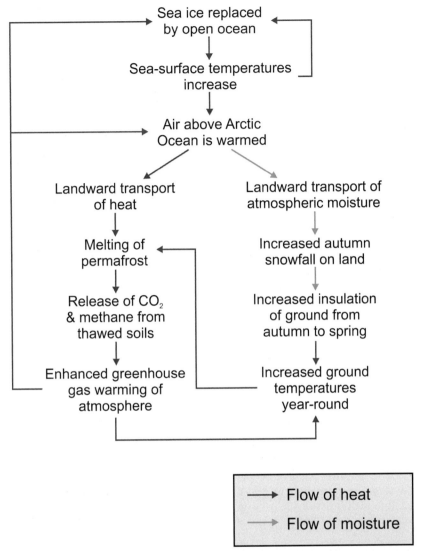

FIG. 1.3 Flowchart of ocean heating spread onto land. (Concepts after Vaks, A., Mason, A. J., Breitenbach, S. F. M., Kononov, A. M., Osinzev, A. V., Rosensaft, M., Borshevsky, A., Gutareva, O. S., & Henderson, G. M., 2020. Palaeoclimate evidence of vulnerable permafrost during times of low sea ice. Nature 577 (7789), 221–225. https://doi.org/10.1038/s41586-019-1880-1.)

The ACSYS project was driven by two principal objectives: to document the present state of Arctic sea ice and to assess the feedback between sea ice and other elements of the climate system. Long-term observations on the extent, concentration, thickness, and velocity of sea ice are essential to the study of natural variations in the marine cryosphere on seasonal, interannual, and decadal scales and to detect anthropogenic influence in the midst of these natural variations. These kinds of data provide standards by which to evaluate and improve sea-ice models.

Pack ice departure from the Arctic Basin. One important aspect of oceanic and atmospheric conditions in the high northern latitudes is the southward movement of pack ice into warmer waters. These movements serve as a measure of sea-ice net production and as a contributor to the freshwater entering the convective gyres of the North Atlantic. As noted earlier, the geography of the Arctic limits the amount of ice that leaves the region. Sea level pressure fields play a role in initiating these pack ice movements and in forcing multiyear ice to move south. These pressure fields generate regional winds that drive the Beaufort Gyre, which keeps sea ice circulating within the central Arctic Basin, while the Transpolar Drift Stream transports sea ice out of the Arctic Basin. Sea-ice drift speeds have increased since 1979, both in the Arctic Basin and through Fram Strait (Oppenheimer et al., 2019 and references therein). This is attributed to thinner ice and changes in wind forcing. The volume of sea ice exiting the Arctic Ocean through Fram Strait is estimated between 600,000 and 1 million km^2 of ice per year. This represents approximately 10% of the ice in the Arctic Basin. Other points of sea ice departure from the Arctic basin include the Barents Sea (<1% of total ice export), the Canadian Arctic Archipelago (5%−20% of total ice export), and Bering Strait (up to 22% in recent years).

THE EFFECTS OF THE ARCTIC OSCILLATION

Over years to decades, the dominant driver of atmospheric variability around the North Pole is the Arctic Oscillation (AO). The AO is an atmospheric oscillation in which air masses shift between the Arctic and the midlatitudes (NASA Earth Observatory, 2016). The shifting intensifies, weakens, and shifts the location of the dominant low- and high-pressure systems. These changes influence the strength of the prevailing westerly winds and storm tracks. The "positive" phase of the AO sees increased intensity of Arctic Ocean winds. Increased winds tend to open up leads in the ice pack, breaking up the surface in conjunction with ocean currents (Fig. 1.4).

The thin, young ice that forms in these leads is more likely to melt in the summer. The strong winds also tend to flush ice out of the Arctic through Fram Strait, between Greenland and Svalbard. "Negative" phases of the oscillation are associated with weaker winds. Fewer leads of open water develop during these years, preserving more multiyear pack ice, which is less likely to be swept south of the Arctic basin into warmer waters.

However, in recent years, the relationship between the AO and the extent of summer sea ice has weakened. For example, a strong negative phase in the winters of 2009 and 2010 was insufficient to maintain high levels of ice cover. Clearly some other factors can override the relationship.

FIG. 1.4 Strong winds caused sea ice to crack and buckle off the coast of Greenland. (Photo by Andy Mahoney, National Snow and Ice Data Center, in the public domain; From National Snow and Ice Data Center, 2013. Pack Ice Break Up. https://nsidc.org/cryosphere/quickfacts/seaice.html.)

Historical Records from the Little Ice Age to World War II

Historical climatologists have worked tirelessly over the past 30 years to reconstruct a history of Arctic sea ice that predates satellite and aerial surveys. Their work has been driven by the following research questions, for which long-term data are essential (Walsh et al., 2016):

(1) Has Arctic sea ice cover been this small since the start of the industrial revolution?

(2) Has sea ice ever declined this rapidly in the historical record?

(3) How is sea ice affected by natural fluctuations over multiple decades?

The reconstruction of sea ice margins in the North Atlantic in the presatellite era (i.e., between 1850 and 1978) have been compiled from a wide range of historic observations of sea ice. These include ship observations, newspaper articles, airplane surveys, compilations by naval oceanographers, diaries and analyses by national ice services and meteorological offices. Sea-ice concentration data from regular aerial surveys of ice in the eastern Arctic have been made by the Arctic and Antarctic Research Institute, St. Petersburg, Russia, beginning in 1933. Sea-ice edge positions for Newfoundland and the Canadian Maritime Region come from observations kept by the Canadian government, covering the interval from 1870 to 1962. Detailed charts of sea ice for Alaska for 1954 to 1978 were compiled by US Navy ice observer William H. Dehn. He collected 6896 paper ice charts of Alaska, the western Canadian Arctic and

Bering Sea waters. Arctic-wide maps of ice cover have been kept by the Danish Meteorological Institute, covering the period from 1901 to 1956. Finally, whaling ship logbooks dating back to the 1850s were examined for entries that noted ship position and whether the ship encountered ice.

As shown in Fig. 1.5, sea-ice cover in March (near the end of the Arctic winter) hovered around 15 million km^2 from 1850 through about 1990 and then declined to levels between 14 and 15 million km^2 since then.

Sea-ice cover in September (the end of the Arctic summer) hovered around 8–9 million km^2 from 1850 through about 1930 and then declined to levels between 6 and 7 million km^2 from the 1930s through about the year 2000. The National Snow and Ice Data Center (2019a,b,c) reports that the September sea-ice cover has fallen between 4 and 5 million km^2 in the 21st century.

Changes in sea-ice cover since the 1950s
Between 1979 and 2015, the average monthly sea-ice extent for September declined by 13.4% per decade. In every Arctic region, in every month, and during every season, Arctic sea ice extent is lower today than it was during the 1980 and 1990s.

Natural variability and global warming both appear to have played a role in this decline. The AO's strongly positive mode through the mid-1990s flushed thicker, older ice out of the Arctic, replacing multiyear ice with first-year ice more prone to melting. After the mid-1990s, the AO was often neutral or negative, but sea ice nevertheless failed to recover to previous levels. Instead, a new pattern began in 2002—one of steep Arctic sea ice decline. The AO likely triggered a phase of accelerated melt that continued into the next decade because of unusually warm Arctic air temperatures (NASA Earth Observatory, 2016).

As sea ice retreats from Arctic coastlines, wind-driven waves—combined with thawing permafrost—are leading to increased coastal erosion. Other potential impacts include changes in regional weather patterns. This is an area of active research, as scientists try to tease out the possible links between sea ice loss and midlatitude weather patterns.

Some researchers have hypothesized that melting sea ice could interfere with established patterns of ocean circulation. In the Arctic, ocean circulation is driven by the sinking of dense, salty water. Fresh meltwater, coming primarily from the Greenland Ice Sheet, could slow ocean circulation at high latitudes. Changes in the location and timing of sea ice growth—where the dense salty waters are formed and then sink to the bottom—may also be an important factor.

Declines in multiyear sea ice
The proportion of Arctic sea ice at least 5 years old declined from 30% to 2% between 1979 and 2018 (IPCC, 2019). Integration of data from submarines, moorings, and earlier satellite radar altimeter missions shows ice thickness declined across the central Arctic by 65%, from 3.59 to 1.25 m between 1975 and 2012 (Lindsay and Schweiger, 2015; Kwok and Rothrock, 2009). It appears that this loss of sea ice volume may be unprecedented over the past century. The thinning of seasonal sea-ice cover contributes to the overall reduction of sea-ice cover through summer melting and the thin ice cover is vulnerable to fragmentation from the passage of intense Arctic cyclones in summer and increased ocean swell conditions.

Current rapid loss of sea ice
Arctic sea-ice cover varies seasonally, in recent decades between 6 million km^2 in the summer and 15 million km^2 in the winter. The summer ice cover is mostly confined to the Arctic Ocean basin and the Canadian Arctic Archipelago, but winter sea ice reaches south as far as 44°N, into the adjacent seas. In September, at the end of Northern Hemisphere summer, the Arctic sea-ice cover consists of the ice that survived the summer melt period (Combal and Fischer, 2016).

The record lowest Arctic sea ice cover occurred in September 2012 when it fell to 3.41 million km^2. That was 20% less than the previous record of 4.17 million km^2 set in 2007, when Arctic sea ice extent broke all prior records by mid-August, more than a month before the end of the melt season (NASA Earth Observatory, 2016). Since the mid-2000s, low minimum sea ice extents in the Arctic have become the "new normal." The extent of Arctic sea ice at the end of the summer of 2019 was effectively tied with 2007 and 2016 for second-lowest since modern record-keeping began in the late 1970s. An analysis of satellite data by NASA and the NSIDC (Fig. 1.6) shows that the 2019 minimum extent, which was likely reached on September 18, measured 4.15 million km^2.

From 2016 to 2018, Arctic winters were extremely warm: nearly double (+6°C) those of previous record highs (Overland et al., 2019). As we have seen, sea ice extents in four successive winters (2015–18) were at record low levels. Thin sea ice grows faster than thick sea ice, and before the 1990s, the winter growth of sea ice rapidly returned to previous values. But in the past decade, there has been a marked decrease in multiyear sea ice. Multiyear sea ice coverage, extent, and thickness are an integrator of climate over years to decades. Its loss is a clear indicator of Arctic climate change. In fact, the extent of multiyear old, thick sea ice is now 60% smaller than that of the 1980s (Fig. 1.7).

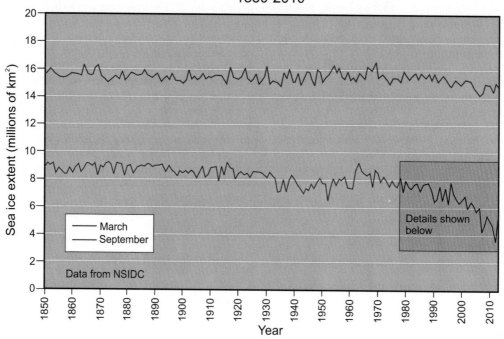

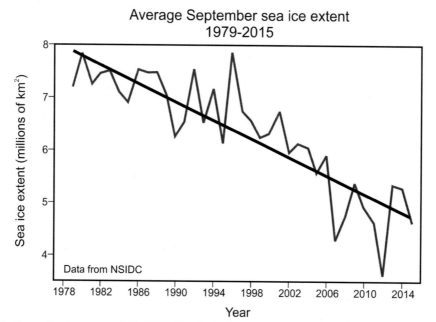

FIG. 1.5 Above: Sea-ice cover, 1850 to 2010. *Blue line* Indicates sea ice extent in March; the *red line* indicates extent in September. Below: Details of sea ice cover, 1978–2015. (Data from National Snow and Ice Data Center, in the public domain; From National Snow and Ice Data Center. 2016. State of the Cryosphere: Sea Ice. https://nsidc.org/cryosphere/sotc/sea_ice.html.)

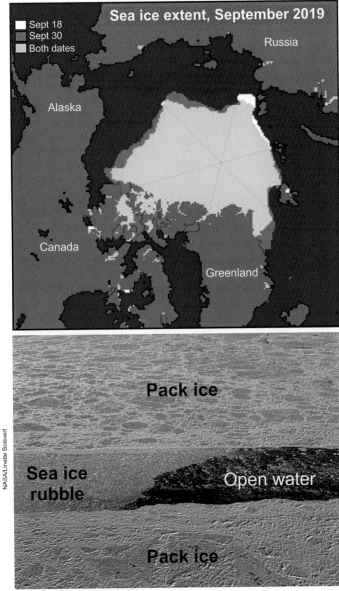

FIG. 1.6 Above: Map of the high Arctic, showing the extent of Arctic sea ice in September 2019. Below: An opening in the sea ice cover north of Greenland is partially filled in by much smaller sea ice rubble and floes. (After National Snow and Ice Data Center, 2019a. Arctic Science News/Thermodynamics: Albedo. https:// nsidc.org/cryosphere/seaice/processes/albedo.html; National Snow and Ice Data Center, 2019b. State of the Crysophere: Sea Ice. https://nsidc.org/cryosphere/sotc/sea_ice.html; Photo by Linette Boisvert, NASA, 2020. Both images in the public domain.)

Shrinking Ice Shelves

Ice shelves are thick, floating ice masses that are attached to the Arctic coast. They form as glaciers expand from the land into the near-shore sea, or they may be derived from accumulations of multiyear land-fast sea ice. They increase in thickness through the surface accumulation of snow and superimposed ice and by the growth of ice from the water beneath them (Dowdeswell and Jeffries, 2017). Glacier tongues are relatively narrow floating ice margins. Unlike

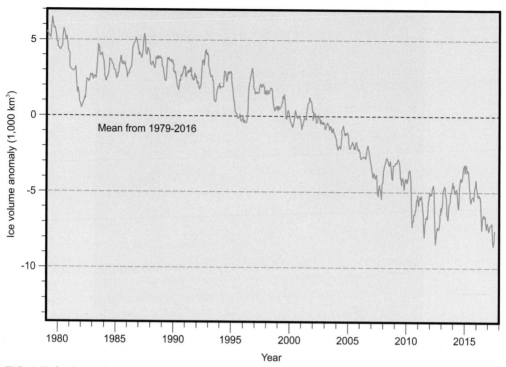

FIG. 1.7 Arctic sea ice volume, 1979–2016. (Data from National Snow and Ice Data Center, in the public domain; From Vizcarra, N. 2016. Late Summer in the Arctic, Sea Ice Melt Continues. http://nsidc.org/arctics eaicenews/author/vizcarrn/page/3/.)

Antarctica, where ice shelves are large and numerous, Arctic ice shelves are restricted to several island groups in the High Arctic and a few Greenland fjords. These are ice masses >20 m thick. They float on the sea but are frozen to the shore. The Ellesmere Ice Shelf developed about 5500 years ago in response to Holocene cooling. During the Little Ice Age, this shelf was about 9000 km^2 along the coast of northern Ellesmere Island. Since then, a series of calving events have broken the once single, continuous ice-shelf fringe into a number of small and isolated remnants. For instance, the Ayles Ice Shelf (66 km^2) broke free from the coast of Ellesmere Island in August 2005 (Fig. 1.8). The Ward Hunt Ice Shelf on the north coast of Ellesmere Island is the largest modern remnant of the Ellesmere Ice shelf (400 km^2), representing only 4% of the area of the former ice shelf.

Floating glacier tongues of the Greenland Ice Sheet have also begun to fragment. Large, often tabular icebergs calve from these Greenland ice shelves. Ice islands are a form of tabular iceberg in the Arctic Ocean. They have a characteristic undulating surface. Icebergs drift mainly under the influence of currents, sometimes

being trapped for years in the Arctic Ocean circulation pattern before eventually drifting south along the east coast of Greenland (Fig. 1.9). Despite the presence of about 4000 km of marine-terminating glaciers and ice caps in the Eurasian Arctic, there are few floating ice shelves here.

PREDICTED LOSS OF SEA ICE WITH GLOBAL WARMING

Will the Arctic lose its sea-ice cover because of global warming? A study by Notz and Stroeve (2016) suggests that the observed rate of Arctic sea-ice loss will not change substantially in the foreseeable future. They developed a numerical model of sea-ice loss, driven by atmospheric CO_2 concentration levels. They estimated that the remainder of Arctic summer sea ice will be lost when an additional 1000 Gt of CO_2 is emitted to the atmosphere. This estimate is based on observations indicating that roughly 3 m^2 of September sea ice is lost for every ton of anthropogenic CO_2 emissions. This value is based on a 30-year running mean of monthly averages. It is therefore a very conservative

FIG. 1.8 **(A)** The Ayles Ice Shelf on the northern coast of Ellesmere Island, Canada. The ice shelf was roughly 66 square kilometers, slightly larger than Manhattan Island, New York. **(B)** On August 13, 2005, the ice shelf broke free in less than an hour and began drifting out to sea. (Both photos courtesy of NASA Earth Observatory, in the public domain; From NASA Earth Observatory. 2007. Ayles Ice Shelf, Ellesmere Island. https://earthobservatory.nasa.gov/images/7282/ayles-ice-shelf-ellesmere-island.)

estimate of the level of CO_2 emissions required to decrease the annual minimum sea-ice area below critical threshold of 1 million km^2. The internal variability of the sea-ice system adds an uncertainty of around 20 years in the prediction of the first year of a near-complete loss of Arctic sea ice. For current emissions of 35 Gt CO_2 per year, the limit of 1000 Gt will be reached before midcentury. However, the results of Notz and Stroeve (2016) also suggest that measures taken to slow global CO_2 emissions will directly slow the ongoing loss of Arctic summer sea ice. Specifically, if CO_2 emissions can be kept at a level associated with global warming of 1.5°C (i.e., cumulative future

emissions well below the 1000 Gt level), then Arctic summer sea ice has a chance of survival in the long term, at least in some parts of the Arctic Ocean.

Using numerical modeling, Askenov et al. (2017) developed a series of detailed, high-resolution projections of Arctic sea ice conditions to the end of the 21st century. They employed an Ocean General Circulation Model (OGCM) to examine changes in ocean circulation and Arctic Ocean sea ice cover for the years 2000–99. The model (NEMO-ROAM025) is a configuration developed in the Regional Ocean Acidification Modeling project (ROAM). The forcing factor in their numerical models was atmospheric CO_2 levels. They

FIG. 1.9 Icebergs off the coast of Greenland. Icebergs can develop into a variety of shapes as they break apart. (Photo by Ted Scambos, National Snow and Ice Data Center, in the public domain; From National Snow and Ice Data Center, 2020a. Quick Facts on Icebergs. https://nsidc.org/cryosphere/quickfacts/icebergs.html.)

used two CO_2 emission scenarios: the RCP8.5 IPCC (Intergovernmental Panel on Climate Change) emission scenario and the RCP 2.6 scenario. The IPCC developed a set of representative concentration pathways (RCPs) that represent different climate change scenarios. These include time series of emissions and concentrations of the full suite of GHGs. RCP2.6 is a pathway in which radiative forcing peaks at approximately 3 W per M^2 before 2100 and then declines to a

low level of emissions after 2100. This could be considered a "best-case scenario" among the IPCC climate projections. In contrast to this, the RCP8.5 is a pathway in which radiative forcing reaches greater than 8.5 W per M^2 by 2100 and continues to rise thereafter, reaching constant, very high concentrations after 2250. This pathway has been termed the "business as usual" model by the IPCC. It assumes that GHG emissions will continue to increase at their current rate for the rest of the 21st century. This can be considered the "worst-case scenario." Under RCP8.5 conditions, by the 2090s, sea surface temperatures in the Arctic Ocean and Siberian seas will increase by about 2°C in the winter and by about 7°C in the summer. As shown in Fig. 1.10 under the RCP2.6 scenario, September sea-ice cover declines until about 2050 and then begins to stabilize around 4 million km^2, approximately the level it is today. However, under the RCP8.5 scenario, September sea-ice cover falls sharply by about 2050 and disappears completely a few years thereafter.

Peng et al. (2020) ran 12 different climate models, using the RCP 4.5 and 8.5 scenarios to predict the first ice-free Arctic summer year (FIASY). They defined FIASY as the first year in which sea-ice cover falls below 1×10^6 km^2. As shown in Table 1.1, the average of the 12 model predictions for the FIASY under the RCP4.5 scenario was the year 2054. The average for the RCP 8.5 scenario was 12 years earlier, in 2042. However, the model predictions for the RCP 4.5 scenario

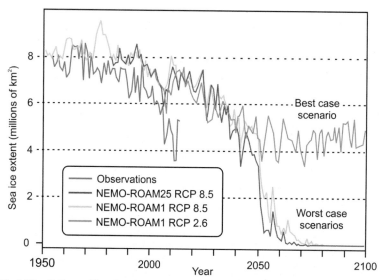

FIG. 1.10 Model Predictions of Sea-Ice Extent Through 2100, Showing the Best- and Worst-Case Scenarios. After Aksenov et al., 2017, On the future navigability of Arctic sea routes: High-resolution projections of the Arctic Ocean and sea ice. Marine Policy 75, 300–317. https://doi.org/10.1016/j.marpol.2015.12.027

TABLE 1.1
Predicted first ice-free Arctic summer year from climate models run using the IPCC's Representative Concentration Pathways (RCP) 4.5 and 8.5.

Representative Concentration Pathways	RCP 4.5	RCP 8.5
Mean of 12 model results	2054	2042
Minimum (absolute value)	2023	2023
Maximum (absolute value)	>2100	2065
Spread	>77	42

Data from Peng et al., 2020.

had a spread of 77 years, and the model predictions for the RCP 8.5 scenario had a spread of 42 years. The authors concluded that there is still room for improvement for future sea-ice prediction models within the global climate model systems.

A university statistics professor of mine once said, "All models are lies, but some are instructive." By this, he meant, of course, that the models themselves do not represent reality but are some mathematical approximation of it. In examining the outcomes of the various climate models, it becomes apparent that the real world of Arctic sea ice and polar climate is too complicated to be accurately modeled, even by our most sophisticated computer models. However, the similarity between the results of the various attempts to predict the timing of the first ice-free Arctic summer year shows that the models are instructive. If climate change carries on at its current pace, then it seems very likely that summer Arctic sea ice will be gone by the middle of this century. If we manage to scale back the emission of gases substantially in the very near future, then perhaps summer sea ice will eventually stabilize at approximately its current extent in the latter half of the 21st century.

Sea Ice and Extreme Weather Events

Stroeve et al. (2019) discussed the global importance of multiyear pack ice, not just to the Arctic but also to the planet as a whole. Multiyear ice is a key element of the Arctic system because it is thicker and stronger than first-year ice. Multiyear ice not only controls the dynamics of the rest of the Arctic sea, but it also influences the Arctic climate and marine ecosystem by limiting the amount of light entering the water column. Changing the light intensity reaching the water disrupts the breeding cycle of marine organisms such as zoo- and

phytoplankton. This adds stress to the organisms in the regional food web. Multiyear ice used to persist through the entire Arctic year, including the summer melt season. But now, the old ice is melting. The only sea-ice cover on the Arctic Ocean is now first-year ice, formed in winter. This thinner ice cannot survive the summer melt. The loss of multiyear ice is a key factor in the process of Arctic amplification. As discussed earlier, when the Arctic Ocean is covered in ice, 80% of the sunlight gets reflected back to space, but with open water, the ocean absorbs 80% of the sunlight. Therefore when the Arctic lacks sea ice, or has only very thin ice, the Arctic Ocean absorbs the sun's heat, which inhibits ice formation the following winter—forming a negative feedback loop (a downward spiral). At the end of the winter, the ice is thinner and more susceptible to melting, hence the massive retreats we are seeing. As Stroeve et al. (2019) told the World Economic Forum, the gradual disappearance of Arctic sea ice cover could dramatically accelerate global climate change. This, in turn, will very likely affect food and water security elsewhere, as well as causing rising sea levels and extreme weather events. The frequency and impact of extreme events increases as the world warms. Extreme weather events in the United States cost $323 billion in 2019 and about $398 billion in 2020. Research is revealing that many extreme events are indeed related to changes in the Arctic. So, when climate scientists identify extreme weather as the highest priority global risk, they are implicitly talking about the dangers of Arctic change. Research shows a strong and direct correlation with CO_2 emissions. Rather than looking at the global emissions data, it is perhaps more helpful to examine emissions by country, especially some of the larger contributors to the problem. Countries such as the United States, Canada, Australia, and Saudi Arabia are some of the largest contributors to sea-ice loss per capita, with Qatar topping the list. Here are the CO_2 emissions and corresponding sea-ice losses from the top four emitters in 2017: China: 10.0 Gt CO_2 = 30,000 km^2 sea-ice loss, the United States: 5.3 Gt CO_2 = 16,000 km^2 sea ice loss, the European Union: 3.5 Gt CO_2 = 10,500 km^2 sea ice loss, and India: 2.5 Gt CO_2 = 7500 km^2 sea ice loss. As of September 2020, the total area of Arctic sea ice was 3.92 million km^2 (National Snow and Ice Data Center, 2020a,b). Just based on the carbon emissions of the top four emitters shown earlier, if sea-ice losses were to continue at this pace, then all Arctic sea ice would be gone in the year 2081. However, as we have noted, this is not a linear process, but rather a self-amplifying feedback loop, so that the loss of sea ice stimulates increasing losses through each annual cycle.

CONCLUDING THOUGHTS

From the time when Europeans first became aware of its existence until the past few decades, Arctic sea ice was mostly thought of as a barrier to ships seeking the Northwest Passage across the Canadian Arctic. As painfully demonstrated by multiple 19th-century naval expeditions, Arctic sea-ice boundaries are constantly shifting in response to winds and ocean currents. Multiyear ice, up to several meters thick, could easily crush the wooden-hulled boats sent into Arctic waters. Boats that were not crushed outright were most frequently trapped in the pack ice during the autumn freeze-up and sometimes remained trapped for years.

It is only in the past 30−40 years that oceanographers and climatologists have come to realize the global importance of Arctic sea-ice cover. Its influence stretches far beyond the Arctic latitudes. We now know that the presence of a largely white surface (sea ice mantled with snow) over the Arctic Ocean helps maintain cold temperatures over a wide region and that the replacement of white surfaces with dark, open ocean drastically lowers Arctic albedo, thereby contributing significantly to amplified warming in the Arctic. Climate scientists are only just now starting to refine their models of the interplay between Arctic sea ice, regional wind patterns, ocean currents, and changes in regional oceanic water columns (Domingues et al., 2008). They have identified positive feedback loops that contribute to the current amplified warming trends in the northern high latitudes.

Ironically, these recent discoveries have taken place while Arctic sea ice is in the process of disappearing. In a sense, we are conducting a climate experiment revealing the regional and global impacts of melting Arctic sea ice. Only with the strictest adherence to reductions in GHG emissions can Arctic sea ice survive the next few decades. This would require the world to adopt GHG emission reductions to the level described in the RCP 2.6 scenario, which the IPCC considers its "best-case" scenario. Specifically, the RCP 2.6 scenario assumes that global annual GHG emissions peak by 2020 and then decline substantially afterward. The opportunity to achieve this goal by 2020 has slipped away. The world community started paying attention to climate change in the 1990s, leading to the Rio Earth Summit (1992), the Kyoto Protocol (1997), the Copenhagen Accord (2009), the Cancun Agreements (2010), the Durban Platform for Enhanced Action (2011), and the Paris Agreement (2015). The good intentions of the political leaders who signed onto the various climate change accords were not backed by any attempts to force countries to meet specific GHG reduction targets. These failures to act in the political realm are having real, immediate consequences for the planet, including the complete loss of Arctic sea-ice cover during the summer months within the next few decades.

The Arctic is the poster child for the importance of the +1.5°C target agreed upon at the 2015 UN Climate Conference in Paris (COP21). If the world can maintain a ceiling of a +1.5°C warmer world, we can save the Arctic summer sea ice and likely prevent some of the most serious global climate impacts. But once global temperatures go beyond this limit, we will almost certainly lose the summer ice, triggering serious consequences for our economies and societies worldwide, including increases in extreme weather in the midlatitudes. From an Arctic sea-ice perspective, upholding the Paris Agreement means the difference between having sea ice or not. Why would these matters concern the World Economic Forum? Preserving Arctic sea ice makes good financial sense. Economic studies suggest that worst-case scenario acceleration of climate change driven by thawing Arctic permafrost and melting sea ice could cause the world's economies $130 trillion in economic losses. On the other hand, if global warming is limited to 1.5°C, the additional costs can be kept under $10 trillion. So, this is where the arcane world of climate science meets the reality of our day-to-day lives. As concluded by Stroeve et al. (2019), Arctic scientists can bring us the facts and projections on Arctic change and global risks, but solutions must ultimately come from politics, business, and civil society.

REFERENCES

Aksenov, Y., Popova, E.E., Yool, A., Nurser, A.J.G., Williams, T.D., Bertino, L., Bergh, J., 2017. On the future navigability of Arctic sea routes: high-resolution projections of the Arctic Ocean and sea ice. Marine Policy 75, 300−317. https://doi.org/10.1016/j.marpol.2015.12.027.

Combal, B., Fischer, A., 2016. 2: model projections of ocean warming under \business as usual\ and \moderate mitigation\ scenarios. In: UNESCO IOC and UNEP, the Open Ocean: Status and Trends, pp. 70−89.

Domingues, C.M., Church, J.A., White, N.J., Gleckler, P.J., Wijffels, S.E., Barker, P.M., Dunn, J.R., 2008. Improved estimates of upper-ocean warming and multi-decadal sea-level rise. Nature 453 (7198), 1090−1093. https://doi.org/10.1038/nature07080.

Dowdeswell, J.A., Jeffries, M.O., 2017. Arctic ice shelves: an introduction. In: Copland, L., Mueller, D. (Eds.), Arctic

Ice Shelves and Ice Islands. Springer. https://doi.org/10.1007/978-94-024-1101-0.

IPCC Fifth Assessment Report, 2013. Climate Change 2013: The Physical Science Basis. https://www.ipcc.ch/report/ar5/wg1/.

Kwok, R., Rothrock, D.A., 2009. Decline in Arctic sea ice thickness from submarine and ICESat records: 1958-2008. Geophysical Research Letters 36 (15), L15501. https://doi.org/10.1029/2009GL039035.

Lindsay, R., Schweiger, A., 2015. Arctic sea ice thickness loss determined using subsurface, aircraft, and satellite observations. The Cryosphere 9, 269−283. https://doi.org/10.5194/tc-9-269-2015.

Melling, H., 2012. Chapter 3. Sea-ice observation: advances and challenges. In: Lemke, P., Jacobi, H.W. (Eds.), Arctic Climate Change: The ACSYS Decade and beyond. Springer. https://books.google.com/books?id=bJy9nCs-fgsC&dq=Atmospheric+and+Oceanographic+Sciences+Library+43&source=gbs_navlinks_s.

NASA Earth Observatory, 2007. Ayles Ice Shelf, Ellesmere Island. https://earthobservatory.nasa.gov/images/7282/ayles-ice-shelf-ellesmere-island.

NASA Earth Observatory, 2016. Arctic Sea Ice. NASA. https://earthobservatory.nasa.gov/features/SeaIce/page3.php.

NASA, 2020. Ocean acidification. https://www.noaa.gov/education/resource-collections/ocean-coasts-education-resources/ocean-acidification. (Accessed 23 July 2020).

National Snow and Ice Data Center, 2013. Pack Ice Break up. https://nsidc.org/cryosphere/quickfacts/seaice.html.

National Snow and Ice Data Center, 2016. State of the Cryosphere: Sea Ice. https://nsidc.org/cryosphere/sotc/sea_ice.html.

National Snow and Ice Data Center, 2019a. Arctic Science News/Thermodynamics: Albedo. In: https://nsidc.org/cryosphere/seaice/processes/albedo.html.

National Snow and Ice Data Center, 2019. State of the Cryosphere: Sea Ice. https://nsidc.org/cryosphere/sotc/sea_ice.html.

National Snow and Ice Data Center, 2019. Arctic Sea Ice News and Analysis October. http://nsidc.org/arcticseaicenews/2019/10/falling-up/.

National Snow and Ice Data Center, 2020. Quick Facts on Icebergs. https://nsidc.org/cryosphere/quickfacts/icebergs.html.

National Snow and Ice Data Center, 2020. Arctic Sea Ice News and Analysis. September. https://nsidc.org/arcticseaicenews/.

Notz, D., Stroeve, J., 2016. Observed Arctic sea-ice loss directly follows anthropogenic CO_2 emission. Science 354 (6313), 747−750. https://doi.org/10.1126/science.aag2345.

Oppenheimer, M., Glavovic, B.C., Hinkel, J., 2019. In: Pörtner, H.-O., Roberts, D.C., Masson-Delmotte, V. (Eds.), IPCC Special Report on the Ocean and Cryosphere in a Changing Climate. Intergovernmental Panel on Climate Change.

Overland, J., Dunlea, E., Box, J.E., Corell, R., Forsius, M., Kattsov, V., Olsen, M.S., Pawlak, J., Reiersen, L.O., Wang, M., 2019. The urgency of Arctic change. Polar Science 21, 6−13. https://doi.org/10.1016/j.polar.2018.11.008.

Peng, G., Matthews, J.L., Wang, M., Vose, R., Sun, L., 2020. What do global climate models tell us about future arctic sea ice coverage changes? Climate 8 (15). https://doi.org/10.3390/cli8010015.

Stroeve, J., Whiteman, G., Wilkinson, J., 2019. The shrinking Arctic ice protects us all. It's time to act. In: Annual Meeting. World Economic Forum, Davos, Switzerland. https://www.weforum.org/agenda/2019/01/the-shrinking-arctic-ice-protects-all-of-us-its-time-to-save-it/.

Tarnocai, C., Canadell, J.G., Schuur, E.A.G., Kuhry, P., Mazhitova, G., Zimov, S., 2009. Soil organic carbon pools in the northern circumpolar permafrost region. Global Biogeochemical Cycles 23 (GB2023). https://doi.org/10.1029/2008gb003327 n/a-n/a.

Timmermans, M.L., Ladd, C., 2018. Arctic Report Card: Update for 2018; Sea Surface Temperature. https://arctic.noaa.gov/Report-Card/Report-Card-201.

Vaks, A., Mason, A.J., Breitenbach, S.F.M., Kononov, A.M., Osinzev, A.V., Rosensaft, M., Borshevsky, A., Gutareva, O.S., Henderson, G.M., 2020. Palaeoclimate evidence of vulnerable permafrost during times of low sea ice. Nature 577 (7789), 221−225. https://doi.org/10.1038/s41586-019-1880-1.

Vizcarra, N., 2016. Late Summer in the Arctic, Sea Ice Melt Continues. http://nsidc.org/arcticseaicenews/author/vizcarrn/page/3/.

Walsh, J.E., Fetterer, F., Stewart, J.S., Chapman, W.L., 2016. A database for depicting Arctic sea ice variations back to 1850. Geographical Review 107, 89−107.

CHAPTER 2

Rising Sea Surface Temperatures

Just as the loss of sea-ice cover in the Arctic Ocean is contributing to amplified warming of the northern high latitudes, so also is the rise in sea surface temperatures (SSTs) in arctic waters. SST data collated by NOAA show that arctic surface waters began warming in the mid-1990s, and SSTs have been increasing ever since. Summer SSTs in the Arctic Ocean are driven mainly by the amount of insolation absorbed at the surface. As discussed in Chapter 1, Arctic Ocean SSTs are also influenced by the extent of sea ice, with greater warming of the open ocean, in addition to variations in cloud cover, water color, and upper-ocean stratification. The situation is more complicated in the Barents and Chukchi seas, where additional heat comes from the advection of warm water from the North Atlantic and North Pacific oceans, respectively. In addition to the physical properties of Arctic SSTs, regional marine ecosystems are also greatly influenced by SSTs, as surface water temperature affects the timing and developmental cycles of primary and secondary production as well as available habitat for individual species. NOAA presents their SST data based on calculations made using the NOAA Optimum Interpolation (OI) SST Version 2 product (OISSTv2), which is a blend of in situ and satellite measurements from December 1981 to present. Their research has shown that this interpolation method yields results that correlate about 80% with in situ temperature measurements (Timmermans and Ladd, 2018). August SSTs provide the most appropriate representation of Arctic Ocean summer SSTs because they are not affected by the cooling and subsequent sea-ice growth that typically takes place in the latter half of September.

RISING SEA SURFACE TEMPERATURES IN ARCTIC WATERS

People have been measuring and recording SSTs for centuries, albeit with varying degrees of accuracy. For instance, SSTs helped sailors verify their course, find their bearings, and predict stormy weather. NOAA and the National Center for Atmospheric Research (NCAR) maintain collections of SST readings dating back to the early 19th century. The database contains more than 155 million observations from fishing, merchant, research, and navy ships from all over the world. There are many problems with older data sets, however. Until the 1960s, most SST measurements were taken from buckets of water dipped into the ocean. Some water temperatures were measured with Fahrenheit thermometers, and others with Celsius. Subsequent conversion from one scale to the other led to rounding errors. The temperatures of some samples were measured immediately after the buckets were brought on board; other samples sat on board for various lengths of time before being measured. After a great deal of data analyses and quality assessment, NOAA and NCAR have developed standardized SST data sets. NOAA has estimated annual SST data for ocean waters between 60°N and 90°N, from 1880 to 2019 (Fig. 2.1).

The interval from 1880 to 1925 included SSTs that were as much as 0.8°C colder than the long-term average. This cold SST interval was followed by a warm interval from 1925 to about 1962 when SSTs averaged about 0.5°C warmer than the long-term average. SSTs oscillated around the long-term average for the next 30 years and then began climbing after 1992. In the past 30 years, Arctic SSTs have consistently risen and are now more than 1°C warmer than the long-term average. Mean August SSTs from 1982 to 2019 show warming trends over much of the Arctic Ocean; statistically significant (at the 95% confidence interval) linear warming trends of up to +1°C per decade are observed. These warming trends coincide with declining trends in summer sea-ice extent (including late-season freeze-up and early melt), increased solar absorption, and increased vertical ocean heat transport (Timmermans and Ladd, 2018). A marked exception to the prevalent August SST warming trend is the cooling trend (-0.06 ± 0.03°C/yr) documented from the northern Barents Sea. A statistically significant cooling trend in the northern Barents Sea is only observed for the months of August and September, with most other months (January to June) characterized by statistically significant warming trends ($+0.03 \pm 0.01$°C/yr). The result is that annually averaged northern Barents Sea SSTs show a warming trend, in spite of the late summer cooling trend. A record of SSTs in the Chukchi Sea

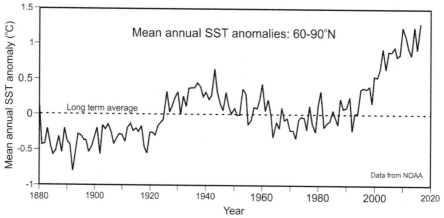

FIG. 2.1 Mean annual SST anomalies for northern seas during the period 1880 to 2018, shown as departures from the long-term average. *SST*, sea surface temperature. (Data from Timmermans, M. L., & Ladd, C., 2018. Arctic Report Card: Update for 2018; Sea Surface Temperature. https://arctic.noaa.gov/Report-Card/Report-Card-2018/ArtMID/7878/ArticleID/779/Sea-Surface-Temperature.)

starting in 1982 shows a gradually increasing trend in temperatures, but with many oscillations above and below the long-term average (Fig. 2.2).

In 2007, Chukchi SSTs rose to almost 3°C above the average, while in 2012, they fell 0.5°C below average (Timmermans and Ladd, 2018). The summer of 2019 saw SSTs on the fringe of the pack ice in the Arctic Ocean that averaged about 2°C. August 2019 mean SSTs ranged from 8 to 9°C in the southern Chukchi and Barents Seas to approximately 1°C in the interior Arctic Ocean near the mean sea-ice edge for that month. These SSTs represent warming of 1.5−2°C above the 1982−2010 average for these seas. Average August 2019 SSTs were likewise 1−7°C warmer than the long-term average in the Beaufort and Laptev Seas, as well as Baffin Bay (Fig. 2.3). The anomalously warm August 2019 SSTs near the edge of the sea ice are linked to an anomalously low sea-ice extent. The reduced ice cover allowed larger areas of the Arctic Ocean to be exposed to solar heating.

However, not every Arctic sea had warmer than average surface temperatures in 2019. The complexity of Arctic SSTs is reflected in the 2019 August data from some of the seas north of Russia. As shown in Table 2.1, SSTs in the Kara Sea and the East Siberian Sea showed little or no change from the long-term average.

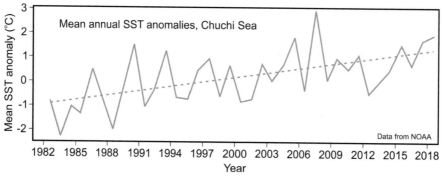

FIG. 2.2 Mean annual SST anomalies for the Chukchi Sea from 1982 to 2018, shown as departures from the long term average. *SST*, sea surface temperature. (Data from Timmermans, M. L., & Ladd, C., 2018. Arctic Report Card: Update for 2018; Sea Surface Temperature. https://arctic.noaa.gov/Report-Card/Report-Card-2018/ArtMID/7878/ArticleID/779/Sea-Surface-Temperature.)

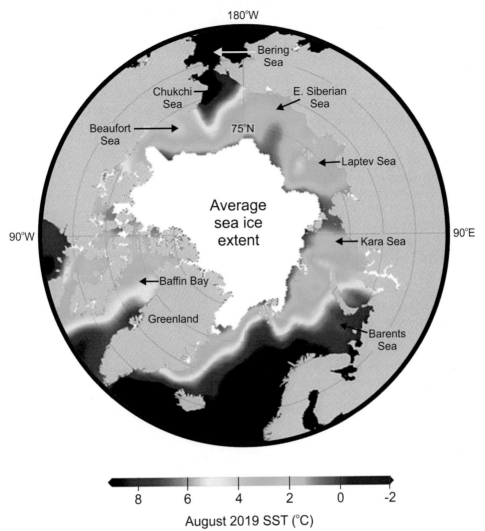

FIG. 2.3 Map of the northern high latitudes, showing August 2019 SSTs for the Arctic Ocean and adjacent seas. *SST*, sea surface temperature. (From National Park Service (US). 2019. Arctic Report Card 2019: Sea Surface Temperatures. https://www.nps.gov/articles/arcticreportcard2019.htm.)

TABLE 2.1
August SSTs Averaged From 1955 to 2012 (Data From World Ocean Atlas, 2013) Compared With 2019 August SSTs (Data From NOAA 2019).

Ocean region	Long-term August SST (°C)	2019 August SST (°C)	Departure from average
Arctic Ocean	−1 in basin; 0 at margins	1 in basin	2
Chukchi Sea	6−7 in south	8−9 in south	+1 to +2
Baffin Bay	1−2 at center; 4 at margins	5	0 to +3
Barents Sea	2 in north; 9 near coast	0 in north; 8−9 near coast	0 to −2
Kara Sea	3.5−4	3.5−4	No change
Laptev Sea	0−1	2−3	0 to +3
East Siberian Sea	0−2	1−2	0 to +1

SST, sea surface temperature.

The Barents Sea had SSTs that were 0.5–2°C colder than the long-term average. As the NOAA (2019) report suggests, the cooler August SSTs in the Barents Sea region require further study. There are aspects of the interplay between sea-ice cover, solar absorption, and lateral ocean heat transport that remain poorly understood. The rise in SSTs is not restricted to the oceans of the Arctic or sub-Arctic. It is, in fact, part of a global problem. Much of the world's attention concerning global warming has focused on increasing GHG levels and their effect on air temperatures. Greater air temperatures warm the ocean surface, generating higher SSTs across much of the world's oceans.

AMPLIFICATION OF TEMPERATURE RISES IN HIGH LATITUDES

As Arctic SSTs rise, the warmer waters melt the sea ice, acting in concert with increasing air temperatures. Thus, increasing SSTs figure into the positive feedback loop illustrated in Fig. 1.2. The greater the loss of sea ice cover, the greater the reduction in albedo over the Arctic Ocean. As more solar energy is absorbed by ice-free ocean waters, the warming process becomes amplified in comparison with lower latitude ocean waters that remain ice-free year-round.

Influx of Warm Waters

Ocean waters are far from static. The Arctic Ocean and adjacent seas are the recipients of warm waters from more southerly latitudes. Major ocean currents move huge volumes of warm water from the tropics to the poles. For instance, the Gulf Stream transports nearly 113,000,000 m^3 of water per second, an amount greater than that carried by all the world's rivers, combined. The Barents Sea is the final recipient of the warm waters carried north by the Gulf Stream. In the Pacific Ocean, the Chukchi Sea receives warm waters carried north from the Pacific through Bering Strait (Timmermans and Ladd, 2018). Once they reach the Arctic, these warm waters are cooled and can take up more CO_2 from the atmosphere. Second, deep waters are formed in high latitudes. This is because cold water is denser than warm water, so the waters cooled by exposure to Arctic conditions increase in density and sink into the ocean's depths, taking with them the CO_2 accumulated at the surface.

PREDICTED ARCTIC SEA SURFACE TEMPERATURES FOR THE NEXT CENTURY

As shown in Fig. 2.4, Arctic SSTs are projected to increase during the rest of this century. Depending on the

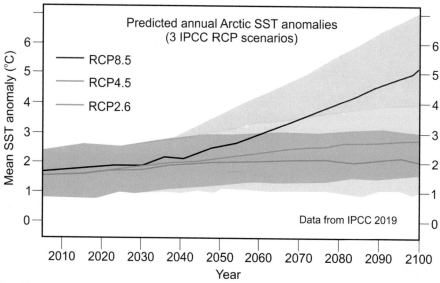

FIG. 2.4 Model predictions of changes in Arctic Ocean SSTs during the 21st century, based on three future greenhouse gas scenarios (RCP 2.6, RCP 4.5, and RCP 8.5). *RCP*, representative concentration pathway; *SST*, sea surface temperature. (Modified from Dahlman, L., & Lindsey, R., 2020. Climate Change: Ocean Heat Content. https://www.climate.gov/news-features/understanding-climate/climate-change-ocean-heat-content Original work published.)

Intergovernmental Panel on Climate Change (IPCC) CO_2 emissions scenario, the model projections show increases in SSTs ranging from about 2°C (RCP 2.6 scenario) to about 5°C (RCP 8.5 scenario) (IPCC, 2019).

The RCP 8.5 scenario presents a substantial increase in global and Arctic surface air temperatures (SATs) as well as Arctic SSTs. Furthermore, increased SSTs are not predicted to be uniform throughout the year. Between the 2000 and 2090s, SSTs in the Arctic Ocean and Siberian seas are predicted to increase by about 2°C in the winter and by about 7°C in the summer, reaching averaged values of about 2–3°C and 5–8°C in the winter and summer respectively.

CHANGES IN THE WATER COLUMN

The modern Arctic Ocean is stratified into three distinct vertical layers: (1) a fresh, cold surface layer that includes a halocline where salinity increases sharply with depth; (2) a warm, saline layer with water of Atlantic provenance; and (3) a cold, saline layer in the abyssal Arctic Ocean. A large portion of the freshwater found in the upper layer of the Arctic Ocean comes from river runoff from adjacent continents and from regional precipitation. The relatively low-saline Pacific water entering the Arctic Basin through the Bering Strait also contributes to the degree of freshness of this upper layer. The warm and saline layer is derived from the North Atlantic. This water enters the Arctic Ocean through Fram Strait and the Barents Sea. However, as discussed by Pemberton and Nilsson (2015, and references therein), climate change is causing a new stratification regime in the upper Arctic Ocean. For instance, recent observations show an increased freshwater content over both the Canadian Basin and the central Arctic Ocean. This is thought to be driven by intensified, large-scale anticyclonic winds over the region, in addition to a freshening of the upper layer from increased river runoff into the Arctic Ocean. River runoff has reportedly increased by about 10% over the past 40 years. The temperature of the Atlantic water (AW) entering the Arctic Ocean has increased in recent decades. This is related to increases in more southerly Atlantic Ocean temperatures. A concurrent increase in both freshwater content and AW temperature has been taking place particularly during the past decade. Freshwater inputs to the Arctic Ocean are expected to increase with about 7% per 1°C of global mean temperature increase. This prediction broadly agrees with future climate projections of increased freshwater inputs linked with temperature increases. These changes may cause a restructuring of the halocline depth. This, in turn, may bring changes in the

depth of the warm Atlantic layer below. If these changes occur, they will alter vertical temperature gradients and heat fluxes, exerting effects on sea-ice cover.

CHANGES IN OCEAN HEAT ENERGY

As reviewed by Dahlman and Lindsey (2020), the oceans play a central role in stabilizing Earth's climate system, because of their ability to absorb large amounts of heat without a large increase in temperature. Thus, ocean water has a tremendous ability to store and release heat over long periods of time. The main source of ocean heat is sunlight. Additionally, clouds, water vapor, and greenhouse gases emit heat that they have absorbed, and some of that heat energy enters the ocean. However, rising levels of greenhouse gases are preventing heat radiated from Earth's surface from escaping into space as freely as it used to. Most of this excess atmospheric heat is being absorbed by the ocean. As a result, upper ocean heat content has increased significantly over the past two decades (Fig. 2.5). In fact, more than 90% of the planetary warming that occurred between 1971 and 2010 has been in the ocean. Averaged over the full depth of the ocean, the 1993–2018 heat gain rates are 0.57–0.81 W per square meter. Shallow ocean waters are heating up more than deep waters. As shown in Fig. 2.5 during the past 25 years, seawater from the surface to 700 m depth has increased in ocean heat content by almost 160×10^{21} J above a 1993 baseline level (Dahlman and Lindsey, 2020).

Waters between 700 and 2000 m depth have also increased in ocean heart content, but only by half as much as shallower waters ($\sim 80 \times 10^{21}$ J). Of course, ocean waters are far from static. They are constantly being mixed by waves, tides, and currents. These actions serve to transfer ocean heat from warmer to colder latitudes and from shallow to deeper levels in the water column. Although there is a great deal of movement of heat within the oceans, it does not disappear. The heat energy eventually reenters the rest of the Earth system by melting polar ice shelves, evaporating surface water, and directly reheating the atmosphere. Thus, heat energy in the ocean can warm the planet for decades after it was initially absorbed. If the ocean absorbs more heat than it releases, its heat content increases. Increasing ocean heat content is contributing to sea-level rise, ocean heat waves, and the melting of ocean-terminating glaciers and ice sheets around Greenland and Antarctica. Also, as discussed earlier, the Arctic Ocean and adjacent seas are not immune to the effects of global ocean warming. On the contrary, the AW, which enters the Arctic Ocean by way of Fram Strait

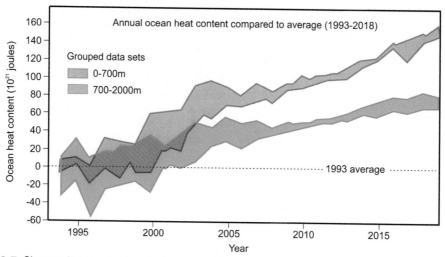

FIG. 2.5 Changes in ocean heat content, averaged for the world's oceans, between 1993 and 2018. (Data from NOAA/NCEI, 2018. Reporting on the State of the Climate in 2017: Sea Surface Temperatures Near-Record High. https://www.ncei.noaa.gov/news/reporting-state-climate-2017.)

and the Barents Sea, and the Pacific waters, which enter through the Bering Strait, are now transporting greater ocean heat energy to the Arctic. From the Pacific Ocean, a difference in mean sea level of about 0.4 m between the Bering Sea and the Arctic Ocean drives a net northward transport of water at about 0.8×10^6 m per second, or 0.8 Sverdrup, through Bering Strait and into the Beaufort Sea (Stabeno and Schumacher, 1999). A Sverdrup is a volumetric flow rate equal to 1 million m^3 per second. The Barents Sea branch of AW loses most of its heat to the atmosphere, as the water travels over the shallow continental shelves of the Barents and Kara Seas, whereas the Fram Strait branch supplies oceanic heat to the warm AW layer (about 200−700 m depth) of the Arctic Ocean. Also, the sill connecting the Arctic and Fram Strait is 2545 m deep, so Fram Strait is the only route by which deep water can be exchanged between the Atlantic and Arctic Oceans. A study of the rates of AW transport through Fram Strait from 1997 to 2010 by Beszczynska-Moeller et al. (1997) shows that the long-term average volume of ocean water transported northward through Fram Strait is about 6.6 Sverdrups (sv) (Fig. 2.6, upper). This flow delivers an average of about 3.0 sv of AW. The mean temperature of this AW inflow is about 3.1°C (Fig. 2.6, lower).

This added warmth may cause Arctic Ocean warming at middepth by intensifying the upward trend of ocean heat supply to the Arctic Ocean through Fram Strait. The

northward transport rates of water through Fram Strait vary as much as 5 sv in any given year, and annual average transport rates vary from 2 to 4 sv. The flow of AW is greatest during the winter months (ca. 4 sv) and least during the summer (ca 2 sv). The Nordic Seas play the role of gatekeeper for these processes. As depicted in Fig. 2.7, the decline in sea ice extent in the Arctic Ocean causes a reduction in the southward export of sea ice through Fram Strait.

This reduction in freshwater input to the Greenland Sea causes increased salinity there. Consequently, the sea surface height decreases, and the gyre circulation strengthens in the Nordic Seas. These changes then drive an increase in the flow of AW into the Nordic Seas and the Arctic Ocean, bringing about significant warming in the AW layer of the Arctic Ocean. The changes in the Arctic Ocean heat budget may accelerate sea-ice decline through basal melting (Wang et al., 2020). There is independent evidence of this warming of Arctic Ocean waters. Data from NOAA (2018a) indicate that most of the Arctic Ocean basin gained 2−3 GJ (10^9 J) of heat energy per square meter in the upper 700 m of the water column during 2018 when compared with the long-term average.

THE BEAUFORT GYRE

The Beaufort Gyre (BG) is an anticyclonic (clockwise) sea ice-ocean circulation system, driven by the

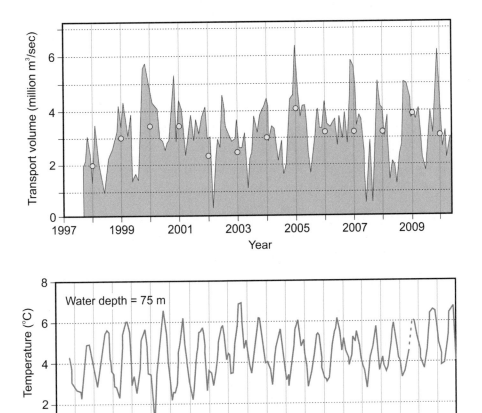

FIG. 2.6 Upper: Graph of seasonal changes in AW transport through Fram Strait, 1997–2010. *Yellow dots* represent the average of lowest and highest values for each year. Lower: Temperature of warm North Atlantic Water at a depth of 75 m as it passes through Fram Strait. (Modified from Wang, Q., Wekerle, C., Wang, X., Danilov, S., Koldunov, N., Sein, D., Sidorenko, D., von Appen, W. J., & Jung, T., 2020. Intensification of the Atlantic water supply to the Arctic Ocean through Fram Strait induced by Arctic sea ice decline. Geophysical Research Letters, 47(3). https://doi.org/10.1029/2019GL086682.)

semipermanent Beaufort Sea high-pressure system. It is the dominant sea ice and ocean surface circulation feature of the western Arctic Ocean. A recent study by Armitage et al. (2020) has revealed the effects of recent Arctic Ocean sea-ice loss on the BG. For the past few decades, the BG has rotated in a clockwise direction, driven by the wind. When ice covers the surface of the ocean, the effects of wind on the gyre are diminished. As the extent and thickness of sea ice cover have declined, wind energy has strengthened the rotation of the gyre. The energized gyre is more effective at corralling and retaining fresh water at the surface. Thus, the gyre has become a major reservoir of freshwater. Since the 1990s, the BG has accumulated around 8000 km³ of additional freshwater, compared with the

long-term average (Armitage et al., 2020). When ocean currents spin, they draw surface waters toward their center. As discussed earlier, freshwater enters the Arctic Ocean basin from melting ice (both terrestrial and sea ice), river runoff, and precipitation. When freshwater reaches the middle of the gyre, it is forced downward. As more freshwater moves toward the center of the gyre, the halocline, the interface between the fresh surface water and the salty water beneath, should be driven deeper in the water column. But the study by Armitage et al. (2020) shows that despite additional freshwater being forced down at the center of the gyre, the depth of the halocline is only slightly increasing. Armitage et al. (2020) propose that an increase in eddy activity could account for the discrepancies in both the

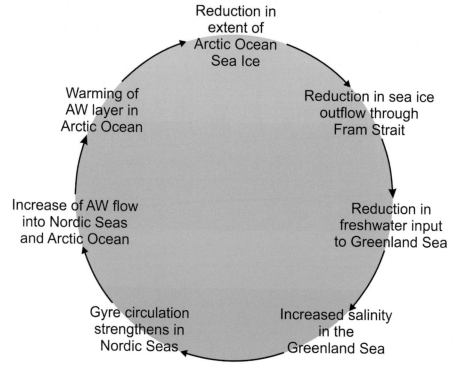

FIG. 2.7 Flowchart showing positive feedback loop of declining sea ice leading to increased input of AW into Arctic Ocean. *AW,* Atlantic water.

freshwater budget and the energy budget. Eddies are circular movements of water, spinning counter to the main current and causing small whirlpools. Eddies may help release the extra freshwater as well as dissipate energy from the gyre. Increased eddy activity has important implications for the hydrology of the Arctic Ocean because increased eddy activity will cause additional mixing of heat, salt, and nutrients throughout the water column. In other words, the Arctic Ocean may become less stratified if eddies stir up the layers. Average temperatures of the Arctic Surface Water range from −1.9 to −1°C. AW entering the Arctic Basin has an average temperature of 3°C, and Arctic Deep Water has a temperature range of −0.8−2°C. If these deep waters are brought to the surface, the extra heat they carry would lead to additional melting of the ice pack. Also, additional mixing of the water column would bring new and different mixtures of nutrients to the surface. This would disrupt marine ecosystems because of changes in the annual timing of nutrient release and in the quantities of nutrients available near the surface. Finally, changes in the BG could alter the amount of cold, freshwater transported out of the Arctic by various currents. This reduction of cold freshwater might disrupt the

warm North Atlantic current, causing a major cooling event in Europe. This is one of the main topics of the next chapter.

CONCLUDING THOUGHTS

Probably the most important "take-home message" from this chapter is that the Arctic Ocean, despite its remote location, is not isolated. Rather, it forms the most northerly part of an interconnected system of the world's oceans. Most of the ocean regions of the world are getting demonstrably warmer; they are holding increasing amounts of heat. This heat energy starts at the surface and has slowly worked its way down to depths of at least 2000 m. This heat must go somewhere. Because of the system of oceanic currents, warm water tends to flow toward the poles. As we have seen in this chapter, warm oceanic water enters the Arctic Ocean from the Pacific side through Bering Strait. The Arctic Ocean receives warm water from the Atlantic side by way of Fram Strait and the Barents Sea. Once this warm water has entered the Arctic Ocean, it begins to melt sea ice and it also undermines the margins of glaciers that reach the sea. This becomes part of a positive feedback system wherein ocean heat from the South contributes to the warming of the Arctic Ocean to the extent that Arctic waters become ice-free and warm-up further because of reduced albedo.

REFERENCES

Armitage, T.W.K., Manucharyan, G.E., Petty, A.A., Kwok, R., Thompson, A.F., 2020. Enhanced eddy activity in the Beaufort Gyre in response to sea ice loss. Nature Communications 11 (1). https://doi.org/10.1038/s41467-020-14449-z.

Beszczynska-Moeller, A., Fahrbach, E., Schauer, U., Hansen, E., 1997. Variability in Atlantic water temperature and transport at the entrance to the Arctic Ocean. ICES Journal of Marine Science 69.

Dahlman, L., Lindsey, R., n.d. Climate Change: Ocean Heat Content. https://www.climate.gov/news-features/understanding-climate/climate-change-ocean-heat-content (Original work published 2020).

IPCC, 2019. In: Shukla, P.R., Skea, J., Calvo Buendia, E., Masson-Delmotte, V., Pörtner, H.O., et al. (Eds.), Climate Change and Land: an IPCC special report on climate change, desertification, land degradation, sustainable land management, food security, and greenhouse gas fluxes in terrestrial ecosystems.

NOAA, 2019. Arctic Sea Surface Temperatures, 2019. https://arctic.noaa.gov/Report-Card/Report-Card-2019/ArtMID/7916/ArticleID/840/Sea-Surface-Temperature (Original work published 2019).

Timmermans, M.L., Ladd, C., 2018. Arctic Report Card: Update for 2018; Sea Surface Temperature. https://arctic.noaa.gov/Report-Card/Report-Card-2018/ArtMID/7878/ArticleID/779/Sea-Surface-Temperature.

Wang, Q., Wekerle, C., Wang, X., Danilov, S., Koldunov, N., Sein, D., Sidorenko, D., von Appen, W.J., Jung, T., 2020. Intensification of the Atlantic water supply to the Arctic Ocean through Fram Strait induced by Arctic sea ice decline. Geophysical Research Letters 47 (3). https://doi.org/10.1029/2019GL086682.

World Ocean Atlas, 2013. Ocean Data View — August Sea Surface Temperatures. Original work published. https://odv.awi.de/en/data/ocean/world_ocean_atlas_2013/.

Oppenheimer, M., Glavovic, B.C., Hinkel, J., 2019. sea level rise and implications for low-lying Islands, Coasts and communities. In: Pörtner, H.-O., Roberts, D.C., Masson-Delmotte, P. (Eds.), IPCC Special Report on the Ocean and Cryosphere in a Changing Climate.

Stabeno, P.J., Schumacher, J.D., 1999. The physical oceanography of the Bering Sea: a summary of physical, chemical, and biological characteristics, and a synopsis of research on the Bering Sea. In: Dynamics of the Bering Sea: A Summary of Physical, Chemical, and Biological Characteristics, and a Synopsis of Research on the Bering Sea, vol. 22.

FURTHER READING

NOAA, 2018. Data Snapshot of Ocean Heat Anomaly, 2018. https://www.climate.gov/maps-data/data-snapshots/heat contentanomaly-annual-ncei-2018-00-00?theme=Oceans (Original work published 2018).

NOAA, 2018. State of the Climate, 2018. https://www.ncdc.noaa.gov/sotc/global/201813.

NOAA, 2018. Arctic Report Card: Update for 2018; Sea Surface Temperature. https://arctic.noaa.gov/Report-Card/Report-Card-201.

NOAA, 2020. State of the Climate. Retrieved February 8, 2020, from https://www.ncdc.noaa.gov/sotc/global/201813 (Original work published 2018)

NOAA, Timmermans, M.L., Ladd, C., 2020. Arctic Report Card: Update for 2018; Sea Surface Temperature. Retrieved February 2, 2020, from. https://arctic.noaa.gov/Report-Card/Report-Card-2018/ArtMID/7878/ArticleID/779/Sea-Surface-Temperature (Original work published 2018)

NOAA, Dahlman, L., Lindsey, R., 2020. Climate Change: Ocean Heat Content. Retrieved February 6, 2020, from. https://www.climate.gov/news-features/understanding-climate/climate-change-ocean-heat-content (Original work published 2020)

NOAA, Timmermans, M.L., Proshutinsky, A., 2015. Arctic Report Card: Update for 2015: Sea Surface Temperatures. https://arctic.noaa.gov/Report-Card/Report-Card-2015/ArtMID/5037/ArticleID/220/Sea-Surface-Temperature.

NOAA/NCEI, 2018. Reporting on the State of the Climate in 2017: Sea Surface Temperatures Near-Record High. https://www.ncei.noaa.gov/news/reporting-state-climate-2017.

National Park Service (US), 2019. Arctic Report Card 2019: Sea Surface Temperatures. https://www.nps.gov/articles/arcticreportcard2019.htm.

Pemberton, P., Nilsson, J., 2015. The response of the central Arctic Ocean stratification to freshwater perturbations. Journal of Geophysical Research: Oceans 121.

Xie, S.P., Deser, C., Vecchi, G.A., Ma, J., Teng, H., Wittenberg, A.T., 2010. Global warming pattern formation: sea surface temperature and rainfall. Journal of Climate 23 (4), 966—986. https://doi.org/10.1175/2009JCLI3329.1.

CHAPTER 3

Changes in Ocean Circulation Patterns

INTRODUCTION

The Arctic Ocean is getting fresher. In other words, more freshwater is entering the Arctic Basin than had entered it previously. The sources of this freshwater are twofold. One is simply increased precipitation over the Arctic Ocean, both in the form of snow in winter and as rain in the summer. These changes and precipitation are mainly due to global warming which is causing more storminess in the high latitudes. The other source of additional freshwater entering the Arctic Basin is runoff from streams flowing north into the Arctic Basin from the adjacent continents. While the Arctic Ocean represents only about 2% of the global ocean in terms of volume and surface area, it receives a disproportionate amount of freshwater from rivers. In fact, the Arctic Ocean collects over 11% of global river discharge. Some of this additional runoff is coming from increased precipitation over the land, and some of it is due to the melting of permafrost in these regions, which is releasing moisture into the rivers that flow into the Arctic Ocean and adjacent seas. The winds over the Arctic Ocean are getting stronger. In the past, the force of the winds over the surface waters of the Arctic Ocean was weakened by thick layers of pack ice in and around the gyre. Indeed, the seasonal sea-ice cover, blanketing the entire Arctic Ocean, has been one key to the remarkable quietness of the Arctic Ocean. The ice dampens surface and internal waves and decreases the transfer of wind momentum to the water. But as the ice thins and the margins retreat toward the North Pole, the effect of the winds is increasing. For instance, prior to the past 20 years of diminishing sea-ice cover, the Beaufort Gyre (BG) in the Western Arctic Ocean (Fig. 3.1) stored about 20,000 km^3 of freshwater in the upper few hundred meters of the water column. However, global change has ushered in a new regime of increased freshwater storage in this region, accompanied by increasing wind speeds. Fig. 3.2 shows the changes in mean monthly wind speeds before and after 2007.

The general trend of low wind speeds in winter and high winds from August to November is seen in both sets of data, but the post-2007 record shows marked increases in wind speeds in both summer and autumn

(Armitage et al., 2020). Because of the rapid loss of sea ice, especially the older and thicker perennial ice, the sea surface in the western Arctic has risen by around 15 cm and now holds 8000 km^3 of additional freshwater at or near the surface. This represents around 10% of all the freshwater in the Arctic Ocean and a 40% increase over freshwater levels in the Western Arctic Ocean in the 1970s. The increase in freshwater content of the western Arctic Ocean during the past 25 years is shown in Fig. 3.3 (Giles et al., 2012; Proshutinsky et al., 2019).

This additional lid of freshwater is associated with a tendency for deepening of the halocline, as shown in Fig. 3.4. In the 1970s, the layer of relatively fresh water was confined to the upper 75 m of the water column.

By 2017, freshwater (salinity <32 parts per 1000) extended to depths of 125 m (Proshutinsky et al., 2019). This Arctic freshwater cannot keep accumulating indefinitely. Sooner or later, the oceanic system will act to balance itself by other processes, as discussed in the following.

Changes in the Beaufort Gyre

Why is so much freshwater pooling in the western Arctic BG? Some models suggest that wind-driven convergence drives this freshwater accumulation. Giles et al. (2012) used continuous satellite measurements taken from 1995 to 2010 to show that the dome in sea surface height associated with the BG has been steepening, indicating the spin-up of the gyre (Fig. 3.5).

Modeling experiments suggest that freshwater accumulates in the BG during anticyclonic regimes and that it is forced to the Arctic Ocean margins during cyclonic regimes, from which it may then be released to the North Atlantic. Giles et al. (2012) found that the trend in wind field curl exhibits a corresponding spatial pattern, suggesting that wind-driven convergence does indeed control freshwater variability. Wind field curl is a measure of spatial gradients in winds that lead to water convergence or divergence. The authors concluded that a future reversal in the wind field (i.e., from counterclockwise to clockwise) could lead to a spin-down of the BG, causing the release of this

Threats to the Arctic. https://doi.org/10.1016/B978-0-12-821555-5.00001-2
Copyright © 2021 Elsevier Inc. All rights reserved.

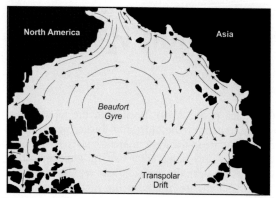

FIG. 3.1 Map of the Arctic Ocean region, showing the location of the Beaufort Gyre and other Arctic Ocean currents. (Modified from Fig. 3.29 in the Arctic Monitoring and Assessment Program, 1998. Arctic Council, 1998, in public domain.)

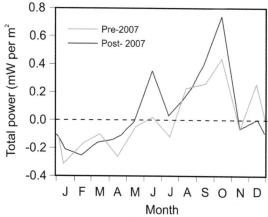

FIG. 3.2 Beaufort Gyre wind speeds, pre- and post-2007. (Modified from Armitage, T.W., Manucharyan, G.E., Petty, A.K., Kwok, R., Thompson, A.F., 2020. Enhanced eddy activity in the Beaufort Gyre in response to sea ice loss. Nature Communications 11, 761. https://doi.org/10.1038/s41467-020-14449-z.)

freshwater to the Arctic Ocean. Using continuous satellite radar altimetry data from the Earth Remote Sensing (ERS-2) (1995–2003) and Envisat (2002–10) satellites, Giles et al. (2012) analyzed the trends in sea surface height (SSH) between 1995 and 2010, from which they calculated both the change in freshwater storage and the corresponding trends in the wind field curl. The BG region of the Arctic Ocean shows a rise in sea level of about 2 cm per year, in comparison with the geoid (the hypothetical shape of the earth,

coinciding with mean sea level and its imagined extension in relation to land areas). On the other hand, Morison et al. (2012) argue that other factors must be affecting the freshwater content of the western Arctic Ocean. They used observations to show that during a time of record reductions in ice extent from 2005 to 2008, the dominant freshwater content changes were an increase in the Canada basin balanced by a decrease in the Eurasian Basin. They drew their observations from satellite data (sea surface height and ocean-bottom pressure) and from in situ data. They concluded that the freshwater changes were due to a cyclonic (counterclockwise) shift in the ocean pathway of Eurasian runoff forced by strengthening of the west-to-east Northern Hemisphere atmospheric circulation characterized by an increased Arctic Oscillation (AO) index. Their results reinforce the importance of runoff as an influence on the Arctic Ocean. They also concluded that the spatial and temporal manifestations of the runoff pathways are modulated by the AO, rather than the strength of the wind-driven BG circulation.

Arctic Oscillation

As summarized by Dahlman (2009), the AO is a large-scale mode of climate variability. It is also called the Northern Hemisphere annular mode. The AO is a climate pattern characterized by winds circulating counterclockwise around the Arctic, centered around 55°N latitude. The most obvious indicator of the phases of this oscillation is the north-to-south location of the storm-steering, midlatitude jet stream. Fig. 3.6 (NOAA Climate Prediction Center, 2020) shows the history of the AO over the interval 1995–2000.

The length of the dark blue columns reflects the relative strength of the positive phase of the AO; the turquoise columns show the relative strength of the negative phase. As shown in the figure, the AO switched modes 58 times in the past 25 years. When the AO is in its positive phase, this ring of strong winds circling the North Pole acts to restrict colder air from leaving the polar regions. The AO's positive phase is characterized by lower-than-average air pressure over the Arctic, paired with higher-than-average pressure over the north Pacific and Atlantic Oceans. The jet stream is farther north than average under these conditions, and storms are typically shifted north of their usual paths. Thus, the midlatitudes of North America, Europe, Siberia, and East Asia generally see fewer cold air outbreaks than usual during the positive phase of the AO. The AO's negative phase has higher-than-average air pressure over the Arctic and lower-than-average pressure over the north Pacific and Atlantic

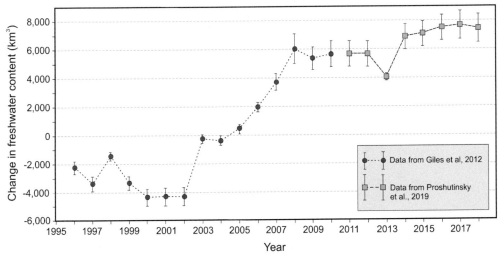

FIG. 3.3 Change in Western Arctic freshwater content 1995–2010. Error bars are the 1δ uncertainty. Data for 1995–2010 (*red circles*) from Giles et al., 2012. Data for 2011–18 (*green boxes*) from Proshutinsky et al. (2019).

Oceans. The jet stream shifts toward the equator under these conditions, so the globe-encircling river of air is south of its average position. The midlatitudes of the Northern Hemisphere are more likely to experience outbreaks of frigid, polar air during winters when the AO is negative. The contention of Morison et al. (2012) is that these shifts in the jet stream and their associated impacts on Arctic climatology are at least as important in the accumulation of freshwater in the western Arctic Ocean as wind-driven convergence.

Effects of Ekman Transport and Pumping

To make sense of the effects of winds on the speed and direction of ocean currents, we must delve briefly into the world of physical oceanography. As reviewed by Chereskin and Price (2019, and references therein), winds on the surface of the ocean exert forces that set the oceans in motion, producing both currents and waves. Winds that exert a normal force (i.e., winds striking the ocean surface at right angles) create waves. Tangential wind forces cause frictional stresses on ocean water, as the winds pull across the sea surface. These tangential winds help drive ocean currents. As described in NASA (2020), if Earth were stationary (i.e., if it did not rotate on its axis), frictional coupling between moving air and the ocean surface would push a thin layer of water in the same direction as the wind. This surface layer would

then drag the layer beneath it, putting it into motion. This motion would propagate down through the water column, with each successive layer moving forward at a slower speed than the layer above it. However, because of Earth's rotation, the shallow surface water set in motion by the wind is deflected to the right of the wind direction in the Northern Hemisphere and to the left of the wind direction in the Southern Hemisphere. This deflection is called the Coriolis effect. This effect causes each layer of water put into motion by the layer above to shift direction because of Earth's rotation (Fig. 3.7), creating what is called the Ekman spiral.

Due to the combined effects of wind force and the Coriolis effect, these frictional stresses may flow in directions quite different from the wind direction. These wind-driven currents are called Ekman layer currents. Ekman layer currents are quite far-reaching but rather weak in comparison with other types of currents. Ekman layer currents typically move water only 0.05–0.1 m per second. These currents are small in magnitude and in vertical extent, typically acting on only the upper ocean water layers. The Ekman spiral dictates that each moving layer is deflected to the right of the movement in the overlying layer. Therefore, the direction of water movement changes with increasing depth. In an ideal case, a steady wind blowing across an ocean of unlimited depth and extent would cause

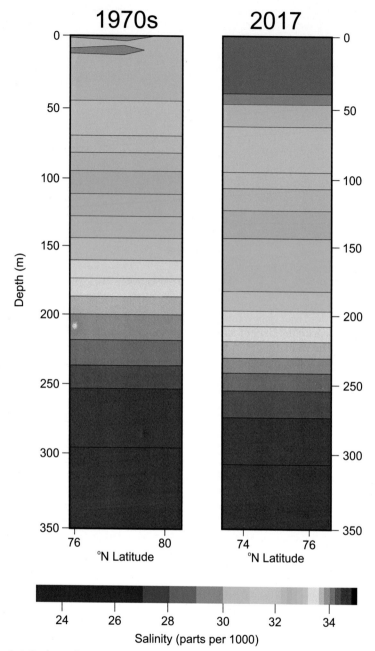

FIG. 3.4 Depth-latitude sections of salinity measurements taken along a transect from 140 to 150°W in the western Arctic Ocean in the 1970s and 2017, data from Proshutinsky et al. (2019).

surface waters to move at an angle of 45° to the right of the wind in the Northern Hemisphere. Each successive layer moves more toward the right and at a slower speed (Fig. 3.7). At depths between 100 and 150 m, the Ekman spiral has gone through less than half a turn. Yet the water at these depths moves so slowly (about

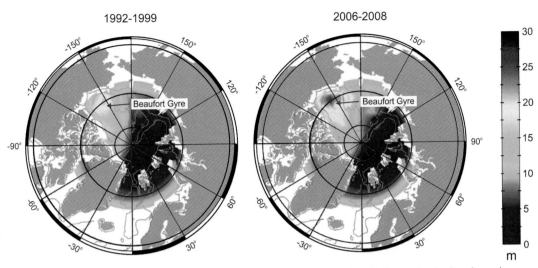

FIG. 3.5 The dome in sea surface height associated with the Beaufort Gyre (yellow and red colored areas), as observed in 1992–1999 versus 2006–08. The maximum height in recent years has been about 27 m. (Figure modified from Aksenov, Y., Karcher, M., Proshutinsky, A., Gerdes, R., De Cuevas, B., Golubeva, E., Kauker, F., Nguyen, A.T., Platov, G.A., Wadley, M., Watanabe, E., Coward, A.C., Nurser, A.J.G., 2016. Arctic pathways of Pacific water: arctic ocean model intercomparison experiments. Journal of Geophysical Research: Oceans 121 (1), 27–59. https://doi.org/10.1002/2015JC011299, in public domain.)

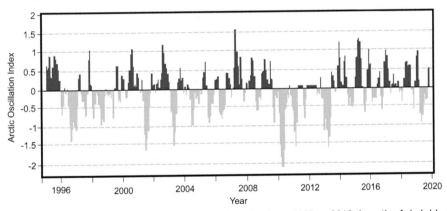

FIG. 3.6 Monthly values for the Arctic Oscillation (AO) index from 1995 to 2019. Length of dark blue bars reflects the strength of positive AO values; length of turquoise bars reflects the strength of negative AO values. (Data from NOAA Climate Prediction Center, 2020. Arctic Oscillation. https://www.cpc.ncep.noaa.gov/products/precip/CWlink/daily_ao_index/ao.shtml. (Original work published 2020).)

4% of the speed of the surface current) against the direction of the surface wind that the surface wind energy dissipates at these depths, setting the lower limit of the wind's influence on ocean movement. Another factor that limits the operational depth of the Ekman spiral is the pycnocline. This is an ocean layer in which water density increases rapidly with depth. A stable pycnocline inhibits the transfer of kinetic energy to deeper waters, helping to contain wind-driven currents to the mixed layer. Thus, the pycnocline acts as a boundary for Ekman transport and surface currents. The net transport of water due to coupling between wind and surface

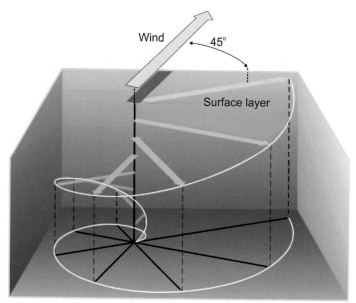

FIG. 3.7 The Ekman spiral describes how the horizontal wind sets surface waters in motion. As represented by horizontal vectors, the speed and direction of water motion change with increasing depth. (Image modified from NASA, 2020. Ocean Motion and Surface Currents. http://oceanmotion.org/html/background/ocean-in-motion.htm, in public domain.)

waters is called Ekman transport. The mass transport of water in the Ekman layer, when integrated across the width of an ocean, is considerable. For instance, the mass transport of water in Ekman layer currents is comparable in scale to the transport of major ocean currents such as the Gulf Stream. Ekman layer currents result in the exchange of surface and deeper ocean waters, a phenomenon known as Ekman pumping. In regions where Ekman transport converges, water is pumped from the surface layer into the ocean interior; in regions where the Ekman transport diverges, water is pumped up to surface from down below. This wind-driven exchange of surficial and deep ocean waters ultimately drives the interior circulation of ocean waters.

EKMAN TRANSPORT

Ekman transport piles up surface water in some areas of the ocean and removes water from other areas, producing variations in SSH. This causes the ocean surface to slope gradually. A sloping ocean surface generates water pressure gradients across ocean surfaces. Just as air flows from regions of high pressure to regions of low pressure,

oceanic pressure gradients give rise to geostrophic flow in which water moves from regions of higher water pressure to regions of lower pressure (NASA, 2020). Ekman pumping also transports nutrients and heat energy from the upper thermocline toward the sea surface. The pattern of converging and diverging Ekman layer currents thus exerts very strong influences on the strength and shape of the major oceanic current gyres, as well as on seasonal biological productivity. The description of these ocean current mechanisms brings us back to the situation in the BG. As we have seen, the BG freshwater content has increased since the 1990s, potentially stabilizing in recent years. As discussed earlier, freshwater cannot simply keep accumulating in the BG indefinitely. Sooner or later, the system will act to balance itself by other processes. The mechanism favored by Armitage et al. (2020) to explain the stabilization of this thick lens of freshwater is based on observational estimates of the BG mechanical energy budget. These estimates indicate that energy dissipation and freshwater content stabilization by eddies increased in the late-2000s. An eddy is a circular current of water (NASA, 2012) (Fig. 3.8).

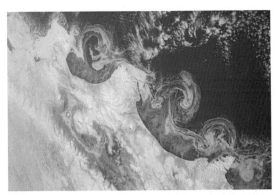

FIG. 3.8 Eddies near the coast of Kamchatka, Eastern Siberia, revealed by ice floes entrained in the currents, photographed by an Expedition 30 crew member on the International Space Station. (Photo courtesy of NASA, 2012. Ice Floes Along Kamchatka Coastline. http://spaceflight.nasa.gov/gallery/images/station/crew-30/html/iss030e162344.html.)

EDDIES AND SEA ICE

Typically, the swirling motion of eddies in the ocean causes colder, deeper waters to come to the surface. The loss of sea ice and the acceleration of ocean currents after 2007 resulted in enhanced mechanical energy input but without corresponding increases in potential energy storage. At least part of this energy must have been dissipated through the formation of eddies in the BG. The authors concluded that eddy dissipation must have increased after 2007. Likewise, they infer that eddies have played an increasing role in BG stabilization. Their results suggest that declining Arctic sea ice cover is contributing to an increasingly energetic BG with eddies playing a greater role in its stabilization. How do eddies affect sea ice distribution? As described by Manucharyan and Thompson (2017), sea ice tends to accumulate in cyclonic eddies (counterclockwise flowing in the Northern Hemisphere). This is because cyclonic eddies are associated with the piling up of water due to winds. These same winds move sea ice, causing it to build up as well. As sea ice is pushed by cyclonic eddy currents from cold water to warmer (previously ice-free) waters, a great deal of heat transfer takes place between the ocean and ice. In contrast to this, anticyclonic eddies typically transport lower concentrations of ice, resulting in less heat transfer. Another aspect of the study by Armitage et al. (2020) deals with modeling the potential outcomes of the lengthening of the ice-free season in the BG region. As discussed in Chapter 2, this part of the Arctic Ocean (like many others) is becoming ice-free earlier in summer and sea ice formation is beginning later in the autumn. Because of the newly developing persistence of open water in the BG for a greater proportion of the year, the current anticyclonic winds will do significantly more work on the ocean surface currents. Meanwhile, year-round dissipation of energy underneath sea ice will also increase as currents speed up. Beyond the confines of the BG, the entire Arctic Ocean is predicted to continue to become more energetic. In this more energetic Arctic Ocean, increases in energy diffusion by eddies, coupled with increased mixing of the water column by eddy activity, are expected to enhance the upwelling of warm Atlantic water in the Arctic Ocean, with consequences for sea ice growth and mixing of biogeochemical tracers.

The Warm Arctic–Cold Continents Pattern

As discussed in the first two chapters, recent decades have seen rapid Arctic warming and sea-ice loss during winter. At the same time, the midlatitude continents have experienced an increase in severe cold winters. This phenomenon has been called the "warm Arctic–cold continents pattern." Some studies have assumed a causal link between reduced sea ice or Arctic warming and cold midlatitude winters (e.g., Tang et al., 2013). This has led to predictions that continued Arctic sea-ice loss could lead to an increased frequency of severely cold winters over the midlatitudes, despite rising global temperatures. As described by Tang et al. (2013), climatic and sea-ice observations from the period 1979–2012 indicate that in some years a high-pressure anomaly, associated with winter sea ice reduction, prevails over the subarctic. In part, this results from fewer cyclonic storms owing to a weakened gradient in sea surface temperatures (SSTs) and lower baroclinicity over sparse sea ice. In this context, baroclinicity relates to the state of ocean waters in which surfaces of constant pressure intersect those of constant density. The results suggest that the winter atmospheric circulation in the Arctic, associated with sea ice loss, favors the occurrence of cold winter extremes at the middle latitudes of the northern continents. Blackport et al. (2019) present evidence from observations and coupled climate model simulations that contradicts this hypothesis. In their view, reduced sea ice is not the cause of cold midlatitude winters. Rather, their study concluded that both phenomena are being driven by the same large-scale atmospheric circulation anomalies.

Freshwater Contributions from Glaciers and the Greenland Ice Sheet

As shown in Table 3.1, observational and modeling studies show that the freshwater budget of the Arctic and sub-Arctic North Atlantic Oceans has been changing over the past few decades (Aagaard and Carmack, 1989; Bamber et al., 2018 and references therein; Carmack et al., 2016; Haine et al., 2015), with possible consequences for the stability of North Atlantic Ocean circulation and for the stratification of the upper ocean. General circulation models (GCMs) of the oceans have thus far disregarded the freshwater entering the system from ice sheets and glacial meltwater.

This is partly due to relatively poor modeling constraints on this meltwater component in the past. To date, land ice has not been an interactive component in Earth system models, further hampering efforts to include it in ocean analyses. However, Bamber et al. (2018) show that from the mid-1990s, there has been a steady increase in the freshwater flux from land ice. This flux peaked around 2010 but has remained well above the long-term average since then. Because of the relative stability of Greenland ice sheet (GrIS) and the glaciers of the Canadian Arctic prior to the 1990s, freshwater flux (i.e., meltwater) from land ice was a relatively small component of the overall Arctic freshwater budget. This is no longer the case. A reconstruction of the surface mass balance (SMB) of the GrIS for the years 1900–2015 indicates only modest variability in runoff for most of the 20th century. However, since 1995, the GrIS has seen unprecedented increases in runoff, exceeding any other interval during the past 115 years, in terms of both absolute magnitude and rate of change (Bamber et al., 2018 and references therein). This increased runoff coincides roughly with an increase in discharge from the GrIS. This mass loss has caused a freshwater flux anomaly in ocean regions close to areas of overturning circulation in the sub-Arctic North Atlantic, and the flux has been increasing steadily since 1995. It is perhaps significant that the Arctic has experienced an atypical period of long-term anticyclonic circulation during this same time frame. This circulation pattern has been associated with changes in the freshwater balance of the Arctic and sub-arctic North Atlantic Oceans. To put things into perspective, consider that while these freshwater anomalies are relatively small in terms of their absolute flux, they may play a central role in the changing climate of the region.

The Younger Dryas Model of North Atlantic Cooling

Some oceanographers predict that future accumulations of freshwater in the Arctic will have large impacts on North Atlantic Ocean circulation. Paleoclimate data

TABLE 3.1
Freshwater Budgets for the Arctic Ocean.

Period	Prior to 1989[1]	1980–2000[2]	2000–10[2]	2018[3]
FRESHWATER RESERVOIRS (KM³)				
Liquid freshwater	80,000	93,000	101,000	
Sea ice	17,300	17,800	14,300	
Total freshwater	97,300	110,800	115,300	
FLUXES (KM³ PER YEAR)				
Runoff	3,300	3,900 ± 390	4,200 ± 420	6,300 ± 316
Bering Strait	1,670	2,540 ± 300	2,640 ± 100	
Total inflow	6,120	8,800 ± 530	9,400 ± 940	
Fram Strait water	−980	−2,700 ± 530	−2,800 ± 420	
Fram Strait ice	−2,790	−2,300 ± 340	−1,900 ± 280	
Davis Strait	−920	−3,360 ± 320	−3,220 ± 190	
Total outflow	−5,520	−8,700 ± 700	−8,250 ± 550	

[1] Aagaard and Carmack, 1989.
[2] Haine et al., 2015.
[3] Bamber et al. (2018).

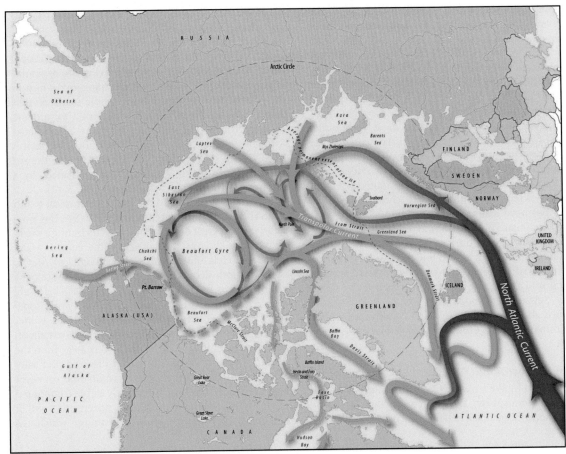

FIG. 3.9 Map of the northern high latitudes, showing the dominant ocean currents of the region. *Red arrows* indicate currents of warm, salty water. *Blue arrows* indicate currents of cold, fresher water. (Image © Woods Hole Oceanographic Institution, Jack Cook.)

show that such large changes to Atlantic Ocean circulation patterns have happened in the past, due to the influx of large volumes of meltwater from continental ice sheets (Overland et al., 2018). There is less certainty about the source of that freshwater and the route it took to get to the North Atlantic. One hypothesis, advanced by Broecker et al. (1989), was that large volumes of freshwater from Lake Agassiz emptied into the North Atlantic in one or more catastrophic floods. Lake Agassiz was an enormous glacial meltwater lake situated south and west of the retreating Laurentide ice sheet. At its largest, it covered over 300,000 km^2, including parts of Manitoba, Ontario, and Saskatchewan in Canada as well as parts of Minnesota and North Dakota in the United States. The lake waxed and waned over several thousand years, undergoing a series of lake

stands, or phases. The start of the Moorhead Phase is synchronous with the start of the Younger Dryas (YD) chronozone. Broecker et al. (1989) proposed a model in which the sudden flooding of the North Atlantic by Lake Agassiz water during the Moorhead Phase sent an estimated 30,000 m^3 per second of cold, freshwater into the North Atlantic via the Gulf of St. Lawrence. This shot of cold freshwater would have acted to turn off the North Atlantic's conveyor belt circulation system. The authors attributed this turnoff to a sudden reduction in surface water salinity and density of the waters in the region where North Atlantic Deep Water (NADW) now forms. As the name indicates, the thermohaline conveyer belt transports both heat and salt northward toward the Arctic Ocean; then the other half of the loop transports cold fresher water southward (Fig. 3.9).

Throughout the Holocene (the past 11,000 years), the North Atlantic conveyor belt has supplied enormous amounts of heat to the atmosphere over the North Atlantic region, keeping the climate of northwest Europe far milder than it would otherwise be. As reviewed by Fisher (2019), numerous investigations during subsequent decades have been performed in various parts of Canada that were formerly on or near the margins of Lake Agassiz, in attempts to find the most likely outlet channel for catastrophic flooding of Lake Agassiz water just at the beginning of the YD interval, about 12,800 years ago. Broecker et al. (1989) proposed that this flooding took place by the eastern drainage route, described earlier, synchronous with the YD chronozone. This model has been assumed by most authors, but Fisher (2019) maintains that a spillway of sufficient size to accommodate the volume of water released by this flooding event has not yet been identified. It is thought that the Moorhead Phase flooding event drained enough water from Lake Agassiz to lower the lake level by about 90 m. Subsequent flooding of glacial Lake Agassiz occurred about 8200 years ago and corresponds with a brief climatic reversal during the early Holocene (Fisher, 2019).

Atlantic Meridional Overturning Circulation

As discussed by Weijer et al. (2019), the climate of the Northern Hemisphere, and particularly that of northwest Europe, is greatly affected by the oceanic transport of heat and salt from the tropics to the subpolar regions. The release of heat from the ocean to the atmosphere in the subpolar North Atlantic makes a major contribution to the relatively mild climate in northwest Europe, which is up to 6°C warmer than the climates of similar latitudes elsewhere. The northward transport of heat in the Atlantic extends through the whole ocean so that South Atlantic Ocean heat is transported toward the equator. This meridional heat transport is associated with the Atlantic Meridional Overturning Circulation (AMOC), an overturning cell that features northward transport of warm upper ocean water, balanced by a southward flow of cooler deep water (Fig. 3.10).

The strength of the AMOC in the subtropical North Atlantic (26.5°N) averages about 17.0 Sv. The time series shows a great deal of variability, including a weakening of 15% over the length of the record. Many climate models indicate that the strength of AMOC varies on both decadal and multidecadal timescales. A key process in the maintenance of the AMOC is likely deep convection in the subpolar North Atlantic, especially in the Labrador and Nordic Seas. Here, winter cooling of relatively salty surface waters leads to convection that drives the newly cooled waters away from the surface, where it eventually forms the cold and salty NADW. As part of the global thermohaline conveyor belt, the AMOC has been a relatively stable component of the greater oceanic circulation system. But will it remain stable despite global warming, which, as

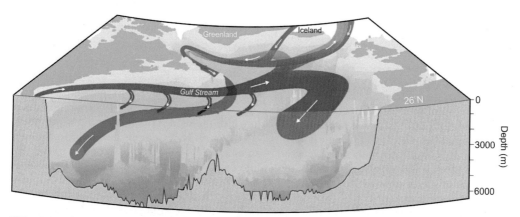

FIG. 3.10 Cross section of the North Atlantic region, view looking north from 29°N, illustrating Atlantic Meridional Overturning Circulation (AMOC), an overturning cell that features northward transport of warm upper ocean water, balanced by a southward flow of cooler deep water. (Image by Louise Bell, Commonwealth Scientific and Industrial Research Organization (CSIRO), in public domain.)

discussed earlier, is introducing substantially colder freshwater into the northern seas while simultaneously reducing the thickness and extent of sea ice cover? As Weijer et al. (2019) point out, a collapse of the AMOC would have far-ranging global consequences and might lead to a drop of surface air temperatures of up to 10°C in the North Atlantic and adjacent continental regions. While this degree of cooling may seem incredible, it is not without precedent in prehistory. Evidence of this scale of temperature change has been clearly identified in both paleoceanographic records from the North Atlantic, as well as terrestrial fossil records from northwest Europe. These multiple lines of proxy evidence strongly support the hypothesis that AMOC collapse events were responsible for rapid climate change during the Pleistocene, especially during the transitions from glacial to interglacial environments (Broecker et al., 1989). The Fifth Assessment Report of the IPCC identifies an AMOC collapse as a potential tipping point in the climate system. The report gave this scenario a low probability of occurrence but noted that its occurrence would have a high impact.

Understanding stability thresholds

Granted that the AMOC has more than one possible state, what are the forces that could potentially drive the system from one state to another? Is the system approaching thresholds of regional oceanic conditions that are about to destabilize the AMOC? As discussed by Weijer et al. (2019), even the possibility that the AMOC is close to surpassing stability thresholds gives us cause for concern, especially in an era of anthropogenic global change, the likes of which we have not previously experienced. Human activity is clearly forcing significant changes to the Earth system. Our understanding of the underlying mechanisms of the AMOC is only now being developed through the application of more sophisticated models. It is clear that more work needs to be done.

Early warning signals?

The search for early warning signals of an impending AMOC collapse is vital but difficult. Even though methods might allow us to detect a change in the AMOC, we do not currently know how to distinguish between the start of a collapse and just longer-term variability or a gradual weakening. Models may help us to detect signals of impending AMOC collapse, but we do not know which models are best suited to address this problem, as they employ a wide range of

variables and not all models are equally adept at capturing the magnitudes and timescales of such changes. Based on the available evidence, the possibility that the AMOC is in, or close to, a regime of multiple equilibria cannot be ruled out. Consider the following facts, put forward by Weijer et al. (2019): The full spectrum of models of the AMOC yields the likelihood of multiple equilibria. These range from relatively simple box models to two-dimensional or zonally averaged models, ocean GCMs, coupled climate models, and quasi-adiabatic models of the overturning circulation. One consistent aspect of the processes responsible for multiple equilibria is salt-advection feedback. This phenomenon is well understood, and as yet, no feedback mechanism has been found that is strong enough to disrupt salt-advection feedback. Important research questions remain, including the following: What temporary perturbation or permanent change in forcing (or combination of both) would it take to trigger an irreversible shutdown of the AMOC? Would the predicted increased rates of GrIS melting release enough cold, freshwater to destabilize the AMOC? Could plausible scenarios of continued anthropogenic forcing change the global heat and freshwater cycles to such an extent that an active AMOC can no longer be sustained? How will receding Arctic sea ice affect the strength and stability of the AMOC? Recent studies indicate that the ongoing Arctic sea ice decline can substantially weaken the AMOC. Liu and Federov (2019) introduced sea ice concentration data from the NSIDC into the Community Earth System Model to predict the effects of Arctic sea-ice changes on the AMOC over the next 200 years. They found different climate responses to Arctic sea-ice decline based on two different timescales: decadal and centennial. In their model runs, within one or two decades after Arctic sea-ice retreat, the global surface air temperature response follows a pattern with a warmer Northern Hemisphere and a colder Southern Hemisphere. This relatively rapid response is driven chiefly by atmospheric processes. Oceanic responses to global change are necessarily slower, as the ocean has a great deal of thermal inertia. In this modeling scenario, the Intertropical Convergence Zone (ITCZ) shifts northward. The ITCZ is a belt of low pressure which circles the Earth, generally near the equator, where the trade winds of the Northern and Southern Hemispheres come together. In a typical Northern Hemisphere summer, the position of the ITCZ drifts north over Asia but remains over the equator in the Atlantic Ocean region.

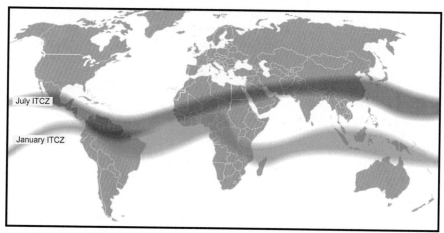

FIG. 3.11 Map of the ITCZ. *ITCZ*, Intertropical Convergence Zone by Mats Halldin. Creative Commons Attribution-Share Alike 3.0 Unported license.

In the Northern Hemisphere winter, the position of the ITCZ usually drifts south over Africa and Southeast Asia but also remains over the equator in the Atlantic Ocean region (Fig. 3.11).

In contrast to the atmospheric-driven short-term response, ocean-mediated impacts gain predominance on multidecadal to centennial timescales. The long-term model run by Liu and Federov (2019) predicts that the AMOC will weaken by 30%–40%. The Southern Hemisphere gradually warms, and the ITCZ shifts back toward the equator, eventually shifting southward from its original position after model runs representing about 40 years. The Arctic remains warm over the multidecadal timescale, but the northern midlatitudes become cooler. The largest cooling takes place over the North Atlantic Warming Hole caused by the AMOC weakening. The North Atlantic Warming Hole, paradoxically a region of reduced warming located in the North Atlantic, significantly affects the North Atlantic jet stream future in climate simulations. This weakening of the AMOC appears in the form of heat and freshwater anomalies that are driven by the contraction of sea-ice cover. These effects accumulate in the Arctic and then spread south in the North Atlantic, suppressing deep convection. The freshening is further amplified by salt-advection feedback. Ironically, the reduction of poleward heat transport due to a weakened AMOC acts to stabilize sea-ice cover. Such two-way interactions (both positive and negative feedback) between the AMOC and Arctic sea ice are an important part of the AMOC response to external forcing, and the global impacts of sea-ice decline critically depend on

whether the AMOC is affected. As discussed earlier, this topic needs a great deal of additional investigation. As of this writing, some of the literature suggests a weakening of AMOC related to Arctic warming, while other work shows that the Arctic component of AMOC has not weakened during the two decades, despite large-scale warming of the Arctic (Overland et al., 2018 and references therein).

Predicted Changes to Pacific Waters in the Arctic

As discussed by Aksenov et al. (2016), Pacific water (PW) enters the Arctic Ocean through Bering Strait, transporting heat, freshwater, and nutrients from the northern Bering Sea. The circulation of PW in the central Arctic Ocean is only partially understood due to the lack of observations. Using computer modeling (regional and global ocean GCMs), the authors posited that PW coming through Bering Strait forms three distinct branches as it enters the Arctic Ocean. One of these starts in Barrow Canyon, bringing PW along the continental shelf of northern Alaska. From there, it traverses the Canadian Straits and then flows into Baffin Bay. The second branch begins in the vicinity of Herald Canyon and transports PW along the continental slope of the East Siberian Sea. From there, this water enters the Transpolar Drift. Then, it passes through Fram Strait on its way to the Greenland Sea. The third branch begins near Herald Shoal and the central Chukchi shelf. It feeds PW into the BG in the western Arctic Ocean. As this third branch of PW crosses the shallow Chukchi Sea, it cools by 0.78°C and loses 0.63×1020 J of energy

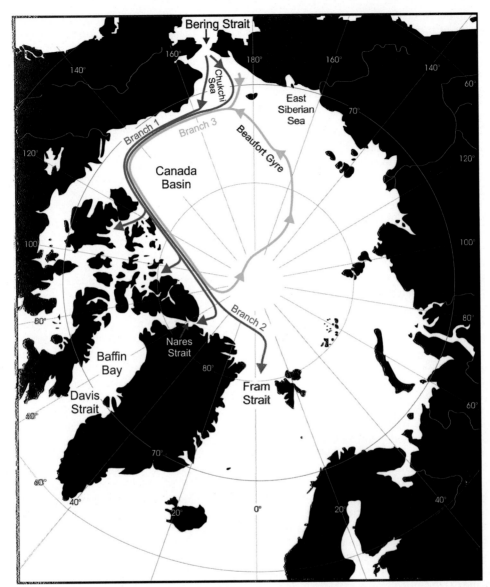

FIG. 3.12 Map of the Arctic Ocean showing the three branches of Pacific water and their routes across the region.

per year, in the form of heat lost to the atmosphere. It then enters the Canada Basin of the Arctic Ocean (Fig. 3.12).

Observations show that the freshwater layer of the BG is dominated by PW (50%–100% of the water column). The freshwater occupies the depth range between the surface and 250 m in the Canada Basin. Between 200 and 800 m, the water mass is mixed with the saltier Atlantic waters. The BG is thought to extend down the water column into the Atlantic water layer. But these freshwater currents apparently change course throughout the year. Different seasonal circulation patterns of PW in the Arctic Ocean have been inferred from observations of their water properties. One proposal has been a cyclonic (counterclockwise) pattern for the winter flow of PW in the Canada Basin with the water mass exiting the Canada Basin along the northern shelf of the Canadian Arctic Archipelago. Another hypothesis suggests an anticyclonic (clockwise) circulation of summer PW in the Canada Basin. These are working hypotheses that need additional field observations. Using a variety of ocean and atmospheric models, Aksenov et al. (2016) reconstructed variations in the storage of PW and associated freshwater content in the Arctic Ocean over the past 50 years. Their results suggest that there was a decrease in the storage of Pacific freshwater in the Arctic Ocean in the 1980s—1990s, followed by an increase in the 21st century. Variations in freshwater content mirrored those of PW content, with a minimum in the 1990s. Observations have revealed persistent flows of relatively fresh PW to the North Atlantic through Fram Strait and the passages through the Canadian Arctic Archipelago (CAA). According to Aksenov et al. (2016), PW outflow through Fram and Davis Straits appears to reflect changes in PW in the Arctic Ocean, with a decrease in the PW export during the two decades. Based on various numerical models, the transit time for PW to cross the Arctic Ocean and enter the North Atlantic was estimated at 10—15 years, and the time for the PW outflow to reach a quasi-equilibrium state was about 20 years. Brown et al. (2019) ran model simulations from a coupled ice-ocean GCM to assess the effects of changes in freshwater input (river runoff and precipitation) on Arctic Ocean freshwater storage. Specifically, they used models to examine responses of freshwater content of the Arctic Ocean to abrupt changes in freshwater input, as are predicted to occur in future global warming scenarios. Their results suggest that mode and footprint of these freshwater inputs play a significant role in determining (1) the fate of the enhanced input in the Arctic Ocean, and (2) the proportion of extra freshwater stored in the Arctic rather than being discharged immediately to the subpolar seas. Climate models predict that river runoff will contribute a larger share of future increases in freshwater input than precipitation. Brown et al. (2019) found that runoff produces a simpler transient response in terms of the storage and export of freshwater. The entry of precipitation into the Arctic Ocean happens in regions where a more complex set of interactions take place, involving freshwater anomalies, sea ice, and circulation patterns. Therefore, the transient responses of freshwater storage in the Arctic Ocean to changes in precipitation are more complex than those associated with changes in runoff. From a modeling standpoint, the nonlinear interactions between ocean, sea ice, and precipitation confound the currently available models. Therefore, accurate projections of the effect of future increases in precipitation will require models that are better able to capture ice-ocean interactions. As discussed by Praetorius et al. (2020), intense warming of the Arctic is predicted by climate modelers, and in fact has already started. We have reviewed possible future scenarios for what may happen in the North Atlantic. But what role, if any, would the waters of the North Pacific play in potentially shutting down AMOC in the North Atlantic? Abrupt climate fluctuations in the Northern Hemisphere during the past glacial and deglacial periods are widely assumed to track changes in freshwater fluxes and sea-ice dynamics of the high-latitude North Atlantic. So far, changes in the North Pacific have received far less attention as a forcing agent of abrupt climate fluctuations, in part due to the lack of deep water formation and the paucity of high-resolution records from the region. Nevertheless, models show that perturbations to the surface waters of the North Pacific have the potential to deliver substantial impacts on Northern Hemisphere climate. The primary elements of these North Pacific impacts are changes in sea-ice extent and poleward moisture transport. At the end of the last glaciation, meltwater from the Cordilleran ice sheet delivered sustained freshwater into the Northeast Pacific. This flow of cold, freshwater is thought to have amounted to 0.02—0.06 Sv, although short-term variations in flux may have been larger. Closure of the Bering Strait when sea level was lower would have impounded an additional 0.03 Sv of freshwater in the Northeast Pacific, based on estimates of modern freshwater flow through Bering Strait. By way of comparison, sustained flows of meltwater into the North Atlantic from the waning Laurentide Ice Sheet are thought to have been 0.05—0.2 Sv. In the model simulation of Pacific currents just after the opening of Bering Strait, a substantial fraction of Cordilleran meltwater flows north through the strait into the Arctic. From here, it flows southward, eventually reaching the western boundary currents of the North Atlantic. Within a decade, this water reaches the Gulf Stream and Atlantic deep water formation regions of the Labrador Sea. This results in the export of Arctic sea ice southward into the North Atlantic. The combination of cold water and additional sea ice

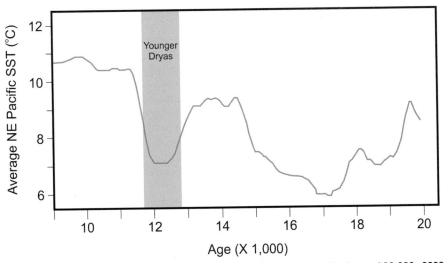

FIG. 3.13 Graph of average SSTs for the northeastern Pacific Ocean during the interval 20,000–9000 years ago, showing the rapid cooling of those temperatures between 13,000 and 12,000 years ago in comparison with the timing of the YD interval in the North Atlantic. *SSTs*, sea surface temperature; *YD*, Younger Dryas. (Data from Praetorius, S.K., Condron, A., Mix, A.C., Walczak, M.H., McKay, J.L., Du, J., 2020. The role of Northeast Pacific meltwater events in deglacial climate change. Science Advances 6, eaay2915.)

brings a cooling of SSTs in the Gulf Stream and Labrador Sea regions. As discussed earlier, massive meltwater input into the North Atlantic has been suggested to have been through the St. Lawrence River and/or via the Arctic during the YD. This cold-water input into the North Atlantic is inferred to have contributed to a decline in the strength of AMOC. But Praetorius et al. (2020) provide evidence for freshening and cooling of the waters along the Northeast Pacific during the YD (Fig. 3.13), indicating that multiple drainage pathways were activated at this time or that closely sequenced meltwater events helped to prolong YD cooling.

The presence of a Pacific bivalve dated to 13,000 years ago in the Arctic suggests a possible YD-age breach of the Bering Land Bridge. This estimate of the timing of PWs breaching the land bridge raises the possibility that Pacific meltwater contributed to the weakening of AMOC at that time.

CONCLUSIONS

It is fair to say that the question of PW input affecting the cooling of the North Atlantic during the YD interval has yet to be solved. This is just one of many thorny problems being examined by scientists who study the Arctic Ocean and surrounding seas. As we have seen

in study after study, more observations need to be made, and more and better ocean data are needed. Of course, this is easy to say, but one must keep in mind that Arctic waters are difficult to work in and very expensive to get to. Obstruction by pack ice and collisions with icebergs have delayed or ruined many a scientific cruise in Arctic waters. There is always an element of danger when working in an extremely cold region where violent storms may bring snow, sleet, and high winds, forcing all work to stop as the scientists hunker down in some kind of shelter. The cold temperatures cause equipment to break down and batteries to go flat very quickly. Flights in and out of remote localities can be greatly delayed, research ships may get stuck in pack ice for days or weeks, and the field season is usually limited to just a few weeks at the height of summer. In many of the research papers reviewed here, the authors concluded that better models need to be developed, for scientific progress to be made. GCMs have improved greatly in the 21st century, yet they still lack the necessary levels of sophistication and precision needed to answer the questions the scientists are now posing about the Arctic. One of the major problems with this is that GCMs need to become even more complicated than they are currently. The major problem here is that the ocean, ice, and atmospheric systems are extremely complex in the Arctic. Modeling their interactions is extremely difficult. Any new models would need to take into account large volumes of data, but these data are not available from every region. Because of the remoteness and great expense of working in the Arctic, only a few regions of the Arctic Ocean and adjacent high-latitude seas have been well studied. This means that the refinement of GCM modeling and the development of new data sets for Arctic seas will have to be done in tandem over the coming decades. Remote sensing data from satellites have revolutionized Arctic oceanography and glaciology. More satellite-based research will supply highly useful data in years to come. Instruments on the satellites have been able to measure such things as the height of sea surfaces and the depth of ocean waters, the course of drifting sea ice and its concentration, and the flux rate of ocean currents. On land, satellites have been measuring the thickness of ice sheets, the rate of ice flow, and the rate of ice discharge into the ocean. They have also been used to measure runoff rates of rivers entering the Arctic Ocean. One gets the sense that now we are entering an era of greater understanding of how the Arctic Ocean works and how it is connected to the rest of the world. Data are being collected at an unprecedented rate, and new kinds of information are contributing to our knowledge of Arctic seas. These advances are exciting, but they are also quite necessary because the Arctic is changing so quickly. We are learning that the slow disappearance of sea ice is affecting more than just the surface of the Arctic Ocean. Rather, the diminution of ice cover on the Arctic Ocean is leading to stronger winds affecting surface waters. These winds are driving oceanic features such as the BG, causing more freshwater to accumulate in patches of the Arctic Ocean. Many compounding factors influence Arctic Ocean conditions, as well as adjacent lands. For instance, we have now come to understand that the climate system called the AO not only affects the Arctic but also extends down to the middle northern latitudes, with effects that vary year to year depending on both the strength and the circulation pattern of the AO. Another aspect of Arctic Ocean research in recent years has been the development of studies of the various water layers between the surface and the seafloor and the recent changes in oceanic circulation that have brought changes in the water column. It appears that global warming is causing the waters of the Arctic Ocean to become more energetic. In this energized environment, increases in energy diffusion by eddies, coupled with increased mixing of the water column by eddy activity, are expected to enhance the upwelling of warm Atlantic water in the Arctic Ocean, with consequences for sea ice growth. We have learned a great about the functioning of AMOC, an overturning cell that transports warm upper ocean water northward, balanced by a southward flow of cooler deep water. The fossil record clearly shows a major disruption of thermohaline circulation in the North Atlantic at the end of the last glaciation. Could such an event happen in modern times? Oceanographers are looking for early warning signals of an impending AMOC collapse. We can currently detect changes in the AMOC, but we cannot currently distinguish between the start of a collapse and just longer-term variability or a gradual weakening. The possible participation of North PWs in the YD cooling of the North Atlantic is now being considered, based on newly obtained data from sentiment cores. A marked drop in SSTs in the North Pacific sediment record aligns with the timing of the YD cooling interval in the North Atlantic. While this scenario is of no direct importance to the modern world, it may serve as a model for what might happen in future disruptions of AMOC when large volumes of cold freshwater enter the North Atlantic. The Arctic Ocean is a highly complex system of currents, sea-ice cover, and freshwater inputs mainly on the Pacific side and outputs on the Atlantic side. The interactions between the atmosphere, ice cover, and water are too complex to be accurately reconstructed by current numeric models. Developing a better understanding of the Arctic Ocean is not just an academic exercise—far from it! We now know enough to say that the Arctic Ocean is changing rapidly today and will keep on changing in the coming decades and centuries. It seems highly likely that at least some of these future changes will never have been seen before by human observers. It is also clear that these changes will have impacts on much of the rest of the world's oceans and climates.

REFERENCES

Aagaard, K., Carmack, E.C., 1989. The role of sea ice and other fresh water in the Arctic circulation. Journal of Geophysical Research 94, 14485–14498.

Aksenov, Y., Karcher, M., Proshutinsky, A., Gerdes, R., De Cuevas, B., Golubeva, E., Kauker, F., Nguyen, A.T., Platov, G.A., Wadley, M., Watanabe, E., Coward, A.C., Nurser, A.J.G., 2016. Arctic pathways of Pacific water: arctic ocean model intercomparison experiments. Journal of Geophysical Research: Oceans 121 (1), 27–59. https://doi.org/10.1002/2015JC011299.

Armitage, T.W., Manucharyan, G.E., Petty, A.K., Kwok, R., Thompson, A.F., 2020. Enhanced eddy activity in the Beaufort Gyre in response to sea ice loss. Nature Communications 11, 761. https://doi.org/10.1038/s41467-020-14449-z.

Bamber, J.L., Tedstone, A.J., King, M.D., Howat, I.M., Enderlin, E.M., van den Broeke, M.R., Noel, B., 2018. Land ice freshwater budget of the arctic and North Atlantic Oceans: 1. Data, methods, and results. Journal of Geophysical Research: Oceans 123 (3), 1827–1837. https://doi.org/10.1002/2017JC013605.

Blackport, R., Screen, J.A., van der Wiel, K., Bintanja, R., 2019. Minimal influence of reduced Arctic sea ice on coincident cold winters in mid-latitudes. Nature Climate Change 9 (9), 697–704. https://doi.org/10.1038/s41558-019-0551-4.

Broecker, W.S., Kennett, J.P., Flower, B.P., Teller, J.T., Trumbore, S., Bonani, G., Wolfli, W., 1989. Routing of meltwater from the Laurentide ice sheet during the younger Dryas cold episode. Nature 341 (6240), 318–321. https://doi.org/10.1038/341318a0.

Brown, N.J., Nilsson, J., Pemberton, P., 2019. Arctic ocean freshwater dynamics: transient response to increasing river runoff and precipitation. Journal of Geophysical Research: Oceans 124 (7), 5205–5219. https://doi.org/10.1029/2018JC014923.

Carmack, E.C., Yamamoto-Kawai, M., Haine, T.W.N., Bacon, S., Lique, C., Melling, H., Polyakov, I.V., Straneo, F., Timmermans, M.L., Williams, W.J., 2016. Freshwater and its role in the Arctic Marine system: sources, disposition, storage, export, and physical and biogeochemical consequences in the Arctic and global oceans. Journal of Geophysical Research G: Biogeosciences 121 (3), 675–717. https://doi.org/10.1002/2015JG003140.

Chereskin, T.K., Price, J.F., 2019. Upper ocean structure: Ekman transport and pumping. In: Encyclopedia of Ocean Sciences. Elsevier, pp. 80–85. https://doi.org/10.1016/B978-0-12-409548-9.11161-3.

Dahlman, L., 2019. Climate Variability: Arctic Oscillation. https://www.climate.gov/news-features/understanding-climate/climate-variability-arctic-oscillation. (Original work published 2008). (Accessed 15 March 2020).

Fisher, T.G., 2019. Megaflooding associated with glacial Lake Agassiz. Earth Science Reviews 201. Article, 102974.

Giles, K.A., Laxon, S.W., Ridout, A.L., Wingham, D.J., Bacon, S., 2012. Western Arctic Ocean freshwater storage increased by wind-driven spin-up of the Beaufort Gyre. Nature Geoscience 5 (3), 194–197. https://doi.org/10.1038/ngeo1379.

Haine, T.W.N., Curry, B., Gerdes, R., Hansen, E., Karcher, M., et al., 2015. Arctic freshwater export: status, mechanisms, and prospects. Global and Planetary Change 125, 13–35. https://doi.org/10.1016/j.gloplacha.2014.11.013.

Liu, W., Federov, A.V., 2019. Global climate impacts of Arctic sea ice loss mediated by the Atlantic Meridional overturning circulation. Geophysical Research Letters 46, 944–952.

Manucharyan, G.E., Thompson, A.F., 2017. Submesoscale sea ice-ocean interactions in marginal ice zones. Journal of Geophysical Research: Oceans 122 (12), 9455–9475. https://doi.org/10.1002/2017JC012895.

Morison, J., Kwok, R., Peralta-Ferriz, C., Alkire, M., Rigor, I., Andersen, R., Steele, M., 2012. Changing Arctic ocean freshwater pathways. Nature 481 (7379), 66–70. https://doi.org/10.1038/nature10705.

NASA, 2012. Ice Floes along Kamchatka Coastline. http://spaceflight.nasa.gov/gallery/images/station/crew-30/html/iss030e162344.html.

NASA, 2020. Ocean Motion and Surface Currents. http://oceanmotion.org/html/background/ocean-in-motion.htm.

NOAA Climate Prediction Center, 2020. Arctic Oscillation. https://www.cpc.ncep.noaa.gov/products/precip/CWlink/daily_ao_index/ao.shtml. (Original work published 2020).

Overland, J., Dunlea, E., Box, J.E., Corell, R., Forsius, M., et al., 2018. The urgency of Arctic change. Polar Science 21, 6–13.

Praetorius, S.K., Condron, A., Mix, A.C., Walczak, M.H., McKay, J.L., Du, J., 2020. The role of Northeast Pacific meltwater events in deglacial climate change. Science Advances 6, eaay2915.

Proshutinsky, A., Krishfield, R., Toole, J.M., Timmermans, M.L., Williams, W., 2019. Analysis of the Beaufort Gyre freshwater content in 2003–2018. Journal of Geophysical Research: Oceans 124, 9658–9689. https://doi.org/10.1029/2019JC015281.

Tang, Q., Zhang, X., Yang, X., Francis, J.A., 2013. Cold winter extremes in northern continents linked to Arctic sea ice loss. Environmental Research Letters 8, 014036, 6 pp. https://doi.org/10.1088/1748-9326/8/1/014036.

Weijer, W., Cheng, W., Drijfhout, S.S., Fedorov, A.V., Hu, A., Jackson, L.C., Liu, W., McDonagh, E.L., Mecking, J.V., Zhang, J., 2019. Stability of the Atlantic meridional overturning circulation: a review and synthesis. Journal of Geophysical Research: Oceans 5336–5375. https://doi.org/10.1029/2019jc015083.

FURTHER READING

Bell, L., 2017. Slowdown of North Atlantic Circulation Caused Sudden Cold Spells Lasting Hundreds of Years. https://watchers.news/2018/07/24/slowdown-of-north-atlantic-circulation-rocked-the-climate-of-ancient-northern-europe/.

Cook, J., Woods Hole Oceanographic Institute, J., 2020. Arctic Ocean Circulation. https://www.whoi.edu/know-your-ocean/ocean-topics/polar-research/arctic-ocean-circulation/.

Halldin, M., 2006. Map of the Average Positions of the ITCZ during Northern Hemisphere Summer and Winter. https://commons.wikimedia.org/wiki/File:ITCZ_january-july.png.

Manucharyan, G.E., Platov, G., Watanabe, E., Kikuchi, T., Nishino, S., Itoh, M., Kang, S.H., Cho, K.H., Tateyama, K., Zhao, J., 2019. Analysis of the Beaufort Gyre freshwater content in 2003–2018. Journal of Geophysical Research: Oceans 124 (12), 9658–9689. https://doi.org/10.1029/2019JC015281.

NASA, 2012. Ice Floes Along Kamchatka Coastline. Retrieved from: http://spaceflight.nasa.gov/gallery/images/station/crew-30/html/iss030e162344.html. (Original work published 2012).

Woodgate, R., Curry, B., Gerdes, R., Hansen, E., Karcher, M., Lee, C., Rudels, B., Spreen, G., de Steur, L., Stewart, K.D., 2015. Arctic freshwater export: status, mechanisms, and prospects. Global and Planetary Change 125, 13—35. https://doi.org/10.1016/j.gloplacha.2014.11.013.

© Woods Hole Oceanographic Institution, Jack Cook, 2012. Retrieved from: https://www.whoi.edu/know-your-ocean/ocean-topics/polar-research/arctic-ocean-circulation/. (Original work published 2012).

CHAPTER 4

Sea Level Change

INTRODUCTION

It is quite clear that sea level is rising around the world. This is not due to any single cause but rather to multiple causes. In this chapter, we will explore those causes and attempt to quantify the various inputs that combine to cause the increases we are seeing. The measurement of global mean sea level (GMSL) has become much more precise in recent years, following the advent of satellite measurements. Prior to this, tide gauges were used to track changes in sea level at thousands of stations around the world. As it turns out, the determination of GMSL based on the new satellite data is quite similar to the estimates based on tide gauges. Sea level does not rise uniformly across the ocean basins. There are regional variations in sea level. However, for this discussion, GMSL provides an excellent metric to assess global sea level change, and studies of the various sources of sea level rise (glacial and ice sheet melting, thermal expansion) use the global mean sea level budget.

Measuring Sea Level

As described by NOAA (2019a), a tide gauge is fitted with sensors that continuously record the height of the surrounding water level. Before the use of computers to record water levels (especially tides), special "tide houses" sheltered permanent tide gauges. The interior of the building housed the instrumentation—including a well and a mechanical pen-and-ink (analog) recorder. A tidal staff was affixed to the outside of the building. This was essentially a large measuring stick that allowed scientists to manually observe tidal levels and then compare them to readings taken every 6 min by the recorder. Tide houses and the data they recorded required monthly maintenance. Scientists would collect the data tapes and mail them to headquarters for manual processing. The computer age led to tide gauges that use microprocessor-based technologies to collect sea level data. While older tide houses used mechanical floats and recorders, modern monitoring stations use advanced acoustics and electronics. Today's recorders send an audio signal down a sounding tube and measure the time it takes for the reflected signal to travel back from the water's surface. Data are still collected every 6 min, and the readings are transmitted to a Geostationary Operational Environmental Satellite (GOES).

The number of tide gauges has increased over the past three centuries, from only a few in northern Europe in the 18th century to more than 2000 today along most of the world's coastlines. Because of their location and limited number, tide gauges sample the ocean sparsely and nonuniformly with a bias toward the Northern Hemisphere. Most tide gauge records are short and have significant gaps. In addition, tide gauges are anchored on land and are affected by the vertical motion of Earth's crust caused by both natural processes (e.g., tectonics and sediment compaction) and anthropogenic activities (e.g., groundwater depletion, dam building).

As described by NOAA (2019b), satellite altimeter radar measurements can be combined with precisely known spacecraft orbits to measure GMSL with a high level of accuracy. A series of satellite missions that started with TOPEX/Poseidon (T/P) in 1992 and continued with Jason-1 (2001–13), Jason-2 (2008–19), and Jason-3 (2016–present) estimate GMSL every 10 days with an uncertainty of 3–4 mm. Jason-3, launched in 2016, is a joint effort between NOAA, NASA, France's Centre National d'Etudes Spatiales (CNES), and the European Organization for the Exploitation of Meteorological Satellites (EUMETSAT).

RECENT CHANGES IN THE GREENLAND ICE SHEET

Although the Greenland ice sheet (GIS) appears to be a monolithic sheet of ice firmly grounded to the bedrock below, the ice sheet is actually divided up in groups of six major drainage basins, with meltwater flowing both on the surface and at the base of the ice (Rignot et al., 2008), as illustrated in Fig. 4.1. Mass balance studies undertake to measure the difference between the snow accumulated in the winter and the snow and ice melted over the summer. If the mass of snow accumulated on an ice sheet or glacier exceeds the mass of snow and ice lost during summer months, the mass

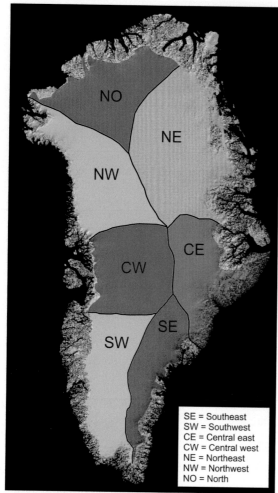

FIG. 4.1 The Greenland ice sheet is divided into six regions based on groups of drainage basins identified by Rignot et al. (2008). (Courtesy of Creative Commons CC BY 4.0.)

SE = Southeast
SW = Southwest
CE = Central east
CW = Central west
NE = Northeast
NW = Northwest
NO = North

measurement of ice sheet mass balance and the associated global sea level contribution (Fig. 4.2). Ice sheets respond to variations in climate, primarily snowfall and temperature, by changing in thickness.

The IMBIE Team (2019) predicts that the GIS holds enough water to raise the mean global sea level by 7.4 m. Greenland ice flows to the oceans through a network of glaciers and ice streams, each with a substantial inland catchment. Over recent decades, ice losses from Greenland have made a substantial contribution to global sea level rise (Fig. 4.2), and model projections suggest that this imbalance will continue in a warming climate. The melting season of 2019 was one of the longest on record, from late April through August (Fig. 4.3, top). It was also exceptionally warm (Fig. 4.3, bottom). At the peak of the warming (late July), upward of 70% of the surface of the GIS was melting (Tedesco and Fettweis, 2020).

Influence of fjords

Greenland's bedrock topography exerts a strong influence on ice flow, grounding line migration, the dynamics of glacial calving, and subglacial drainage (Fig. 4.4).

Recent research has brought to light the vital role of coastal fjords in the melting of tidewater glaciers in Greenland. The bathymetry of these fjords controls the degree of penetration of warm Atlantic water (AW). When this warm water penetrates to the heads of fjords, it rapidly melts and undercuts tidewater glaciers (Fig. 4.5).

Morlighem et al. (2017) present a compilation of Greenland bed topography that assimilates seafloor bathymetry and ice thickness data through a mass conservation approach. Their map reveals recent calving front response of numerous outlet glaciers and points to new pathways by which AW can reach glaciers with marine-based basins. This information is critical to the development of accurate future sea level models, as it highlights the sectors of Greenland that are most vulnerable to future oceanic forcing.

Not all fjords allow the entrance of AW. For example, sills in some fjords block AW from reaching glacier calving fronts. AW is typically found at depths below 200–300 m below sea level. Thus, shallow fjords also block AW from reaching glacier fronts. Likewise, some glacier fronts are grounded on the ocean bed, but at depths above the AW level. When AW reaches the terminus of a glacier, its calving front is exposed to strong ocean-induced melting. This melting may be enhanced by subglacial discharge, leading to glacier undercutting, enhanced calving, ice front retreat, flow acceleration, and glacier thinning (Fig. 4.6).

balance is positive. On the other hand, if summer melting exceeds the mass of snow accumulation of the previous winter, the mass balance of the glacier is negative. Each GIS drainage basin has its own level of mass balance, and these can vary considerably between basins (Table 4.1).

Mass Balance Changes

Starting in 2011, the Ice Sheet Mass Balance Inter-Comparison Exercise (IMBIE) is a collaboration between scientists supported by the European Space Agency (ESA) and NASA. The data collected by the IMBIE team contribute to IPCC assessment reports. IMBIE studies have led to improved confidence in the

TABLE 4.1
Mass Balance and Annual Ice Discharge for the GIS Drainage Basins.

Region	Area (103 km²)	SMB: 1971–88 (Gt per year)	ANNUAL ICE DISCHARGE (GT PER YEAR)							
			1958	1964	1996	2000	2004	2005	2006	2007
North	484.3	43 ± 6	54 ± 6	54 ± 6	59 ± 5	54 ± 3	55 ± 3	59 ± 3	56 ± 3	56 ± 3
East	375.6	154 ± 23	151 ± 34	159 ± 30	168 ± 9	188 ± 10	217 ± 12	262 ± 14	237 ± 13	224 ± 12
Southwest	147.5	32 ± 5	32 ± 5	32 ± 5	32 ± 5	34 ± 5	34 ± 5	36 ± 5	36 ± 5	36 ± 5
West	529.2	111 ± 16	161 ± 17	168 ± 17	145 ± 9	164 ± 9	174 ± 9	176 ± 9	178 ± 9	186 ± 9
SMB per year			272 ± 41	299 ± 45	300 ± 45	277 ± 42	242 ± 36	233 ± 35	234 ± 35	228 ± 34
SMB discharge	1536.5		−119 ± 68	−106 ± 68	−97 ± 47	−156 ± 44	−231 ± 40	−293 ± 39	−265 ± 39	−267 ± 38

GIS, Greenland ice sheet; SMB, surface mass balance.
Data from Rignot et al. (2008).

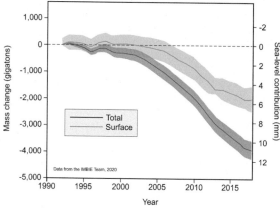

FIG. 4.2 Cumulative anomalies in the total mass (*blue line*) and surface mass balance (*green line*) of the Greenland ice sheet from 1992 to 2018, in relation to the mean values of these records from 1980 to 1990. *Shaded blue* and *green areas* represent one standard deviation from the total mass and surface mass balance values, respectively. (After the IMBIE Team (2019).)

It is therefore critical to determine the locations that are currently exposed to AW and that may be exposed to AW in the future. How far do these glaciers need to retreat before their margins reach higher ground (<200–300 m depth) or terminate on land (bed > 0 m)? It must be pointed out that glaciers, themselves, never retreat. They are, after all, bodies of ice lying on a slope. The ice always flows downhill (albeit slowly), akin to a giant conveyor belt. As the climate warms and/or snowfall near the head of the glacier decreases substantially, the ice will continue to flow downhill, but the downhill margin, or terminus, will begin to retreat upslope.

Narrow and deep fjords have important roles to play in the current and future state of the GIS. They can provide pathways for AW to interact with glacier termini. To investigate the regions that are in contact with the ocean, Morlighem et al. (2017) delineated locations that are continuously below sea level from the continental shelf to the ice sheet bed (light blue area in Fig. 4.7).

As glacier margins along the Greenland coast retreat, these regions will remain in contact with the ocean. Morlighem et al. (2017) also determined the regions that are continuously below a depth of 200 and 300 m, respectively, and are currently connected to the ocean below these depths (pale and dark blue areas in Fig. 4.7, respectively). Glaciers that retreat within these regions will likely remain in contact with warm AW. The mapping of Greenland landscapes beneath the ice cap has shown that submarine bed channels

are widespread and extend far inland. That means that these glaciers will remain vulnerable to ocean warming, even if their margins retreat for hundreds of kilometers.

Connections between global sea level and the Greenland ice sheet. Bindschadler et al. (2013) were the first to develop a process-based case study modeling the response of the GIS to predicted future climate changes. Their models indicated that Greenland is more sensitive than Antarctica to likely atmospheric changes in temperature and precipitation, while Antarctica is more sensitive to increased ice shelf basal melting. They ran a modeling experiment based on the IPCC RCP8.5 (representative concentration pathway; worst-case) scenario. These model runs indicate additional contributions to sea level of 22.3 and 8.1 cm from Greenland and Antarctica, respectively, in the coming century. A 200-year model run projected sea level rise contributions of 53.2 cm for the GIS and 26.7 cm for Antarctica. The estimates for the 21st century are in stark contrast to the most recent IPCC forecast of sea level rise in the next 100 years (Fig. 4.8).

Under the worst-case scenario (RCP8.5), the latest IPCC estimate is that sea level will rise between 84 and 110 cm above its current level by 2100.

Comparisons with the last interglacial period. As discussed by Thomas et al. (2020), the planet's atmosphere, geosphere, hydrosphere, and cryosphere are in a configuration that has been relatively stable through much of the current interglacial interval, the Holocene. However, anthropogenic forcing is putting enormous stress on all of these components of Earth's environmental system. So far, these human-caused stresses have induced a series of incremental changes. These changes are bad enough, but the greater danger is that soon the Earth system will be pushed into a new state of equilibrium. There are several aspects of this scenario that are very troubling. First, we are not sure how far the current environmental system can be pushed before it reaches the tipping point and enters a new, unchartered environmental system. Second, it appears likely that the transition from the old to the environmental system will be quite rapid, permanent, and irreversible.

Tipping points

My favorite analogy of this scenario concerns a group of people paddling a canoe across a lake. As the voyage begins, the canoe is, of course, upright. As long as the occupants sit still in the canoe as its being paddled, it will remain upright. However, imagine that one of the occupants drops a pair of sunglasses over the side. He

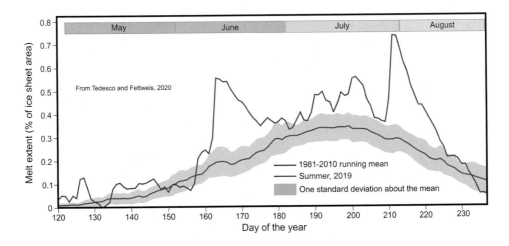

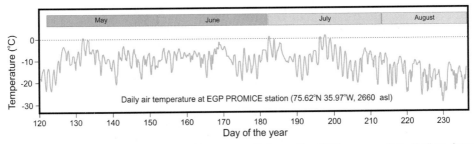

FIG. 4.3 Top: Number of days when melting occurred during the 2019 summer (late April to August) according to space-borne passive microwave observations (*red line*). Note that by the end of July, more than 70% of the surface of the GIS was melting. *Blue line* represents running mean values from 1981 to 2010, with *shaded blue area* showing boundaries of one standard deviation. Bottom: Daily air temperatures during the melting season of 2019 at the EGP PROMICE station (75.62°N 35.97°W, Elevation: 2660 asl). *GIS*, Greenland ice sheet. (Both graphs after Tedesco, M., & Fettweis, X., 2020. Unprecedented atmospheric conditions (1948–2019) drive the 2019 exceptional melting season over the Greenland ice sheet. The Cryosphere 14, 1209–1223. https://doi.org/10.5194/tc-14-1209-2020; Creative Commons Attribution 4.0 License.)

reaches his arm out of the canoe in an effort to grab the sunglasses before they sink, but they are just beyond his reach. He then leans even farther out of the canoe, to the point where the canoe begins to tip to that side. Another occupant of the canoe tries to reach the sunglasses with her paddle, leaning much of her weight over the side. A moment later, all the people join the pair of sunglasses in the lake, as the canoe tips over in the water. Unfortunately for the people, they discover that this new, half-submerged, upside-down condition of the canoe is quite stable and that the canoe is extremely difficult to flip back to its original upright orientation. Humans have been pushing their earthly "canoe" toward that tipping point. How close is our vessel to the tipping point, and what will the new equilibrium look like?

Climate scientists, glaciologists, and oceanographers are trying to come to grips with the real-life "tipping elements" of the Earth system, but these tipping elements remain a major source of uncertainty, hindering our ability to project the timing, intensity, and nature of future climate change. Nowhere are these tipping elements more important than in the high latitudes because, as we have seen, these regions are particularly susceptible to amplification of climate change through atmosphere-ocean-ice interactions. Unfortunately, the short instrumental record fails to capture the full range of past or projected climate scenarios. Returning to our canoeing analogy, the instrumental record would compare to a few seconds of film footage, taken when the canoe was in its stable upright position. This lack of a meaningful instrumental record is particularly acute in the high latitudes since the instrumental records of these regions are the shortest and patchiest of the planet. To gain a better understanding of these potentially radical changes to the Earth System, we must investigate the

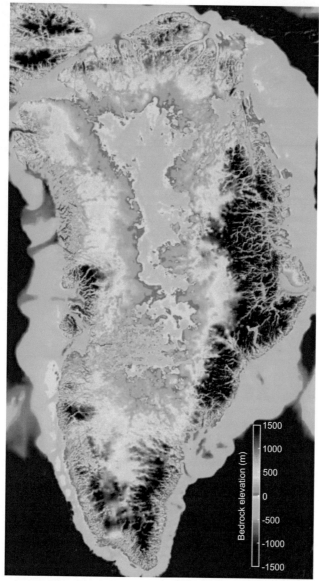

FIG. 4.4 BedMachine v3 bed topography (m), color-coded between −1500 m and +1500 m with respect to mean sea level, with areas below sea level in blue. (After Morlighem, M., Williams, C. N., & Rignot, E., 2017. BedMachine v3: complete bed topography and ocean bathymetry mapping of Greenland from multibeam echo sounding combined with mass conservation. Geophysical Research Letters 44 1, 11051−11061. https://doi.org/10.1002/2017GL074954; Image courtesy of Creative Commons Attribution License, in the public domain.)

past. Natural archives from past periods that were warmer than today can be used to better understand the range of possible tempos and amplitudes of Arctic climate change. Ancient proxy data can be used to test climate models, offering potential insights into the future.

The last interglacial

Geologically speaking, we do not have to look back very far into the past to find a time period that offers a useful warmer-than-modern analog. The Eemian Interglacial (129,000 to 116,000 years before present [yr BP]) was

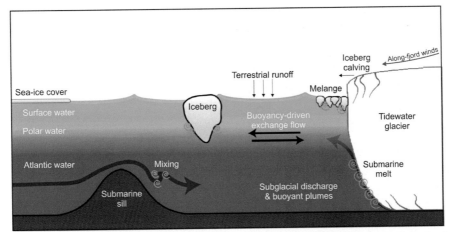

FIG. 4.5 Cross section of a Greenland fjord and adjacent tidewater glacier. (After Straneo, F., Sutherland, D. A., & Stearns, L., 2019. The case for a sustained Greenland ice sheet-ocean observing system (GrIOOS). Frontiers in Marine Science 138, 1–23. https://doi.org/10.3389/fmars.2019.00138, used with permission.)

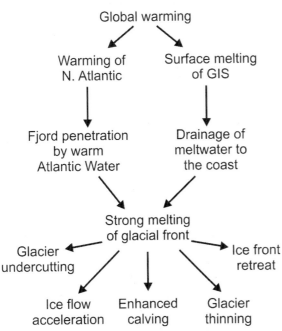

FIG. 4.6 Flowchart illustrating the mechanisms by which the melting of a tidewater glacial front is enhanced.

the warmest interglacial period of the last 800,000 years. As such, it represents the most recent period during which global temperatures were comparable with low-end temperature projections of the IPCC, which point to a world that will be 2°C warmer than today by the end of this century, with even greater

temperature increases in the polar regions. Substantial environmental changes happened during the Eemian interglacial. Thomas et al. (2020) synthesized the available paleo-data on the nature and timing of potential high-latitude tipping elements during the Eemian, including sea ice cover, the extent of boreal forests, permafrost, ocean circulation, and ice sheets/sea level. They also reviewed the thresholds and feedbacks that characterized the environmental changes of the Eemian. Notably, substantial ice mass loss from the GIS, the West Antarctic Ice Sheet (WAIS), and possibly sectors of the East Antarctic Ice Sheet were the major contributors to a 6–9 m rise in global sea level at that time. This was accompanied by reduced summer sea-ice cover, northward expansion of boreal forest, and a reduction in the extent of permafrost.

While it is difficult to date Eemian events precisely, Thomas et al. (2020) found that high latitude tipping elements had undergone abrupt changes and that these changes took place within 1–2 millennia of each other in both the Northern and Southern Hemispheres. In contrast to these rapid changes to new conditions, the time of recovery to prior conditions lasted many thousands of years. Their synthesis also demonstrated important feedback loops between tipping elements, amplifying polar and global change during the Last Interglacial.

The NEEM ice coring project
The most recent of a long line of multidisciplinary studies of ice cores drilled from the GIS was carried out by the members of the North Greenland Eemian Ice Drilling (NEEM) community. Their report (NEEM Members, 2013) details the results of the analysis of a

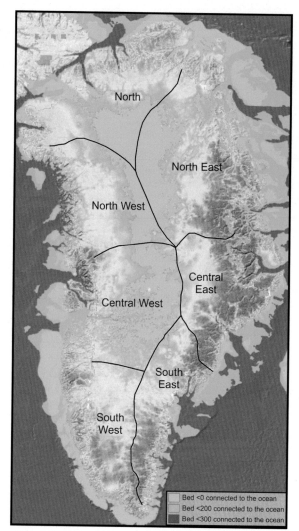

2540-m-long ice core drilled through the ice during field seasons in 2008−12 at a site in northwest Greenland (77.45°N, 51.06°W). They took measurements of stable water isotopes in the ice ($\delta^{18}O_{ice}$) and CH_4, N_2O concentrations and the isotopes $\delta^{15}N$ of N_2 and $\delta^{18}O_{atm}$ of O_2, noble gases, and the air content, all extracted from air bubbles in the ice. The top 1419 m is from the Holocene (the last 11,000 yr), and together with the glacial ice below, it can be matched to existing GIS ice-core timescales down to 108,000 yr BP (108 kyr BP). Below this, the ice was found to be disturbed and folded. The research team concluded that this deeper, folded ice predates the Eemian. This pre-Eemian ice lies just above bedrock, and its low $\delta^{18}O_{ice}$ values suggest that the ice layers most probably represent the glacial period before the Eemian. Northern Hemisphere temperatures are known to vary in parallel with the atmospheric methane (CH_4) concentrations, which increased abruptly at 128 kyr BP, signaling the start of the Eemian period. This indicates that the NEEM ice core recovered the Eemian record intact, without any gaps before the commencement of the interglacial period. The reconstructed Eemian sequence spans the interval 128.5−114 kyr BP. The ice core layers from 127 to 118.3 kyr show clear evidence of melt features, an indication of warmer temperatures at the study site than those of the last millennium. This interpretation is independently substantiated by a decrease of $\delta^{15}N$ in this zone, which indicates that mean annual firn temperatures at the depositional site were 5°C warmer than today. Firn is the term used for granular snow, especially on the upper part of a glacier or ice sheet, that has not yet been compressed into ice. Between 128.5 kyr BP and 126.0 kyr BP, $\delta^{18}O_{ice}$ increased from −35‰ to −31.4‰. This increase in $\delta^{18}O$ values at 126 kyr BP implies that surface temperatures at the NEEM core site were 7.5 ± 1.8 °C warmer than those of the last millennium. Fig. 4.9 (above) illustrates the progression of changes in $\delta^{18}O_{ice}$ values during the Eemian and the corresponding interpreted changes in mean annual temperature at the NEEM site during the same interval (Fig. 4.9, below).

Stable isotope studies

The results of the stable isotope studies agree with an independent paleoclimate reconstruction from the NEEM core, based on an isotope of beryllium (BE). Sturevik-Storm et al. (2014) reported on the ^{10}Be record as a proxy for climate change at the NEEM site. The North Greenland Eemian Ice Coring project (NEEM, 2007−12), retrieved ice from the Eemian period dating from 115.36 to 128.48 kyr BP. To establish the chronologic boundaries of the Eemian, the authors matched $\delta^{18}O_{atm}$ and CH_4 records from the NEEM core with the records of the same isotopes from the NGRIP (North Greenland Ice core Project) and EDML (Epica Dronning Maud Land, Antarctica) cores. They also compared the variability in N_2O, $\delta^{15}N_{atm}$ and air content to match the various ice core records. The study showed that average ^{10}Be concentrations during the Eemian were roughly 70% lower than in the

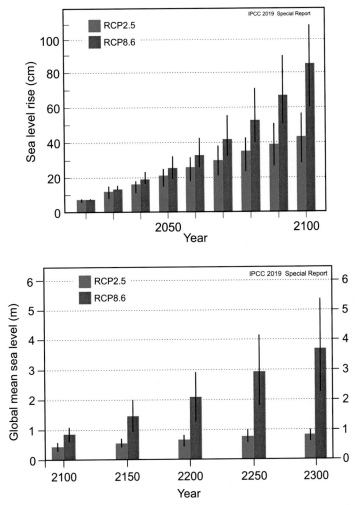

FIG. 4.8 Above: IPCC assessment of the likely range of the sea level rise until 2100 under the conditions associated with RCP2.6 and RCP8.5. Below: IPCC projected sea level rise until 2300. *IPCC, Intergovernmental Panel on Climate Change. (Both graphs based on data from Oppenheimer, Glavovic, B. C., Hinkel, J., Wal, Magnan, A. K., & et al., 2019. Sea level rise and implications for low-lying islands, coasts and communities. In IPCC Special Report on the Ocean and Cryosphere in a Changing Climate (pp. 321–445); in public domain.)*

Holocene. This suggests that Eemian climates in North Greenland were warmer and wetter, as the regional precipitation was reconstructed as 65%–90% higher than today.

Melting of the Eemian Greenland ice sheet
Within 6000 yr, from 128 to 122 kyr BP, the elevation of the ice sheet at the NEEM site is estimated to have decreased by about 80 m below the present surface

elevation. Based on this estimate, the ice thickness at NEEM decreased by an average of 764 cm per year between 128 and 122 kyr BP and stayed at this level until 117−114 kyr BP, long after surface melt ceased, and temperatures fell below modern levels. This is one example of an environmental change that came on rapidly and then returned to its previous state much more slowly.

This amount of ice thinning on top of the GIS during the Eemian represents a reduction of only about

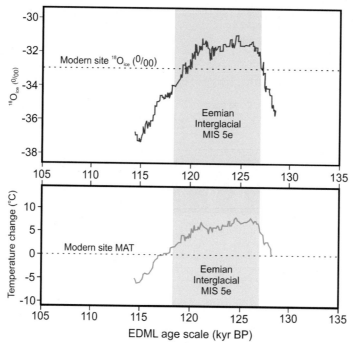

FIG. 4.9 Above: The reconstructed records of $\delta^{18}O_{ice}$ records centering on the Eemian interglacial period, plotted on the EDML timescale. Millennial-scale variations of $\delta^{18}O_{atm}$ are evidenced over MIS 5 both on the Antarctic EPICA Dronning Maud Land (EDML) and Greenland ice cores and are coupled to CH4 profiles to synchronize the NEEM and EDML records. Below: Reconstruction of the temperature history during the Eemian interglacial. (Based on the stable water isotopes ($\delta^{18}O_{ice}$) and air content records; Data from NEEM Members. 2013. Eemian interglacial reconstructed from a Greenland folded ice core. Nature 493, 489—494. https://doi.org/10.1038/nature11789.)

10% compared with the modern ice thickness. However, the NEEM team (2013) points out that there may well have been substantial melting and reduction of ice thickness near the margins of the GIS. Such thinning at the margins would have reduced the overall volume of the GIS. Although the documentation of ice thickness at one location on the GIS cannot constrain the overall ice sheet changes during the Eemian, the NEEM data agree with GIS models that indicate only a modest contribution (2 m) from the melting of the GIS to the observed 4—8m Eemian sea level high stand. This suggests that the melting of Antarctic ice must have contributed far more significantly to the Eemian sea level rise.

Tipping point temperature thresholds

As reviewed by Thomas et al. (2020), recent literature discussing the trajectory of the Earth System to a "hothouse" condition suggests some of the temperature

thresholds above which some elements will "tip" into an alternative state (Table 4.2). It appears that the tipping point for the GIS, WAIS, and Arctic summer sea ice is 1—3°C warming above recent levels. Changes in the boundaries of the world's boreal forests as well as changes in thermohaline circulation are predicted to take place if temperatures rise by 3—5°C. Arctic winter sea ice, permafrost, and the East Antarctic ice sheet will be significantly altered at temperatures >5°C above modern, although some subregions are vulnerable to change below this temperature.

Arctic sea-ice retreat appears to be one of the first elements within the Arctic environment to have tipped during the Eemian. This retreat predated other large-scale environmental changes in the Arctic, starting before 130 ka. Evidence of substantial Greenland ice sheet mass loss, as recorded from the North Greenland Eemian Ice Drilling core site, begins about 128 ka (NEEM Members, 2013).

TABLE 4.2
Tipping Points and Responses of Various High-Latitude Systems.

System	Predicted Tipping Point	System Response	References
(1) Polar ice sheets	1–3°C warming	Substantial melting, thinning, ice loss	Meredith et al. (2019)
(2) Arctic summer sea ice	1–3°C warming	Likely disappearance	Meredith et al. (2019)
(3) Arctic winter sea ice	>5°C warming	Substantial thinning and reduction of extent	Meredith et al. (2019)
(4) Sea level	1–2°C warming	~43 cm sea level rise level by 2100; ~90 cm rise by 2300	Oppenheimer et al. (2019)
	>3°C warming	~84 cm sea level rise by 2100; ~4 m rise by 2300	Oppenheimer et al. (2019)
(5) Mountain glaciers	3–5°C warming	No more than 1 m increase in sea level	Bamber et al. (2018)
(6) Ocean circulation	Atlantic meridional overturning circulation	Rapid centennial-scale reductions of North Atlantic deep-water formation	Thornalley et al. (2018)
(7) Boreal forest	3–5°C warming	Northward extension by 600–1000 km in Eurasia	Otto-Bliesner et al. (2020)
(8) Extent of permafrost	1–3°C warming	By 2100, near-surface permafrost area decrease up to 66% (RCP2.6) or up to 99% (RCP8.5)	Meredith et al. (2019)

Concepts from Thomas, Z. A., Jones, R. T., Turney, C. S., Golledge, N., Fogwill, C., Bradshaw, C. J. A., Menviel, L., McKay, N. P., Bird, M., Palmer, J., Kershaw, P., Wilmshurst, J., & Muscheler, R. 2020. Tipping elements and amplified polar warming during the Last Interglacial. Quaternary Science Reviews 233. https://doi.org/10.1016/j.quascirev.2020.106222.

One of the most important Earth system feedbacks is positive ice-albedo feedback. As discussed in Chapter 1, increased exposure of the dark ocean waters with a lower surface albedo following sea-ice melt causes increased absorption of solar radiation, enhancing surface melt. New areas of open water can increase in temperature by up to 4–5°C during summer, resulting in the delay in the onset of autumn sea-ice formation. Important feedbacks have been identified in Arctic sea ice. For instance, regional warming causes first-year sea ice to grow more slowly and melt more rapidly, allowing increasingly larger areas of ice-free water in each subsequent year. Thus, a potential tipping point can occur when the summer melt rate exceeds the winter growth rate such that multiyear ice cannot establish, and there is only seasonal ice cover.

A similar phenomenon is being observed today on the GIS. Here, there are strong positive feedbacks related to albedo and elevation. As the surface of the ice sheet melts, the local elevation of the ice decreases, effectively causing a rise in temperature of the surface. Ice-albedo feedback also includes the effects of surface melt-ponds in increasing the absorption of solar radiation.

This feedback mechanism further increases melting until a new equilibrium is reached (Thomas et al., 2020).

The sea-ice threshold

Although the Arctic Ocean represents just 1% of the global ocean volume, it receives an order of magnitude more freshwater runoff, relative to its volume, than the other oceans, meaning that the changing freshwater supply is an important determinant of its properties, e.g., salinity and buoyancy. The loss of perennial sea-ice cover in the Arctic influences the thermohaline circulation and regional climate in the sub-Arctic and North Atlantic. Freshwater anomalies from Greenland ice sheet melt propagate into the Nordic seas, North Atlantic, and beyond, enabling a change to the thermohaline structure that influences sea-ice coverage and thickness. Reduction of the thickness of sea ice can also increase heat flux from the ocean through reduced insulation of the thinner and/or absent ice. This also warms the shallow continental shelf, helping to melt offshore permafrost, in turn contributing to the release and decomposition of trapped methane hydrates. While a relationship between an ice-free Arctic and permafrost thaw has been established during interglacials prior to 400,000 years ago, links during the Last Interglacial are less clear, possibly due to the presence of perennial/seasonal sea ice in the Western Arctic. However, if the current trend toward reduced sea-ice extent in the Arctic continues, the thawing of the Siberian permafrost is a likely candidate to join the tipping cascade.

Recent history of GIS melting: In recent decades, the GIS has experienced increased surface melt. Tedesco and Fettweis (2020) report on the extremely high level of SMB loss and runoff of the GIS during the summer of 2019. They used a combination of remote sensing observations, regional climate model outputs, data reanalysis, and artificial neural networks that documented the unprecedented atmospheric conditions that took place that summer over Greenland. The summer of 2019 was characterized by highly persistent anticyclonic conditions. In conjunction with low albedo associated with reduced snowfall in summer, these environmental conditions enhanced melt-albedo feedback by facilitating the absorption of insolation. The conditions also enhanced the advection of warm, moist air along the western portion of the GIS toward the north. This is the region where surface melt has been the highest since 1948 (Fig. 4.10).

Summer of 2019

Since records of the volume of runoff from the GIS began to be kept in 1948, the 2019 runoff was second only to that of 2012. The GIS also suffered the highest negative anomaly of SMB for the hydrological year of 1 September 2018–31 August 2019. Statistical analyses showed that the total number of days with the five most frequent atmospheric patterns that characterized the summer of 2019 was five standard deviations above the 1981–2010 mean, confirming the exceptional nature of the 2019 season. The duration of summer melting in 2019 exceeded the long-term (1981–2010) mean by up to 40 days along the western side of the ice sheet where dark, bare ice is exposed. Over the rest of the ice sheet, the anomaly of the number of melting days during the summer of 2019 was around 20 days. Negative anomalies (i.e., fewer than average melting days) were rare and geographically concentrated over a small area in the southern portion of the ice sheet. Surface melting in 2019 started relatively early, around mid-April (day 105 of the year) and exceeded the 1981–2010 mean for the rate of melting for about 82% of the days from June through August.

Mechanisms of enhanced Greenland ice sheet surface melting

The underlying cause of increased surface melting and how it relates to environmental changes throughout the Arctic remains unclear. Liu et al. (2016) produced evidence indicating that an important contributing factor is a decrease in Arctic sea ice. Reduced summer sea

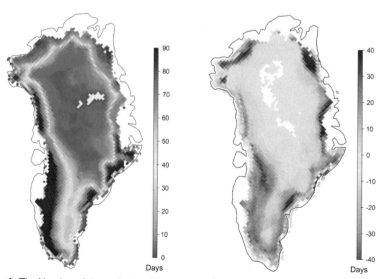

FIG. 4.10 Left: The Number of days when melting occurred during the 2019 summer (June, July, August, JJA) according to spaceborne passive microwave. Right: Anomaly of the Number of Melting Days with Respect to the 1981–2010. (Baseline period obtained from spaceborne passive microwave data shown on the left; After Tedesco, M., & Fettweis, X., 2020. Unprecedented atmospheric conditions (1948–2019) drive the 2019 exceptional melting season over the Greenland ice sheet. The Cryosphere 14, 1209–1223. https://doi.org/10.5194/tc-14-1209-2020.)

ice favors stronger and more frequent occurrences of blocking-high pressure events over Greenland. During Greenland summers, these blocking-high pressure systems, which spin clockwise, stay largely in place and can block cold, dry Canadian air from reaching the island. Blocking highs not only retard the entrance of cold, dry air into Greenland, but they also enhance the transport of warm, moist air, which increases downwelling infrared radiation. They also contribute to the number of extreme heat events and account for most of the observed warming trends. Blocking-high pressure systems over Greenland usually form when Arctic summer weather is dominated by warm air. This has been the case in many recent years. Arctic warmth tends to weaken the jet stream, which typically flows west to east, allowing it to meander more to the north and south.

Modeling sea-ice effects
The findings of Liu et al. (2016) are supported by analyses of observations and data reanalysis, as well as by independent atmospheric model simulations using a state-of-the-art atmospheric model that is forced by varying only the sea-ice conditions. Reduced sea-ice conditions in the model favor more extensive Greenland surface melting. The authors find positive feedback between the variability in the extent of summer Arctic sea ice and melt area of the summer GIS, which affects the GIS mass balance. This linkage may improve the projections of changes in the global sea level and thermohaline circulation. A large pool of abnormally cool ocean water south of Greenland that could slow circulation was first observed in 2013.

Greenland blocking index
Hanna et al. (2013) correlated that Greenland coastal weather station temperature data with the North Atlantic Oscillation (NAO) and the Atlantic Multidecadal Oscillation (AMO) indices for the summer season (when GIS melt and runoff occur) reveal significant temporal variations over the past 100 years, with periods of strongest correlations in both the early 20th century and recent decades. During the mid-20th century, temperature changes at the stations are not significantly correlated with these atmospheric circulation patterns. Greenland coastal summer temperatures and GIS runoff since the 1970s are more strongly correlated with the Greenland Blocking Index (GBI) than with the NAO Index (NAOI), making the GBI a potentially useful predictor of ice sheet mass balance changes (Hanna et al., 2020). The GBI measures the occurrence and strength of atmospheric high-pressure systems,

which tend to remain stationary for long periods of time, causing long runs of relatively stable and calm weather conditions. The high pressure also blocks storm systems from moving into the region.

Their results show that the changing strength of NAOI-temperature relationship that is seen in the boreal regions also extends to Greenland summers. Hanna et al. (2013) found that Greenland temperatures and GIS runoff over the past 30—40 years are significantly correlated with AMO variations, although they are more strongly correlated with GBI changes. They have found an increase in the occurrence of the GBI over Greenland since the 1980s throughout all seasons, which is linked with significantly strong warming of Greenland and the whole Arctic region compared with the rest of the world. The extent of GIS summer melting is less significantly correlated with atmospheric and oceanic index changes than runoff, which we attribute to the latter being a more quantitative index of GIS response to climate change. Moreover, the four recent warm summers of 2007—10 are characterized by unprecedented high pressure (since at least 1948—the start of the NCEP/NCAR reanalysis record) in the tropospheric column. The results of Hanna et al. (2013) suggest complex and changing atmospheric forcing conditions that are not well captured using the NAO alone, and support theories of oceanic influence on the recent increases in Greenland temperatures and GIS runoff.

Complexities of glacial ice-ocean interactions
As discussed by Straneo et al. (2019), the rapid loss of ice mass from the GIS is raising scientific interest in glacier-ocean interactions for three main reasons. First, melting at the ocean margins of Greenland glaciers is now being considered a potential trigger for the observed dynamic ice loss (roughly half of the total ice loss of the GIS) with important consequences for sea level rise. Second, increased freshwater discharge from the GIS has the potential to affect global climate by altering the Atlantic Meridional Overturning Circulation (AMOC). Third, changes in freshwater discharge may impact marine ecosystems along the Greenland margin and potentially farther afield in the North Atlantic by altering nutrient fluxes, productivity, and biogeochemical properties of coastal waters. This will affect organisms using these regions as habitat and feeding grounds as well as human societies relying on these ecosystems for subsistence.

Understanding exchanges of heat, freshwater, and nutrients that occur at the GIS marine margins is key to interpreting ongoing and projecting future ice loss, and its impact on the ocean (Fig. 4.11). The mechanisms

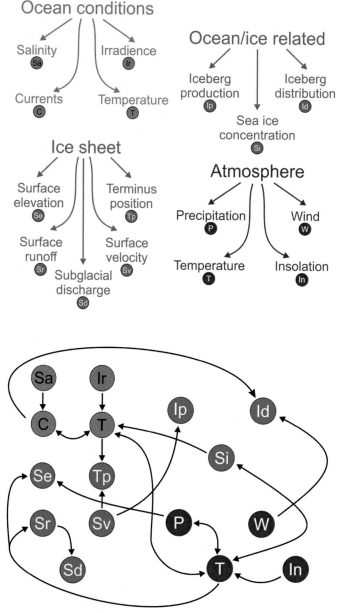

FIG. 4.11 Flowchart illustrating the component variables within ocean conditions, ocean-ice related conditions, ice sheet conditions, and atmospheric conditions, as prescribed by Straneo et al. (2019) for use in the development of a GIS-ocean observing system. Below: The author's concept of most likely interactions between the variables shown in the upper panel. *GIS*, Greenland ice sheet.

involved in these glacier-marine exchanges are not well understood. Only limited observations are available from the regions where glaciers terminate into the ocean. There are also significant challenges in the development of models at the appropriate spatial and temporal scales. Thus far, then, ice sheet/ocean exchanges are either poorly represented or not accounted for in models used for projection studies (Straneo et al., 2019).

The marine-glacial boundary

Processes occurring at the ice sheet-ocean boundary include iceberg calving, sediment-laden turbulent up-welling plumes from subglacial discharge, submarine melting of glacial ice, water circulation in fjords, and strong katabatic winds. All of these processes are intrinsically challenging to observe and quantify because of their complexity and small spatial scales. Changes at the ice sheet-ocean boundary can trigger dynamic glacial responses such as stress perturbations in the ice sheet momentum balance. These result in the thinning of the upstream glacier and further GIS ice loss. In other words, events at the marine-glacial boundary tend to destabilize at least the distal end of the glacier.

A direct consequence of ice loss from the GIS is the increase in the liquid (meltwater) and solid (iceberg) discharge of freshwater into the North Atlantic. As discussed earlier, this may have an impact on AMOC, with potentially important climate consequences beyond the Arctic. To properly assess the impacts of a shrinking GIS on large-scale ocean circulation, we must gain a far better understanding of GIS freshwater discharge boundary conditions and apply this knowledge to develop better-coupled ocean-atmosphere or ocean-only models. The current model boundary conditions quantify the ice and meltwater discharge at the ice sheet-ocean margins (typically at the head of narrow fjords) but neglect what is happening within the fjords themselves (Straneo et al., 2019). We know that there are important within-fjord processes, based on observations of Greenland's fjords. For instance, there is a substantial modification of the glacial meltwater discharge by in-fjord processes. These include iceberg melt, dilution of surface melt by turbulent plumes, and more complex fjord-ocean exchanges of freshwater than would be expected from simple surface export of freshwater. The development of our understanding of the mechanisms of freshwater flux from Greenland fjords to the North Atlantic is a vital step in the development of models to predict the impact of future GIS mass loss on the ocean.

FJORD DYNAMICS

Glacial fjords represent a bottleneck through which oceanic heat is delivered to the ice sheet margins. While warm water enters the fjords, meltwater, icebergs, and nutrients exit the fjords to enter the global ocean. These fjords are typically 100 km long and 5–10 km wide. Outlet glaciers in Greenland are often grounded in hundreds of meters of water. Only a few fjords in Northern Greenland have a floating ice tongue that covers much of the fjord's extent. Fjord topography, including the presence of a sill, regulates the exchange of water masses with the continental shelf, where the waters of Atlantic and Arctic origin coexist. Fjords with sills deeper than 100 m are characterized by a warm, Atlantic-sourced layer underneath a colder, fresher surface layer (Straneo et al., 2019). Certain processes govern the exchanges of heat, freshwater, and nutrients at the ice/ocean boundary, and the upwelling of deep nutrient-rich ocean waters at the glaciers' margins. These processes include localized plume upwelling driven by the subglacial discharge of ice sheet surface melt at glacier grounding lines and distributed melting along the glacier face. The circulation of waters in the fjord is likely driven by a combination of buoyancy, shelf-driven, and wind-driven forcing. This circulation maintains a steady supply of warm water to melt ice as well as regulating the export of the strongly diluted meltwater. Icebergs, commonly found next to calving glaciers, release meltwater throughout much of the fjord water column as they melt. Subglacial discharge represents ice sheet surface melt that is routed to the ice sheet base via an englacial drainage system. This discharge has emerged as a major component of ice sheet-ocean interactions. It amplifies ice sheet-ocean exchanges. Calving of icebergs, which balances most of the ice flux across the grounding line, is another poorly understood process that is likely influenced by climatic conditions and is a key regulator of glacier dynamics. Efforts to reconstruct the history of ice sheet mass loss are receiving considerable research attention.

Global Ocean Warming and Sea Level Rise

The oceans are absorbing more than 90% of the increased atmospheric heat associated with anthropogenic greenhouse gas (GHG) emissions. Not surprisingly, the heat content of the oceans (expressed in units of heat [Joules]) has been steadily rising. The heat content of the upper 700 m of the ocean began rising in about 1990, and as of 2019, the heat content stands at 16×10^{22} J greater than the long-term average. The contribution of oceanic thermal expansion of the upper 2000 m of seawater has likewise been continually rising since 1985, and as of 2019, this contribution to GMSL was 28 mm above the long-term average (Fig. 4.12) (NOAA/NESDIS/NCEI Ocean Climate Laboratory 2019).

Thermal expansion of seawater

The addition of heat causes water to expand in volume, but not in a uniform way. The amount of thermal expansion for a given volume of ocean water depends

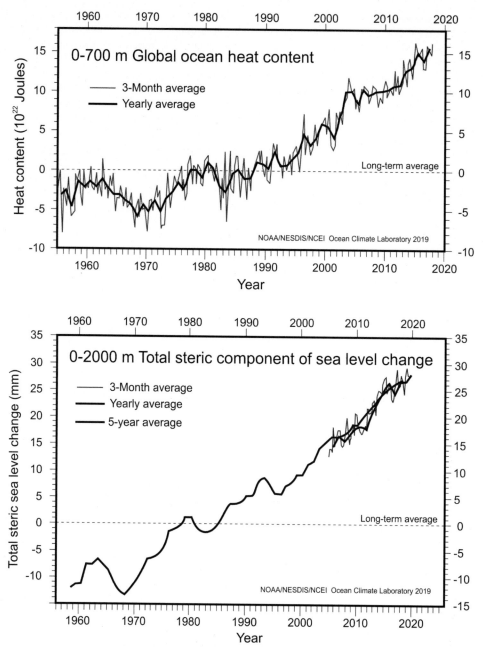

FIG. 4.12 Above: 0–700 m—global ocean heat content, compared with the long-term average. Below: 0–2000 m—total steric component of sea level change. (After NOAA/NESDIS/NCEI Ocean Climate Laboratory. 2019. Global Ocean Heat Content. https://www.nodc.noaa.gov/OC5/3M_HEAT_CONTENT/.)

on depth, salinity, and the existing temperature. Because of this, a 1°C change in ocean water from 10 to 11°C produces a different amount of expansion than a change from 21 to 22°C. Therefore, a fixed

quantity of energy (e.g., 1 GJ, or 1×10^9 J) injected as heat into different parts of the ocean will cause varying amounts of thermal expansion. Thus, even an "average" warming of 1°C across the entire ocean produces a wide

Main causes of recent sea level rise (mm per year)

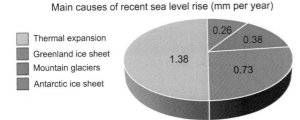

Thermal expansion
Greenland ice sheet
Mountain glaciers
Antarctic ice sheet

FIG. 4.13 Percentage contributions to modern sea level rise from the various sources. (Data from Allison, I., Colgan, W., King, M., & Paul, F., 2015. Ice sheets, glaciers, and sea level. In Snow and Ice-Related Hazards, Risks, and Disasters (pp. 713–747). Elsevier Inc. https://doi.org/10.1016/B978-0-12-394849-6.00020-2.)

range of thermal expansions, depending on the temperature signatures in the various ocean regions.

As reviewed by Allison et al. (2015), the thermal expansion of seawater has contributed about 45% of total sea level rise since 1972, compared with a glacier mass loss contribution of about 40%, with most of

the remaining sea level rise assumed to be sourced from the ice sheets (Fig. 4.13).

For the period since 1972, for which a reasonably good distribution of ocean temperature data is available, warming of the upper 700 m of the ocean is estimated to have resulted in a sea level rise of 0.63 mm/year. More recently, during the post-1993 satellite altimetry period, upper ocean thermal expansion is equivalent to 0.71 mm/year of sea level rise, and over the full ocean depth, the expansion is equivalent to 0.88 mm/year of sea level rise. Data taken from Argo profiling floats show that this warming has caused roughly one-third of the global sea level rise observed by satellite altimeters since 2004. Despite this slight acceleration in the sea level rise associated with thermal expansion, in the past decade, the sea level rise contribution from terrestrial ice has overtaken that due to thermal expansion. The rates of sea level rise have been steadily increasing since the middle of the 20th century and are now somewhere between 3.3 and 3.58 mm per year, based on various estimates (Fig. 4.14 and Table 4.3).

Tide Gauge Estimates

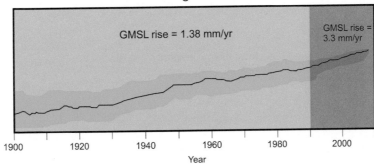

Estimates from Observations and Climate Models

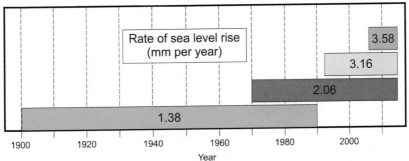

FIG. 4.14 Above: Observed Rates of sea level rise, 1900–2015, based on tide gauge data. Below: Rates of sea level rise (in mm per year) estimated from a combination of climate models and tide gauge data. (Data from Oppenheimer, Glavovic, B. C., Hinkel, J., Wal, Magnan, A. K., & et al. 2019. Sea level rise and implications for low-lying islands, coasts and communities. In IPCC Special Report on the Ocean and Cryosphere in a Changing Climate (pp. 321–445); In the public domain.)

TABLE 4.3
Global Mean Sea Level Budget Over Different Periods From Observations and From Climate Model Base Contributions, Shown in mm per year. Values in Brackets Are Uncertainties Ranging From 5% to 95%.

Source	1901–90	1970–2015	1993–2015	2006–15
OBSERVED CONTRIBUTION TO GMSL RISE				
Thermal expansion		0.89 (0.84–0.94)	1.36 (0.96–1.76)	1.40 (1.08–1.72)
Glaciers except in Greenland and Antarctica	0.49 (0.34–0.64)	0.46 (0.21–0.72)	0.56 (0.34–0.78)	0.61 (0.53–0.69)
GIS including peripheral glaciers	0.40 (0.23–0.57)		0.46 (0.21–0.71)	0.77 (0.72–0.82)
Antarctica ice sheet including peripheral glaciers			0.29 (0.11–0.47)	0.43 (0.34–0.52)
Observed GMSL rise from tide gauges and altimetry	1.38 (0.81–1.95)	2.06 (1.77–2.34)	3.16 (2.79–3.53)	3.58 (3.10–4.06)
MODELED CONTRIBUTIONS TO GMSL RISE				
Thermal expansion	0.32 (0.04–0.60)	0.97 (0.45–1.48)	1.48 (0.86–2.11)	1.52 (0.96–2.09)
Glaciers	0.53 (0.38–0.68)	0.73 (0.50–0.95)	0.99 (0.60–1.38)	1.10 (0.64–1.56)
Greenland surface mass balance	−0.02 (−0.05–0.02)	0.03 (−0.01–0.07)	0.08 (−0.01–0.16)	0.12 (−0.02–0.26)
Total including land water storage and ice discharge	0.71 (0.39–1.03)	1.88 (1.31–2.45)	3.13 (2.38–3.88)	3.54 (2.79–4.29)

GIS, Greenland ice sheet; *GMSL*, global mean sea level.
Data from Oppenheimer, Glavovic, B. C., Hinkel, J., Wal, Magnan, A. K., & et al. 2019. Sea level rise and implications for low-lying islands, coasts and communities. In IPCC Special Report on the Ocean and Cryosphere in a Changing Climate (pp. 321–445).

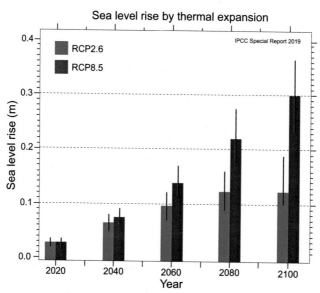

FIG. 4.15 Predicted sea level rise based on thermal expansion under various RCP scenarios. *RCP*, representative concentration pathway. (Based on data from Church, J. A., Clark, P. U., Cazenave, A., Gregory, J. M., Jevrejeva, & et al., 2013. The physical science basis. Contribution of working group I to the Fifth assessment report of the intergovernmental panel on climate change. In Sea Level Change (pp. 1137–1216); in the public domain.)

Predicted sea level rise

The IPCC Fifth Assessment Report (2014) contained a set of predicted sea level rises by the year 2100, based on various levels of GHGs (best-case scenario: RCP2.6; worst-case scenario: RCP8.5).

The graph (Fig. 4.15) indicates that under the best-case scenario, thermal expansion of seawater will cause GMSL to rise by 0.15 ± 0.4 m by the end of the 21st century. Under the worst-case scenario, the IPCC predicts that thermal expansion of seawater will cause GMSL to rise by 0.30 ± 0.6 m (Church et al., 2013).

Conclusions

The melting of the GIS is one of the main components of the predicted sea level rise in the coming centuries. As we have seen, the GIS holds enough water to raise mean global sea level by 7.4 m. It might seem unlikely that the entire GIS will disappear in the foreseeable future, but keep in mind that global warming may reach levels that have not been seen on this planet in millions of years. If we do little or nothing too slow the increase in GHG emissions in the coming decades, then our descendants will be forced to live with the worst-case scenario forecast by the IPCC. As discussed by Oppenheimer et al. (2019), future sea level rise will depend on which RCP emission scenario is followed. Sea level rise at the end of the 21st century will almost certainly be faster under all the RCP scenarios—even those compatible with achieving the long-term temperature goal set out in the Paris Agreement. Beyond 2100, sea level will continue to rise for centuries due to continuing deep-ocean heat uptake and mass loss of the GIS and Antarctic ice sheet. Even if GHG emissions were to end today, sea levels will remain elevated for thousands of years into the future, simply because we have pumped an enormous amount of heat into the world's oceans, by way of GHG warming of the atmosphere.

Human impacts

Impacts of sea level rise on coastal populations are painful to contemplate, although no accurate worldwide estimate of the number of people likely to be affected by coastal inundation or flooding has thus far been published. Certain facts are undeniable. Of the world's 15 largest cities, 11 cities lie along coasts or estuaries. In the United States, around 53% of the population lives near the coast. More than 10% of the world's population (approximately 780 million people) now reside in towns and cities situated at less than 10 m above sea level. These low-lying cities and towns will certainly be inundated by rising sea levels during the next century or two.

FIG. 4.16 Photo of the Thames Barrier by Andy Roberts. (Creative Commons Attribution 2.0 Generic license.)

The wealthiest nations of the world may find ways of minimizing the damage done by high sea levels to big coastal cities through extraordinary feats of engineering. Consider the Thames Barrier (Fig. 4.16), built to protect London from storm surges up the Thames from the North Sea. The barrier spans 520 m near the mouth of the Thames and protects 125 square km of central London from flooding. It has 10 steel gates that can be raised into position across the River Thames. Completed in 1982, it took 12 years to build at a cost of £534 million, the equivalent of £1.9 billion today. Many developing countries that lie close to sea level do not have the financial resources to protect their coasts from rising sea levels. These include Bangladesh (average elevation 8.5 m), Gambia (34 m), Guinea-Bissau (70 m), Maldives (1.8 m), Senegal (69 m), and Trinidad and Tobago (83 m). Relatively poor regions of the United States, such as rural Alaska, are already suffering the effects of record-setting storm surges. As will be discussed in some detail in Chapter 17, these surges and high tides are having some devastating effects on Arctic towns and villages, most of which are coastal. Warming regional temperatures are causing coastal permafrost to melt and soil to collapse into the sea.

Storm surges have caused significant damage because of coastal erosion in several native villages in Alaska (Fig. 4.17).

Some villages are being forced to relocate to higher ground as their buildings are being inundated and the ground is literally washing away beneath their feet. One of the dominant themes of this book is the interconnectedness of the Arctic with the rest of the world.

Coastal erosion near Kaktovik, AK. Photo by M. Torre Jorgenson, USGS

FIG. 4.17 Above: Coastal erosion near Kaktovik, AK. Below: A cabin along Alaska's Arctic coast was recently washed into the ocean because the bluff it was sitting on eroded away. (Photo by M. Torre Jorgenson, USGS, in the public domain; Photo by Benjamin Jones, USGS; in the public domain.)

Thus, global warming, driven by the industrialized countries of the midlatitudes, causes the GIS to melt. The meltwater from the GIS flows on the surface in streams and accumulates in ponds and lakes (Fig. 4.18).

Meltwater also finds its way through cracks in the ice, eventually flowing down to the base, where it travels in streams that emerge at or near glacial termini at the sea.

These accumulated meltwaters, along with meltwater from mountain glaciers and Antarctica, combined to raise sea levels around the world. As if that were not enough, the water in the world's oceans is expanding because of atmospheric warming.

Rising sea level also puts coastal regions in danger from the storm surges and high tides that accompany elevated sea level. A study using models to predict the impacts

NASA

NSF

FIG. 4.18 Above: The Greenland ice sheet (GIS), showing streams of meltwater feeding into a lake, July 19, 2015. Below: A canyon on the surface of the GIS, filled with melt water in summer 2010. (Above: Maria-José Viñas, NASA Earth Science News Team; Below: Photo by Ian Joughin, UW APL Polar Science Center, courtesy of NSF; In the public domain.)

of rising sea levels on future storm surges in the United States (Tebaldi et al., 2012) shows that, by midcentury, some low-lying locations may experience annual high-water levels that today would be considered "once in a century" events. Likewise, today's century level coastal flooding events will become "decade"-level events. In North America, regional storm surges around the Gulf of Mexico are predicted to be the largest, since this region is the most likely to be struck by hurricanes. Storm surges on the Pacific coast will be the smallest, due to the absence of hurricanes and the presence of a wide, shallow continental shelf. These are regional generalizations, but local shoreline characteristics also play a role.

REFERENCES

Allison, I., Colgan, W., King, M., Paul, F., 2015. Ice sheets, glaciers, and sea level. In: Snow and Ice-Related Hazards, Risks, and Disasters. Elsevier Inc, pp. 713−747. https://doi.org/10.1016/B978-0-12-394849-6.00020-2.

Bamber, J.L., Westaway, R.M., Marzeion, B., Wouters, B., 2018. The land ice contribution to sea level during the satellite era. Environmental Research Letters 13, 063008. https://doi.org/10.1088/1748-9326/aac2f0.

Bindschadler, R.A., Nowicki, S., Ouchi, A., Aschwanden, A., Choi, H., Fastook, J., Granzow, G., Greve, R., Gutowski, G., Herzfeld, U., Pollard, D., Price, S.F., Sato, T., Seddik, H., Seroussi, H., Takahashi, K., Walker, R., Wang, W.L., 2013. Ice-sheet model sensitivities to environmental forcing and their use in projecting future sea level (the SeaRISE project). Journal of Glaciology 59 (214), 195−224. https://doi.org/10.3189/2013JoG12J125.

Church, J.A., Clark, P.U., Cazenave, A., Gregory, J.M., Jevrejeva, et al., 2013. The physical science basis. Contribution of working group I to the Fifth assessment report of the intergovernmental panel on climate change. In: Sea Level Change, pp. 1137−1216.

Hanna, E., Jones, J.M., Cappelen, J., Mernild, S.H., Wood, L., 2013. The influence of North Atlantic atmospheric and oceanic forcing effects on 1900−2010 Greenland summer climate and ice melt/runoff. International Journal of Climatology 33, 862−880. https://doi.org/10.1002/joc.3475.

Hanna, E., Pattyn, F., Navarro, F., Favier, V., Goelzer, H., van den Broeke, M.R., Vizcaino, M., Whitehouse, P.L., Ritz, C., Bulthuis, K., Smith, B., 2020. Mass balance of the ice sheets and glaciers − progress since AR5 and challenges. Earth-Science Reviews 201, 102976, 17 pp. https://doi.org/10.1016/j.earscirev.2019.102976.

IMBIE Team, 2019. Mass balance of the Greenland ice sheet from 1992 to 2018. Nature 579, 233−238. https://doi.org/10.1038/s41586-019-1855-2.

IPCC, 2014. Climate Change 2014: Synthesis Report. Contribution of Working Groups I, II and III to the Fifth Assessment Report of the Intergovernmental Panel on Climate Change. IPCC, Geneva.

Liu, J., Chen, Z., Francis, J., Song, M., Mote, T., Hu, Y., 2016. Has arctic sea ice loss contributed to increased surface melting of the Greenland ice sheet? Journal of Climate 29 (9), 3373−3386. https://doi.org/10.1175/JCLI-D-15-0391.1.

Meredith, M., Sommerkorn, M., Cassotta, S., Derksen, C., Ekaykin, A., 2019. Polar regions. In: Pörtner, H.-O., Roberts, D.C., Masson-Delmotte, V. (Eds.), IPCC Special Report on the Ocean and Cryosphere in a Changing Climate. IPCC, p. 118.

Morlighem, M., Williams, C.N., Rignot, E., 2017. BedMachine v3: complete bed topography and ocean bathymetry mapping of Greenland from multibeam echo sounding combined with mass conservation. Geophysical Research Letters 44 (1), 11,051−11,061. https://doi.org/10.1002/2017GL074954.

NEEM Members, 2013. Eemian interglacial reconstructed from a Greenland folded ice core. Nature 493, 489–494. https://doi.org/10.1038/nature11789.

NOAA, 2019. What Is a Tide Gauge? Retrieved February 4, 2019. https://oceanservice.noaa.gov/facts/tide-gauge.html, 04/09/20. Original work published.

NOAA, 2019. Global Ocean Heat and Salt Content. https://www.nodc.noaa.gov/OC5/3M_HEAT_CONTENT/.

NOAA/NESDIS/NCEI Ocean Climate Laboratory, 2019. Global Ocean Heat Content. https://www.nodc.noaa.gov/OC5/3M_HEAT_CONTENT/.

Oppenheimer, Glavovic, B.C., Hinkel, J., Wal, Magnan, A.K., et al., 2019. Sea level rise and implications for low-lying islands, coasts and communities. In: IPCC Special Report on the Ocean and Cryosphere in a Changing Climate, pp. 321–445.

Otto-Bliesner, B.L., Brady, E.C., Zhao, A., Brierley, C., Axford, Y., 2020. Large-scale features of Last Interglacial climate: results from evaluating the lig127k simulations for CMIP6-PMIP4. David Salas Y Melia 17. https://doi.org/10.5194/cp-2019-174.

Rignot, E., Box, J.E., Burgess, E., Hanna, E., 2008. Mass balance of the Greenland ice sheet from 1958 to 2007. Geophysical Research Letters 35, L20502, 5pp. https://doi.org/10.1029/2008gl035417.

Straneo, F., Sutherland, D.A., Stearns, L., 2019. The case for a sustained Greenland ice sheet-ocean observing system (GrIOOS). Frontiers in Marine Science 138, 1–23. https://doi.org/10.3389/fmars.2019.00138.

Sturevik-Storm, A., Aldahan, A., Possnert, G., Berggren, A-M., MuschelerDahl-Jensen, R.D., et al., 2014. ^{10}Be climate fingerprints during the Eemian in the NEEM ice core, Greenland. Scientific Reports 4, 6408. https://doi.org/10.1038/srep06408 1.

Tebaldi, C., Strauss, B.H., Zervas, C.E., 2012. Modeling sea level rise impacts on storm surges along US coasts. Environmental Research Letters 7, 014032, 11pp.

Tedesco, M., Fettweis, X., 2020. Unprecedented atmospheric conditions (1948–2019) drive the 2019 exceptional melting season over the Greenland ice sheet. The Cryosphere 14, 1209–1223. https://doi.org/10.5194/tc-14-1209-2020.

Thomas, Z.A., Jones, R.T., Turney, C.S., Golledge, N., Fogwill, C., Bradshaw, C.J.A., Menviel, L., McKay, N.P., Bird, M., Palmer, J., Kershaw, P., Wilmshurst, J., Muscheler, R., 2020. Tipping elements and amplified polar warming during the Last Interglacial. Quaternary Science Reviews 233, 106222, 17 pp. https://doi.org/10.1016/j.quascirev.2020.106222.

Thornalley, D.J.R., Delia, W., Ortega, P., Robson, J.I., Brierley, C.M., Davis, R., Hall, I.R., Moffa-Sanchez, P., Rose, N.L., Spooner, P.T., Yashayaev, I., Keigwin, L.D., 2018. Anomalously weak Labrador Sea convection and Atlantic overturning during the past 150 years. Nature 556 (7700), 227–230. https://doi.org/10.1038/s41586-018-0007-4.

FURTHER READING

Church, J.A., Clark, P.U., 2013. Sea level change. In: Climate Change 2013: The Physical Science Basis. Fifth Assessment Report of the Intergovernmental Panel on Climate Change. IPCC.

Dahl-Jensen, D., Albert, M.R., Aldahan, A., Azuma, N., Balslev-Clausen, D., Baumgartner, M., Wolff, E.W., Xiao, C., Zheng, J., Waddington, E., Wegner, A., Weikusat, I., White, J.W.C., Wilhelms, F., Winstrup, M., Witrant, E., 2013. Eemian interglacial reconstructed from a Greenland folded ice core. Nature 493 (7433), 489–494. https://doi.org/10.1038/nature11789.

Jones, B., Jorgenson, M.T., 2020. Coastal Erosion, Arctic Alaska. https://www.usgs.gov/media/images/erosion-along-alaskas-arctic-coastline-near-village-kaktovik.

NASA, 2020. Understanding Sea Level. https://sealevel.nasa.gov/understanding-sea-level/key-indicators/global-mean-sea-level/201.

NASA, Viñas, M.J., 2014. Watching the Rivers Flow on Greenland. https://earthobservatory.nasa.gov/images/86508/watching-the-rivers-flow-on-greenland.

NOAA, 2016. Sea Level Variations of the United States 1854–2006. NOAA Technical Report NOS CO-OPS 053. NOAA. Original work published 2016.

NOAA Laboratory for Satellite Altimetry, 2020. Sea Level Rise. Original work published 2020. https://www.star.nesdis.noaa.gov/socd/lsa/SeaLevelRise/.

NOAA. (2019). Global ocean heat and salt content. Retrieved 12020, from https://www.nodc.noaa.gov/OC5/3M_HEAT_CONTENT/ (Original work published 2019).

Pörtner, H.-O., Roberts, D.C., Masson-Delmotte, V., 2019. Sea level rise and implications for low-lying islands, coasts and communities. In: Oppenheimer, M., Glavovic, B.C., Hinkel, J. (Eds.), IPCC Special Report on the Ocean and Cryosphere in a Changing Climate. IPCC.

Rietbroek, R., Brunnabend, S.E., Kusche, J., Schröter, J., Dahle, C., 2016. Revisiting the contemporary sea-level budget on global and regional scales. Proceedings of the National Academy of Sciences of the United States of America 113 (6), 1504–1509. https://doi.org/10.1073/pnas.1519132113.

Roberts, A., 2020. Photo of the Thames Barrier. https://commons.wikimedia.org/wiki/File:Thames_Barrier_03.jpg.

Jorgenson, T.M., U. S. Geological Survey, Jorgenson, M.T., 2020. Coastal Erosion along Alaska's Arctic Coastline Near the Village of Kaktovik. https://www.usgs.gov/centers/eros/science/monitoring-arctic-and-boreal-ecosystems-through-assimilation-field-based?qt-science_center_objects=0#qt-science_center_objects.

CHAPTER 5

Impacts of Ocean Acidification on Arctic Marine Ecosystems

INTRODUCTION

There are many aspects of global change that remain hidden from our view. They have little effect on our daily lives, and most people ignore them: out of sight, out of mind. If this is true of global change impacts on land, then it is doubly true of the impacts on the world's oceans. Although 70% of the Earth's surface is covered by ocean waters, most of us know very little about the oceans. It is perhaps even more difficult to generate public interest in oceanic environmental impacts that are completely invisible. One exception to this is the enormous piles of plastic debris that drift across the world's oceans and wash up on shores around the world. This problem can be photographed, and films can be made about it, showing marine mammals, sea turtles, and sea birds dying because of it. But changes in ocean temperature and chemistry are invisible to the naked eye. They are happening, slowly but surely, as the planet's atmosphere warms, and as atmospheric CO_2 levels rise. This chapter concerns one of these invisible impacts: the rising acidity of the world's ocean waters.

SOURCE OF THE ACIDIFICATION PROBLEM

One property of water, including ocean water, is that it holds dissolved atmospheric gases. In fact, there is a constant exchange of gases between the air and the water. The ocean serves as a vast reservoir for dissolved carbon dioxide. About 94% of the Earth's CO_2 is held in the world's oceans. The remaining few percent is mostly tied up in terrestrial vegetation. The carbon cycle moves CO_2 between the atmosphere, oceans, and biosphere (Fig. 5.1).

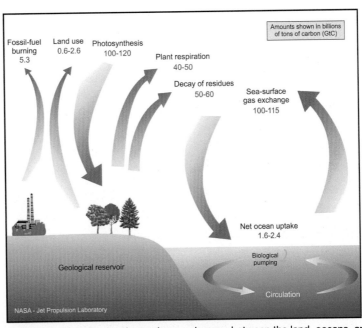

FIG. 5.1 The global carbon cycle, showing carbon exchanges between the land, oceans, and atmosphere. (Image courtesy of NASA Jet Propulsion Laboratory, in the public domain.)

Threats to the Arctic. https://doi.org/10.1016/B978-0-12-821555-5.00003-6

History of CO$_2$ Emissions

Before the Industrial Revolution, the carbon cycle had been relatively stable for the previous 11,000 years. But the burning of fossil fuels has caused an increase of atmospheric carbon dioxide levels from 278 parts per million at the start of the Industrial Revolution to over 416 parts per million today, a 40% increase. The oceans are absorbing much of the increased carbon dioxide being pumped into the atmosphere by the burning of fossil fuels. This is a critical aspect of the global carbon cycle. In 2019, global fossil fuel emissions reached 33.4 billion metric tons of CO$_2$. This level of emissions is an estimated 150 times higher than it was near the start of the Industrial Revolution, in 1850. The COVID-19 pandemic in 2020 caused the first decline in global fossil fuel emissions in more than 50 years, to 31.5 billion metric tons. The ocean's uptake of CO$_2$ is outstripping the buffering capacity of seawater, leading to a tight coupling between pH and the saturation state. These two variables decrease in concert (Gattuso et al., 2011). Tynan et al. (2016) plotted the pH of surface waters from an east-west transect of Arctic and sub-Arctic seas (Fig. 5.2). Their data clearly show the tight linkage between pH and Ωar (aragonite saturation state) across the northern seas.

The ocean is a sink for about one-quarter % of the atmospheric CO$_2$ emitted by human activities, an amount greater than 2 Pg of carbon per year (Pg C yr^{-1}; 1 Pg $= 1 \times 10^{15}$ g). Estimates of global ocean-atmosphere CO$_2$ flux (flow) provide an important

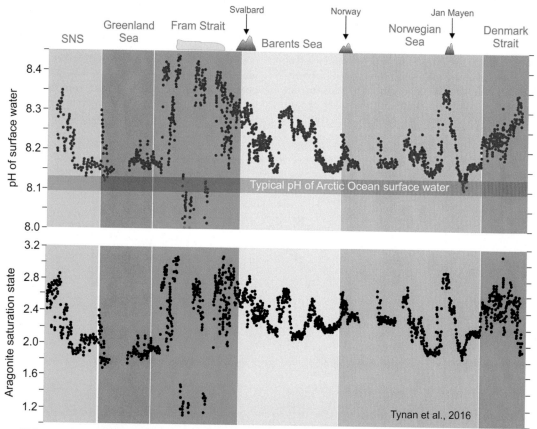

FIG. 5.2 Above: pH of surface water from a transect of sites across the northern seas. SNS stands for the South Norwegian Sea. Below: Aragonite saturation state of surface water from a transect of sites across the northern seas. (Both images after Tynan, E., Clarke, J.S., Humphreys, M.P., Ribas-Ribas, M., Esposito, M., Rérolle, V.M.C., Schlosser, C., Thorpe, S.E., Tyrrell, T., Achterberg, E.P., 2016. Physical and biogeochemical controls on the variability in surface pH and calcium carbonate saturation states in the Atlantic sectors of the Arctic and Southern Oceans. Deep-Sea Research Part II Topical Studies in Oceanography 127, 7–27. https://doi.org/10.1016/j.dsr2.2016.01.001.)

constraint on the global carbon budget. However, previous estimates of this flux were derived from surface ocean CO_2 concentrations. Those estimates were not corrected for temperature gradients between the surface and sampling of ocean waters at a few meters depth, or for the effect of the cool ocean surface skin. Watson et al. (2020) reconstructed the history of ocean-atmosphere CO_2 fluxes from 1992 to 2018 and used a numerical model that corrected for these effects. Their revised estimates increased the net CO_2 flux into the oceans by 0.8–0.9 Pg C per year. Some of the revised flux estimates were double the previous (uncorrected) estimates. Their corrections reconcile figures for the surface uptake of CO_2 in the oceans with independent estimates of the increase in ocean CO_2 inventory and suggest that previous ocean models have underestimated CO_2 uptake. We now understand that the oceans are holding even more CO_2 than we thought. This phenomenon has two obvious consequences: (1) the greater the CO_2 content of the ocean waters, the more carbonic acid is produced, with the potential to lower ocean water pH to a greater extent; (2) the higher concentrations of CO_2 in ocean water also mean that even if atmospheric CO_2 declines in the coming century, additional CO_2 will be drawn out of the ocean into the atmosphere, as these two CO_2 reservoirs balance each other. Global warming is acting to slow the flux rate of atmospheric CO_2 to the oceans. This is because increasing surface water temperatures make it more difficult for wind to mix the surface layers with the deeper layers. Thus, the ocean stratifies (settles into layers). Without an infusion of fresh carbonate-rich water from below, the surface water becomes saturated with carbon dioxide. The stagnant water also supports fewer phytoplankton, and carbon dioxide uptake from photosynthesis slows. In short, stratification cuts down the amount of carbon the ocean can take up.

The rates of change in global ocean pH and Ω are unprecedented. Modern changes are a factor of 30–100 times faster than the rate of changes in the recent geological past, and the perturbations will last many centuries to millennia. The geological record does contain past ocean acidification events, the most recent associated with the Paleocene-Eocene Thermal Maximum 55.8 million years ago (Fig. 5.3). But these events may have occurred gradually enough and under different enough background conditions for ocean chemistry and biology that there is no good analog in the fossil record for the current situation.

Ocean acidification has been called "the other CO_2 problem" (Doney et al., 2009). The enormous reservoir of carbon held in solution by the world's oceans is going to determine the extent of ocean acidification over the coming centuries. Even if anthropogenic CO_2 emissions were to stop today, the impacts of CO_2 emissions on seawater carbonate chemistry will continue to disrupt biogeochemical cycles and marine ecosystems for centuries to come (Gattuso et al., 2011, Table 5.1). Surface ocean $CaCO_3$ saturation states are declining everywhere, and Doney (2010) predicted that polar surface waters would become undersaturated for aragonite when atmospheric CO_2 reached 400–450 ppm for the Arctic (the 400 ppm threshold was passed in 2013; the 2020 atmospheric CO_2 level is 416 ppm).

Dissolved CO_2 forms carbonic acid (H_2CO_3) in oceans, leading to ocean acidification. The CO_2 added to the ocean has already caused a 30% increase in surface ocean acidity (0.1 drop in pH), which is sufficient to stress calcium carbonate ($CaCO_3$) shell-producing organisms and influence the rates of metabolic reactions in ocean biota. Ocean acidification has significant consequences for microbial abundance, marine food webs, ecosystem productivity, and commercial fisheries (Table 5.2).

One physical property of the cold waters of the Arctic region is that they absorb more gases, including CO_2 and O_2 than the warmer waters of lower latitudes. In the world oceans, the Arctic and Southern Oceans act as major sinks of CO_2 because they are colder. Cold water is also denser than warm water, causing it to sink. The carbon dioxide taken up at the ocean surface can be effectively transported to the deeper waters by convection. Hence, CO_2 is stored in the bottom layers of the ocean. However, salty water holds less dissolved gases than freshwater. This aspect of seawater offsets a large part of the temperature effect, but overall, Arctic waters are less salty than ocean waters at lower latitudes. The global average concentration of salt in seawater is 35 parts per thousand. The salinity of waters in the Arctic Ocean ranges from 30 to 34 parts per thousand. Salinity levels vary by region in the Arctic, and areas with strong river input may have salinity levels less than 30 parts per thousand. This is illustrated in Fig. 5.4, based on studies of a north-south transect of sampling sites in the Chukchi Sea (Qi et al., 2020).

Individual ocean basins display large differences in their ability to absorb atmospheric CO_2 and in their vulnerability to ocean acidification. Arctic waters have large inputs of freshwater, and the large continental shelf seas are relatively prone to ocean acidification and CO_2 outgassing. This is due in part to the remineralization of terrestrial organic carbon supplied by rivers and coastal erosion. In the Siberian shelf seas, degradation of terrestrial organic carbon is a dominant control

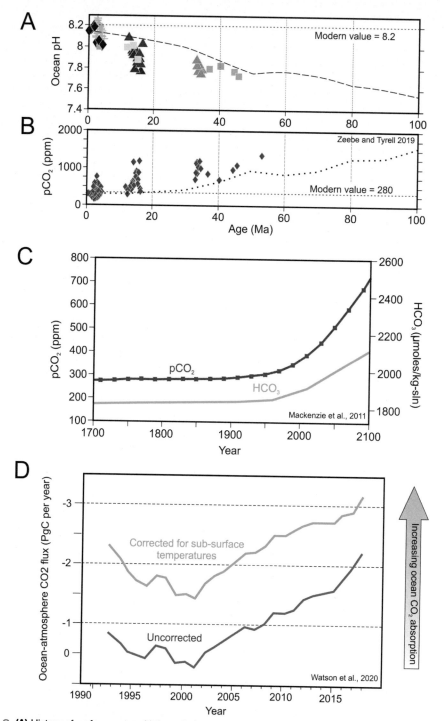

FIG. 5.3 **(A)** History of surface water pH through the past 100 million years, reconstructed from various proxies; **(B)** History of surface water pCO_2 through the past 100 million years. The pCO_2 estimates from the GEOCARB-III model. **(C)** Estimated coastal ocean HCO_3 (carbonate) concentrations and atmospheric pCO_2 from 1700–2100 AD; **(D)** Effect of near-surface temperature corrections on global CO_2 flux between the atmosphere and the oceans. ((**A** and **B**) After Zeebe, R.E., Tyrrell, T., 2019. History of carbonate ion concentration over the last 100 million years II: revised calculations and new data. Geochimica et Cosmochimica Acta 257, 373–392. https://doi.org/10.1016/j.gca. 2019.02.041; **(C)** After Mackenzie, F.T., Andersson, A.J., Arvidson, R.S., Guidry, M.W., Lerman, A., 2011. Land-sea carbon and nutrient fluxes and coastal ocean CO_2 exchange and acidification: past, present, and future. Applied Geochemistry 26, S298–S302. https://doi.org/10.1016/j.apgeochem.2011.03.087; **(D)** After Watson et al. (2020).)

TABLE 5.1
Changes in Various Ocean Parameters Predicted From CMIP5 Models and Several IPCC RCP Emissions Scenarios.

Year and RCP Emission Scenario	SST (°C)	Δ pH	Δ O_2 Content	Sea Level (m)	Volume Ω_{ar} > 1%
CHANGES RELATIVE TO 1990–99					
2090–99 (RCP8.5)	2.73	−0.33	−3.48	0.67	9.4
2090–99 (RCP4.5)	1.28	−0.15	−2.37	0.49	15
2090–99 (RCP2.6)	0.71	−0.07	−1.81	0.41	17.3
1990s (1990–99)	0	0	0	0	24
Preindustrial (1870–99)	−0.44	0.07	−	−	25.6
Preindustrial (1870–79)	−0.38	0.07	−	−	25.6
CHANGES RELATIVE TO 1870–99 (EXCEPT SEA LEVEL, RELATIVE TO 1901)					
2090–99 (RCP8.5)	3.17	−0.40	−	0.86	−
2090–99 (RCP4.5)	1.72	−0.22	−	0.68	−
2010s (2010–19)	0.83	−0.11	−	−	−
Past 10 years (2005–14)	0.72	−0.10	− 0.19*	−	−
1990s (1990–99)	0.44	−0.07	−	−	−
Preindustrial (1870–99)	0	0	−	0	−

*Value for 2010 obtained from instrumental records.
Data from Gattuso, J.-P., Magnan, A., Billé, R., Cheung, W.W.L., Howes, E.L., Joos, F., Allemand, D., Bopp, L., Cooley, S.R., Eakin, C.M., Hoegh-Guldberg, O., Kelly, R.P., Pörtner, H.-O., Rogers, A.D., Baxter, J.M., Laffoley, D., Osborn, D., Rankovic, A., Rochette, J., Turley, C., 2015. Oceanography. Contrasting futures for ocean and society from different anthropogenic CO_2 emissions scenarios. Science (New York, N.Y.) 349(6243). https://doi.org/10.1126/science.aac4722

TABLE 5.2
Assessment of Taxon Sensitivity to Future Ocean Acidification Under three RCP Scenarios for the Year 2100.

	RCP 2.6		RCP 6		RCP 8.5	
pCO$_{2atm}$	**426 PPM**		**670 PPM**		**936 PPM**	
Taxon	Evidence	Sensitivity	Evidence	Sensitivity	Evidence	Sensitivity
Pteropods[b]	Robust	Moderate	Robust	Very high	Robust	Very high
Echinoderms[a]	Medium	Moderate	Medium	Low	Robust	Moderate
All mollusks[a]	Medium	Moderate	Limited	Moderate	Robust	High
Bivalves[b]	Medium	High	Medium	Very high	Medium	Very high
Crustaceans[a]	Limited	Low	Limited	Low	Medium	Low
Krill[b]	Limited	Moderate	Limited	High	Limited	High
Fin fish[b]	Medium	Moderate	Medium	Very high	Medium	Very high
All fish[a]	Limited	Low	Limited	High	Limited	High

[a] Data from Wittmann and Pörtner (2013).
[b] Data from Gattuso et al. (2015).

of acidification and CO_2 accumulation in surface waters. In the Arctic outflow shelf sea Hudson Bay, in northern Canada, strong links between $CaCO_3$ saturation (a measure of the tendency for $CaCO_3$ to dissolve, CO_2 air-sea flux, and freshwater abundance) have prompted speculation that CO_2 flux and ocean

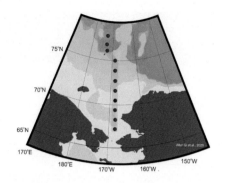

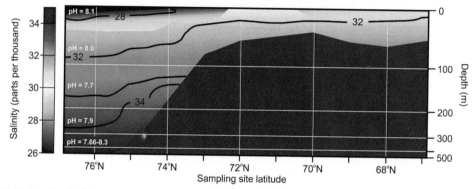

FIG. 5.4 Latitudinal distributions of salinity and pH measured in the upper 500 m along a shelf-slope-basin section from the Chukchi Sea to the Chukchi Abyssal Plain in the western Arctic Ocean. (After Qi, D., Chen, B., Chen, L., Lin, H., Gao, Z., Sun, H., Zhang, Y., Sun, X., Cai, W., 2020. Coastal acidification induced by biogeochemical processes driven by sea-ice melt in the western Arctic ocean. Polar Science 23. https://doi.org/10.1016/j.polar.2020.100504.)

acidification are influenced by terrestrial carbon delivery and remineralization. Additional factors are also important, such as the delivery of fresh water from rivers and ice melt, which lowers the capacity of seawater to resist changes in pH. Understanding the key drivers of ocean acidification and CO_2 flux in high-latitude shelf seas is urgent given how rapidly these areas are responding to global warming, with many effects that potentially impact carbon cycling, including loss of permafrost, snow, and sea ice. These various drivers will determine the effects of ocean acidification and future air-sea CO_2 exchanges in the high-latitude northern seas (Table 5.3).

CARBON BUDGETS OF ARCTIC WATERS

Capelle et al. (2020) developed a carbon budget model using published and recently collected measurements to estimate carbon inputs, transformations, and losses

in Hudson Bay, Canada. They estimated the annual effects of terrestrial carbon remineralization on aragonite saturation (ΩAr, a proxy for ocean acidification) and the partial pressure of CO_2 (pCO_2, a proxy for air-sea CO_2 flux), as well as the effects of marine primary production, marine organic carbon remineralization, and terrestrial calcium carbonate dissolution. They found that the remineralization of terrestrial dissolved organic carbon (DOC) is the main driver of CO_2 accumulation and aragonite undersaturation in coastal surface waters, but this is largely offset by marine primary productivity. Their study indicates that marine organic carbon remineralization is the largest contributor to CO_2 below the surface mixed layer in Hudson Bay.

Overall, the annual delivery and processing of carbon reduces the ΩAr of the water flowing through Hudson Bay by up to 0.17 units and raises the pCO_2 by up to 165 µatm. The similarities between Hudson Bay and other Arctic shelf seas suggest these areas are also

TABLE 5.3
Key Drivers Causing Direct Effects on Arctic Marine Ecosystems.

	CONDITION[a]			
	Loss of multiyear sea ice	**Acidification**	**Ocean stratification**	**Ocean warming**
Organisms	Whales↓	Crabs↓	Bowhead whales↓	Crabs↔
	Polar bears↓	Pteropods↓	Benthic organisms↔	Subarctic cod↔
	Seals, walrus↓	Phytoplankton↔	Phytoplankton↓	Subarctic flatfish↔
	Phytoplankton↑			Arctic char↔
	Ice-algal bloom↔			Plankton feeders↓
				Large zooplankton↓

[a] ↓ indicates negative effects; ↔ indicates mixed effects; ↑ indicates positive effects.
Data from Meredith et al. (2019).

significantly influenced by terrestrial carbon inputs and transformation.

Capelle et al. (2020) also compared delivery rates of terrestrial organic carbon from rivers and coastal erosion along a transect of major Arctic shelf seas (Fig. 5.5). They divided the carbon sources into (1) particulate organic carbon (POC) from rivers; (2) POC from coastal erosion; and (3) dissolved organic carbon from rivers. As shown in the figure, the Beaufort Sea receives the most organic carbon inputs of all studied regions, with a total of 35 kg of carbon per km^2 per year. Several Arctic seas receive less than 6 kg of carbon per km^2 per year.

THE PH OF OCEAN WATERS

The oceans are naturally alkaline, with an average pH of around 8.2 (Fig. 5.6). Ocean water pH varies up to 0.3 units depending on location and season. The term "pH" is an acronym from the Latin phrase *potentia Hydrogenii*,

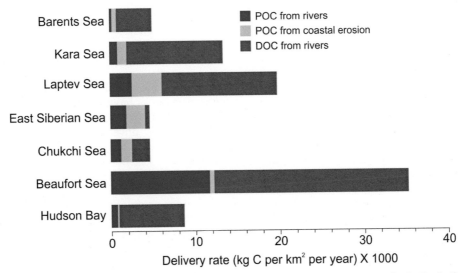

FIG. 5.5 Delivery rates of terrestrial organic carbon from rivers and coastal erosion to major Arctic shelf seas. *DOC*, dissolved organic carbon; *POC*, particulate organic carbon. (After Capelle, D.W., Kuzyk, Z.Z.A., Papakyriakou, T., Guéguen, C., Miller, L.A., Macdonald, R.W., 2020. Effect of terrestrial organic matter on ocean acidification and CO$_2$ flux in an Arctic shelf sea. Progress in Oceanography 185. https://doi.org/10.1016/j.pocean.2020.102319.)

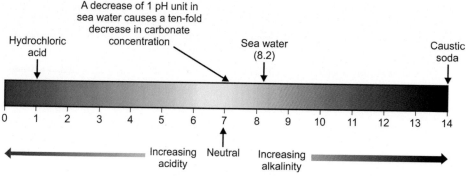

FIG. 5.6 The pH scale.

meaning "the power of hydrogen." The pH scale is thus used to indicate hydrogen ion activity. On the pH scale, pH values below 7.0 represent acidic solutions (hydrogen ion activity greater than hydroxide ion activity), while values above 7.0 represent basic solutions. A solution with a pH of 7.0 has equal hydrogen ion and hydroxide ion activity. Ocean water, a basic solution, has hydroxide ion activity greater than hydrogen ion activity. The pH scale is logarithmic. The possible range of hydrogen (H+) and hydroxide (OH−) ion activity spans 14 orders of magnitude.

The degree of acidification of Arctic Ocean waters is neither uniform through the entire region, nor is it consistent throughout the water column, nor is it constant throughout the year. An example of this variability in Arctic Ocean pH was presented by Qi et al. (2020). They made a series of pH, salinity, and temperature measurements along a shelf-slope-basin transect from the Chukchi Sea shelf to the Chukchi Abyssal Plain (CAP) in the western Arctic Ocean on a research cruise during the summer of 2010 (Fig. 5.4). They observed low pH values in the Chukchi Sea shelf bottom waters (from about 30 m to the bottom) and the upper halocline layer (UHL) (100–200 m) of CAP. A halocline is a strong, vertical salinity gradient within a body of water. Because salinity, in combination with temperature, affects the density of seawater, it can play a role in its vertical stratification. In the shelf bottom waters, the pH values were 7.66–8.13, about 0.07–0.68 pH units lower than the surface values of 8.20–8.24. In the CAP subsurface waters, the pH values were 7.85–7.98, about 0.08–0.31 pH units lower than the surface value (Fig. 5.4). Biogeochemical model simulations suggest that remineralized CO_2 driven by sea-ice loss is primarily responsible for the low pH values in the bottom waters of the Chukchi Sea (shelf) and the UHL waters of the CAP (basin). Another driver of pH change in Arctic

waters comes from enhanced organic matter production at the surface, triggered by sea-ice melt. The melting ice drops its load of organic matter, and this, in turn, enhances the growth and reproduction of algae and other phytoplankton at the ocean surface. The biogenic particles generated by this activity sink to the bottom where they decompose. This process consumes a large amount of dissolved oxygen and produces quantities of CO_2. Moreover, low pH bottom waters were flushed into the UHL during winter, which sustains the low pH characteristics in the subsurface basin layers. Their model suggests that the thermodynamic effect of pH is small. However, increasing temperature significantly increased aragonite saturation (Ωar), which slowed down the speed of acidification.

The CO_2 liberated by bacterial decomposition on the seafloor reacts with alkaline CO^{2-}_3 in seawater, as follows:

$$CO_2 + CO_3^{2-} + H_2O \leftrightarrow 2HCO_3-$$

Thus, the seawater becomes more acidic during the remineralization of biomass particles. In contrast, the pH increases during photosynthesis.

Another strong driver of pH in the western Arctic Ocean is the inflow of Pacific Ocean water from the south (Fig. 5.7). The inflow pathway is split into three branches: the Alaskan Coastal Current, Bering shelf water, and the Anadyr Current.

The residence time of the water on the Chukchi shelf is typically less than 6 months. It appears that the combined effects of Bering Sea throughflow variation and biological processes are the main controlling factor of pH variation in the Chukchi Sea. Some physical processes also contribute to the decrease in pH, including the mixing of low pH sea-ice meltwater and river water with seawater, ocean warming in summer, and oceanic uptake of CO_2 from the atmosphere after sea-ice retreat.

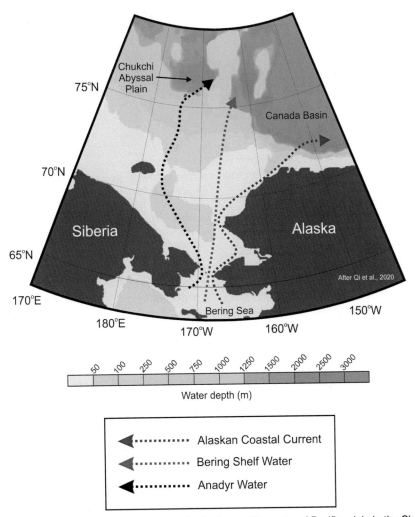

FIG. 5.7 Map of the study area in the western Arctic Ocean. The water of Pacific origin in the Chukchi Sea follows three main transport pathways: the Alaskan Coastal Current, Bering Shelf Water, and Anadyr Water. (After Qi, D., Chen, B., Chen, L., Lin, H., Gao, Z., Sun, H., Zhang, Y., Sun, X., Cai, W., 2020. Coastal acidification induced by biogeochemical processes driven by sea-ice melt in the western Arctic ocean. Polar Science 23. https://doi.org/10.1016/j.polar.2020.100504.)

In short, the driving mechanisms of pH changes in the western Arctic Ocean are more complex than other coastal regions and open oceans in the world.

Ocean surface waters are significantly supersaturated with calcium carbonate ($CaCO_3$). This means that ocean water, which is constantly in motion, holds more calcium carbonate in solution than it would if it were left in a stable environment for a long time. The solubility of calcium carbonate depends strongly on pH. If the pH of sea water drops below 8.2, calcium carbonate becomes more soluble. Bicarbonate and carbonate are different forms of the same ion. At lower pH, the bicarbonate form (HCO_3^-) predominates. At higher pH, the carbonate form (CO_3^{--}) dominates. Because the pH scale is logarithmic, small changes in pH values bring about major changes in ocean water carbonate chemistry. A 0.3 pH unit drop below about pH 9 causes a twofold drop in carbonate concentration. A decrease of one full pH unit brings a tenfold decrease in carbonate concentration.

The Effects of Melting Sea Ice

The mixed layer of water immediately under Arctic sea ice is typically low in salinity and pCO_2. When Arctic sea ice melts in summer, the surface water that was previously under the ice layer is exposed to the atmosphere and is also diluted by freshwater from sea ice melt. This dilution has become more prominent in Arctic wasters since multiyear sea ice has begun to disappear, because it reduces surface seawater calcium and carbonate ion concentrations, as well as lowering the pH. The combination of these processes is expected to drive the expansion of areas of carbonate mineral undersaturation in the Arctic Ocean over the next decades.

Future CO_2 Level Predictions

Since the early 19th century, the pH of surface ocean waters has fallen by 0.1 pH unit. This represents about a 30% increase in acidity. A model by ocean scientists at NOAA (NOAA-PMEL, 2019) predicts that if humans continue pumping carbon dioxide into the atmosphere at the current rate, by the end of this century the surface waters of the ocean could be nearly 150% more acidic than the preindustrial level. This is a level of ocean acidity that has not existed since the early Miocene, more than 20 million years ago.

Climate modelers (e.g., Gattuso et al., 2015) are confident that future projections of ocean acidification are reliable because the chemical mechanisms of carbon exchange are well understood. The International Panel on Climate Change (IPCC) scenarios give reductions in average global surface pH of between 0.14 and 0.35 units over the 21st century, so under this scenario, surface pH may drop as low as 7.8 by the year 2100. As can be seen, the estimates of the level of acidification predicted for the year 2100 vary somewhat, but the trend toward dramatic acidification is the same in all these scenarios (Fig. 5.8). Especially in the worst-case scenario (RCP8.5), there are elevated risks for keystone aragonite shell-forming species due to crossing an aragonite stability threshold year-round in the Arctic and sub-Arctic oceans by 2081–2100. For the best-case scenario (RCP2.6), these conditions will likely be avoided this century, but some eastern boundary upwelling systems are projected to remain vulnerable (Bindoff et al., 2019).

Several anthropogenic inputs amplify the effects of ocean acidification along coastal margins where human impacts are greatest. These impacts include nitrogen and phosphate runoff from agricultural, industrial, urban, and domestic sources. These additional nutrients cause eutrophication, triggering population spikes of algae and some other plankton. This artificial fertilization causes algal blooms. Eventually, these population

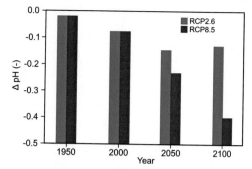

FIG. 5.8 Simulated global changes in ocean surface water pH from 1900 to 2100. Lines represent mean values. (From Bindoff, N.L., Cheung, W.W.L., Kairo, J.G., Arístegui, J., Guinder, V.A., et al., 2019. Changing ocean, marine ecosystems, and dependent communities. In: IPCC Special Report on the Ocean and Cryosphere in a Changing Climate.)

explosions die out, and when the organic matter decays, it generates CO_2, further acidifying seawater (Gattsuo, 2015).

MINERAL SATURATION IN ARCTIC WATERS

A large fraction of the global net CO_2 uptake during recent decades has occurred over the relatively small surface area of the Arctic Ocean. This global uptake amounts to ~2200 Tg of carbon per year (2.2 trillion kg). From the estimated net CO_2 uptake rates of the Arctic Ocean (65–175 Tg C per year), as much as 7.5% of global oceanic CO_2 uptake may occur in the Arctic Ocean, which comprises only 3.9% of the global ocean's surface (Robbins et al., 2013). The uptake occurs predominantly in seasonally ice-free areas, which, for now, are just a fraction of the Arctic Ocean surface. One driving factor behind the disproportionate uptake of CO_2 in the Arctic Ocean, and the resulting acidification of Arctic waters, is the relatively cold surface water, which absorbs more CO_2 than warmer seawater. Furthermore, over the past decade, summer sea-ice extent has rapidly declined, reaching a record low in 2012. Multiyear sea ice has undergone the greatest decline. This loss of multiyear ice has exposed the surface mixed layer (typically, 50 m thick), which is undersaturated with respect to atmospheric CO_2. This results in oceanic CO_2 uptake. Prior to the 21st century, summer ice cover significantly constrained CO_2 exchange between surface waters and the atmosphere. This melt-associated exposure of undersaturated waters is unique to the Arctic, where a steep vertical density gradient exists. This gradient is due primarily to the cold halocline underlying the surface mixed layer, which inhibits upward mixing of CO_2-rich deep waters, even during winter when the mixed layer temperature is lowest, and salinity is highest due to sea ice formation.

The mixed layer immediately under Arctic sea ice remains low in salinity and pCO_2. Thus, when Arctic sea ice melts in summer, the underice layer is exposed to the atmosphere and is also diluted by freshwater from sea ice melt. This dilution is an important consequence of multiyear sea ice loss because it reduces surface seawater calcium, carbonate ion concentrations, and pH. The combination of these processes is expected to drive the expansion of areas of carbonate mineral undersaturation in the Arctic Ocean over the coming decades (Robbins et al., 2013).

The uptake of CO_2 by the oceans alters seawater chemistry in several ways (Tynan et al., 2016). It increases the concentrations of dissolved inorganic carbon (DIC) and bicarbonate ions and reduces carbonate ion concentrations, calcium carbonate mineral saturation states (Ω), and pH. High-latitude oceans have naturally lower pH, Ω, and buffering capacity due to a higher solubility of CO_2 in their cold waters. As such, polar waters are expected to be the first to experience undersaturation ($\Omega < 1$) for calcium carbonate minerals. In fact, surface waters of the Arctic Ocean already experience seasonal undersaturation due to increased sea-ice melt, river runoff, and Pacific water intrusion. Undersaturated waters can be corrosive to calcifying organisms that lack protective mechanisms or rely on seawater pH for calcification (Cross et al., 2018). Pteropods (sea snails and sea slugs) living in polar waters are predicted to be the most affected by a decrease in aragonite saturation state (Comeau et al., 2012), and shell dissolution of living pteropods has been reported in the Southern Ocean. However, the response of marine organisms and ecosystems to ocean acidification also depends on other factors, such as nutrient availability, species interactions, and the previous exposure history of an organism to high pCO_2 waters.

Although ocean acidification results in long-term trends in mean ocean chemistry, it can also influence seasonal cycles. Observations indicate that the seasonal cycle of global surface-ocean pCO_2 increased in amplitude by 2.2 ± 0.4 µatm between 1982 and 2014 (Landschützer et al., 2018). CMIP5 models and data-based products similarly project consistent future increases in the seasonal cycle of surface-ocean pCO_2 under the RCP8.5 emissions scenario, with enhanced amplification in high-latitude waters (McNeil and Sasse, 2016). The amplitude of the seasonal cycle of global surface-ocean free acidity ($[H+]$) is projected to increase by

71%–91% over the 21st century under RCP8.5, also with greater amplification in the high latitudes (Kwiatkowski and Orr, 2018). Conversely, models project a 12%–20% reduction (across 90% confidence intervals) in the seasonal amplitude of surface-ocean pH, as changes in pH represent relative changes in $[H+]$ due to their logarithmic relationship, and there are typically greater projected increases in annual mean state $[H+]$ than the seasonal amplitude of $[H+]$. Models also project a 4%–14% reduction in the seasonal amplitude of global mean surface-ocean aragonite saturation state under RCP8.5. The contrasting changes in the seasonal amplitudes of ocean carbonate chemistry variables derive from different sensitivities to atmospheric CO_2, climate change and diverging trends in the seasonal cycles of DIC, alkalinity, and temperature. Overall, alongside the strong mean state changes, it is very likely that the amplitude of the seasonal cycle in free acidity will increase by 71%–91%, while it is very likely that the seasonal cycles of pH and aragonite saturation will decrease by 12%–20% and 4%–14%, respectively.

It is virtually certain that by 2081–2100, surface ocean pH will decline by 0.036–0.042 for the RCP2.6 scenario, and by 0.287–0.29 pH units for the RCP8.5 scenario, relative to 2006–15. Under the RCP8.5 scenario, these pH changes are very likely to cause the Arctic Ocean, and adjacent sectors of the Pacific and Atlantic Oceans to become corrosive for the major mineral forms of calcium carbonate. However, these changes are very likely to be avoided under the RCP2.6 scenario. There is increasing evidence of an increase in the seasonal exposure to acidified conditions in the future, and model runs indicate a very likely increase in the amplitude of seasonal pH cycle by 2100, relative to 2000 for the RCP8.5 scenario, especially at high latitudes.

Calcium Carbonate Saturation

These are three forms of carbonates used by marine organisms to build skeletons and shells, namely, aragonite, calcite, and magnesian calcite (Mg-calcite, formula $(MgCa)CO_3$. Aragonite and calcite have the same chemical composition but differ in mineral structure (Fig. 5.9). Calcite and Mg-calcite have the same mineral structure, but in Mg-calcite, some calcium ions are replaced by magnesium ions. These differences result in somewhat different chemical and physical properties. For example, Mg-calcite is more soluble in water than aragonite, which is more soluble than calcite.

The carbonate system in polar oceans has high natural variability and strong spatial gradients, which makes

BYU Geology Dept displays

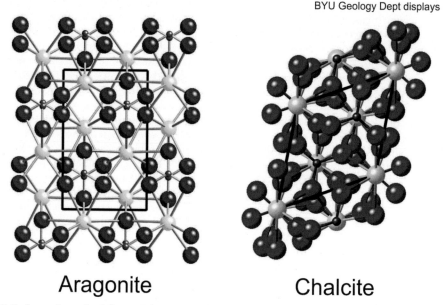

Aragonite   Chalcite

FIG. 5.9 Aragonite and calcite crystal structures. (© 2020, Eric Tingey, BYU Geology Department Displays. Used with permission.)

it challenging to distinguish natural processes from perturbations resulting from a gradual uptake of anthropogenic CO_2. To predict the future impact of climate change on the polar carbonate system, we must, therefore, understand the interactions between the physical and biogeochemical processes that drive the spatial variability. Polar oceans are particularly vulnerable to ocean acidification due to their low temperatures and reduced buffering capacity. Increasing acidification of polar waters will bring reduced carbonate mineral saturation states (Ω) in the near future.

Tynan et al. (2016) investigated the carbonate chemistry in the Atlantic sector of the Nordic Seas and Barents Sea in the Arctic Ocean, to determine the physical and biogeochemical processes that control surface pH and Ω. Their observations showed large gradients in surface pH (changes of 0.10–0.30 pH units) and aragonite saturation state (Ω_{ar}) (0.2–1.0) over small spatial scales, and these were particularly strong in sea ice–covered areas (up to 0.45 in pH and 2.0 in Ω_{ar}). In the Arctic, sea-ice melt facilitates algal bloom initiation in light-limited and iron-replete (dFe > 0.2 nM) regions, such as the Fram Strait. This results in more alkaline waters (pH = 8.45) and Ω_{ar} (3.0) along the sea-ice edge. Strong local gradients in pH and Ω may mean that organisms thriving in favorable conditions (high pH

and Ω_{ar}) can easily become exposed to lower pH and Ω_{ar}. This may have detrimental effects on calcifying organisms that lack sufficient protective mechanisms. Alternatively, it may be the case that polar ecosystems are already adapted to large variations in pH and Ω_{ar} and are therefore more resilient than expected to anthropogenic ocean acidification. We still do not understand the functioning of arctic marine life sufficiently well, either at the organismic level or at the ecosystem level, to accurately predict their responses to future acidification.

Robbins et al. (2013) examined calcite saturation rates and the role of freshwater inputs in Arctic sea basins (Fig. 5.10). Their data show aragonite undersaturation in about 20% of the surface waters of the Canada and Makarov basins and the Beaufort Sea. These regions have undergone a recent acceleration in the loss of sea-ice cover. Conservative tracer studies using stable oxygen isotopic data from 307 sites show that while the entire surface of this area receives abundant freshwater from meteoric sources, freshwater from sea-ice melt is most closely linked to the areas of carbonate mineral undersaturation. These data link the Arctic Ocean's largest area of aragonite undersaturation to sea-ice melt and atmospheric CO_2 absorption in areas of low buffering capacity. Some relatively supersaturated areas

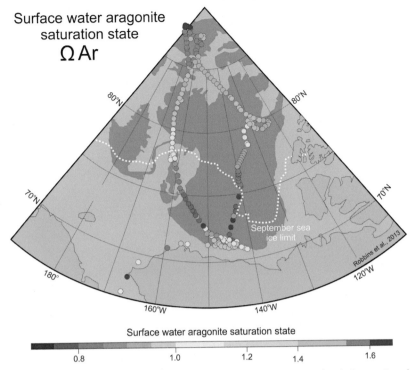

FIG. 5.10 Surface water aragonite saturation state for a transect of sampling sites in the western Arctic seas. (After Robbins, L.L., Wynn, J.G., Lisle, J.T., Yates, K.K., Knorr, P.O., Byrne, R.H., Liu, X., Patsavas, M.C., Azetsu-Scott, K., Takahashi, T., 2013. Baseline monitoring of the western Arctic ocean estimates 20% of Canadian basin surface waters are undersaturated with respect to aragonite. PloS One 8(9). https://doi.org/10. 1371/journal.pone.0073796.)

can be linked to localized biological activity. Collectively, these observations can be used to project trends of ocean acidification in higher-latitude marine surface waters where inorganic carbon chemistry is largely influenced by sea-ice meltwater. These areas exhibit the lowest aragonite saturation state (Ω_{ar}) values during late summer when sea ice is at its minimum annual extent. Undersaturated water extended poleward from the Beaufort Sea to nearly 80°N between the Canadian Archipelago and the Chukchi Cap, an ocean region more than 37,600 km² in area. In contrast, waters supersaturated with aragonite were observed by Robbins et al. (2013) in the Makarov Basin and Sever Spur areas ($1.14 < \Omega_{ar} < 1.61$) in association with nearly total ice cover and relatively low pCO₂ (Fig. 5.11). Beaufort The coastal waters had Ω_{ar} values significantly lower than Makarov Basin waters. Sea-ice melt made the greatest contribution to water masses in the Beaufort Sea and Canada Basin and comparatively low contributions in the Makarov Basin and Sever Spur area.

STRESSES OF CHANGING pH ON PLANKTON AND MOLLUSKS

Changes in the saturation state of ocean waters have both immediate and long-term consequences. The excess carbon dioxide in the oceans is making seawater more acidic. This trend was first discussed by Caldeira and Wickett (2003), who predicted, based on a combination of a general circulation model of the atmosphere and a geochemical model, that the pH of ocean water may decrease by 0.7 units if CO₂ emissions from fossil fuel burning continue unabated. This would reduce the pH of oceans from 8.2 to 7.5. This may not sound significant, but because the pH scale is logarithmic, this translates into water that is five times as acidic than the preindustrial level.

The chemical reaction that causes this increase in acidity is as follows:

$$CO_2 + H_2O + CO_3 \rightarrow 2HCO_3$$

In this reaction, carbon dioxide combines with water and a carbonate ion to form two bicarbonate

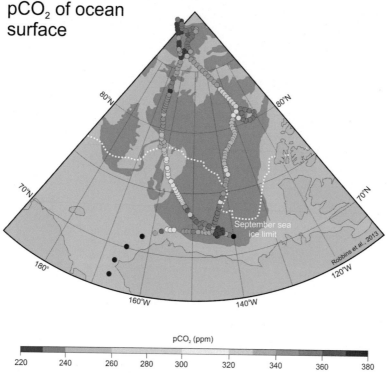

FIG. 5.11 Surface pCO_2 for a transect of sampling sites in the western Arctic seas. (After Robbins, L.L., Wynn, J.G., Lisle, J.T., Yates, K.K., Knorr, P.O., Byrne, R.H., Liu, X., Patsavas, M.C., Azetsu-Scott, K., Takahashi, T., 2013. Baseline monitoring of the western Arctic ocean estimates 20% of Canadian basin surface waters are undersaturated with respect to aragonite. PloS One 8(9). https://doi.org/10.1371/journal.pone.0073796.)

molecules. This process takes carbonate ions out of ocean water. The lowering of carbonate ion concentration acidifies the water, and it also lowers the saturation states of calcium carbonate minerals. As discussed earlier, these minerals are essential to many marine organisms because they form the building blocks for their skeletons and shells. In areas rich in marine life, the seawater is supersaturated with calcium carbonate minerals. But the current trend of ocean acidification is causing many parts of the ocean to become undersaturated with these minerals, which affects the ability of a huge variety of marine organisms to produce and maintain shells or, in the case of corals, to produce and maintain calcium-carbonate skeletons.

Importance of Carbonates to Marine Invertebrates

The changes in Arctic seawater pH, Ω_{carb}, and pCO_2 affect calcifying organisms (especially marine invertebrates) at or near the base of the food chain, creating effects that propagate through higher trophic levels. Thus,

the expansion of acidification into areas previously isolated from atmospheric contact by multiyear ice cover is likely to have ecological and economic consequences. There are three commonly occurring carbonate mineral phases used by marine organisms to form shells and skeletons: aragonite, calcite, and magnesium calcite. Some organisms build their shells or skeletons from one of these mineral phases, while others use more than one, either in different parts of the body or during different phases of life. These patterns of calcite mineral usage are linked to the mechanism and control of the calcification process, which likewise differs in various organisms (Andersson et al., 2011).

The effects of ocean acidification on marine life are many and varied, but two are prominent. First is the role that pH plays in ocean acidification, which has been called "the osteoporosis of the sea." Ocean acidification creates conditions that dissolve the minerals used by marine invertebrates to build their shells and skeletons (Fig. 5.12). Both CO_2 and bicarbonate (HCO_3) are used in photosynthesis and carbonate is a building block

Dissolution of a pteropod shell

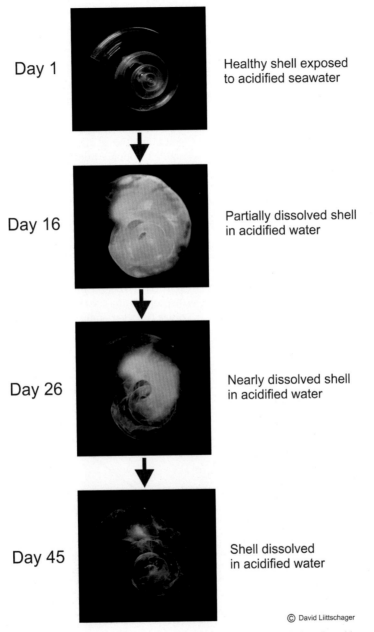

Day 1 — Healthy shell exposed to acidified seawater

Day 16 — Partially dissolved shell in acidified water

Day 26 — Nearly dissolved shell in acidified water

Day 45 — Shell dissolved in acidified water

© David Liittschager

FIG. 5.12 Progressive dissolution of the shell of the pteropod *Limacina helicina* placed in an experimental tank with acidified water over a period of 45 days. (Photos courtesy of David Liittschwager, used with permission.)

of shells and skeletons made of calcium carbonate. Ocean acidification can stimulate the growth of marine plants (mainly algae); hence, it causes primary production of the surface oceans to increase. As with so many aspects of biological response to a stressor, there are additional complications to consider. One of these is that within the same species, there may be different levels of sensitivity to ocean acidification during different life stages. For instance, there may be enhanced acid sensitivity among larval stages (Gattsuo, 2015). The larval and juvenile stages of marine organisms are typically more sensitive to environmental conditions and can suffer extremely high mortality in acidified environments (Kroeker et al., 2010). As discussed earlier, acidification typically decreases an organism's ability to construct and maintain calcium carbonate shells and skeletons. This suggests that highly calcium carbonate–dependent ecosystems—such as oyster and mussel beds—will be particularly vulnerable to increased acidification. However, the deposition of shells, tests, and skeletons by marine organisms is also quite variable. The organisms that make hard parts for their bodies are called calcifiers. Many of these calcifiers deposit hard parts that contain significant concentrations of Mg-calcite. A report by Andersson et al. (2008) warns that increasing ocean acidity is putting calcifying organisms living in high-latitude and/or cold-water environments at immediate risk

because of decreasing seawater carbonate saturation. Today these calcifiers live in seawater that is just slightly supersaturated in the carbonate phases they secrete to make their hard parts (Fig. 5.13). If CO_2 emissions remain at the current level or increase, the model developed by Andersson et al. (2011) indicates that high-latitude ocean waters may become undersaturated in aragonite in just a few decades. High-latitude waters will become more acidic than tropical waters because cold water can hold more dissolved gases (including CO_2) than warm water. Cold waters will become undersaturated in Mg-calcite minerals, because of their higher solubility. The relative proportion of calcifiers depositing stable carbonate minerals, such as calcite and low Mg-calcite, will increase, and the average magnesium content of carbonate sediments will decrease. A summary of the calcifiers that deposit hard parts with calcite, Mg-calcite, and aragonite is shown in Table 5.4.

Changes in community structure

At the community level, differential sensitivities to acidification have important implications for marine ecosystems where individual species often play disproportionately strong roles in structuring communities (Kroeker et al., 2010). For instance, the lowering of Ω_{carb} in the Arctic marine environment affects calcifiers at the base of the food chain, creating effects that

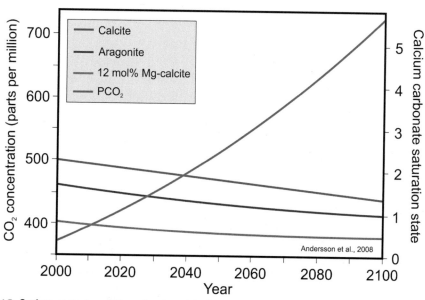

FIG. 5.13 Surface seawater pCO_2 and carbonate Ω for calcite, aragonite, Mg-calcite for high latitude ocean environment during the 21st century. (Modified from Andersson, A.J., Mackenzie, F.T., Bates, N.R., 2008. Life on the margin: implications of ocean acidification on Mg-calcite, high latitude and cold-water marine calcifiers. Marine Ecology Progress Series 373, 265–273. https://doi.org/10.3354/meps07639.)

TABLE 5.4
List of Major Marine Invertebrate Groups That Deposit Hard Parts With Calcite, Aragonite, and Mg-Calcite.

	Calcite	Aragonite	Mg-Calcite
Planktonic organisms	Coccolithophores, foraminifera	Sea snails (pteropods)	
Benthic organisms	Brachiopods, sponges, oysters	Corals, some mollusks	Coraline algae, echinoderms, bryozoans, foraminifera, sponges, some mollusks

propagate through higher trophic levels. Thus, the expansion of acidification into areas previously isolated from atmospheric contact by multiyear ice cover is likely to have ecological and economic consequences (Robbins et al., 2013).

Aragonite saturation state is influenced by not only air-sea CO_2 exchange but also by biological activities such as carbon fixation and respiration (Robins et al., 2013). Photosynthesizing plants (especially phytoplankton, such as diatoms) remove CO_2 from the water, thereby tending to increase Ω_{ar}. On the other hand, marine plant respiration adds CO_2 to seawater, causing a decrease in Ω_{ar}.

Role of pH in animal physiology
pH also plays a vital role in many aspects of animal physiology. The pH of bodily fluids and organs is critical to the physiology of organisms ranging from one-celled plankton to fish. Metabolic processes are closely linked to pH, and when pH falls outside the normal biochemical boundaries, these physiological systems can be overwhelmed (Gattsuo, 2015). In other words, a decrease in marine water pH changes the concentration of biomolecules that are key to animals' physiology. Wittmann and Pörtner (2013) identified the variability of the effects of ocean acidification among the different kinds of marine animals (Table 5.1 and Fig. 5.14). Phyto- and zooplankton, echinoderms, mollusks, crustaceans, and fish all have different responses to acidification. Crustaceans appear to suffer the least impacts because their exoskeletons are a mixture of biologically synthesized material (chitin) and calcium carbonate. As discussed in the following, fish do not build calcium carbonate shells, but they

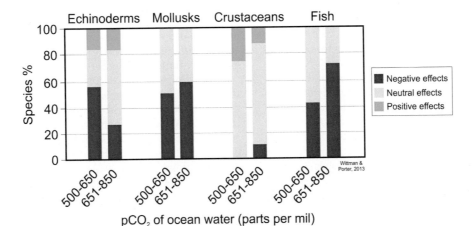

FIG. 5.14 Sensitivities of animal taxa to ocean acidification. Fractions (%) of coral, echinoderm, mollusk, crustacean and fish species exhibiting negative (red bars), neutral (yellow bars), or positive (green bars) effects on performance indicators reflecting individual fitness in response to the respective levels of atmospheric CO_2 levels (expressed as the partial pressure of CO_2 in ocean waters). (After Wittmann, A.C., Pörtner, H.O., 2013. Sensitivities of extant animal taxa to ocean acidification. Nature Climate Change 3(11), 995–1001. https://doi.org/10.1038/nclimate1982.)

must maintain suitable pH levels in their soft tissues, to enable the chemical reactions that sustain life and facilitate reproduction.

Gattuso et al. (2011) reckon that increasing ocean acidity will have a negative and significant impact on calcifying algae, coccolithophores, and mollusks. Increased acidity may have a positive effect on crustaceans and convey a slight advantage to some echinoderms (for instance, starfish, sea urchins, sand dollars, and sea cucumbers). Some organisms may have the innate ability to boost their metabolism and calcification abilities to compensate for lower calcite saturation levels. These assessments generally agree with those of Kroeker et al. (2010), who described negative effects on survival, calcification, growth, and reproduction of most calcifiers. However, they also discussed the significant variation in the sensitivity of marine organisms to increasing ocean water acidity. They found that calcifiers that deposit one of the more soluble forms of calcium carbonate (Mg-calcite) can be more resilient to ocean acidification than those that deposit the less soluble forms of calcite and aragonite. The life histories of these organisms play a role in calcifier response to acidification. Kroeker et al. (2010) discussed variations in the sensitivities of the different developmental stages of these organisms. This is another aspect that varies between taxonomic groups. Overall, their analyses suggest that the biological effects of ocean acidification are generally large and negative. However, the variation in sensitivity between different groups of organisms will have important implications for the response of marine ecosystems to increasing acidification.

Moreover, responses of planktonic taxa to ocean acidification end up having consequences for organisms higher up the food chain. Elevated CO_2 levels have been shown to alter the biochemical composition of phytoplankton species. These changes lead directly to modifications in the productivity and nutritional value of zooplankton. These responses are in addition to the negative effects that have been observed in critical zooplankton species that serve as the forage base for early life stages of larger-bodied marine fishes. These indirect food web-mediated impacts have the greatest effect on fish larvae because they have limited energy reserves and are highly sensitive to their foraging environment during early life stages (Hurst et al., 2019).

Barnacles

Research into marine biotic response to increasing acidification is advancing from relatively simple, short-term experiments to more complex and comprehensive studies that include long-term experiments, multiple life history stages, multipopulation, and multitrait approaches. One example of this more recent approach was published by Pansch et al. (2018), who used a population of the barnacle *Balanus improvisus*, a species known to be sensitive to short-term acidification, to determine its potential for long-term acclimation to acidification (Fig. 5.15). They placed a combination of laboratory-bred and field-collected barnacles in tanks with water of pH 8.1 and 7.5 (about 400 and 1600 µatm pCO_2, respectively) for up to 16 months. They found that acidification caused significant mortality and reduced growth rates. After 6 months of exposure to reduced pH, acidification suppressed the respiration rates of the barnacles and induced a higher level of feeding activity. However, normal respiration rates were reestablished after 15 months. Laboratory-bred barnacles developed mature gonads but failed to produce fertilized embryos. Field-collected barnacles reared in the laboratory for 8 months at the same pH levels developed mature gonads, but only those in pH 8.1 (i.e., the normal pH of seawater) produced viable embryos and larvae. This reproductive failure following long-term acidification demonstrates that *B. improvisus* will likely not survive long-term acidification in the wild.

Coccolithophores

Coccolithophores are a group of one-celled marine algae clothed in disk-shaped scales made of calcite. These scales are called coccoliths. Each coccolithophore is surrounded by at least 30 scales. These coccoliths are shed into the water when the coccolithophores multiply asexually, die, or make too many scales. During algal blooms, trillions of coccolithophores turn ocean waters an opaque turquoise (Fig. 5.16).

It is estimated that coccolithophores release more than 1.4 billion kilograms of calcite a year, making them the leading calcite producers in the ocean (NASA, 2016). Studies on the effects of increasing ocean acidity on coccolithophores have yielded some contradictory results. Iglesias-Rodriguez et al. (2008) did laboratory studies on the widespread coccolithophore species *Emiliania huxleyi*. This species is the world's single most important calcifying organism. They found that calcification and net primary production in this species are significantly increased by high CO_2 partial pressures in seawater. The inference of this study is that increasing ocean acidity actually benefits *E. huxleyi*. In contrast to this, Beaufort et al. (2011) measured the calcite mass of dominant coccolithophores in the modern ocean and over the past 40,000 years. They found a marked pattern of decreasing

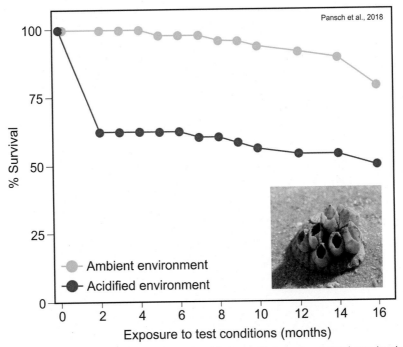

FIG. 5.15 Survival of laboratory-bred barnacles. Amphibalanus (*Balanus improvises*) was incubated under two different pH conditions for 16 months. The pH of the ambient environment was 8.1; the pH of the acidified environment was 7.5. (Data from Pansch, C., Hattich, G.S.I., Heinrichs, M.E., Pansch, A., Zagrodzka, Z., Havenhand, J.N., 2018. Long-term exposure to acidification disrupts reproduction in a marine invertebrate. PloS One 13(2). https://doi.org/10.1371/journal.pone.0192036. Barnacle photo by Andrew Butko, Creative Commons Attribution-Share Alike 3.0 Unported license.)

calcification with increasing partial pressure of CO_2, linked with decreasing concentrations of carbonate. Their study showed that species with different levels of calcification and varying morphotypes (individuals of a species having different physical appearances) are distributed in the ocean according to their carbonate chemistry. Because coccolithophores are the most important biological source of marine carbonates, their reduction due to increasing acidity would have a substantial impact on the marine carbon cycle. Beaufort et al. (2011) also discovered a heavily calcified morphotype of *E. huxleyi* in waters with low pH. This heavy-bodied variety is apparently able to withstand higher levels of acidity than other coccolithophores. This discovery highlights the complexity of phytoplankton assemblage responses to environmental impacts.

The solution to this seeming contradiction of results may lie in rapid evolutionary response to increasing ocean acidity. Lohbeck et al. (2013) examined the same species, *E. huxleyi*, and its ability to evolve in response to ocean acidification. This is a rapidly reproducing organism that employs asexual reproduction,

and the research team was able to raise cultures of this species over two 500-generation selection experiments. They exposed *E. huxleyi* populations founded by single or multiple clones to increased CO_2 concentrations. Compared with populations kept at normal levels of seawater CO_2, those exposed to ocean acidification conditions showed higher growth rates. When populations grown in normal seawater were exposed to the acidified water environment, their calcification rates were up to 50% lower than the acid-adapted populations. The authors suggest that contemporary evolution could help to maintain the functionality of microbial processes at the base of marine food webs in the face of global change.

Pteropods

Pteropods (sea snails and sea slugs) are tiny creatures, mostly less than 1 cm in length, but they occur in such great numbers in Arctic waters that they form a major component near the bottom of the food web in this region. There are three species of pteropods that range throughout the Arctic: *Limacina helicina*, *Limacina*

FIG. 5.16 NASA SeaWiFS image taken on April 25, 1998, showing a coccolithophore bloom off the Alaskan coast in the Bering Sea.

retroversa, and *Clione limacine*. They have been documented occurring in great numbers. Kacprzak et al. (2017) discovered a population of *L. retroversa* in waters near the Norwegian coast with a population density of nearly 52,000 individuals per 1000 m^{-3}. Many different organisms eat them, from tiny krill to fish to whales. Other animals, such as seals, rely on the fish that eat the pteropods. These "sea butterflies" are also a major food source for North Pacific juvenile salmon. Sediment trap studies have shown that pteropods are the major source of the carbonate flux (>50%) into the ocean's interior in the polar regions. These pelagic snails contribute to the vertical flux of carbon through the production of fecal pellets, mucous flocs, and rapid settling of aragonite shells upon their death.

Bednarsěk et al. (2014) examined the effects of acidified ocean water on the shells of the pteropod *L. helicina* in a transect of sites off the coast of the Pacific Northwest. They found that large portions of the shelf waters are currently corrosive to pteropods. They also demonstrated a strong positive correlation between the proportion of pteropod individuals with severe shell dissolution damage and the percentage of undersaturated water in the top 100 m with respect to aragonite. More than half of individuals from shallow water

and 24% of offshore individuals suffered severe dissolution damage (Fig. 5.12). The extent of undersaturated waters in the top 100 m of the water column has increased sixfold in these waters, compared with preindustrial CO_2 concentrations. These results demonstrate that habitat suitability for pteropods in the North Pacific is declining.

STRESSES OF CHANGING pH ON FISH

As discussed in the following, fish do not need to build a calcium carbonate shell, but they must maintain suitable pH levels in their soft tissues, to enable the chemical reactions that sustain life and facilitate reproduction. This makes fish vulnerable to acidification of their environment.

Acidification of ocean waters does not have a uniformly negative effect on all fish. As discussed in the following, many, but not all kinds of fish, suffer various kinds of physiological damage when exposed to lower pH in their environment, especially in the egg and larval stages. Some of these effects carry on through the life of the fish so that as adults, their senses may be dulled, and their behavior may be altered (Fig. 5.17). Furthermore, their growth, reproductive fitness, and even life span may be reduced.

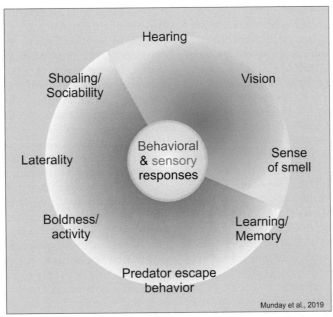

FIG. 5.17 A summary on the results obtained from studies investigating the effects of elevated CO_2 on the senses and behavioral responses of marine fish. (Data from Munday, P.L., Jarrold, M.D., Nagelkerken, I., 2019. Ecological effects of elevated CO_2 on marine and freshwater fishes: from individual to community effects. In: Fish Physiology, vol. 37. Elsevier Inc., pp. 323–368. https://doi.org/10.1016/bs.fp.2019.07.005.)

There is one common theme that runs throughout the literature on the effects of acidification on marine fish. Virtually every author concludes their work by saying that there are large gaps in our understanding of this problem and that we do not know enough to be able to predict future responses of fish to the acidification of pH have ocean waters that are linked with increased atmospheric emissions of carbon dioxide.

Interactions Between Temperature and pH

Many studies have examined the interactions of elevated ocean temperatures and elevated CO_2 on fishes. Not surprisingly, these studies show that the effects of elevated CO_2 are affected by raised temperatures, with higher temperatures having the potential to exacerbate, mitigate, or even reverse the effects of elevated CO_2 on a broad range of traits, including growth, development, survival, and behavior. The senses (vision, hearing, sense of smell) and various behavioral adaptations can be affected by acidification (Fig. 5.16). In addition, new studies are considering how food availability could interact with elevated CO_2. A vitally important piece of the puzzle concerns the interactions between the various species in marine biological communities, starting with the phyto- and zooplankton, working up through predatory fish at the top of the food chain. About the only thing we can say with certainty is that these marine ecosystems will change in response to lowered pH.

Munday et al. (2019) evaluated the current understanding of the effects of acidification on the autecology and synecology of marine fish communities, focusing on interactions with other stressors such as global warming, and assessed the uncertainties and major gaps in our knowledge. Laboratory experiments have shown that elevated CO_2 can affect ecologically important traits of some species, but fish biologists are not yet able to predict which species are sensitive and which are tolerant to higher CO_2 levels. Moreover, the impacts of elevated CO_2 on ecologically relevant traits do not occur in a vacuum. Rather, these impacts can be influenced in unexpected ways by food availability and interactions with temperature and other stressors.

Impacts on immature stages

Fish were initially thought to be unaffected by rising CO_2 levels because of their well-developed ability to regulate acid-base chemistry in their bodies. However, this assumption was overturned by studies showing that elevated PCO_2 (1000–2000 ppm) affects behaviors, causes damage to major organs, and reduces growth and survivorship, especially during early life stages. The effects of elevated CO_2 on fish early life stages (larvae and juveniles) are highly variable and species-specific, with negligible, negative, and positive effects on survival, growth, and development all reported. Most studies to date have focused on larval stages, as these are expected to be the most vulnerable to elevated CO_2 because their acid-base balance system may not be fully developed, and their smaller size increases the cost of homeostasis. The large surface-to-volume ratios of small larvae not only increase their exposure to environmental stressors, but their less developed anatomy also limits their capacity to buffer these stressors. Only a few studies have evaluated the potential of fish to acclimate over longer-term periods, and they are mainly based on tropical species. A study by Rodriguez-Dominguez et al. (2018) showed that the detrimental effects of ocean acidification on embryonic fish are not only irreversible but also show a lack of acclimation after a 3-month exposure.

Based on laboratory experiments, Hurst et al. (2019) published a multifaceted analysis of the sensitivity of Pacific cod (*Gadus macrocephalus*) larvae to elevated CO_2. Fish behavior in a horizontal light gradient was used to evaluate the sensitivity of behavioral phototaxis in 4- to 5-week-old cod larvae. Fish raised at elevated CO_2 levels (~ 1500 and 2250 µatm) exhibited a stronger phototaxis (moved more quickly to regions of higher light levels) than fish at ambient CO_2 levels (~ 600 µatm). In an independent experiment, they examined the effects of elevated CO_2 levels on the growth of larval Pacific cod over the first 5 weeks of life under two different feeding treatments. Fish exposed to elevated CO_2 levels (~ 1700 µatm) were smaller and had lower lipid levels at 2 weeks of age than fish at low (ambient) CO_2 levels (~ 500 µatm). However, by 5 weeks of age, this effect had reversed, so that fish reared at elevated CO_2 levels were slightly larger and had higher total lipid levels and storage lipids than fish reared at low CO_2. Fatty acid composition differed significantly between fish reared at high and low CO_2 levels after 2 weeks of feeding, but this effect diminished by week five. Effects of CO_2 on the fatty acid composition of the larvae differed between the two diets, an effect possibly related more to dietary equilibrium and differential lipid class storage than a fundamental effect of CO_2 on fatty acid metabolism. These experiments point to the stage-specific sensitivity of Pacific cod to the effects of ocean acidification.

Interference with enzyme function

The Atlantic halibut is a benthic marine fish widely distributed in the northern regions of the Atlantic Ocean and parts of the Arctic Ocean. Juveniles stay in coastal areas at depths of 20–60 m before migrating to deeper waters. Almroth et al. (2019) conducted an experiment in which Atlantic halibut were exposed to a range of temperatures and water pH for 3 months. They analyzed physiological changes through adjustments in the functioning of several antioxidant enzymes. Antioxidants are substances that remove potentially damaging oxidizing agents in a living organism. The treatments resulted in oxidative stress, and the damage was evident in the form of protein carbonyls, which were consistently higher in the elevated CO_2-treated fish at all temperatures. Oxidative stress is an imbalance toward the prooxidant side of prooxidant/antioxidant homeostasis. The presence of high levels of protein carbonyl (CO) groups has been used as a marker of oxidative stress in many different organisms, including humans. Analyses of antioxidant enzymes did not show the same results, suggesting that the exposure to elevated CO_2 increased reactive oxygen species formation but not defenses. The antioxidant defense system was insufficient, and the resulting oxidative damage could affect the physiological functions of the halibut on a cellular level.

Interference with acid-base balance

Brauner et al. (2019) discuss at length the acid-base physiology of fish. Acid-base balance is one of the most tightly controlled physiological processes. Accumulation of metabolic CO_2 produced in fish tissues causes extra- or intracellular acidosis that can disrupt cellular processes. Acidosis is an excessively acidic condition of the body fluids or tissues. Consequently, most fish have a well-developed system for CO_2 transport and excretion. However, the system varies considerably in different fish taxa. Elevated environmental CO_2 induces a rapid acid-base disturbance in fish. Fish respond to the resulting acidosis by increasing plasma HCO_3^- in exchange for Cl^-, primarily through the gills. The rate and success of acid-base compensation during fish exposure to increased CO_2 levels are affected by water ion composition. At high CO_2 levels, there appears to be an upper limit to the increase in plasma HCO_3^-. Fish that naturally live in such high CO_2 environments appear to have an exceptional capacity for intracellular pH regulation. While it has long been thought that fish would not be affected by climate change relevant CO_2 levels, negative physiological effects are seen. The effect of fluctuating CO_2 levels in seawater may be especially problematic, an area where more research is required.

Virtually every physiological process in fish is affected by the acid-base condition of their tissues and bodily fluids. As discussed earlier, fish have developed homeostatic responses to compensate for acid-base disturbances that arise from metabolic and environmental sources. Tresguerres et al. (2019) discuss the ability of fish to sense acid-base disturbances, which is a common requirement for all homeostatic responses. In a broad sense, this means sensing acidosis and alkalosis in comparison with a specific (optimal) pH level. Maintaining acid-base homeostasis in physiological fluids is essential for life because the concentration of H^+ and the pH of internal fluids in fish bodies greatly affect protein folding and function. Additionally, acid-base conditions affect various physiological processes such as metabolism, pH buffering, biomineralization, neurotransmission, oxygen (O_2) delivery, feeding, and digestion. Fish acid-base status can also be affected by environmental factors, the most prominent being elevated CO_2 (hypercapnia) and the associated reduction in pH. Environmental hypercapnia often is associated with hypoxia (oxygen starvation).

CONCLUSIONS

The world is on a collision course with elevated levels of atmospheric CO_2. Even if we were to stop burning fossil fuels today, we would still be living with the consequences of many decades of uncontrolled CO_2 emissions. This is particularly true of the world's oceans, which have been soaking up excess CO_2 and storing it as carbonic acid. About 94% of the Earth's CO_2 is being held in the world's oceans. It will take many centuries for the bulk of this oceanic CO_2 to leave the water and enter the atmosphere. In the meantime, as well as so many other aspects of global change, the Arctic regions are particularly susceptible to CO_2 enrichment of ocean water. This has nothing to do with any higher concentration of CO_2 in the atmosphere of high northern latitudes. Studies have shown that the concentration of atmospheric CO_2 is essentially the same throughout the world because of mixing in the atmosphere. The reason that Arctic waters absorb more CO_2 is the same reason they absorb more of any other atmospheric molecules. This is the simple property that cold water can hold more dissolved gases than warm water. This is one reason why the commercial fishing fleets of the

world tend to focus their efforts in the higher latitudes of both hemispheres. Cold waters, holding more oxygen and CO_2, foster growth of phytoplankton and the zooplankton that forms the basis of all marine food webs. Tropical waters are poor in dissolved gases and therefore support less abundant and diverse marine life than high-latitude waters.

The current rates of change in ocean pH and Ω are unprecedented, as much as 30 to 100 times faster than the rates of change in the past 50 million years. The most recent time of extreme ocean was during the Paleocene-Eocene Thermal Maximum, 55.8 million years ago. It should therefore not be a surprise that the modern marine biota do not appear to have the mechanisms to effectively combat acidification. Modern animals that build shells, tests, and skeletons from calcium carbonate are under threat as the pH of ocean water decreases. Modern ocean water is already 30% more acidic than the waters of the preindustrial era. Under a worst-case scenario, scientists of the IPCC predict that by the end of the century, ocean pH will have dropped from 8.1 to 7.8. This level of acidification over just a few decades is truly unprecedented, and it remains doubtful whether most of the modern marine biota will be able to adapt before they go extinct. What appears likely is that those species that are best adapted to more acidic waters will flourish at the expense of poorly adapted species. This will undoubtedly bring wholesale changes to marine ecosystems around the world and perhaps especially so in the Arctic. Species with more genetic plasticity are more likely to adapt to a variable and changing environment than species with narrow environmental tolerances and strict physiology or life history. For example, copepods and krill in the Barents Sea show marked trophic plasticity, shifting from herbivory during the algal bloom to omnivory when fresh material is less abundant (Michel et al., 2013). Predator fishes such as Atlantic cod also show high feeding plasticity, shifting their prey from fishes to zooplankton in response to changes in abundance. Such flexibility in feeding strategies may provide an advantage in highly variable environments such as the Barents Sea. Phenotypic plasticity is also expected to dominate the responses of marine mammals to environmental change in the short term. Accordingly, biodiversity can offer functional redundancy and increase the resilience of marine systems to multiple stressors. However, this resilience ultimately depends on the response of each species to individual and combined stressors and the resulting trophic interaction.

Arctic marine ecosystems are not as species-diverse as lower latitude marine ecosystems. Of the more than 20,000 species of fish known from the world's oceans, there are only 400 species living in Arctic seas and adjacent waters. Just as a cord of many strands is more difficult to break, more species-rich ecosystems are more resilient to environmental change than those with fewer species. This makes polar marine ecosystems particularly vulnerable to disturbance because the elimination of just a few species may cause the collapse of the entire food chain.

As discussed by Johannessen and Miles (2011), the Arctic and sub-Arctic marine food webs are fundamentally influenced by the coupled ocean-ice-atmosphere system. For example, the food web of the Barents Sea can be described in a simplified manner as consisting of phytoplankton (first level), zooplankton (second level), capelin and herring (third level), cod (fourth level), and seals and whales (fifth level). It is worth noting that phytoplankton, capelin, seals, and whales are all closely linked to the sea-ice edge. Moreover, all members of the ecosystem are highly dependent on each other.

One of the greatest impediments to understanding the ongoing changes in the functioning and biodiversity of Arctic marine ecosystems is the fragmented nature of the existing knowledge and the lack of consistent and regular long-term monitoring programs in most Arctic marine regions, including unique or vulnerable ecosystems. A commitment to long-term studies is essential in this regard. The effects of disturbances and stressors on the Arctic marine ecosystems are not well understood. The lack of baseline information in many areas, the wide range of ecosystems, and the impact of cumulative effects make it difficult to predict the direction of changes. The multiple stressors currently affecting Arctic marine ecosystems operate simultaneously at various temporal and spatial scales, emphasizing the need for local and concerted ecological assessment and species monitoring. There is also a need to develop indicators that properly reflect the unique characteristics of Arctic marine ecosystems. Similarly, unique habitats associated with sea ice and ice shelves are poorly understood, and their biodiversity is largely unknown. Deep basins in the Arctic Ocean were largely inaccessible until quite recently. However, some of these regions are becoming ice-free in summer, bringing new opportunities for research and exploration. As one of the last frontiers on Earth, the marine Arctic still holds many discoveries concerning the species that inhabit them including their ecophysiology, biodiversity, and ecosystem functioning.

REFERENCES

Almroth, B.C., Souza de, B., Jönsson, E., Sturve, J., 2019. Oxidative stress and biomarker responses in the Atlantic halibut after long term exposure to elevated CO_2 and a range of temperatures. Comparative Biochemistry and Physiology. Part B, 238, 110321, 8 pp.

Andersson, A.J., Mackenzie, F.T., Bates, N.R., 2008. Life on the margin: implications of ocean acidification on Mg-calcite, high latitude and cold-water marine calcifiers. Marine Ecology Progress Series 373, 265–273. https://doi.org/10.3354/meps07639.

Andersson, A.J., Mackenzie, F.T., 2011. Effects of ocean acidification on benthic processes, organisms, and ecosystems. In: Ocean Acidification, Oxford University Press, pp. 122–173.

Beaufort, L., Probert, I., de Garidel-Thoron, T., Bendif, E.M., Ruiz-Pino, D., et al., 2011. Sensitivity of coccolithophores to carbonate chemistry and ocean acidification. Nature 476, 80–83.

Bednaršek, N., Feely, R.A., Reum, J.C.P., Peterson, B., Menkel, J., Alin, S.R., Hales, B., 2014. Limacina helicina shell dissolution as an indicator of declining habitat suitability owing to ocean acidification in the California current ecosystem. Proceedings of the Royal Society B: Biological Sciences 281 (1785). https://doi.org/10.1098/rspb.2014.0123.

Bindoff, N.L., Cheung, W.W.L., Kairo, J.G., Arístegui, J., Guinder, V.A., et al., 2019. Changing ocean, marine ecosystems, and dependent communities. In: IPCC Special Report on the Ocean and Cryosphere in a Changing Climate.

Brauner, C.J., Shartau, R.B., Damsgaard, C., Esbaugh, A.J., Wilson, R.W., Grosell, M., 2019. Acid-base physiology and CO_2 homeostasis: regulation and compensation in response to elevated environmental CO_2. In: Fish Physiology, vol. 37. Elsevier Inc., pp. 69–132. https://doi.org/10.1016/bs.fp.2019.08.003

Caldeira, K., Wickett, M.E., 2003. Anthropogenic carbon and ocean pH. Nature 425 (6956), 365. https://doi.org/10.1038/425365a.

Capelle, D.W., Kuzyk, Z.Z.A., Papakyriakou, T., Guéguen, C., Miller, L.A., Macdonald, R.W., 2020. Effect of terrestrial organic matter on ocean acidification and CO_2 flux in an Arctic shelf sea. Progress in Oceanography 185, 102319, 14 pp. https://doi.org/10.1016/j.pocean.2020.102319.

Comeau, S., Gattuso, J.P., Nisumaa, A.M., Orr, J., 2012. Impact of aragonite saturation state changes on migratory pteropods. Proceedings of the Royal Society B: Biological Sciences 279 (1729), 732–738. https://doi.org/10.1098/rspb.2011.0910.

Cross, J.N., Mathis, J.T., Pickart, R.S., Bates, N.R., 2018. Formation and transport of corrosive water in the Pacific Arctic region. Deep-Sea Research Part II 152, 67–81. https://doi.org/10.1016/j.dsr2.2018.05.020.

Doney, S.C., 2010. The growing human footprint on coastal and open-ocean biogeochemistry. Science 328 (5985), 1512–1516. https://doi.org/10.1126/science.1185198.

Doney, S.C., Fabry, V.J., Feely, R.A., Kleypas, J.A., 2009. Ocean acidification: the other CO_2 problem. Annual Review of Marine Science 1, 169.

Gattuso, J.-P., Hansson, L., 2011. Ocean acidification: background and history. In: Gattuso, J.-P., Hansson, L. (Eds.), Ocean Acidification. Oxford University Press, pp. 1–20.

Gattuso, J.-P., Magnan, A., Billé, R., Cheung, W.W.L., Howes, E.L., Joos, F., Allemand, D., Bopp, L., Cooley, S.R., Eakin, C.M., Hoegh-Guldberg, O., Kelly, R.P., Pörtner, H.-O., Rogers, A.D., Baxter, J.M., Laffoley, D., Osborn, D., Rankovic, A., Rochette, J., Turley, C., 2015. Oceanography. Contrasting futures for ocean and society from different anthropogenic CO_2 emissions scenarios. Science (New York, N.Y.) 349, 46–58. (6243). https://doi.org/10.1126/science.aac4722.

Hurst, T.P., Copeman, L.A., Haines, S.A., Meredith, S.D., Daniels, K., Hubbard, K.M., 2019. Elevated CO 2 alters behavior, growth, and lipid composition of Pacific cod larvae. Marine Environmental Research 145, 52–65. https://doi.org/10.1016/j.marenvres.2019.02.004.

Iglesias-Rodriguez, M.D., Halloran, P.R., Rickaby, R.E.M., Hall, I.R., Colmenero-Hidalgo, E., Gittins, J.R., Green, D.R.H., Tyrrell, T., Gibbs, S.J., von Dassow, P., Rehm, E., Armbrust, E.V., Boessenkool, K.P., 2008. Phytoplankton calcification in a high-CO_2 world. Science (New York, N.Y.) 320 (5874), 336–340. https://doi.org/10.1126/science.1154122.

Johannessen, O.M., Miles, M.W., 2011. Critical vulnerabilities of marine and sea ice-based ecosystems in the high Arctic. Regional Environmental Change 11 (1), 239–248. https://doi.org/10.1007/s10113-010-0186-5.

Kacprzak, P., Panasiuk, A., Wawrzynek, J., Weydmann, A., 2017. Distribution and abundance of pteropods in the western Barents Sea. Oceanological and Hydrobiological Studies 46 (4), 393–404. https://doi.org/10.1515/ohs-2017-0039.

Kroeker, K.J., Kordas, R.L., Crim, R.N., Singh, G.G., 2010. Meta-analysis reveals negative yet variable effects of ocean acidification on marine organisms. Ecology Letters 13 (11), 1419–1434. https://doi.org/10.1111/j.1461-0248.2010.01518.x.

Kwiatkowski, L., Orr, J.C., 2018. Diverging seasonal extremes for ocean acidification during the twenty-first century. Nature Climate Change 8 (2), 141–145. https://doi.org/10.1038/s41558-017-0054-0.

Landschützer, P., Gruber, N., Bakker, D.C.E., Stemmler, I., Six, K.D., 2018. Strengthening seasonal marine CO_2 variations due to increasing atmospheric CO_2. Nature Climate Change 8 (2), 146–150. https://doi.org/10.1038/s41558-017-0057-x.

Lohbeck, K.T., Riebesell, U., Collins, S., Reusch, T.B.H., 2013. Functional genetic divergence in high CO_2 adapted Emiliania huxleyi populations. Evolution 67 (7), 1892–1900. https://doi.org/10.1111/j.1558-5646.2012.01812.x.

Mackenzie, F.T., Andersson, A.J., Arvidson, R.S., Guidry, M.W., Lerman, A., 2011. Land-sea carbon and nutrient fluxes and coastal ocean CO_2 exchange and acidification: past, present, and future. Applied Geochemistry 26, S298–S302. https://doi.org/10.1016/j.apgeochem.2011.03.087.

McNeil, B.I., Sasse, T.P., 2016. Future ocean hypercapnia driven by anthropogenic amplification of the natural CO_2 cycle. Nature 529 (7586), 383–386. https://doi.org/10.1038/nature16156.

Meredith, M., Sommerkorn, M., Cassotta, S., Derksen, C., Ekaykin, A., 2019. In: Roberts, D.C., Masson-Delmotte, V. (Eds.), Polar Regions. IPCC Special Report on the Ocean and Cryosphere in a Changing Climate. H.-O. Pörtner. IPCC, pp. 203–320.

Michel, C., Bluhm Ford, V., Gallucci, V., Gaston, A.J., et al., 2013. Chapter 14, Marine ecosystems. In: Arctic Biodiversity Assessment Status and Trends in Arctic Biodiversity. Conservation of Arctic Flora and Fauna, pp. 29–31.

Munday, P.L., Jarrold, M.D., Nagelkerken, I., 2019. Ecological effects of elevated CO_2 on marine and freshwater fishes: from individual to community effects. In: Fish Physiology, vol. 37. Elsevier Inc., pp. 323–368. https://doi.org/10.1016/bs.fp.2019.07.005.

NASA, 2016. Canadian waters teem with phytoplankton. NASA Earth Observatory. https://earthobservatory.nasa.gov/images/88687/canadian-waters-teem-with-phytoplankton.

NOAA-Pacific Marine Environmental Laboratory, 2019. The future of ocean acidification. https://www.pmel.noaa.gov/news-story/future-ocean-acidification.

Pansch, C., Hattich, G.S.I., Heinrichs, M.E., Pansch, A., Zagrodzka, Z., Havenhand, J.N., 2018. Long-term exposure to acidification disrupts reproduction in a marine invertebrate. PloS One 13 (2), e0192036. https://doi.org/10.1371/journal.pone.0192036.

Qi, D., Chen, B., Chen, L., Lin, H., Gao, Z., Sun, H., Zhang, Y., Sun, X., Cai, W., 2020. Coastal acidification induced by biogeochemical processes driven by sea-ice melt in the western Arctic ocean. Polar Science 23, 100504, 8 pp. https://doi.org/10.1016/j.polar.2020.100504.

Robbins, L.L., Wynn, J.G., Lisle, J.T., Yates, K.K., Knorr, P.O., Byrne, R.H., Liu, X., Patsavas, M.C., Azetsu-Scott, K., Takahashi, T., 2013. Baseline monitoring of the western Arctic ocean estimates 20% of Canadian basin surface waters are undersaturated with respect to aragonite. PloS One 8 (9), e73796. https://doi.org/10.1371/journal.pone.0073796.

Rodriguez-Dominguez, A., Connell, S.D., Baziret, C., Nagelkerken, I., 2018. Irreversible behavioural impairment of fish starts early: embryonic exposure to ocean acidification. Marine Pollution Bulletin 133, 562–567. https://doi.org/10.1016/j.marpolbul.2018.06.004.

Tresguerres, M., Milsom, W.K., Perry, S.F., 2019. CO_2 and acid-base sensing. In: Fish Physiology, vol. 37. Elsevier Inc., pp. 33–68. https://doi.org/10.1016/bs.fp.2019.07.001

Tynan, E., Clarke, J.S., Humphreys, M.P., Ribas-Ribas, M., Esposito, M., Rérolle, V.M.C., Schlosser, C., Thorpe, S.E., Tyrrell, T., Achterberg, E.P., 2016. Physical and biogeochemical controls on the variability in surface pH and calcium carbonate saturation states in the Atlantic sectors of the Arctic and Southern Oceans. Deep-Sea Research Part II Topical Studies in Oceanography 127, 7–27. https://doi.org/10.1016/j.dsr2.2016.01.001.

Watson, A.J., Schuster, U., Shutler, J.D., Holding, T., Ashton, I.G., et al., 2020. Revised estimates of ocean-atmosphere CO_2 flux are consistent with ocean carbon inventory. Nature Communications 11, 4422. https://doi.org/10.1038/s41467-020-18203-3.

Wittmann, A.C., Pörtner, H.O., 2013. Sensitivities of extant animal taxa to ocean acidification. Nature Climate Change 3 (11), 995–1001. https://doi.org/10.1038/nclimate1982.

Zeebe, R.E., Tyrrell, T., 2019. History of carbonate ion concentration over the last 100 million years II: revised calculations and new data. Geochimica et Cosmochimica Acta 257, 373–392. https://doi.org/10.1016/j.gca.2019.02.041.

FURTHER READING

Krumhardt, K.M., Lovenduski, N.S., Iglesias-Rodriguez, M.D., Kleypas, J.A., Landschützer, P., Gruber, N., Bakker, D.C.E., Stemmler, I., Six, K.D., (n.d.). Strengthening seasonal marine CO_2 variations due to increasing atmospheric CO_2. Progress in Oceanography, 159, 146–150.

Rehm, E., Armbrust, E.V., Boessenkool, K.P., Hall, I.R., Colmenero-Hidalgo, E., Gittins, J.R., Green, D.R.H., Tyrrell, T., Gibbs, S.J., Von Dassow, P., 2008. Phytoplankton calcification in a high-CO_2 world. Science 320 (5874), 336–340. https://doi.org/10.1126/science.1154122.

CHAPTER 6

Impacts of Chemical Pollution on Marine Ecosystems

INTRODUCTION

The oceans cover more than 70% of the Earth's surface, but oceanic ecosystems are arguably the poorest known and least studied, until very recently. This is especially true of the polar seas, the least accessible and most hazardous seas of the world. Oceanic ecosystems do not behave like terrestrial ecosystems. While terrestrial biological communities fall within discrete geographic regions, the world's oceans are all interconnected. The gradients between oceanic environments are all very gradual; there are no firm ecological boundaries. This means that the ocean biota are not as geographically differentiated as the terrestrial biota. Ocean water is continually moving through the action of currents. Coldwater flows into warm regions and vice versa. Currents aid the dispersal of marine organisms. Unlike terrestrial ecosystems, the oceans display distinct vertical patterns of biotic distribution on daily to seasonal timescales; these also interact with the horizontal patterns of biotic distribution.

Researchers cannot set up field camps, survey the regional flora and fauna of a certain district, or observe interactions between the biota and their physical environment without organizing very costly ocean voyages, lasting many weeks, and persisting through years and decades of observations. The polar seas are the most difficult and costly regions for scientific investigation. Research cruises cost between $10,000 and $20,000 per day at sea, depending on the size and capabilities of the ship. Polar research is mostly done onboard the few ice-breaking ships converted for scientific use.

Studies of life below the photic zone (the water sufficiently shallow to allow sunlight to penetrate) have to overcome the logistical problems of darkness and increasing water pressure with depth. Until recently, much of the Arctic Ocean remained inaccessible because of drifting pack ice, even during the brief Arctic summer. The Arctic Ocean and adjacent seas cover large areas of the continental shelf, and the shelf regions are easier to explore because of the shallow water. The deep water in the Arctic remains virtually unstudied.

Because of the enormous logistical problems, large parts of the Arctic Ocean and adjacent seas remain largely unexplored. Therein lies an enormous problem, because the adage, "out of sight—out of mind" has meant that the world's oceans, including the polar seas, have remained unstudied. Indeed, conservation of the oceans and their biota only started in earnest in recent decades. Contrary to most people's way of thinking, the oceans, although enormous, do not have an infinite capacity to absorb waste material, chemicals, and an overabundance of atmospheric carbon dioxide. The rise of the chemical industry since the 1950s brought newly created poisons (pesticides, industrial compounds, and chemical fertilizers) into the oceans in ever-increasing amounts. As discussed in the following, one of the most ubiquitous and damaging of these chemical pollutants are the many kinds of plastics that have been invented in the past 60 years.

PLASTIC PRODUCTION AND DISPERSAL

Plastics are long-chain synthetic molecules. All types of plastic were created in a laboratory—there are no natural plastics. Between 1960 and 2000, the world production of plastic resins increased 25-fold (Moore, 2008). Annual global plastic resin production reached 359 million metric tons in 2019. The volume of plastics entering the oceans in the year 2017 was estimated to be more than 33 times the total plastics that accumulated in the oceans up to 2015. Up to 90% of trash floating in the ocean and littering our shores is plastic (National Park Service Oceans, Coasts & Seashores, 2018)

The MacArthur Foundation predicts that plastic production will double again over the next 20 years, as plastics come to serve an increasing number of applications. The plastics industry is a huge consumer of petroleum. Over 90% of the plastic products made are derived from virgin petroleum. Plastic manufacture uses petroleum at a rate equivalent to the oil consumption of the global aviation sector (Ellen MacArthur Foundation, 2016).

Threats to the Arctic. https://doi.org/10.1016/B978-0-12-821555-5.00010-3

Plastics have been grouped into several types, as follows:

Thermoplastics (Fig. 6.1, Top)

These plastics can be remelted and essentially returned to their original state—much like the way an ice cube can be melted and then cooled again. Thermoplastics usually are produced first in a separate process to create small pellets; these pellets then are heated and formed to make all sorts of consumer and industrial products. See the discussion of nurdles in the following. Thermoplastics include many familiar varieties, such as polyethylene (PE), polypropylene (PP), polyvinyl chloride (PVC), polystyrene (PS), nylon, and polycarbonates.

Thermosets (Fig. 6.1, Bottom)

Thermosetting plastics are a group of polymers that are irreversibly hardened by curing from a soft solid or viscous liquid prepolymer or resin. Curing is induced by heat or suitable radiation and may be promoted by high pressure or mixing with a catalyst. Because of this, these plastics are usually produced and formed into products at the same time. They cannot be returned to their original state. Thermosets include vulcanized synthetic rubber, acrylics, polyurethanes, polyester resin, melamine, silicone, and epoxies.

Plastic fibers

Plastic fibers are plastics that have been spun into fibers or filaments and used to make fabrics, string, ropes, and cables, even optical fibers. Some of the most recognizable plastic fibers are polyester, nylon, rayon, acrylic, and spandex, although there are many more. The name polyester is short for polyethylene terephthalate (PET) and is also used to make beverage bottles.

Other types of plastic are used in coatings, adhesives, and elastomers. Types of artificial rubber include fluoroelastomer rubber, ethylene-propylene-diene rubber, nitrile rubber, and polychloroprene.

Geyer et al. (2017) estimated that 8300 million metric tons of virgin plastics had been produced by 2017. As of 2015, approximately 6300 million metric tons of plastic waste had been generated, but only about 9% of this had been recycled (Fig. 6.2). These are numbers that are very difficult to visualize. For comparative purposes, 1 million metric tons of water would fill a cube with height, depth, and width each 100 m in size. The biggest producer of plastic waste in recent years has been China, followed by the rest of Asia, Europe, and North America (Jambeck et al., 2015) (Fig. 6.3).

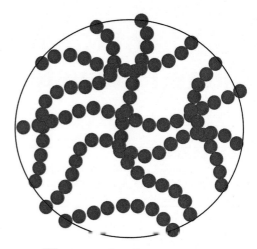

Thermoplastic polymer

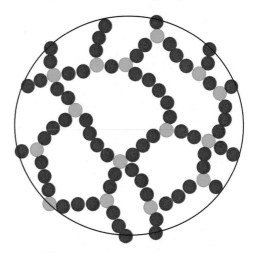

Thermosetting polymer

FIG. 6.1 Top: Linear chains of macromolecules (a thermoplastic polymer). Bottom: A three-dimensional macromolecule (a thermosetting polymer). The tri- and tetravalent cross-linking nodes are represented in green. The dimension of circles equals about one ångström (Å). (Modified from Structures of Macro Molecules, c. 2018. https://commons.wikimedia.org/wiki/File:Structures_of_macromolecules.png.)

Nearly all plastic ocean debris comes from coastal regions. Continental plastic litter enters the ocean largely through stormwater runoff, flowing into watercourses or directly discharged into coastal waters. Around 50% of plastic products are buoyant and 60%—64% of the terrestrial load of floating plastic to the sea are

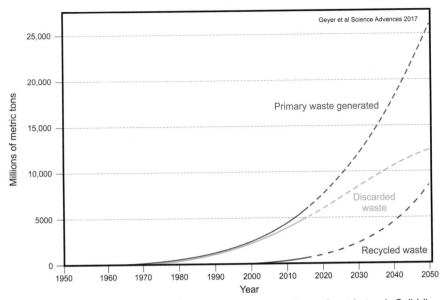

FIG. 6.2 Cumulative plastic waste generation and disposal (in millions of metric tons). *Solid lines* show historical data from 1950 to 2015; *dashed lines* show projections of trends to 2050. (Modified from Geyer, R., Jambeck, J., Law, K., 2017. Production, use, and fate of all plastics ever made. Science Advances 3 (7), e1700782. https://doi.org/10.1126/sciadv.1700782.)

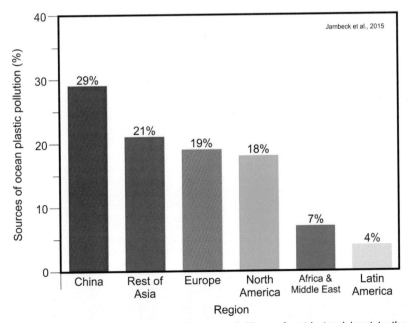

FIG. 6.3 Estimated mass of mismanaged plastic waste (millions of metric tons) input to the ocean by populations living within 50 km of a coast in 192 countries, plotted as a cumulative sum from 2010 to 2025. (Modified from Jambeck, J., Geyer, R., Law, K., 2015. Plastic waste inputs from land into the ocean. Science 347, 768–771. https://doi.org/10.1126/science.1260352.)

estimated to be exported from coastal waters to the open ocean (Cózar et al., 2014). Unlike natural organic materials, the synthetic polymers in plastics are extremely durable and may persist, virtually unaltered, for centuries to millennia. Once plastic materials enter the ocean, their breakdown becomes even slower than on land (Fig. 6.4). Because plastics in the ocean are not exposed to thermal oxidation as they are on land, plastics in water degrade much slower than they do on land. This problem is further exacerbated in the Arctic where cold-water and freezing-air temperatures greatly retard chemical reactions. Eventually, plastic objects in seawater are broken down into smaller particles through mechanical abrasion, hydrolysis, or photodegradation through UV from sunlight. Increasing brittleness promotes the breakdown of the polymer structure, but degradation rates are slow, especially under cold arctic temperatures.

Jambeck et al. (2015) published current and future predictions of the cumulative quantities of plastic waste in the world, showing three different scenarios (Fig. 6.5). Under the worst-case scenario, the world will have generated 250 million tons of plastic waste by 2025. Even in the best-case scenario, the world will have generated 100 million tons of plastic waste by 2025. The estimates are based on conversion rates of mismanaged plastic waste to marine debris (high, 40%; mid, 25%; low, 15%). Only a small fraction of this plastic waste enters the oceans, but even the small amount has delivered 150 million metric tons of plastic into the oceans, with an additional 8 million metric tons being added every year. Plastic pollution exists almost everywhere in the ocean, from the remote seas of the Arctic to the floor of the deep sea. As discussed in the following, this enormous volume of plastic has become one of the most important threats to marine biota.

Kühn and van Franeker (2020) did an inventory of data from 747 studies concerning the effects of marine plastic debris on marine birds, mammals, turtles, fish, and invertebrate species. They reported that this debris has affected 914 species through entanglement and/or ingestion. Ingestion was recorded for 701 species; entanglement was documented for 354 species. More specifically, marine plastic pollution has affected 100% of marine turtles, 59% of whales, 36% of seals, and 40% of seabirds of those examined (Plastic in the Ocean Statistics, 2020). Sea turtles mistakenly eat plastic bags that they confuse with jellyfish. Sea turtles have up to 74% (by dry weight) of their diets composed of

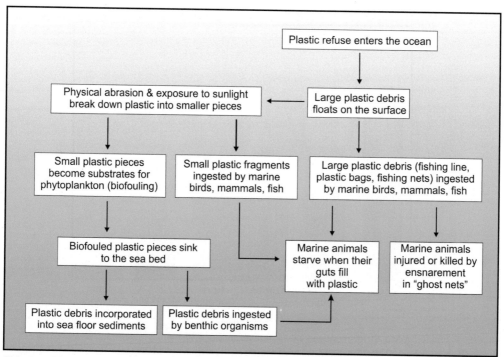

FIG. 6.4 Flowchart tracing the path of plastic debris from its entrance into the ocean through ingestion by marine wildlife.

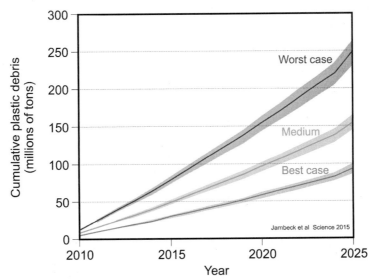

FIG. 6.5 Graph showing rise in plastic debris in the oceans, 2010 to 2025, after Jambeck et al. (2015). Best-case, medium-case, and worst-case scenarios depicted, with 95% confidence intervals shown in shaded regions surrounding *lines*. (Data from Oceans, Coasts & Seashores: Ocean Plastics. 2018. https://www.nps.gov/subjects/oceans/ocean-plastics.htm.)

ocean plastics. In recent years, all of the baby sea turtles examined by biologists throughout the world were found to have plastic in their stomachs (Plastic in the Ocean Statistics, 2020). Sea birds, whales, dolphins, and other marine animals often turn up dead with stomachs full of plastic or get caught in abandoned plastic fishing nets (Table 6.1).

Plastic ocean debris does not remain intact and unaltered in the water. Part of the breakdown process comes from physical abrasion. Eventually, most large plastic objects are abraded into smaller pieces. Hammer et al. (2012) divide plastic ocean debris into three classes: macrodebris, mesodebris, and microdebris. Macrodebris consists of the larger parts of plastic debris (>20 mm to several meters). This large debris includes plastic chairs, shoes, car/plane/boat parts, buoys, footballs, etc. Nearly all kinds of objects larger than 20 mm that have ever been made from plastic have found their way into the oceans.

MACRODEBRIS

One important component of plastic macrodebris is the "ghost net." A ghost net is an abandoned or lost fishing net that drifts across the ocean. These nets persist for years, traveling with the currents and tides, continually catching animals and other macrodebris in their strands, until they become filled with other floating debris—mainly other plastic objects. Ghost nets can grow to masses of 6 tons, making it virtually impossible to remove them from the water. Experts have estimated that there are roughly 640,000 metric tons of these nets currently in the world's oceans, accounting for 10% of the total plastic waste in the sea. More than 650,000 marine animals, including dolphins, whales, seals, turtles, and seabirds, are killed or injured in fishing nets each year (Fig. 6.6, below). In many cases, animals get caught and die in nets that are being actively used by fishermen. Pinnipeds are particularly susceptible to entanglement, especially in abandoned, lost, or discarded fishing gear and packaging straps (Table 6.2). As of 2019, 67% of pinniped species are known to be susceptible to entanglement.

Entanglement in oceanic plastic pollution is a threat for at least 243 marine species, fewer species than for plastic ingestion, although entanglement is responsible for more known mortalities (Jepsen and Nico de Bruyn, 2019). Most of the plastics that entangle marine wildlife are fishing nets, monofilament lines, ropes, and other fishing-related gear that enter the water near coastal cities, via commercial fishery activities and in shipping corridors. Before the middle of the 20th century, fishing gear was made from natural fibers such as hemp or cotton that disintegrate over time. However, the advent of synthetic material in its stead resulted in long-lasting plastics that withstand disintegration. Much of this

TABLE 6.1

Percent of Marine Birds and Mammals Affected by Plastic Entanglement or Ingestion.

Taxa	No. of Species in Taxon	% of Affected Species
SEABIRDS		
Marine ducks	6	23%
Grebes	6	30%
Tubenoses (albatrosses, petrels, fulmars)	91	63%
Gannets, cormorants	24	49%
Gulls, terns, skuas, auks	77	57%
All seabirds	409	55%
MARINE MAMMALS		
Polar bears	1	100%
Sea otters	2	100%
Eared seals	14	86%
Walrus	1	0%
True seals	16	62%
All seals	31	71%
Baleen whales	14	86%
Toothed whales	72	65%
All whales	86	69%
Manatees, dugongs	3	67%
All marine mammals	123	70%
Sea turtles	7	100%

Data from Kühn, S., van Franeker, J.A., 2020. Quantitative overview of marine debris ingested by marine megafauna. Marine Pollution Bulletin 151. https://doi.org/10.1016/j.marpolbul.2019.110858.

material (especially ropes and nets) is abandoned, lost, or purposefully discarded and continue to entangle through ghost-fishing.

MESODEBRIS

Mesodebris, also called microplastics (MPs), range in size from 2 to 20 mm in diameter. It often consists of plastic resin pellets, also known as nurdles. Nurdles are small granules that have the shape of a cylinder or disk and have a diameter of 2–6 mm (Wu et al., 2019) (Fig. 6.7).

Nurdle production began in the 1940s and 1950s (NOAA, 2020). These pellets are made as a raw industrial material and are sent to manufacturers for melting and molding into plastic products. Nurdles end up in the marine environment through accidental spillage during transport and handling, not as litter or waste. Nurdles are the most economical way to transfer large amounts of plastic to end-use manufacturers around the globe, and the United States produces about 60 billion pounds of them annually. One-half pound (227 g) of high-density PE plastic is made from 22,000 nurdles. They are sufficiently small to pass through many water filtration systems. Only a small fraction of the trillions of nurdles produced every year make their way into the world's oceans, but they still constitute a major source of marine plastic debris. Nurdles are accumulating on the world's beaches (Fig. 6.8), having been first washed out to sea, and then carried by waves to the shore. Researchers have found upward 40,000 plastic items per square meter on some island beaches—mostly nurdles (Hammer et al., 2012). The durability of these industrial plastic pellets in the marine environment is still uncertain, but they seem to last from 3 to 10 years, and chemical additives used in many nurdles can probably extend this period to 30–50 years. These additives are used to give them nurdles certain characteristics, such as color, reduced flammability, or increased flexibility. In some cases, the additives weigh more than the nurdles. For example, additives constitute up to 75% of the weight of PVC nurdles. Some additives are toxic to marine life, such as phthalates, bisphenol A (BPA), and organotins. These chemicals may leach out of the plastic once the nurdles enter the marine environment.

Nurdles enter marine food chains in several ways. Ubiquitous throughout the world's oceans, these tiny plastic particles often enter the food chain as they are consumed by zooplankton, mistaken for the organic detritus that forms a natural part of their diet. The particles also accumulate in marine mollusks and other benthic filter feeders. Zooplankton are then consumed by crustaceans and fish, and by some whales. Mollusks, especially clams, are consumed by walrus. The crustaceans and fish are consumed by marine birds; marine mammals such as seals, sea lions, and killer whales, and humans. One in three fish caught for human consumption contains plastic (Plastic in the Ocean Statistics, 2020). Indigenous peoples of the North (Yupik and Inupiat Eskimos and Inuit) harvest whales as well as fish.

Once at sea, nurdles and other plastics also attract oceanic chemical contaminants to their surfaces. These are persistent bioaccumulating toxins (PBTs), industrial chemicals that can accumulate in animal and human

US Fish and Wildlife Service

NOAA Marine Debris Program 2014

FIG. 6.6 Above: Plastic debris found in dead albatross chick, North Pacific Gyre region. Below: Dead shearwater trapped in a fishing net. (Above: Photo courtesy of bluebird-lectric.net; Below: NOAA, 2014. The Great Pacific Garbage Patch. National Ocean Service. https://oceanservice.noaa.gov/podcast/june14/mw1.)

tissue causing long-term damage (Great Nurdle Hunt, 2020). Plastics act like a sponge for these toxins and can concentrate them to levels millions of times higher than the surrounding seawater. The manufacture of most PBTs has been banned in recent years, but not in all countries. Although current production has been reduced, these chemicals continue to persist in the oceans.

Bioaccumulation and Biomagnification

Bioaccumulation is the process by which toxins enter a food web by building up in individual organisms. Thus, it can be particularly harmful to species with great longevity, including several marine mammals since they accumulate toxins from their diets throughout their lives. As shown in Table 6.3, long-lived marine mammals, such as the striped dolphin (50–60 years)

TABLE 6.2
Number of Pinnipeds (by Species) Entangled in Ghost Nets Since 1979.

Species	Common Name	Deaths since 1979	Range in Northern Seas
Zalophus californianus	California sea lion	2170	Southeast Alaska
Arctocephalus gazella	Antarctic fur seal	2063	None
Arctocephalus pusillus	Brown fur seal	1179	None
Callorhinus ursinus	Northern fur seal	823	Chukchi Sea
Neomonachus schauinslandi	Hawaiian monk seal	447	None
Arctocephalus forsteri	New Zealand fur seal	280	None
Phoca vitulina	Harbor seal	261	Circumpolar in the low Arctic
Mirounga angustirostris	Northern elephant seal	251	Aleutian Islands, Alaska
Eumetopias jubatus	Steller sea lion	251	Sub-Arctic Siberia and Alaska
Halichoerus grypus	Grey seal	145	Subarctic North Atlantic
All 12 other species, combined		277	N/A
Total deaths do to net entanglement		2735	

Data from Jepsen, E.M., de Bruyn, P.J.N., 2019. Pinniped entanglement in oceanic plastic pollution: a global review. Marine Pollution Bulletin 145, 295–305. https://doi.org/10.1016/j.marpolbul.2019.05.042.

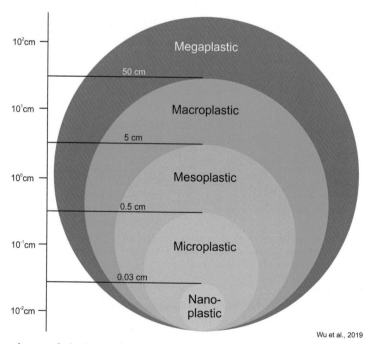

FIG. 6.7 Size classes of plastic debris: nanoplastic (<0.03 cm), microplastic (0.05–0.5 cm), mesoplastic (0.5–5 cm), macroplastic (5–50 cm), and megaplastic (>50 cm). (Modified from Wu, P., Huang, J., Zheng, Y., Yang, Y., Zhang, Y., 2019. Environmental occurrences, fate, and impacts of microplastics. Ecotoxicology and Environmental Safety, 184, 109612. https://doi.org/10.1016/j.ecoenv.2019.109612.)

Microplastics (<2 mm)

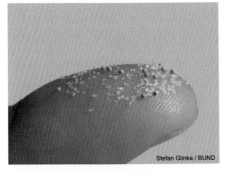

Stefan Glinka / BUND

Mesoplastics (2-20 mm)

Photo by Jace Tunnell,
published by NOAA 2020

Macroplastics (>20 mm)

FIG. 6.8 Top: Microplastic particles. Photo by Stefan Glinka, BUND, used with permission. Middle: Plastic pellets (nurdles) found on an ocean beach. Bottom: Macroplastic debris on a beach in Svalbard. (Top: From Glinka, Bund, S., 2020. Top: Micro plastic particles. https://www.bund.net/service/presse/pressebilder/aktionen/#c3479. Middle: Photo by Jace Tunnell, Mission Aransas National Estuarine Research Reserve, published by NOAA, in the public domain. Bottom: Photo by Ashley Cooper, Getty Images, used with permission.)

and humpback whale (80 years), accumulate dangerously large quantities of mercury in their bodies, especially concentrated in their livers. In contrast to this, the polar bear (life span of 20 years) accumulates only a small fraction of the mercury found in the dolphins and whales. This relationship between longevity and bioaccumulation of toxins is quite variable because the different animals involved have quite different diets. This is where biomagnification comes in.

Biomagnification is the process by which toxins are passed from one trophic level to the next (and thereby increase in concentration) within a food web. As shown in Fig. 6.9, the concentration of mercury increases dramatically from the bottom to the top of the food chain in the Arctic marine ecosystem. Mercury is a useful tracer for all contaminants that enter at the bottom of the food chain and work their way up, becoming more concentrated with each trophic level. Two examples typify this process. Beluga whales eat fish that have eaten other fish and zooplankton. The concentrations of polychlorinated biphenyls (PCBs) in beluga blubber in the Canadian arctic range from about 3000 to 5000 ng per gram of blubber, a concentration about 150 times that found in zooplankton. Polar bears eat mainly ringed seals in these waters. The seals eat fish that have eaten other fish and zooplankton. The concentration of PCBs in polar bear fat is about 1000—2500 ng per gram of fat. Walruses eat mainly bivalves, such as clams. These prey items have very low concentrations of pollutants in their bodies, so walruses likewise have lower concentrations of pollutants than fish-eating marine mammals.

Studies have demonstrated the ubiquity of MPs in marine ecosystems. These present serious threats to the health of fish, birds, and marine animals, causing symptoms such as malnutrition, inflammation, chemical poisoning, growth thwarting, decreased fecundity, and death due to damages at individual, organ, tissue, cellular, and molecular levels (Peng et al., 2020). The damage done to marine wildlife is shocking. More than one million seabirds and 100,000 marine animals die from plastic pollution every year. Over 90% of all seabirds examined have plastic in their stomachs (Plastic in the Ocean Statistics, 2020) (Fig. 6.6, above).

The information on nanoplastics (NPs) in the marine ecosystem has been scarce due to the challenges in sampling and detecting these nano-scaled entities. In vitro and in vivo experiments have demonstrated that NPs have the potential to penetrate different

TABLE 6.3
Longevity of Arctic Marine Species Compared With Mercury Concentrations in Liver Tissue.

Species	Scientific Name	Life Span	Typical Mercury Concentration in Liver Tissue
Humpback whale	*Megaptera novaeangliae*	80 years	300–370 µg/g dry weight
Striped dolphin	*Stenella coeruleoalba*	50–60 years	514 µg/g dry weight
Beluga	*Delphinapterus leucas*	35–50 years	2–17 µg/g dry weight
Ringed seal	*Pusa hispida*	Up to 43 years	11 µg/g dry weight
Narwhal	*Monodon monocerus*	30–40 years	2–6 µg/g dry weight
Walrus	*Odobenus rosmarus*	40 years	<1 µg/g dry weight
Polar bear	*Ursus maritimus*	30 years	2–20 µg/g dry weight
Blue whiting	*Micromesistius poutassou*	7–10 years	1.3 µg/g dry weight
Arctic cod	*Boreogadus saida*	6 years	<0.09 µg/g dry weight

Data from Kershaw, J.L., Hall, A.J., 2019. Mercury in cetaceans: exposure, bioaccumulation and toxicity. Science of the Total Environment 694. https://doi.org/10.1016/j.scitotenv.2019.133683.

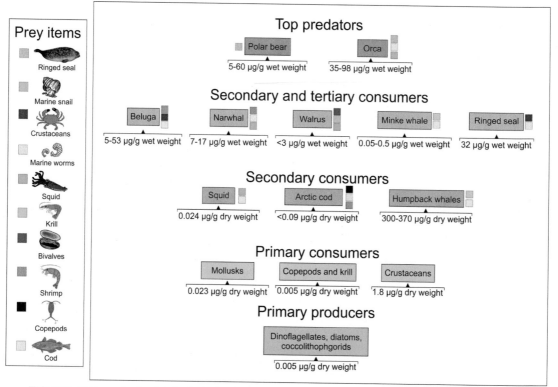

FIG. 6.9 Arctic marine food web, displaying the biomagnification of mercury from the bottom to the top of the food chain. Each species is shown with the concentrations of mercury found in its body, as well as symbols indicating the three most important items in their diet. The key to the identification of food items is shown to the left of the figure.

biological barriers including the gastrointestinal barrier and the brain-blood barrier and have been detected in many important organs such as brains, the circulation system, and livers of sampled animals.

Through the actions of tides and currents, huge amounts of plastic nurdles may wash up on a given beach. The densest site on the coast of Alaska had an estimated 278,000 nurdles per square meter of material (Das, 2020). Whether on beaches or floating at sea, mesodebris presents many hazards to animal life.

MICRODEBRIS

MPs come from a variety of sources, including from larger plastic debris that degrades into smaller and smaller pieces. In addition, microbeads, a type of microplastic, are very tiny pieces of manufactured PE plastic that are added as exfoliants to health and beauty products, such as some cleansers and toothpaste. These tiny particles easily pass through water filtration systems and end up in the ocean, posing a potential threat to marine life. As an emerging field of study, not a lot is known about microplastics and their impacts (NOAA, 2020).

MPs and NPs are of special concern to conservationists because of their minute size, the propensity to float in seawater, and persistence in the ocean for multiple decades. These qualities facilitate their long-distance transport across oceans, across trophic levels in food webs, and across different organs within contaminated marine fish, birds, and mammals (Fig. 6.10) (Peng et al., 2020).

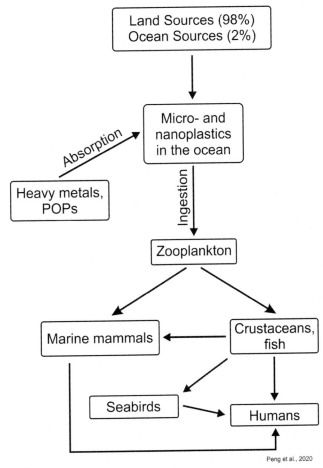

Peng et al., 2020

FIG. 6.10 Flowchart illustrating the pathways of microplastics in the ocean. (Data from Peng, L., Fu, D., Qi, H., Lan, C.Q., Yu, H., Ge, C., 2020. Micro- and nano-plastics in marine environment: source, distribution and threats — a review. Science of the Total Environment 698. https://doi.org/10.1016/j.scitotenv.2019.134254.)

Microbeads are not a new problem. Plastic microbeads first appeared in personal care products about 50 years ago, with plastics increasingly replacing natural ingredients. As recently as 2012, this issue was still relatively unknown, despite an abundance of consumer products containing microbeads. Eventually, the environmental damage done by microbeads was discovered, and in 2015, the US government banned their use in cosmetics and personal care products. However, because of their persistence in the environment, there are now 51 trillion MPs littering our oceans—twice the number of macroplastics (Plastics in the Ocean Statistics, 2020).

MPs enter the Arctic Ocean and adjacent northern seas region via long-distance transport by both winds and ocean currents, and through local input from arctic communities, although the exact origins and pathways bringing MPs to the Arctic remain unclear.

Microplastics in Arctic Sea Ice and Snow

Zhang et al. (2020) recently documented the fact that MPs float in the atmosphere surrounding the globe, from the tropics to the poles. The abundance of aerial MPs and their deposition on land and sea varies by as much as three orders of magnitude across different regions and localities. Fibers and fragments of larger plastic objects are the most frequently reported types in the MP size range. In a study by Bergmann et al. (2019), snow deposited on ice floes drifting in Fram Strait and on snowbanks on Svalbard was sampled to assess the level of atmospheric MP fallout in the Arctic. These snow samples yielded MPs with the following size ranges: 80% of the detected MPs were less than or equal to 25 μm; 98% of all particles were less than 100 μm (Bergmann et al., 2019). They found that the amount of MP particles was inversely proportional to their size (Bergmann et al., 2019). In fact, they found large numbers of particles below the standard detection limit of 11 μm. The MPs they recovered from Arctic snowfall included a wide variety of polymers, with varnish (acrylates), plasticized rubber, and polyamides among the most common varieties (Table 6.4).

As it forms, sea ice scavenges and concentrates particulates from the water column. With rising quantities of MPs entering Arctic seas, these human-made particles have become trapped in the forming sea ice, along with natural grains of sand, silt, and clay. Obbard et al. (2014) documented the fact that Arctic sea ice from remote locations not only contains MPs but that their concentration in sea ice is several orders of magnitude greater than those reported from highly contaminated surface waters, such as those of the Pacific Gyre.

Their findings indicate that MPs accumulating in Arctic sea ice constitute a major global sink of human-made particulates. As inferred by Lusher et al. (2015), the study by Obbard et al. (2014) documents the problem that melting Arctic sea ice will release substantial quantities of chemically contaminated MP particles into the ocean as the ice, with as-yet-unknown physical and toxicological effects on marine life.

The polymers identified by Obbard et al. (2014) from sea ice samples have a wide range of domestic and industrial uses. The relative proportions of different polymer types were like those reported in other ocean sediments and biota, including polyester, acrylic, PP, PE, and polyamide. If current trends continue, in the next decade 2.04 trillion m^3 of Arctic sea ice will melt. Even if this ice were to contain the lowest quantity found in the sea ice samples (38 particles per cubic meter), the melting sea ice would release over one trillion pieces of plastic into the northern seas.

Microplastics in Arctic Waters

Lusher et al. (2015) performed the first investigation of MPs in Arctic waters, working in transects of sites to the south and southwest of Svalbard. MPs were found in surface (top 16 cm) and subsurface (6 m depth) samples. The vast majority (95%) of MP particles found in their samples were plastic fibers. This suggests that they may either result from the breakdown of larger items (e.g., discarded fishing nets, transported over large distances by prevailing currents, or derived from local fishing vessel activity), or input in sewage and wastewater from coastal areas.

In addition to microbeads and microfibers, other kinds of MPs find their way into Arctic seas in the form of granules, pellets, and powders used in the production of larger plastic products and from the washing of synthetic clothing. In addition to the MP floating on the surface, plastic litter quantities have increased significantly on the deep Arctic seafloor over the past 15 years, as highlighted in a time-series study by Tekman et al. (2017).

Because of their minute size, MPs are difficult to mechanically remove from the ocean and their distribution and fate have still to be studied in depth. Recently, MP particles were reported from ice cores from remote areas of the Arctic Ocean, at levels greater than those reported for Pacific Gyre surface waters. Ocean circulation models propose the transport and a potential accumulation zone of MPs at higher northern latitudes, including the Barents Sea. However, field studies have not yet revealed the geographic distribution of MP in polar waters.

TABLE 6.4
Plastic Polymer Compounds Found in Fram Strait Ice Floes and in Svalbard Snowpack.

Compound	Abundance in Fram Strait Ice Floes	Abundance in Svalbard Snow Pack	Uses of Compounds
Polyethylene	Common	Common	Packaging film, trash and grocery bags, cable insulation, housewares
Oxidized polyethylene (wax)	Rare	Not found	Printing inks, rubber, adhesives, sealants
Polyester	Very common	Very common	Fabrics, bottles, films, filters, film insulation for wire
Acrylates/PUR/varnish	Very common	Not found	Cosmetics, adhesives/glue/wood finishing
Fluoroelastomer rubber	Rare	Not found	Tubing, hoses, seals, and gaskets
Ethylene-propylene-diene rubber	Very common	Very common	Roofing membranes, gaskets, shoe soles, cable insulation
Nitrile rubber	Very common	Very common	Medical gloves, automotive belts, gaskets and hoses, synthetic leather
Ethylene-vinyl-acetate (EVA)	Very common	Not found	Foam rubber used in packaging and seat cushioning; adhesives, sealants
Polyamide	Very common	Very common	Textiles, automotive parts, fishing nets, carpets, kitchen utensils
Polyvinylchloride	Rare	Not found	Door and window frames, water pipes, cable insulation, medical devices
Polypropylene	Very common	Very common	Fabrics, cling-film, storage and grocery bags, food containers
Chemically modified cellulose	Common	Common	Fabric finishing
Polychloroprene	Not found	Rare	Gaskets, cable jackets, tubing, seals, tires, hoses, orthopedic braces

Data from Bergmann, M., Mützel, S., Primpke, S., Tekman, M.B., Trachsel, J., Gerdts, G., 2019. White and wonderful? Microplastics prevail in snow from the Alps to the arctic. Science Advances 5 (8). https://doi.org/10.1126/sciadv.aax1157.

In the study by Lusher et al. (2015), MPs were found across the survey area of the Barents Sea in more than 90% of samples, both in the surface waters and at 6 m depth. MP abundance values in surface waters were of the same order of magnitude as those found in the North Pacific and North Atlantic and were greater than those of the Californian current system and south and equatorial Atlantic.

Arctic waters support a large quantity of filter-feeding organisms, from copepods to fish and baleen whales, which could actively target or passively ingest MPs floating in the surrounding water. The potential effects of MP ingestion include reduced ability to feed, energy depletion, injury, and death due to toxicological responses to contaminants associated with the plastics.

Previous studies have not focused on the organisms that live just beneath the surface of Arctic waters. However, a high abundance of zooplankton in the surface waters of the Barents Sea suggests a high probability of encounter between MPs and marine fauna just beneath the surface. Further fragmentation of MPs into NPs might increase potential interaction with zooplankton that mistake plastics for prey (Lusher et al., 2015). These extremely small objects range in size from 1 to 1000 nm (nm). One nanometer is one-billionth of a meter (1×10^{-9} M).

Secondary impacts may occur at higher trophic levels if ingested NPs are transferred within the prey items of fish, birds, and mammals. Animals such as baleen whales usually feed on aggregations of planktivorous fish, crustacea, and cephalopods, both in the water column and at the surface. This habit greatly increases the likelihood of primary and secondary ingestion of MPs while feeding. If, as seems likely, sea ice is holding large quantities of MP particles, then such ecological interactions will be exacerbated as the sea ice melts and releases its particulate contents (Lusher et al., 2015).

As we have seen, despite an exponential increase in global data from studies on marine plastic debris, information on levels and trends of plastic pollution (especially MPs) in the Arctic is based on just a handful of pioneering studies. The available literature points to a ubiquitous distribution of plastic particles in all environmental compartments, including sea ice, snow, and ocean waters (Halsband and Herzke, 2019). The number of MPs in the world's oceans likely exceeds the numbers of plastic pieces large enough to be collected and counted. It now appears that the proportion of MPs now exceeds 90% of the total plastic debris in the world's oceans (Zarfl and Matthies, 2010). Presumably, a similar level of MP dominance occurs in Arctic seas.

TOXIC EFFECTS OF MARINE PLASTICS

Polar waters support highly productive marine food webs and ecosystems that are demonstrably vulnerable to marine plastic pollution. Prevailing ocean currents, wind currents, and migratory species transport containments to the Arctic. For instance, the North Atlantic branch of the Thermohaline Circulation delivers plastics of all sizes to the Arctic (Cózar et al., 2014, 2017). Plastic particles are swallowed by accident by various marine organisms and accumulate up the food chain from prey to predator. Filter feeders such as krill (the basic source of food for many marine organisms), forage fish (herrings, sardines, and others), jellyfish, sharks, whales, and sea birds are particularly affected by ingested plastics due to their modes of food intake. Thus, plastic particles are pollutant vectors that can be transported directly into the food chain. As of this writing, the quantification of plastic particles transported to the Arctic is insufficient to assess the long-range transport potential of toxic chemicals in the Arctic.

In addition to the mechanical impacts of plastics on marine organisms, such as suffocation and starvation due to entanglement or ingestion of plastic pieces or bags, a relationship between toxic chemicals and plastic ingestion has been reported. In particular, the accumulation of persistent organic pollutants (POPs) has been well documented. Toxic chemicals that are persistent in the environment and bioaccumulate in animals are called persistent bioaccumulative toxic chemicals, or PBTs. POPs are halogenated organic compounds that are resistant to environmental degradation through chemical, biological, or photolytic processes. These toxic chemicals have been wreaking havoc on the Arctic biota for decades. For instance, during a research cruise, Gao et al. (2019) identified 16 different chemical compounds that form parts of organophosphorus flame retardants (OBFRs) in Arctic water samples. MPs absorb PCBs and other POPs from the surrounding seawater. When they adhere to pieces of plastic, their concentration is far higher than their concentration in the surrounding seawater. For instance, MP pellets have been found to hold two orders of magnitude higher concentrations of PCBs than natural particles suspended in seawater. Therefore, plastic is acting as not only a conduit for POPs but also as a powerful concentrator. Organisms ranging from zooplankton to marine mammals and birds consume these plastic particles, either accidentally or because the plastics are mistaken by them for food items. When this happens, the toxic chemicals attached to the plastic pallets are released into these animals. Besides sorption from surrounding seawater, several toxic chemical additives are contained in the plastic matrix. Colorants, UV stabilizers and matting agents, brominated flame retardants (BFRs), phthalate plasticizers, bisphenol A, and antimicrobial agents are additives of particular concern (Thompson et al., 2009). Bound to the plastic matrix, these chemical compounds escape rapid degradation and may thus become "persistent" and subject to long-range transport. When ingested by organisms, these substances may follow the "biomagnification" route of organic chemicals sorbed to or contained in the plastics, as shown by investigations of polybrominated diphenyl ethers (PBDEs), polycyclic aromatic hydrocarbons (PAHs), triclosan, and nonylphenol (NP).

CHEMICAL POLLUTION IN ARCTIC SEAS

The flora and fauna of the Arctic seas are swimming in a pool of poisons, practically none of which originated in the Arctic. But even this aspect is starting to change. Pollution of Arctic Ocean waters used to be caused solely by long-distance transport of pollutants via ocean currents, atmospheric deposition (from long-distance transport), or from rivers draining into the Arctic Basin. However, as Arctic sea ice is becoming less of a threat to shipping and offshore petroleum drilling, local pollution is becoming a serious problem in Arctic waters.

The Stockholm Convention on POPs is a global treaty created to protect human health and the environment through the reduction, restriction, or elimination of the production and use of specified POPs. The Convention came into force in 2004 with a list of 12 POPs, the "Dirty Dozen" with the subsequent addition of another 16 POPs (UNEP, 2018c). Their criteria for chemical compounds to be included in their list are as follows:

(a) Chemical identity: the compound must have a name, including trade names and synonyms. It must have a Chemical Abstracts Service (CAS) Registry Number, International Union of Pure and Applied Chemistry (IUPAC) name, and its chemical structure must be identified.

(b) Persistence: There must be evidence that the half-life of the chemical in water is greater than 2 months, or that its half-life in soil is greater than 6 months, or that its half-life in sediment is greater than 6 months. If these specific half-lives are not yet identified, then there must be evidence that the chemical is otherwise sufficiently persistent to justify its consideration within the scope of the Stockholm Convention.

(c) Bioconcentration: There must be evidence that the bioconcentration factor or bioaccumulation factor in aquatic species for the chemical is greater than 5000. Or there must be evidence that a chemical presents other reasons for concern, such as high bioaccumulation in other species, high toxicity, or ecotoxicity; or there must be monitoring data in biota indicating that the bioaccumulation potential of the chemical is sufficient to justify its consideration within the scope of the Convention.

(d) Potential for long-range environmental transport: There must be measured levels of the chemical in locations distant from the sources of its release that are of potential concern. There must be monitoring data showing that long-range environmental transport of the chemical, with the potential for transfer to a receiving environment, may have occurred via air, water, or migratory species. For a chemical that migrates significantly through the air, its half-life in air should be greater than 2 days.

(e) Adverse effects: There must be evidence of adverse effects to human health or to the environment that justifies consideration of the chemical within the scope of the Convention, or there must be toxicity or ecotoxicity data that indicate the potential for damage to human health or the environment.

As a result of these initiatives, concentrations of many POPs have decreased in the Arctic environment (Fig. 6.11). Although new emissions of these chemicals have been virtually eliminated, large quantities remain in environmental reservoirs (oceans and forests). These POPs will eventually break down, but some compounds (e.g., PCBs) will persist for a long time (Munthe et al., 2018). Meanwhile, scientists continue to discover previously unknown contaminants in the Arctic; many of these newer POPs will continue to increase in the environment until they become regulated on a global basis.

Table 6.5 provides a list of chemicals found in Arctic environments that are currently being regulated (or being considered for regulation) and those considered by the Arctic Monitoring and Assessment Program to be of emerging Arctic concern (AMAP, 2017).

Local Sources

Although most toxic substances in the Arctic come from long-distance transport, there are several examples of regional sources of pollution in the Arctic. These include PCBs from decommissioned "distant early warning line" sites in Canada and Alaska, dioxins (PCDDs and PCDFs) from smelters in Norway, heavy metals from industrial activities in the Kola Peninsula and the Norilsk industrial complex in Russia (Burkow and Kallenborn, 2000).

Increased human access to the Arctic Ocean is enabling the development of new oil and gas fields on the continental shelves surrounding the Arctic Ocean. Drilling of new offshore oil and gas wells is already being planned in these waters, as sea ice retreats. Such activities will alter habitats, pollute the water, and disturb marine life. These topics are discussed at length in Chapter 15 (Oil and Mineral Extraction).

Long-Distance Transport of Chemical Pollutants

Burkow and Kallenborn (2000) mapped multiple sources of pollution entering the Arctic. These pollutants arrive in the North via air currents, ocean currents, on drifting sea ice, and from rivers emptying into the Arctic basin. Some toxic chemicals and heavy metals are produced either in or near the Arctic, in Scandinavia and northern Russia. Other pollutants travel thousands of kilometers before arriving in the Arctic. For instance, it is thought that ongoing PCB emissions affecting polar bears on the Greenland coast come from urban sources in the midlatitudes of North America and Europe.

Among the various anthropogenic contaminants, POPs are relatively more toxic, persistent, bioaccumulative, and prone to long-range atmospheric transport than other toxins. These characteristics

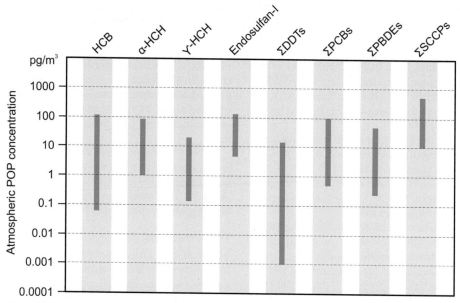

FIG. 6.11 The current concentration ranges of atmospheric POPs in the Arctic. *DDT*, dichloro-diphenyl-trichloroethane; *HCB*, hexachlorobenzene; *HCH*, hexachlorocyclohexane; *PBDEs*, POPs, persistent organic pollutants; polybromodiphenyl ethers; *PCBs*, polychlorinated biphenyls; *SCCPs*, short-chain chlorinated paraffins (Wang, X., Wang, C., Zhu, T., Gong, P., Fu, J., Cong, Z., 2019. Persistent organic pollutants in the polar regions and the Tibetan Plateau: a review of current knowledge and future prospects. Environmental Pollution 248, 191–208. https://doi.org/10.1016/j.envpol.2019.01.093.)

promote their widespread distribution across the globe, including remote regions where they have never been used (Wang et al., 2019).

Nearly all POPs are manufactured in cities and towns in the midlatitudes and then are accidentally transported from densely populated areas at lower latitudes via the atmosphere to the Arctic, with subsequent deposition and volatilization. This has been termed "the grasshopper effect" and explains how POPs travel extremely long distances and arrive in remote and cold areas. The trajectory of the winds carrying POPs is heavily influenced by seasonal weather patterns. For example, winter high-pressure systems in Siberia push air masses north into the Arctic. This wind trajectory accounts for much of the Eurasian contaminant transportation to the Arctic. In the Pacific, the El Niño Southern Oscillation (ENSO) likewise plays an important role in the atmospheric transportation of POPs to the Arctic. Long-term regional climate patterns, such as the Arctic Oscillation and the North Atlantic Oscillation, also influence the strength and direction of regional wind currents, especially the westerly winds

in the Arctic and North Atlantic regions (Wang et al., 2019). Changes in wind patterns also affect the levels of POP deposition in the Arctic.

The quantification of transported mass fluxes is not sufficient to assess the long-range transport potential of the various chemical pollutants listed in Table 6.5. Atmospheric scientists need to identify aerial transport routes that may directly serve as pollutant vectors into arctic organisms and the food chain (Zarfl and Matthies, 2010).

Reactivation of in situ persistent organic pollutants

A secondary, but important, source of POPs in the Arctic comes from recycled pollutants that were lodged in glacial ice, sea ice, and snow. These contaminants may have been in the Arctic for decades, trapped in frozen water. Warming climates will very likely enhance the release and reactivation of POPs out of these frozen repositories. Global POP transport models indicate that the Arctic receives a relatively small fraction of POPs but that climate change will probably increase the total

TABLE 6.5
Persistent Organic Pollutants Found in Arctic Waters and Biota, Listed in the Stockholm Convention (UNEP, 2018a), the Minimata Convention (UNEP, 2018b), and Chemicals of Emerging Arctic Concern (AMAP, 2017).

Compound	Abbreviation/Common Name	Primary Uses	Status
Aldrin	HHDN	Insecticide	Banned 1990s
Brominated flame retardants	BFRs, PBDEs	Flame retardants	Banned 2004
Chlordecone hydrate	CLDh	Insecticide	Banned 1975
Chlorinated camphene	Toxaphenes	Insecticides	Banned 2001
Chlorinated flame retardants	CFRs	Flame retardants	Banned 2010
Current-use pesticides	CUPs	Pesticides	Still in use
Cyclodiene chlorinated hydrocarbon	Heptachlor	Insecticide	Banned 1988
Cyclodiene organochlorine	Endosulfan Chlordane	Insecticide Insecticide (kills termites)	Being phased out Banned 1988
Dichloro-diphenyl-trichloroethane	DDT	Insecticide	Banned 2001
Dieldrin	HEOD	Insecticide	Banned 1974
Endosulfan-sulfate	Endosulfan	Insecticide	Banned 2011
Endrin aldehyde	Endrin	Insecticide, rodenticide	Banned 1984
Halogenated natural products	HNPs	Flame retardants	Banned 2017
Hexabromocyclododecane	HBCDD	Flame retardants	Banned 2014
Hexachlorobenzene	HCB	Pesticides	Banned 1966
Hexachlorobutadiene	HBCD	Synthetic rubber, lubricants	Banned 2017
Hexachlorocyclohexane	HCH	Insecticide	Banned 1976
Monochlorobenzene	Dicofol	Acaricide (kills mites)	Still in use
Organophosphate-based flame retardants	OBFR	Flame retardants	Some banned 2005
Organic derivatives of tin	Organotins	Antifouling of ship's hulls	Banned 2008
Pentachloroanisole	PCA	Herbicide, insecticide, fungicide	Banned in Europe (2015
Pentachlorobenzene	PeCB	Pesticide, fungicide, flame retardant	Banned 2009
Pentachlorophenol	PCP	Pesticide	Partial ban in 2015
Per- and polyfluoroalkyl substances	PFOS	Fabric treatments, cleaning products	Partial ban in 2019
Perchloropentacyclodecane	Mirex	Insecticide	Banned 1978
Perfluorohexane sulfonic acid	PFHxS	Food packaging, water-resistant materials	Banned 2019
Perfluorooctane sulfonic acid	PFOS	Metal plating and semiconductor industries	Partial ban in 2019

Continued

TABLE 6.5
Persistent Organic Pollutants Found in Arctic Waters and Biota, Listed in the Stockholm Convention (UNEP, 2018a), the Minimata Convention (UNEP, 2018b), and Chemicals of Emerging Arctic Concern (AMAP, 2017).—cont'd

Compound	Abbreviation/Common Name	Primary Uses	Status
Perfluorooctanoic acid	PFOA	Fabric treatments, nonstick cookware	Partial ban in 2019
Phthalic anhydride esters	Phthalates	Solvents, adhesives, soap, shampoo	Partial ban in 2018
Polybromodiphenyl ethers	PBDEs	Building materials, fabrics, polyurethane	Partial ban in 2007
Polychlorinated biphenyls	PCBs	Flame retardant	Banned 1977
Polychlorinated dibenzo-p-dioxins/furans	PCDD/PCDF	Electrical transformers (source of dioxins)	Banned 1979
Polychlorinated naphthalenes	PCNs	Wood preservatives, oil additives	Banned 2013
Polychlorocamphene	Toxaphene	Insecticide	Banned 1990
Polycyclic aromatic hydrocarbons	PAHs	Dyes, plastics, pesticides	Partial ban in 2014
Short-chain chlorinated paraffins	SCCPs	Lubricants in metal processing, sealants	Banned 2009
Siloxanes	Silicone	Cosmetics, medical implants, sealants	Partial ban in 2019

mass of all toxic compounds in the Arctic (Wang et al., 2019). For instance, when Arctic sea-ice extent reached a record low in 2012, the deposition of high concentrations of β-HCH was observed throughout the year at the Alert station in the Canadian High Arctic. β-HCH is a by-product of the manufacture of the insecticide lindane (γ-HCH). In addition to sea ice and snowbanks, Arctic glaciers may also rerelease pollutants into the environment. One example of this concerns perfluorinalkyl substances (PFASs), found in a wide range of consumer products, including cookware, pizza boxes, and stain repellants. PFASs have been identified in arctic glacial meltwater that feeds into streams draining into the Arctic Ocean.

Effects of Chemical Pollution on Arctic Marine Ecosystems

Most polymers used in common consumer products, such as PP, PE, and PS, have lower densities than water and thus float on the surface of the ocean. This provides a pathway for entrainment of these chemicals in Arctic sea ice. Higher-density polymers such as PVC and PET more easily become deposited in terrestrial soils and sink to the seafloor in marine systems. As discussed earlier, plastics can adsorb and concentrate POPs. If such chemicals are present in the surrounding seawater, even surface-active hydrophobic contaminants such as PCBs and BFRs can be transported by MPs into the Arctic where they are ingested by marine wildlife.

Some toxic substances are particularly prone to bioaccumulate in marine organisms and biomagnify through marine food webs. POPs such as dioxins and PCB and heavy metals such as mercury tend to accumulate in the tissues of many marine organisms, from zooplankton to fish and marine mammals (UNEP, 2018a; Wenning et al., 2011). These pollutants are soluble in lipids rather than water, so they tend to accumulate in the fatty tissues of marine animals. Because of biomagnification, their concentrations become far higher in second- and third-order consumers than those found in organisms lower down the food chain.

The biological effects of POPs on Arctic wildlife are challenging to quantify. Laboratory studies are the usual method of ascertaining cause-and-effect relationships between contaminants and their effects on animal health. However, due to the sheer number of POPs in the Arctic environment, these studies are virtually incapable of duplicating what is happening in nature. Scientists have opted to investigate biomarkers (e.g., hormones, vitamins, immune system activation, liver enzyme activity) in wildlife, as indicators of the biological effects of the actual toxins. Specific effects associated with POPs have been reported for several Arctic species. PCBs are the dominant group of compounds that affect the top predators in the Arctic marine food web. Killer whales (*Orcinus orca*) are the species most at risk from PCB contamination (Levin et al., 2018), while long-finned pilot whales from the Faro Islands and several populations of birds of prey, such as white-tailed eagles (*Haliaeetus albicilla*) and peregrine falcons (*Falco peregrinus*) are also at risk. Genotoxicity has also been shown for polar bears and some birds and fish, mainly driven by PCB exposure (AMAP, 2018). Genotoxicity is the property of a chemical agent that damages the genetic information within a cell. This kind of genetic damage can cause mutations that may lead to cancer.

Steroid hormones play a vital role in numerous physiological processes of vertebrates. These include regulation of reproduction, growth, and development. The regulation of steroid hormones is linked to biological factors such as sexual maturity and body size, but toxic chemicals are known to suppress mammalian endocrine systems (Routti et al., 2019). Thus, contaminants acting on steroid hormone homeostasis may pose a threat to the population dynamics in Arctic animals, including polar bears. Because of slow maturity and small numbers of cubs born only every-other-year, a female polar bear can probably only produce six surviving cubs over her entire life span. Therefore, even small declines in fecundity may cause polar bear populations to irreparably decline.

Physiological Effects of Persistent Organic Pollutants and Mercury on Marine Vertebrates

Dietz et al. (2018) identified quantifiable effects of POPs and mercury on the health of various arctic marine mammals. They found that these pollutants interfere with many aspects of vertebrate physiology. In marine mammals (the best-studied group), the pollutants disrupted or damaged vitamin metabolism, immune function, thyroid and steroid hormone balances, oxidative stress, tissue pathology, and reproduction capacity.

Polychlorinated biphenyls

Until their manufacture was banned in 1977, the estimated global production of PCBs was 1.5 million metric tons. About 10% of these highly toxic compounds remain in the environment today (Reddy et al., 2019). PCBs occur in 209 different forms or congeners. Each congener has two or more chlorine atoms located at specific sites on the PCB molecule. One of the challenges in dealing with PCBs is their great longevity in the environment. Of all the POPs, PCBs may have the longest residence time, particularly in the Arctic where chemical reactions are slowed by cold temperatures (Gamberg, 2020). Due to their excessive chemical stability, PCBs are persistent in both biotic and abiotic environments. The breakdown of PCBs occurs mainly from sunlight, followed by an attack from microorganisms. Sunlight plays a key role in the breakdown of PCBs in water, air, and soil. Generally, the persistence of PCBs in the environment increases with the number of chlorine atoms in the molecule. In temperate environments, molecules of PCB that contain seven chlorine atoms (heptachlorobiphenyl) may persist in water for up to 54 years, and they may persist in soil and sediments upward of 75 years (Reddy et al., 2019). At the other end of the spectrum, PCB molecules with only three chlorine atoms (trichlorobiphenyl) persist just 120 days in water and up to 6 years in soils and sediments. Even though PCBs were banned more than 40 years ago, chemists predict that they will persist in the Arctic for the next 100 years. Under a global warming scenario, atmospheric modeling of PCB composition and behavior predicts some increases in environmental PCB concentrations in a warmer Arctic, but the general decline in PCB levels is still the most prominent feature (Carlsson et al., 2018). The levels of PCBs in Arctic human populations have declined since the 1980s and are predicted to decline further.

PCBs were first detected in the Arctic more than 30 years ago, in 1988. PCBs are probably the best understood compounds in the POP group in terms of physical and chemical properties, emissions, pathways, and observed concentrations in the global environment.

PCBs have a high tendency to transform into persistent metabolites, which are capable of accumulating in specific tissues and body fluids. Furthermore, PCBs can enter tissues through blood and then rapidly transform into water-soluble substances. Table 6.6 shows levels of PCBs measured from animals in various levels of the Arctic food chain, from plankton to polar bears. This table illustrates the degree of biomagnification of PCBs from the bottom to the top of the food chain.

Polychlorinated biphenyls in Rorqual whales

An interesting example of the differences in POP contamination of different species was presented by Tartu et al. (2020). They examined pollutant concentrations in blue whales (*Balaenoptera musculus*) and fin whales (*Balaenoptera physalus*) (Fig. 6.12). Blue whales feed almost exclusively on krill, whereas fin whales have a more varied diet that includes krill, amphipods, copepods, shrimps, small fish, and squid. POP concentrations were determined by blubber biopsies and through the use of stable isotopes of nitrogen (δ^{15}N) and carbon (δ^{13}C), measured using skin biopsies from whales sampled in waters around Svalbard.

The POPs were dominated by DDTs, PCBs, and toxaphenes. The median concentrations in blue whales for these three compounds were 208, 127, and 133. The concentrations in fin whales were 341, 275, and 233 ng/g lipid weight, respectively. Thus, pollutant concentrations were 1.6–3 times higher in fin whales than in blue whales. This is likely due to the fact that fin whales are higher up the food chain than blue whales. This higher trophic position is also indicated by higher ratios of δ^{15}N. In contrast, fin whales had lower δ^{13}C than blue whales, suggesting that fin whales feed farther north than blue whales. Another interesting aspect of this study was the revelation that pollutant levels were approximately twice as high in males as in females of the same species. The authors concluded that the female whales offload pollutants to their offspring during gestation and lactation, like many other mammals. Pollutant concentrations in these two species of whales from Svalbard waters were generally much lower than in the same species of whales that live in the Mediterranean Sea or the Gulf of California.

Persistent organic pollutants and mercury in polar bears

According to the World Wildlife Fund (2019), there are an estimated 22,000–31,000 polar bears (*Ursus maritimus*) in today's Arctic. Four subpopulations of polar bears in the Canadian Arctic are in decline (Fig. 6.13). Five subpopulations from locations around the Arctic are currently stable in numbers. Two subpopulations are growing: one in northwest Greenland and one in the Canadian Arctic. The numbers of polar bears in the nine other subpopulations are not sufficiently well known to yield population estimates.

Polar bears are an integral part of the Arctic ecosystem and the diet of Indigenous peoples, who have hunted polar bears sustainably for millennia. But, beginning in the 1700s, large-scale hunting by Europeans, Russians, and North Americans brought marked declines in polar bear numbers, especially since the end of World War II. Unregulated hunting continued until the 1970s. Then, in 1973, Canada, the United States, Denmark, Norway, and the former USSR signed the International Agreement on the Conservation of Polar Bears and their Habitat, strictly regulating commercial hunting. That same year, the US government classified the Polar Bear under its Endangered Species Act (ESA). In 2005, the IUCN changed their designation of polar bears from "Least Concern" to Vulnerable.

Today, polar bears are among the few large carnivores still found in their original habitat and range. In some regions, their numbers have returned to historic levels. Although most of the world's 19 subpopulations have returned to healthy numbers, there are differences between them. Some are stable, some seem to be increasing, and some are decreasing due to various pressures (Fig. 6.13). Climatologists predict that only a fringe of ice will remain in Northeast Canada and Northern Greenland by 2040 and that all the other large areas of summer ice will be gone. This "Last Ice Area" is likely to become vital for polar bears, arctic seals, and other life that depends on ice. If these predictions prove accurate, then global polar bear numbers are projected to decline by 30% by 2050 (World Wildlife Fund, 2020).

As if sea-ice habitat loss were not enough, polar bears are among the most chemically contaminated of Arctic species because of their position at the top of the food web and their lipid-rich diet. Polar bears range throughout the Arctic and hunt in the continental shelf regions that are covered by sea ice for much of the year. They particularly prey on ringed seals (*Pusa hispida*) and bearded seals (*Erignathus barbatus*). Polar bears are exposed to a wide range of bioaccumulative contaminants. High levels of POPs have been found in their tissues, including PCBs, organochlorine pesticides (DDT, Lindane, Dieldrin, etc.), BFRs, perfluoroalkyl substances (PFASs), and mercury (Hg) (Routti et al., 2019). The chemical toxins listed before are termed "legacy POPs," because they stopped being produced many

TABLE 6.6
PCB Levels Measured From Arctic and Sub-Arctic Marine Wildlife.

Species	Scientific Name	Region	PCB Level[1]
INVERTEBRATES			
Copepods[2,3]	Copepoda	Near Axel Heiberg Island White Sea	0.04 to 0.27 2.3−23.6
Amphipods[2]	*Gammarus wilkitzkii*	Near Axel Heiberg Island	1.1−129.9
Spider crab[3]	*Hyas araneus*	White Sea	8.5−19.2
Whelk[3]	*Buccinum undatum*	White Sea	64.1
FISH			
Glacial eelpout[2]	*Lycodes frigidus*	Arctic Ocean (2000 m)	5.7
Arctic cod[2]	*Boreogadus saida*	Barrow Strait	1.3
Atlantic cod[3]	*Gadus morhua*	White Sea	6.80−25.7
Rose fish[4]	*Sebastes marinus*	East Greenland	1.2
Thorny skate[4]	*Amblyraja radiata*	Arctic and sub-Arctic waters	0.7
European plaice[4]	*Pleuronectes platessa*	Greenland	0.92
Atlantic halibut[4]	*Hippoglossus hippoglossus*	Arctic and sub-Arctic waters	1.45
PINNIPEDS			
Harp seal[3,5]	*Pagophilus groenlandicus*	Davis Strait East Greenland White Sea	1.81−77.7 0.14−8.9 367−2020
Hooded seal[5]	*Cystophora cristata*	Davis Strait East Greenland	61.5−358.2 14.1−100.6
Walrus[6,7]	*Odobenus rosmarus*	Pechora Sea Svalbard coast	40.9−50,400 26.9−31,617
Ringed seal[5,8]	*Pusa hispida*	East Greenland Arctic Bay Western Hudson Bay Prince of Wales Island Pond Inlet, Baffin Island South Hudson Bay NW Svalbard	1−37.2 1.7−58 0.1−254.9 0.2−40 0.43−34.3 0.6−75.9 336.7−624.8
Bearded seal[3,8]	*Erignathus barbatus*	White Sea Svalbard coast Southern Svalbard	2980−5320 159.1−247.6 83−422
CETACEANS			
Humpback whale[9,10]	*Megaptera novaeangliae*	Gulf of St. Lawrence Bering Sea	897.2 160
Blue whale[10]	*Balaenoptera musculus*	Gulf of St. Lawrence	2017.9
Narwhal[5]	*Monodon monoceros*	Qaanaaq, NW Greenland East Greenland	0.2−132.1 0.4−18.1
Stellar sea lion[11]	*Eumetopias jubatus*	Gulf of Alaska	889-26,107
Pilot whale[5]	*Globicephala* sp.	Faroe Islands	17.5−574
Sperm whale[12]	*Physeter macrocephalus*	North Atlantic	19,500
Beluga whale[5,13]	*Delphinapterus leucas*	Beaufort Sea[5] Beaufort Sea[13] South Jameson Land	0.3−143.7 3093 16−64

Continued

TABLE 6.6
PCB Levels Measured From Arctic and Sub-Arctic Marine Wildlife.—cont'd

Species	Scientific Name	Region	PCB Level[1]
Killer whale[4,5,13]	*Orcinus orca*	East Greenland	26.7–199.8
		Kamchatka coast	72–3300
		Alaskan coast	5,000–14,000
		Greenland	40,000–113,000
Minke whale[9]	*Balaenoptera acutorostrata*	North Pacific	620–3100
CARNIVORA: URSIDAE			
Polar bear[5]	*Ursus maritimus*	Baffin Bay	34.1–106.8
		Chukchi Sea	3.93–12.7
		Lancaster Sound	24.7–95
		N. Beaufort Sea	35–414.3
		East Greenland	6.5–186.8

[1] ng/g wet weight.
[2] Hargrave et al. (1992).
[3] Muir et al. (2003).
[4] Atkinson et al., 2019.
[5] AMAP, 2018.
[6] Boltunov et al., 2019.
[7] Scotter et al., 2019.
[8] Bang et al. (2001).
[9] Aono et al. (1998).
[10] Metcalfe et al. (2004).
[11] Keogh et al. (2020).
[12] Holsbeek et al. (1999).
[13] Desforges et al., 2018.
Data from the sources listed above.

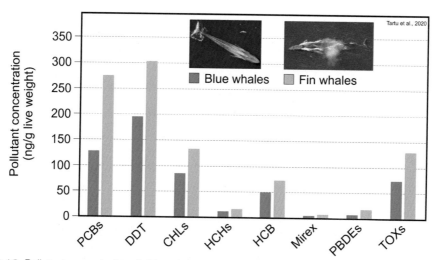

FIG. 6.12 Pollutant concentrations in blue whales and fin whales from the Svalbard Archipelago. (Data from Tartu, S., Fisk, A.T., Götsch, A., Kovacs, K.M., Lydersen, C., Routti, H., 2020. First assessment of pollutant exposure in two balaenopterid whale populations sampled in the Svalbard Archipelago, Norway. The Science of the Total Environment 718. https://doi.org/10.1016/j.scitotenv.2020.137327.)

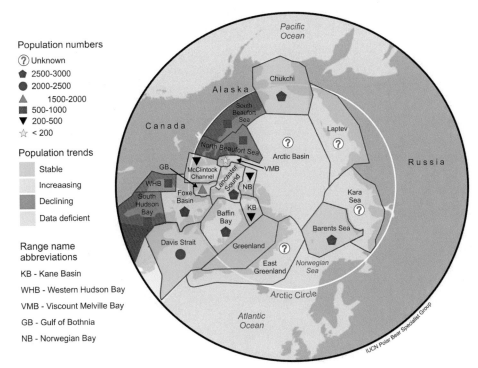

FIG. 6.13 Distribution map of the subpopulations of polar bears in the Arctic. Subpopulation boundaries from Environment and Climate Change Canada (2018). Polar bear subpopulation numbers from World Wildlife Fund (2019).

years ago. The concentrations of several of these legacy POPs have declined in polar bears in recent years. However, it appears that there are certain hot spots where the concentrations of these contaminants are considerably higher than average, and recent studies indicate that some populations of polar bears are being exposed to higher concentrations of POPs. For instance, Routti et al. (2019) note that concentrations of PCBs in the polar bear populations living in the Kara Sea and East Greenland regions were about twice the level of bears in the Hudson Bay and Barents Sea populations. Also, considerably lower POP concentrations were found in the bears of Chukchi Sea and southern Beaufort Sea regions (Fig. 6.14). The higher levels of PCBs in the Kara Sea population may have been caused by the discharge of PCBs into rivers in the Russian Arctic. A different spatial pattern of concentrations was found for oxychlordane. Its concentrations in polar bear tissues were found to increase westward from the Chukchi Sea and Kara Sea toward the Barents Sea and East

Greenland. The highest oxychlordane concentrations were found in bears in the Hudson Bay and Southern Beaufort Sea populations (Fig. 6.14). This geographic trend corresponds with the higher usage of chlordanes in the United States compared with other parts of the world. The impact on the health of polar bears likewise varies in accordance with the concentrations of these dangerous chemicals.

The various POPs discussed before are known to cause a wide range of diseases and disorders, including disruption of the endocrine and metabolic systems, immunotoxicity, neurotoxicity, and direct damage to various tissues. Steroid hormones play a critical role in many physiological processes, including regulation of reproduction, growth, and development. Toxic chemicals are known to suppress mammalian endocrine systems, affecting such life processes as the age of sexual maturity and body size. Thus, these contaminants pose a threat to the population dynamics of polar bears and other mammals.

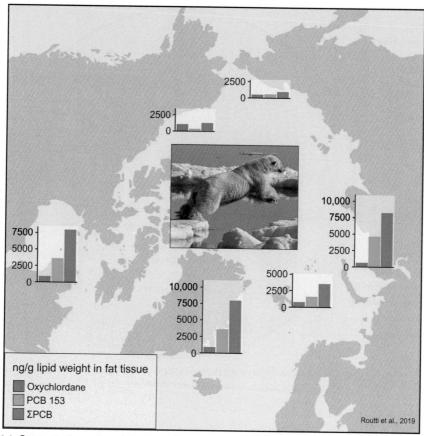

FIG. 6.14 Concentrations of persistent organic pollutants in adipose tissue of polar bears of various regions in the Arctic. (Data from Routti, H., Atwood, T.C., Bechshoft, T., Boltunov, A., Ciesielski, T.M., et al., 2019. State of knowledge on current exposure, fate and potential health effects of contaminants in polar bears from the circumpolar Arctic. Science of The Total Environment 664, 1063–1083. https://doi.org/10.1016/j.scitotenv.2019.02.030.)

Pharmacokinetics is the branch of pharmacology concerned with the movement of drugs within the body. Using physiologically based pharmacokinetic modeling, physiologists suggest that the concentrations of POPs in 11 of the 19 polar bear subpopulations (from Alaska to Svalbard) exceed threshold levels for various aspects of polar bear health. These include impacts on the reproductive and immune systems and greatly increased susceptibility to cancers. The level of these health risks has waxed and waned over the past 50 years. For instance, two studies on East Greenland polar bears suggest that this subpopulation has been at risk for contaminant effects for several decades and that the risk quotients peaked in the early 1980s and again in 2013 (Dietz et al., 2018). In both studies, PCBs were the main contributors to the risks.

The relationship between fertile males and females is crucial for the maintenance of the population size of virtually all mammals. Research on polar bear fertility by Pavlov et al. (2016) has uncovered a pernicious and potentially devastating effect of PCBs polar bear reproduction. The biochemical evidence shows that PCBs are damaging the reproductive capability of male polar bears. These bears accumulate large amounts of this toxin in their bodies. Large amounts of PCB in the bodies of male polar bears cause their testes to shrink and their penis bones to weaken, thus lowering reproductive success. Male polar bears reach the peak of sexual maturity at 10–14 years. When seeking mates, fully grown bears generally outcompete younger, smaller, or older males. But because the largest males also have the highest PCB levels, many of them suffer from reduced fertility. Their attempts at mating are likely to fail, and they keep other, potentially more reproductively fit males from gaining access to receptive females. It appears that the small number of pregnant females seen in the Arctic today can be explained by high levels of PCB in the dominant males.

Mercury (Hg) is a naturally occurring element. Since the Industrial Revolution of the mid-19th century, Hg has been released into the atmosphere through coal combustion and through an extraction process used in gold mining. The total amount of mercury released to the air each year from anthropogenic sources is estimated to be about 2000 metric tons. A further 3000–4000 metric tons are released to the air either from natural sources or as a result of reemission of mercury that has previously been deposited onto surfaces, back into the air (Gamberg, 2020). It has been estimated that Arctic Ocean seawater currently accumulates about 25 metric tons of mercury each year (AMAP, 2011).

Once released into the ocean, mercury bioaccumulates and shows stronger biomagnification in Arctic marine food webs relative to those in warmer regions, resulting in levels of concern in high trophic position marine species, particularly the apex predator: the polar bear. Indeed, concentrations of total Hg in polar bear liver tissues may be at least an order of magnitude greater than the concentrations reported in the liver of their main prey, the ringed seal.

Mercury emissions have recently been subject to regulation at the international level through the 2017 ratification of the Minamata Convention on Mercury. However, geochemical modeling suggests that global emissions of Hg will have to be aggressively reduced, to just keep oceanic Hg at the current levels. This is because of the release of decades-old mercury deposits into the oceans. Within Arctic marine ecosystems, river discharge into the Arctic Ocean is considered the dominant pathway for Hg entrance to the Arctic, as it is with PCBs, although atmospheric, terrestrial, and oceanic inputs also occur (Routti et al., 2019). Mercury is mostly deposited from the air in inorganic forms which are not very bioavailable. However, bacteria can convert these forms of Hg through the process of methylation. This process yields methylmercury, a toxic organic form that is readily bioavailable, bioaccumulates within animals (and people), and biomagnifies up the food chain. Methylation can occur in aquatic environments, sediments, and wetlands. Virtually all the mercury present in fish is in the toxic methylmercury form, allowing biomagnification through the Arctic marine food chain to seals and fish-eating seabirds, and onto polar bears

and people. As a result, mercury levels tend to be higher in marine ecosystems than in terrestrial ecosystems. Marine mammals are more exposed to mercury (Hg) than any other animals in the world.

Mercury tends to accumulate in muscle and liver tissue of vertebrates and can have toxic effects on the nervous, digestive, and immune systems, particularly in the unborn fetus. Mercury is considered by the World Health Organization as one of the top 10 chemicals of major public health concern (WHO, 2017). As with many trace elements, Hg can impair brain function, leading to population declines. Specifically, mercury interferes with several neurotransmitters in polar bear brains and brain stems, although the exact interactions between the contaminants and brain chemistry remain unknown (López-Berenguer et al., 2020) (Fig. 6.15).

As mentioned earlier, bacteria and phytoplankton are the main entry points for the uptake of Hg into marine food webs. Methyl mercury is then biomagnified to higher trophic levels, to marine top predators including marine mammals and seabirds. As discussed by Kershaw and Hall (2019), the biomagnification rate for MeHg in the Arctic marine food chain is estimated at about sixfold increase with each trophic level. Once MeHg is absorbed, it enters the bloodstream and is distributed quickly to various tissues and organs. It binds to cysteine in fluids, mimicking methionine, which makes it easily transported across cell membranes by amino acid transporters. First, it is distributed to the liver, kidney, and spleen. Later it enters muscle and brain tissues.

There is increasing concern about the impacts of mercury on the Arctic marine ecosystem because of ongoing changes in water temperatures, ocean currents, and prey availability. All of these global changes are predicted to affect the exposure of arctic marine life to mercury. The accumulation of mercury in various tissues has been linked to renal and hepatic damage as well as reported neurotoxic, genotoxic, and immunotoxic effects (Kershaw and Hall, 2019).

Persistent organic pollutants and mercury in whales, dolphins, and seals

Cetaceans have a limited ability to metabolize and eliminate or excrete Hg, so it becomes sequestered in their tissues. As a result, Hg concentrations in cetaceans are between 10 and 100 times higher than those measured in other predators at the same trophic level that have a similar average life span and dietary intake, such as tuna (Kershaw and Hall, 2019). A study by Simond et al. (2017) reported that the concentration

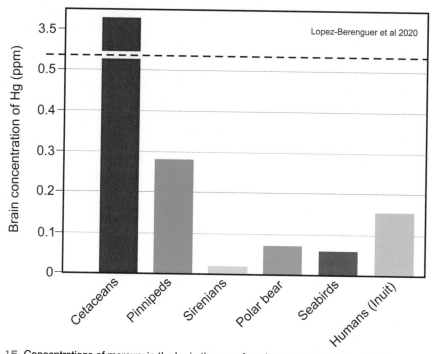

FIG. 6.15 Concentrations of mercury in the brain tissues of marine mammal species, polar bears, seabirds, and humans living in the Arctic. (Modified from López-Berenguer, G., Peñalver, J., Martínez-Lóez, E., 2020. A critical review about neurotoxic effects in marine mammals of mercury and other trace elements. Chemosphere 246, 125688. https://doi.org/10.1016/j.chemosphere.2019.125688.)

of PBDEs, one group of POPs, in the liver of Canadian Arctic belugas was 53 ng/g live weight (lw). The blubber of minke whales (*Balaenoptera acutorostrata*) from Norway, Greenland, and Iceland contained PBDEs at a level of 71−126 ng/g lw. Minke whales in the Arctic contained 1.5 to 7 times higher levels of POPs than minke whales living in lower latitudes. Blubber concentrations of PBDEs were reported in this study in belugas from the seas off the Svalbard coast, ranging from 161 to 314 ng/g lw. These levels were generally higher than those found in some beluga populations from the Canadian Arctic (range: 1.29−160 ng/g lw) and Alaska (range: 4.33−45.6 ng/g lw). Other marine mammals from the Canadian Arctic were also found to be contaminated with PBDEs. These include ringed seals (*Pusa hispida*) (range: 6−28.8 ng/g lw) and narwhals (*Monodon monoceros*) (range: 18 ng/g lw). The greater blubber concentrations of PBDEs that accumulated in Canadian Arctic belugas from the present study relative to previous Arctic beluga studies suggest continuous inputs of PBDEs to the Arctic, despite their partial production ban in 2007. Taken together, these

results highlight the marked geographical differences in marine mammal POP contamination in the circumpolar regions.

Dolphins are particularly susceptible to the accumulation of PCBs as it amasses through the fish that they eat (which also have high levels of PCB) and it gets stored in their fatty tissues. Female dolphins draw on their fat reserves—where PCB compounds are stored—to produce milk. In dolphins with high PCB toxicity, their first offspring usually dies as the mother passes the PCBs to them in its milk (Featherstone, 2012). Subsequent calves have a higher rate of survival but are prone to birth defects. High levels of PCBs in the tissues of dolphins and other marine mammals weaken their immune systems. PCBs and other contaminants suppress the immune system in animals, making it more difficult for them to fight off infections from viruses and bacteria.

POPs and other toxins are particularly problematic for cetaceans in years of limited prey availability. POPs that have accumulated in their blubber are mobilized in years of food shortage. These fat stores are metabolized for energy. However, once these fat tissues are broken down, they release the POPs stored in fats and lipids, causing higher concentrations of toxins within the body. As discussed earlier, female marine mammals often mobilize their fat stores during lactation, causing high contaminant loads to be passed from female to calf, especially the first-born calf (Featherstone, 2012).

CONCLUSIONS

Marine wildlife, from plankton to birds and marine mammals, are being exposed to a veritable witch's brew of toxic chemicals, MPs, and heavy metals, especially mercury. As we have seen in this chapter, Arctic marine food chains are experiencing some of the worst levels of contamination from these toxins. When I was a child, a major chemical manufacturer advertised the slogan, "Better living through chemistry." Little did they or anybody know in the 1960s was that the explosion of newly created halogenated chemical compounds and plastics would leave a semipermanent legacy of plastic pollution and chemical poisons that would permeate the entire surface of the planet and its oceans in the coming decades.

The 37 POPs listed in Table 6.5 represent some of the most damaging toxins that affect marine life, but these are just the tip of the iceberg. Beyond the well-known POPs identified as dangerous by the Stockholm Convention, Scheringer et al. (2012) discussed an additional 510 chemicals that exceed the screening criteria defined in Annex D of the Stockholm Convention. These 510 new contenders probably merit inclusion in the list of POPs. Ninety-eight percent of them are halogenated; specifically, they represent mostly halogenated aromatic compounds. The nonhalogenated substances in contention for the Stockholm Convention POP listing are highly branched alkanes and nitroaromatic compounds.

Sadly, one of the most pervasive legacies of the modern era is the trillions of pieces of plastic that have found their way into the ocean and subsequently spread throughout the world from pole to pole. The Plastics in the Ocean website (2020) reports that an estimated 25 trillion macroplastics and 51 trillion MPs litter our oceans. A large proportion of this plastic litter floats and 88% of the sea's surface is polluted by plastic waste. These pervasive pollutants have negative effects on virtually all forms of marine life. Over 1 million seabirds and 100,000 marine mammals are killed by ocean plastic every year. Over 90% of all seabirds have plastic in their stomachs. Sea turtles have up to 74% (dry weight) of their diets composed of ocean plastics. More than two-thirds of the world's fish stocks are suffering from plastic ingestion, and 100,000 marine animals are killed by plastic bags every year. MP pieces also attract and concentrate POPs from seawater. These chemically laden plastics drift to the most remote regions of the Arctic and become part of the food chain as they are consumed by everything from zooplankton to fish and birds. Once ingested, the plastic bits release their toxic chemical contents into the animals' digestive systems, sometimes delivering a fatal dose of toxic chemicals.

All of these life-threatening hazards are going to persist in the environment for many decades or centuries to come. Long-chain plastic polymers and halogenated organic compounds last for a very long time before eventually breaking down. This is a human-caused legacy that threatens the very existence of marine life in the Arctic and elsewhere.

REFERENCES

AMAP, 2018. A map assessment 2018: arctic ocean acidification. Arctic Monitoring and Assessment Programme, Tromso, Norway, p. 187.

AMAP Assessment 2011: Mercury in the Arctic, 2011. Arctic Monitoring and Assessment Program. AMAP), p. 193.

Aono, S., Tanabe, S., Fujise, Y., Kato, H., Tatsukawa, R., 1998. Persistent organochlorines in minke whale (*Balaenoptera acutorostrata*) and their prey species from the Antarctic and the North Pacific. Environmental Pollution 98, 81–89.

Arctic Monitoring and Assessment Program (AMAP), 2017. AMAP Assessment 2016: Chemicals of Emerging Arctic Concern. Arctic Monitoring and Assessment Program.

Atkinson, S., Branson, M., Burdin, A., Boyd, D., Ylitalo, G.M., 2019. Persistent organic pollutants in killer whales (*Orcinus orca*) of the Russian Far East. Marine Pollution Bulletin 149, 110593. https://doi.org/10.1016/j.marpolbul.2019.110593.

Bang, K., Jenssen, B.M., Lydersen, C., Skaare, J.U., 2001. Organochlorine burdens in blood of ringed and bearded seals from north-western Svalbard. Chemosphere 44 (2), 193–203. https://doi.org/10.1016/S0045-6535(00)00197-1.

Bergmann, M., Mützel, S., Primpke, S., Tekman, M.B., Trachsel, J., Gerdts, G., 2019. White and wonderful? Microplastics prevail in snow from the Alps to the arctic. Science Advances 5 (8), eaax1157. https://doi.org/10.1126/sciadv.aax1157.

Boltunov, A., Semenova, V., Samsonov, D., Boltunov, N., Nikiforov, V., 2019. Persistent organic pollutants in the Pechora Sea walruses. Polar Biology 42 (9), 1775–1785. https://doi.org/10.1007/s00300-019-02457-9.

Burkow, I.C., Kallenborn, R., 2000. Sources and transport of persistent pollutants to the Arctic. Toxicology Letters 112–113, 87–92. https://doi.org/10.1016/S0378-4274(99)00254-4.

Cózar, A., Echevarría, F., González-Gordillo, J.I., Irigoien, X., Úbeda, B., 2014. Plastic debris in the open ocean.

Proceedings of the National Academy of Sciences 111, 10239–10244. https://doi.org/10.1073/pnas.1314705111.

Cózar, A., Martí, E., Duarte, C.M., García-de-Lomas, J., Van Sebille, E., Ballatore, T.J., Eguíluz, V.M., González-Gordillo, J.I, Pedrotti, M.L., Echevarría, F., Troublè, R., Irigoien, X., 2017. The Arctic Ocean as a dead end for floating plastics in the North Atlantic branch of the Thermohaline Circulation. Science Advances 3, e1600582. https://doi.org/10.1126/sciadv.1600582.

Das, A., 2020. Alaskan "Sweet Spots" for Nurdles. https://nurdlepatrol.org/Forms/News/article.php?article_id=51.

Desforges, J.-P., Hall, A., McConnell, B., Rosing-Asvid, A., Barber, A.L., et al., 2018. Predicting global killer whale population collapse from PCB pollution. Science 361, 1373–1376. https://doi.org/10.1126/science.aat1953.

Dietz, R., Desforges, J.P., Gustavson, K., Rigét, F.F., Born, E.W., Letcher, R.J., Sonne, C., 2018. Immunologic, reproductive, and carcinogenic risk assessment from POP exposure in East Greenland polar bears (Ursus maritimus) during 1983–2013. Environment International 118, 169–178. https://doi.org/10.1016/j.envint.2018.05.020.

Ellen MacArthur Foundation, 2016. The New Plastics Economy: Rethinking the Future of Plastics. Ellen MacArthur Foundation. http://www.ellenmacarthurfoundation.org/publications.

Featherstone, A., 2012. How PCBs Continue to Affect the Marine Environment. Conservation Jobs. https://www.conservationjobs.co.uk/articles/.

Gamberg, M., 2020. Threats to Arctic Ecosystems. Encyclopedia of the World's Biomes. Elsevier BV, pp. 532–538.

Gao, X., Huang, P., Huang, Q., Rao, K., Lu, Z., Xu, Y., Gabrielsen, G.W., Hallanger, I., Ma, M., Wang, Z., 2019. Organophosphorus flame retardants and persistent, bioaccumulative, and toxic contaminants in Arctic seawaters: on-board passive sampling coupled with target and non-target analysis. Environmental Pollution 253, 1–10. https://doi.org/10.1016/j.envpol.2019.06.094.

Geyer, R., Jambeck, J.R., Law, K.L., 2017. Production, use, and fate of all plastics ever made. Science Advances 3 (7), e1700782. https://doi.org/10.1126/sciadv.1700782.

Great Nurdle Hunt, 2020. Plastic - Toxic Combination. https://www.nurdlehunt.org.uk/whats-the-problem/toxic-combination.html.

Halsband, C., Herzke, D., 2019. Plastic litter in the European Arctic: what do we know? Emerging Contaminants 5, 308–318. https://doi.org/10.1016/j.emcon.2019.11.001.

Hammer, J., Kraak, M.H.S., Parsons, J.R., 2012. Plastics in the marine environment: the dark side of a modern gift. Reviews of Environmental Contamination and Toxicology 220, 1–44. https://doi.org/10.1007/978-1-4614-3414-6_1.

Hargrave, B.T., Harding, G.C., Vass, W.P., Erickson, P.E., Fowler, B.R., Scott, V., 1992. Organochlorine pesticides and polychlorinated biphenyls in the Arctic Ocean food web. Archives of Environmental Contamination and Toxicology 22 (1), 41–54. https://doi.org/10.1007/BF00213301.

Holsbeek, L., Joiris, C.R., Debacker, V., Ali, I.B., Roose, P., Nellissen, J.P., Gobert, S., Bouquegneau, J.M., Bossicart, M., 1999. Heavy metals, organochlorines and polycyclic aromatic hydrocarbons in sperm whales stranded in the southern North Sea during the 1994/1995 winter. Marine Pollution Bulletin 38 (4), 304–313. https://doi.org/10.1016/S0025-326X(98)00150-7.

Jambeck, J.R., Geyer, R., Law, K.L., 2015. Plastic waste inputs from land into the ocean. Science 347, 768–771. https://doi.org/10.1126/science.1260352.

Jepsen, E.M., de Bruyn, P.J.N., 2019. Pinniped entanglement in oceanic plastic pollution: a global review. Marine Pollution Bulletin 145, 295–305. https://doi.org/10.1016/j.marpolbul.2019.05.042.

Keogh, M.J., Taras, B., Beckmen, K.B., Burek-Huntington, K.A., Ylitalo, G.M., Fadely, B.S., Rea, L.D., Pitcher, K.W., 2020. Organochlorine contaminant concentrations in blubber of young Steller sea lion (Eumetopias jubatus) are influenced by region, age, sex, and lipid stores. Science of the Total Environment 698, 134183. https://doi.org/10.1016/j.scitotenv.2019.134183.

Kershaw, J.L., Hall, A.J., 2019. Mercury in cetaceans: exposure, bioaccumulation and toxicity. Science of the Total Environment 694, 133683. https://doi.org/10.1016/j.scitotenv.2019.133683.

Kühn, S., van Franeker, J.A., 2020. Quantitative overview of marine debris ingested by marine megafauna. Marine Pollution Bulletin 151, 110858. https://doi.org/10.1016/j.marpolbul.2019.110858.

Levin, M., Ross, P.S., Samarra, F., Víkingson, G., Sonne, C., Dietz, R., De Guise, S., Eulaers, I., Jepson, P.D., Letcher, R.J., 2018. Predicting global killer whale population collapse from PCB pollution. Science 361 (6409), 1373–1376. https://doi.org/10.1126/science.aat1953.

López-Berenguer, G., Peñalver, J., Martínez-López, E., 2020. A critical review about neurotoxic effects in marine mammals of mercury and other trace elements. Chemosphere 246, 125688. https://doi.org/10.1016/j.chemosphere.2019.125688.

Lusher, A.L., Tirelli, V., O'Connor, I., Officer, R., 2015. Microplastics in Arctic polar waters: the first reported values of particles in surface and sub-surface samples. Scientific Reports 5, 14947. https://doi.org/10.1038/srep14947.

Metcalfe, C., Koenig, B., Metcalfe, T., Paterson, G., Sears, R., 2004. Intra- and inter-species differences in persistent organic contaminants in the blubber of blue whales and humpback whales from the Gulf of St. Lawrence, Canada. Marine Environmental Research 57 (4), 245–260. https://doi.org/10.1016/j.marenvres.2003.08.003.

Moore, C.J., 2008. Synthetic polymers in the marine environment: a rapidly increasing, long-term threat. Environmental Research 108 (2), 131–139. https://doi.org/10.1016/j.envres.2008.07.025.

Muir, D., Savinova, T., Savinov, V., Alexeeva, L., Potelov, V., Svetochev, V., 2003. Bioaccumulation of PCBs and chlorinated pesticides in seals, fishes and invertebrates from the

White Sea, Russia. Science of the Total Environment 306 (1–3), 111–131. https://doi.org/10.1016/S0048-9697(02)00488-6.

Munthe, J., MacLeod, M., Odland, J.Ø., Pawlak, J., Rautio, A., Reiersen, L.O., Schlabach, M., Stemmler, I., Wilson, S., Wöhrnschimmel, H., 2018. Polychlorinated biphenyls (PCBs) as sentinels for the elucidation of Arctic environmental change processes: a comprehensive review combined with ArcRisk project results. Environmental Science and Pollution Research 25 (23), 22499–22528. https://doi.org/10.1007/s11356-018-2625-7.

National Park Service Oceans, Coasts & Seashores, 2018. Ocean Plastics. National Park Service. https://www.nps.gov/subjects/oceans/ocean-plastics.htm.

NOAA, 2014. The Great Pacific Garbage Patch. National Ocean Service. https://oceanservice.noaa.gov/podcast/june14/mw1.

NOAA, 2020. The Nurdle Patrol. https://oceanservice.noaa.gov/podcast/jan20/nurdle-patrol.html.

Obbard, R.W., Sadri, S., Wong, Y.Q., Khitun, A.A., Baker, I, Thompson, R.C., 2014. Global warming releases microplastic legacy frozen in Arctic Sea ice. Earth's Future 2, 315–320. https://doi.org/10.1002/2014EF000240.

Pavlova, V., Nabe-Nielsen, J., Dietz, R., Sonne, C., Grimm, V., 2016. Allee effect in polar bears: a potential consequence of polychlorinated biphenyl contamination. Proceedings of the Royal Society B: Biological Sciences 283 (1843). https://doi.org/10.1098/rspb.2016.1883.

Peng, L., Fu, D., Qi, H., Lan, C.Q., Yu, H., Ge, C., 2020. Micro- and nano-plastics in marine environment: source, distribution and threats — a review. Science of the Total Environment 698, 134254. https://doi.org/10.1016/j.scitotenv.2019.134254.

Plastics in the Ocean, 2020. Plastic in the Ocean: Statistics. https://www.condorferries.co.uk/plastic-in-the-ocean-statistics.

Reddy, A.V.B., Moniruzzaman, M., Aminabhavi, T.M., 2019. Polychlorinated biphenyls (PCBs) in the environment: recent updates on sampling, pretreatment, cleanup technologies and their analysis. Chemical Engineering Journal 358, 1186–1207. https://doi.org/10.1016/j.cej.2018.09.205.

Routti, H., Atwood, T.C., Bechshoft, T., Boltunov, A., Ciesielski, T.M., et al., 2019. State of knowledge on current exposure, fate and potential health effects of contaminants in polar bears from the circumpolar Arctic. Science of The Total Environment 664, 1063–1083. https://doi.org/10.1016/j.scitotenv.2019.02.030.

Scheringer, M., Strempel, S., Hukari, S., Ng, C.A., Blepp, M., Hungerbuhler, K., 2012. How many persistent organic pollutants should we expect? Atmospheric Pollution Research 3 (4), 383–391. https://doi.org/10.5094/APR.2012.044.

Scotter, S.E., Tryland, M., Nymo, I.H., Hanssen, L., Harju, M., Lydersen, C., Kovacs, K.M., Klein, J., Fisk, A.T., Routti, H., 2019. Contaminants in Atlantic walruses in Svalbard part 1: relationships between exposure, diet and pathogen prevalence. Environmental Pollution 244, 9–18. https://doi.org/10.1016/j.envpol.2018.10.001.

Simond, A.E., Houde, M., Lesage, V., Verreault, J., 2017. Temporal trends of PBDEs and emerging flame retardants in belugas from the St. Lawrence Estuary (Canada) and comparisons with minke whales and Canadian Arctic belugas. Environmental Research 156, 494–504. https://doi.org/10.1016/j.envres.2017.03.058.

Tartu, S., Fisk, A.T., Götsch, A., Kovacs, K.M., Lydersen, C., Routti, H., 2020. First assessment of pollutant exposure in two balaenopterid whale populations sampled in the Svalbard Archipelago, Norway. Science of the Total Environment 718, 137327. https://doi.org/10.1016/j.scitotenv.2020.137327.

Tekman, M.B., Krumpen, T., Bergmann, M., 2017. Marine litter on deep Arctic seafloor continues to increase and spreads to the North at the HAUSGARTEN observatory. Deep-Sea Research Part I Oceanographic Research Papers 120, 88–99. https://doi.org/10.1016/j.dsr.2016.12.011.

Thompson, R.C., Moore, C.J., vom Saal, F.S., Swan, S.H., 2009. Plastics, the environment and human health: current consensus and future trends. Philosophical Transactions of the Royal Society of London Series B 364, 2153–2166. https://doi.org/10.1098/rstb.2009.0053.

UNEP, 2018a. Minamata Convention on Mercury.

UNEP, 2018b. Stockholm Convention. UNEP. http://chm.pops.int/.

UNEP, 2018c. The Stockholm Convention. http://www.pops.int/TheConvention/Overview/tabid/3351/Default.aspx.

Wang, X., Wang, C., Zhu, T., Gong, P., Fu, J., Cong, Z., 2019. Persistent organic pollutants in the polar regions and the Tibetan Plateau: a review of current knowledge and future prospects. Environmental Pollution 248, 191–208. https://doi.org/10.1016/j.envpol.2019.01.093.

Wenning, R.J., Martello, L., Prusak-Daniel, A., 2011. Dioxins, PCBs, and PBDEs in aquatic organisms. In: Environmental Contaminants in Biota: Interpreting Tissue Concentrations, vol. 13, 978–1.

World Health Organization (WHO), 2017. World Health Organization Fact Sheets: Mercury and Health. http://www.who.int/news-room/fact-sheets/detail/mercury-and-health.

World Wildlife Fund, 2019. Polar Bear Populations. https://arcticwwf.org/species/polar-bear/population/.

World Wildlife Fund, 2020. Timeline of Polar Bear Conservation. WWF. https://arcticwwf.org/species/polar-bear/population/.

Wu, P., Huang, J., Zheng, Y., Yang, Y., Zhang, Y., 2019. Environmental occurrences, fate, and impacts of microplastics. Ecotoxicology and Environmental Safety 184, 109612. https://doi.org/10.1016/j.ecoenv.2019.109612.

Zarfl, C., Matthies, M., 2010. Are marine plastic particles transport vectors for organic pollutants to the Arctic? Marine Pollution Bulletin 60 (10), 1810–1814. https://doi.org/10.1016/j.marpolbul.2010.05.026.

Zhang, Y., Kang, S., Allen, S., Allen, D., Gao, T., Sillanpää, M., 2020. Atmospheric microplastics: a review on current status and perspectives. Earth-Science Reviews 203, 103118. https://doi.org/10.1016/j.earscirev.2020.103118.

CHAPTER 7

Impacts of Overfishing in Arctic and Sub-Arctic Waters

INTRODUCTION

Humans have a seemingly insatiable appetite for fish. Since World War II, most of the world's oceans have been stripped of commercially desirable fish to help feed the burgeoning human population, which has more than tripled during this interval. According to the FAO (2020), 90% of fish stocks were within biologically sustainable levels in 1990. By 2017, the sustainability level had dropped to 66%. The fisheries statistics include some staggering figures. For instance, in 2018, total global capture fisheries production reached the highest level ever recorded at 96.4 million metric tons (mt). This equals 96.4 billion kilograms of fish caught. With the human population at 7.9 billion in 2021, this record-setting global marine catch, if evenly distributed, would provide each person on the planet with just over 14 kg (31 lbs.) of fish per year. The increase in 2018 was mostly driven by marine capture fisheries, with production from marine areas increasing to 84.4 million mt. People are particularly fond of predatory fish species, which puts them particularly at risk of overfishing, to satisfy world markets. These predatory species include cod, herring, sardines, tuna, haddock, and salmon (FAO, 2020).

The countries that brought in the largest number of fish in 2020, ranked in order, are as follows: China, India, Bangladesh, and Myanmar. The commercial fishing industry employed 30,768,000 people in Asia, 5,021,000 people in Africa, and 2,455,000 people in the Americas. Altogether, the global fishing industry employed 38,976,000 people in 2018. These workers went to sea in an estimated 4.6 million fishing vessels. These boats range in size from small canoes to the world's largest trawler that is 144 m long and can catch, process, and freeze 400 mt of fish per day.

Two of the most important types of modern fishing vessels are pelagic trawlers and purse seiners (Fig. 7.1). Pelagic trawlers often work in pairs to maximize their catch. In this fishing method, a trawling net is towed in midwater between two vessels to target pelagic fish (NOAA, 2019b; Seafish, 2019). The height of the net

in the water column can be changed by altering vessel speed and length of wire fed out into the sea. The nets can be very large: as big as 240 m wide and 160 m deep.

Sea turtles are at risk of being captured in midwater trawling nets as they come up from the bottom, where they rest and forage, to the surface, where they breathe. Because turtles breathe air, the likelihood of their drowning increases the longer a turtle is held underwater. When hauled aboard a fishing boat, turtles can also be crushed by the weight of the catch on top of them, resulting in broken appendages or shell. Injury may also occur when the net is emptied onto the fishing vessel and turtles are dropped onto the hard deck along with the fish catch. The stress of being captured can leave turtles exhausted and barely alive when they are tossed back into the water.

Many species of marine mammals forage and swim at midwater depths, putting them at a high risk of being captured or entangled in trawling nets. They can become trapped in netting and tow lines or become disoriented by vessel noise. Pilot whales and dolphins are particularly susceptible to being caught in midwater trawls in nearshore areas. Killer whales and Steller sea lions are occasionally captured in Alaskan waters.

Sea turtle mortality in trawling gear was once very high. Turtle excluder devices (TEDs) have greatly reduced mortality in some trawl fisheries. If small and juvenile cetaceans reach the narrow end of a tapered trawl net, they may also be able to escape through the TED if not entangled. Mitigation measures are like those recommended for bottom trawling, such as reducing the number of turns per tow and tow duration. Ongoing research is investigating how animals might behave differently when in the vicinity of fishing vessels and assessing potential gear modifications that might lead to animals avoiding trawls.

The purse seine is used mainly for catching dense, mobile schools of pelagic fish (NOAA, 2019a; Seafish, 2011). The schools of fish are surrounded and impounded by means of large pursed surround nets

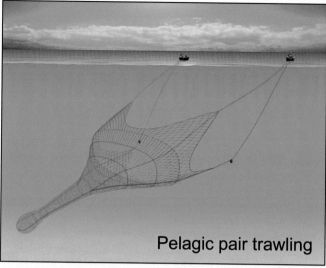

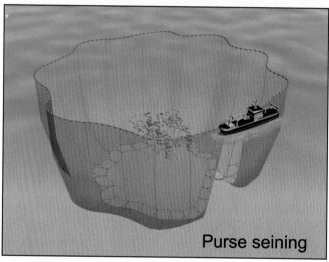

FIG. 7.1 Above: Pelagic pair trawling. This fishing method is where one trawl is towed in midwater between two vessels to target pelagic fish. Illustration by Orchid Information Systems Limited, courtesy of Seafish (www. seafish.org), used with permission. Below: Purse seining. The purse seine is used mainly for catching dense, mobile schools of pelagic fish. Illustration from the FAO, in the public domain. Pelagic trawler illustration by Orchid Information Systems Limited, courtesy of Seafish (www.seafish.org), used with permission.

called either ring nets or purse seines according to design. A purse seine is a wall of netting with a mesh size to suit the target species and a headrope carrying numerous floats to keep the net on the surface. The net is equipped with rings (purse rings) along its lower edge through which a special cable (purse wire) is passed to enable the fisherman to close off the space surrounded by the purse seine from below, preventing the fish from escaping downwards and forming a bowl-like shape of the net in the water containing the fish. Purse seines can reach more than 2000 m (6500 ft) in length and 200 m (650 ft) in depth, varying in size according to the vessel, mesh size, and target species.

Purse seining captures everything that it surrounds, including sea turtles and marine mammals. Sea turtles become entangled in the net mesh as the net is hauled in, often sustaining injuries. In large catches, turtles may be crushed to death under the weight of the net contents. Captured turtles can be released alive if they are quickly retrieved and removed from the net.

Purse seines easily encircle marine mammals. Historically, dolphin pods were used by fishermen to locate abundant schooling fish (called "setting on dolphins"). Once the netting has been set, encircled marine mammals become entangled, injured, or stressed. Even with quick retrieval, marine mammals' sensitive bodies and internal organs cannot usually withstand the weight of the catch or the impact of being dropped onto the deck of the vessel. In US fisheries, species most captured include bottlenose dolphins and humpback whales.

Bottom trawling is a fishing practice that herds and captures the target species, like flatfish or crabs, by towing a net along the ocean floor (NOAA, 2019c). Floats are attached to the headrope (top of trawl opening), while weights and special gear are attached to the footrope (bottom of trawl opening), to keep the net open as it moves through the water across the ocean floor. The mesh is designed to confine fish inside the net, trapping them in the narrow end as the trawl is hauled to the surface. A sweep attached to the net's footrope collects marine animals as they lay on the bottom or gather before the trawl opening. The trawling gear may be constructed and rigged for various target species over different types of bottom surfaces. Sweep types include chain sweep for smooth surfaces, cookies (small steel or rubber rollers) for soft and irregular ocean floors, and rockhoppers (large molded rubber rollers) for rocky bottoms. Raised foot ropes reduce groundfish bycatch.

Overfishing is defined as the removal of a fish species from a body of water at a rate that the species cannot replenish, resulting in the species becoming underpopulated or extirpated in that area. Extirpation is the dying out of the populations of a species in a particular region. This contrasts with extinction, which is a global dying out of a species, leaving no populations alive. Extirpation may be remedied by the reintroduction of a species into regions where it formerly lived. Extinction, on the other hand, is irreversible.

The middle- and high-latitude seas are most at risk for overfishing because these regions are the most biologically productive. It all starts with the physics of water. Cold water holds more dissolved gases than warm water. Thus, the world's cold ocean regions hold both more oxygen and carbon dioxide than do tropical waters. These dissolved gases, especially in regions with rich nutrient supply, support the abundant growth of phytoplankton. This, in turn, supports a rich invertebrate fauna of zooplankton, which forms the bulk of the primary consumers in ocean ecosystems. Rich plankton populations support large numbers of fish and marine mammals, from small phytoplankton consumers to the largest predators.

NATURE OF THE ARCTIC MARINE ECOSYSTEM

The central basins of the Arctic Ocean (AO) are surrounded by 16 ocean regions, of which 12 are true Arctic seas and four are gateways between the Arctic and the Atlantic or the Pacific Ocean (Aune et al., 2018). The distribution of sea ice largely determines the distribution of species in Arctic waters. Some Arctic shelf seas (e.g., the Barents Sea) are not entirely covered by sea ice, even in winter. Other areas (e.g., the Bering Sea) are covered with seasonal sea ice that extends well south of the Arctic Circle. The bathymetry of the AO and adjacent seas (NOAA, 2004) is shown in Fig. 7.2. As discussed in Chapter 1, the summer sea-ice extent has declined steadily since satellite records started in 1979, with a record minimum recorded in 2012. This is observed particularly in the Marginal Ice Zone (MIZ), defined as that part of the ice cover which is close enough to the open ocean boundary to be affected by its presence. This zone often coincides with the area between the summer minima and winter maxima of ice extent. The MIZ covers most of the Arctic shelf and the shelf break. The increase in open water not only occurs around the MIZ. Open water is now exposed in increasing numbers of leads in the pack ice throughout the AO.

Changes in Oceanic Circulation

As discussed by Waga et al. (2020), climate changes are affecting polar marine ecosystems more rapidly and with greater intensity than those in other regions. The rapidity of these changes is making it difficult for marine scientists to assess the responses of marine biota in the AO. Based on current observations, oceanographers predict that in the future, the AO will be warmed by the increased northward transport of Pacific Water through the Bering Strait. This warming, in turn, may facilitate the northward invasion of warm-adapted biota from the North Pacific into the AO. This may be happening now, but because of the logistic difficulties of research in Arctic waters, well-documented examples of this remain scarce, particularly for benthic organisms. One might assume that because the benthic macrofauna (i.e., mollusks and echinoderms) are normally

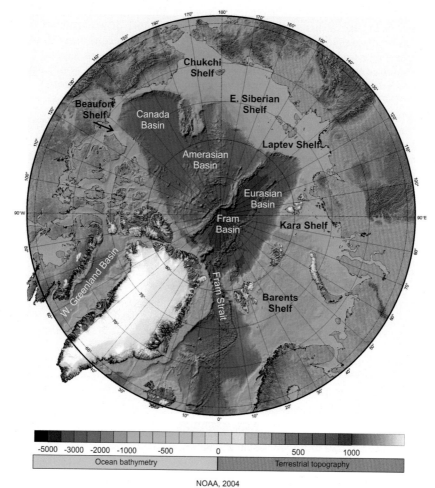

NOAA, 2004

FIG. 7.2 Bathymetry of the Arctic Ocean and adjoining seas, after NOAA, 2004. Continental shelf regions labeled in black, and basins labeled in white.

sessile, such organisms would be much less likely to invade the AO as the water warms. However, this fails to take into account that many of these less mobile, benthic macrofauna have immature life stages that are pelagic and thus could easily float northward, allowing their species ranges to expand (Renaud et al., 2015). Pelagic fish live in ocean waters that are neither close to the bottom nor near the shore (the pelagic zone). In addition, the dispersal distance of marine larvae in cold water is much greater than those in warmer water, because temperature is inversely related to larval duration.

Biodiversity and Ecology of the Arctic Ocean

Bluhm et al. (2015) provide an excellent review of the physical and biological attributes of the deep AO. Benthic organisms that live in the AO basins show a level of biodiversity that was not thought possible until quite recently. They also show surprising dispersal abilities, given the isolation of the individual basins and low level of carbon flux reaching the seafloor. Larval dispersion is aided by the large-scale current flows originating in the North Atlantic and the North Pacific. In contrast to this, zooplankton diversity is low in the AO basins (Haug et al., 2017), but surprisingly, zooplankton faunal assemblages are equally distributed between the Eurasian Basin and the Amerasian Basin. Sea ice is now retreating past the shelf break every summer, exposing the open basin waters to increased sunlight and wind forcing. Winds that promote upwelling generally flow from east to west. These winds are now driving upwelling along the shelf breaks, drawing nutrients from subsurface basin waters onto the shelf.

The Arctic Marine Food Web

Of course, all ecosystems are founded on the energy derived from plants through photosynthesis. In the AO, the primary producers are almost all phytoplankton (algae and diatoms) (Rey, 2004). The bulk of Arctic phytoplankton appears only during the relatively short interval of spring/early summer. Nevertheless, this short-lived food resource represents the most important high-quality food for zooplankton grazers and higher trophic level marine animals during the year. Hence, the timing of this production pulse, relative to the timing of other ecological key processes, such as reproduction, is critical for the fate of the produced biomass and the efficiency of trophic pathways (Aune et al., 2018).

Much of the algal growth comes on or near the bottom of the ice pack. The bloom of these bottom sea-ice algae (Fig. 7.3) is usually the first algal bloom of the year in Arctic waters, marking the transition from winter to springtime. Light availability early in the season controls the timing of its initiation and development. For example, underice algal blooms have been recorded in both Svalbard and the Amundsen Gulf from the end of March. Arctic phytoplankton blooms are normally limited by nutrient availability, and these blooms end when the available inorganic nutrients are depleted. Nitrate is the nutrient that is typically the first to be depleted (Randelhoff et al., 2020). After that, less vigorous primary production continues throughout the summer and may persist as late as September (Aune et al., 2018).

The contribution of these ice algae to the total primary production ranges from 4% to 25% on the Arctic shelves, as much as 50% or more in the basins, and up to 90% in the Canada Basin. Due to nutrient limitations, both sea-ice algal and pelagic primary production in the two basins are far less than on the adjacent continental shelves. In addition to the oligotrophic (nutrient-poor) status of the AO basins, biological productivity here is also limited by the stratification of these waters, combined with light limitation due to snow and ice cover and extreme sun angle. Across the AO, there is a great diversity of primary producers, including 1874 known species of phytoplankton and 1027 known species of ice algae. Most of these species

FIG. 7.3 A block of overturned sea ice reveals the growth of algae underneath. (Photo by S. Falk-Petersen, in Aune, M., 2018. Seasonal ecology in ice-covered Arctic seas - considerations for spill response decision making. Marine Environmental Research 141, 275–288. https://doi.org/10.1016/j.marenvres.2018.09.004.)

form large cells (>20 μm). In addition to algae, the phytoplankton flora is dominated by diatoms and dinoflagellates.

Sea-ice algae represent a high nutritional quality food source early in the season in sea ice−covered areas. In and below the sea ice, they are grazed upon by meiofauna (small benthic invertebrates), and herbivorous zooplankton, such as the specialized pelagic grazer *Calanus glacialis*. This copepod is the key grazer in Arctic shelf sea areas and can account for up to 70% of the total mesozooplankton biomass (Aune et al., 2018). Mesoplankton live in the middle depths of the ocean, below the penetration of photosynthetically effective light. *C. glacialis* females remain close to the underside of the pack ice, where they actively graze upon the sea ice algae at the ice-water interface. Ice amphipods constitute another important link between the ice algae and upper trophic levels. For instance, *Apherusa glacialis* is a typical herbivore, whereas *Gammarus wilkitzkii*, *Onisimus glacialis,* and *Onisimus nanseni* are typical omnivores and carnivores.

When ice algae are released from melting sea ice, they are partly fed upon by pelagic grazers. Since ice algae often form large colonies that sink rapidly, a substantial amount of this production reaches the seafloor and represents an important food supply for benthic organisms.

As a result of low in situ primary productivity in the AO basins and planktonic consumption of freshly produced carbon while particles are sinking, vertical carbon flux to the deep-sea floor is comparatively low. Particles advected from productive shelves such as the Barents, Chukchi, and Kara Sea shelves add substantial amounts of carbon to the vertical supply. While the areas around the basin's perimeter receive this allochthonous input (of external origin), biota in the central Arctic Basins away from the shelves do not. As a result, biomass tends to decrease with water depth and/or distance from the shelf, both in the water column and on the seafloor. In fact, biomass drops to a third or half of the slope values toward the deep basins. Elevated biomass at the basin periphery roughly coincides with the marginal ice zone and seasonally ice-free waters.

Based on modeling and fieldwork in the Barents Sea, Reigstad et al. (2011) estimated that annual gross primary productivity for ice-free, Atlantic water masses in the southwest Barents Sea is about 160 g per cm^2 per year. They reckoned that annual gross primary production in seasonally ice-covered Arctic waters further north was 60 g per cm^2 per year, less than 40% of the productivity of the ice-free waters. As much as 53% of the primary production in Arctic waters sinks toward

the bottom, a rate considerably higher than that found in the Atlantic Ocean at lower latitudes.

This implies that Arctic waters are more strongly governed by benthic than pelagic processes and that the degree of ice coverage has a direct influence on primary production rates (Reigstad et al., 2011). As such, interannual and long-term variations in ice and water mass conditions have consequences for species distributions and ecosystem functioning. As noted by Aune et al. (2018), many species of the AO fish fauna migrate seasonally, and their larvae have parts of the year in which they are particularly vulnerable to predation by larger fish (Fig. 7.4).

As shown in Table 7.1, annual primary productivity varies greatly across the Arctic seas, ranging from less than 30 g per cm^2 per year in the central basins of the AO to 127 g per cm^2 per year in the Eurasian Arctic. Of the eight marine regions for which trends in productivity had been recorded, five regions have experienced downturns in productivity since the early 21st century, and three regions have experienced increases.

The zooplankton community of the Arctic consists of about 300 species that spend their entire life cycle drifting in the plankton. Numerous benthic invertebrate and fish species have pelagic larval stages which join the zooplankton community for parts of the year. Brine channels in the sea ice sustain species-rich food webs throughout the year, but these communities are typically most abundant and diverse in the spring and summer seasons. While many of these species are unique to the sea ice environment, other species originate from the benthic or pelagic realms and visit the sea ice to feed or hide from predators. These species include bacteria and protists, as well as species from higher trophic levels such as cnidarians, copepods, amphipods, euphausiids, and arthropods. Copepods dominate the Arctic invertebrate fauna, in terms of species number (>50% of all Arctic holoplankton [plankton that spend their entire life cycle as free-swimming organisms]), abundance, and biomass.

The few historical assessments of the zooplankton biomass in the Arctic basins reported very low biomass with less than 0.2−3.0 g dry weight (DW) per m^2 in the 0−1500 m water layer. More recent studies suggest that the zooplankton stock of the AO basins may previously have been significantly underestimated. The newer studies report greater zooplankton biomass and production. Recent comprehensive zooplankton assessments have been made at more than 80 locations scattered over the Eurasian and Amerasian basins, using consistent methods. Based on these more recent studies,

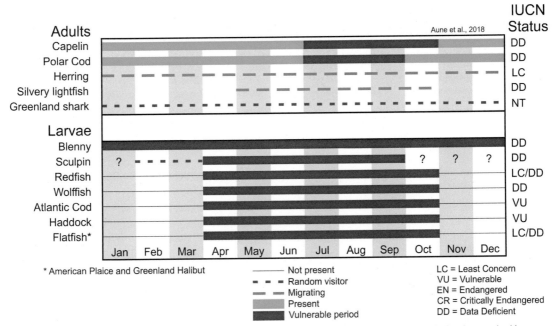

FIG. 7.4 Seasonal calendar showing the presence of fish in the upper water layers of the Arctic marginal ice zone. (After Aune, M., 2018. Seasonal ecology in ice-covered Arctic seas - considerations for spill response decision making. Marine Environmental Research 141, 275–288. https://doi.org/10.1016/j.marenvres.2018. 09.004.)

there is now an estimated zooplankton biomass integrated over the entire water column at ca. 2–24 g DW m² dry mass (Bluhm et al., 2015).

Almost everywhere in the AO basins, vertical profiles of zooplankton abundance and biomass in summer are characterized by maximum concentrations in the 0–50 m water layer with a noticeable decrease, by several orders of magnitude, at greater depths. As shown in Table 7.2, 75% of mesozooplankton biomass occurs in the upper 500 m of the water column, and 82% of macrobenthic biomass occurs in the upper 2000 m (Bluhm et al., 2015).

Although Arctic marine ecosystems are structurally more complex than previously thought, they remain characterized by a relatively simple ecosystem structure and a high degree of specialization among species. This lack of functional redundancy renders them more vulnerable than less specialized systems with higher biodiversity. In an ecological context, functional redundancy occurs when multiple species representing a variety of taxonomic groups share similar, if not identical, roles in the ecosystem (e.g., algae consumers, scavengers, and top predators). The structure of Arctic marine ecosystems is unusual in comparison with sub-Arctic or boreal marine ecosystems. The AO

ecosystem appears to be more strongly dominated by benthic than pelagic processes (Bluhm et al., 2015).

Arctic marine fish diversity

Aune et al. (2018) describe the fish fauna of the Arctic, as follows. There are 242 marine fish species known from Arctic waters, a number verified by an independent assessment from Mecklenburg et al. (2010), who classified these species within 45 different families of fish. Most of these fish species live on the Arctic shelves (Fig. 7.5). In the deep, central Arctic basin (average depth 2418 m), only 13 fish species have been recorded. The distribution, abundance, ecology, and life history of 90% of these species are poorly understood. The three most species-rich families are the snailfish (Liparidae), eelpout (Zoarcidae), and sculpins (Cottidae). Ongoing phylogenetic studies (tracing the evolutionary origins of species) suggest that eelpout, sculpins, and several other groups of Arctic fish are more closely related than previously thought. The geographic distributions of Arctic marine fish species are incompletely known. This is due to insufficient sampling, especially in the western Arctic, as well as unresolved taxonomic issues, and difficulties in species identification (Mecklenburg et al., 2010).

TABLE 7.1
Annual Primary Productivity for Various Ocean Regions in the Arctic.

Region	Prevailing Conditions	Annual Primary Productivity (g Carbon per m² per year)	2018 Departures from 2003 to 2017 Averages	References
Central Arctic Ocean	Perennial sea ice dominant	<30	N/A	Haug et al., 2017
Outer Arctic Ocean	Seasonally ice-covered	30–100	N/A	Reigstad et al. (2011)
Northern Barents Sea	Seasonally ice-covered	40	↓ 1 g C per m² per year	Dalpadado et al. (2012)
Southern Barents Sea	Typically ice-free	113	↓ 1 g C per m² per year	Ray and McCormick-Ray (2004)
Eurasian Arctic	Seasonally ice-covered	127	↑ 4 g C per m² per year	Frey et al. (2018)
Sea of Okhotsk	Seasonally ice-covered	20.8	↓ 5.8 g C per m² per year	Frey et al. (2018)
Bering Sea	Seasonally ice-covered	24.5	↑ 1.1 g C per m² per year	Frey et al. (2018)
Hudson Bay	Seasonally ice-covered	54.6	↓ 2.8 g C per m² per year	Frey et al. (2018)
Greenland Sea	Seasonally ice-covered	60	↓ 2.1 g C per m² per year	Frey et al. (2018)
Baffin Bay	Seasonally ice-covered	56.7	↑ 2.1 g C per m² per year	Frey et al. (2018)

TABLE 7.2
Average Biomass Distribution, by Water Depth, of Mesozooplankton and Macrobenthos Across Arctic Ocean Basins.

Water Depth (m)	Mesozooplankton Biomass (mg C per m²)	Water Depth (m)	Macrobenthic Biomass (mg C per m²)
0–100	950	500–1000	450
100–500	700	1000–2000	150
500–1000	350	2000–3000	100
>1000	200	3000–4000	20
		>4000	10

Data from Bluhm, B.A., Kosobokova, K.N., Carmack, E.C., 2015. A tale of two basins: an integrated physical and biological perspective of the deep Arctic Ocean. Progress in Oceanography 139, 89–121. https://doi.org/10.1016/j.pocean.2015.07.011.

Polar cod: a keystone species

Two cryopelagic (i.e., living and spawning in association with sea ice) fish species live in the AO: polar cod (*Boreogadus saida*) and ice cod (*Arctogadus glacialis*). Both species live throughout the Arctic and are endemic there (Aune et al., 2018). Polar cod is a highly abundant species in the Arctic ecosystem, while ice cod is seldom seen and less closely tied to the sea-ice environment. Young polar cod are commonly observed both underneath Arctic sea ice and in the pelagic zone. Immature

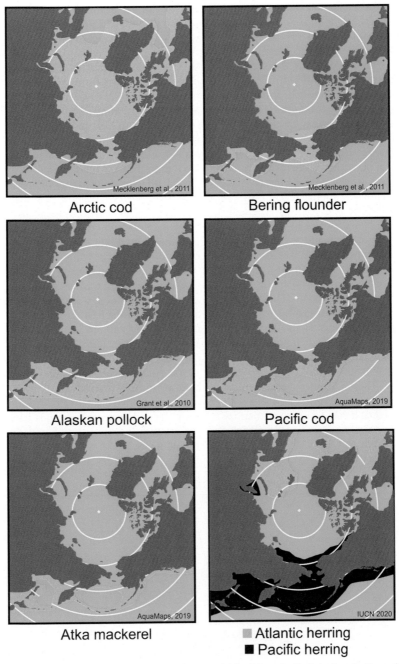

Atlantic herring
Pacific herring

FIG. 7.5 Distribution maps for species of fish with potential for future Arctic fisheries. Arctic cod and Bering flounder maps. Walleye Pollock map. Pacific cod map. (Modified from Mecklenberg, C.W., 2011. Biodiversity of arctic marine fishes: taxonomy and zoogeography. Marine Biodiversity 41, 109–140. https://doi.org/10.1093/icesjms/fsq079; Grant, W.S., Spies, I., Canino, M.F., 2010. Shifting-balance stock structure in North Pacific walleye pollock (*Gadus chalcogrammus*). ICES Journal of Marine Science 67 (8), 1687–1696. https://doi.org/10.1093/icesjms/fsq079; AquaMaps, 2019. Computer Generated Native Distribution Map for *Gadus macrocephalus* (Pacific Cod), with Modelled Year 2050 Native Range Map Based on IPCC RCP8.5 Emissions Scenario. https://www.aquamaps.org/receive.php?type_of_map=regular#.)

stages remain close to the ice and are separated vertically from the larger congeners who reside in pelagic waters. In the Barents Sea, polar cod spawn near the edge of the pack ice from November to March, either in the southeastern Barents Sea or in the Svalbard area. From these areas, the larvae drift with the ocean currents in the surface layers. Research shows that healthy polar cod larvae stay in the upper 15 cm of the water column, whereas larvae that do not remain close to the surface do not reach maturity. Polar cod is considered a keystone species in the ice-associated Arctic marine food web, because it is one of just a few species that link the lower trophic levels (i.e., zooplankton) and the higher trophic levels, such as other fish, marine mammals, and seabirds. The early life history of polar cod is tied closely to the presence of ice. The females spawn eggs under the ice. The eggs can develop in subfreezing temperatures. The cod larvae feed on zooplankton specific to seasonal ice-melt algal blooms. Finally, the larvae survive well in the near-freezing temperatures of Arctic waters (Huserbråten et al., 2019). Bouchard and Fortier (2011) examined hatching patterns for polar cod across the Arctic. The study tested the theory that early hatching occurs in underice river plumes, where the input of freshwater keeps temperatures just below 0°C, compared with −1.8°C under the ice. These "warmer" temperatures accelerate the development of embryos and facilitate feeding when the larvae first hatch. To test this "thermal refuge" hypothesis, the researchers studied six oceanographic regions of the Arctic that are heavily influenced

by rivers, as well as some polynyas (areas surrounded by sea ice that remain open water throughout the year) with little freshwater input. Results showed that the duration of the hatching period became shorter and shifted from winter to summer in regions with decreasing levels of freshwater input. For example, hatching started as early as December and January in areas where freshwater input is high, including the Laptev Sea, Hudson Bay, and the Beaufort Sea. In contrast, hatching was delayed until spring (April or May) in Baffin Bay and the Northeast Water polynya system, situated over the continental shelf region of northeast Greenland, where freshwater input is minimal. The results of this study suggest that the effects of global warming on AO environments, including earlier ice breakup, more numerous polynyas, and increased temperatures caused by increased river discharge, promote conditions that favor winter hatching.

The Barents Sea polar cod stock has been monitored annually by a joint Norwegian-Russian survey since 1986, and the total stock biomass (TSB) has varied enormously in the past three decades, between a minimum of 127,000 mt in 1990, to a maximum of 1,941,000 mt in 2006. Yet there is no clear trend in the population size of this stock (Fig. 7.6). In the 1990s, polar cod were spawning in regions east of Svalbard at latitudes from about 75 to 78°N. As the outer margin of sea ice has retreated toward the North Pole, the polar cod spawning grounds have likewise shifted progressively northward and now lie between about 78 and 81°N.

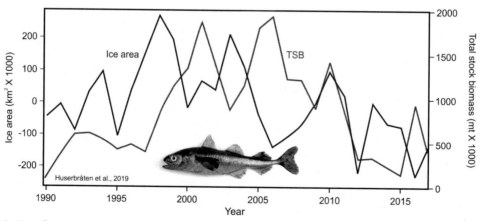

FIG. 7.6 Comparison of extent of sea ice cover with Arctic cod biomass sampled from the seas east of Svalbard and in the Pechora Sea. TSB is measured in thousands of metric tons and represented by the red line. The blue line represents the annual maximum marginal ice cover. Diagram after Huserbråten et al. (2019). Arctic cod photo courtesy of Norwegian Institute of Marine Research, Creative Commons Attribution 4.0 International license.

In general, many Arctic demersal fish species have pelagic juveniles (i.e., past the larvae and postlarvae stages) before they embark on a demersal lifestyle. The larvae are pelagic, and make use of the elevated biological production in the summer season. However, in the Barents Sea, some species have prolonged pelagic larval stages that may last for several annual cycles, including wintertime. For many fish larvae, the pelagic zone is probably a less exposed and therefore safer habitat, with fewer predators and higher food availability.

The incursion of subarctic species

Regional variability in Arctic marine fish diversity is tied closely to the proximity of the Atlantic and Pacific gateway areas. On the Atlantic side, populations of the Atlantic copepod *Calanus finmarchicus* are being carried by currents into the Arctic basin, causing a buildup of biomass near the boundary of the Atlantic and AO. Similarly, locations in the Amerasian Basin close to the Pacific inflow also show higher biomass at all depths. A study by Waga et al. (2020) demonstrated that an increased northward influx of relatively warm Pacific Water is likely to contribute to the expansion of subarctic taxa into the AO and adjacent seas. One of the logical predictions for a future Arctic characterized by warmer waters and reduced sea ice is that new taxa will expand or invade Arctic seafloor habitats (Renaud et al., 2015; Waga et al., 2020). Specific predictions regarding where this will occur and which taxa are most likely to become established or excluded will remain moot until some basic questions are answered (Bluhm et al., 2015):

Under a global warming scenario, will primary production in the AO basins increase or decrease in response to reduced ice cover?

Will the Arctic food web of the future provide more or less energy to the higher trophic levels, with potential implications for the development of commercial fisheries?

Historically, a warmer Arctic is more easily invaded by taxa associated with the Boreal or sub-Arctic zones of the Atlantic and Pacific. Climate models predict that Arctic marginal seas may warm by 2−3°C or more by the year 2100 (Waga et al., 2020). AO warming may well reduce the physiological barriers that have previously kept subarctic taxa out of the AO. This will have significant implications for the northward expansion of subarctic taxa, ranging from benthic organisms with pelagic life stages to demersal (bottom-feeding) and pelagic fishes.

Several physical and biological factors will determine the success or failure of the invasion of more warm-adapted biota into the AO in the coming years. Among these factors are the oceanographic characteristics of a future ice-free AO, the life history traits of the invading taxa, and the availability of suitable habitat. It is difficult to generalize about the specific groups of invading organisms or locations into which they are likely to invade and expand. However, the autecology of species and perhaps individual populations will ultimately determine the success or failure of invasion. The autecology of a species encompasses its interactions with the biological and physical factors of its environment. It speaks to the ecological requirements of a species.

Dietary differences between Arctic and Atlantic cod

Atlantic cod and haddock are extending further into Arctic waters, potentially competing with native Arctic cod for food resources. However, a study by Renaud et al. (2012) indicates little competition for food between these two groups at present. However, increased levels of climate change would certainly increase the geographic overlap between the invading North Atlantic species and Arctic cod, potentially ramping up competition between the species. As mentioned earlier, polar cod is considered a keystone species in the AO. For instance, they are an important food source for seabirds, marine mammals, and other fish. Population declines in polar cod driven by increased competition with invasive species would cause damage and instability throughout the AO ecosystem. Other predatory fish are also expanding north into the Arctic. Haug et al. (2017) note the recent northern expansion of mackerel (*Scomber scombrus*) and capelin (*Mallotus villosus*). These stocks are currently found as far north as the shelf break north of Svalbard. Greenland halibut (*Reinhardtius hippoglossoides*), redfish (*Sebastes* spp.), and shrimp (*Pandalus borealis*) are also present on the slope between the Barents Sea and the AO.

Renaud et al. (2012) studied the stomach contents of young cod and haddock collected in fjords near Svalbard, Norway, over three years. They found that Arctic cod (*Boreogadus saida*) ate mainly krill and crustaceans, such as species of the copepods *Pseudocalanus*, *Calanus*, and *Thermisto* (Fig. 7.7). Atlantic cod and haddock fed on some of the same species, copepods, and krill. However, in regions where Atlantic cod, haddock, and Arctic cod inhabit the same waters, there is less than a 40% overlap in their diets. Isotopic studies of the stomach

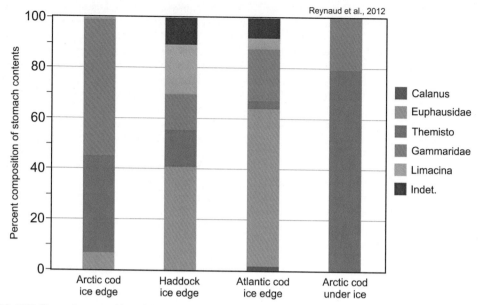

FIG. 7.7 Percent composition of the diets of Arctic cod, haddock, and Atlantic cod found together in waters off the coast of Svalbard, based on their stomach contents. (Bar chart after Reynaud, P.E., 2012. Is the poleward expansion by Atlantic cod and haddock threatening native polar cod, *Boreogadus saida*? Polar Biology 45, 401–412. https://doi.org/10.1007/s00300-011-1085-z.)

contents of these fish likewise revealed distinct diets for all three species. Nevertheless, the authors suggest that there will be increased competition between Arctic cod, Atlantic cod, and haddock if the two North Atlantic species continue their invasion of Arctic waters. One of the dangers for Arctic cod is that the adults of the predatory, invasive species will prey heavily on immature Arctic cod. Taken together, the potential population changes could reduce Arctic cod numbers to the point where a trophic cascade takes place, effecting many different members of the Arctic marine food web. A trophic cascade begins with a significant disruption to the populations of a predator in an ecosystem. This disturbance of the system propagates downward through food webs across multiple trophic levels. The disturbance can be anything that negatively affects the predator, such as increased competition with other predators, or the introduction of a potentially fatal disease.

Disruption of Arctic marine communities

The Arctic marine biotic response to global warming will likely vary from one region to another. On the continental shelf regions adjacent to the AO basins, removal of ice cover beyond the shelf break will enhance wind and ice-forced shelf-break upwelling.

This will result in increases in both nutrient fluxes and the amount of insolation reaching open waters, both of which will increase biological productivity (Bluhm et al., 2015). The current distribution of faunal biomass, from the benthos to zooplankton and fish, indicates that energy transfer from producers (photosynthesizers) to consumers (animals of all kinds) is concentrated along the perimeters of the AO basins. In the basins themselves, biological productivity quickly decreases with depth. Peaks of productivity occur in areas receiving an influx of Atlantic and Pacific waters. These are the sources of nutrients that facilitate productivity and help the spread of both allochthonous and autochthonous (of local origin) fauna. Along the slopes where the continental shelves give way to the deep basins, for instance in regions where the Atlantic layer is in contact with the seafloor, the increased productivity may improve habitat conditions for some species during all or some portion of their life cycle. Mechanisms of connection between the pelagic and benthic zones are critical to this process. For example, concentrations of Arctic cod appear to peak in areas where the upper portion of Atlantic Water meets the seafloor. If this distribution pattern holds for the basin perimeter in the coming decades, these fish stocks may provide an easy and concentrated target for future commercial fisheries at the edge of the Arctic basins. To avoid overfishing of these regions, new management policies will need to be established in the nations most involved in fishing northern high-latitude waters.

Several examples of biotic expansion into the Arctic have been observed, but these have been based on the results of just a few Arctic biological investigations. The available data suggest that inflow shelves on the Barents and Chukchi Seas, as well as West Greenland and the western Kara Sea, are the most likely locations for expansion (Renaud et al., 2015). Temperature thresholds have been identified for some characteristic sub-Arctic and Boreal benthic fauna. Under global warming scenarios, these temperature thresholds will tend to foster the invasion of Boreal or sub-Arctic marine fauna into the AO, at the expense of the more cold-adapted native fauna.

Invasions of warmer adapted species into the AO will certainly cause some ecological disruption to AO marine communities. There will likely be serious consequences to such invasions, many of which we cannot now predict. Shifting species have the potential to seriously affect an ecosystem by outcompeting native species, resulting in disruptions of the existing biological interactions and food web structure (Waga et al., 2020). Since the end of the last glaciation, sub-Arctic and Boreal species have withstood a greater range of environmental conditions than their counterparts in the AO. It is therefore likely the future AO ecosystem will come to resemble the sub-Arctic systems that we see today. However, Alabia et al. (2018) rightly point out that individual marine taxa each respond to environmental changes in their own way, exhibiting different paces and extents in their distributional responses to ocean warming. Their research underpins the importance of incorporating species-specific climatic sensitivity and exposure to climatic changes when predicting range shift responses and evaluating species vulnerability. These insights are critical for conservation and future management of Arctic fisheries resources. A series of maps showing predicted range shifts of several Arctic species by 2050 are presented in Fig. 7.8.

HISTORY OF THE ARCTIC FISHERY

For most of human history, fishing in Arctic waters was too dangerous and the locations too remote for all but the indigenous people of the North. Fishing boats from Europe did make the dangerous crossing of the North Atlantic to the Grand Banks off the coasts of southeastern Canada and the northeastern United States, to

fish for cod. In fact, some historians have suggested that Columbus learned of the possibility of safe trans-Atlantic voyaging from Spanish, Portuguese, or English fishermen who had already discovered the Grand Banks before 1492. Be that as it may, in 1497, John Cabot led the first well-documented transatlantic crossing that reached the Grand Banks. Some 15th-century European texts refer to a land called *Bacalao*, "The land of the codfish," which likely refers to Newfoundland. Shortly after Cabot's voyage, the existence of fishing grounds on the Grand Banks became well known throughout Europe, but these fishing expeditions kept well south of Arctic waters. The wooden-hulled boats of renaissance Europe were not built to withstand collisions with sea ice. So the only people fishing in Arctic waters until much later were native inhabitants of the Arctic region, gathering fish for their subsistence. Their fishing had little or no impact on regional fish stocks.

Many things have changed since those days. As mentioned earlier, the world's appetite for marine fish has increased many times, especially since the end of World War II. Most importantly, our ability to gather huge quantities of fish from the seas changed radically in the middle of the 20th century with the invention of large trawlers, able to stay at sea for months at a time, sweeping up tons of fish in their giant nets and then keeping the catch frozen on board ship until their eventual return to port. Even these ships with their enormous steel hulls have stayed away from the ice-choked waters of the Arctic. But as discussed in previous chapters, the edge of the Arctic sea ice is retreating, and the ice pack is getting thinner each year in response to global warming. As we will explore in the following, this thawing of the AO and adjacent seas is offering an opportunity for fishing boats to venture farther north than ever before. In addition to that, there is now mounting evidence that many of the fish that currently inhabit waters south of the Arctic are moving north. The Atlantic and Pacific waters in the boreal latitudes have been fished so heavily in the past 60 years that the stocks of the most commercially attractive species, such as Atlantic Cod, have been depleted to the point of near-extinction (Table 7.3). Having decimated the fish stocks of the North Atlantic and Pacific, the commercial fishing fleets are very likely to head north into Arctic waters if they are allowed to do so, especially now that the sea ice is retreating. However, the Arctic marine ecosystem is distinctly different from marine ecosystems further south. In many ways, it is more fragile than adjacent (warmer) seas. Fewer species of marine organisms live here, so the food web is much smaller than marine food webs in warmer seas. The lack of

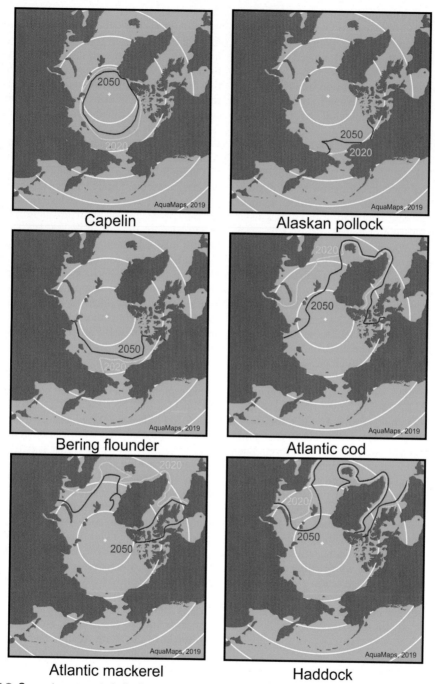

FIG. 7.8 Computer generated distribution maps for several northern fish species of commercial value, projected for the year 2050, based on IPCC RCP8.5 emissions scenario. (After maps from Aquamaps.org.)

TABLE 7.3
Fish Stocks in the North Atlantic Identified by the FAO (2020) as Falling Into Its "Worst" Category, "Depleted."

Region	Species	Common Name	Status
Northwest Atlantic FAO Area 21	*Gadus morghua*	Atlantic Cod	Depleted
	Melanogrammus aeglefinus	Haddock	Depleted
	Clupea harengus	Atlantic Herring	Underexploited to recovering
Northeast Atlantic FAO Area 27	*Salmo salar*	Atlantic salmon	Fully exploited to depleted
	Pleuronectes platessa	European plaice	Overexploited
	G. morghua	Atlantic cod	Overexploited to depleted
	Micromesistius poutassou	Blue whiting	Overexploited
	M. aeglefinus	Haddock	Overexploited to depleted
	Pollachius virens	Pollock	Fully exploited to overexploited
	Merlangius merlangus	Whiting	Fully exploited to depleted
Northeast Pacific (FAO area 67)	*Oncorhynchus tshawytscha*	Chinook Salmon	Fully exploited to overexploited
	Oncorhynchus kisutch	Coho Salmon	Fully exploited to overexploited
	Merluccius productus	North Pacific Hake	Underexploited to depleted
	Clupea pallasii	Pacific Herring	Moderately exploited to overexploited

diversity in Arctic marine biota makes the ecosystem easier to disrupt and more likely to collapse under external pressures such as excessive commercial fishing.

COLLAPSE OF NORTH ATLANTIC COMMERCIAL FISHING STOCKS

The fish stocks of the AO have yet to be exploited. To predict what might happen if the fishing trawlers head north into Arctic waters, we need to look no further than the history of the North Atlantic cod fishery. The most dramatic collapse of a fishery in modern times has been that of North Atlantic cod. In 1497, John Cabot's crew reported that "the sea there is full of fish that can be taken not only with nets but with fishing-baskets." This nutrient-rich shallow water region of the North Atlantic supported enormous stocks of cod. This fishery was incredibly productive for centuries, with enormous fish caught regularly (Fig. 7.9, upper left). The fishing was largely done from sailboats by people using relatively small nets (Fig. 7.9, lower panel). This kind of fishing was sustainable until the second half of the 20th century when factory fishing began. Factory ships, as the name suggests, represent a combination of industrial-scale fishing and a factory where the fish are gutted and flash-frozen. The nets reeled out by these huge ships are extremely large (Fig. 7.1). The largest trawler nets are large enough to hold 13 Boeing 747 jumbo jet airplanes (Clover, 2004). By severely overfishing the Grand Banks, the

supertrawlers brought an end to a fishery that had thrived for four centuries. One of the richest of the world's fisheries was brought to the point of total collapse in less than 40 years.

The North Atlantic cod catch peaked at 810,000 tons in 1968 (Fig. 7.10). This catch represents three times the maximum yearly catch before factory fishing. It is estimated that 8 million tons of cod were caught from 1647 to 1750, representing between 25 and 40 cod generations. The factory trawlers removed the same amount of cod from the North Atlantic in just 15 years. In 1993, Canada banned all cod fishing on the Grand Banks because 90%—95% of the regional cod populations had collapsed. Cod have just barely returned to the Grand Banks since then. However, the reestablishment of cod in this region is made more difficult by the fact that the marine environments are not the same as they were before the cod fishery collapse. The ecosystem has changed because of the loss of one of its top predators. Capelin, a smaller fish previously eaten by cod, now prey on the few immature codfish in the region, keeping cod from becoming reestablished. The waters of the Grand Banks are now dominated by crab and shrimp. However, the Canadian ban on cod fishing on the Grand Banks has started to pay dividends. An assessment by the Canadian government in 2010 concluded that cod numbers are back to 10% of their historical levels. A cod fishing moratorium was extended to include New England coastal waters in 2013. The number of cod hatchlings on the Grand Banks has been rising for the

FIG. 7.9 Top right: Cod depicted on a 1932 Newfoundland postage stamp; Top left: Enormous cod were caught on the Labrador coast until the second half of the 20th century. This young boy was the son of the fisherman who caught these fish in 1902. (Photo from National Archives, Canada; Bottom: Charles Napier Hemy - The Fisherman, 1888. https://commons.wikimedia.org/wiki/File:Charles_Napier_Hemy_-_The_Fisherman_1888.jpg. Photographic reproduction in the public domain.)

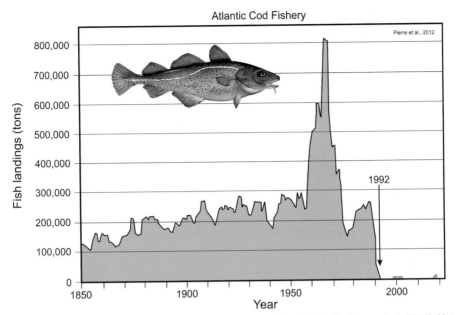

FIG. 7.10 History of the annual catch of Atlantic Cod from 1850 to 2010, after Pierre et al. (2012). Note the collapse of the fishery 1992, which led to an international moratorium on Atlantic cod fishing that is still partially in force today.

past several years, although the Canadian government has noted that the species is not yet out of danger (Miel, 2019).

As of 2019, the number of Atlantic cod in Canadian waters remained at only half the level needed for the species to move out of the critical zone and into the cautious zone. The northwestern Atlantic cod stocks remain far from being healthy. There are now positive signs of a fragile, tentative recovery, but a fish population that has been depleted for as long and as hard as the Atlantic cod will take a great deal of time and effort to recover. The Atlantic fishing fleets of Canada in the eastern United States started calling for increased cod fishing quotas in 2019, even though these, of all people, should have known that Atlantic cod populations had barely started to recover from a near-regional extinction. The Department of Fisheries and Oceans Canada (DFO) has acknowledged that when a fish stock is in the critical zone, the commercial fishing for this population should either be banned or kept to the lowest possible level. Despite this, commercial fisheries in Canada harvested over 18,000 mt of Atlantic cod in 2018 (DFO, 2018). History has clearly shown that increasing fishing pressure on very small, fragile fish populations can lead to their regional extirpation,

thwarting all hopes of recovery. The southern Newfoundland and Labrador population of Atlantic cod is a good example of this. As discussed by Miel (2019), prematurely ramping up the fishery comes at the expense of rebuilding a sustainable population that can provide much greater benefits to fishing communities and the ocean, in the long term.

Premature Reopening of the Cod Fishery

History has clearly shown that increasing fishing pressure on a stock before it is ready can thwart recovery. The cod population of the Canadian fisheries district of southern Newfoundland and Labrador provides an object lesson of this principle. This fishery was reopened in 1997, just 4 years after the cod fishing moratorium. At that time, the government quota was set at 10,000 mt. Thus, the quota allowed about 440,000 cod to be caught that year in this district. However, intense lobbying by the fishing industry pressured the Canadian government into tripling this quota to 30,000 mt by 1999, allowing 1.3 million cod to be taken. This higher quota represents a level of cod fishing not seen since before the moratorium began. Predictably, the small regional cod population off the shores of southern Newfoundland and Labrador immediately began to collapse again, and the cod population has not recovered in the intervening 20 years. Nevertheless, the government of Canada has continued to allow cod fishing in this region. The fishing quotas have fluctuated between 30,000 mt in 1999—6000 mt in 2000 and then back up to 20,000 mt in 2001 and then to 15,000 mt until 2005. In 2018, Canadian fisheries landed just over 18,000 mt of Atlantic cod (DFO, 2018). As of 2019, the quota was set at 6000 mt.

Fishing Pressures on Atlantic Cod Biology

There have been centuries of population pressure placed on Atlantic cod from commercial fishing. This is especially true of the past 70 years when enormous trawlers have plied their trade. All this commercial fishing put tremendous biological stress on North Atlantic cod populations. The most obvious change in Atlantic Cod in the past 175 years has been a drastic reduction in size. In 1850, the average mature adult cod was 78 cm in length and weighed nearly 6 kg. By 2005, the size of an average mature cod was 68 cm, and it weighed 3.6 kg—a 40% reduction (McCrea-Strub and Pauly, 2011). The largest cod in the record books weighed 96 kg (212 pounds) and was taken in 1895 off the coast of Massachusetts. Atlantic cod is a relatively long-lived, slow-growing species whose reproductive capacity could not keep up with the increasing rate of mortality due to fishing. McCrea Strub and Pauly (2011) assembled the following comparison between Atlantic cod populations in 1850 and 2005. The 1850 estimates for Atlantic cod near the New England and southeast Canadian coasts were as follows:

Total biomass: 10.2 million mt
Average density: 8.8 mt per km^2
Total abundance: 3.4 billion mt
The 2005 estimates for Atlantic cod from the same fishing grounds were as follows:

Total biomass: 360,000 mt (<4% of the 1850 biomass)
Average density: 0.3 mt per km^2 (<4% of the 1850 density)
Total abundance: 285 million (8% of the 1850 abundance)
These statistics refer to the Grand Banks, one of the richest historic fishing grounds ever known, turned into a barren region by overfishing. Given these radical changes in the biomass, density, and abundance of cod in the northwest Atlantic, it is not surprising that this species suffered enormous selection pressures. These pressures affected their population structure, longevity, physical appearance, and genetic makeup, as discussed in the following.

Loss of Atlantic cod phenotypic diversity

Assessing historic measurement records of Atlantic cod fisheries, Olsen et al. (2004) showed that, until the Canadian cod fishery moratorium, the life history of Atlantic cod continually shifted toward maturation at earlier ages and smaller sizes. This study was followed up by Olsen et al. (2009), who reported on changes in phenotypic variability in natural Atlantic cod populations. "Phenotype" refers to the physical properties of an organism. These include physical appearance, development, and behavior. An organism's phenotype is determined by its genotype, which is the set of genes the organism carries, as well as by environmental influences upon those genes. Diversity in these observable traits, or a lack of diversity, may have impacts on adaptive evolutionary changes and population stability. The authors examined two data sets to illuminate complex trait changes in Atlantic cod along the Norwegian Skagerrak coast. The first of these were data collected during an annual beach seine survey, starting in 1919. Through the years, the survey participants recorded juvenile cod body size and abundance of cod stocks. Olsen et al. (2009) also gathered capture-mark-recapture data from which they estimated juvenile body size and growth. Their research showed that the variability of juvenile size has steadily decreased across the past 90 years. Their report also noted negative effects from commercial fishing and hence negative selection pressures, on both small, slow-growing fish as well as large, fast-growing fish. Taken together, these results suggest long-term stabilizing selection pressures on Atlantic cod populations. Phenotypic diversity is thus declining in these fish, which may ultimately lead to their decreased ability to cope with environmental change. Further studies are needed to evaluate the full complexity of trait changes in wild Atlantic cod populations.

Eikeset et al., 2013 examined the evolutionary changes in Atlantic cod brought on by many decades of commercial fishing. They found that there is a consistent trend toward faster growth and earlier maturation, even in cod stocks that are optimally managed. In their model (Fig. 7.11), selection pressure from long-term commercial fishing triggered genetic responses in the fish that affected the species' growth and maturation rates, their reproduction, and their length of life. Eventually, genetic selection pressures lead to phenotypic changes.

The authors then proceeded to model past and predicted future changes in Atlantic cod biomass, length at maturity, and age of maturation, under a variety of modeling scenarios that included historic levels of fishing and optimal harvest control methods (Fig. 7.12).

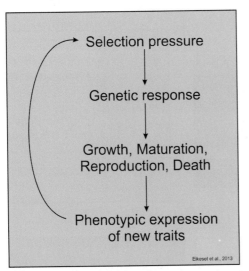

FIG. 7.11 Flowchart illustrating genetic and phenotypic responses of fish to selection pressures such as overfishing. (After Eikeset, A.M., 2013. Economic repercussions of fisheries-induced evolution. Proceedings of the National Academy of Sciences of the United States of America 110, 12259–12264. https://doi.org/10.1073/pnas.1212593110.)

Cod length at maturation was more than 10% greater for the OHC method predictions than for the predictions based on fishing at historic levels. Also, cod fished under the OHC method are predicted to mature at about 7 years of age, while predictions based on fishing at historic levels showed cod reaching maturity at 6–6.5 years.

Kindsvater and Palkovacs (2017) used numerical modeling to predict the eco-evolutionary impacts of commercial fishing on Atlantic cod body size and trophic role. Fishing has caused changes in abundance and demography in the exploited populations, in part due to rapid decreases in age and size at maturation. They contrasted purely demographic effects with those that include adaptive responses to fishing. Examining the age demographic models of fished and unfished cod populations (Fig. 7.13), they predicted that unfished populations tend to live longer, with a substantial proportion of the population living up to 15 years. In contrast to this, the projections for heavily fished populations suggest that most of these individuals do not live longer than 10 years.

Fishing clearly decreases cod abundance under both scenarios, but the mean trophic level decreases more when there is an adaptive response in maturation. In other words, smaller cod end up lower on the food chain than large cod. As found by Eikeset et al., 2013,

phenotypic adaptation also resulted in fish that were smaller at maturity. Some larger fish persisted in the model runs, even with heavy fishing. Because of their size in comparison with other North Atlantic fish species, they occupy a higher trophic position. These modeling results suggest that adaptations in heavily fished species can bring about alterations in food web dynamics.

Loss of Atlantic cod genetic diversity

Genomic analyses have increasingly revealed chromosome structural rearrangements (e.g., inversions and translocations) underpinning ecologically variable traits, supporting an important role for genomic architecture in promoting intraspecific ecological diversity (Colella et al., 2020). However, Kess et al. (2019) argue that the relationship between genomic architecture and key ecological traits remains largely unknown in most exploited species. The Atlantic cod has an extensive history of exploitation, and there is a growing body of evidence of genomic structural variation in this species, associated with ecological adaptation. As discussed earlier, the cod population in Northwest Atlantic waters around Newfoundland and Labrador has undergone multiple population declines, the most recent and drastic of which have been driven largely by overfishing following the adoption of industrial-scale fishing in the 1950s, in tandem with climate shifts. Individual and regional genomic structural diversity that differentiates Atlantic cod populations based on their environmental requirements and migratory behavior has been identified across the species' range in the North Atlantic. The migratory behavior of Atlantic cod may have increased their vulnerability to overfishing.

A large suite of variable sites within this rearrangement with predicted effects on gene function in more than 300 genes were previously shown to be differentiated between coastal and migratory ecotypes in Norwegian waters. Several of these variants are suggested to affect swim bladder function and muscular efficiency, facilitating vertical movement at variable depths in offshore sites and enhancing muscular capacity for strenuous migration.

The genetic evidence indicates that the Atlantic cod population crash resulted in more than a 90% reduction in population size. The study by Kess et al. (2019) revealed different demographic histories between individuals with different genotypes, with the differentiation starting in the North Atlantic population as early as 1860. After this time, populations with different genotypes began to appear, suggesting different selective histories, especially within the past century. Subsequent shifts in population genotypes indicate that the different

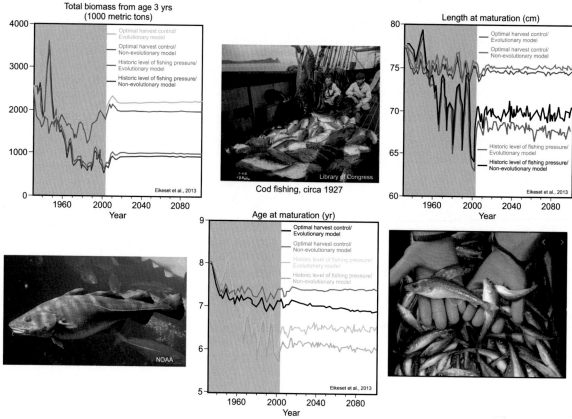

FIG. 7.12 Graphs showing changes in total biomass at age 3 years, length at maturation, and age at maturation of Atlantic cod. Two scenarios are shown for each graph: one based on optimal harvest control and the other based on continued fish mortality rates from commercial fishing from 1946 to 2005 (shaded area on each graph). Lower left: Photograph of mature Atlantic cod courtesy of NOAA, 2020. Lower right: Photo of immature Atlantic cod, courtesy of Northeastern University, College of Science (2017). Upper center: Photograph of cod fishing before World War II, and the subsequent invention of commercial trawlers, courtesy of the Library of Congress. (Adapted from Eikeset, A.M., 2013. Economic repercussions of fisheries-induced evolution. Proceedings of the National Academy of Sciences of the United States of America 110, 12259–12264. https://doi.org/10.1073/pnas.1212593110.)

demographic histories of these populations may have resulted in distinct genetic consequences from the bottleneck imposed by overfishing. However, additional factors shaping demographic history such as a change in climate have also potentially played a role.

As discussed by Miel (2019), the collapse of Atlantic cod stocks has resulted in a massive restructuring of coastal shelf ecosystems of eastern North America. Continued overfishing may lead to loss of the migration-associated genotype, possibly resulting in further changes in the ecological function and composition in the Northwest Atlantic marine ecosystem because of changing cod distribution and site persistence. Removal of this genomic diversity through

overfishing may also reduce the buffering effect provided by phenotypic diversity within these populations, further increasing the risk of future collapse.

Flatfish Depletion in North American Atlantic Waters

Flatfish are bottom dwellers that prey on smaller fish, mollusks, and other invertebrates on the seafloor. In US waters, this group includes four species of flounder, American plaice (*Hippoglossoides platessoides*), and Atlantic halibut (*Hippoglossus hippoglossus*) (NOAA Fisheries, 2019). Bottom trawling from huge factory ships has also taken its toll on this group of fish (Fig. 7.14). The US catch of flatfish peaked at 80,000 mt in 1983.

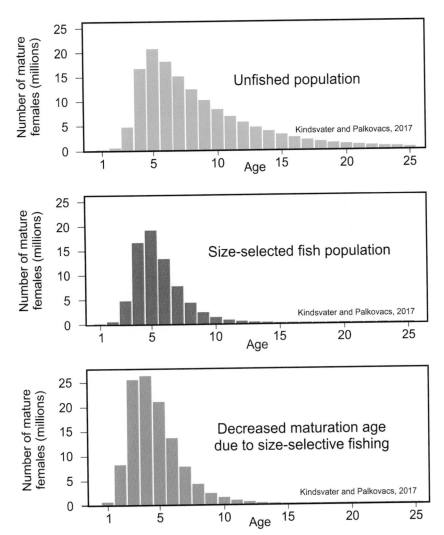

FIG. 7.13 Above: Age distribution of mature females in the unfished population. Middle: Age distribution of mature females after size-selective fishing mortality. Below: Age distribution of mature females when maturation age decreases in response to size-selective fishing. (After Kindsvater, H.K., Palkovacs, E.P., 2017. Predicting eco-evolutionary impacts of fishing on body size and trophic role of Atlantic cod. Copeia 105 (3), 475–482. https://doi.org/10.1643/OT-16-533.)

Then the fishery started to collapse. Only 30,000 mt of flatfish were taken in 1989. 21st-century flatfish catches have dropped even further. In 2007, 12,000 mt of flatfish were taken, and this dropped to just 3000 mt in 2018. That same year, Canadian fisheries landed over 5000 mt of Atlantic Halibut and 9698 mt of other flatfish (DFO, 2018).

American plaice and Atlantic halibut are large, slow-growing fish that are highly sensitive to overfishing.

Atlantic halibut were already considerably depleted in the 19th and early 20th centuries, and their numbers have never recovered (Pierre et al., 2012). Fishing levels have fluctuated around the level projected to allow the stock to rebuild by 2056, but there is significant uncertainty in this estimate. American plaice abundance is of low conservation concern, and current fishing pressure is at a sustainable level. Summer flounder (*Paralichthys dentatus*) is inherently vulnerable to fishing pressure,

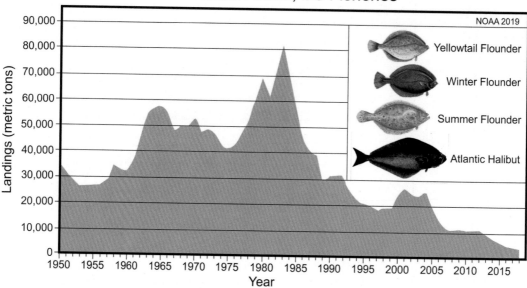

FIG. 7.14 History of the annual catch of flatfish from 1950 to 2018. (Data from NOAA - National Marine Fisheries Service, 2020. Annual Catch of Flatfish from 1950–2018. https://foss.nmfs.noaa.gov/apexfoss/f? p=215:200:0:::::&tz=-6:00.)

even though its numbers have been rebuilding. Windowpane flounder (*Scophthalmus aquosus*) is somewhat vulnerable to fishing pressure. There are two regional stocks of this species. The northern stock suffers from low abundance and high fishing mortality, and so is of high conservation concern. Because of higher abundance and lower fishing mortality, the southern stock is of lower conservation concern. Winter flounder (*Pseudopleuronectes americanus*) has a low natural vulnerability to fishing pressure, but stock abundance varies considerably near the shores of New England and southeastern Canada. Witch flounder (*Glyptocephalus cynoglossus*) and yellowtail flounder (*Pleuronectes ferruginea*) are both somewhat vulnerable to fishing pressure, and the abundance and fishing mortality of the stocks of both species are of high conservation concern (NOAA Fisheries, 2019).

Failure of EU Member States to Meet Quota Targets

After lengthy discussions in December 2019, it became clear that European member states failed to meet a legal obligation to end overfishing by 2020. This is because the ministers in charge of deciding annual fishing quotas for their respective countries declined to agree to the targets set by concerned scientists. Some progress

was made toward ensuring fish stock sustainability. For instance, the quotas for Atlantic cod fishing were cut in half. But the agreed quotas for several other species, such as haddock and sole, were allowed to increase.

The results of the meeting were succinctly summarized by Andrew Clayton of The Pew Charitable Trusts. He said, "The limits set by ministers suggest that progress to end overfishing has stalled or even reversed, a disappointing outcome for the year when overfishing was supposed to become a thing of the past" (Gregory, 2019).

Meanwhile, poleward fish species range shifts have also commenced on the Pacific side of the AO. The gateway of this intrusion of boreal and sub-Arctic biota into the Arctic is the Bering Sea, which is being transformed by global warming from an Arctic to a sub-Arctic system. Several fish species have been observed farther north than previously reported and in increasing abundances. For instance, Pacific cod (*Gadus macrocephalus*) has recently been making inroads into the AO (Aquamaps, 2019). Using genetic markers, Spies et al. (2020) assessed whether these fish migrated from stock in the eastern Bering Sea, the Gulf of Alaska, or the Aleutian Islands, or whether they represent a separate population. Genetic data indicate that nonspawning cod collected in the northern

Bering Sea (NBS) were closely related to spawning stocks of cod in the eastern Bering Sea (EBS). This result suggests an increasing northward movement of the large EBS stock during summer months. Whether the cod observed in the NBS migrate south during winter to spawn or remain in the northern Bering Sea as a sink population remains unknown.

The North Pacific Fishery Management Council (2009) approved a new Fishery Management Plan for Fish Resources of the Arctic Management Area (Arctic FMP), in the waters north and northwest of Alaska. The plan was in response to the changing environmental conditions of the Arctic, including warming trends in ocean temperatures, the loss of seasonal ice cover, and the potential long-term effects of these changes on the Arctic marine ecosystem. The plan covers all US waters of the Chukchi and Beaufort seas. Since 2009, all Arctic waters under US Federal jurisdiction have been closed to commercial fishing for all species of fish, mollusks, crustaceans, and all other forms of marine animal and plant life (Koenig, 2019).

ANTHROPOGENIC STRESSORS ON NORTHERN MARINE ECOSYSTEMS

Andersen et al. (2017) examined human activities that place a heavy strain on the marine environments of Greenland. The marine environment in general and marginal seas in particular have experienced ecosystem regime shifts and altered food web structures. These have come about through a combination of human activities including commercial fishing and marine mammal hunting by Native peoples, pollution, physical modification, the introduction of nonnative species, and climate change brought on by greenhouse gas (GHG) emissions. Further north, in Arctic waters, the environmental stressors include artisanal fishing, commercial trawling and shipping intensity, ocean acidification, and ocean warming. In addition to rising sea surface temperatures, climate change is also causing warming of the seawater throughout the water column, as well as changes in salinity.

The overall pattern of marine ecosystem diversity of the mid- and south-western coast of Greenland features a year-round concentration of species (Fig. 7.15). The lowest stressor index values were seen to occur in the northern part of the research area—the least inhabited part of the study region. In contrast, the stressor index was found to be greatest in the southern part of the study area: from the coastal and shelf area of Qeqertarsuaq (Disko Island) to the southern tip of Greenland. The combined impacts of multiple stressors in high-

diversity marine ecosystem regions increase the likelihood of complex cumulative impacts especially along the more southern parts of Greenland's west coast, and in some offshore waters.

Lotze et al. (2019) modeled the effects of global change on marine animal biomass (including fish stocks), using a variety of GHG emission scenarios devised by the IPCC (Fig. 7.16). In a model run without fishing, they found that mean global marine animal biomass decreased by 5% under the best-case scenario (RCP2.6: low emissions). Biomass decreased by 17% under a worst-case scenario (RCP8.5: high emissions) by 2100. Their model predicted an average 5% decline in fish stocks for every 1°C of warming. These projected declines were mainly driven by increasing temperatures, leading to decreasing primary productivity. As discussed earlier, warm waters hold less dissolved gases than cold waters. These dissolved gases are essential to the growth and development of the primary producers, mainly marine algae and diatoms. Furthermore, these losses were more pronounced in animals occupying higher trophic levels (i.e., top predators). This process is known as trophic amplification (Stock et al., 2019). Trophic amplification of the climate signal from primary producers to higher trophic levels arises from multiple factors that include changes in phytoplankton size composition, lengthening of food chains, reduced trophic efficiencies, and higher metabolic costs with increased body size.

This model predicted considerable regional variation. For instance, the model predicted substantial decreases in marine animal biomass in the ocean waters of the middle to low latitudes. These results indicate that global ocean animal biomass will consistently decline with global warming and that these impacts will be magnified at higher trophic levels. Actual increases in marine animal biomass are projected for high latitude seas. However, Lotze et al. (2019) warn that future projections based on global models are often less certain for polar regions. Future model refinements will be made through regional downscaling to incorporate higher resolution climate and ecosystem features.

In the coming years, if there are substantial increases in polar fish stocks, these will likely be viewed by the commercial fishing industry as a new set of opportunities for exploiting hitherto untapped Arctic fisheries. This scenario demonstrates an urgent need for protecting sensitive species and rapidly changing ecosystems.

When the impacts of commercial fishing are added to the model, the projections become direr. In a natural environment, the predicted warming would enhance both fish growth and predation rates. But predation rates will be reduced if commercial fishing is allowed,

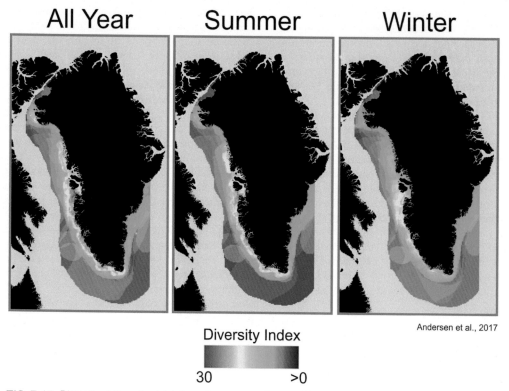

Andersen et al., 2017

FIG. 7.15 Diversity of Greenland fish fauna in summer, winter, and year-round. Note that the highest diversity of fish species consistently occurs close to the Greenland coast. (From Andersen, J.H., Berzaghi, F., Christensen, T., Geertz-Hansen, O., Mosbech, A., Stock, A., Zinglersen, K.B., Wisz, M.S., 2017. Potential for cumulative effects of human stressors on fish, sea birds and marine mammals in Arctic waters. Estuarine, Coastal and Shelf Science 184, 202–206. https://doi.org/10.1016/j.ecss.2016.10.047.)

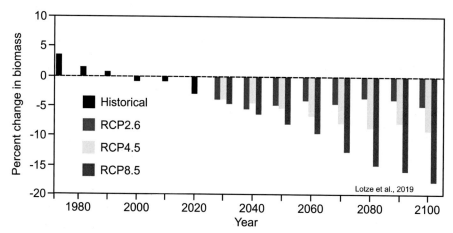

Lotze et al., 2019

FIG. 7.16 Multimodel mean change in total biomass without fishing for historical and four future emission scenarios (RCPs) relative to 1990–99. *RCP*, representative concentration pathway. (Data from Lotze, H.K., Tittensor, D.P., Bryndum-Buchholz, A., 2019. Global ensemble projections reveal trophic amplification of ocean biomass declines with climate change. Proceedings of the National Academy of Sciences of the United States of America 116, 12907–12912. https://doi.org/10.1073/pnas.1900194116.)

due to selective fishing of larger species and lower predator abundance. The reduction of predators from the ocean may indirectly enhance prey biomass and weaken the relative climate change effect, but this is a relatively small effect compared with the large direct effect of fishing itself. The model predicts that there will be a 16%−80% decrease in biomass for fish and other commercially fished marine animals (for instance, lobsters and crabs) that have body lengths greater than 10 cm. Compared with unfished conditions under RCP2.6, biomass losses will be even greater (48%−92%) by 2100 for animals greater than 30 cm in length.

However, the results of the model runs strongly suggest that limiting future warming to 1.5−2.0°C above preindustrial levels would limit marine animal biomass declines to 4%−6% by 2100, underscoring the potential impact of climate change mitigation in accord with the Paris Agreement (Lotze et al., 2019).

CONCLUSIONS

This chapter has dipped its toe in many topics related to overfishing and the marine ecology of the AO. The take-home message of this chapter is that commercial fishing of Arctic waters may well lead to a collapse of the Arctic marine ecosystem. It all starts with the ongoing global warming, which has many ramifications. A few of these may be "good" for the AO ecosystem, but there are many that are bad (Fig. 7.17, above). Global warming is causing rising ocean temperatures in both the North Atlantic and North Pacific. The change in temperature regime is bringing about shifts in oceanic circulation patterns, strengthening warm water currents from the North Atlantic and Pacific to the extent that they are now penetrating the boundaries of the AO for the first time in recorded history (Polyakov et al., 2020). This increase in warm water input to the AO facilitates the invasion of warm-adapted biota, ranging from microbes to fish and marine mammals. These changes have not been witnessed before, and they are very likely to produce unpredictable outcomes for AO ecosystems. Because of its remoteness and logistic difficulties, we know far less about the AO ecosystem and its functionality than we know about most other ocean regions of the world.

Invasive Species

The arrival of invasive fish species that are adapted to warmer water than their current Arctic counterparts is certain to have negative impacts on several aspects of the AO ecosystem. First, there will be new competition for food resources from subarctic predatory fish, such as Atlantic cod, haddock, halibut, and flounder. There is a substantial size difference between Arctic Cod and the larger, more aggressive species that will probably invade the AO in the coming decades. A fully mature Arctic cod may attain a length of 325 mm. Adult Atlantic cod are more than three times that size, reaching lengths of 1000−1070 mm. However, there is some evidence that dietary differences between these species may help maintain Polar cod populations, despite the new competition. A more sinister threat to Arctic cod is that the potentially invasive species, including Atlantic cod, are known to prey on its immature stages.

Loss of Sea-Ice Cover

As discussed in Chapter 1, global warming is bringing about decreased sea-ice cover at a remarkably rapid rate. Aune et al. (2018) traced the history of latitudinal shifts in the northern sea-ice margin from 1580 onward (Fig. 7.18). This margin extended south as far as 76°N in the late 17th century and the 1780s. Today the sea-ice margin lies about 83°N, the furthest north it has been in the past 500 years.

Given the present rate of sea-ice shrinkage, climate modelers predict that the most likely date for the permanent disappearance of Arctic sea ice is 2034. The loss of sea ice will usher in many changes to the AO ecosystem. Again, some of these changes will be positive and some will be negative. On the plus side, the newly open waters of the AO will receive increased exposure to sunlight throughout the growing season, stimulating the productivity of marine primary producers (diatoms and algae). Warmer Arctic waters will facilitate the growth of immature polar cod because warmer water temperatures will facilitate winter hatching of Arctic cod eggs, giving them a head start on the growing season. The increase in open water will also promote upwelling on the continental shelf regions of the AO, drawing nutrients up from basin waters and onto the shelves, thereby boosting the biological productivity of the shelf regions. On the negative side, the warming of the AO will likely disrupt the Arctic marine food web by eliminating a key resource: sea-ice bottom algae. As discussed earlier, this kind of algae provides a substantial proportion of the food for zooplankton, and by extension, the entire animal food chain. Also, the loss of sea ice will disrupt the life cycle of Arctic cod because the immature forms of this fish live beneath sea-ice cover. Likewise, sea-ice cover is an essential aspect of polar cod spawning grounds. As we have seen in this chapter, Arctic cod have been shifting northward year by year, tracking the southern edge of the sea ice. What will happen when the sea ice disappears?

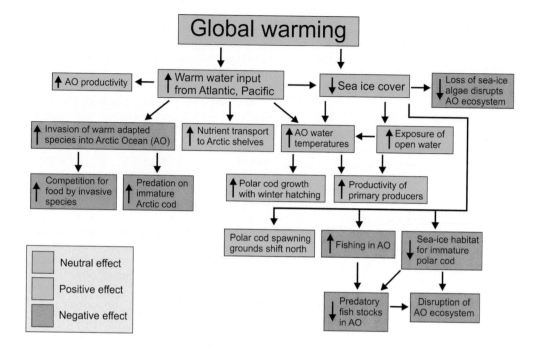

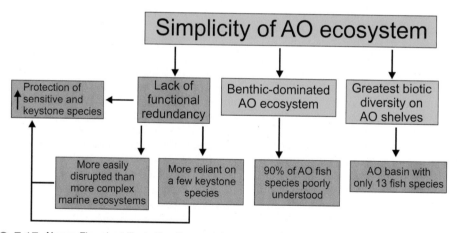

FIG. 7.17 Above: Flowchart illustrating the neutral, positive, and negative effects of global warming on the AO ecosystem. Below: Flowchart illustrating the weaknesses of the AO ecosystem because of its simplicity. *AO*, Arctic Ocean.

A groundbreaking study by Lewis et al. (2020), based on a long time series of satellite observations, indicates that net primary production (NPP) in the AO has increased 57% since 1998, a far greater change than was indicated by earlier estimations. The largest increases were eastern Arctic shelf regions of the Siberian, Laptev, and Kara seas, the inflow shelves of the Chukchi and Barents subregions, and the central basin. The inflow shelves together contributed 70% of the NPP increase. From 1998 to 2008, increased AO NPP was likely due to regional sea ice loss. However, Lewis et al. (2020) conclude that the rise in NPP since 2008 has been driven primarily by increased phytoplankton biomass in the AO (Fig. 7.19). Their evidence indicates that these biomass

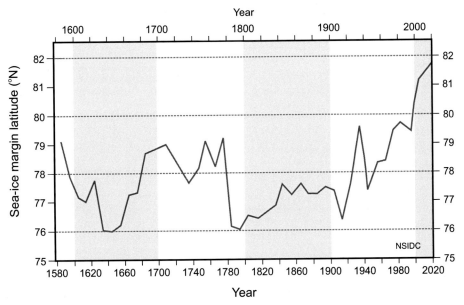

FIG. 7.18 The position of the ice edge in August between Svalbard and Franz Josef Land for the period 1553–2017, given by its mean latitude in the sector 20–45°E. (Modified from Aune, M., 2018. Seasonal ecology in ice-covered Arctic seas - considerations for spill response decision making. Marine Environmental Research 141, 275–288. https://doi.org/10.1016/j.marenvres.2018.09.004, with more recent data from National Snow and Ice Data Center (NSIDC), 2020. Arctic Sea Ice News and Analysis. http://nsidc.org/arcticseaicenews/category/analysis/.)

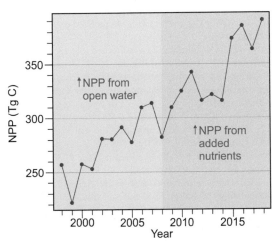

FIG. 7.19 Annual time series of mean net primary production (NPP) of the AO. Blue shaded area represents an interval when increased open waters brought increased NPP. Green shaded area represents an interval when increased nutrients brought increased NPP. After Lewis et al., 2020. AO, Arctic Ocean. (Modified from Lewis, K., van Dijken, G., Arrigo, K., 2020. Changes in phytoplankton concentration now drive increased Arctic Ocean primary production. Science 369, 198–202. https://doi.org/10.1126/science.aay8380.)

increases were sustained by increases in nutrient flux into the AO from Atlantic and Pacific waters. As discussed earlier, the increased advection of Atlantic waters comes via the Barents Sea. Pacific waters enter the AO via the Chukchi Sea. The combination of these two oceanic inputs may supply enough additional nutrients to sustain the higher biomass observed on the inflow shelves. According to Lewis et al. (2020), the Pacific inflow through the Bering Strait increased by about 50% from 1999 to 2015. The "Atlantification" of the Barents Sea is likely associated with increased advection of phytoplankton and nutrients. But the incoming warm Atlantic water does more than add nutrients to the AO. Because of the temperature differential between the Atlantic and AO waters, this warm inflow also reduces stratification in the AO, which has remained stable since 1979. The better-mixed AO waters support increased nutrient availability, which fosters substantial increases in phytoplankton biomass and production. Finally, the decreased sea-ice cover has led to warmer sea surface temperatures in the AO. The warmer water facilitates increased frequency and intensity of regional storms. These storms stimulate increased internal wave mixing and storm-induced upwelling throughout the shallow shelf regions. This, in turn, results in periodic injections of nutrients that collect on the AO seafloor. If these current trends in nutrient availability persist into the coming years, then the AO may become more biologically productive at all trophic levels.

The potential collapse of a simple ecosystem

Finally, I would like to speak to the issues surrounding the simplicity of the Arctic marine food web. Very few species of plants or animals are adapted to live in the extremely cold waters of the AO. The small number of species in this ocean causes a lack of functional redundancy in the ecosystem. To paraphrase an admonition from King Solomon, a rope with just a few strands is much more likely to break than a rope of many strands. For instance, if there were 50 predatory fish species in the AO, then the removal of one species would have a little impact on the system. Unfortunately, there are just a few kinds of predatory fish in the AO, and only 13 fish species of all trophic levels living the AO basins. So, the removal of even one species could cause catastrophic damage to the system. As we have seen, the AO ecosystem relies heavily on the presence of a few keystone species, such as Arctic cod, ringed seal, and the polar bear. The loss of sea ice cover will certainly have devastating effects on both the ringed seal and polar bear, both of which rely on this habitat for their survival.

One of the most serious problems in AO fish management is our lack of knowledge. Sadly, about 90% of AO fish species are poorly understood. This might seem a shocking state of ignorance in the 21st century, but keep in mind that the AO and the Southern Ocean surrounding Antarctica are the two most remote oceanic regions of the world. Research in polar seas is extremely costly, logistically difficult, and made dangerous by ever-changing pack ice conditions. Perhaps after the sea ice disappears completely, oceanographers and marine biologists will have an easier time doing fieldwork in the AO. However, this easier access comes at a price: a greatly depleted and disrupted marine ecosystem that may be on the point of collapse before the 22nd century.

REFERENCES

Alabia, I.D., García Molinos, J., Saitoh, S.I., Hirawake, T., Hirata, T., Mueter, F.J., 2018. Distribution shifts of marine taxa in the Pacific Arctic under contemporary climate changes. Diversity and Distributions 24 (11), 1583–1597. https://doi.org/10.1111/ddi.12788.

Andersen, J.H., Berzaghi, F., Christensen, T., Geertz-Hansen, O., Mosbech, A., Stock, A., Zinglersen, K.B., Wisz, M.S., 2017. Potential for cumulative effects of human stressors on fish, sea birds and marine mammals in Arctic waters. Estuarine, Coastal and Shelf Science 184, 202–206. https://doi.org/10.1016/j.ecss.2016.10.047.

AquaMaps, 2019. Computer Generated Native Distribution Map for *Gadus macrocephalus* (Pacific Cod), with Modelled Year 2050 Native Range Map Based on IPCC RCP8.5 Emissions Scenario. https://www.aquamaps.org/receive.php?type_of_map=regular#.

Aune, M., Aniceto, A.S., Biuw, M., Daase, M., Falk-Petersen, S., Leu, E., Ottesen, C.A.M., Sagerup, K., Camus, L., 2018. Seasonal ecology in ice-covered Arctic seas - considerations for spill response decision making. Marine Environmental Research 141, 275–288. https://doi.org/10.1016/j.marenvres.2018.09.004.

Bluhm, B.A., Kosobokova, K.N., Carmack, E.C., 2015. A tale of two basins: an integrated physical and biological perspective of the deep Arctic Ocean. Progress in Oceanography 139, 89–121. https://doi.org/10.1016/j.pocean.2015.07.011.

Bouchard, C., Fortier, L., 2011. Circum-arctic comparison of the hatching season of polar cod *Boreogadus saida*: a test of the freshwater winter refuge hypothesis. Progress in Oceanography 90 (1–4), 105–116. https://doi.org/10.1016/j.pocean.2011.02.008.

Clover, C., 2004. The End of the Line: How Overfishing is Changing the World and What We Eat. Ebury press, London.

Colella, J.P., Talbot, S.L., Brochmann, C., Taylor, E.B., Hoberg, E.P., Cook, J.A., 2020. Conservation genomics in a changing arctic. Trends in Ecology and Evolution 35 (2), 149–162. https://doi.org/10.1016/j.tree.2019.09.008.

Dalpadado, P., Ingvaldsen, R.B., Stige, L.C., Bogstad, B., Knutsen, T., Ottersen, G., Ellertsen, B., 2012. Climate effects on Barents Sea ecosystem dynamics. ICES Journal of Marine Science 69 (7), 1303–1316. https://doi.org/10.1093/icesjms/fss063.

Department of Fisheries and Oceans Canada, 2018. Sea Fisheries Landed Quantity by Region, 2018. Department of Fisheries and Oceans Canada (DFO). https://www.dfo-mpo.gc.ca/stats/commercial/land-debarq/sea-maritimes/s2018aq-eng.htm.

Eikeset, A.M., Richter, A., Dunlop, E.S., Dieckmann, U., et al., 2013. Economic repercussions of fisheries-induced evolution. Proceedings of the National Academy of Sciences of the United States of America 110, 12259–12264. https://doi.org/10.1073/pnas.1212593110.

FAO, 2020. The State of World Fisheries and Aquaculture 2020. Sustainability in Action. Food and Agriculture Organization of the United Nations. https://doi.org/10.4060/ca9229en.

Frey, K.E., Comiso, J.C., Cooper, L.W., Grebmeier, J.M., Stock, L.V., 2018. Arctic Ocean Primary Productivity: The Response of Marine Algae to Climate Warming and Sea Ice Decline. NOAA Arctic Report Card: Update for 2018. https://arctic.noaa.gov/Report-Card/Report-Card-2018/ArtMID/7878/ArticleID/778/Arctic-Ocean-Primary-Productivity-The-Response-of-Marine-Algae-to-Climate-Warming-and-Sea-Ice-Decline.

Grant, W.S., Spies, I., Canino, M.F., 2010. Shifting-balance stock structure in North Pacific walleye pollock (*Gadus chalcogrammus*). ICES Journal of Marine Science 67 (8), 1687–1696. https://doi.org/10.1093/icesjms/fsq079.

Gregory, A., 2019. EU Ministers Vote to Allow Overfishing Despite Legal Deadline to Heed Scientists' Advice. The Independent, 18 December. https://www.independent.co.uk/environment/eu-fishing-quota-overfishing-common-fisheries-policy-cod-legal-2020-a9252276.html.

Haug, T., Bogstad, B., Chierici, M., Gjøsæter, H., Hallfredsson, E.H., et al., 2017. Future harvest of living resources in the Arctic Ocean north of the Nordic and Barents Seas: A review of possibilities and constraints. Fisheries Research 188, 38–57. https://doi.org/10.1016/j.fishres.2016.12.002.

Hemy, C.N., 1888. The Fisherman. https://commons.wikimedia.org/wiki/File:Charles_Napier_Hemy_-_The_Fisherman_1888.jpg.

Huserbråten, M.B.O., Eriksen, E., Gjøsæter, H., Vikebø, F., 2019. Polar cod in jeopardy under the retreating Arctic sea ice. Communications Biology 2 (1). https://doi.org/10.1038/s42003-019-0649-2.

Kess, T., Bentzen, P., Lehnert, S.J., Sylvester, E.V., Lien, S., et al., 2019. A migration-associated supergene reveals loss of biocomplexity in Atlantic cod. Science Advances 5 (6), eaav2461. https://doi.org/10.1126/sciadv.aav2461.

Kindsvater, H.K., Palkovacs, E.P., 2017. Predicting eco-evolutionary impacts of fishing on body size and trophic role of Atlantic cod. Copeia 105 (3), 475–482. https://doi.org/10.1643/OT-16-533.

Koenig, R., 2019. Prospect of Commercial Fishing in Central Arctic Ocean Poses Big Questions for Science. https://www.alaskapublic.org/2019/03/05/prospect-of-commercial-fishing-in-central-arctic-ocean-poses-big-questions-for-science/.

Lewis, K.M., van Dijken, G.L., Arrigo, K.R., 2020. Changes in phytoplankton concentration now drive increased Arctic Ocean primary production. Science 369, 198–202. https://doi.org/10.1126/science.aay8380.

Lotze, H.K., Tittensor, D.P., Bryndum-Buchholz, A., 2019. Global ensemble projections reveal trophic amplification of ocean biomass declines with climate change. Proceedings of the National Academy of Sciences of the United States of America 116, 12907–12912. https://doi.org/10.1073/pnas.1900194116.

McCrea Strub, A., Pauly, D., 2011. Atlantic Cod: Past and Present. http://www.seaaroundus.org/atlantic-cod-past-and-present.

Mecklenburg, C.W., Møller, P.R., Steinke, D., 2010. Biodiversity of arctic marine fishes: taxonomy and zoogeography. Marine Biodiversity 41, 109–140. https://doi.org/10.1007/s12526-010-0070-z.

Miel, K.A., 2019. Oceana Blog: Northern Cod Recovery is Still Fragile, Proceed with Caution to Avoid Past Mistakes. https://oceana.ca/en/blog/.

National Snow and Ice Data Center (NSIDC), 2020. Arctic Sea Ice News and Analysis. http://nsidc.org/arcticseaicenews/category/analysis/.

NOAA - National Marine Fisheries Service, 2020. Annual Catch of Flatfish from 1950-2018. https://foss.nmfs.noaa.gov/apexfoss/f?p=215:200:0:::::&tz=-6:00.

NOAA, 2004. International Bathymetric Chart of the Arctic Ocean. National Oceanographic and Atmospheric Administration (NOAA). https://www.ngdc.noaa.gov/mgg/fliers/04mgg03.html.

NOAA, 2004. Bathymetry of the Arctic Ocean and Adjoining Seas. https://www.ngdc.noaa.gov/mgg/fliers/04mgg03.html.

NOAA, 2019. Fishing Gear: Pelagic and Mid-water Trawling. https://www.fisheries.noaa.gov/national/bycatch/fishing-gear-midwater-trawls.

NOAA, 2019. Fishing Gear: Purse Seines. https://www.fisheries.noaa.gov/national/bycatch/fishing-gear-purse-seines.

NOAA, 2019. Fishing Gear: Bottom Trawls. https://www.fisheries.noaa.gov/national/bycatch/fishing-gear-bottom-trawls.

NOAA Fisheries, 2019. Landings by NOAA Marine Fisheries Service Regions. https://foss.nmfs.noaa.gov/apexfoss/f?p=215:200:0:::::&tz=-6:00.

North Pacific Fishery Management Council, 2009. Arctic Fishery Management. https://www.npfmc.org/arctic-fishery-management/.

Olsen, E.M., Heino, M., Lilly, G.R., Morgan, M.J., Brattey, J., Ernande, B., Dieckmann, U., 2004. Maturation trends indicative of rapid evolution preceded the collapse of northern cod. Nature 428 (6986), 932–935. https://doi.org/10.1038/nature02430.

Olsen, E.M., Carlson, S.M., Gjøsæter, J., Stenseth, N.C., 2009. Nine decades of decreasing phenotypic variability in Atlantic cod. Ecology Letters 12 (7), 622–631. https://doi.org/10.1111/j.1461-0248.2009.01311.x.

Pierre, J., Siple, M., Simon, R., 2012. Seafood Watch: Atlantic Flatfish [stock status updated, 2016]. Monterey Bay Aquarium, Monterrey, California, 212 pp.

Polyakov, I.V., Alkire, M.B., Bluhm, B.A., Brown, K., Carmack, C.E., Melissa, C., Danielson, S.L., Ingrid, E., Elizaveta, E., Katarina, G., Randi, B.I., Pnyushkov, A.V., Dag, S., Paul, W., 2020. Borealization of the Arctic Ocean in response to Anomalous advection from sub-arctic seas. Frontiers in Marine Science 7, 491–523. https://doi.org/10.3389/fmars.2020.00491.

Randelhoff, A., Holding, J., Janout, M., Sejr, M.K., Babin, M., Tremblay, J.É., Alkire, M.B., 2020. Pan-Arctic Ocean primary production constrained by turbulent nitrate fluxes. Frontiers in Marine Science 7, 150. https://doi.org/10.3389/fmars.2020.00150.

Ray, G.C., McCormick-Ray, J., 2004. Coastal-Marine Conservation: Science and Policy. Blackwell Science, Oxford.

Reigstad, M., Carroll, J.L., Slagstad, D., Ellingsen, I., Wassmann, P., 2011. Intra-regional comparison of productivity, carbon flux and ecosystem composition within the northern Barents Sea. Progress in Oceanography 90 (1–4), 33–46. https://doi.org/10.1016/j.pocean.2011.02.005.

Renaud, P.E., Berge, J., Varpe, O., Lønne, O.J., Nahrgang, J., Ottesen, C., Hallanger, I., 2012. Is the poleward expansion by Atlantic cod and haddock threatening native polar cod,

Boreogadus saida? Polar Biology 35 (3), 401−412. https://doi.org/10.1007/s00300-011-1085-z.

Renaud, P.E., Sejr, M.K., Bluhm, B.A., Sirenko, B., Ellingsen, I.H., 2015. The future of Arctic benthos: expansion, invasion, and biodiversity. Progress in Oceanography 139, 244−257. https://doi.org/10.1016/j.pocean.2015.07.007.

Rey, F., 2004. Phytoplankton: the grass of the sea. In: Skjoldal, H.R. (Ed.), The Norwegian Sea Ecosystem. Tapir Academic Press, pp. 97−136. https://doi.org/10.5194/bg-12-3273-2015.

Seafish, 2011. Gear Technology Note − Encircling Gear. https://seafish.org/gear-database/fishing-gear/options/encircling-gear/.

Seafish, 2019. Gear Technology Note − Pelagic Pair Trawl. https://seafish.org/gear-database/gear/pelagic-pair-trawl/.

Stock, C.A., Verley, P., Volkholz, J., Walker, N.D., Worm, B., Fernandes, J.A., Schewe, J., Shin, Y.J., Silva, T.A.M., Steenbeek, J., 2019. Global ensemble projections reveal trophic amplification of ocean biomass declines with climate change. Proceedings of the National Academy of Sciences of the United States of America 116 (26), 12907−12912. https://doi.org/10.1073/pnas.1900194116.

Spies, I., Gruenthal, K.M., Drinan, D.P., Hollowed, A.B., Stevenson, D.E., Tarpey, C.M., Hauser, L., 2020. Genetic evidence of a northward range expansion in the eastern Bering Sea stock of Pacific cod. Evolutionary Applications 13 (2), 362−375. https://doi.org/10.1111/eva.12874.

Waga, H., Hirawake, T., Grebmeier, J.M., 2020. Recent change in benthic macrofaunal community composition in relation to physical forcing in the Pacific Arctic. Polar Biology 43 (4), 285−294. https://doi.org/10.1007/s00300-020-02632-3.

FURTHER READING

Loeng, H., Naustvoll, L.J., Røttingen, I., Sunnanå, K., Hallfredsson, E.H., Høines, Å.S., Hoel, A.H., Ingvaldsen, R.B., Jørgensen, L.L., Knutsen, T., 2017. Future harvest of living resources in the Arctic Ocean north of the Nordic and Barents Seas: a review of possibilities and constraints. Fisheries Research 188, 38−57. https://doi.org/10.1016/j.fishres.2016.12.002.

CHAPTER 8

Impacts of Global Shipping to Arctic Ocean Ecosystems

INTRODUCTION

As discussed in the introduction, the search for a North-west Passage (NWP) from the Atlantic to the Pacific was a 400-year endeavor, beset by failure and disaster. Since the voyages by Columbus to the Caribbean islands, which he mistakenly took to be part of Asia, most Europeans had the false impression that the distance between Europe and Asia, traveling west was relatively small. As the crow flies, traveling eastward from Lisbon to Hong Kong is a journey of about 11,000 km. By comparison, traveling westward from Lisbon to Hong Kong is a journey of more than 18,500 km. The length of the NWP itself is roughly equal to the distance between Nice and Amsterdam. The NWP spans roughly 1450 km from the North Atlantic entrance north of Baffin Island in the east to the Beaufort Sea north of Alaska in the west. The route is located entirely within the Arctic Circle, less than 1900 km from the North Pole (Fig. 8.1).

Native Navigation of the Canadian Arctic Archipelago

The Inuit, natives of Arctic Canada and Greenland, have always been maritime people. Their principal mode of sea transportation has been the kayak. They traditionally used kayaks to hunt marine mammals and sea birds, both on the open water of summer and through the pack ice in winter and spring. The top of the kayak is completely covered, allowing the user to stay dry, even in stormy seas. Since they weigh only about 12–15 kg, kayaks can easily be carried and launched from the edge of the land or sea ice and maneuvered through moving ice. Most European explorers never even considered enquiring whether the native peoples of the eastern Canadian Arctic knew of an NWP, but they could have learned a great deal about it from the Inuit. The Inuit utilized what is now referred to as the NWP for thousands of years to migrate across their island chain of homelands, called Inuit Nunangat.

Two World Views

The world views of European explorers and the Inuit of the Canadian High Arctic could not have been more different. To the explorers, all naval officers, and seamen, the Arctic waters were highly treacherous, the conditions unpredictable, and Arctic nature itself seemed completely hostile. Like a late 20th-century lunar mission, they brought along everything they thought they needed. In the case of the Franklin expedition, this included enough dried and canned food to last

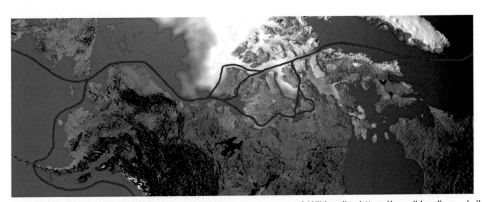

FIG. 8.1 Current Northwest Passage routes. (Image courtesy of Wikipedia. https://en.wikipedia.org/wiki/Northwest_Passage)

Threats to the Arctic. https://doi.org/10.1016/B978-0-12-821555-5.00008-5

3 years. When their ships were trapped in the ice for years, this food ran out, and they starved. They endured incredible hardships to conquer these natural foes, find the NWP, and win glory for king and country. For the Inuit, these same land- and seascapes represented their homeland. The Arctic is a harsh, unforgiving region for all human beings, but the natives found ways to make it work for them. The seas were navigable, especially in small kayaks. Local knowledge, reinforced by many generations of observations of ice, sea, and sky, informed Inuit hunters of both the dangers and opportunities of traversing ice-choked waters. Openings in the pack ice could be found and exploited. Polynyas, areas of Arctic seas that remain open water throughout the year, were well known to the Inuit. They hunted sea birds and fished there. Marine mammals were abundant and supplied most of the calories in their diet. As an example, consider icebergs. For European explorers, icebergs represented potentially fatal danger to be avoided at all costs. For the Inuit of Canada and Greenland, winter icebergs can be a valuable resource when they are frozen into the pack ice. They provide a source of freshwater for people who are far from land. They also provide a place to rest and shelter from the wind. They are used as meeting places or landmarks when traveling on wide expanses of flat sea ice. One remarkable exception to this neglect of Native knowledge concerns American explorer Charles Francis Hall. Hall came to the Arctic to find evidence of what happened to the lost Franklin Expedition. Rather than disparaging Native stories, Hall sought them out. His attitude was summed up in the following statement, "Why not then, be able to ascertain from the same natives — that is, of the same Inuit race — all those particulars so interesting and many of them important to science, concerning the Lost Polar Expedition?" (Hall, 1865: 280).

Hall led two expeditions to find evidence of Franklin's expedition between 1860 and 1869. He lived among the Inuit near Frobisher Bay on Baffin Island and later at Repulse Bay on the Canadian mainland. Inuit guides helped Hall find camps, graves, and relics on the southern coast of King William Island. In 1869, local Inuit took Hall to a shallow grave on King Edward Island containing well-preserved skeletal remains and fragments of clothing. These remains were brought to London and buried beneath the Franklin Memorial at Greenwich Old Royal Naval College.

The Search for the Northwest Passage

For over 400 years, European explorers attempted to find the NWP, from the earliest voyage of John Cabot

(1497) to Roald Amundsen (1906). The search was commercially motivated. Europeans developed an interest in alternative routes to Asia after the Ottoman Empire monopolized major overland trade routes between Europe and Asia (e.g., the Silk Road) in the 15th century.

John Cabot expedition 1497–1498

Originally Giovanni Caboto, Cabot was a Venetian navigator living in England and became the first European to explore the NWP. He sailed from Bristol in May 1497 with a crew of 18 men and made landfall somewhere in the Canadian Maritime islands the following month. Cabot thought he had reached the shores of Asia. King Henry VII authorized a second, larger expedition for Cabot in 1498. This expedition included five ships and 200 men. Cabot and his crew were lost at sea—probably shipwrecked in a severe storm in the North Atlantic.

Jacques Cartier expeditions 1534–41

In 1534, King Francis I of France sent Jacques Cartier to explore the New World in search of riches and a new route to Asia. With two ships and a crew of 61 men, he explored the coast of Newfoundland and the Gulf of St. Lawrence. He discovered Prince Edward Island, but his voyage was well south of the NWP. On Cartier's second voyage, he navigated the St. Lawrence River to modern-day Quebec. Faced with hostile natives (Iroquois) and a crew weakened by scurvy, Cartier returned to France, bringing captured Iroquois chiefs with him. The Iroquois told King Francis I that another great river in their homeland led west to riches and, perhaps, Asia. Cartier made a third voyage to Canada in 1541 that failed to find the NWP. However, unlike most of his fellow seekers of the passage, Cartier returned to France and died aged 66, in his own bed. His explorations formed the basis for later French claims to land in North America.

Robert McClure search expeditions 1850–54

In 1850, Irish Arctic explorer Robert McClure and his crew set sail from England in search of the lost Franklin expedition. In 1854, McClure became the first to traverse the NWP, both by ship and over the sea ice on sleds.

Roald Amundsen expedition 1903–06

Norwegian explorer Roald Amundsen was the first European to successfully traverse the NWP in 1903–06. He did not find an ice-free route across the Canadian Arctic. Rather, he allowed his ship to be frozen into the pack ice and then drift West with the current until he reached the coast of northern Alaska. After a 3-year expedition, Amundsen and his crew, aboard a small fishing ship called Gjøa, reached Nome on Alaska's Pacific coast in 1906. In 1911, Roald Amundsen defeated the British in their quest to be the first to reach the South Pole.

Freeing up the Northwest Passage

Until quite recently, traversing the frozen NWP was an extremely hazardous journey, requiring the avoidance of thousands of giant icebergs rising 100 m above the surface of the water while seeking small channels of open water surrounded by constantly shifting masses of sea ice that could seal the passage and trap ships for months at a time. Through the 20th century, the passage was not a commercially viable shipping route due to the constant danger of sea ice. That has now changed. As discussed in Chapter 1, the amplified warming of the Arctic is causing sea ice to melt, creating greater access to the waters. In the summer of 2007, the entire NWP route was ice-free for the first time in recorded history. Wilson et al. (2020) examined the impacts of sea-ice loss on the Arctic seals and walrus that depend on sea ice as a platform on which to rest and have their young.

Of the seven Arctic species, the walrus, bearded seal, and ringed seal are most ubiquitous, occurring in nearly all of the Arctic sea regions (Fig. 8.2). Ringed seals rely most on stable sea ice to survive. The ice is where they rest, molt, and reproduce. But that ice is disappearing. Seal and walrus populations are widespread in the Arctic (Fig. 8.3), but their future is insecure because of sea-ice loss, increasing industrial and cargo ship traffic, and increasing fishing boat traffic. Traffic through the Arctic sea route has increased in the past decade. In 2012, a record 30 ships made the transit. Crystal Serenity, a luxury cruise ship, made headlines in 2016 when it became the first tourist cruise ship to navigate the NWP.

The Northeast Passage

The Northeast Passage is the eastward shipping route from the Atlantic to the Pacific Ocean, along the Arctic coasts of Norway and Russia (Fig. 8.4) From west to east, the Northeast Passage crosses the Barents Sea, Kara Sea, Laptev Sea, East Siberian Sea, and Chukchi Sea. As with the NWP, the motivation to find and navigate the Northeast Passage was mainly economic. The idea of a possible sea route connecting the Atlantic and Pacific was first proposed by the Russian diplomat Gerasimov in 1525. The western parts of the passage

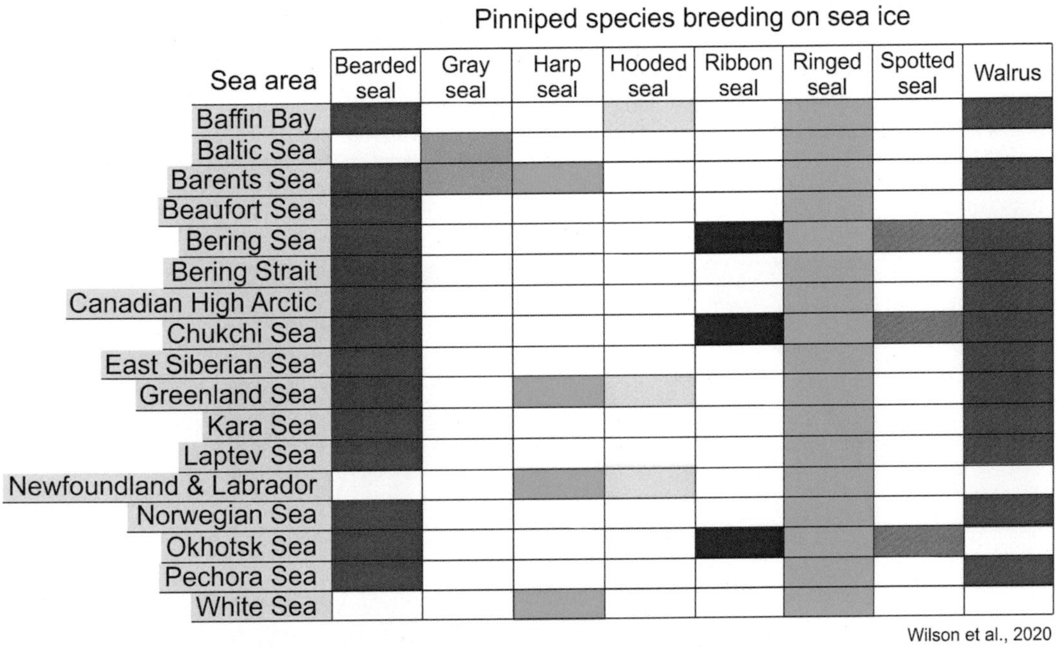

Pinniped species breeding on sea ice

Sea area	Bearded seal	Gray seal	Harp seal	Hooded seal	Ribbon seal	Ringed seal	Spotted seal	Walrus
Baffin Bay	■			■		■		■
Baltic Sea		■				■		
Barents Sea	■		■			■		■
Beaufort Sea	■					■		■
Bering Sea	■				■	■	■	■
Bering Strait	■					■	■	■
Canadian High Arctic	■					■		■
Chukchi Sea	■				■	■	■	■
East Siberian Sea	■					■		■
Greenland Sea	■		■	■		■		
Kara Sea	■					■		■
Laptev Sea	■					■		■
Newfoundland & Labrador	■		■	■		■		
Norwegian Sea	■					■		■
Okhotsk Sea	■				■	■	■	■
Pechora Sea	■					■		■
White Sea	■					■		

Wilson et al., 2020

FIG. 8.2 List of seas where ice breeding seals are found. (After Wilson, S.C., Crawford, I., Trukhanova, I., Dmitrieva, L., Goodman, S.J., 2020. Estimating risk to ice-breeding pinnipeds from shipping in Arctic and sub-Arctic seas. Marine Policy 111, 103694. https://doi.org/10.1016/j.marpol.2019.103694.)

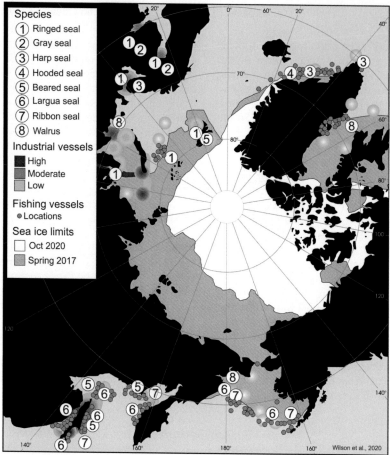

FIG. 8.3 Map of the high northern latitudes, showing regions occupied by eight species of native Arctic pinnipeds. Regions marked in red, amber, and green indicate high, medium, and low levels of industrial vessel traffic. *Blue dots* represent regions with substantial numbers of fishing vessels. (Modified from Wilson, S., Crawford, I., Trukhanova, I., Dmitrieva, L., Goodman, S., 2020. Estimating risk to ice-breeding pinnipeds from shipping in Arctic and sub-Arctic seas. Marine Policy 111, 103694. https://doi.org/10.1016/j.marpol.2019.1036945.)

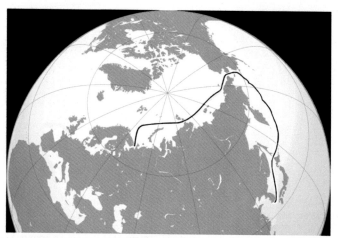

FIG. 8.4 Map of the Northeast Passage by Collin Knopp-Schwyn and Turkish Flame. Creative Commons Attribution 4.0 International license. (Modified from Knopp-Schwyn, C., 2020. Map of the Northeast Passage. https://en.wikipedia.org/wiki/Northeast_Passage.)

were explored by northern European countries such as England, the Netherlands, Denmark, and Norway, looking for an alternative seaway to China and India. Although these expeditions failed, new coasts and islands were discovered.

The Northeast passage

Dutch navigator Willem Barents, after whom the Barents Sea is named, led three voyages in search of a Northeast Passage. During these voyages, he discovered Spitsbergen and Bear Island. His was also the first European voyage into the Kara Sea. On his last voyage in 1596, he and his crew explored the northern coast of Novaya Zemlya. Barents and his second-in-command Heemskerck were 81°N at their highest latitude, beyond any point previously reached. By November, however, the ice had grown thick, and it finally imprisoned the ship. Still close to Novaya Zemlya, realizing that they must build a solid shelter ashore to survive, they made one of logs and driftwood and moved into this "Safe House" in October (Fig. 8.5). They lived there until June 1597, suffering but at first in good spirits, calling themselves "burghers of Novaya Zemlya." Conditions deteriorated when the firewood gave out and their ship was finally crushed by the pack ice. The men began to construct two small boats. They provisioned the longboats as best they could and started on the 2575-km journey home. Sadly, Barents did not survive the trip and died at sea, succumbing to scurvy, which had weakened everyone on the expedition. Barents died at the end of June, soon after asking for a friend to lift his head up for a final look at Novaya Zemlya. Heemskerck and the other survivors reached the Kola Peninsula. Here they were rescued by the third captain of the expedition, Rijp, who had argued with Barents, returned to Holland, and came back for trade.

Later expeditions

By the 17th century, traders had established a continuous sea route from Archangel (Arkhangelsk) to the Yamal Peninsula, where they were forced to portage their boats to the Gulf of Ob. This route, known as the Mangazeya seaway, after its eastern terminus, the trade depot of Mangazeya, was an early precursor to the Northern Sea Route (NSR). East of the Yamal, the route north of the Taimyr Peninsula, proved impossible or impractical. East of the Taimyr, from the 1630s, Russians began to sail the Arctic coast from the mouth of the Lena River to a point beyond the mouth of the Kolyma River. A series of attempts to complete the Northeast Passage were made in the 18th and 19th centuries, including expeditions led by Vitus Bering in 1725—30, under the patronage of Russian czar Peter the Great. Both Vitus Bering (in 1728) and James Cook (in 1778) entered the Bering Strait from the south and sailed some distance northwest, but until 1879 no one is known to have sailed eastward between the mouth of the Kolyma River and Bering Strait.

FIG. 8.5 The hut in which Willem Barents and his crew overwintered on Novaya Zemlya in 1596—97. Author and date unknown. In the public domain.

Nordenskiöld's voyage 1879—79

In 1878, Adolf Nordenskiöld led an expedition that sailed around the north coast of Asia, exiting by way of the Bering Strait. His was the first voyage to complete the sailing of the entire Northeast Passage. This he accomplished in his ship the *Vega*. Starting from Karlskrona on June 22, 1878, the *Vega rounded Cape Chelyuskin* the following August, and was frozen in at the end of September near Bering Strait. Nordenskiöld successfully completed the voyage the following summer.

SEA-ICE GROWTH AND MELTING SEASONS

The processes of sea ice formation and melting have been succinctly described by the NSIDC (2020c and 2020d). As ocean water begins to freeze, frazil forms. This is composed of small needle-like ice crystals. These crystals are typically 3—4 mm in diameter. Because salt does not freeze, the crystals expel salt into the water; thus, frazil crystals consist of nearly pure freshwater. Sheets of sea ice form when frazil crystals float to the surface, accumulate, and bond together. Depending upon the climatic conditions, sheets can develop from grease and congelation ice, or from pancake ice. In calm waters, frazil crystals form a smooth, thin form of ice, called grease ice for its resemblance to an oil slick. Each of these types of sea ice has a set of distinguishing characteristics, as described by Timco and Johnston (2003) (Table 8.1). Grease ice develops into a continuous, thin sheet of ice called nilas. Initially, the sheet is very thin and dark, becoming lighter as it thickens. Currents or light winds often push the sheets of nilas around so that they slide over each other, a process called rafting. Eventually, the ice thickens into a more stable sheet called congelation ice. This has a smooth bottom surface. Frazil ice cannot form in the relatively still waters under sea ice, so only congelation ice developing under the ice sheet can contribute to the continued growth of a congelation ice sheet. Congelation ice crystals are long and vertical because they grow much slower than frazil ice. If the ocean is rough, the frazil crystals accumulate into slushy disks, called pancake ice, because of their shape. A signature feature of pancake ice is raised edges or ridges on the perimeter, caused by the pancakes bumping into each other in the ocean waves. If the motion is sufficiently strong, rafting occurs. If the ice is thick, ridging occurs, where the sea ice bends or fractures and piles on top of itself, forming lines of ridges on the surface. Each ridge has a corresponding structure, called a keel, that forms on the underside of the ice. Ridges up to 20 m (60 feet) thick can form in the Arctic when thick ice deforms. Eventually, the pancakes freeze together and consolidate into a coherent ice sheet. Unlike the congelation process, sheet ice formed from consolidated pancakes has a rough bottom surface. Once sea ice forms into sheet ice, it continues to grow through the winter.

When temperatures increase in spring and summer, the first-year ice begins to melt. If the ice does not thicken sufficiently during the winter, it will completely melt during the summer. If the ice grows enough during the winter, it thins during the summer but does not completely melt. In this case, it remains until the following winter, when it grows and thickens and is classified as multiyear ice. The growth and preservation of multiyear sea ice in the Arctic is becoming very rare. First-year ice dominates most parts of the Arctic Ocean. Sea ice melts during the summer when solar radiation heats the ice surface. The amount of solar radiation absorbed by the ice depends on its surface albedo. Sea ice reflects about half of the insolation (incoming solar radiation). This reflection prevents the ice from warming up as quickly as open water, but insolation nonetheless heats the ice sufficiently to initiate melting. Thick sea ice covered with snow reflects even more insolation, approximately 90%. As a result, sea ice with snow takes even longer to melt. After the snow starts to melt, melt ponds form. As the melt ponds grow and deepen, the albedo continues to decrease, leading to higher absorption of solar radiation and an increased rate of melting. Energy to melt ice can come from sources besides direct solar energy. Water that is under the ice and that has a temperature above the freezing point causes the ice to melt from the bottom up. Warm surface waters cause the edges of the ice to melt, particularly in leads and polynyas.

MODERN SHIPPING IN THE ARCTIC

Those who think long and hard about the Arctic Ocean, admittedly not a huge number of people, are pretty much divided into two camps. There are conservationists who look on the Arctic Ocean as a precious ecosystem to be preserved. Standing with them are climatologists who warn that the melting of polar ice and snow will send shock waves throughout the global climate system. On the other hand, there are industrialists and shipping magnates who are starting to look on the Arctic Ocean as a developing resource to be exploited. If the loss of sea ice is considered a tragedy to conservationists, it is considered a boon to the shipping industry. Arctic seas never have had much of a role in global commerce until quite recently. The ice-covered oceans of the North were remote and dangerous. The endlessly shifting pack ice was liable to crush the ships

TABLE 8.1
Stages of Development of Sea Ice.

Sea Ice Type	Abbreviation	Thickness (cm)	Distinguishing Characteristics
New ice	(N)	<10 cm	Sea ice that is in early stages of formation. This ice forms in small platelets or lumps, or ice in a soupy-looking layer. New ice is usually subdivided into frazil, grease ice, slush, or shuga.
Nilas	NI	10 to 30	A thin elastic crust of floating ice that easily bends from waves and swells. It has a matte surface appearance.
Young ice	YN		
Gray young ice	G	0 to 15	Gray ice is less elastic than nilas. It often breaks from swells. In a compression regime, gray ice will raft.
Gray-white young ice	GW	15 to 30	Gray-white sea ice is more rigid. In a compression regime, gray-white ice will ridge.
Thin first-year ice	FY	30 to 70	Sea ice that grows from young ice. This is separated into stage 1 (30–50 cm thick) and stage 2 (50–70 cm thick)
Medium first-year ice	MFY	70 to 120	First-year ice that is 70–120 cm thick
Thick first-year ice	TFY	>120	First-year ice that is >120 cm thick
Second-year ice	SY	Variable	Sea ice that has survived one melt season. It stands higher out of the water than first-year ice. Summer melting has often smoothed and rounded it. Melt water puddles in the summer are often greenish-blue.
Multiyear ice	MY	Variable	Sea ice that has survived more than two melt seasons. This ice can be more than 3m thick and is very strong. It has a characteristic bluish color and a weathered, undulating surface.
Glacial ice		Variable	Ice that has originated from a glacier that calved into the sea.

Stages of sea ice decay, data from Timco and Johnston (2003).

Condition	Season	Description
No melt	Winter	Sea ice intact
Snow melt	Spring	Snow on surface of sea ice melts away
Ponding	Spring and summer	Meltwater from snow and sea ice pools on the sea ice surface
Thaw holes	Spring and summer	Meltwater ponds penetrate to bottom of sea ice, speeding the melting process
Rotten ice	Summer	Remaining ice cover collapses

Data from Timco, G., Johnston, M., 2003. Arctic Ice Regime Shipping System: Pictorial Guide. Canadian Hydraulics Centre, National Research Council of Canada and Transport Canada.

that were sent there. Also, shipping companies have little reason to visit the Arctic for its own sake. There were no large towns or cities where mariners could engage in trade. The small groups of Arctic natives were largely nomadic, and remarkably self-sufficient. So the Arctic was never a destination for the shipping industry. Sailing in and around the Arctic, even during the height

of summer, was simply a means to an end, namely, to get to some other parts of the world more quickly. Until recent decades, only the fringes of the Arctic Ocean were sufficiently safe for navigation, and then only in the summer. But now that the sea ice is retreating, the Arctic Ocean is becoming progressively safer for mariners. Ironically, the same profit-driven world economy that has produced enormous quantities of greenhouse gases hopes to profit from that atmospheric pollution as the driver of Arctic warming. In other words, they hope to take advantage of the disappearance of sea ice to increase access to shipping routes across the Arctic. The nature of shipping in the Arctic is rapidly evolving, chiefly due to the progressive loss of sea-ice cover. The ice-free season is growing in length, and the thickness of the ice has decreased throughout most of the Arctic Basin. Until the 21st century, most Arctic shipping was restricted to a few months of the summer and only ships with the most robust, heavily reinforced hulls attempted lengthy Arctic voyages. Even these reinforced ships often needed icebreaker escorts to safely get across large stretches of Arctic water (Table 8.2). That has changed in recent years (Fig. 8.6). With the threat of ice being reduced year by year, shipping companies are starting to make plans for expanded use of the NWP, the northeast passage, and even a route directly across the North Pole. A research paper by Reeves et al. (2014) summarizes the current state if knowledge of the distribution and movement patterns of the three ice-associated whale species that reside year-round in the Arctic, the narwhal (*Monodon monoceros*) (Fig. 8.7), beluga (*Delphinapterus leucas*) (Fig. 8.8), and bowhead whale (*Balaena mysticetus*) (Fig. 8.9). It maps their current distribution and identifies areas of seasonal aggregation, particularly focusing on high-density occurrences during the summer. Routes used for commercial shipping in the Arctic (Fig. 8.10) are compared with the distribution patterns of the whales, with the aim of highlighting areas of special concern for conservation. The authors made a series of recommendations on ways of keeping whales safe in shipping lanes. Measures should be taken to mitigate the impacts of human activities on these Arctic whales and the Native peoples who depend on them for subsistence. These measures include (1) careful planning of ship traffic lanes (rerouting if necessary); (2) ship speed restrictions in high-traffic marine mammal areas; (3) temporal or spatial closures of specified areas. These include areas where critical whale activities such as calving, calf rearing, resting, or intense feeding take place; (4) strict regulation of seismic surveys and other sources of loud underwater noise; and (5) close and sustained monitoring of whale populations to track their responses to environmental disturbance.

Longer Operational Windows, Shorter Transit times

Khon et al. (2010) compared the simulation of Arctic sea ice concentration in 1980–99 from 21 global climate models with remote sensing data. By the end of the 21st century, under the global greenhouse emission A1B scenario, they predicted that the length of time for safe navigation of the Arctic Northeast Passage would increase from 49 days per year in 1997–2008 to about 100 days per year in 2080–99. Between 1981 and 2010, Arctic sea ice extent remained above 8 million km^2 except for a few weeks from August to early October (Fig. 8.6). The summer of 2012 was an exceptionally warm summer and sea ice cover dropped well below 4 million km^2, with the open water season lasting from mid-July to late October (NSIDC, 2020a). The year 2020 has seen an exceptionally long open water season. Open waters began in early July and remained open past the end of October in many regions, notably the Laptev Sea (Fig. 8.11). The average navigation time for the Arctic Northeast Passage (North Sea Route) shrank from 20 days in the 1990s to 11 days during 2012–13. With the help of model outputs, Melia et al. (2016) estimated that European routes to Asia will become 10 days faster via the Arctic Northeast Passage than the alternatives by midcentury, and 13 days faster by late in the century. While transit of the NWP will become 4 days faster, Liu et al. (2017) estimated the navigable windows of the various legs of the NWP (Fig. 8.12). Three legs of the voyage are currently safe to navigate only in the 50-day window from August to the first week in October.

Table 8.3 tallies the number of voyages using the North Sea Route from 2016 to 20, based on statistics kept by the Northern Sea Route Information Office (2020). The number of voyages has consistently increased, with 1705 voyages in 2016 building to 2694 in 2019. For the North Pole Route, Aksenov et al. (2017) predicted that during the 2030s, only four types of vessels (CAC 3–4 and types A–B) will be able to safely traverse the North Pole Route, while during the 2050s, all seven types of vessels (CAC 3–4 and types A–E) will likely be able to safety traverse this route (Zhang et al., 2019).

Expansion of Other Kinds of Vessels

Hong (2012) assessed the evolving opportunities for various kinds of vessels to use Arctic waters. First, Arctic warming is transforming the fishing industry. As sea-ice margins retreat, open waters are expanding, and the fishing industry is moving into these new regions. Some fishers are already beginning to move up the western Greenland coast to pursue both turbot and shrimp. The expansion of open water and lengthening of the ice-free season is also providing access to new cruise ship destinations. Cruise vessels have increased their visits

TABLE 8.2
Ship Ice Class Systems.

Ship Type	Canadian Classification	Polar Class	Maximum Allowable Ice Type	Allowable Ice Thickness (cm)	NUMBER IN NATIONAL FLEET							
					Russia	Finland	Canada	Sweden	Kazakhstan	United States	Denmark	China
Icebreaker	CAC1	1	No limit	No limit	26	6	2	5	5	4	4	2
Icebreaking commercial ship	CAC2	2	Multiyear	No limit	2		1	2				
Commercial cargo ship with heavily reinforced hull	CAC3	3	Second-year	No limit			1			1		1
Commercial cargo ship with reinforced hull	CAC4	4	Thick first-year	≤120	4	4	3					
Type A		5	Medium first-year	70–20								
Type B		6	Thin first-year (stage 2)	50–70	9							
Type C		7	Thin first-year (stage 1)	30–50	5							
Type D			Open water/gray	10–15								

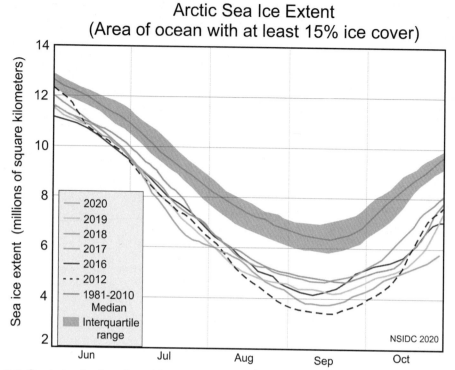

FIG. 8.6 Graph showing the extent of Arctic sea ice (area of ocean with at least 15% ice cover). Gray line and the surrounding gray area represent the median values and one standard deviation from 1981 to 2010. (From NSIDC, 2020. Arctic sea ice news and analysis, November 4, 2020. http://nsidc.org/arcticseaicenews/.)

to the Arctic region, and it seems likely that the novel experience of cruising in the Arctic will continue to increase. There are two types of cruise destinations in northern North America: the Alaska panhandle route and northern Canada. For instance, in 2016, the cruise ship *Crystal Serenity* made an unprecedented voyage through the NWP. It was the largest cruise ship to ever pass through the NWP. The ship has a gross tonnage of 68,870; length of 250 m beam of 32.2 m, and a draught of 8 m. The cruise ship is rated ice class: 1C, so it had to be escorted by a Polar-Grade ship that could break the ice through which the cruise ship could pass. Crystal Cruises proudly took the *Serenity* through the NWP on multiple voyages for "true explorers" and is expecting the delivery of a Polar-Grade ship for more cruises (Maddox, 2019). During the 1960s and 1970s, nuclear-powered submarines of the American, Soviet/ Russian, British, and French navies have cruised Arctic waters, diving beneath the pack ice as needed. They are the only type of vessel that can enter High Arctic waters year-round. After the Cold War, the number of submarines operating in these waters has significantly decreased. However, naval interest in operating surface

vessels in Arctic waters has increased. Canada and the United States have stated their intentions to improve their Arctic surface vessel fleets. Thus, steps are now being taken that will result in an increasing surface naval presence in Arctic waters.

The North Sea Route

Askenov et al. (2017) summarized the history and potential future of shipping via the Northeast Passage, which Russians call the NSR. Arctic summer sea ice reduction has accelerated in the past 10 years, leading to debates in the shipping industry on the viability of increased cargo shipping on this route. Average sailing times on the North Sea Route have decreased 20 days in the 1990s to 11 days in 2012—13, because of the easing of sea-ice conditions along the Siberian coast. However, the use of this shipping route still carries substantial risks, which Askenov et al. (2017) explore in their article. They made detailed projections of ocean and sea ice conditions along the Siberian coast, forecasting conditions to the end of the 21st century. Their model used the worst-case IPCC emission scenario, RCP8.5. Their results show that in future summers, large

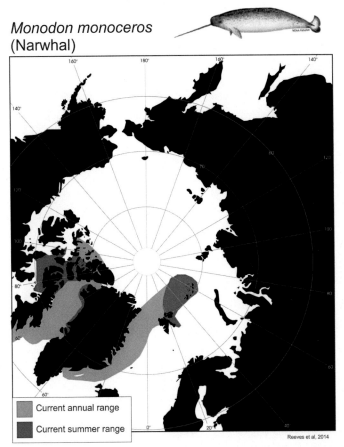

Monodon monoceros
(Narwhal)

FIG. 8.7 Modern range of Narwhal. (Modified from Reeves, R., Ewins, P., Agbayani, S., Heide-Jørgensen, M., Kovacs, K., 2014. Distribution of endemic cetaceans in relation to hydrocarbon development and commercial shipping in a warming Arctic. Marine Policy 44, 375–389. https://doi.org/10.1016/j.marpol.2013.10.005.)

areas of open water will develop in the Arctic Ocean, in regions previously covered by pack ice. The future Arctic Ocean features more fragmented, thinner sea ice, stronger winds, ocean currents, and waves. They also considered the viability of the shipping route across the North Pole. According to their model, by the mid-21st century, North Pole route summer season sailing times are estimated to be 13–17 days, making this route as fast as the NSR. Fuglestvedt et al. (2014) concur that the savings and time and fuel for the NSR would be substantial when compared with southern roots. Sailing routes between Europe and East Asian ports through the Arctic Ocean along the NSR are about 11,100 km shorter than the routes around the Cape of Good Hope and are about 5000 km shorter than the Europe to East Asia routes via Suez Canal. The NSR route is also shorter than the Panama Canal route by about 10,000 km. The

use of the NSR also eliminates the cost of using these canals. The cost for an oil tanker using either the Suez Canal or the Panama Canal is currently about $450,000 per trip. The savings would be even greater for the megaships that are too large to pass through the Panama and Suez Canals and so are currently forced to sail around the Cape of Good Hope and Cape Horn.

Sea-ice Dynamics and Models

According to simulations run by Fuglestvedt et al. (2014), before the 2030s sea-ice conditions will be the principal factor affecting navigation and ship safety on the NSR. Currently, mariners planning to transit the NSR must rely on sea-ice models that rely on static information such as sea-ice concentration and sea-ice types. Such estimates are adequate for the navigation of pack ice areas (albeit, with the aid of an icebreaker).

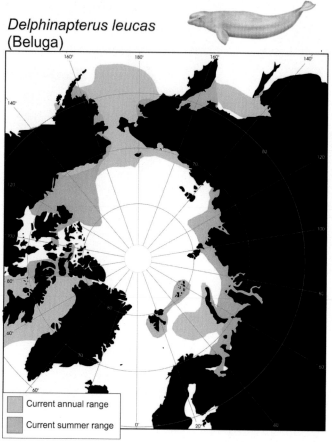

Delphinapterus leucas
(Beluga)

☐ Current annual range
☐ Current summer range

FIG. 8.8 Modern range of Beluga whales. (From Reeves, R., Ewins, P., Agbayani, S., Heide-Jørgensen, M., Kovacs, K., 2014. Distribution of endemic cetaceans in relation to hydrocarbon development and commercial shipping in a warming Arctic. Marine Policy 44, 375–389. https://doi.org/10.1016/j.marpol.2013.10.005.)

However, sea ice drift has become much more rapid in the last decade, and mariners now need models that account for sea-ice dynamics. These dynamics include such factors as ice drift and ice internal pressure due to sea-ice convergence and compression. An analysis of shipping accidents on the NSR occurring between the 1930s and 1990s concluded that ice drift and compression caused about half of the shipwrecks. Ice jets were a particular danger. These are rapid sea ice flows between drifting and land-fast ice. They are typically generated by storm surges and constitute the most dangerous hazard to shipping along the NSR. A route optimization algorithm is needed to estimate sailing times and accessibility projections for more detailed short-term forecasting. Such an algorithm should be extended by developing a route optimization tool to estimate the fastest trans-Arctic route under various ice conditions and the season of the year. Optimal route simulations should come about more easily because of recent developments in coupled atmosphere-ocean general circulation models (GCMs). These GCMs are continually adding more advanced models of sea ice, ocean, and atmospheric physics. Of these elements, shipping forecasts are most in need of ice ridging and ice thermal decay projections. After the 2030s, the rapid decline in sea-ice cover and the development of more extensive marginal ice zone type sea-ice provinces will necessitate new approaches to forecasting. The new transport and accessibility models will require forecasts of winds, currents, and waves. More detailed sea-ice data will also be required, such as sea-ice floe sizes and ice drift parameters. Definitions of sea-ice mechanical properties may need revisiting, as sea ice will be thinner, more saline, and weaker.

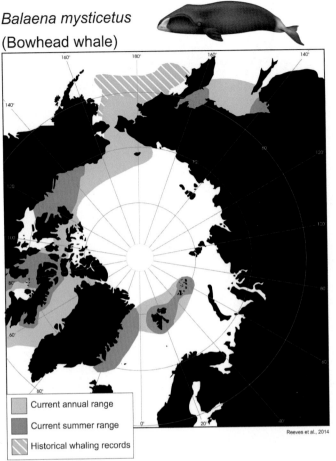

FIG. 8.9 Modern range of bowhead whale. (From Reeves, R., Ewins, P., Agbayani, S., Heide-Jørgensen, M., Kovacs, K., 2014. Distribution of endemic cetaceans in relation to hydrocarbon development and commercial shipping in a warming Arctic. Marine Policy 44, 375—389. https://doi.org/10.1016/j.marpol.2013.10.005.)

Strengths and Weaknesses of the North Sea Route

The development of shipping activities along the NSR for transiting vessels and those operating in the Arctic has recently been subject to much attention (Faury and Cariou, 2016). The cooperation among Arctic countries since the end of the cold war has allowed the development of this economic area, facilitating transiting along this route. Another boost has come from the decline in sea-ice extent due to climate change. Finally, the scientific and technical advancements in shipping navigation have reduced the unpredictability of sailing conditions along this route. The use of the Russian NSR brings a 40% reduction in sailing distance and a saving of 6.5—14 transit days compared with the Suez Canal Route. Despite these advantages, the climatic, technical, and economic conditions of the NSR limit the number of transits. For instance, only 124 transits took place in 2013 and 2014, combined. These numbers are extremely low compared with the 15,917 oil tankers that transited via the Suez Canal in 2013. Table 8.4 provides estimates of the potential savings to be had by tankers of various classes to take either the Northeast Passage or NWP, rather than using the Suez Canal. The continued unpredictability of the weather and sailing conditions along the NSR contribute significantly to the relatively low competitiveness of this route. This, despite a general decrease in sea-ice thickness and extent, facilitates increased vessel speeds and decreased costs. Another limitation of this route is its bathymetry. Areas of shallow water limit the size of vessels deployed on this route.

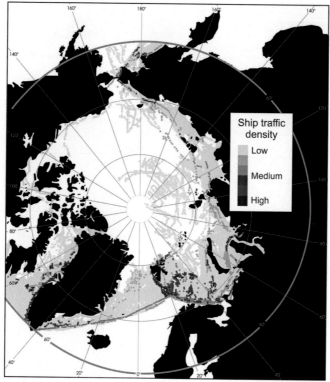

FIG. 8.10 Arctic shipping in 2012, including the activity of all vessels with AIS transponders onboard. Ship traffic density colored as follows: Yellow, low; red, medium; blue, high. Data from the Norwegian Coastal Administration and Det Norske Veritas. (Modified from Reeves, R., Ewins, P., Agbayani, S., Heide-Jørgensen, M., Kovacs, K., 2014. Distribution of endemic cetaceans in relation to hydrocarbon development and commercial shipping in a warming Arctic. Marine Policy 44, 375–389. https://doi.org/10.1016/j.marpol.2013.10.005.)

Faury and Cariou (2016) applied a model to a specific NSR route and two vessel types. They found that the NSR is commercially viable from July to November, even with conservative assumptions on ice thickness.

SAFETY AND AFFORDABILITY OF TRANS-ARCTIC SHIPPING

As discussed by Fuglestvedt et al. (2014), the economic viability of Arctic shipping is still being debated. The challenges involve a balancing act between economic drivers in Europe, the Americas, and far-eastern countries, such as China, Malaysia, Singapore, Taiwan, the economic risks of high latitude navigation, and the costs of Arctic infrastructure development. If cargo ships and oil tankers are going to make major use of Arctic sea routes, they will need to have access to modern port facilities, warehouse space, and roads. Such facilities are extremely scarce in Arctic villages of today, and it

is not at all clear that the inhabitants of those villages want this kind of development in their "backyard." Some people involved in the shipping industry also take an interest in the impact of large-scale shipping on Arctic communities and ecosystems (both marine and terrestrial).

Fuel Savings and CO_2 Emission Reductions

The exploitation of shipping routes in the Arctic Ocean should reduce the length of voyages between Europe and Asia by about 40%, saving fuel and reducing CO_2 emissions. Schøyen and Bråthen (2011) analyzed the reduction in sailing time, fuel, and CO_2 emission savings for two types of bulk cargo vessels sailing along the NSR, in comparison with travel via the Suez Canal. They concluded that the main advantage of shipping operations using an ice-free NSR would be the reduction of sailing time, cutting fuel use by more than 50%, and reducing CO_2 emissions by 49%–78%. This

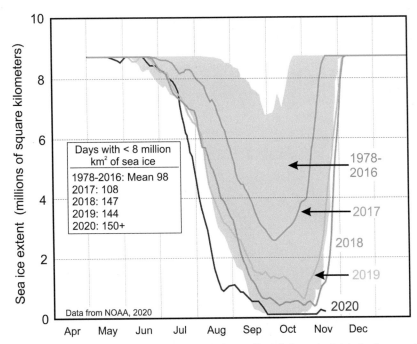

FIG. 8.11 Arctic sea ice extent as of October 4, 2020, along with daily ice extent data for four previous years and the record low year. 2020 is shown in red, 2019 in yellow, 2018 in gold, 2017 in blue. The 1978 to 2016 median is in dark gray. The gray areas around the median line show the interquartile range of the data. (Modified from NSIDC, 2020. Arctic Sea Ice News and Analysis - Lingering Seashore Days. https://nsidc.org/arcticseaicenews/2020/10/.)

route would be of primary interest for bulk shipping. While this might be the case for individual cargo ships, Schøyen and Bråthen (2011) warned that this would not necessarily be the case for liner shipping. They gave several reasons for caution, including the uncertainty of schedule reliability for the NSR, and added costs because of other factors. They specified higher construction costs for ice-classed ships, service irregularity and slower speeds, navigation difficulties, greater safety risks, and, most importantly, fees for icebreaker services. The recent six-day blockage of the Suez Canal by the *Ever Given*, one of the world's largest container ships, cost an estimated $400 million per hour in international trade. The accident showed the vulnerability of this canal to blockages by today's giant ships. The Kremlin is hoping to use this shipping catastrophe to promote the NSR.

Insuring Arctic Voyages

The complexities of insuring ships sailing in Arctic waters are discussed by Saul et al. (2020). As climate change opens new sea routes, shipping companies are looking for captains with lengthy experience in Arctic

waters. Such experience, mainly dealing with sea-ice conditions, may serve to lower the price of insurance for such voyages. Nevertheless, many risks accompany such voyages, especially when ships begin using new routes, such as the transpolar route. The insurance companies must deal with a fundamental question: If something goes wrong, who pays? So far, it is unclear that the cost of a major accident would be completely covered by insurance. Damages from a ship spilling oil, hitting an iceberg, or becoming marooned can run into the hundreds of millions of dollars. Helle Hammer is chair of the policy forum with the International Union of Marine Insurance (IUMI), the leading association for the global marine insurance market. She points out that "It's all very much new territory" (Saul et al., 2020). Without years' worth of data on the number of casualties, accidents, collisions, or oil spills, she said, "it's impossible to do the risk modeling." More than 10 insurance companies or brokers have complained that they still have too little knowledge of the region to resolve liability questions. According to IUMI data, most marine insurers have paid out more for ship damage than they have collected in insurance premiums in

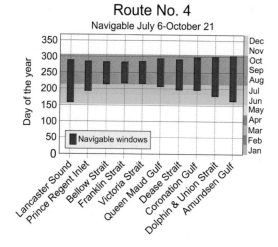

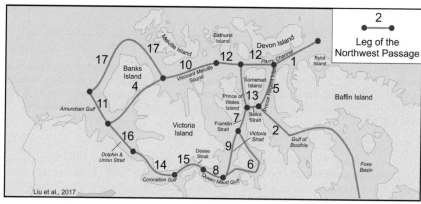

FIG. 8.12 Above: The navigable windows of route No. 4 of the NWP and its legs. The Y-axis refers to the day of the year (left side) and corresponding months of the year (right side). Below: The legs of NWP voyages. (Modified from Liu, X., Ma, L., Wang, J., Wang, Y., Wang, L., 2017. Navigable windows of the Northwest Passage. Polar Science 13, 91—99. https://doi.org/10.1016/j.polar.2017.02.001.)

the past few years. Some marine insurers are quitting the market completely. There is little appetite to underwrite risks in the Arctic market among marine insurance companies. And without insurance coverage, no one is going to send a freighter carrying millions of dollars of goods across Arctic seas.

Shipwrecks

Despite all safety precautions, there will always be shipwrecks, and most of them occur in relatively familiar waters. However, since navigation in and around the Arctic Ocean has only been possible in recent decades, only about 6% of the Arctic Ocean is charted. No wonder the insurance companies are nervous. Smaller ships are less costly to insure because their greater maneuverability makes it easier to avoid such dangerous obstacles as icebergs. On the other hand, cargo ships are much more at risk of collision with ice. Complete with cargo, crew, fuel, and ballast water, such a ship is much harder to maneuver or tow than a trawler. At the top of the scale, a loaded supertanker could take as much as 15 min to come to a full stop, in the meantime covering 3 km of ocean. Turning to avoid collisions is likewise extremely cumbersome for these huge vessels. Many have turning diameters of about 2 km. So far, the most common Arctic shipping problems have not involved hitting icebergs. Much more typical is the problem of equipment freezing and seizing up. Out of 512 incidents reported through 2019, machinery damage or failure accounted for

TABLE 8.3
The Number of Voyages Using the North Sea Route, 2016–20.

Year	Month	No. of Voyages	Year	Month	No. of Voyages	Year	Month	No. of Voyages	Year	Month	No. of Voyages	Year	Month	No. of Voyages
2016	Jan	58	2017	Jan	65	2018	Jan	93	2019	Jan	140	2020	Jan	157
	Feb	61		Feb	75		Feb	92		Feb	131		Feb	156
	Mar	63		Mar	93		Mar	93		Mar	152		Mar	169
	Apr	71		Apr	90		Apr	104		Apr	160		Apr	171
	May	65		May	100		May	97		May	153		May	156
	Jun	87		Jun	107		Jun	93		Jun	126		Jun	14
	Jul	221		Jul	261		Jul	205		Jul	292		Jul	299
	Aug	262		Aug	339		Aug	344		Aug	439		Aug	448
	Sep	345		Sep	410		Sep	399		Sep	447		Sep	N/A
	Oct	261		Oct	192		Oct	256		Oct	346		Oct	N/A
	Nov	117		Nov	101		Nov	123		Nov	159		Nov	N/A
	Dec	94		Dec	74		Dec	123		Dec	149		Dec	N/A
Totals		1705			1907	2022		2694			2694			2671 (estimated)

Data from Northern Sea Route Information Office, 2020. NSR Shipping Traffic - Activities in January—June 2020. https://arctic-lio.com/nsr-shipping-traffic-activities-in-january-june-2020/.

TABLE 8.4
Tanker Cargo Sizes and Draughts When Loaded.

Ship Size	Ship Size in This Study (DWT Metric Tons)	Cargo Size in This Study (metric tons)	Overall Length (m)	Maximum Speed (Knots)	Draught when Loaded (meters)	Operating and Capital Costs per day ($US) (Non–ice-Class Vessel)	Time and Financial Savings Using the Northeast Passage	Time and Financial Savings Using the Northwest Passage
MR—Handymax	50,000	35,000	183	16	10	16,020	10 days; $160,200	10 days; $160,200
LR1—Panamax	75,000	60,000	225	16	12.5	18,454	10 days; $184,540	10 days; $184,540
LR2—Aframax	115,000	90,000	250	16	13	19,938	10 days; $199,380	10 days; $199,380
LR3—Suezmax	150,000	120,000	274	16	14.2	23,353	10 days; $233, 530	10 days; $233, 530

Data from Theocharis, D., Sanchez Rodrigues, V., Pettit, S., Haider, J., 2019. Feasibility of the Northern Sea Route: The role of distance, fuel prices, ice breaking fees and ship size for the product tanker market. Transportation Research Part E 129, 111–135. https://doi.org/10.1016/j.tre.2019.07.003.

almost half, according to a 2020 shipping and safety report by insurer Allianz Group (Saul et al., 2020). Other incidents included a crack in the hull, onboard explosions, and sinking. A heavy burden for insurance companies is the cost of towing damaged ships back to port from remote locations. Tugboat fees may amount to millions of dollars, far exceeding the cost of ship repairs in some cases. Accordingly, insurance companies that cover Arctic voyages add up to a 40% surcharge to their basic premiums, which range from $50,000 to $125,000 per Arctic voyage, depending on the class of ship, the chosen route, and the proximity of an icebreaker. In 2020, Swedish icebreaker fees were about 900 Euros ($1050 US) per hour. One example of an Arctic ship disaster occurred in 2010 when a passenger vessel, the *Clipper Adventurer*, ran aground in 3 m water in the Canadian Arctic carrying around 200 passengers and crew. The cruise ship operator claimed that the ship ran aground on an "uncharted rock." Fortunately, a Canadian Coast Guard mapping vessel responded to the radioed SOS signal. The Coast Guard vessel was just 500 nautical miles away (a relatively small distance by Arctic standards) and arrived some 40 h later. Fortunately, the sea was calm, and the Coast Guard crew was able to get all the passengers safely off the *Clipper Adventurer*. The ship had to return to Europe for repairs. The Canadian

government sued the ship's owners for costs incurred for pollution control and was awarded some $445,000. Four salvage vessels were employed to refloat the vessel, which was subsequently towed to Poland for repairs. The cost of temporary and permanent repairs, payment to the salvors, business interruption, and related matters resulted in damages of US$13.5 million. The UN's International Maritime Organization (IMO) introduced new standards for Arctic navigation in 2017, including ship design and equipment, search and rescue protocols, and special training for captains.

Complexities of Ship Emissions

While the shift to shorter Arctic shipping routes may decrease fuel use and lower CO_2 emissions, there are other climate impacts to be considered. The use of Arctic shipping routes will lead to increased emissions of non-CO_2 gases, aerosols, and particulates into the Arctic atmosphere. These other pollutants change regional radiative forcing. For instance, the incomplete burning of ship fuel contributes to the deposition of black carbon on sea ice and snow, producing more complex regional warming/cooling effects. Modeling of these aspects of Arctic shipping suggests that they may initially cause a net global warming effect before the long-term decrease in CO_2 emissions takes effect and climates begin to cool (Fuglestvedt et al., 2014.)

STANDARDS OF SHIP SAFETY IN ICY WATERS

The Polar Code was recently implemented by the IMO to improve the safety of Arctic vessel traffic. The Polar Code applies to all ships certified under SOLAS (International Convention for the Safety of Life at Sea), which includes cargo vessels 500 gross tons or more, and all passenger vessels with greater than 12 passengers. The Polar Code does not apply to pleasure craft, fishing vessels, military vessels, and any other vessels not covered by SOLAS. The Polar Code strives to ensure that vessels traveling in the Arctic meet certain standards and make appropriate voyage plans. The code states that mariners should take into account current information and measures to be taken, relevant routing systems, speed recommendations, and vessel traffic services relating to areas with higher densities of marine mammals, including seasonal migration areas. Mariners are to follow national and international laws and guidelines related to reducing the impacts of vessels on marine mammals.

Arctic Ice Regime Shipping System

Transport Canada has implemented a Regulatory Standard that is intended to minimize the risk of pollution in the Arctic due to damage of vessels by ice. It is called "AIRSS," for Arctic Ice Regime Shipping System. AIRSS combines information on the ice conditions and a ship's capability to assess the potential for ice damage. It uses the concept of an "ice regime" to characterize the ice. An ice regime is a region of ice with more-or-less consistent ice conditions. From a navigation standpoint, it is an area in which there is the likelihood of encountering certain ice types while maintaining a constant mode of navigation. Basically, the ice regime is the ice that the ship will likely encounter. The ice regime considers several important factors of the ice: concentration, thickness, age, state of decay, and roughness. AIRSS also takes into account a vessel's ability to travel safely in all types of ice conditions. Because different vessels have different capabilities in ice-covered waters, each vessel is assessed and assigned to a vessel class. This rating reflects the strength, displacement, and power of the vessel. The relative risk of damage to a vessel by different types of ice is reckoned using "weighting" factors, called "ice multipliers." In the Ice Regime System, a simple calculation relates the strength of the ship to the danger presented by different ice regimes. The calculation gives an "ice numeral." Ice regimes that are not likely to be hazardous have zero or "positive" ice numerals. Those regimes that could be dangerous have "negative" ice numerals. As always, however, the safety of the ship is the responsibility of the master.

Ice concentration

Ice coverage in an area is determined by its total concentration, expressed in "tenths." AIRSS uses the partial concentration of each ice type in determining the ice numeral, as follows: open water: less than 10% covered with ice; very open drift: 10%–30% ice cover; open drift: 40%–60% ice cover; close pack/drift: 70%–80% ice cover; very close pack: 90% ice cover; compact/consolidated ice: 100% ice cover (Fig. 8.13).

Inuit Input on Arctic Shipping Lanes

The rapid increase in Arctic shipping, propelled by climate change and the prospect of heightened global maritime trade through the Arctic, poses significant threats to a region with fragile and limited biodiversity. It also threatens Inuit livelihoods, as they continue to rely on access to a functioning and intact maritime environment. Over the past decade, shipping in Arctic Canada has nearly tripled, and further growth is expected as open water seasons lengthen. It is expected that by 2030 the NWP will be ice-free in the summertime enabling the possibility of international trade and increased proportional traffic volumes. In response, the Canadian government is developing Low Impact Shipping Corridors to support safety and sustainability in this rapidly changing environment. These shipping corridors represent a promising adaptation strategy that responds to climate change–induced increases in Arctic shipping activity. The initial development of these Low Impact Shipping Corridors was done with little consultation of Inuit Arctic communities. As described by Dawson et al. (2020), the Arctic Corridors and Northern Voices (ACNV) project was established to fill this knowledge gap. Dawson et al. (2020) discuss 13 Canadian Arctic communities across the Inuit homelands that became involved in the ACNV project through a series of participatory community mapping workshops. The findings of the study emphasize the vital need for meaningful inclusion of northern voices in the development of Arctic shipping policy and governance. Some of the most meaningful impacts of climate change in the Arctic are the observed reduction in extent and thickness of sea ice. Changing sea ice has broad-ranging impacts on the marine environment, animals, and human communities. Sea ice plays a central part in Inuit culture. It enables Inuit to travel and hunt for much of

Open water   (Photo by A eith, CCA-SA4.0)

Very open drift   (Source: NOAA, 2019)

Open drift   (Source: NSIDC, 2020)

Close pack/drift   (Source: NPR, 2019)

Very close pack   (Source: Aweith, CCA-SA4)

Research vessel Akademik Fedorov, November, 2019

Consolidated Ice   (Source: NASA, 2020)

FIG. 8.13 Sea ice conditions. Upper left: Open water; AWeith. Upper right: Very open drift; NOAA, 2019. Middle left: Open drift; NSIDC, 2021. Middle right: Close pack/drift; AWeith. Lower left: Very close pack; NSIDC. Lower right: Consolidated Ice; NASA, 2020.

the year by acting as an extension of land as an "ice highway." However, changing sea ice conditions have reduced local people's ability to hunt, fish, and travel, and as a result "Canadian Inuit are experiencing challenges to their food security, with consequences for nutritional composition in their diets and other aspects of their well-being" (Wesche and Chan, 2010).

THE TRANS-POLAR SEA ROUTE

A recent meeting of the Organization for Economic Cooperation and Development (OECD)'s International Transport Forum focused much attention on Arctic shipping routes. While the Northwest Pass and the NSR were the focus of some attention, one participant reported that the NSR's potential was starting to seem passé in the face of rapid climate change in the Arctic. Because the shrinking and thinning of sea ice are happening faster than scientists thought possible, the hot topic at the meeting was the trans-Arctic passage cutting straight across the North Pole (Bennett, 2019). As the Arctic continues to warm, summer sea ice will continue to recede, and a greater expanse of Arctic waters will become navigable. These changes may result in an increase in vessel traffic to the region, including via the Transpolar Sea Route (TSR), through the high

seas area of the central Arctic Ocean (Stevenson et al., 2019) (Fig. 8.14). The Arctic Ocean is covered most of the year by sea ice and serves as important habitat for several regionally endemic species and may help stabilize global climate systems. There are many seamounts, ridges, and other important geomorphic features in the Arctic Ocean that also serve as important habitat for sustaining existing, and presumably future, biodiversity in the region. Legally, the central Arctic Ocean (CAO) is defined as the region of the Arctic Ocean beyond the 200-mile exclusive economic zones (EEZs) of the five Arctic coastal states, and thus, any management or governance actions concerning human activity in the CAO require international agreement.

Sea-ice Modeling

Ten different global climate models predict that the central section of TSR (i.e., near the North Pole) will mostly be ice-clogged, limiting vessel accessibility until 2035. However, the TSR will gradually replace the NSR by 2050. The TSR allows vessels to avoid the Russian EEZ, avoiding high-cost tariffs associated with the NSR. This presents an economic argument for shipping via the TSR route, but there will remain hidden costs associated with this route. These include risks inherent to the region's remoteness, the unpredictability of

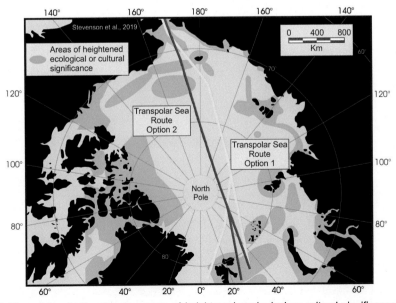

FIG. 8.14 Map of the Arctic, showing areas of heightened ecological or cultural significance (*in green*), Transpolar Sea Route Option 1 (*yellow lines*); Transpolar Sea Route Option 2 (*red lines*). (Modified from Stevenson, T., Davies, J., Huntington, H., Sheard, W., 2019. An examination of trans-Arctic vessel routing in the Central Arctic Ocean. Marine Policy 100, 83–89. https://doi.org/10.1016/j.marpol.2018.11.031.)

shifting sea ice, and extremely poor emergency response infrastructure (Stevenson et al., 2019). There are clear indications that the Arctic Sea Ice is fading fast. Russia's Arctic Laptev Sea finally froze in December, 2020. This is the latest freeze-up since records began, according to the NSIDC (2020b). The sea, which is known as the "birthplace of ice," thawed much earlier in 2020 than in any previous year since 1979 (Fig. 8.15).

Scientists attribute the lack of autumn ice to early summer warming and an extreme heat wave in Siberia, as well as warm Atlantic currents flowing into the Arctic. Climate change has reduced sea-ice coverage in the Arctic Ocean in recent decades, with 2019 tying for second-lowest in recorded history. Ocean temperatures have risen by more than 5°C in the area. If all Arctic sea ice disappears even for just one summer, this would

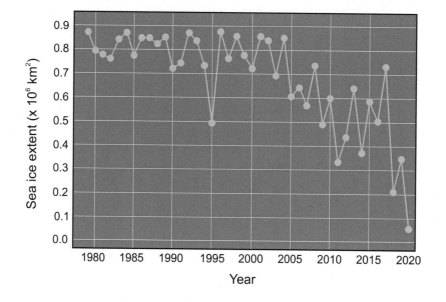

Vladislav / Wicicommons

FIG. 8.15 Above: Sea-ice cover graph from NSIDC, 2020d. Below: Laptev Seashore with open water at the end of October 2020. (Above: NSIDC, 2020d. All about Sea Ice Formation. https://nsidc.org/cryosphere/seaice/characteristics/formation.html. Below: Photo by Vladislav Kyamyarya, Wikimedia Commons.)

cause the disappearance of multiyear sea ice in the central Arctic Ocean. This thick ice near the North Pole blocks all commercial shipping except with the aid of an icebreaker. In the absence of thick multiyear ice, which can be up to 5 meters deep, any water that refreezes would take the form of much thinner, more navigable seasonable ice. In other words, if the thick ice melts away, the relatively thin, pliable first-year ice offers far less resistance to shipping. Such a change in sea-ice conditions would trigger changes in the design, construction, and operational standards of future Arctic ships. Icebreakers might even become obsolete. Within the next few decades, summer voyages of freighters, tankers, and other commercial vessels may traverse the top of the Earth regularly, even if insurance companies and the Polar Code still mandate Polar-class, ice-resistant ships.

The Case for Shipping via the North Pole

For journeys between Europe and Asia, the NSR along the Siberian coast is already cutting 2—3 weeks from voyages using the Suez Canal. By cutting straight across the Arctic, the Transpolar Passage could save a further 2 days. Speed is not everything, of course. Shipowners must also consider risks and costs, and polar shipping remains more dangerous and expensive because of the need for advanced types of ships, required insurance costs, and, for the time being, icebreaker escort fees. Most modern shipping ventures follow the pendulum model, with vessels stopping along the way at ports between their origin and destination to make deliveries. Also known as the hub and spoke model, this logistics chain requires markets, which are lacking in the middle of the Arctic Ocean. However, there may be a way to work around this problem. As described by Bennett (2019), ports situated at the Pacific and Atlantic gateways to the Arctic, like Dutch Harbor, Alaska, or the soon-to-be-developed deep-water port at Finnafjord in north-eastern Iceland, could become hubs for the Transpolar Passage. Cargo could be fast-tracked between Europe and Asia and North America via Polar-class shuttles sailing across the Arctic Ocean. These ships would not have the size restrictions imposed elsewhere, because the CAO is far deeper than the relatively shallow waters that form parts of the NSR and the NWP. Cargo from the polar passage ships would then be transshipped to ports further south via Dutch Harbor and Finnafjord.

Native Concerns for the Bering Sea

Bennett (2019) notes, however, that these gateway areas, especially around the Bering Sea, are also where the negative impacts of increased shipping may be felt. This is because subsistence hunting and fishing still play important parts in Native diet and culture. The Bering Strait is essentially where the NSR, NWP, and Transpolar Passage all meet, making this the busiest part of the Bering Sea. The Alaskan ports of Nome, Teller, and Dutch Harbor will likely see substantial increases in ships docking by 2050. For the residents of these towns and the surrounding regions, this potential boost in shipping activity is not being greeted with unbridled enthusiasm. While the increase in the extent and season of open water will facilitate the shipping increases, that same open water threatens to undermine food security and permanently alter a way of life. Since time immemorial, the Yupik and Inupiat peoples have relied on the presence of sea ice. The ice attracts beluga whales, walruses, and seals, which are all hunted along the northern Alaskan shores. To counteract these losses, in a 2014 workshop called "Bering Strait Voices on Arctic Shipping," one participant suggested that the ships that are already docking could contribute to a fund that helps protect Native food security. The money might be helpful, but what the Bering Strait communities need most is the pack ice, and that does not seem likely to return.

Growth in Arctic Shipping

Arctic sea ice retreat is expected to make new shipping routes available for crossing the Arctic Ocean. Most commercial shipping takes place in the midlatitudes, but some routes in the Arctic are attracting interest, particularly for destination shipping, that is, voyages to and from Arctic ports. As reviewed by Stevenson et al. (2019), Arctic vessel traffic projections suggest this pattern will continue. For example, the US Committee on the Marine Transportation System (2019) projects that in the US Arctic, vessel traffic will grow by as much as 500% by 2025 (Fig. 8.16). Shipping along the NSR has likewise experienced substantial traffic increases. The number of voyages that used all or part of the NSR along Russia's Arctic coast has increased dramatically in recent years. From 2015 to 2019, traffic increased by 58% to 2694 voyages (NSR, 2020). What is more, vessels are transiting Arctic waters faster and earlier than in previous years. For example, in 2017, a Russian tanker sailed the NSR in record time without icebreaker assistance. Vessel traffic is expected to grow in the NSR and the NWP, albeit under different climate-forcing scenarios. Now that attention has started shifting toward the TSR, it is time to assess the potential environmental effects of vessel traffic in the CAO. It is not too early to start considering potential

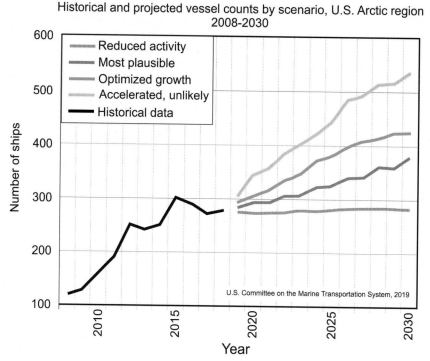

FIG. 8.16 Historical and projected vessel counts by scenario, US Arctic region, 2008–30. (Modified from U.S. Committee on the Marine Transportation System, 2019. A 10-year projection of maritime activity in the US Arctic region. vol. 118.)

risks and available options for reducing those risks. For example, as part of the recommendations of the Arctic Council Arctic Marine Shipping Assessment, *Det Norske Veritas* produced a report evaluating the potential for specially designated marine areas in the Arctic High Seas. The report recommends several options for a more comprehensive approach, centering on the concept that some or all of the CAO should be designated a Particularly Sensitive Sea Area by the IMO and that sea ice and other important habitats should be considered as Areas to Be Avoided. Designation as a Particularly Sensitive Sea Area provides both a higher level of recognition of the unique ecological, socioeconomic, or cultural values of an area and a means of developing Associated Protective Measures as a single package of management procedures to prevent or mitigate impacts from shipping through these areas.

OCEAN POLLUTION

As with all industrial-scale operations, global shipping is responsible for polluting the air and water through which ships travel. Over 90% of global trade is transported across the world's oceans by some 90,000 marine vessels. Ship engines burn fossil fuels, such as heavy fuel oil, marine gas oil, and natural gas. They emit significant levels of carbon dioxide into the atmosphere, as well as a variety of other pollutants that contribute to the problem. The shipping industry is responsible for a significant proportion of the global climate change problem. In fact, if global shipping were a country, it would be the sixth-largest producer of greenhouse gas emissions. More than 3% of global carbon dioxide emissions can be attributed to ocean-going ships. This amount continues to grow rapidly as shipping increases. Only the United States, China, Russia, India, and Japan emit more carbon dioxide than the world's shipping fleet. But because the shipping industry is not a country, CO_2 emissions from ocean-going vessels remain unregulated. In addition to CO_2, seafaring vessels emit methane (CH_4), nitrogen oxides (NOx), sulfur oxides (SOx), carbon monoxide (CO), and varieties of particulate matter such as organic carbon (OC) and black carbon (BC). A recent modeling study showed that emission controls, fuel-type restrictions (e.g., heavy fuel oil ban), and mechanical controls (e.g., exhaust cleaning systems) on seafaring vessels can significantly reduce BC and SOx (Stevenson et al., 2019). Developing smart regulatory policies to improve vessel-related air pollutants is achievable, but this will require considerable cooperation at the international level.

Black Carbon

As reviewed by Oceana (2017), along with CO_2, ships emit various global warming pollutants, including black carbon (BC), nitrogen oxides (NOx), and nitrous oxide (N_2O). These pollutants all contribute to global climate change. They either act as agents that trap heat in the atmosphere or aid the creation of additional greenhouse gases. Black carbon, more commonly known as soot, is made up of fine particles created by the incomplete combustion of hydrocarbons such as oil or coal. Aging engines and poor engine maintenance contribute to incomplete combustion. Unlike greenhouse gases, black carbon is a solid and it warms by absorbing sunlight, rather than absorbing infrared or terrestrial radiation Black carbon contributes to warming in two ways: as it floats through the atmosphere, it directly absorbs solar heat in the top of the atmosphere. When it comes down in the Arctic, it is often deposited on snow and ice. Here, it warms the surface by lowering the albedo, or reflectivity. Reducing black carbon from ships could slow warming, buying time for further steps to reduce carbon dioxide emissions.

Oil Spills and Shipwrecks

Vessel oiling is a major concern with projected vessel traffic growth in the central Arctic Ocean. Spilled fuel and other oil products cause harm to marine life. Vessel collisions with sea ice or other vessels are the most likely causes of oil spills in the Arctic Ocean and adjacent seas. Oil pollution is particularly problematic in the Arctic marine environment because the breakdown and volatilization of oils are much slower in near-freezing Arctic waters and sea ice-oil interactions are complex and unpredictable. While such spills are a major environmental threat, collisions between vessels or between vessels and sea ice may be quite serious. Of course, the worst-case scenario is a collision with ice in regions far from port, leading to the vessel sinking and loss of life. This is the scenario played out by the Titanic in April 1912, when a collision with an iceberg sank this ocean liner, with the loss of 1522 passengers and crew. Another Arctic shipping catastrophe occurred in 1959, when the freighter *Hans Hedtoft* sank on its return maiden voyage, from Godthabb in Greenland to Copenhagen. Like the Titanic, which was hailed as

"unsinkable" before its maiden voyage, the *Hans Hedtoft*, a 2857-ton freighter, was called the "safest ship afloat" by its owners. As discussed by Dasgupta (2019a), the ship hit an iceberg on January 30, about 20 nautical miles (37 km) southeast of Cape Farewell. An SOS signal was picked up by the German trawler *Johannes Kruss*, but when it reached the last reported position, there was no sign of the *Hans Hedtoft*. Rescue operations followed up for many days, but the only identified piece of wreckage was a lifebuoy which washed ashore on Iceland, 9 months later. The sinking of the *Hans Hedtoft* took the lives of all on board: 40 crew and 55 passengers. Less serious collisions with ice reduce the seaworthiness of the vessel or vessels involved, potentially requiring rescue, such as happened with the *Clipper Adventurer*. Existing regulatory mechanisms, such as those under the International Maritime Organization Polar Code, focus on maintaining technical aspects of vessel structure, stability, and integrity that help prevent unintended oil spills from occurring in the CAO. While these regulations offer de facto protections, improvements, such as making protection of the marine environment a priority within the Polar Code, have been advocated.

Vessel Discharges

Vessels are currently allowed to discharge raw sewage at distances more than 22 km from the nearest land; however, vessels greater than 400 gross tonnage and carrying more than 15 passengers are required to have onboard sewage treatment capabilities. Most vessels transiting the central Arctic Ocean would presumably have onboard sewage treatment capabilities. As vessel traffic increases over time, concerns around raw sewage and gray water discharge may become more pertinent and issues around decomposition rates of raw sewage and impacts on nutrient cycling should be considered. At present, vessel discharge in the Arctic Ocean is not a major concern. As described by Parks et al. (2019), ships discharge, either accidentally or deliberately, oil, trash, sewage, and gray water into Arctic waters. Marine collisions, groundings, and sinking in northern waters significantly increase environmental risks, especially due to the scarcity of response resources available in these remote regions. Sewage discharge may release zoonotic pathogens that are responsible for transmitting diseases between humans and animals. As shipping increases in the northern Bering Sea, the frequency of harmful algal blooms and paralytic shellfish poisoning may likewise increase. A 2016 study found elevated levels of harmful algal bloom toxins in marine mammals from the region. In the fall of 2017, 39 walruses washed ashore in western Alaska, and of the four walruses sampled, all tested positive for algal biotoxins (Parks et al., 2019). Vessel traffic through the Bering Sea increased by 145% between 2008 and 2015 (Parks et al., 2019). This trend is projected to increase in the future. As the number of ships and passengers continue to grow, so will the amount of waste discharged into the water. Granted, the amount of actual vessel traffic in the polar north is relatively low compared with other major waterways, but this increased activity throughout the Arctic poses risks to people, cultures, and the environment. Vessel traffic in the region is diverse, with vessels primarily engaged in transiting between markets or supporting the extraction of natural resources, including oil, minerals, and fish. Cargo ships, both refrigerated bulk and container vessels, transport seafood to global markets. Tankers, general cargo ships, and barges travel the northern Bering Sea, providing goods and supplies to coastal and inland communities. Cruise ships are beginning to transit through the region in greater numbers. The increasing pollution risk posed by vessels is causing great concern to the Indigenous people of the region. Their culture and food security is based upon the harvest of living marine resources that are untainted by pollutants. Though the issue of vessel-generated waste in this region has not yet reached crisis proportions, preventative actions now to maintain the status quo will be much less costly and difficult than future efforts to undo the damage.

Cruise Ships

As mentioned earlier, cruise ships are starting to cross the NWP as well as traversing the coasts of adjacent seas, such as the Bering Sea. This has been made possible by the great reduction in sea ice extent, as well as the lengthening of the summer season when the ice has retreated northward. Halliday et al. (2018) discussed the increase in cruise ship traffic across the western Canadian Arctic and the potential effects of this traffic on whale populations of the region. Cruise vessels have increased their visits to the Arctic region, and it seems likely that the novel experience of cruising in the Arctic will only continue to increase in popularity. The number of cruise ships navigating Arctic waters has already begun to increase in recent years. There are two types of cruise destinations in northern North America: the Alaska panhandle route and northern Canada. Next to the Caribbean, cruises along the Alaska Panhandle and southern Alaska have emerged as one of the largest cruise markets almost on par with the trade in the Mediterranean (Hong, 2012).

Arctic village infrastructure

The increase in Arctic tourism facilitated by cruise ship voyages might seem a genuine boon to the Native villages of the region. However, there is some concern among the Indigenous peoples about what Arctic tourism will really bring to their communities. For one thing, these communities are all small villages. As such, they do not have the infrastructure to deal with the influx of people and garbage. In the Canadian Arctic territory of Nunavut, the largest town, Iqaluit, has 7740 residents. All the other settlements in the territory are considered hamlets, which range in population size from 129 in Grise Fiord to 2842 residents in Rankin Inlet. Except in Iqaluit, these hamlets have no hotels, restaurants, or other tourism-related infrastructure. Hamlets can be overwhelmed even by limited groups of tourists visiting. The likelihood of a catastrophic oil or fuel spill from cruise ships rises with increased ship traffic.

Eco-tourism

Starting in 1966, Lars-Eric Lindblad, the "father of eco-tourism," took the very first "citizen explorers" to Antarctica, a region never before visited by nonscientists (Lonsdale, 2019). This was the start of a new sector of the travel market: expedition cruising. For over 50 years, small ships specifically designed to travel in high-latitude waters have transported small groups of tourists to remote locations. Expedition team members generally include experts on regional history, geology, botany, and zoology. For many tourists, this is a once-in-a-lifetime experience. The average prices for such cruises are about $6000 per person. The NWP route has become more and more accessible in recent years, but the cruise ship companies do not guarantee that the cruises will get through. One cruise ship had to turn back in the summer of 2018. More recently, the Russian authorities have seen the financial opportunities for western passenger ships to attempt the Northeast Passage (NSR). The first commercial passenger cruise on this route was made in 2014, aboard the *Hanseatic*, an ice-strengthened ship.

Northwest versus the Northeast Passage

As far as tourist attractions are concerned, the Northwest and Northeast passage trips are quite different. On the Northeast Passage, the passengers are given demonstrations of Russian Folklore, they view Chuckchi and Inuit bone carvings, and a Paleo-Eskimo camp dating back 3400 years. NWP customers visit the Native hamlets along the route. Special side trips are arranged for small-group visits to the Beechy Island site where some of Franklin's crew are buried. Those traveling the Northeast Passage will become better acquainted with Soviet military bases and nuclear installations. One of the most popular aspects of these cruises is wildlife spotting. Passengers are almost guaranteed to see polar bears. During the first Northeast Passage voyage on the *Hanseatic*, passengers spotted 86 bears. Bear sightings are unlikely to be so numerous on the NWP cruise, but passengers may see narwhal in the water, a rare sight in Russian waters (Lonsdale, 2019). Passengers looking toward the land on either journey can expect to see musk oxen, reindeer, and Arctic hare. Out to sea, chances are good that passengers will see walrus, a variety of seals, and humpback, bowhead, and beluga whales. Six different cruise ship companies offer journeys on these two routes. The trips are scheduled to make the passages in about 24 days. Most trips depart in mid- to late August when the ice has receded to its limits. Passengers are required to sign a disclaimer indicating their understanding that the voyage may not be possible, as ice or political conditions may cause last-minute changes.

PROTECTING WHALES FROM SHIPS

As discussed by Halliday et al. (2018), vessel traffic can pose serious threats to whales. Vessels may injure or kill whales through collisions. The frequent occurrence of the vessels in whale habitat may cause behavioral disturbance and increase stress levels. Ships also pollute the waters where whales live. Many populations of whales live in continuous proximity with vessels, putting them constantly at risk. Populations in more remote areas may only experience these threats on a seasonal basis. Although contact with vessels is less frequent in remote Arctic regions, these whales are not as acclimated to vessels, which increases their risk of injury. As reviewed by Dasgupta (2019b), collisions with ships seem to be a greater threat than noise pollution for some species of whales, particularly the North Atlantic right whales, whose primary habitat happens to be the high-traffic waters off the east coast of the United States and Canada. Most whales flee from oncoming large ships, but for some reason, North Atlantic right whales seldom make any effort to avoid large ships, often with fatal consequences. Vessel operators sometimes do not bother to slow down their ships in high-density areas and also fail to heighten their visual awareness for possible whale presence. Apart from right whales, the whale species most affected by ship collisions are the fin whales, humpback whales, sperm whales, and gray whales. Whale injuries from ship collisions include fractured skulls, jaws, vertebrae, and other bones. Deep cuts from propellers are also common, as are long, parallel cut marks on the dorsal aspect, penetrating the insulating blubber. Many of these injuries are fatal, if not immediately then within a few days or weeks as infections take hold.

Management Measures

Three different management measures are typically used for decreasing the risks of vessels to whales: (1) keeping vessels away from whales, either through rerouting ships or establishing exclusion zones; (2) restricting vessel speed, which reduces risks of ship strikes and can lower noise pollution; (3) using marine mammal observers or other vessel behavior if whales are nearby (e.g., changing course, stopping the engine). For example, adjusting the vessel corridor in the Roseway Basin of Canada to avoid the Right Whale Conservation Area was implemented to reduce the risk of ship strikes for North Atlantic right whales. This has resulted in a 62% decrease in ship strikes in this region. The Port of Vancouver (Canada) recently enacted an 11-knot slowdown zone in Haro Strait, reducing the impacts of ship traffic on killer whale foraging in this region (Halliday et al., 2018). Multiple measures are usually more effective than single ones, for instance, the establishment of an exclusion zone surrounded by a slowdown zone. Additional vessel traffic in Arctic waters will certainly bring more vessels into close contact with whales, seals, and walrus. Management measures should be put in place for the protection of these marine mammals, but any measure requiring enforcement may fail because enforcement vessels are faced with a much greater distance to patrol.

Canadian Arctic Ship Traffic

Vessel traffic has been increasing in the Canadian Arctic over the past 30 years and is triple the level it was in the 1980s. The most rapidly expanding vessel types are pleasure craft and passenger vessels, such as cruise ships and expedition-style tour vessels. These are defined under Canadian law as vessels carrying 12 or more passengers. Pleasure craft include the full spectrum of privately owned vessels used for pleasure, but most are private yachts that can range in size from very small to quite large. Both of these vessel classes often head to specific Arctic destinations and may spend time exploring and seeking out areas with more marine wildlife. Such exploration may cause greater disturbance to whales than freighters and tankers, as the latter do not spend time searching for marine wildlife.

Native Reliance on Whales

As discussed by Bennett (2019), whales are also a primary subsistence food source for Indigenous people in Arctic communities. Native whale hunting crews want commercial ships to stay out of their vital whaling areas. The remoteness of the Arctic also means that the distribution and abundance of whales are not as well understood. Even if mariners try to avoid key whale areas, they may not have access to information on the locations of these whale areas. Finally, severe, unpredictable weather and sea conditions often force ships to change course to keep afloat in the storm. In such circumstances, avoidance of whale areas may become impossible.

NOISE POLLUTION EFFECTS ON MARINE MAMMALS

Marine mammals spend most of their time below the photic zone in the oceans. This means that they are navigating in the dark and scarcely rely on vision to accomplish their usual tasks, such as food acquisition, finding other members of their species, looking after infants, and keeping watch for predators. Their world is an acoustic world where sounds are far more important than sights. For millions of years, whales, dolphins, porpoises, seals, and sea lions lived in relatively quiet oceans. The only background noises they had to contend with were natural sounds such as rain and wind on the surface of the water and volcanic eruptions and earthquakes on the seafloor. Up through the time of seafaring by sailing ships, the oceans remained relatively quiet. However, the Industrial Revolution of the mid-19th century brought engines onboard ships. Other technological advances introduced extremely loud noises into ocean waters from sonar, underwater explosions, and acoustic air guns. Within a few short decades, marine mammals were faced with noise levels they had never heard before and to which they were not physiologically, behaviorally, or psychologically adapted. The peaceful acoustic world of marine mammals and other sea life was shattered by human-caused noise. In a sense, marine mammals were acoustically blinded by this unceasing noise. They have been trying, with varying levels of success, to negotiate life with acoustic blinders on. From a technical standpoint, this severe acoustic impairment can be quantified. At the 2004 meeting of the US Marine Mammal Commission's Advisory Committee on Sound, research was presented that suggests human noise can shrink the area in which whales can communicate with each other by two to four orders of magnitude. In other words, when the sea is especially noisy, the effective communication area for whales shrinks down to a space between one hundredth and one ten-thousandth the size that it would be in the absence of human noise.

Loudness Scales

The loudness of a sound is measured in decibels (dB). Decibel levels indicate how loud a sound is relative to some reference (represented in units of amplitude called micropascals, or μPa). For instance, in air, measurements are referenced to 20 micropascals (μPa). In water, a different reference is used: 1 μPa. As such, dB levels are always reported as "dB re 1 μPa" for a sound in water or "dB re 20 μPa" for a sound in the air. The difference in reference levels used, combined with the difference between the density of air and water (the latter being denser), means that the volume in dB of a sound transmitted through the air cannot be directly compared with its volume in water (Stafford, 2013). It is important to keep in mind at the decibel scale is logarithmic. If we compare the sound level associated with 70 dB with a sound of 60 dB, we find that the 60 dB sound is only half as loud. Furthermore, a sound produced at 80 dB is twice as loud as the 70 dB sound, and a 100 dB sound is 8 times as loud as a 70 dB sound. Thus, the acoustic energy expressed by the decibel sound scale ranges from very weak below about 40 dB but climbs rapidly. Exposure to sounds at or above 100 dB can cause serious damage to human hearing (Fig. 8.17). As we shall see in this section, the extremely loud underwater noises to which marine life is exposed is causing a wide variety of health problems, especially in marine mammals.

Commercial Shipping Contribution to Ocean Noise

As discussed by Rako-Gospić and Picciulin (2018), commercial shipping contributes significantly to overall ambient noise in the sea. Generated noise is related mainly to propeller cavitation and turbulence which is known to peak at 50−150 Hz but can extend up to 10,000 Hz. In the period 1950−2000, an increase of 3 dB in the overall background noise has occurred every decade. During this period, the world's fleet of ships tripled, and the average gross tonnage of commercial ships increased. The noise generated by marine vessels varies according to ship type. The volume of sound generated by ship engines varies from 178 to 188 dB re 1 μPa at 1 m, in a wide range of frequencies (20−1000 Hz).

Clashing Sound Frequencies

Stafford (2013) has assessed the problem of anthropogenic sound and marine mammals in the Arctic. As a

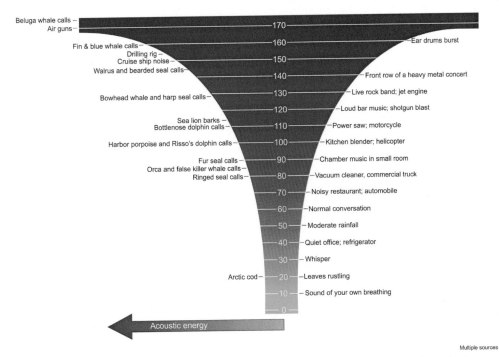

FIG. 8.17 Decibel scale with examples. (Data from multiple sources; figure by the author.)

group, marine mammals hear and produce sounds over a broader range of frequencies than do humans, with each species generally using quasi-specific frequency bands to communicate (Fig. 8.18). For instance, bowhead whales produce sounds that are relatively low frequency (usually less than 1000 Hz), bearded seals use midfrequencies (500–6000 Hz), and beluga whales produce sounds in high frequencies (from a few thousand hertz to more than 100,000 Hz). Low-frequency sounds can travel great distances underwater. Owen et al. (2019) note that the transmission of humpback whale songs between populations may involve great distances. Scientists have documented the transmission of humpback whale song that extended across the South Pacific, spanning 6000 km from eastern Australia in the west to French Polynesia in the east. Many marine mammals have developed vocal adaptations that allow them to cope with highly variable ambient sound conditions. In noisy conditions, animals may call more or less often, alter the pitch of their vocalizations, change their acoustic communication strategy from vocal to nonvocal, or adopt a response known as the Lombard effect. This is when an animal alters the amplitude (loudness) of calls in response to the loudness of their acoustic environment. These strategies have been documented for several marine mammal species, such as North Atlantic right whales (*Eubalaena glacialis*), beluga

whales (*D. leucas*), and humpback whales (*Megaptera novaeangliae*), indicating that these animals can adapt their behavior in response to changing noise conditions in their acoustic habitat (Fournet et al., 2018). Particularly during the spring when many species produce elaborate vocal displays as part of mating behavior, Arctic marine mammals create such a cacophony of sounds that even marine mammal experts find it difficult to distinguish individual animals or even species. This is a critical season of the year when it appears to be vital for marine mammals to communicate with each other. It is also the season when they may be most sensitive to background noises.

Sound frequency versus amplitude

Both the frequency and amplitude (loudness) of sound waves need to be considered when studying the potential consequences of noise pollution on marine species. Sounds in the same range as those produced or heard by animals are considered more likely to affect them than sounds outside of their hearing range. This is because those sounds within the same range can interfere with signals that are important to animals, making it harder for them to hear each other or detect acoustic cues in the environment (Fig. 8.18). The sound pressure levels exerted on the ear are directly proportional to the distance one is from the sound source. However, sound

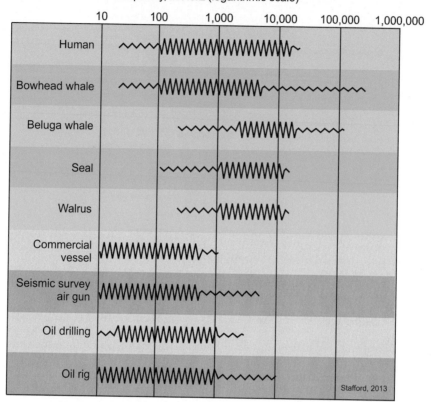

Frequency ranges of marine mammals and industrial activities

Frequency, in Herz (logarithmic scale)

FIG. 8.18 Frequency ranges of marine mammals and industrial activities. Frequency, in Hertz (logarithmic scale). (Modified from Stafford, K., 2013. Anthropogenic Sound and Marine Mammals in the Arctic. Pew Charitable Trust, U.S. Arctic Program. https://www.pewtrusts.org/~/media/Assets/2013/06/07/arcticnoise_final_web.pdf.)

travels much farther and faster in water than in air, and low-frequency sounds travel farther than higher frequency sounds. Therefore, the area of possible impact for low sounds can be very large.

WHALE RESPONSE TO OCEAN NOISE

There are many examples of whale responses to ocean noise. For instance, beluga whales are known to flee from icebreaker noise, leaving an area with active ice breaking and remaining away for as long as 2 days. Marine mammal response to anthropogenic marine noise is not always so easy to assess, but one way of examining these responses is to observe what happens when the human-caused noise stops. A recent study of northern right whales, a close relative of the bowhead whale, found that the sharp decrease in low-frequency underwater sound due to the near cessation of ship traffic immediately after 9/11 was significantly correlated with a reduction in stress hormones in the whales. These data suggest that northern right whales, at least, may be chronically stressed from high levels of ship sound. Similarly, the last time researchers were able to listen to humpback whales in a quiet ocean in Alaska was in 1976, when commercial whale watching began. When the COVID-19 pandemic broke out, tourism in Alaska ground to a halt. Cruise ships did not visit Alaskan waters in the summer of 2020, and whale watching boats stayed in port. For once, the waters of south-eastern Alaska were devoid of anthropogenic noise. Marine mammal researchers have not completed their analyses of humpback whale responses to the quiet ocean, but Dr. Heidi Pearson of the University of Alaska Southeast is quoted as saying "Based on my observations, it does seem that whales are exhibiting more resting behavior this year … I have also observed larger groups and more social behavior than I have in previous years" (Pennington, 2020).

Studies in Glacier Bay, Alaska

In previous years, most whale watching trips in Alaska start from Juneau's Auke Bay. Over 350,000 tourists pack the whale watching boats each year. On a busy day, 10 or more boats cluster around a whale or group of whales, prompting concerns about vessel overcrowding (Pennington, 2020). On any given day during recent summers, as many as 65 whale-watching boats zip around Auke Bay and nearby Glacier Bay National Park. In response, the humpbacks have changed their behavior in two ways: they call more loudly and less frequently. There is concern that this change in whale communication may be altering their social structure. Fewer calls decrease the likelihood of a humpback finding a companion. The International Whaling Commission has established a set of guidelines to protect whales from being overwhelmed by whale watching boats (Table 8.5). Fournet et al. (2018) studied humpback whale responses to acoustic noise in Glacier Bay, Alaska. They found that controlling for ambient sound levels, the probability of a humpback whale calling in the survey area was 31%−45% lower when vessel noise contributed to the soundscape than when only natural sounds were present. They also found that cruise ships and tour boats, roaring harbor seals (*Phoca vituline*),

and weather events were primary drivers of ambient sound levels and that the sound levels vary both seasonally and diurnally. As ambient sound levels increased, humpback whales responded by increasing the volume of their calls (nonsong vocalizations) by 0.81 dB for every 1 dB increase in ambient sound. Cruise ship noise is pervasive during the summer months at Glacier Bay. Daily ambient sound levels at the Beardslee Island Complex of Glacier Bay National Park are shown in Table 8.5. This noise is louder at the source than harbor seal roars. Cruise ship engines generate noise at about 171−188 dB, and noise is continuous for as long as the vessel is within acoustic range, which is up to 40 min.

The Behavioral Context of Marine Mammal Response to Noise

As discussed by Southall (2007), 30 years of research has failed to find any consistent and predictable relationships between anthropogenic ocean noise and the responses of marine mammals. Certain species such as beaked whales and harbor porpoises clearly respond to anthropogenic noise at low sound thresholds, while other species such as bottlenose dolphins and humpback whales not only do not avoid loud noises from sound-producing sources, they actually swim up to them for various reasons. Southall et al. (2017) found that within certain contexts (e.g., feeding vs. traveling bowhead whales), there was a generally increasing likelihood of response as the noises increase, especially if the sound is recurrent.

Behavioral changes in narwhal exposed to air guns

Narwhals spend their lives in the Arctic waters of Canada, Greenland, Norway, and Russia (Fig. 8.19). The

TABLE 8.5
Daily Ambient Sound Levels, Beardslee Island Complex of Glacier Bay National Park, Southeast Alaska.

Month	5th Percentile	Median (dB)	95th Percentile	Whale Call Volume (dB)
May	88	95	105	145
June	90	99	108	149
July	93	103	108	153
August	86	94	105	144
September	86	92	105	142
Overall	87	97	107	147

Data from Fournet, M.E.H., Matthews, L.P., Gabriele, C.M., Haver, S., Mellinger, D.K., Klinck, H., 2018. Humpback whales Megaptera novaeangliae alter calling behavior in response to natural sounds and vessel noise. Marine Ecology Progress Series 607, 251−268. https://doi.org/10.3354/meps12784.

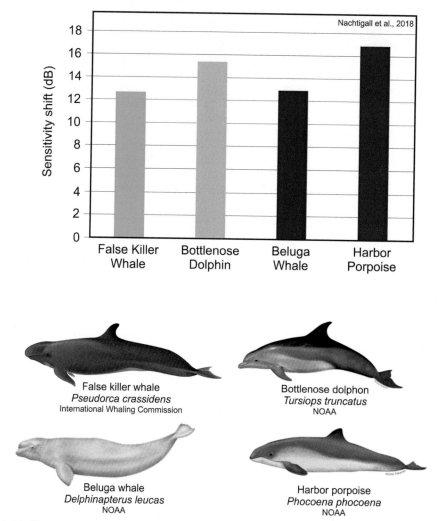

FIG. 8.19 The maximum sensitivity shift demonstrated for four marine mammal species that occurred following simply pairing a loud sound with a warning sound. (Modified from Nachtigall, P., Supin, A., Pacini, A., Kastelein, R., 2018. Four odontocete species change hearing levels when warned of impending loud sound. Integrative Zoology 13 (2), 160–165. https://doi.org/10.1111/1749-4877.12286.)

majority of the world's narwhals winter for up to 5 months under the sea ice in the Baffin Bay—Davis Strait area (between Canada and western Greenland). Cracks in the ice allow them to breathe when needed, especially after dives, which can be up to a mile and a half deep. They feed mainly on Greenland halibut, along with other fish, squid, and shrimp. Narwhals, like most toothed whales, communicate with "clicks," "whistles," and "knocks." As most toothed whales, narwhals use sound to navigate and hunt for food. Narwhals may also adjust the duration and pitch of their pulsed calls to maximize sound propagation in varying acoustic environments. Other sounds produced by narwhals include trumpeting and squeaking door sounds. The narwhal vocal repertoire is like that of the beluga whale. Narwhals may be particularly sensitive to this noise, but as yet no studies have addressed the question of direct effects of high-energy air gun pulses on these animals. Heide-Jørgensen et al. (2013) studied the response of narwhals to seismic exploration. They postulated that seismic noise is increasing the risk of ice entrapment by narwhals. Their concern was prompted by recent events. Three recent large-scale ice entrapments were linked to seismic survey activities. On these occasions, narwhals remained in their coastal summering areas far longer than usual—well into early winter. This delay in their annual offshore migration led the narwhals to be trapped by rapidly forming fast ice. About 1000 narwhals died in an ice entrapment in Canada in 2008 and about 100 died in two entrapments in Northwest Greenland in 2009–10. Studies of the direct effects of seismic surveys on narwhals are urgently needed and should ideally precede further seismic surveys in narwhal habitats. Considering specific responses to noise exposure, Ellison et al. (2012) identified the central importance of exposure context in determining response type and probability. The authors proposed that a variety of external and internal factors greatly affect the responses of marine mammals to low-to-moderate noise. They termed these factors "exposure scenarios." These include the animal's behavioral state and its spatial orientation in relation to the source of the noise (i.e., is it distant or close, above the animal in the water, or below it). A marine mammal's familiarity with a sound source can make a big difference. For example, dolphins and porpoises kept in captivity can be conditioned to ignore loud noises to which they have been repeatedly exposed. Finally, if the noise in the water is similar in pitch and strength to sounds emitted by a species' predators, they will tend to flee, either because the anthropogenic sound has tricked them into thinking a predator is near, or because they realize that such noises are so similar to the predator's sounds, they will not be able to detect when predators use the human-induced sound to mask an impending attack. Ellison et al. (2012) also proposed that within the context of their common behavior patterns, a general principle is that the louder the noise, the greater the likelihood of animal response. Subsequent research has demonstrated these context dependencies in marine mammal behavioral responses to noise and has added additional resolution and clarity to Ellison et al.'s (2012) findings. Goldbogen et al. (2013) exposed blue whales (*Balaenoptera musculus*) at sea to sounds simulating military sonar. They found a much greater probability of response in whales feeding at depth than in whales engaged in feeding or traveling in shallow water. Their study likewise demonstrates a strong link between the noise exposure context of the whales and the likelihood of their response.

Ship Strikes

Loud underwater noises also confuse marine mammals that rely on their own sonar to identify large objects in the water. Whales and dolphins use echolocation or "biosonar" systems to locate predators and prey. In such circumstances, collisions may result from marine mammals, especially whales, being unable to locate and avoid ships because ships emit sounds over a wide range of frequencies and in all directions. This has yet to become a serious problem in Arctic waters. Only about 1% of bowhead whales from the Arctic Ocean exhibit scarring from ship collisions. This problem is more prevalent in the very busy waters of the North Atlantic. Here, about 7% of North Atlantic right whales show signs of ship collisions. In fact, such collisions are often fatal. Nearly half of recently documented deaths of North Atlantic right whales were caused by ship strikes. Right whale habitat is exposed to higher levels of vessel traffic than that occurring in the Arctic Ocean. However, as shipping and vessel traffic in the Arctic Ocean is projected to increase substantially in the near future, the number of strikes is almost certain to increase.

Masking of Sounds

Masking occurs when a sound is obscured because of added noise. When this happens, a marine mammal's behavior may be affected because it is not able to detect, interpret, and respond to biologically relevant sounds (Marine Mammal Commission, 2007). For example, masking may adversely affect reproduction if a female cannot hear potential mates vocalizing at a distance. Mother-offspring bonding and recognition may be more difficult if the pair cannot communicate effectively. Foraging may be disrupted for marine mammals that use echolocation to detect prey. Species that hunt cooperatively, such as orca, humpback whales, and dolphins cannot coordinate attacks on prey if they cannot effectively communicate. Finally, sound masking may endanger survival if an animal cannot detect predators or other threats. Nevertheless, any marine mammals demonstrate remarkable abilities to overcome this problem. For instance, Beluga whales can detect the return of their echolocation signals when they are only 1 dB above the background noise, and gray whales can detect the calls of predatory killer whales at less than 1 dB above the background. Masking may affect marine mammal hearing even when the sound levels are below the threshold needed to cause observable responses by scientists. For instance, masking sounds may have traveled far greater distances from the sound source than those at which the animal shows a behavioral response (Marine Mammal Commission, 2007). Anthropogenic ocean noise is not the sole source of masking in the world's oceans. Natural sounds such as rain, waves, sea-ice cracking, and vocalizations of other marine mammals can also mask important signals.

Marine mammals have developed ways of overcoming certain levels of masking by increasing the level, changing the temporal pattern, and shifting the frequency of their vocalizations.

Physiological Reactions

Exposure of marine mammals to high-intensity sound may cause a temporary threshold shift or a temporary loss of hearing sensitivity. A reduction in sensitivity is the usual response of a mammalian sensor exposed to an intense or prolonged stimulus and, within limits, is reversible (Marine Mammal Commission, 2007). Nonetheless, because of the importance of sound in the daily lives of marine mammals, even temporary threshold shifts have the potential to increase an animal's vulnerability to predation, reduce its foraging efficiency, or impede its communication. Nachtigall et al. (2018) studied changes in hearing sensitivity when cetaceans are first exposed to a warning sound, followed by a loud sound. They examined the responses of the false killer whale (*Pseudorca crassidens*), the bottlenose dolphin (*Tursiops truncatus*), the beluga whale (*D. leucas*), and the harbor porpoise (*Phocoena phocoena*) (Fig. 8.19). Hearing sensitivity was measured using pip-train test stimuli and auditory evoked potential recording. When the test/warning stimuli preceded a loud sound, hearing thresholds before the loud sound increased relative to the baseline by 13−17 dB. Experiments with multiple frequencies of exposure and shift provided evidence of different amounts of hearing change depending on frequency, indicating that the hearing sensation level changes were not likely due to a simple auditory reflex of closing off the inner ear.

Physical Injury

Permanent threshold shifts, or permanent loss of hearing sensitivity, can result when animals are exposed even briefly to very intense sounds, over a longer duration to moderately intense sounds, or intermittently but repeatedly to sounds sufficient to cause temporary threshold shifts. Permanent threshold shifts result in the loss of sensory cells and nerve fibers. In terrestrial animals, temporary reductions in sensitivity of about 40 dB have been required to cause permanent threshold shifts. New et al. (2015) noted that the noise generated day after day by whale-watching boats in southeast Alaska can cause masking and temporary threshold shifts in the hearing of resident whales. To date, temporary shifts of only about 20 dB have been induced experimentally in marine mammals, which is much less than required for a permanent shift if marine mammals respond similarly to terrestrial animals.

Variable Response to Vessel Noise

Working in the Pacific Arctic region, (Stafford, 2013) examined the effects of vessel noise on beluga and bowhead whales. The authors suggested that masking may be more of an issue for bowhead whales than for toothed whales, because the hearing of baleen whale (the bowhead) should be most sensitive at low frequencies, such as are generated by ship engines. Near these vessels, however, the acoustic disturbance overlaps with the communication channel of both species. On the other hand, beluga hearing has been shown to be very sensitive, such that even low-level exposure to distant shipping noises disturbs their acoustic space. Noise-triggered changes in vocal behavior have also been documented in bowhead whales. They increase their calling rate in the presence of distant air guns up to a certain level of noise (127 dB re 1 $\mu Pa^2/s$, over 10 min intervals), but calling ceased altogether when noise levels exceeded 160 dB re 1 $\mu Pa^2/s$. The responses of both species may be seen as a mechanism to compensate for the loss of acoustic space.

Cumulative Effects of Noise Pollution

The long-term, cumulative significance of repeated sublethal effects is an important topic of debate and central to the concerns of many with respect to anthropogenic sound. Effects that are individually insignificant may become significant when repeated over time or combined with the effects of other sound sources. Baleen whales, for example, use low-frequency sound for communication and therefore may be affected by both seismic air guns and shipping noise. Similarly, the effects of sound may interact additively or synergistically with the effects of other risk factors. Beluga whales enjoy a wide distribution through much of the Arctic (Fig. 8.8), but their abundance may be compromised. Their ability to survive and reproduce under climate change scenarios may alter the distribution and availability of their prey, and persistent organochlorine contaminants have already altered their immune function and made them susceptible to disease and parasites, and noise from oil and gas operations, icebreakers, or commercial vessels has caused them to abandon important habitat.

Noise Effects on Migrating Whales

Another aspect of this problem is that many whale species migrate across oceans every year. A recent assessment of the cumulative impact of ship noise and seismic signals on migrating bowheads in the Beaufort Sea during the open water season to estimate the aggregated sound pressure levels to which the whales would be exposed. This study found that bowheads would experience noise levels well above the natural ambient baseline throughout much of their migration. Here, as elsewhere in the world, the noise contribution from all human activities occurring in the Pacific Arctic should be considered in the context of the cumulative impact on the already changing acoustic habitat (Stafford et al., 2018).

SOURCES OF DAMAGING OCEAN SOUNDS

The "background noise" in the ocean has been steadily building from the 20th century onward, mostly from ship engine noise. However, there are other sources of ocean noise that are not part of the background noise to which marine mammals have become accustomed. These are short bursts of extremely loud noise that cause damage to marine mammal hearing and tissue damage to vital organs and in some cases lead to the death of marine mammals that happen to be too close to the source of the noise. These include high powered military sonar, seismic surveys using air guns, underwater explosions, and another extremely loud anthropogenic noise.

Development of Military Sonar

D'Amico and Pittenger (2009) provided a concise history of active naval sonar (sounding navigation and ranging). The most important echo-ranging system to emerge between the two world wars was the ultrasonic ASDIC (Allied Submarine Detection Investigation Committee), a cooperative effort by the British and French Navies. The first ASDIC shipboard systems were installed in 1919. Operating frequencies varied from 20 to 50 kHz. During the 1920s and early 1930s, ASDICs were developed for use on destroyers for antisubmarine warfare (ASW). One key discovery during this period was that the amplitude of higher frequencies of underwater sounds is more rapidly attenuated (i.e., they do not carry as far) than lower frequencies as they pass through seawater. Based on this observation, the frequency range for a newer version ASDIC was eventually set to 14−22 kHz. This refined sonar was designed to help surface ships detect the presence of submarines. Between the late 1940s and 1960, in response to improvements in submarine technology and the increased threat that this represented, surface ship active sonars were developed for the US Navy. The major Cold War active sonar technology development was the advent of scanning sonar to compensate for faster submarine speeds and the need to switch rapidly from long-range to short-range detection of an attacking submarine. Scanning sonar

provides directional search capability via sending and receiving focused sound energy in multiple directions simultaneously with different ping intervals. The detection range of these sonars was improved by the development of lower-frequency active sonars that increased detection ranges through minimizing attenuation loss. The US Navy continued its quest for lower-frequency sonars, developing a series of lower active sonar frequencies, culminating in the 1950s with a tactical midfrequency active sonar (MFAS) unit operating at 3.5 kHz. Modern naval sonar instruments operate at 2.6 and 3.3 kHz. This in itself is not hazardous to marine life, but these sonars generate sound signals at 235 dB. This level of sound energy is the highest ever generated, far exceeding any natural sounds. Human eardrums are known to burst when exposed to 160 dB. Because decibel levels are measured on a logarithmic scale, 235 dB is more than 128 times louder than 160 dB. Is it any wonder that whales are injured or killed by exposure to such high-energy sound blasts? This unintended effect of sonar was not linked to marine mammal distress until the following decade. In the meantime, because of their submarine detection capabilities, MFAS has become ubiquitous, employed by virtually every navy in the world (D'Amico and Pittenger, 2009).

Sonar Effects on Whales

Mass stranding events (MSEs) of cetaceans are defined as those in which two or more individuals strand alive at approximately the same place and time. The mass stranding of beaked whales is most often associated with sound. The use of military-grade sonar has unfortunately contributed significantly to beaked whale injuries and deaths by stranding. Naval MFAS was developed in the 1950s to detect submarines, using frequencies of 8 kHz or higher. However, the beaked whale atypical mass stranding events, mainly of Cuvier's beaked whales, did not occur until MFAS shifted to lower frequency ranges of 4.5−5.5 kHz in the 1960s (Bernaldo de Quirós et al., 2019). Mass stranding of beaked whales (BWs) were extremely rare before the 1960s (15 reported cases involving five species), and none were atypical mass stranding events. However, between 1960 and 2004, 121 beaked whale mass strandings took place and the number of species involved increased. The first of these mass strandings occurred in Corsica, followed by others in Italy, the United States, and the Bahamas during the 1960s. At least 37 of these events were atypical mass stranding events that involved three or more individuals and took place during naval exercises, or they occurred in naval training areas where US and NATO fleets were based. All 121 strandings occurred in the Northern Hemisphere, and 61 involved Cuvier's beaked whale.

What is killing the beaked whales?

Necropsies performed on 10 stranded beaked whales revealed symptoms consistent with decompression sickness. In humans, decompression sickness occurs in divers when dissolved gases (mainly nitrogen) come out of solution in the bloodstream in the form of bubbles. These bubbles can damage many parts of the body, including joints, lungs, heart, skin, and brain. The gas bubbles are released from the bloodstream in response to rapid changes in pressure during scuba diving. Nitrogen and other gases in the body increase in pressure as a diver descends. For every 10 m descent in ocean water, the pressure due to nitrogen goes up by 80,000 Pa (11.6 pounds per square inch). As the pressure due to nitrogen increases, more nitrogen dissolves into the tissues. The longer a diver remains at depth, the more nitrogen dissolves. This nitrogen gas is not utilized by the body and so it builds up in body tissues over time. As a diver swims to the surface, the pressure decreases. The nitrogen, which has dissolved in tissues, changes back to its gaseous form because the body can hold only a certain amount based on that nitrogen pressure. If a diver surfaces too fast, the excess nitrogen will come out rapidly as gas bubbles. Depending on which organs are involved, these bubbles produce the symptoms of decompression sickness. The nitrogen bubbles are released when the diver returns to the surface, blocking blood flow and disrupting blood vessels and nerves by stretching or tearing them. They may also cause emboli, blood coagulation, and the release of vasoactive compounds. Although this phenomenon is much more common in human divers that attempt to rise to the surface too quickly, under certain conditions it is possible for marine mammals to suffer the same problem: the development of gas emboli. Several mechanisms have been proposed to explain how sonar might lead to stranding and/or death of beaked whales and decompression sickness features in two of them (Bernaldo de Quirós et al., 2019). It could be that beaked whales simply swim away from the sound source into shallower waters, where they become stranded. A scenario that fits better with the necropsy results involves a behavioral response that disrupts their normal diving behavior, for instance, if the whales swim far too rapidly to the surface, perhaps in an attempt to escape from the noise. Their rapid ascent to the surface results in nitrogen bubble formation and tissue damage. A third hypothesis concerns physiological changes that are a follow-on result of nitrogen

accumulation, bubble formation, and tissue damage. One other scenario is that direct tissue damage is caused by exposure. The most recent scientific evaluation suggests that gas embolism following behavioral and physiological responses to sound is the most plausible explanation. Whatever the mechanism, some animals that are severely injured or even dead wash up on the shore, but most of these whales die unnoticed in the open sea and sink (Safety4Sea, 2020). Beaked whales dive deeper and longer than other cetaceans, often diving to more than 1000 m multiple times per day. In general, there is a trimodal dive pattern composed of respiration (i.e., ventilation) dives, long-duration and deep-foraging dives (typically greater than 400 m and 30 min), and shorter and shallower (intermediate) nonforaging dives that come between foraging dives. Following deep foraging and intermediate dives, beaked whales typically ascend more slowly than they previously descended, thus avoiding decompression sickness (Bernaldo de Quirós et al., 2019).

Mass Strandings

Mass strandings are becoming increasingly common across the globe. The strandings of 600 whales in New Zealand (February 2017), 150 in Western Australia (March 2018), and 140 whales again in New Zealand (November 2018) stand testimony to this. More recently, an estimated 470 pilot whales were stranded along ashore on the western coast of Tasmania in September 2020. This is believed to be the largest mass stranding event in Australia's history. Hundreds more whales were discovered the following day, on another section of Tasmania's west coast. By Wednesday, an additional 200 whales were discovered further into the harbor. Since the 1960s, the frequency of such events has continuously increased. Scientists are now convinced that human-caused noise is a reason behind this. Noise from military sonar, oil and gas exploration, ships, and seismic surveys is often unbearable. The sonars, for instance, can emit sound equivalent to a rocket launch (Fig. 8.17).

Seismic Surveys Using Air Guns

Seismic surveys utilize arrays of air guns to produce powerful sound waves. These devices quickly release pressurized air bubbles to create the sound source, with up to 20 guns fired simultaneously, while batches of hydrophones in the water record echoes. Using sophisticated acoustic processing, these echoes can provide information about geological structures up to 40 km below the seafloor. At sea, seismic surveys are used by the oil and gas industry to find hydrocarbon deposits. Air gun arrays generate sound at 200–260 dB (in water) (Rako-Gospić and Picciulin, 2018). There is a difference of about 60 dB when converting the sound level from water to air, so in air, the air gun sound level would be about 140–200 dB. A typical seismic air gun array towed by a ship might fire its compressed air bubbles into the ocean five or six times a minute, around the clock, amounting to more than 7000 sound blasts in 24 h. Marine mammal noise exposure criteria for injury and behavior are summarized in Table 8.6. Because sound can travel thousands of km under water, the noise from seismic air guns can be heard at great distances by marine mammals. In 2004, marine mammal acoustics experts reported that air gun noise from seismic surveys along the coast of South America can constitute the dominant background noise in the mid-Atlantic region, at times making it difficult or impossible to hear the whales they are trying to study. Air gun noise is generated at over 200 dB (often 230 dB) at the source but diminishes quickly to under 180 dB (usually within 50–500 m) and continues to drop more gradually over the next few km, until leveling off at somewhere near 100 dB. At this still-high level, the sound may travel for thousands of km. When the air guns raise the background noise to this level, it potentially masks local biological calls and signals.

TABLE 8.6
Marine Mammal Noise Exposure Criteria for Injury (Top) and Behavior (Bottom) for Five Groups of Marine Mammals.

LOW-FREQUENCY CETACEANS (BALEEN WHALES: BLUE, SEI, HUMPBACK, GRAY, MINKE, BOWHEAD, NORTH ATLANTIC RIGHT, AND NORTH PACIFIC RIGHT WHALES)

SOUND EXPOSURE LEVEL (BEHAVIORAL RESPONSE)

183 dB*	N/A	N/A
Single pulses	Multiple pulses	Nonpulses
198 dB*	198 dB*	215 dB*

MIDFREQUENCY CETACEANS (TOOTHED WHALES: SPERM WHALE, BAIRD'S BEAKED WHALE, CUVIER'S BEAKED WHALE, NORTHERN BOTTLENOSE WHALE, NARWHAL, BELUGA, ORCA)

SOUND EXPOSURE LEVEL (BEHAVIORAL RESPONSE)

183 dB*	N/A	N/A
Single pulses	Multiple pulses	Nonpulses
198 dB*	198 dB*	215 dB*

HIGH-FREQUENCY CETACEANS (KILLER WHALE, LONG-FINNED PILOT WHALE, ATLANTIC WHITE-SIDED DOLPHIN, WHITE-BEAKED DOLPHIN)

SOUND EXPOSURE LEVEL (BEHAVIORAL RESPONSE)

183 dB*	N/A	N/A
Single pulses	Multiple pulses	Nonpulses
198 dB*	198 dB*	215 dB*

PINNIPEDS IN WATER (WALRUS, SEA LION, HARP SEAL, HOODED SEAL, RINGED SEAL, BEARDED SEAL, SPOTTED SEAL, RIBBON SEAL)

SOUND EXPOSURE LEVEL (BEHAVIORAL RESPONSE)

171 dB*	N/A	N/A

PINNIPEDS IN AIR (WALRUS, SEA LION, HARP SEAL, HOODED SEAL, RINGED SEAL, BEARDED SEAL, SPOTTED SEAL, RIBBON SEAL)

SOUND EXPOSURE LEVEL (BEHAVIORAL RESPONSE)

100 dB«	N/A	N/A

* re: 1 µPa²s (MIf)
« re: 20 µPa²s (MPa)

Data from Rako-Gospić, N., Picciulin, M., 2018. Underwater noise: sources and effects on marine life. In: World Seas: An Environmental Evaluation Volume III: Ecological Issues and Environmental Impacts, Elsevier. p. 367–389. https://doi.org/10.1016/B978-0-12-805052-1.00023-1.

Such effects have been noted at ranges from 1300 to 3000 km from active surveys. These sounds are primarily in the low-frequency range, so at long distances, the effects are most pronounced for larger species such as the great whales and some fish that use low-frequency sounds.

Other air gun impacts

In addition to marine mammal impacts, the auditory assault from seismic surveys has been found to damage or kill fish eggs and larvae and to impair the hearing and health of fish, making them vulnerable to predators. In marine mammals, these disturbances can disrupt and displace important migratory patterns, pushing marine life away from suitable habitats such as nurseries and foraging, mating, spawning, and migratory corridors. This is especially dangerous for critically endangered species like the North Atlantic right whale, with an extremely low population (only 411 individuals left in the world). For this whale, just one death can have population-wide impacts. Sadly, in November 2018, as part of the America First Energy Strategy, the Trump administration authorized five seismic surveys in the Atlantic. This is the first step in expanding new offshore oil drilling off the US coasts.

Pros and cons of seismic surveys

A well-established body of research indicates that marine mammals seek to avoid seismic survey vessels, responding at ranges of 5–30 km. On the other hand, it is not uncommon for whales or dolphins to approach vessels operating air guns, whether out of curiosity or because of a biological need to be where they are. In 2002, two beaked whales were found dead along a shoreline near where an academic seismic survey was underway. The whales were too decomposed to determine a cause of death, but the incident became the first case of a survey being stopped by the courts due to animal safety concerns. Since then, seismic surveys have continued to be implicated in whale and dolphin beaching and stranding incidents. The passage of an energy bill in the United States in the summer of 2005 called for a comprehensive inventory to be made of oil and gas reserves on the Outer Continental Shelf. Public hearings were conducted in April 2012 to receive public input on seismic surveys for the mid-Atlantic and south-Atlantic regions. According to a report issued by Oceana in April 2013, seismic testing in the mid-Atlantic and south-Atlantic regions would cause 138,000 injuries to whales and dolphins and 13.5 million disturbances to vital behaviors in marine mammals such as breathing, feeding, mating, and communicating—injuries to nine critically endangered North Atlantic right whales and disruption of their critical habitat, when only roughly 500 right whales remain—widespread displacement of whales. Disruption of loggerhead sea turtles as they head to nesting beaches that are soon to be designated as critical habitats—death of fish eggs and larvae—potential disruption of fish migration and spawning.

Underwater Explosions

The armed forces of many countries frequently set off underwater explosions. This is often done to eliminate unexploded ordinance from modern or historic marine battles. As discussed by Keevin and Hempen (1997), gas-containing organs (e.g., lungs, intestinal tract) in marine mammals can be ruptured when the animals are in close proximity to underwater detonations (Fig. 8.20). For instance, injuries have been reported in humpback whales exposed to the pressure waves from underwater explosions.

EFFECTS OF ANTHROPOGENIC NOISE ON PINNIPEDS

Mikkelsen et al. (2019) examined the effects of anthropogenic noise on marine pinnipeds (seals, sea lions, walrus). Like cetaceans, pinnipeds have sensitive underwater hearing; their full hearing range extends from a few hundred Hz to 70–80 kHz (Fig. 8.18). They rely on sound for communication, predator detection, and possibly also for navigation and listening for prey. Pinnipeds in captivity have been found to respond strongly to underwater tone pulses at 8–45 kHz and to sounds from seismic surveys and pile driving in the wild. Ship noise may be especially relevant to coastal seals that rely on periodic land-based resting (hauling out) and therefore spend much of their lives in coastal habitats that strongly overlap with marine traffic. Conservation of pinnipeds has mainly been focused on their haul-out sites, which is due to the limited knowledge of how underwater noise may affect the animals at sea. As reviewed by Jones et al. (2017), injury due to collisions with vessels is widely recognized as a serious risk for cetaceans. Trauma from ship strikes has also been identified in a proportion of both live-stranded and dead-stranded seals in the United States, suggesting that mortality resulting from these collisions may pose a risk, albeit lower, for pinnipeds. However, difficulties in observing these unpredictable events mean that mortality rates are still poorly understood. The phocid (earless) seals found in the Arctic include the harp, hooded, ribbon, bearded, ringed, and harbor seals. These seals

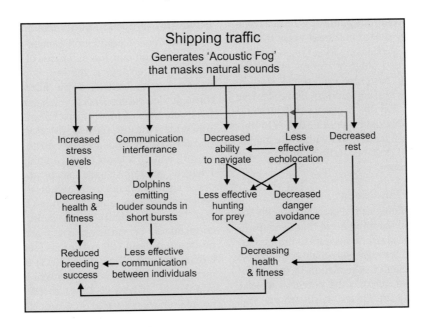

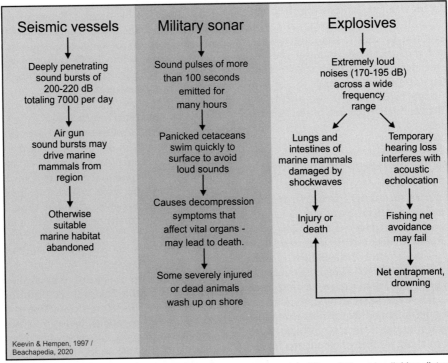

FIG. 8.20 Potential consequences of disturbance conceptual model of the relationships linking disturbance to changes in behavior and physiology, health, vital rates, and population dynamics. (Concepts from Keevin, T.M., Hempen, G.L., 1997. The Environmental Effects of Underwater Explosions with Methods to Mitigate Impacts. Environmental Protection Agency.)

all rely on sound for communication, as well as navigation and predator-prey detection. Vessel noise is likely to be audible to seals at relatively long ranges and has the potential to lead to a range of chronic effects. For pinnipeds exposed to nonpulse underwater sounds, sound exposure levels causing auditory tissue damage were predicted to be 203 dB. A permanent threshold shift in pinniped hearing occurs at 183 dB. It is important to identify areas of greatest risk within the marine environment and to develop techniques to assess long-term sound exposure.

Sound Masking in Pinnipeds

Auditory masking of biologically significant sounds is a potential risk for seals. Shipping noise that overlaps in frequency with seal vocalizations appears to inhibit harbor seal reproduction, through male-male competition or advertisement to females. A reduction in the ability of seals to detect these calls has the potential to lead to biologically significant effects. Furthermore, seal behavioral responses to persistent anthropogenic sound in specific regions of the ocean may include avoidance of important foraging habitats, limiting caloric intake. However, there have been very few empirical studies on behavioral responses by seals to shipping noise, so impacts associated with avoidance have not yet been quantified. This remains a clear data gap when considering the potential risks posed by shipping to seal populations. Where populations seal are also affected by other stressors, the cumulative impacts of the stressors may be substantial.

REMEDIATION OF ANTHROPOGENIC SOUND POLLUTION

As the Arctic becomes increasingly attractive to shipping, its previously undisturbed oceanic regions will certainly become noisier and therefore more dangerous for marine mammals and other marine life forms. There are several approaches to the remediation of this problem. As discussed in the following, one of these solutions is simply for humans to reduce the noise level of their ships, military sonar, seismic surveys, and other sources of anthropogenic sound that penetrate the water column. Another approach to this problem is to reroute shipping to avoid regions where marine mammals either spend much of their year or migrate through on a regular basis. Of course, to do this, we need to know where these animals spend their time so that we can direct shipping away from those areas. Such knowledge is fragmentary at this time, but there are hopeful signs of progress with some species of whales.

Bowhead Whales: Follow the Food

Identification of population centers of marine mammals can be difficult, but in the case of bowhead whales, their feeding ecology is the key. Bowhead whales are baleen whales and feed by filtering zooplankton through their exceptionally long baleen. They consume large quantities of small crustaceans, especially calanoid copepods (mostly *Calanus hyperboreus* and *Calanus glacialis*), euphausiids (mostly Arctic krill, *Thysanoessa raschii*), and, to a lesser extent, amphipods and mysids. Energetic models suggest that bowhead whales need dense aggregations of zooplankton to meet their energetic requirements, and research in Greenland suggests that bowhead whales target dense aggregations of zooplankton. Scientists estimate that a bowhead whale needs to eat about 100 metric tons (100,000 kg) of crustaceans per year. Citta et al. (2015) tracked the movements of the Bering-Chukchi-Beaufort (BCB) population of bowhead whales (*B. mysticetus*). This population ranges across the seasonally ice-covered waters of the Bering, Chukchi, and Beaufort seas (Fig. 8.21). They used locations from 54 bowhead whales, obtained by satellite telemetry between 2006 and 2012, to define areas of concentrated use, termed "core-use areas." In so doing, they identified six primary core-use areas and describe the timing of use and physical characteristics (oceanography, sea ice, and winds) associated with these areas.

Bowhead seasonal migrations

In spring, most bowhead whales migrate from their wintering grounds in the Bering Sea to the Cape Bathurst polynya, Canada (Area 1), and spent most of their time foraging for food in the vicinity of the halocline at depths <75 m. This range of water depths falls within the euphotic zone, where calanoid copepods ascend following winter diapause. Peak use of the polynya occurred between 7 May and 5 July; whales generally left in July, when the copepods typically descend to greater depths. Between 12 July and 25 September, most tagged whales swam east to the shallow shelf waters adjacent to the Tuktoyaktuk Peninsula, Canada (Area 2). Here, wind-driven upwelling promotes the concentration of copepods. Between 22 August and 2 November, whales also congregated near Point Barrow, Alaska (Area 3). East winds promote upwelling that shifts zooplankton populations first onto the Beaufort shelf, but the subsequent relaxation of these winds promotes the zooplankton aggregations to drift toward Point Barrow. Between 27 October and 8 January, this population of bowheads congregated along the northern shore of Chukotka, Russia (Area 4), where

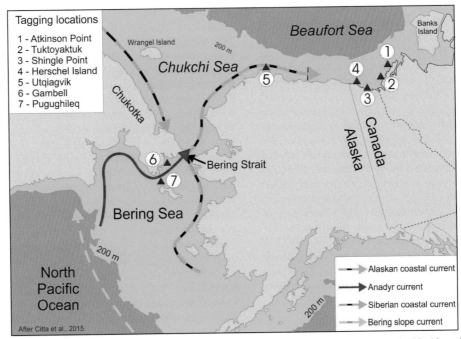

FIG. 8.21 Bowhead whale tagging locations, generalized currents in the western Arctic. (Modified from Citta, J., Quakenbush, L., Okkonen, S., Druckenmiller, M., Maslowski, W., Clement-Kinney, J., George, J., Brower, H., Small, R., Ashjian, C., Harwood, L., Heide-Jørgensen, M., 2015. Ecological characteristics of core-use areas used by Bering-Chukchi-Beaufort (BCB) bowhead whales, 2006—12. Progress in Oceanography 136, 201—222. https://doi.org/10.1016/j.pocean.2014.08.012.)

zooplankton likely concentrate along a coastal front between the southeastward-flowing Siberian Coastal Current and northward-flowing Bering Sea waters. The two other core-use areas are in the Bering Sea: Anadyr Strait (Area 5), where peak use occurs between 29 November and 20 April, and the Gulf of Anadyr (Area 6), where peak use falls between 4 December and 1 April. During this study (2006—12), both areas had fractured sea ice. Whales near the Gulf of Anadyr spent much of their time diving to depths between 75 and 100 m, usually near the seafloor. Here, a subsurface front between cold Anadyr Water and warmer Bering Shelf Water likely supports zooplankton aggregations. The amount of time whales spend near the seafloor in the Gulf of Anadyr, where copepods (in diapause) and perhaps also Arctic krill likely aggregate strongly indicates that these whales feed through the winter months. The timing of bowhead spring migration corresponded with the timing of zooplankton ascent to shallow water in April. The core-use areas identified by Citta et al. (2015) support the results of previous research. Throughout the year, high densities of whales occupy these core-use

areas. Of course, the boundaries of these core-use areas will shift over time, in response to changing ocean conditions.

Multispecies Studies

Following their landmark research on bowhead whales, Citta et al. (2018) expanded their research to track the seasonal movements of six species of ice-associated marine mammals in the Pacific Arctic: ringed seals (*Pusa hispida*), bearded seals (*Erignathus barbatus*), spotted seals (*Phoca largha*), Pacific walruses (*Odobenus rosmarus divergens*); bowhead whales (*B. mysticetus*), and five Arctic and sub-Arctic stocks of beluga whales (*D. leucas*). This study was based on satellite telemetry data for the various species. They also included one seasonal resident, eastern North Pacific gray whales (*Eschrichtius robustus*). This chapter summarized the distribution of daily locations from satellite-linked transmitters during summer (May—November) and winter (December—April) and then examined the overlap among species. Six multispecies core-use areas were identified during the summer period (Fig. 8.22): (1)

Chukotka/Bering Strait; (2) Norton Sound; (3) Kotzebue Sound; (4) the northeastern Chukchi Sea; (5) Mackenzie River Delta/Amundsen Gulf, and (6) Viscount Melville Sound. During the winter period, they identified four multispecies core-use areas: (1) Anadyr Gulf/Strait; (2) central Bering Sea; (3) Nunivak Island, and (4) Bristol Bay. During the summer period, four of the six areas were centered on the greater Bering Strait region and the northwestern coast of Alaska and included most of the species examined. The two remaining summer areas were in the western Canadian Arctic and were characterized by the presence of Bering-Chukchi-Beaufort stock bowhead whales and Eastern Beaufort Sea stock beluga whales, whose distribution overlapped during both summer and winter periods. During the winter period, the main multispecies core-use area was located near the Gulf of Anadyr and extended northward through Anadyr and Bering Straits. This area is contained within the Bering Sea "green belt," an area of enhanced primary and secondary productivity in the Bering Sea.

Movements Tied to the Bering Sea Green Belt

The Green Belt in the Bering Sea, as discussed by Springer et al. (1996), is based on observations of high primary productivity and biomass at numerous trophic levels in the region of the southern edge of the continental shelf in the Bering Sea (Fig. 8.23). The mixing of ocean currents at the shelf edge brings nutrients into the euphotic zone. These nutrients prolong high primary productivity after the spring bloom over the shelf and basin is finished. The prolonged blooming period maintains primary productivity that supports key mesozooplankton species, especially the copepods that are critical to the diet of bowhead whales, ribbon seals, and polar cod. The shelf edge also sustains abundant stocks of the marine fish, birds, and mammals of the Bering Sea. The delineation of areas within the greater Bering Strait region (areas 1–4) is somewhat arbitrary in that the entire region could be considered a single area; however, there are clearly different zones within this area. For example, Norton Sound is quite distinct from Kotzebue Sound. This region has elevated densities of marine mammals partly because it is a migratory corridor connecting the Bering Sea to more northern waters and many animals pass through in spring and fall. It is also known to have elevated benthic and pelagic biomass. In the spring, primary production from ice algae is known to fall to the seafloor in this area and supports a rich benthic environment.

Summer movements in the Bering and Chukchi seas

According to Springer et al. (1996), the Chukotka/Bering Strait area includes a large proportion of daily locations (>20%) of ringed, spotted, and bearded seals, bowhead whales, gray whales, walruses, and belugas. The Norton Sound area was largely defined by belugas; 52% of beluga daily locations were in Norton Sound. Norton Sound was also used by 20% of tagged bearded seals and 33% of tagged spotted seals. The Kotzebue Sound area was visited by many species, including some that were not tagged locally, such as beluga whales. However, except for bearded seals, most species spent little time in Kotzebue Sound. The northeastern Chukchi Sea area was the most diverse of the summering areas and contained >20% of daily locations for bearded seals, gray whales, walruses, and belugas. In addition, many species used the area: >20% of tags for all species were located within the northeastern Chukchi Sea area, except those for belugas, which never entered the Chukchi Sea. The Tuktoyaktuk/Amundsen Gulf area was predominantly important for ringed seals, bowhead whales, and belugas. Viscount Melville Sound was predominantly important for belugas, although it was also visited by a ringed seal and a bowhead whale.

Winter movements in the Bering and Chukchi Seas

Citta et al. (2018) identified four main areas where species overlap during the winter period (December–April): (1) Anadyr Gulf/Strait; (2) central Bering Sea; (3) Nunivak Island, and (4) Bristol Bay. The largest and most diverse area was Anadyr Gulf/Strait. This area extended from the west side of Bering Strait, south through Anadyr Strait, spanned the entrance into the Gulf of Anadyr, and branched east toward St. Matthew Island. The Anadyr Gulf/Strait core area included >20% of all locations for bowhead whales, Pacific walruses, and beluga during the winter period. Additionally, more than 50% of all individuals for all species used this area during the winter period. The Anadyr Gulf/Strait winter core-use area loosely follows what is referred to as the Bering Sea "green belt." The Bering Slope Current flows westward along the shelf break and, upon reaching Cape Navarin, most flows south as the Kamchatka Current. A northern branch, named the Anadyr Current, flows northward into the Gulf of Anadyr and passes through Anadyr and Bering Straits. Shelf breakwaters are characterized by periodic eddies and upwelling, which moves nutrient-rich waters from

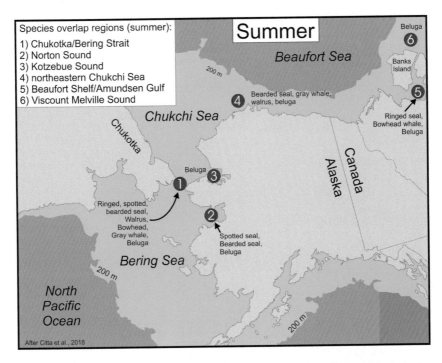

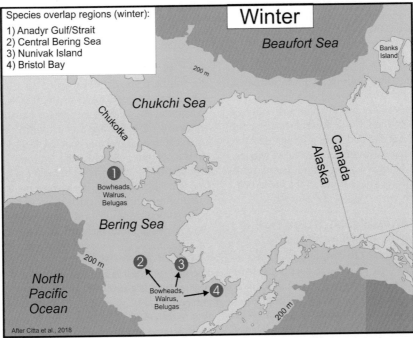

FIG. 8.22 Above: Multispecies core-use areas during the summer (May–November). Below: Multispecies core-use areas during the winter period (December–April). (Modified from Citta, J., Lowry, L., Quakenbush, L., Kelly, B., Fischbach, A., London, J., Jay, C., Frost, K., Crowe, G., Crawford, J., Boveng, P., Cameron, M., Von Duyke, A., Nelson, M., Harwood, L., Richard, P., Suydam, R., Heide-Jørgensen, M., Hobbs, R., et al., 2018. A multi-species synthesis of satellite telemetry data in the Pacific Arctic (1987–2015): Overlap of marine mammal distributions and core use areas. Deep-Sea Research Part II: Topical Studies in Oceanography 152, 132–153. https://doi.org/10.1016/j.dsr2.2018.02.006.)

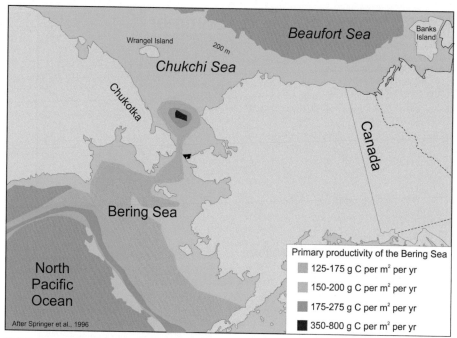

FIG. 8.23 A generalized pattern of primary production in the Bering Sea. The Green Belt is the region of high productivity at the shelf edge, with branches to the northwest and southwest. Elevated production along the Aleutian Arc completes a circle around the basin. (Modified from Springer, A., Peter McRoy, C., Flint, M., 1996. The Bering Sea Green Belt: Shelf-edge processes and ecosystem production. Fisheries Oceanography 5 (3–4), 205–223. https://doi.org/10.1111/j.1365-2419.1996.tb00118.x.)

the Aleutian Basin into the euphotic zone near the shelf break. These nutrient-rich waters result in a green belt of high primary and secondary productivity that extends along the Bering Sea shelf break toward Russia and northward through Anadyr and Bering Straits. During the winter period, the distribution of bowhead whales, Pacific walrus, and beluga whales, in addition to all three seal species, falls within portions of the green belt.

VESSEL RISK MANAGEMENT STRATEGIES

Halliday et al. (2018) discussed three different management strategies that are used to minimize the risks of vessels to whales. The first of these is keeping vessels away from whales, either through ship routing measures or exclusion zones. The second is restricting vessel speed, which reduces risks of ship strikes and can lower noise pollution. The third is monitoring shipping lanes using marine mammal observers when whales are nearby. The observers could tell ship's captains to change course, slow down, or stop engines if a collision with a marine mammal is imminent. For example, adjusting the vessel corridor in the Roseway Basin off

the southern tip of Nova Scotia in Canada to avoid the Right Whale Conservation Area was assessed to reduce the risk of ship strikes for North Atlantic right whales by 62%. For another example, the Port of Vancouver (Canada) recently enacted an 11-knot slowdown in Haro Strait, reducing the amount of time when foraging by southern resident killer whales would be impacted by about 10%. Management schemes that use multiple measures, such as an exclusion zone with a slowdown zone around it, may be more effective than any single management measure.

Canadian High Arctic

Keeping marine mammals safe in the Canadian Arctic Archipelago is a difficult and complicated task. As discussed by Halliday et al. (2018), the Polar Code states that mariners should respect national and international rules related to areas with high densities of marine mammals, but information on the exact location of these areas is not readily available to mariners traveling through the Inuvialuit Settlement Region (ISR) (Fig. 8.24). The Code applies to passenger vessels but does not apply to pleasure craft. Unfortunately, the

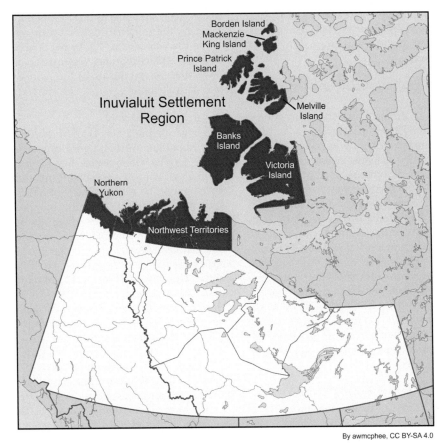

FIG. 8.24 Map of Inuvialuit Settlement Region by awmcphee, CC BY-SA 4.0. (From Map of Inuvialuit Settlement Region, 2020. https://commons.wikimedia.org/w/index.php?curid=79588984.)

two whale areas are very close to Herschel Island and the Smoking Hills, two common destinations for tourists. Herschel Island is just west of the beluga area, and all vessels traveling to the island from the east pass directly through this area. The Smoking Hills are on the west side of Franklin Bay, directly south of the bowhead area near Cape Bathurst, so vessels following the coastline to see the Smoking Hills pass directly through this area. Given the proximity of marine mammal areas to tourist destinations, it is very unlikely that management interventions could fully exclude tourist vessels from these marine mammal areas. A corridor would not stop vessels from visiting sites such as the Smoking Hills or Herschel Island, but it might reduce the impact for all other vessels traveling through the ISR. For vessels traveling outside of the corridor, a speed restriction of 10 knots could be an effective management tool for reducing impacts in whale concentration areas. Slowdowns are typically used to reduce ship strikes and to

reduce acoustic disturbance. Slowdowns are effective for both of these reasons, but they also add travel time to Arctic voyages, leading to conflicts with the shipping industry.

Remediation of Ship Noise in the Western Canadian Arctic

Halliday et al. (2018) suggest that excluding vessels from the two marine protected areas in the western Canadian Arctic region would have little or no impact on whales within concentration areas since vessels would likely just be displaced to adjacent areas with similar whale concentrations. Restricting vessels to the Canadian government's proposed low-impact corridor may reduce impact slightly, but creating a corridor completely outside of the known whale area could more significantly reduce the potential impact of vessels on whales in those areas. Restricting vessel speed within whale areas would also reduce the impact of passenger vessels but would not likely reduce the impact of plea-sure craft. Overall, a combination of management mea-sures may be the best way to reduce impacts on whales in concentration areas.

Canadian Inuit Input on Arctic Shipping Regulation

Dawson et al. (2020) discussed the Canadian govern-ment's development of Low Impact Shipping Corridors as an adaptation strategy that supports safety and sus-tainability under rapidly changing environmental con-ditions. The corridors are voluntary maritime routes where services and infrastructure investments are prior-itized. While various experts were consulted to establish corridor locations, Indigenous knowledge from Arctic communities was not considered in much detail. The Arctic Corridors and Northern Voices (ACNV) project was established to address this omission. Altogether, 13 Canadian Arctic communities across the Inuit home-lands were involved in the project. A summary of the recommendations includes the following: (1) establish-ment of preferred corridors, (2) establishment of areas to avoid, (3) seasonal route restrictions, (4) modifica-tion of vessel operation, and (5) identification of areas where charting is needed.

Regulating Traffic in the Bering Strait

Huntington et al. (2019) discuss the problem of increasing ship traffic in the Bering Strait region, which is home to a spectacular abundance of seabirds, marine mammals, and marine productivity. The confluence of expanding maritime commerce, remoteness, vibrant Indigenous cultures, and extraordinary biological rich-ness requires robust governance to promote maritime safety, cultural protection, and environmental conserva-tion. The use of areas to be avoided (ATBAs) offers one mechanism to help achieve this goal, and three have already been adopted by the IMO, along with shipping routes through the Bering Strait. The Bering Strait pro-vides the only marine connection between the Pacific and Arctic Oceans. Huntington et al. (2019) define the larger Bering Strait region as comprising the north-ern Bering Sea and the Chukchi Sea from St. Matthew Island in the south to Wrangel Island in the northwest and Point Barrow in the northeast (Fig. 8.25). The re-gion is home to the Chukchi, Iñupiaq, St. Lawrence Is-land, and Siberian Yupik peoples. These groups have maintained their lifeway of hunting, fishing, traveling, and trading on the sea. This is a good place to hunt ma-rine mammals—a long-standing tradition of all these peoples. All marine mammal migrations between the Bering Sea in the Pacific Ocean and the Chukchi Sea

in the Arctic must pass through the Bering Strait. Unfor-tunately, the same is true for ocean vessel traffic be-tween the two oceans. The Bering Strait is thus a focal point for Arctic shipping. Vessel traffic in the region is increasing, as are threats to the marine environment and native peoples.

Areas to be avoided

The larger Bering Strait region includes many ecologi-cally important areas and great cultural importance to native peoples. Commercial shipping poses a threat to these areas, but this can be mitigated through sensible governance. The establishment of ATBAs can make a substantial contribution to the management of ship traffic in the strait. In parts of the Bering Strait region, there is sufficient information about sea ice, seabirds, and marine mammals to support the definition of fixed ATBAs in addition to those that have already been desig-nated. Huntington et al. (2019) recommend that the designation of additional ATBAs should be done in close consultation and active participation of indige-nous residents (Fig. 8.26). Their knowledge of marine mammal behavior and ecology, as well as their knowl-edge of culturally important sites, areas, and practices, will make significant contributions to the process. Mar-iners also have needs that must be taken into consider-ation, including safety and freedom of navigation. Compliance with staying outside ATBA boundaries must be monitored effectively, which will be difficult in this remote region. ATBAs cannot address all the needs for the governance of commercial vessel traffic in the region, but they can make a valuable contribution.

LIMITING WASTEWATER POLLUTION FROM SHIPS

Parks et al. (2019) assessed the impacts and possible remediation of wastewater pollution from increasing ship traffic along the coast of northwestern Alaska. The Native people of this region rely upon maritime food resources. They have deepening concerns over vessel waste and its potential impact upon people and animals. Subsistence use areas are large, comparable with the size of some US states, and therefore, impacts could be felt over large areas. Consequently, the following concerns are expressed by people throughout the region. Alaska Natives are not prepared for a large-scale oil spill to this pristine environment, nor can they deal with the cleanup of polluted wastewater from ships passing their coastal settlements. Climate change is threatening to Alaska Natives' way of life. In

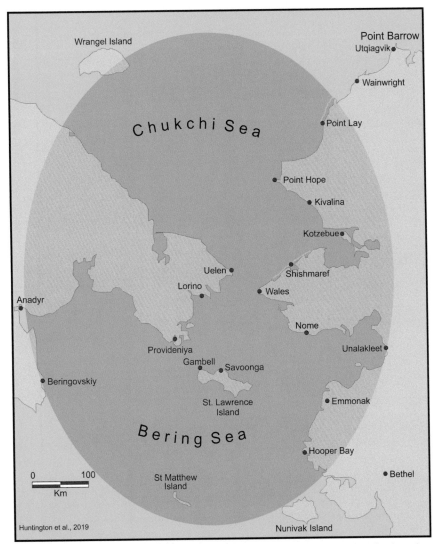

FIG. 8.25 Map of the Bering and Chukchi sea regions, showing the greater Bering Strait region (shaded region) as defined by Huntington et al., (2019). Illustration by the author.

the face of these threats, Indigenous people must have a hand in determining their own destiny.

Biological Pathogens Promoted by Wastewater

Ocean pollution from shipping includes such things as discharges of oil, trash, sewage and gray water, emissions from engines, and noise. Such discharges will increase linearly as vessel traffic continues to grow. Marine accidents involving collision, grounding, and sinking significantly increase the risk to the environment. This problem is intensified due to the lack of available response resources in the region. It often takes many days for rescue vessels to arrive at the scene of a marine accident in the Arctic. Discharges, including sewage, may expose maritime resources to zoonotic pathogens, which are responsible for transmitting diseases between humans and animals. In the northern Bering Sea, harmful algal blooms and paralytic shellfish poisoning may start occurring more often. A 2016 study found elevated

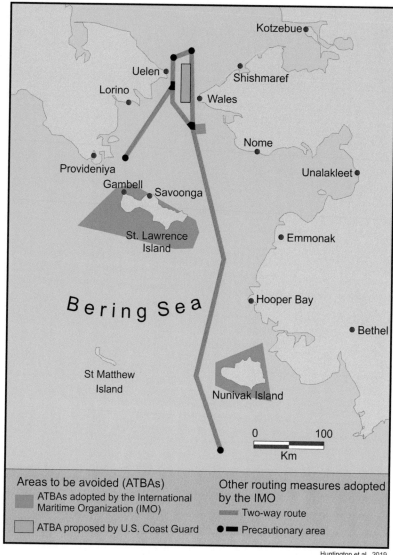

FIG. 8.26 Map of existing areas to be avoided (ATBAs) and routing measures in the Bering Strait area, as adopted by the International Maritime Organization (*deep blue shaded areas*), plus an additional ATBA proposed by the US Coast Guard as part of its Port Access Route Study for the Bering Strait (*Green shaded areas*). Two-way routes show in pink. Precautionary areas are shown by *black dots*. (Modified from Huntington, H., Bobbe, S., Hartsig, A., Knight, E., Knizhnikov, A., Moiseev, A., Romanenko, O., Smith, M., Sullender, B., 2019. The role of areas to be avoided in the governance of shipping in the greater Bering Strait region. Marine Policy 110. https://doi.org/10.1016/j.marpol.2019.103564 4.)

levels of algal bloom toxins in marine mammals from the region. In the fall of 2017, 39 walruses washed ashore in western Alaska. Four walruses underwent necropsy, and all tested positive for biotoxins from algae. As pointed out by Parks et al. (2019), the issue of vessel-generated waste in this region has not yet reached crisis proportions, but preventative actions now to maintain, as near as possible, the status quo will be cheaper and easier to implement than corrective measures to undo a deteriorated situation in the future.

OCEAN NOISE REMEDIATION

Attempts to limit ocean noise have been carried out at local, regional, national, and international levels. While awareness of the problem goes back several decades, attempts at remediation have mostly come in the 21st century. First, it was necessary to gain an understanding of the kind of damage that undersea noise is inflicting on marine life, from plankton to blue whales. This process is still under way, and it is going to take a great deal of additional research. In the meantime, we know that marine mammals are being hurt by acoustic pollution of their habitats and so efforts are under way to limit that damage, especially from very loud sources of sound.

Remediation of Naval Sonar Impacts

Limiting the damage done to marine mammals from naval sonar has been discussed by Bernaldo de Quirós et al. (2019). Given that tactical MFAS is used by many nations, it is important to find ways to mitigate its impact. If the likelihood or degree of an individual's response to MFAS depends on its prior exposure to sonar, one option to mitigate acute effects of frequency active sonar exposure on beaked whales would be to restrict the use of such sonar to areas where training and testing are already regularly dome, with no associated mass strandings of whales, and prohibiting use in areas where mass strandings have occurred, or beaked whale populations are known or suspected to occur, and where MFAS is rarely or never used.

US government plan for noise reduction

The NOAA Ocean Noise Strategy Roadmap is a document designed to support the implementation of an agency-wide strategy for addressing ocean noise from 2016 to 2026 (Gedamke et al., 2016). The roadmap highlights the expansion of NOAA's historical mission to protect specific species by addressing noise impacts in acoustic habitats of special interest. It is clear that sound pollution and other environmental stressors lead to a combination of physiological and behavioral changes (Fig. 8.27). These changes damage marine mammal health. Weakened animals have poorer adult survival rates, lower fecundity, and lower offspring survival. The document presents a series of goals and recommendations that would improve NOAA's ability to manage both species and the places they inhabit in the context of a changing acoustic environment. The recommendations include actions to be taken by individual programs within NOAA, while others call for partnerships among multiple NOAA programs. The roadmap attempts to balance the competing needs of commercial, economic, scientific, national defense and security activities, protected species, and natural acoustic habitats. This is a delicate balancing act that is certain to present NOAA with significant challenges over the coming decade. Staff and leadership from NOAA Fisheries' Offices of Protected Resources and Science and Technology and the National Ocean Service's Office of National Marine Sanctuaries identified four overarching goals the Strategy aims to achieve the following: (1) Science: NOAA and federal partners are filling shared critical knowledge gaps and building an understanding of noise impacts over ecologically relevant scales. (2) Management: NOAA's actions are integrated across the agency and minimizing the acute, chronic, and cumulative effects of noise on marine species and their habitat. (3) Decision support tools: NOAA is developing publicly available tools for assessment, planning, and mitigation of noise-making activities over ecologically relevant scales. (4) Outreach: NOAA is educating the public on noise impacts, engaging with stakeholders, and coordinating with related international efforts.

Reduction of emissions by Arctic shipping fleets

The ocean conservation group, Oceana (2020), has put forward the following recommendations to reduce global ship emissions: Shipping fleets should implement technical and operational measures to reduce global warming pollution immediately. Such measures include speed reductions, weather routing, fuel switching, and specialized hull coatings. Also, these fleets should begin to implement longer-term measures to reduce global warming pollution, such as fuel-efficient design of new ships and engines created specifically for slow steaming. Furthermore, the IMO should set international emission standards to reduce global warming pollutants from the shipping industry. Overall, speed reductions are a quick, easy, and effective way to achieve emissions reductions from ocean-going vessels. Given the recent increases in oil prices, speed reduction makes sense not only environmentally but also economically. Emissions, especially those of carbon dioxide, are directly proportional to fuel consumption. Greater speeds require increased fuel consumption. Consequently, slowing down, even by a small amount, can result in significant fuel savings and emissions reductions. The IMO calculated that a speed reduction of just 10% across the global fleet would result in a 23.3% reduction in emissions. Research has shown that slowing some of large cargo ships by just five knots, or 20%, resulted in savings of around 50% on fuel costs. Restrictions on vessel speed

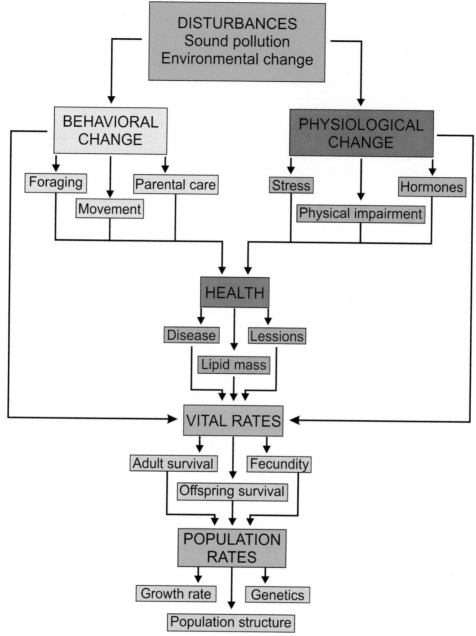

New et al., 2014

FIG. 8.27 A diagram of the "Population Consequences of Disturbance" framework linking disturbance to changes in behavior and physiology, health, vital rates, and population dynamics. (After New, L.F., Clark, J.S., Costa, D.P., Fleishman, E., Hindell, M.A., Klanjšček, T., Lusseau, D., Kraus, S., McMahon, C.R., Robinson, P.W., Schick, R.S., Schwarz, L.K., Simmons, S.E., Thomas, L., Tyack, P., Harwood, J., 2014. Using short-term measures of behaviour to estimate long-term fitness of southern elephant seals. Marine Ecology Progress Series 496, 99–108. https://doi.org/10.3354/meps10547.)

would reduce emissions of carbon dioxide, black carbon, nitrogen oxides, and nitrous oxide. Compared with other forms of transport, ships traveling at slow speeds have been found to be far more efficient and less polluting—roughly 10 times more efficient than trucks and at least a hundred times more efficient than air transport. As ship speeds increase, much of this efficiency is lost. Ships traveling at very high speeds have been found to have similar energy demands to those of airplanes.

CONCLUSIONS

Marine mammals live in an oceanic soundscape. The underwater world used to be a relatively quiet realm. Then the Industrial Revolution came along, and the sound coal- and oil-fired ship engines began to build year by year. The number and size of marine ships grew concurrently with the power of marine engines. Human awareness of the impacts of the noise we make at sea was much slower in developing. Research since the 1980s and 1990s has begun to raise that awareness, but so far little or nothing has been done to dampen the noise.

How Loud is It?

As discussed by Science of Sound (2020), sound levels in the ocean differ from location to location and change with time. Different sources of sound contribute to the overall noise level, including shipping, breaking waves, marine life, and other anthropogenic and natural sounds. The amplitude (volume level) of ocean noise varies with the frequency of the sound. For instance, in a study of ocean noise off Point Sur, California, sounds emitted at 50 Hz have an average volume level of 91 dB (re 1 μPa^2/Hz) (Fig. 8.28). Shipping noise is the dominant contributor to background noise at this frequency. Sounds in the 100 Hz range are emitted at an average of 81 dB. The sound volume gradually diminishes, so that at 400 Hz, the average volume is 71.6 dB.

Cumulative Effects

Rako-Gospić and Picciulin (2018) assessed mitigation procedures for sound pollution in the oceans. In their view, effective and responsible mitigation measures are particularly important where critical habitats are inhabited by small, vulnerable populations. Current mitigation measures and regulations are mainly designed to address exposure to a single event, not to a series of noises or a continuous barrage of noise. We still do not understand the effects of cumulative

exposure to anthropogenic sounds on marine mammals, nor do we fully comprehend how repeated exposure is accumulated by an animal. Much more research needs to be done to better our understanding of the range of sea noise levels and the source of the dominant noise in each frequency range associated with a particular habitat. This will not be a quick process. We need long-term data sets to monitor trends in soundscapes, especially in crucial areas, such as reproductive or territorial sites, migration routes, biodiversity hot spots, and areas where anthropic activities are anticipated in the future. This information represents the base for local management and the development of proper marine spatial planning, site evaluation, and impact assessments.

Out of Sight, Out of Mind

As with so many other human-caused problems in the oceans and the Arctic, most people are unaware of the damage we are doing because we do not see it. Few of us will ever appreciate the damage we are doing to marine mammals as we fill the ocean with our ever-increasing noise. This is a form of pollution that is invisible to our eyes, but it has serious impacts on marine mammals, contaminating the soundscapes upon which they rely for communication, prey detection, and predator evasion. The only visual evidence most of us have concerning the plight of whales is when a number of them wash up on a beach, either already dead or very close to it. In most cases, the news media treat these events as unsolved mysteries, rather than human-caused fatalities, such as when beaked whales die from decompression sickness in their attempts to evade loud sonar.

Jurisdiction Problems

Once again, we find ourselves asking whether we have the political will to do something about the problems discussed in this chapter. It would certainly help if countries that have ships that sail in Arctic waters would agree to help remediate marine life damage caused by their vessels, especially as the decline in sea ice fosters increased Arctic Ocean voyages. But these problems are made more complex by the mixed governance of northern waters. The Arctic can be divided geographically into two sets of regions. Some waters fall within the auspices of national governments because they are in national territorial waters. Most of the Arctic Ocean lies in international waters where national governments have no jurisdiction. In recent years, governments and international organizations have issued sets of guidelines meant to alleviate environmental problems caused

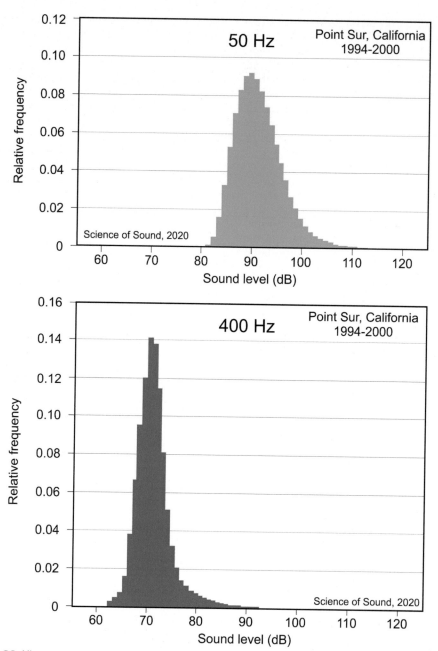

FIG. 8.28 Histograms showing the distribution of sound levels (dB re 1 μPa^2/Hz) at 50 Hz (above) and 400 Hz (below) as observed at a receiver on the continental slope off Point Sur, California, from November 1994 to December 2000. (Modified from Science of Sound, 2020. Ocean Noise Variability and Noise Budgets. https://dosits.org/science/advanced-topics/ocean-noise-variability-and-noise-budgets/.)

by shipping in the Arctic. Some excellent recommendations have been made by the UN IMO, the International Whaling Commission, and the Arctic Council. However, for these guidelines to succeed, these international organizations must rely on the voluntary compliance of ship's captains and shipping company management. The Arctic Ocean is simply too remote from "civilization." It would be impossible, extremely expensive, and even dangerous to attempt to send fleets of ships to enforce international guidelines in Arctic waters. And when all is said and done, guidelines are not laws. No one has the jurisdiction to enforce guidelines. Increasing the public's awareness of shipping damage to marine life in the Arctic would be very helpful, but it is a slow process at best. In the meantime, ship's captains have deadlines to meet, and shipping companies keep an eye on their bottom line.

REFERENCES

Aksenov, Y., Popova, E.E., Yool, A., Nurser, A.J.G., Williams, T.D., Bertino, L., Bergh, J., 2017. On the future navigability of Arctic sea routes: high-resolution projections of the Arctic Ocean and sea ice. Marine Policy 75, 300–317. https://doi.org/10.1016/j.marpol.2015.12.027.

Bennett, M., 2019. The Arctic shipping route no one's talking about. Cryopolitics.

Bernaldo de Quiros, Y., Fernandez, A., Baird, R.W., Brownell, R.L., Aguilar de Soto, N., et al., 2019. Advances in research on the impacts of anti-submarine sonar on beaked whales. Proceedings of the Royal Society B: Biological Sciences 20182533. https://doi.org/10.1098/rspb.2018.2533.

Citta, J.J., Quakenbush, L.T., Okkonen, S.R., Druckenmiller, M.L., Maslowski, W., Clement-Kinney, J., George, J.C., Brower, H., Small, R.J., Ashjian, C.J., Harwood, L.A., Heide-Jørgensen, M.P., 2015. Ecological characteristics of core-use areas used by Bering-Chukchi-Beaufort (BCB) bowhead whales, 2006-2012. Progress in Oceanography 136, 201–222. https://doi.org/10.1016/j.pocean.2014.08.012.

Citta, J.J., Lowry, L.F., Quakenbush, L.T., Kelly, B.P., Fischbach, A.S., London, J.M., Jay, C.V., Frost, K.J., Crowe, G.O.C., Crawford, J.A., Boveng, P.L., Cameron, M., Von Duyke, A.L., Nelson, M., Harwood, L.A., Richard, P., Suydam, R., Heide-Jørgensen, M.P., Hobbs, R.C., et al., 2018. A multi-species synthesis of satellite telemetry data in the Pacific Arctic (1987–2015): overlap of marine mammal distributions and core use areas. Deep-Sea Research Part II Topical Studies in Oceanography 152, 132–153. https://doi.org/10.1016/j.dsr2.2018.02.006.

Dasgupta, S., 2019a. How Shipping has Become a Great Threat to Whales. https://www.marineinsight.com/environment/how-shipping-has-become-a-great-threat-to-whales/.

Dasgupta, S., 2019b. 10 Ships Sunk by Accidents with Icebergs. https://www.marineinsight.com/maritime-history/10-ships-sunk-by-accident-with-iceberg/.

Dawson, J., Carter, N., van Luijk, N., Parker, C., Weber, M., Cook, A., Grey, K., Provencher, J., 2020. Infusing inuit and local knowledge into the low impact shipping corridors: an adaptation to increased shipping activity and climate change in Arctic Canada. Environmental Science and Policy 105, 19–36. https://doi.org/10.1016/j.envsci.2019.11.013.

D'Amico, A., Pittenger, R., 2009. A brief history of active sonar. Aquatic Mammals 35. https://doi.org/10.1578/AM.35.4.2009.426.

Ellison, W.T., Southall, B.L., Clark, C.W., Frankel, A.S., 2012. A new context-based approach to assess marine mammal behavioral responses to anthropogenic sounds. Conservation Biology 26 (1), 21–28. https://doi.org/10.1111/j.1523-1739.2011.01803.x.

Faury, O., Cariou, P., 2016. The Northern Sea Route competitiveness for oil tankers. Transportation Research Part A: Policy and Practice 94, 461–469. https://doi.org/10.1016/j.tra.2016.09.026.

Fournet, M.E.H., Matthews, L.P., Gabriele, C.M., Haver, S., Mellinger, D.K., Klinck, H., 2018. Humpback whales Megaptera novaeangliae alter calling behavior in response to natural sounds and vessel noise. Marine Ecology Progress Series 607, 251–268. https://doi.org/10.3354/meps12784.

Fuglestvedt, J.S., Dalsøren, S.B., Samset, B.H., Berntsen, T., Myhre, G., Hodnebrog, Ø., Eide, M.S., Bergh, T.F., 2014. Climate penalty for shifting shipping to the Arctic. Environmental Science and Technology 48 (22), 13273–13279. https://doi.org/10.1021/es502379d.

Gedamke, J., Harrison, J., Hatch, L., Angliss, R., Barlow, J., et al., 2016. Ocean Noise Strategy Roadmap.

Goldbogen, J.A., Southall, B.L., DeRuiter, S.L., Calambokidis, J., Friedlaender, A.S., Hazen, E.L., Falcone, E.A., Schorr, G.S., Douglas, A., Moretti, D.J., Kyburg, C., McKenna, M.F., Tyack, P.L., 2013. Blue whales respond to simulated mid-frequency military sonar. Proceedings of the Royal Society B: Biological Sciences 280 (1765). https://doi.org/10.1098/rspb.2013.0657.

Hall, C.F., 1865. Arctic researches, and life among the Esquimaux. Harper, New York.

Halliday, W.D., Têtu, P.L., Dawson, J., Insley, S.J., Hilliard, R.C., 2018. Tourist vessel traffic in important whale areas in the western Canadian Arctic: risks and possible management solutions. Marine Policy 97, 72–81. https://doi.org/10.1016/j.marpol.2018.08.035.

Heide-Jørgensen, M.P., Hansen, R.G., Westdal, K., Reeves, R.R., Mosbech, A., 2013. Narwhals and seismic exploration: is seismic noise increasing the risk of ice entrapments? Biological Conservation 158, 50–54. https://doi.org/10.1016/j.biocon.2012.08.005.

Hong, N., 2012. The melting Arctic and its impact on China's maritime transport. Research in Transportation Economics 35 (1), 50–57. https://doi.org/10.1016/j.retrec.2011.11.003.

Huntington, H.P., Bobbe, S., Hartsig, A., Knight, E.J., Knizhnikov, A., Moiseev, A., Romanenko, O., Smith, M.A., Sullender, B.K., 2019. The role of areas to be avoided in the governance of shipping in the greater Bering Strait region. Marine Policy 110. https://doi.org/10.1016/j.marpol.2019.103564.

Map of Inuvialuit Settlement Region, 2020. https://commons.wikimedia.org/w/index.php?curid=79588984.

Jones, E.L., Hastie, G.D., Smout, S., Onoufriou, J., Merchant, N.D., Brookes, K.L., Thompson, D., 2017. Seals and shipping: quantifying population risk and individual exposure to vessel noise. Journal of Applied Ecology 54 (6), 1930–1940. https://doi.org/10.1111/1365-2664.12911.

Keevin, T.M., Hempen, G.L., 1997. The Environmental Effects of Underwater Explosions with Methods to Mitigate Impacts. Environmental Protection Agency.

Khon, V.C., Mokhov, I.I., Latif, M., Semenov, V.A., Park, W., 2010. Perspectives of Northern Sea route and northwest passage in the twenty-first century. Climatic Change 100 (3), 757–768. https://doi.org/10.1007/s10584-009-9683-2.

Knopp-Schwyn, C., 2020. Map of the Northeast Passage. https://en.wikipedia.org/wiki/Northeast_Passage.

Liu, X.h., Ma, L., Wang, J.y., Wang, Y., Wang, L.n., 2017. Navigable windows of the northwest passage. Polar Science 13, 91–99. https://doi.org/10.1016/j.polar.2017.02.001.

Lonsdale, E., 2019. Northwest Passage or Northeast Passage: Which Cruise to Choose? Travel Advice. Original work published 2019. https://www.mundyadventures.co.uk/adventure-news/travel-advice/northwest-northeast-passage-cruise.

Maddox, L., 2019. The New Northwest Passage: tourism through the Arctic. American Geographical Society Library. https://agslibraryblog.wordpress.com/2019/09/17/the-new-northwest-passage-tourism-through-the-arctic/.

Marine Mammal Commission of the United States, 2007. Marine Mammals and Noise: A Sound Approach to Research and Management. Report to Congress from the Marine Mammal Commission.

Melia, N., Haines, K., Hawkins, E., 2016. Sea ice decline and 21st century trans-Arctic shipping routes. Geophysical Research Letters 43 (18), 9720–9728. https://doi.org/10.1002/2016GL069315.

Mikkelsen, L., Johnson, M., Wisniewska, D.M., van Neer, A., Siebert, U., Madsen, P.T., Teilmann, J., 2019. Long-term sound and movement recording tags to study natural behavior and reaction to ship noise of seals. Ecology and Evolution 9 (5), 2588–2601. https://doi.org/10.1002/ece3.4923.

Nachtigall, P.E., Supin, A.Y., Pacini, A.F., Kastelein, R.A., 2018. Four odontocete species change hearing levels when warned of impending loud sound. Integrative Zoology 13 (2), 160–165. https://doi.org/10.1111/1749-4877.12286.

New, L.F., Clark, J.S., Costa, D.P., Fleishman, E., Hindell, M.A., Klanjšček, T., Lusseau, D., Kraus, S., McMahon, C.R., Robinson, P.W., Schick, R.S., Schwarz, L.K., Simmons, S.E., Thomas, L., Tyack, P., Harwood, J., 2014. Using short-term measures of behaviour to estimate long-term fitness of southern elephant seals. Marine Ecology Progress Series 496, 99–108. https://doi.org/10.3354/meps10547.

New, L.F., Hall, A.J., Harcourt, R., Kaufman, G., Parsons, E.C.M., et al., 2015. The modelling and assessment of whale-watching impacts. Ocean & Coastal Management 115, 10–16. https://doi.org/10.1016/j.ocecoaman.2015.04.006.

Northern Sea Route Information Office, 2020. NSR Shipping Traffic - Activities in January-June 2020. https://arctic-lio.com/nsr-shipping-traffic-activities-in-january-june-2020/.

NSIDC, 2020a. Arctic Sea Ice News and Analysis - Lingering Seashore Days. https://nsidc.org/arcticseaicenews/2020/10/.

NSIDC, 2020b. All about Sea Ice Formation. https://nsidc.org/cryosphere/seaice/characteristics/formation.html.

NSIDC, 2020c. Arctic Sea Ice News and Analysis, November 4, 2020. http://nsidc.org/arcticseaicenews/.

NSIDC, 2020d. Arctic Sea Ice News and Analysis. Ocean Waves in November—In the Arctic. https://nsidc.org/arcticseaicenews/2020/.

NSIDC, 2020e. All about Sea Ice: Thermodynamics: Melt. In: https://nsidc.org/cryosphere/seaice/processes/thermodynamic_melt.html.

Oceana, 2020. Shipping Pollution. https://europe.oceana.org/en/shipping-pollution-1.

Owen, C., Rendell, L., Constantine, R., Noad, M.J., Allen, J., Andrews, O., Garrigue, C., Poole, M.M., Donnelly, D., Hauser, N., Garland, E.C., 2019. Migratory convergence facilitates cultural transmission of humpback whale song. Royal Society Open Science 6 (9). https://doi.org/10.1098/rsos.190337.

Parks, M., Ahmasuk, A., Compagnoni, B., Norris, A., Rufe, R., 2019. Quantifying and mitigating three major vessel waste streams in the northern Bering Sea. Marine Policy 106. https://doi.org/10.1016/j.marpol.2019.103530.

Pennington, E., 2020. Humpback Whales Seize Chance to Sing in Alaska's Cruise-free Covid Summer. The Guardian.

Rako-Gospić, N., Picciulin, M., 2018. Underwater noise: sources and effects on marine life. World Seas: An Environmental Evaluation Volume III: Ecological Issues and Environmental Impacts. Elsevier, pp. 367–389. https://doi.org/10.1016/B978-0-12-805052-1.00023-1.

Reeves, R.R., Ewins, P.J., Agbayani, S., Heide-Jørgensen, M.P., Kovacs, K.M., Lydersen, C., Suydam, R., Elliott, W., Polet, G., van Dijk, Y., Blijleven, R., 2014. Distribution of endemic cetaceans in relation to hydrocarbon development and commercial shipping in a warming Arctic. Marine Policy 44, 375–389. https://doi.org/10.1016/j.marpol.2013.10.005.

Reeves, R.R., Ewins, P.J., Agbayani, S., Heide-Jørgensen, M.P., Kovacs, K.M., 2014. Distribution of endemic cetaceans in relation to hydrocarbon development and commercial shipping in a warming Arctic. Marine Policy 44, 375–389. https://doi.org/10.1016/j.marpol.2013.10.005.

Current Northwest Passage routes, 2020. Image courtesy of Wikipedia. https://en.wikipedia.org/wiki/Northwest_Passage.

Safety4Sea, 2020. Underwater Noise Problems Come on the Surface. https://safety4sea.com/cm-underwater-noise-problems-come-on-the-surface/.

Saul, J., Baldwin, C., Adomaitis, Solsvik, T., Cohn, C., 2020. Arctic Headache for Ship Insurers as Routes Open up. MSN News.

Schøyen, H., Bråthen, S., 2011. The Northern Sea route versus the Suez canal: cases from bulk shipping. Journal of Transport Geography 19 (4), 977–983. https://doi.org/10.1016/j.jtrangeo.2011.03.003.

Science of Sound, 2020. Ocean Noise Variability and Noise Budgets. https://dosits.org/science/advanced-topics/ocean-noise-variability-and-noise-budgets/.

Southall, B.L., 2017. Noise. In: Würsig, B., Thewissen, J.G.M., Kovacs, K. (Eds.), Encyclopedia of Marine Mammals, third ed. Academic Press, pp. 637–645.

Southall, B.L., Bowles, A.E., Ellison, W.T., Finneran, J.J., Gentry, R.L., 2007. Marine mammal noise exposure criteria: initial scientific recommendations. Aquatic Mammals 33, 411–521.

Springer, A.M., Peter McRoy, C., Flint, M.V., 1996. The Bering Sea green belt: shelf-edge processes and ecosystem production. Fisheries Oceanography 5 (3–4), 205–223. https://doi.org/10.1111/j.1365-2419.1996.tb00118.x.

Stafford, K., 2013. Anthropogenic Sound and Marine Mammals in the Arctic. Pew Charitable Trust. U.S. Arctic Program. https://www.pewtrusts.org/~/media/Assets/2013/06/07/arcticnoise_final_web.pdf.

Stafford, K.M., Castellote, M., Guerra, M., Berchok, C.L., 2018. Seasonal acoustic environments of beluga and bowhead whale core-use regions in the Pacific Arctic. Deep-Sea Research Part II 152, 108–120. https://doi.org/10.1016/j.dsr2.2017.08.003.

Stevenson, T.C., Davies, J., Huntington, H.P., Sheard, W., 2019. An examination of trans-Arctic vessel routing in the Central Arctic Ocean. Marine Policy 100, 83–89. https://doi.org/10.1016/j.marpol.2018.11.031.

Theocharis, D., Sanchez Rodrigues, V., Pettit, S., Haider, J., 2019. Feasibility of the Northern Sea Route: The role of distance, fuel prices, ice breaking fees and ship size for the product tanker market. Transportation Research Part E 129, 111–135. https://doi.org/10.1016/j.tre.2019.07.003.

Timco, G., Johnston, M., 2003. Arctic Ice Regime Shipping System: Pictorial Guide. Canadian Hydraulics Centre, National Research Council of Canada and Transport Canada.

U.S. Committee on the Marine Transportation System, 2019. A Ten-Year Projection of Maritime Activity in the U.S. Arctic Region, vol. 118.

Wesche, S.D., Chan, H.M., 2010. Adapting to the impacts of climate change on food security among inuit in the Western Canadian Arctic. EcoHealth 7 (3), 361–373. https://doi.org/10.1007/s10393-010-0344-8.

Wilson, S.C., Crawford, I., Trukhanova, I., Dmitrieva, L., Goodman, S.J., 2020. Estimating risk to ice-breeding pinnipeds from shipping in Arctic and sub-Arctic seas. Marine Policy 111, 103694. https://doi.org/10.1016/j.marpol.2019.103694.

Zhang, Z., Huisingh, D., Song, M., 2019. Exploitation of trans-Arctic maritime transportation. Journal of Cleaner Production 212, 960–973. https://doi.org/10.1016/j.jclepro.2018.12.070.

Arctic Ice

This section concerns the cryosphere—the frozen world of glaciers and ice sheets, snow, sea ice, and permafrost. This realm of frozen water constitutes around 40% of the Earth's land area. The cryosphere received little attention from the world at large until very recently, when it started to melt. Other than the Antarctic ice sheets, the temperature of the world's snow and ice cover has been rising in recent decades. This warming has reached the point where most of the modern cryosphere is very close to the freezing point of water (0°C), making it increasingly vulnerable to melting as the planet grows warmer. The Greenland ice sheet (GrIS) covers about 1.7 million km², covering most of the island of Greenland. Dated ice cores from various parts on the island indicate that the current ice sheet formed more than 100,000 years ago. While the ice sheet still covers the vast majority of Greenland, it is not in a healthy condition. the temperature of the ice is warming, and the ice is getting thinner. Beyond the polar regions, the rest of the cryosphere this confined to high elevations where temperatures remain cold year-round. Just as the Arctic regions are getting warmer and the cryosphere is melting, so likewise is the part of the cryosphere that resides near mountaintops. The melting of Alpine glaciers affects millions of people who live in the tropical and temperate regions. Millions of people live in or near the high mountain regions. They rely heavily on alpine resources, especially melt-water from snowfields and glaciers.

CHAPTER 9

Decline in Mountain Glaciers

INTRODUCTION

To put mountain glaciers in their proper perspective, it is useful to step back from the topic and consider their contribution to the larger realm of the cryosphere. The cryosphere consists of the frozen world of glaciers and ice sheets, snow, sea ice, permafrost, and lake and river ice. Changes in the cryosphere affect the lives of hundreds of millions of people and many ecosystems in various direct and indirect ways. Seasonal or year-round snow covers around 45 million km^2, and the combination of the world's glaciers and the Greenland and Antarctic ice sheets adds 15 million km^2. Together, the cryosphere constitutes around 40% of the Earth's land area. Most of the world paid little attention to the cryosphere until quite recently, when it started to melt. Outside of the polar ice sheets, much of the cryosphere today is very close to the freezing point of water (0°C), making it increasingly vulnerable to melting as the planet grows warmer. Although this book is about the Arctic, it is important to consider the rest of the cryosphere because all frozen regions share these vulnerabilities. Furthermore, very few people are directly affected by conditions of the Greenland and Antarctic ice sheets, but millions of people live in the high mountain regions of the world and rely heavily on alpine resource regions, especially meltwater from snowfields and glaciers.

The alpine realm is small. The combined total of all alpine regions of the world amounts to about 4 million km^2. The World Glacier Monitoring Service (WGMS) keeps track of approximately 32,500 glaciers around the world. These are scattered along the various mountain ranges of Asia, Europe, Greenland, North and South America, and New Zealand, with just a few rapidly shrinking glaciers and ice fields remaining in Africa (Fig. 9.1). These glaciers represent only about 2.6% of the global land area outside Antarctica (Körner, 2017). About one-quarter of the alpine regions (~1 million km^2) consist of glaciers, snowfields, bare rock, rock fields, and scree slopes. This tiny fraction of global real estate is far more important to Earth's ecosystems, climate, and freshwater supply than its size would indicate.

The "Third Pole"

The region that encompasses the Himalaya-Hindu Kush mountain range and the Tibetan Plateau is widely known as the Third Pole because its ice fields contain the largest reserve of fresh water outside the polar regions. This region is the source of the 10 major river systems that provide irrigation, power, and drinking water for over 1.3 billion people in Asia—nearly 20% of the world's population. The Qinghai-Tibet Plateau is the highest in the world with an average elevation of around 4500 m. Hence, the plateau is known as "the roof of the world." This extreme elevation is due to the tectonic forces that have driven the Indian plate into the Eurasian plate. Because both these continental landmasses have about the same rock density, one plate could not be subducted under the other. The pressure of the impinging plates could only be relieved by thrusting skyward, contorting the collision zone, and forming the jagged Himalayan peaks. The Himalayas and the Tibetan Plateau to the north have risen very rapidly. In just 50 million years, the tallest peaks in the world have risen in the Himalayan Range. The impinging of the two landmasses has yet to end. The Himalayas continue to rise more than 1 cm a year. This astonishing growth rate represents 10 km per million years.

This chapter considers the Third Pole, but also the other glaciated mountain regions of the world. Unfortunately, they are all suffering the same fate. No mountain region is getting colder. On the contrary, as we shall see in this chapter, all of them are getting warmer and most glaciers are in the process of melting.

WHAT MAKES THE ALPINE DIFFERENT FROM THE ARCTIC?

At high latitudes (>65–70° N), the alpine zone merges with the Arctic. Despite similarities in vegetation and the overwhelming influence of low mean temperature, the Arctic is very different from the alpine zone in terms of climate, land surface structure, and vegetation. The most characteristic feature of the alpine life zone is its fragmentation into a multitude of microhabitats created by relief, exposure, and slope. These topographic

Threats to the Arctic. https://doi.org/10.1016/B978-0-12-821555-5.00007-3

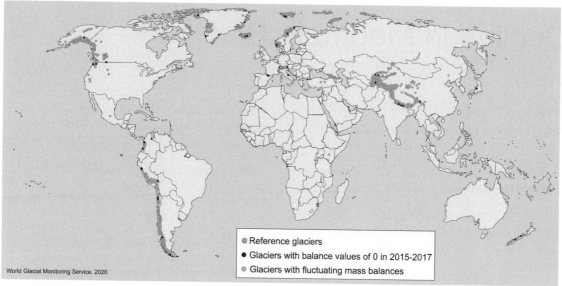

World Glacial Monitoring Service, 2020

Reference glaciers
Glaciers with balance values of 0 in 2015-2017
Glaciers with fluctuating mass balances

FIG. 9.1 Location of the 32,500 glaciers for which fluctuation data or special events are available from the WGMS. This overview includes 163 glaciers with reported mass balance data for the observation periods 2015/16 and 2016/17, and 41 "reference" glaciers with well-documented and independently calibrated, long-term mass balance programs based on the glaciological method. *WGMS*, World Glacier Monitoring Service. (Modified from World Glacier Monitoring Service, 2020. Global glacier change bulletin, 2016−2017. In: Zemp, M., Gärtner-Roer, I., Nussbaumer, S.U., Bannwart, J., Rastner, P., et al. (Eds.), Global Glacier Change Bulletin (vol. 3). ISC(WDS)/IUGG(IACS)/UNEP/UNESCO/WMO, World Glacier Monitoring Service.)

features interact with solar radiation and wind to cause large variations in soil moisture, temperature, and substrate quality over very short distances. These diverse microhabitats, shaped by the steepness of the terrain, make the alpine zone very different from the arctic tundra, most of which is characterized by relatively flat, undulating landscapes with a few outstanding topographic features. Another difference between Arctic and alpine climates is their precipitation patterns. Many high mountain regions receive large amounts of precipitation, while most regions of the Arctic receive less than 500 mm per year. Finally, average wind speeds are quite low in the Arctic, except for occasional storms. As discussed in the following, the alpine regions are some of the windiest in the world.

Wind Speeds

Many tall mountains jut up into parts of the atmosphere where high winds are the norm, not the exception. For instance, during the winter months of November through March, the jet stream normally flows in the vicinity of the Himalayas, if not directly over them (around 28°N latitude). February is the windiest month of the year on Mt Everest, with a record wind speed estimated at 282 km per hour, based on meteorological data collected at the South Col of Everest (elevation 7896 m above sea level). It is estimated that between October and the end of February, the summit of Everest experiences an average of three out of four days with hurricane-force (120 km per hour) or stronger winds. The highest point in North America is Denali in Alaska, at 6190 m. Although the top of Denali is more than 2000 m below the summit of Mt Everest, computer models predict that maximum wind gusts on Denali exceed 482 km per hour (kph), although there is no anemometer on the peak. One mountain top where there is a weather station is Mount Washington, New Hampshire. Here, average wind speeds range from about 75 kph in the winter months, tapering to 40−50 kph in the summer. But this is not the whole story, because peak gusts on Mount Washington range between about 220 and 290 kph through the year, and a record peak gust of 372 kph occurred there one April in the 1930s. Long-term meteorological data from Niwot Ridge, Colorado, show average monthly wind speeds ranging from about 20 kph in summer to 50 kph in winter. Monthly average high wind speeds range from a low of about 58 kph in July to a high of about 90 kph in winter. A little further north along the Colorado Front Range, Longs Peak has a record wind gust of 323 kph in 1981, but the station only lasted a short time. Anemometers tend to get blown over at such wind speeds.

Snow Accumulation Patterns

Alpine regions may receive large quantities of snow during mountain storms, but because of the persistent windiness at high elevations, the snow gets rapidly redistributed, leaving some surfaces bare of snow, while filling depressions with deep snowbanks. An examination of snow persistence (SP) spatial patterns in three representative areas revealed three dominant patterns (Wayand et al., 2018). Over the coldest regions and highest elevations, the ridges that surround glaciers are scoured free from snow. In nonglacial alpine areas, two patterns dominate (1) deep snow accumulation and high persistence of snow in gullies perpendicular to main ridges, and (2) high persistence of snow in areas that are parallel and offset of the main ridge. Both features are influenced by a combination of variable patterns of ablation rates and of snow redistribution (blowing snow and avalanching). On exposed ridges, snow depths rarely exceeded 0.25 m before being wind-scoured, while at sheltered sites a maximum snow depth of 1.5 m was measured by Wayand et al. (2018) in their study. A similar study was done by Litaor et al. (2008) on Niwot Ridge in the Colorado Front Range (Table 9.1).

Atmospheric Pressure

In the alpine zone, atmospheric pressure is only 40% −60% of that at sea level (Fig. 9.2). Low atmospheric pressure has a wide range of effects on alpine environments. The thin air is highly transparent to incoming solar radiation (insolation). However, the air holds less heat than the denser air closer to sea level, so the alpine

climate is cold. The thin air also holds less moisture, making alpine air quite dry. Water vapor is the largest contributor to the greenhouse effect (American Chemical Society, 2020). On average, it accounts for about 60% of the warming effect. Therefore, the lack of moisture in high mountain environments also reduces the greenhouse effect. Water vapor, carbon dioxide, methane, and other trace gases in Earth's atmosphere absorb the longer wavelengths of outgoing infrared radiation from Earth's surface. These gases then emit infrared radiation in all directions, both outward toward space and downward toward Earth. The lack of water vapor in the rarified alpine atmosphere thus leads to large differences between daytime and nighttime temperatures. The cold air temperatures cause most of the alpine precipitation to fall as snow. The thin air also does a poor job of screening out ultraviolet radiation, so UV levels are high in the alpine.

Lapse Rates

The decrease in air temperature due to increasing elevation shapes many aspects of alpine climate. The temperature lapse rate (TLR) is mainly controlled by the topography and climatology of the region. The rate of decrease varies in different mountain chains. It is 0.63°C per 100 m of elevation gain in the dry mountains of the Western United States and 0.26°C per 100 m in the moister mountains of the Eastern United States. Lapse rates also have seasonality. For instance, the mean lapse rate in the Himalayas varies from 0.43°C per 100 m in January to 0.61°C per 100 m in April (Kattel et al., 2013). An average lapse rate for mountainous regions in the northern midlatitudes is 0.6°C per 100 m elevation gain.

Of course, there is also cooling of regional temperatures associated with increasing latitude. A 1000 m increase in altitude occurring over a short horizontal distance roughly corresponds to the same temperature

TABLE 9.1
Snow Depths at Windblown and Sheltered Sites, Niwot Ridge, Colorado.

Site	Snow-Free Days per Year	Mean Snow Depth (cm)	Maximum Snow Depth (cm)
Extremely wind blown	>300	7.2 ± 1.5	24 ± 3.4
Wind blown	200−300	14.6 ± 2.4	50 ± 7
Minimal snow cover	150−200	102 ± 10	244 ± 17
Early melting snow bank	100−150	215 ± 15	391 ± 15
Late melting snow bank	50−100	298 ± 47	475 ± 40

After Litaor, M.I., Williams, M., Seastedt, T.R., 2008. Topographic controls on snow distribution, soil moisture, and species diversity of herbaceous alpine Vegetation, Netwot Ridge, Colorado. Journal of Geophysical Research: Biogeosciences 113 (2). https://doi.org/10.1029/2007JG000419.

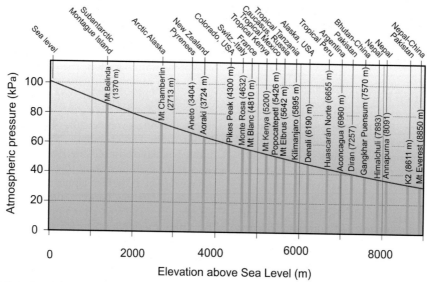

FIG. 9.2 In the alpine zone, atmospheric pressure is only 40%–60% of that at sea level. (Figure by the author.)

change as a 1000 km increase in latitude. Two-thirds of the global alpine area is situated in the temperate and subtropical zones. Only 10% occurs in the tropics. Tropical mountains need to be at least 3600–4000 m to support an alpine zone, and glaciated peaks in the tropics are generally 5000 m or more in elevation, whereas mountains in the subpolar regions need only to be 600–800 m to support alpine habitat (Körner et al., 2011) (see Fig. 9.2).

Temperature Regime

The climate of the alpine zone is as cold as some Arctic regions. High mountains in the temperate zone have summer afternoon temperatures that average about 5–9°C, and many nighttime temperatures fall below freezing. The wettest, least windy sites are dominated by snow beds, which persist into late summer, leaving behind soils saturated with meltwater for a few brief weeks before snowfall resumes. The meteorological station on Mount Evans, Colorado, is situated just above treeline. Here, maximum summer temperatures reach 11°C in July, but minimum temperatures remain below freezing during all months of the year and fall to −22°C in December and January. The diurnal temperature changes are substantial: 10–15°C difference between day and night.

Not all high mountain regions experience freezing temperatures in the summer. A weather station at the base camp on Annapurna in the high Himalayas (4130 m above sea level) records minimum temperatures above freezing from May to early September. This pattern of temperature regime is like that found on top of Mount Washington, New Hampshire, United States, even though the elevation is only 1917 m above sea level. A weather station at 4794 on Mount Xixibangma in Tibet shows differences between monthly high and low temperatures of 12–14°C. Here, summer temperatures rise to 17°C, and winter temperatures fall as low as −18 in February. Climate data from 3488 m above sea level on the slopes of the Matterhorn in the Swiss Alps show a different pattern. Here, the difference between maximum and minimum temperatures is only 4–5°C throughout the year. Maximum summer temperatures are only 3°C. One of the coldest alpine regions of the world is the upper slopes of Denali in Alaska. The highest peak in North America has bitterly cold temperatures, year-round. Temperatures as low as −59.7°C and wind chills as low as −83.4°C have been recorded by an automated weather station located at 5700 m, which is still almost 500 m below the summit. According to the National Park Service, a science expedition to Denali in 1932 recovered a self-recording minimum thermometer left at about 4600 m on Denali by a previous group of scientists in 1913. The spirit thermometer was calibrated down to −71 °C, and the lowest recorded temperature was below that point. Another thermometer was placed at the 4600 m level by the US Army Natick Laboratory and was there from 1950 to 1969. The coldest temperature recorded during that period was −73°C.

ALPINE THERMAL ZONES

Körner et al. (2011) reported the results of a Global Mountain Biodiversity Assessment that divided the alpine regions of the world into three thermal zones. The nival (snow) zone, the highest of the alpine zones, has mean annual air temperatures of less than 3.5°C. This zone is home to the alpine glaciers of the world, although the ice may extend down to the lower alpine zone. It is estimated that the nival zone occupies about 3% of total mountainous areas. The upper alpine zone experiences mean annual air temperatures greater than 3.5°C and occupies about 4.5% of total mountainous areas. The lower alpine zone has mean annual air temperatures less than 6.4°C and occupies about 13.75% of total mountainous areas. Thus, the combined zones of the alpine comprise about 21.5% of total mountainous areas.

HOW AND WHEN MOUNTAIN GLACIERS FORMED

Glaciers and ice fields dominated the highlands of mid- and high-latitude terrain during much of the Pleistocene, and especially during the last million years when glaciations became longer and more intense. Outside of the ice sheets of Antarctica and Greenland, nearly all the glacial ice on mountaintops formed during the Holocene (the last 11,000 years), especially during the Little Ice Age (LIA), as discussed in the following.

Equilibrium Line Altitudes

Mountain glaciers are built by repeated snowfalls that accumulate year after year without completely melting. To have this buildup of snow, which eventually becomes glacial ice, two ingredients are required: cold summers and heavy winter snow. The cold summers allow the snows of previous winters to remain in high mountain valleys. Heavy snowfalls insure that there is enough frozen water to create and maintain a glacier. Many regions of the world are sufficiently cold in summer to maintain snowpack, but they lack sufficient winter moisture to form glaciers. Much of the southern Rocky Mountains are characterized by this kind of climate. During glacial intervals of the Pleistocene, they were rarely glaciated because of the lack of moisture in this dry region. Other regions receive a great deal of winter snow, but their summers are too warm to preserve the previous winter's snowpack. The upslope retreat of mountain glacial margins can be caused by either increasing temperatures or decreasing snowfall, or a combination of the two.

The boundary on a glacier between regions where winter snows melt off in summer and regions of permanent snow is called the equilibrium line altitude or ELA. Above this line, snow continues to accumulate. It compresses under its own weight and forms ice, adding to the mass of the glacier. Below the line, snow melts during the summer months, so it does not accumulate year after year, and ice does not build up.

ELAs tend to fall on a climatic boundary that corresponds to regions in which summer temperatures average 0°C—the freezing point of water. However, different climatic conditions seem to control ELAs in different regions. Obviously, precipitation plays a major role. No matter how cold it is, if there is no new snow, new ice will not form. During glacial intervals, temperatures were depressed to the extent that snow accumulated not just on mountaintops, but across much of the high-latitude regions, eventually forming ice sheets that covered most of the northern regions and spread south to the midlatitudes. In mountains, glaciers advanced when the amount of ice buildup above the ELA was greater than the amount of ice melt below the ELA. In the Rocky Mountains, for instance, ELAs during the Last Glaciation were as much as 1000 m lower than they are today.

Late Pleistocene

Based on dated polar ice cores, the oldest glacial ice in Antarctica may well exceed 2,700,000 years. The oldest glacial ice in Greenland is more than 100,000 years old. There are a few remnants of Late Pleistocene-age ice in the high latitudes. For instance, the age of the oldest Alaskan glacial ice thus far recovered is about 30,000 years old. Glacial flow moves newly formed ice through the entire length of a typical Alaskan valley glacier in a few hundred years or less. For instance, based on flow rates, it takes less than 400 years for ice to journey the entire 225 km length of Bering Glacier, Alaska's largest and longest glacier (US Geological Survey, 2020a). There are patches of glacial ice in the Northwest Himalayan region that are thought to have formed in the early Holocene but remain undated (Shukla et al., 2020). With these and a few other rare exceptions, all the mountain glaciers that remain today formed during the LIA.

The Little Ice Age

The period between about AD 1500 and 1900 has been called the "Little Ice Age" by historians, geologists, and climatologists because much of the Northern Hemisphere experienced temperatures colder than the Holocene average (i.e., the warm interval during the last

11,000 years). The LIA cooling brought on a buildup of glacial ice in many mountain ranges and in the Arctic. Pack ice clogged sea lanes around Iceland and Greenland. By 1400, the settlement on Greenland's west coast had been abandoned, and by 1450, the inhabitants in the Eastern Settlement on the island's southern tip were gone as well. During the 14th century, as the cold climate phase continued, ice apparently clogged the seas farther south and remained longer each year, disrupting voyages to and from Europe. As discussed in Chapter 1, Franklin's attempt to find a Northwest Passage in 1845 was made virtually impossible by the LIA conditions that caused pack ice to clog the passages in the Canadian Arctic Archipelago.

How cold was the LIA, in comparison with the rest of the late Holocene? Ljungqvist (2010) used a suite of paleotemperature proxy records to reconstruct mean annual temperatures for Northern Hemisphere regions from 30 to 90°N for the past 2000 years. His reconstruction (Fig. 9.3) shows a distinct Roman Warm Period c. 1−300 AD, reaching up to the late 20th-century average temperature level, followed by an Early Medieval Cold Period c. 300−800 AD. The Late Medieval Warm Period falls within c. 800−1300 AD, and the LIA is clearly visible c. 1300−1900 AD, followed by a rapid temperature increase in the 20th century. The highest average temperatures in the reconstruction were encountered in the mid- to late 10th century, and the lowest were in the LIA, during the late 17th century. Decadal mean temperatures seem to have reached or exceeded the 1961−90 mean temperature level during substantial

parts of the Roman Warm Period and the Medieval Warm Period.

Glaciers in the Alps and most other Northern Hemisphere mountain ranges advanced dramatically, overrunning alpine meadows. In the Rocky Mountains, glaciers advanced out of their cirques and spilled down mountainsides, forming large moraines. Coincidentally, the fields of glacial geology and paleoclimatology were just getting underway toward the end of the LIA. When Louis Agassiz described the glaciers of the Swiss Alps and their origins, he was looking at glaciers that were larger than they had been during nearly all the Holocene.

The impressive glaciers that existed in northwestern Montana at the turn of the 20th century, along with the impressive mountain scenery and wildlife, led to the creation of Glacier National Park. In some ways, the way in which the park got its name is quite fortuitous. Not many people realize that when the park was founded, the glaciers there were beginning the process of thinning and melting back from the maximum extent they reached in about 1850. At that time—the culmination of more than three centuries of LIA cooling—the glaciers were bigger than they had been in the previous 10,000 years.

Cause(s) of the little ice age cooling

Oosthoek (2015) reviewed the probable causes of LIA cooling. While the exact cause of the LIA remains unresolved, there is a striking coincidence in the sunspot cycle and the timing of the LIA. During the LIA, there was

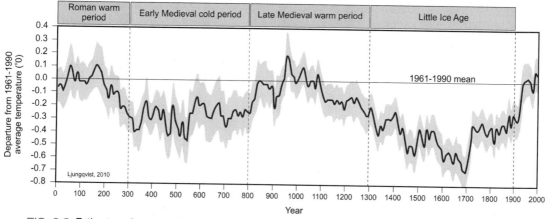

FIG. 9.3 Estimates of extratropical Northern Hemisphere (30–90°N) decadal mean temperature variations (*red line*) AD 1–1999, relative to the 1961–90 mean instrumental temperature record (*black line*) with 2 standard deviation error bars (light red shading). (Modified from Ljungqvist, F., 2010. A new reconstruction of temperature variability in the extra-tropical northern hemisphere during the last two millennia. Geografiska Annaler, Series A: Physical Geography 92 (3), 339–351. https://doi.org/10.1111/j.1468-0459.2010.00399.x.)

a minimum in sunspots, indicating an inactive and possibly cooler sun. This interval of reduced solar activity is called the Maunder Minimum. The Maunder Minimum occurred during the coldest period of the LIA between 1645 and 1715 AD when the number of sunspots was very low. It is named after British astronomer E.W. Maunder who discovered the dearth of sunspots during that period. The lack of sunspots meant that solar radiation was probably lower at this time, but models and temperature reconstructions suggest this would have reduced average global temperatures by 0.4°C at most, so this phenomenon cannot fully explain the regional climate cooling in Europe and North America.

North Atlantic oscillation

Another likely candidate that at least contributed to LIA cooling—a drop of up to 2°C in winter temperatures—is the North Atlantic Oscillation (NAO), as discussed at length in Chapter 4. The North Atlantic is one of the most climatically unstable regions in the world, caused by complex interactions between the atmosphere and the ocean. The main feature of the NAO is a see-saw of atmospheric pressure between a persistent high over the Azores and an equally persistent low over Iceland. Sometimes the pressure cells weaken, which has severe consequences for the weather in Europe. When the Azores high pressure grows exceptionally strong and the Icelandic low becomes deeper than normal, the result is warm and wet winters in Europe and cold and dry winters in northern Canada and Greenland. This also pushes the North Atlantic storm track northward, causing more frequent and severe storms over northern Europe. This situation is called a Positive NAO Index. When both pressure systems are weak, cold air invades Northern Europe more easily, resulting in cold winters. The North Atlantic storm track shifts south, bringing wet weather to the Mediterranean region. This set of conditions reflects a Negative NAO Index. Climatologists now consider that the NAO Index was more persistently negative during the LIA, making the North Atlantic region colder than average.

Volcanic eruptions

Miller et al. (2012) observed that intervals of land ice expansion in the North Atlantic region coincide with two of the most volcanically active episodes of the past millennium (Fig. 9.4). They examined dated records of ice-cap growth in Iceland and Arctic Canada. These records show that LIA summer cold and ice growth began abruptly between 1275 and 1300 AD, followed by a substantial increase in glacial growth from 1430 to 1455. Backed by climate modeling and the

results of other regional proxy climate reconstructions, their results suggest that episodes of explosive volcanism acted as a climate trigger, allowing Arctic Ocean sea ice to expand. Volcanic eruptions can inject huge quantities of particulates or chemical aerosols into the atmosphere, reducing insolation—sometimes globally—for weeks to years. Such strong reductions in insolation could easily explain the evidence for immediate regional cooling (e.g., rapid snowline lowering on Arctic and subarctic islands), but a single eruption could not sustain centuries of depressed temperatures, as occurred during the LIA. Of course, a series of volcanic eruptions every few decades would produce greater cooling than a single large eruption if the recurrence interval is shorter than the upper ocean temperature relaxation time of decades. But volcanism was not the only factor driving the LIA cooling. The volcanic eruptions took place at a time when Earth's orbital configuration brought a centuries-long interval of low summer insolation to the Northern Hemisphere. Increased sea ice export to the North Atlantic from the Arctic Ocean and adjacent seas may have engaged a self-sustaining sea-ice/ocean feedback system in the northern North Atlantic. This sea ice–driven phenomenon suppressed summer air temperatures in the Northern Hemisphere for centuries after the aerosols from volcanic eruptions were removed from the atmosphere. Miller et al. maintained that the coincidence of repeated explosive volcanism with centuries of lower-than-modern solar irradiance indicates that volcanic impacts were likely reinforced by external forcing, but that an explanation of the LIA does not require a solar trigger.

The great sea-ice anomaly

Following on the findings of Miller et al. (2012) that highlighted the importance of sea ice in the LIA cooling, Miles et al. (2020) reconstructed the history of sea-ice exportation from the Arctic Ocean to the North Atlantic over the past 1400 years, using a spatial network of proxy records. They found strong evidence for extreme export of sea ice commencing abruptly around 1300 AD and ending in less than a century (i.e., the late 1300s). The authors termed this phenomenon the "Great Sea-Ice Anomaly" because of its exceptional magnitude and duration. During this interval, they found evidence that a series of pulses of sea ice passed through Fram Strait and then along the East Greenland coast. This resulted in downstream effects in the North Atlantic, brought on by increases in polar waters and ocean stratification. It must be remembered that sea ice does not only passively respond to Arctic change—it is also an active driver of climate change on seasonal to decadal timescales. Numerical model experiments have

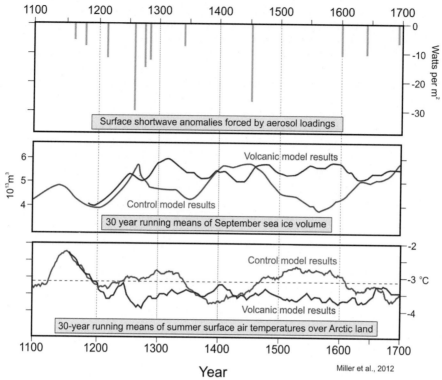

Miller et al., 2012

FIG. 9.4 Climate model results for 1000 AD control models *(blue)*, and volcanically perturbed transient models *(red)*. Top: Monthly global surface shortwave radiation anomalies (in watts per m²) forced by volcanic aerosols; Middle: Mean Northern Hemisphere sea ice volume modeled for September (sea-ice minimum of the year) with the volcanically perturbed transient *(red line)* compared with the control model (blue); Bottom: 30-year running mean of average summer (JJA) surface air temperature over Arctic lands in the North Atlantic region (>60°N and 90°W to 30°E) with the volcanically perturbed transient *(red line)* compared with the control model (blue). (Modified from Miller, G., Geirsdóttir, A., Zhong, Y., Larsen, D., Otto-Bliesner, B., Holland, M., Bailey, D., Refsnider, K., Lehman, S., Southon, J., Anderson, C., Björnsson, H., Thordarson, T., 2012. Abrupt onset of the Little Ice Age triggered by volcanism and sustained by sea-ice/ocean feedbacks. Geophysical Research Letters 39 (2). https://doi.org/10.1029/2011GL050168.)

highlighted the importance of sea-ice cover in paleoclimate reconstructions, noting that feedbacks in the climate system associated with sea-ice cover are essential for explaining the onset and/or sustainment of longer climate anomalies, during cold periods such as the LIA as well as in warm periods, such as the current global warming. The Great Sea Ice Anomaly culminated about 1400 AD, but the North Atlantic conditions it fostered were more-or-less self-sustaining during subsequent centuries, chilling the region and maintaining sea ice at unusually low latitudes (e.g., along the southern coast of Iceland).

Miles et al. (2020) went one step further than Miller et al. (2012). The latter group had suggested the necessity of external forcing from volcanic eruptions in the development of the LIA, but Miles et al. (2020)

suggested that such an external trigger was not necessary. They noted that the onset and development of the LIA were markedly similar to spontaneous abrupt cooling episodes predicted by some climate models. In these models, the sustained cooling is driven only by sea-ice input. Their results provide evidence that marked climate changes may not require an external trigger. This concept may stimulate a lot of new thinking in the climate science community.

The little ice age in the Himalayas

As reviewed by Rowan (2017), most previous investigations of recent advances of Himalayan glaciers assumed that these events were synchronous with LIA advances known from Europe. These inferences were based on the appearance and position of Himalayan moraines, but they lacked numerical age control. In recent decades, such methods as cosmogenic nuclide dating have allowed researchers to date these young moraines and clarify the timing of glacial maxima. Rowan (2017) reviewed the geochronological evidence for the last advance of Himalayan glaciers. Cosmogenic nuclide dating of 138 samples collected from glacial landforms demonstrates that regional glacial advances centered on the interval between 1300 and 1600 AD, slightly earlier than the coldest period of Northern Hemisphere air temperatures, but otherwise agreeing with the timing of the onset of glacial advances in the Alps (Larsen et al., 2013) (Fig. 9.5). For instance, the Aletsch and Gorner glaciers in Switzerland both advanced from 1300–1420 AD, and then advanced again from 1550 to 1680 (Nussbaumer et al., 2011).

The timing of Himalayan LIA advances varied spatially, influenced by topographic setting and meteorological conditions along the mountain range. According to Rowan (2017), paleoclimate proxies indicate cooling air temperatures starting at 1300 AD and leading to a southward shift in the Asian monsoon. Increased Westerly winter precipitation and generally wetter conditions dominated the Himalayas from about 1400–1800 AD. The combination of colder temperatures and increased moisture supply caused glaciers to advance in this region. This set of Himalayan

glacial advances was fed by increased snowfall and sustained by cold climate as a result of atmospheric circulation reorganization during the LIA.

Ages for some Holocene moraines in the Himalaya were recently obtained as part of larger glaciological studies. Two sets of recent moraines occur in front of several Himalayan glaciers, such as Khumbu Glacier in Nepal, Pasu Glacier in Pakistan, and Batal Glacier in the Lahual region of northern India and are assigned to the LIA and a preceding late-Holocene (Neoglacial) advance. The LIA moraines are likely constrained by the preexisting moraines, producing additional crests inside the older ones.

The moraines representing the two advances look markedly different and thus can be distinguished from each other in the field. The late-Holocene advance is marked by low relief moraines that are laterally extensive. Once established, these moraines formed a barrier to the subsequent flow of ice during the LIA advance. The LIA moraines are marked by higher relief. They formed directly within the older landforms by glaciers that thickened rather than expanded laterally as they gained mass.

The glacial landform ages presented in the study by Rowan (2017) span the last 2000 years. The samples were nearly all taken from glaciers in the Central Himalaya. Based largely on ^{210}Be dating, the timing

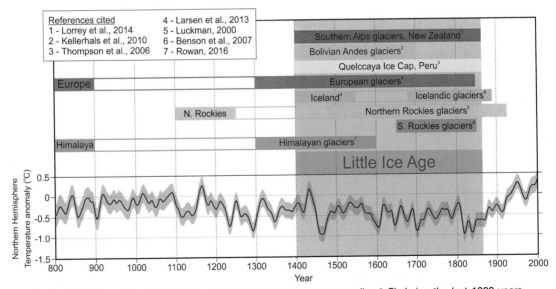

FIG. 9.5 Diagram showing Northern Hemisphere temperature anomalies (oC) during the last 1200 years, showing the timing of LIA glacial advances in major mountain ranges of the world, from multiple sources. *LIA*, Little Ice Age. (Data from Lorrey, A., Fauchereau, N., Stanton, C., Chappell, P., Phipps, S., et al., 2014. The Little Ice Age climate of New Zealand reconstructed from Southern Alps cirque glaciers: a synoptic type approach. Climate Dynamics 42. https://doi.org/10.1007/s00382-013-1876-8.)

of the two sets of glacial advances discussed earlier has been clarified. The first was a late-Holocene advance around 800–900 AD; the second (LIA advance) occurred around 1300–1600 AD. The latter part of the LIA appears to have extended almost until modern times. The end of the LIA is difficult to define, particularly as these glaciers are debris-covered, and this cover thermally insulates the underlying ice. Therefore, the areal extent of these Himalayan glaciers has changed little, and the glacial termini still occupy their LIA moraines. It appears that most Himalayan glaciers remained close to their LIA maxima until the end of the 19th century.

Multiple ice cores were recovered by Kaspari et al. (2007) from East Rongbuk Glacier in the Mount Everest region at altitudes near 6500 m. A 108-m ice core spanning the years 1534–2001 showed that the average glacial ice accumulation rate was 0.8 m per year between 1500 and 1600; then it decreased to 0.3 m per year around 1850. It increased again from 1880 to 1970 and finally decreased from 1970 to 2001. Since the 20th century, monsoon precipitation seems to have been the main driver of regional glacial expansion and contraction. The evidence from the East Rongbuk core indicates an abrupt southward shift of the South Asian monsoon around 1400 AD, driven by a reorganization of Northern Hemisphere atmospheric circulation. This change in monsoonal circulation was synchronous with a reduction in solar irradiance and the onset of the LIA (Kaspari et al., 2007).

Three cores up to 160 m long were recovered by Thompson et al. (2000) from Dasuopu Glacier, Tibet, at elevations of 7000–7200 m. These cores span the interval from 1440 to the present. Decadal-averaged snow accumulation from two of these cores varied between the years 1400 and 1600, giving values like those of 0.5–1.2 m water equivalent (w. e.) recorded from the years 1600–1817. Accumulation rates increased significantly between 1817 and 1880 and then gradually decreased to modern values. Again, it is likely that glacier accumulation rates were driven by monsoon intensity and not redistribution by ice flow. Like other ice cores from the Tibetan Plateau, the oxygen isotope evidence from Dasuopu suggests a large-scale 20th-century warming trend across the Tibetan Plateau that appears to be amplified at higher elevations (Fig. 9.6).

The little ice age in the Southern Alps, New Zealand

Lorrey et al. (2014) reconstructed departures from modern temperatures for LIA austral summers, based on paleo-ELAs at 22 cirque glacier sites across the Southern Alps of New Zealand. Modern analog seasons

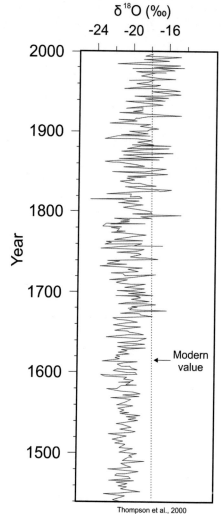

FIG. 9.6 Annual averages of δ18O values from ice core samples taken from Dasuopu, Tibet, are shown from 1440 AD to present. (Modified from Thompson, L., Yao, T., Mosley-Thompson, E., Davis, M., Henderson, K., Lin, P., 2000. A high-resolution millennial record of the South Asian Monsoon from Himalayan ice cores. Science 289 (5486), 1916–1919. https://doi.org/10.1126/science.289.5486.1916.)

with temperature anomalies like the LIA reconstructions were selected and then applied in a sampling of high-resolution gridded New Zealand climate data and global reanalysis data to generate LIA climate composites at local, regional, and hemispheric scales. The temperature anomaly patterns they reconstructed improve our understanding of atmospheric circulation contributions to the LIA climates of New Zealand, placed in a hemispheric context of past conditions. An LIA summer temperature anomaly of −0.56°C (±0.29 °C) for the Southern Alps, based on paleoequilibrium lines, compares well with local tree-ring reconstructions of austral summer temperature. The study by Lorrie et al. (2014) suggests that enhanced southwesterly flow across New Zealand during the LIA was required to generate this level of climatic cooling. Associated land-based temperature and precipitation proxy reconstructions suggest both colder- and wetter-than-normal conditions dominated New Zealand during the LIA, accompanied by colder-than-normal Tasman Sea SSTs (sea surface temperatures). The evidence suggests that LIA summer climate and atmospheric circulation over New Zealand was driven by increased frequency of weak El Niño-Modoki in the tropical Pacific and negative Southern Annular Mode activity. The term El Niño Modoki has been used to identify the phenomenon of ocean-atmosphere coupling that records SST warming anomalies in the central tropical Pacific and SST cooling anomalies in the eastern and western tropical Pacific.

In an independent assessment, Carrivick et al. (2020) determined volume changes for a series of mountain glaciers across the Southern Alps, New Zealand, for three intervals: LIA to 1978, 1978 to 2009, and 2009 to 2019. Glacier ice in the Southern Alps has become restricted to higher elevations and large debris-covered ablation tongues terminating in lakes. They found that the most widespread and best-preserved moraines and trim lines across the Southern Alps represent the former extent of mountain glaciers between the 14th and 19th centuries, which is contemporaneous with the Northern Hemisphere LIA.

Overall, Carrivick et al. (2020) mapped LIA outlines of 400 glaciers. They determined that most of these represent the coalescence of many, now smaller, glaciers that have fragmented greatly since the 19th century. The total area they mapped as LIA-age ice, combined with estimates of unmapped glaciers, was at least 1932 ± 135 km^2. Compared with a modern inventory of glaciers in the Southern Alps (1463 km^2), Carrivick et al. (2020) found a loss of at least 24% in glacier ice areal extent since the LIA. Furthermore, using newly generated glacier outlines for the year 2019, they computed a modern ice coverage of 1021 km^2, which is 53.8% of the LIA area. The authors described a major recession of ice termini between 2009 and 2019, especially where ice is flowing and calving into proglacial lakes. The difference in terminus outlines between 2009 and 2019 far exceeds the 1978 to 2009 changes.

In terms of Southern Hemisphere comparisons, the Southern Alps has a much larger proportion of ice volume loss than do the glaciers in Patagonia, but the rate of ice loss is slower in New Zealand. There is no inherent reason why these two widely separated Southern Hemisphere mountain ranges should experience synchronous glacial episodes, especially at centennial and finer timescales. More refined glacial chronologies, combined with additional earth surface process studies (e.g., landscape, hydrology, mass balance), would allow glacial geologists to contextualize modern conditions and predict future glacial responses in a warming world.

Across the Southern Alps, former LIA glacial ablation zones are now almost completely ice-free. Only 121 km^2 (12%) of glaciated areas in the Southern Alps in 2019 fall within the LIA ablation areas. These exceptional regions are debris-covered ablation tongues that terminate on valley floors at relatively low elevations. These large ablation tongues are nearly stagnant, and often terminate in proglacial lakes, which can exacerbate ice loss. The remaining high-elevation ice is becoming increasingly fragmented. These patches of glacial ice are thinning and have become disconnected from their original ablation areas. Not surprisingly, local conditions (e.g., angle of exposure, steepness of slopes, prevailing wind direction) drive the exceptionally dynamic proglacial geomorphology, sedimentology, and net changes in outwash plain elevation. Proglacial lakes play a large role in the modern activity of mountain glaciers, affecting glacier evolution, interrupting the flux of meltwater and sediments down valley, and controlling long-term landscape evolution. Carrivick et al. (2020) suggest that peak glacial meltwater production in the Southern Alps may already have occurred or will occur soon.

The little ice age in the Alps
The European Alps are very sensitive and vulnerable to climate change. Alpine temperatures dropped below the average of the preceding millennia during the LIA, causing glaciers to extend several hundred meters beyond the modern ice margin. Although the LIA and corresponding glacier responses were heterogeneous across time and space, prominent phases of glacier advance are documented during the 14th, 17th, and 19th centuries in the Alps.

Nussbaumer et al. (2011) reevaluated older Alpine glacier length records in and around the Swiss Alps, and they performed an independent climate reconstruction based on analysis of annually laminated sediments of Lake Silvaplana in southeast Switzerland (Fig. 9.7). They found good agreement between the two different

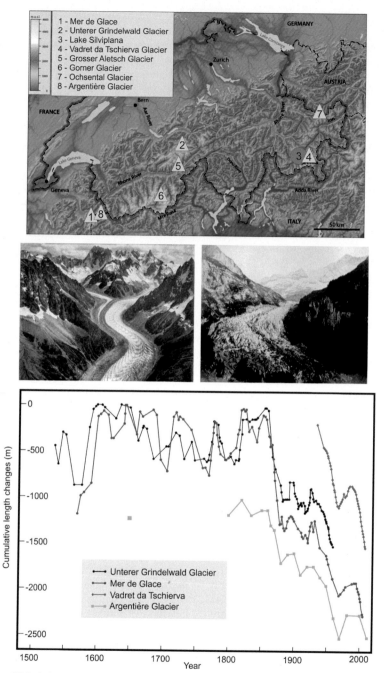

FIG. 9.7 Top: Digital elevation model (DEM) of Switzerland, showing locations discussed in the text. After Pfiffner, 2021; Middle: Historic alpine glacier photographs. Left: Mer de Glace in 1909, photographed from a balloon by Eduard Spelterini. Right: Lower Grindelwald Glacier in 1855/56, photographed by the Bison brothers. Photos in the public domain. Bottom: Comparison of Alpine glacier length changes during the past 500 years. The Vadret da Tschierva glacial record is available only back to AD 1934. After Nussbaumer et al., 2011. (Modified from Pfiffner, O., 2021. Chapter 2. The Geology of Switzerland. In: Reynard, E. (Ed.), World Geomorphological Landscapes, Springer, Switzerland. https://doi.org/10.1007/978-3-030-43203-4_28.)

FIG. 9.8 Oblique view of Morteratsch Glacier, Swiss Alps. IRS-1C satellite image of September 20, 1997 (10 m resolution, black and white), fused with a Landsat image of August 31, 1998 (25 m resolution, color), and draped over a digital terrain model. Marked glacier extents are 1850 *(red)* and 1973 *(blue)*. (From Haeberli, W., Huggel, C., Paul, F., Zemp, M., 2020. The response of glaciers to climate change: observations and impacts. In: Elias, S. (Ed.), Reference Module in Earth Systems and Environmental Sciences. Elsevier. https://doi.org/10.1016/B978-0-12-818234-5.00011-0.)

climate archives during the past millennium. Mass accumulation rates and biogenic silica concentration from Lake Silvaplana sediments are largely in phase with the glacier length changes of Mer de Glace and Unterer Grindelwald Glacier and with the records of glacier length of Grosser Aletsch Glacier and Gorner Glacier.

As seen in Fig. 9.7 (bottom), Swiss alpine glaciers advanced up to 2 km or more downslope from their current termini during the interval from 1560 to 1875. The two glaciers for which long, unbroken records were available were Unterer Grindelwald Glacier in the Bernese Alps and Mer de Glace in the French Alps. As can be seen in the figure, the frequency of waxing and waning of these two ice bodies throughout this 300-year interval is remarkably similar. The amplitude of the two glacial "signals" is not always the same, an aspect of glacial dynamics governed more by local conditions, such as the timing and intensity of snowstorms.

To help visualize the dynamics of an alpine glacier, Haeberli et al. (2020) combined an IRS-1C satellite image of the Swiss alpine Morteratsch Glacier from 1997 (10 m resolution, black and white) with a Landsat image of 1998 (25 m resolution, color) and draped the combined images over a digital terrain model. The glacier boundaries in 1850 (red) and 1973 (blue) are marked in the image (Fig. 9.8).

The study by Nussbaumer et al. (2011) also compares two sets of paleoclimate records which align with that of concurrent solar activity. Solar (Hallstatt) minima are identified around 500, 2500, and 5000 years ago. As discussed earlier, the Maunder Minimum in solar activity took place between 1645 and 1715 and coincided with the timing of glacier advances, taken as indications of climatic cooling in the Alps. During solar (Hallstatt) maxima around 0, 2000, and 4500 years ago, Alpine glacier margins generally retreated, indicating a warmer climate.

The importance of precipitation regime in determining the size of Alpine glaciers was born home in a study by Protin et al. (2019), who observed that the Argentière Glacier in the French Alps was larger in 1820 than in 1850. They argue that the French Alps received greater precipitation in 1820 than in 1850, thus explaining the earlier glacial advances in France. Given that the 19th-century maximum of the largest glaciers in the Swiss Alps occurred in 1850–60 rather than in 1820, local variations in precipitation during the LIA could explain the differences in the behavior of Argentière glacier compared with glaciers elsewhere in the Alps. The results of the Argentière glacier study demonstrate the impact of local precipitation on mountain glacier dynamics.

Braumann et al. (2020) developed a new Holocene mountain glacier chronology from the Eastern Alps in Austria. The study area, the Ochsental Glacier in the Silvretta Massif, features pronounced Holocene moraine ridges. They used ^{10}Be exposure ages of bedrock outcrops and boulders, combined with historical records, correlated with climate proxy records, to reconstruct ice margin positions at different times during the Holocene (Fig. 9.9). The Ochsental chronology spans the time interval from 12.9 ka to the present. Glaciers in the Eastern Alps were much smaller during the Mid-Holocene and readvanced during the LIA to positions close to the elevations of Early Holocene moraines. Based on the ^{10}Be age chronology, the evidence for LIA glacial advances falls between 1250 and 1850 AD in the Silvretta Massif. ^{10}Be ages range from 390 ± 20 years ago to 135 ± 5 years ago (1630–1885 AD) and point to multiple advances within this period. The most robust glacial evidence indicates that the greatest extent of LIA glaciation in the Austrian Alps occurred in the 18th century.

The ^{10}Be record and the historical glacier records overlap and are remarkably consistent, which demonstrates that ^{10}Be surface exposure dating produces reliable ages even for young glacial deposits. Within the last c. 170 years, Ochsentaler glacier has retreated c. 2.3 km, highlighting the impact of recent regional warming.

Little ice age in the Andes

The Andes form the longest mountain chain in the world, stretching over 6900 km from Venezuela to Tierra del Fuego. The northern Andes lie in the tropical zone of the Northern Hemisphere. The Southern Andes range from the tropical zone of the Southern Hemisphere in the North to the sub-Antarctic zone in the South. Not surprisingly then, the history of mountain glaciation during the late Holocene varies considerably

from one region of the Andes to another. Because of their higher latitude, the southern Andes were more heavily glaciated, both in the Pleistocene and in the late Holocene.

Southern Andes glaciers

Davies et al. (2020) reviewed the glacial history of the Patagonian region of southern South America (Fig. 9.10). They affirmed that regional glacial advances took place in the mid-Holocene (6–5 ka), the late Holocene (2–1 ka), and at the same time as the LIA in Europe (1500–1800 AD) (Fig. 9.11A), although they emphasize that the climate drivers for the LIA of Europe and the concurrent cooling in Patagonia were likely different. They also suggest that 20th- and 21st-century glacial recession has been more rapid than at any other time during the Holocene.

The timing of most recent glacial advances in this region has been established through historical documents, dendrochronology, lichenometry, and ^{10}Be dating. On satellite imagery, the most recent moraines and trim lines, generally assumed to date from the period 1500–1800 AD, are distinctively different than older moraines in terms of vegetation and degradation. In the Isla de Chiloé and Archipiélago de los Chonos sector, the dating of Glaciar Torrecillas (Fig. 9.10, No. 1) using lichenometry suggests a series of consecutively stratigraphically younger moraines that date from 1735 to 1934 AD. Lichenometry is a geomorphic dating method that uses lichen growth to determine the age of exposed rock, based on a presumed specific rate of increase in radial size over time.

Multiple lines of evidence from the Northern Patagonian Ice Field (NPI) revealed a series of advances and retreats of the margins of Glaciar San Rafael (Fig. 9.10, No. 2) centered on 1675 and 1766 AD, followed by a final advance into a lagoon in 1871 AD. Dendrochronological records indicate an advance of Glaciar Benito of the NPI took place in 1860 AD. On the eastern NPI, Glaciar Nef (Figure 9.10, No. 3) advanced in 1863 AD, and moraine crests around Lago Arco date an advance of a previously unnamed glacier (NPI-24) in 1881 AD. For Monte San Lorenzo, lichenometry indicates ice-margin stabilization at ca. 1925 AD (Glaciar Rio Lacteo) and 1795–1955 AD (Glaciar San Lorenzo—Fig. 9.10, No. 4).

For the Southern Patagonian Ice Field (SPI), historical documents provide evidence of a substantial advance of Glacier Jorge Montt (Fig. 9.10, No. 5) relative to the present at 1898 AD. Glacier O'Higgins (Fig. 9.10, No. 6) likewise advanced into Lago O'Higgins in 1896 AD. On the eastern SPI, dendrochronology

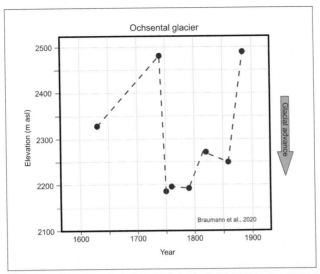

FIG. 9.9 Top: Graph illustrating the elevational position of the terminal moraines of Ochsental Glacier since about 1650 AD, showing major advance from 1750 to 1800, then a major retreat after 1850. Bottom: Photograph of the Ochsental Glacier valley, Vorarlberg, Austria. Photo by Basotxerri, Creative Commons Attribution-Share Alike 4.0 International license. The dotted line shows the approximate position of the LIA ridgeline. *LIA*, Little Ice Age. (Data from Braumann, S.M., Schaefer, J.M., Neuhuber, S.M., Reitner, J.M., Lüthgens, C., Fiebig, M., 2020. Holocene glacier change in the Silvretta Massif (Austrian Alps) constrained by a new 10Be chronology, historical records and modern observations. Quaternary Science Reviews 245, 106493. https://doi.org/10.1016/j.quascirev.2020.106493; Modified from Miles, K., Hubbard, B., Irvine-Fynn, T., Miles, E., Quincey, D., Rowan, A., 2020. Hydrology of debris-covered glaciers in High Mountain Asia. Earth-Science Reviews 207, 103212. https://doi.org/10.1016/j.earscirev.2020.103212.)

indicates glacier advance to prominent moraines in 1626−1850 AD. Moraines near Huemul Glaciar (Fig. 9.10, No. 7) are dated from 1481 to 1886 AD by lichenometry.

On the Herminita Peninsula, several moraines, just inside the mid-Holocene and last two millennia moraines described earlier, have been dated by [10]Be to ∼1400−1800 AD. Moraines just inside those of

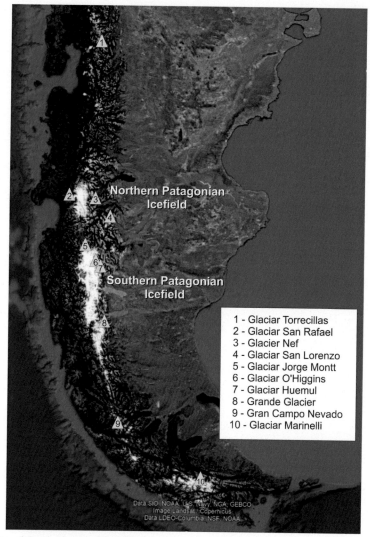

FIG. 9.10 Map of the Northern and Southern Patagonian Ice Fields, showing locations of sites discussed in the text.

Mid-Holocene age at Glacier Torres have [10]Be exposure ages of 0.5 ka. In the south-western Lago Argentino basin, moraines inside the mid-Holocene limits document an advance of Grande Glacier (Fig. 9.10, No. 8) at 1500–1800 AD. A Late Holocene advance of the Gran Campo Nevado ice cap (Fig. 9.10, No. 9) is recorded by dendrochronology on glacier moraines dating from 1628 to 1886 AD. An advance of glaciers at Cordillera Darwin is recorded at Glaciar Marinelli (Fig. 9.10, No. 10), where two advances are dated to ~1270 and 1540 AD. The most recent of these reached Narrows Moraine, where it remained until historical times.

Patagonian glacial moraines contemporaneous with the LIA are typically large and well defined, surrounding ice-scoured bedrock along the edges of the valley glaciers. These advances reached similar extents to the mid-Holocene glacial advances (5 ka). Morphological similarities between these moraines allowed Davies et al. (2020) to extrapolate the 1800 AD ice margin across the NPI. This time is associated with a persistent cold/wet, negative Southern Annular Mode (SAM) interval, which produced conditions favorable for glacier growth. The SAM describes the low-frequency variability of the Southern Hemisphere climate outside of the tropical zone. A positive index (lower polar

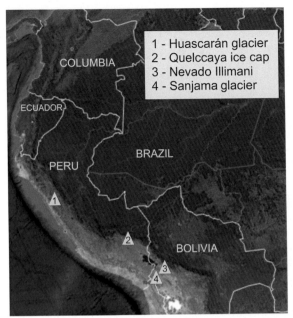

1 - Huascarán glacier
2 - Quelccaya ice cap
3 - Nevado Illimani
4 - Sanjama glacier

COLUMBIA

ECUADOR

PERU

BRAZIL

BOLIVIA

FIG. 9.11 Map of the northern Andes, showing sites discussed in the text. (Figure by the author.)

Interpretation of $\delta^{18}O$ records from low-latitude mountain glaciers

The first long tropical ice cores were retrieved in 1983 from the Quelccaya ice cap, the Earth's largest tropical ice cap, which is 5670 m above sea level in the southeastern Andes of Peru. Because of its dome shape, Quelccaya is very sensitive to the current increase in the elevation of the 0°C isotherm, and it loses more mass per unit rise in the equilibrium line than steeper alpine glaciers.

As shown in Fig. 9.12, Thompson et al. (2006) compared the $\delta^{18}O$ records (AD 1540−1983) from the 1983 core (blue) and 2003 core (red), confirming that the timescales are virtually identical. This agreement confirms the timescale for the 1983 core, and it demonstrates excellent preservation of the $\delta^{18}O$ record of the ice cap even though its surface has been warming during the intervening 20 years. Thus, the Quelccaya $\delta^{18}O$ record puts the modern accelerating glacier retreat within a 1500-year context. The $\delta^{18}O$-air temperature (Ta) relationship is reasonably well established for ice core samples from Antarctica and Greenland, but the relationship is not as straightforward for high mountain regions in mid-to-low latitudes. Few specific empirical studies exist. As discussed earlier, Thomson et al. (2000) examined the $\delta^{18}O$-Ta relationship for discrete precipitation events at three sites on the Tibetan Plateau. Although dynamic processes dominate $\delta^{18}O$ on short timescales (weeks to months), they concluded that air temperature is the primary control on longer-term (more than annual) $\delta^{18}O$ variations at those sites.

In the Andes, the controls on $\delta^{18}O$ in recent snowfall have been investigated by using observations and models. Strong linkages have been demonstrated between SSTs across the equatorial Pacific Ocean and $\delta^{18}O$ in ice cores from the tropical Andes and the Himalaya (Dasuopu glacier). More recently, it has been suggested that $\delta^{18}O$ ultimately records changes in the atmospheric circulation that integrate all aspects of transport from source to sink. Although not fully resolved, the weight of available empirical evidence strongly suggests that on longer timescales, $\delta^{18}O$ variations in low-latitude mountain precipitation reflect variations in air temperature more strongly than variations in weather patterns.

Based on the assertion of Thompson et al. (2006) that $\delta^{18}O$ variations in low-latitude mountain precipitation reflect variations in air temperature, the following paleotemperature approximations may be made. Note that these are the author's estimates—not those of Thompson et al. (2006).

pressure) is associated with weaker, zonal winds. A negative value is associated with stronger zonal winds.

Northern Andes

The northern Andes lie in the tropical zone of the Northern Hemisphere (Fig. 9.11). All of the Holocene paleoclimate research for the region has been done through the analysis of stable isotopes from ice cores taken from mountaintop glaciers.

Kellerhals et al. (2010) provided a reconstruction of paleotemperatures and glacial history over the past ∼1600 years, based on a highly resolved and carefully dated ice core taken from Nevado Illimani in the eastern Bolivian Andes (Figure 9.11, No. 3). Their reconstruction reveals that Medieval Warm Period−type and Little Ice Age−type episodes are distinguishable in the high Andes of tropical South America. This region has had only very limited temperature proxy data. For the interval from about 1050 to 1300 AD, their reconstruction shows relatively warm conditions, followed by cooler conditions from the 15th to the 18th century, when temperatures dropped as much as 0.6°C below the 1961−90 average. The last decades of the 20th century are characterized by temperatures than those reconstructed for the last ∼1600 years.

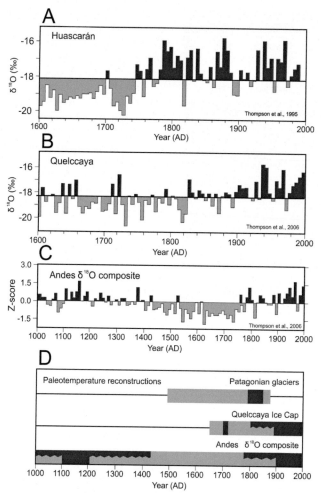

FIG. 9.12 Oxygen isotope records from the northern Andes and paleoclimate reconstructions from Patagonia and combined for South America. (A) The oxygen isotope record for the last 500 years from Huascarán, Peru. (B) The oxygen isotope record for the last 500 years from the Quelccaya Ice Cap, Peru. (C) Composite oxygen isotope record for the last 1100 years from ice cores taken from the Andes. (D) The author's comparison of paleotemperature reconstructions from sites throughout the Andes. (After Thompson, L.G., Mosley-Thompson, E., Brecher, H., et al., 2006. Abrupt tropical climate change: past and present. Proceedings of the National Academy of Sciences of the United States of America 103, 10536–10543. https://doi.org/10.1073/pnas.0603900103.)

The $\delta^{18}O$ record from the Quelccaya ice cap is considered by Thompson et al. (2006) to be the only reliable regional record. The ice layers at Sajama Glacier and Huascarán are rapidly thinning (see Thompson et al., 1995). Flow models were required, constrained by time-stratigraphic horizons, to construct their time-scales. Net mass balance records could not be constructed for these two sites. Thus, Quelccaya provides the only reliable ice core—derived net mass balance history for the past 400 years in the Andes.

A summary of inferred paleotemperature reconstructions for the Quelccaya ice cap is shown in Fig. 9.12B. The interval from 1660 to 1710 was dominated by colder-than-modern conditions. There was a brief warming from 1710 to 30, followed by almost a century of colder conditions (1730–1812). Between 1812 and 1900, there were a series of temperature regime oscillations, approximately on a decadal scale. Since 1900, nearly all $\delta^{18}O$ data indicate warm conditions at Quelccaya.

The paleotemperature reconstructions for the Huascarán site are not as reliable because of difficulties with the ice core chronology. In rough approximation, then, the interval from 1600 to 1745 at Huascarán saw colder-than-modern conditions. Since 1745, conditions have mostly been warm, with only brief (<5 years) cooling events, and one cold decade from 1890 to 1900.

Thompson et al. (2006) also developed a composite δ¹⁸O record for three glaciers in the Andes (Quelccaya, Huascarán, and Sajama) (Fig. 9.12C). Because the three sites are at different latitudes and elevations, the only useful way to combine the data from all sites was to standardize them, using z-scores for each data point. A z-score gives an indication of the distance of a data point from the mean of the entire data set (in this case, mean values from the individual ice-coring sites). It is a measure of the number of standard deviations a particular score falls below or above the population mean. Z-scores are thus a way to compare results to a "normal" population. The Andes composite δ¹⁸O record shows a lengthy cold interval from 1400 to 1790 AD. This is followed by some warm and cold oscillations throughout the 19th century, and then the composite record indicates nearly all warming in the 20th century.

Summary of South American little ice age
As in other parts of the world, high mountain climates cooled in the Andes during the time interval of the LIA. Since the teleconnections between the Andes and climate drivers of the LIA in the Northern Hemisphere are not well understood, the authors I have quoted in this section are, for the most part, reluctant to equate the Late Holocene climatic history of the Andes with the LIA history of Europe and the North Atlantic. Be that as it may, the principal reason for discussing LIA mountain glaciers in this book is to set a baseline from which to compare the extent and thickness of modern glaciers around the world. So, for our purposes, the climatic trigger for glacial advances in the last 1000 years does not matter. It is more important to discuss the location and size of these mountain glaciers by the mid-19th century, as the Industrial Revolution started pumping greenhouse gases (GHGs) into the atmosphere.

Unfortunately, despite a great deal of work, as described by Davies et al. (2020), we still do not know the limits of late Holocene Patagonian glaciers with much certainty. We know that glaciers advanced well beyond their modern elevations in the last 500–1000 years, but it is difficult to distinguish glacial features from this time interval and features from 5000 years ago. It is clear that the climate of the Andes of southern South America was colder than modern from 1500–1790 AD, followed by warmer conditions through about 1850. From 1850 to 1900, this region experienced another cold interval, but then from 1900 to the present day, the climate appears to have been too warm to foster the growth of regional glaciers.

The composite δ¹⁸O record from ice cores from Peruvian glaciers and ice caps provides a thousand years of climate history for the northern Andes. A warm interval dominated the region between 1100 and 1200 AD. This was followed by more than 200 years of oscillating temperature regimes, and then a lengthy cold period began about 1420 AD. From 1420 to 1780 AD, regional climates were colder than modern and regional glaciers expanded. Much of the 19th-century record from the Peruvian oxygen isotope samples suggests oscillations between cold and warm climate, followed by continuous warming of the region since about 1900 AD.

MOUNTAIN GLACIER DYNAMICS
With a few notable exceptions, mountain glaciers appear to be stable accumulations of ice, clinging to high mountain regions. But they are not static, they simply move more slowly than generally can be observed in a day or a week, or even a month. Frozen water is a unique substance in that it is a solid, but it is also plastic, that is it changes shape in response to pressures such as gravity. Therefore, the ice in all glaciers and ice sheets is constantly moving from higher elevations to lower. When regional temperatures fall and there is sufficient moisture, glaciers and ice sheets expand. When regional temperatures climb or when there is a lack of moisture, the margins of glaciers and ice sheets contract. This is putting glacial dynamics in its simplest form. Let us examine a few particular aspects of this topic.

Glacial Mass Balance
The mass balance of a glacier, also called its surface mass balance, is the difference between the snow accumulated in the winter and the snow and ice melted over the summer. If the mass of snow accumulated on a glacier exceeds the mass of snow and ice lost during the summer months, the mass balance is positive. If summer melting exceeds what was gained the previous winter, the mass balance of the glacier is negative. Mass balance is a measure of a glacier's health. Over time, mass balance data reveal glacial responses to climate change.

As reviewed by the US Geological Survey (2020b), nearly all Earth's alpine glaciers have been in negative mass balance conditions in the 21st century, so they are losing ice. Rates of mass loss for North American glaciers are among the highest on Earth and shrinking glaciers are often the most visible indicators of mountain ecosystem response to climate change. World Glacier Monitoring Service analysis of glacial mass balance data collected from a set of reference glaciers around the world from 1976 to 2005 showed glacier mass loss of over 20 m of w.e. (water equivalent), with average annual melting rates doubling after the year 2000 Lindsey (2020). 2018 was the 30th year in a row of mass loss of mountain glaciers worldwide. This trend continued through 2019 (NSIDC, 2019).

Mass balance measurement

Traditionally, determining glacier mass balance involves measuring the change in mass at specific sites with snow stakes and snow pits and then extrapolating these point values over the entire glacier surface. This field method of measuring mass balance is labor-intensive, requiring repeated visits to sites dispersed along the glacier's elevational gradient to capture differences in snow depth and density. Measurements are generally made twice annually: once to measure the minimum mass of the season (autumn) and a second time to measure the maximum mass of the season (spring). Based on these two sets of measurements, seasonal mass gains and losses, as well as the net change over the entire year, may be calculated.

Glaciers are made of several kinds of snow and ice, of various densities, expressed in units of kg per m^3. For instance, the snow that fell in the most recent winter has densities from 200 to 500 kg per m^3 (heavy, wet snow has a greater density than light, dry powder snow). Firn, which is snow that accumulated from previous winters, has densities ranging from 500 to 900 kg per m^3. Finally, ice has a density of 900 kg per m^3. Because of these highly variable densities, average mass balance estimates for a glacier are presented in water-equivalent (w.e.) units. To do this, the densities and thicknesses of snow, firn, and ice layers are each converted to w.e. units. Then changes in mass balance are represented as a uniform w.e. thickness over the entire glacier. Water equivalent is the amount of liquid water that would result if a given amount of ice or snow melted and spread out over the surface of a glacier.

More recently, various types of digital measurements and modeling have been used to calculate glacial mass balance. Digital elevation models (DEMs) accurately determine the size and shape of a glacier at a specific time. Comparisons of DEM data taken from different times can be used to estimate glacier volume change over the interval between DEM acquisitions.

Recent advances in remote sensing from satellites and airplanes provide new opportunities to quantify how glaciers are changing. As discussed in Chapter 11, image- and radar-based DEMs provide three-dimensional representations of glacier surfaces with meter-scale precision. DEMs allow field measurements to be extrapolated over the entire glacier. In addition, repeat DEM acquisitions over several years can be compared to determine the total volume change of a glacier. This can be used to improve traditional stake and pit estimates of long-term mass balance trends.

Glacial Debris

As glaciers creep down mountainsides, they grind up the bedrock beneath and alongside their path. Rock debris that is eroded and plucked up at the base of the glacial ice is slowly entrained in the flowing ice until it reaches the terminus of the glacier, where it helps to form a terminal moraine. Likewise, rocks that are scoured from the sidewalls of a glacier can be entrained in the ice and move to the terminus. This kind of rocky debris can also be pushed to the sides of the moving ice, where it forms lateral moraines. The third destination for glacial debris is the surface of the glacier. Some glaciers are more-or-less covered by a mantle of rocky debris, which ranges in size class from the finest clay particles to boulders.

As discussed by Herreid and Pellicciotti (2020), the structure of a debris cover is unique to each glacier and sensitive to climate. A layer of rock debris on the surface of a glacier causes the glacial melt rate to decrease. The debris acts as a layer of insulation, limiting or blocking sunlight from reaching the ice beneath. Glacial debris varies in thickness from ~2 cm to several meters. The structure of the debris present on any one glacier is a unique and complex function of many local factors, which include the surrounding geology, topographic relief, and glacier dynamics. A consensus of studies from around the world shows a spatial expansion of debris cover and links this change to global warming. The authors showed that over half of glacial debris is concentrated in three regions: Alaska (38.6% of the total debris-covered area), Southwest Asia (12.6%), and Greenland (12%) (Table 9.2). Of Earth's glaciers, 20% have a substantial percentage of debris cover.

As described by Miles et al. (2020), the hydrology of debris-covered glaciers is complex: rugged surface topography typically directs the paths of meltwater

TABLE 9.2
Percent Cover of Rock Debris on Glaciers and Ice Sheets, by Region.

Region	% Debris Cover
New Zealand	19
Southeast Asia	18.6
North Asia	17
Central Europe	16.2
Caucasus	15.9
Alaska	14.3
Southwest Asia	13.7
Central Asia	10
Iceland	10
Western Canada and the United States	7.2
Low latitudes	5.4
Russian Arctic	3.6
Scandinavia	3.6
Southern Canadian Arctic	3.5
Southern Andes	3.3
Svalbard	2
Greenland	1.5
Northern Arctic Canada	0.8

Data from Herreid, S., Pellicciotti, F., 2020. The state of rock debris covering Earth's glaciers. Nature Geoscience 13 (9), 621–627. https://doi.org/10.1038/s41561-020-0615-0.

through a series of supraglacial-englacial systems (above and within the glacier) including channels and ponds (Fig. 9.13). Low-gradient tongues of glacial debris may extend several km from the glacial terminus, slowing meltwater flow and trapping water in englacial storage. Englacial conduits are constantly shifting; channels are frequently abandoned, new channels form, and old channels become reactivated. New cracks of permeability in the sediments are exploited, capturing drainage as rates of surface and subsurface runoff ebb and flow.

Herreid and Pellicciotti (2020) found that the percentage of debris cover on glaciers increases exponentially with increased distance from the poles (Table 9.2). This suggests that warm regions are more conducive to supraglacial debris production. While this relationship between glacial debris cover and regional temperatures is generally reliable, such factors as mountain-range age, relief, and lithology may cause exceptions to the rule, as exemplified by glaciers in Alaska.

Herreid and Pellicciotti (2020) used multispectral satellite imagery and a map of glacier area as input to a simple image segmentation approach to map debris cover. The strong color contrast between rocks and glacier ice, firn and snow enables a robust and repeatable method to map debris cover if the glacier map is accurate and the imagery is controlled for such factors as seasonal snow and cloud cover. With the recent release of the global Randolph Glacier Inventory, it became feasible for the authors to map debris cover globally by this method.

Proglacial Lakes

Proglacial lakes are masses of water impounded at the edge of a glacier or on the margin of an ice sheet. The term "proglacial lake" has been used to refer to ice-contact or ice-marginal lakes, which are physically attached to an ice margin, as well as lakes detached from a contemporary ice margin. The term proglacial lake, therefore, includes all lakes that are or have been directly influenced by either a glacial ice margin or by

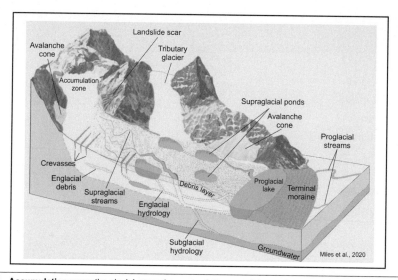

Accumulation zone: the glacial area where snowfall accumulates and exceeds the losses from ablation (melting, evaporation, and sublimation).
Avalanche cone: material deposited at the base of an avalanche track, including ice, snow, rock and debris.
Crevasse: a deep, open crack in a glacier.
Debris layer: a deposit of unstratified sediment, mainly boulders, gravel, sand, and clay left on the surface by a glacier.
Englacial debris: debris dispersed in the interior of a glacier, originating either as buried surface debris or as basal debris is raised from the bed by thrusting or folding.
Englacial hydrology: structures in the ice produced by tension, such as crevasses, that allow the water to penetrate from the surface into the ice
Proglacial lake: a lake formed when a glacier margin retreats and meltwater is impounded in the topographic low between the ice front and the abandoned moraine ridges.
Proglacial stream: streams that form immediately adjacent to a glacier, often fed by glacial meltwater.
Subglacial hydrology: streams and pools of water that form beneath the surface of a glacier.
Supraglacial pond: ponds that form on the surface of a glacier, fed by meltwater carried by supraglacial streams.
Terminal moraine: a moraine deposited at the point of furthest advance of a glacier or ice sheet.
Tributary glacier: a small glacier that flows into a larger glacier.

FIG. 9.13 Conceptual illustration of the main landscape and hydrological features of a typical debris-covered glacier. (From Miles, K., Hubbard, B., Irvine-Fynn, T., Miles, E., Quincey, D., Rowan, A., 2020. Hydrology of debris-covered glaciers in High Mountain Asia. Earth-Science Reviews 207, 103212. https://doi.org/10.1016/j.earscirev.2020.1032122.)

surface glacial meltwater. The formation of moraine-dammed lakes is quasi-periodic and is usually associated with periods of glacier retreat or downwasting following a previous ice advance.

Proglacial lakes are widespread in the Quaternary record and can provide an abundance of paleoenvironmental information. The current global deglaciation is increasing the number and size of proglacial lakes. Carrivick and Tweed (2013) show how various geomorphological, sedimentological, chemical, and biological characteristics can be used to distinguish between "ice-marginal" and "ice-contact" lakes and other distal proglacial lakes.

Proglacial lakes are an active and dynamic part of glacial systems. In some situations, they intensify mountain glacier and ice sheet margin ablation via mechanical and thermal stresses. Very large lakes, on the other hand, can moderate summer air temperatures and relatively retard summer ice ablation. Proglacial lakes interrupt the flow of glacial meltwater and are very efficient sediment traps. The sudden drainage of proglacial lakes and resultant glacial lake outburst floods cause the rerouting of streams. In consequence, geomorphological activity can be radically modified. The drainage of exceptionally large proglacial lakes during the Pleistocene affected global ocean circulation and climate. Overall, analyses of proglacial lakes can provide valuable insights into the patterns, character, and behavior of mountain glaciers, ice sheets, and glaciations. This historic information can be used to assess the potential impacts of present and future deglaciation.

Glacial lake outburst floods, which are a type of jökulhlaup, are ubiquitous within the Quaternary record of deglaciation. They have a modern geological imprint and are modern natural hazards. These floods can be powerful agents of landscape change through erosion and sediment deposition.

Jökulhlaups (an Icelandic word) are glacial outburst floods. They occur when a lake fed by glacial meltwater breaches its dam and drains catastrophically. These lakes can take several different forms. They may be ice-dammed lakes that are held in by the glacier ice itself, or moraine, rock, or sediment dammed lakes, or even lakes that lie beneath glaciers. For a jökulhlaup to occur, the lake water levels must reach a critical point such that the lake causes its ice dam to float. Alternatively, the lake can overtop its dam, causing rapid incision into the sediment, rock, or ice that contains it. These outburst floods carve large meltwater channels beneath the glacier ice that allow for rapid drainage. Jökulhlaups may occur with somewhat regular periodicity, but others drain without warning.

The present deglaciation is producing an increased number and size of proglacial lakes around the world, for example, in the European Alps, the Caucasus, Iceland, South America, and across the Himalaya and specifically within the Mt Everest region, in Bhutan, and Tibet. It is very important to understand the effects that these lakes could have on modern and future geological and environmental systems.

Thermal structure of proglacial lakes

Calving glaciers are rapidly retreating in many regions under the influence of ice-water interactions at the glacier front. Many studies have been conducted on fjords in front of tidewater glaciers, but very few studies have been done on freshwater calving glaciers that terminate in lakes. Sugiyama et al. (2016) measured lake water temperature, turbidity, and bathymetry near three large calving glaciers in the SPI. The thermal structures of these lakes are significantly different from those found in glacial fjords (Fig. 9.14). For instance, there was no indication of upwelling subglacial meltwater. The authors found that turbid and cold glacial water discharge filled the region near the lake bottom. This was because water density is controlled by suspended sediment concentrations rather than by water temperature. Near-surface wind-driven

circulation reaches a depth of ~180m, forming a relatively warm isothermal layer. Mean temperatures of this layer ranged from 3–4°C to 6–7°C at the study sites, sufficient to convey heat energy to the ice-water interface. However, the deeper part of the glacier front is in contact with stratified cold water, implying a limited amount of melting there. For instance, in the proglacial lake of Glaciar Viedma, Argentina, water deeper than 120 m was turbid and very cold. The water temperature of proglacial lakes has been reported from sites in Alaska, New Zealand, the Himalayas, and Patagonia. Most of the study results agree that proglacial lakes are filled with cold water, often colder than 1°C. In Alaska, the temperature of the lakes was colder than 3°C in front of Mendenhall, Yakutat, and Bering Glaciers, except for near-surface shallow layers. These temperatures are significantly lower than that of seawater in front of tidewater glaciers in the same region.

Freshwater calving glaciers are not subject to the continual wave action and currents of ocean water—factors that one might assume would enhance subaqueous melting in comparison with glaciers terminating in lake water. However, this is not always the case. For instance, some of the freshwater lake-calving glaciers in Alaska have shown more rapid changes than tidewater glaciers in the same region. Because of proglacial lake formation following glacier retreat, the number of freshwater calving glaciers is increasing around the world. Recent studies have shown that meltwater lakes are increasing in number and expanding in size at the margins of the Greenland ice sheet, potentially contributing to ice sheet mass loss. Formation and expansion of moraine-dammed lakes due to glacier retreat are also of great concern in the high mountain regions of Asia.

The circulation and thermal structure of water near freshwater calving glaciers are significantly different from those at tidewater glaciers. The heat source for melting in freshwater is different from the one for submarine melting. Solar energy is the most important heat source in a closed lake, whereas abundant heat is available in fjords via convection from the open ocean.

Evidence from Patagonia

As discussed by Glasser et al. (2011), of the 64 major outlet glaciers in the NPI and SPI, more than 80% of them terminate in water. Most glaciers in the NPI and those on the eastern side of the SPI flow into freshwater lakes. These lake-calving glaciers account for ~70% of the calving glaciers in Patagonia. Frontal ablation plays a key role in calving glacier variations, which are

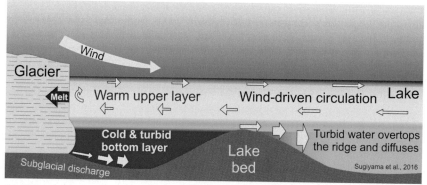

FIG. 9.14 Schematic diagram showing the thermal structure of a lake in front of a calving glacier in Patagonia. (Modified from Sugiyama, S., Minowa, M., Sakakibara, D., Skvarca, P., Sawagaki, T., Ohashi, Y., Naito, N., Chikita, K., 2016. Thermal structure of proglacial lakes in Patagonia. Journal of Geophysical Research: Earth Surface 121 (12), 2270–2286. https://doi.org/10.1002/2016JF004084.)

generally more rapid than those of land-terminating glaciers. Frontal ablation is accomplished through a combination of subaqueous melting and iceberg calving at the termini of marine—or freshwater-terminating glaciers. It was previously thought that frontal ablation occurs mostly through calving, but recent studies on tidewater glaciers in Greenland and Alaska have shown that the contribution of subaqueous melting is substantially larger than previously assumed. The margins of freshwater calving glaciers in the SPI and NPI are rapidly retreating, contributing substantially to the recent ice mass loss in Patagonia. The authors estimated that subaqueous melting accounts for about 5%–30% of the frontal ablation at Glaciar Perito Moreno. In the case of Glaciar Upsala and Glaciar Viedma, it accounts for only a limited portion (<10%) because of the cold lake water and the greater amount of ice discharge into the lake.

THE CURRENT DECLINE OF MOUNTAIN GLACIERS

As we have seen in this chapter, many of the mountain glaciers of the world benefitted greatly from the LIA. In some cases, these alpine glaciers had not been as big since the end of the last glaciation, 11,000 years ago. It is just a quirk of history that the LIA took place at the same time as the development of the sciences of glaciology and climatology, in the late 18th and 19th centuries. If these sciences had developed in Medieval times, then the practitioners of these disciplines would have had less to talk about. But the scientists who lived during the 18th and 19th centuries looked upon

Alpine landscapes populated by massive glaciers that extended down into valleys, blocked rivers, and dominated landscapes. Those of us alive in the 21st century have the unenviable duty of watching these glaciers nearly disappear. Repeat photography (Figs. 9.15 and 9.16) shows us in quite dramatic terms just how much glacial ice has been lost within the last century or so.

Li et al. (2019) assessed regional differences in glacial retreat across all regions from 1980 to 2015, based on two primary features: area change, and mass balance (Table 9.3). The data were compiled from published glacial records. They found the following large-scale trends. The average rate of global glacier area shrinkage was 0.18% per year, equating with a global glacier mass loss of 0.25 m w.e. per year. Mountain glaciers in low and middle latitudes suffered severe area shrinkage and mass loss. In contrast to this, Arctic deglaciation was characterized by ice thinning rather than high area reduction, but the thinning has resulted in relatively high rates of mass loss. The survey concluded that glaciers in high southern latitudes were relatively stable. High Mountain Asia exhibited the lowest rate of area shrinkage and mass loss among glaciers situated in the low and middle latitudes. These glaciers also had a slower rate of mass loss than the global average. Glaciers in the Tropical Andes exhibited the most rapid glacier area shrinkage (−1.6% per year), whereas Antarctic and sub-Antarctic glaciers had the lowest rate (−0.11% per year). The glaciers with the most negative mass balance occurred in the Southern Andes (−0.81 m w.e. per year), followed by Alaska (−0.74 m w.e. per year). Li et al. (2019) concluded

FIG. 9.15 Upper left: Muir Glacier, Alaska in 1941 (photo by William Field) and 2004 (photo by Bruce Molina, USGS) (both photos in the public domain). Upper right: Grinnell Glacier, Montana in 1910 (photo by Morton Elrod) and 2016 (photo by Lisa McKeon, USGS) (both photos in the public domain). Middle left: Kilimanjaro, Tanzania in 1993 and 2000 (photos by NASA, in the public domain). Right: Torres del Paine, Chile in 1916 (photo by Alberto de Agostini, in public domain) and 2016 (photo by Fabio Ventura, used with permission). Bottom: Mendenhall Glacier, Alaska in 1958 (photo by M.T. Millett) and 2011 (Photo by M. J. Beedle, USGS) (both photos in the public domain).

that surface climate variability in the glacial environment is a key driver of regional differences in glacier retreat.

Asia

The WGMS (2020) estimates that the glaciers of Central Asia have lost an average of about 22% of their mass between 1950 and 2020. This level of ice mass loss is essentially the same as the average value for all mountain regions (Fig. 9.17). The largest accumulations of ice outside of the polar regions occur on the Tibetan Plateau (Fig. 9.18). This high plateau is known as the world's Third Pole because of the amount of glacial ice held there, but since the 1950s, average temperatures have risen 1.5°C, with no end to warming in sight. Therefore,

the outlook is grim for glaciers of the Third Pole. Some of the best studies of the regional glaciers concern the 2684 glaciers in the Qilian Mountain region.

As discussed by Pollard (2020), glacier retreat in the Qilian Mountains was 50% faster in 1990–2010 than it was from 1956 to 1990, as revealed by data from the China Academy of Sciences (Fig. 9.19). The flow of water in a stream near the terminus of the Laohugou No. 12 glacier has doubled in the past 60 years.

Cao et al. (2019) studied changes in glacial mass from the Lenglongling Mountains (LLM), an eastern part of the Qilian Mountains. Their analyses employed multitemporal satellite images and the study of DEMs based on topographic maps and ZY-3 data. The climate of the LLM region is influenced by the East Asian

Qori-Kalus Glacier, Peru

Vernagtferner Glacier, Austria

Main Rongbuk Glacier, Tibet

FIG. 9.16 Top: Qori-Kalus Glacier, Peru in 1978 and 2004. Photos by permission of Lonnie Thompson. Bottom: Main Rongbuk Glacier, Tibet, in 1921 (photo by George Mallory, in the public domain) and 2020. (Photo courtesy of Glacierworks.)

Monsoon, the South Asian Monsoon, and the Westerlies. In 1972, 244 glaciers were documented in the LLM, covering a total area of 103 km². This area included 103 glaciers on the southwest-facing slopes of the LLM, with total coverage of 39.1 km²; these glaciers feed the Datong River drainage basin (the

Datong River is a tributary of the Yellow River). The remaining 141 glaciers, covering a total area of 64.1 km², were identified on the northeast-facing slopes of the LLM; they feed the drainage basins of the Shiyang River. The results of the study showed that from 1972 to 2016, glacier coverage in these mountains decreased by

TABLE 9.3
Rate of Retreat of Glaciers From Mountain Regions of the World.

Major Region	Research Region	Study Interval	Rate (% per year)
Alaska	Chugach Mountains	1952–2007	−0.42
	Harding Ice Field	1986–2000	−0.16
Western Canada and the United States	Western Canada	1985–2005	−0.55
	Yukon	1959–2007	−0.46
	North Cascades	1958–2006	−0.34
	Wind River Range	1966–2006	−1.1
Arctic Canada North	Queen Elizabeth Islands	1960–2000	−0.066
Arctic Canada South	Barnes Ice Cap	1958–2000	−0.048
	Penny Ice Cap	1959–2000	−0.046
	Terra Nivea	1958–2000	−0.33
	Grinnell Ice Cap	1958–2000	−0.26
	Baffin Island	1975–2000	−0.16
	Bylot Island	1959–2001	−0.12
Greenland	Central East Greenland	2000–05	−0.019
Iceland	Most Iceland Glaciers	1990s–2000s	−0.3
Svalbard	Svalbard	1980–2010	−0.23
Scandinavia	Norway	1966–2013	−0.29
North Asia	Ural Mountains	1956–2000	−0.51
	Kodar Mountains	1995–2010	−2.68
	Altai Mountains in Russia	1952–2004	−0.39
	Altai Mountains in China	1960–2009	−0.75
Central Europe	Alps	1985–2005	−1.2
Caucasus Middle East	Georgian Caucasus	1960–2014	−0.7
	Turkey Glacier	1970–2013	−1.3
Central Asia	Tianshan Mountains	1961–2012	−0.35
	Inner Tibet	1970s–2000	−0.27
	Pamir	1964–2001	−0.21
	Qilian Mountains	1956–2003	−0.46
	West Kunlun	1970s–2001	−0.01
	Gongga Mountains	1966–2009	−0.2
South Asia West	Mapam Yumco Basin	1974–2003	−0.23
	Himachal Pradesh	1962–2001	−0.54
	Ten basins mean	1962–2002	−0.38
	Kang Yatze	1991–2010	−0.09
	Gharwal Himalaya	1968–2006	−0.12
	Naimona'Nyi region	1976–2003	−0.31
South Asia—East	Koshi Basin Nepal	1976–2000	−0.15
	Mt. Qomolangma Preserve	1976–2006	−0.5
	Boshula Mountain Range	1975–2001	−0.28
	South-Eastern Tibet	1980–2001	−0.9
	Tisa	1997–2004	−0.36
	Mt. Everest	1962–2011	−0.27
Low latitudes	Tropical Andes	1980–2000	−1.6

Continued

TABLE 9.3			
Rate of Retreat of Glaciers From Mountain Regions of the World.—cont'd			
Major Region	**Research Region**	**Study Interval**	**Rate (% per year)**
Southern Andes	Subtropical Andes of Chile	1955–2007	−0.56
	Central Chile and Argentina	1955–2013	−0.51
	Monte San Lorenzo region	1985–2008	−0.81
	Patagonian region	1986–2011	−0.17
New Zealand	Southern Alps	1978–2002	−0.69

Data from Li, Y.J., Ding, Y.J., Shangguan, D.H., Wang, R.J., 2019. Regional differences in global glacier retreat from 1980 to 2015. Advances in Climate Change Research 10 (4), 203–213. https://doi.org/10.1016/j.accre.2020.03.003.

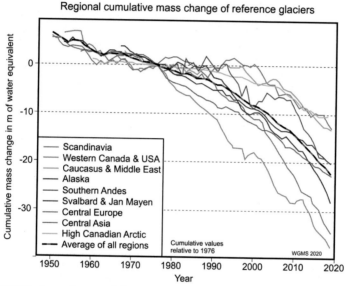

FIG. 9.17 Cumulative mass change relative to 1976 for regional and global means based on data from reference glaciers. Cumulative values are given on the y-axis in the unit meter water equivalent (m w.e.). The mass balance estimates considered here are based on a set of global reference glaciers with more than 30 continued observation years for the time period, which are compiled by the World Glacier Monitoring Service (WGMS) in annual calls-for-data from a scientific collaboration network in more than 40 countries worldwide. Regional values are calculated as arithmetic averages. Global values are calculated using only one single value (averaged) for each region with glaciers to avoid a bias to well-observed regions. Values before 1960 and in 2019 need to be taken with caution due to the limited sample size. (Data from World Glacier Monitoring Service, 2020. Global glacier change bulletin, 2016–2017. In: Zemp M., Gärtner-Roer, I., Nussbaumer, S., Bannwart, J., Rastner, P., et al. (Eds.), Global Glacier Change Bulletin (vol. 3). ISC(WDS)/IUGG(IACS)/UNEP/UNESCO/WMO, World Glacier Monitoring Service.)

40% (41.1 km²), leaving only 62.1 km² of ice cover in 2016.

A study by Farinotti et al. (2019) used an ensemble of up to five models to provide a consensus estimate for the ice thickness distribution of all the about 215,000 glaciers outside the Greenland and Antarctic ice sheets (Table 9.4). They found that High Mountain Asia hosts about 27% less glacier ice than previously thought. This

FIG. 9.18 Map of the glaciers of central and eastern Asia. (Modified from NSIDC, 2018. Impact of High Asian glaciers and snowpack on water resources. Research Project: Contribution to High Asia. https://nsidc.org/charis/project-summary/.)

implies that the date by which the region is expected to lose half of its present-day glacier area must be moved forward by about one decade. Previously, researchers had estimated that the area covered by glaciers in this region would halve by the 2070s. This is now expected to happen by the 2060s. This rapid decline in glacier volume will eventually diminish meltwater volumes, greatly restricting the water supply of regional rivers. Low river levels will have serious impacts on arid regions such as the Andes or central Asia, which depend on this water source for agriculture and human consumption. The glaciers of High Asia supply the headwaters of several large rivers, including the Indus, the Tarim, and rivers feeding into the Aral Sea. Hundreds of millions of people depend on these water supplies (Ruegg, 2019).

Gardelle et al. (2013) studied the region-wide glacier mass balances over the Pamir-Karakoram-Himalaya regions for the interval 1999–2011. During that time, they found only moderate mass losses in the eastern and central Himalaya (−0.22 ± 0.12 m w.e. per year to −0.33 ± 0.14m w.e. per year) and larger losses in the western Himalaya (−0.45 ± 0.13m w.e. per year).

Following on from the work of Gardelle et al. (2013), Brun et al. (2017) computed the mass balance for more than 90% of the glaciers in 12 Asiatic high mountain regions from the inner Tibetan Plateau in the east to the Pamir Mountains in the west (Fig. 9.20 and Table 9.5), using time series of DEMs derived from satellite stereoimagery. They calculated a total mass change between 2000 and 2016 of about −16.3 ± 3.5 Gt of ice per year, the equivalent of about −0.18 ± 0.04m w.e. per year. These figures are less negative than most previous estimates. Region-wide mass balances vary from ∼4.0 ± 1.5 Gt per year (∼0.62 ± 0.23m w.e. per year in

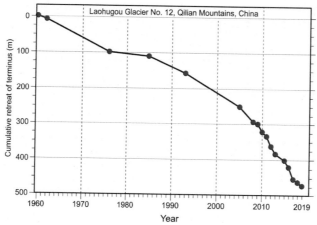

FIG. 9.19 Cumulative retreat (m) of the terminus of Laohugou Glacier No. 12, Qilian Mountains, China, since 1960. (Modified from Pollard, M., 2020. The thaw of the third pole: China's glaciers in retreat. Reuters. https://widerimage.reuters.com/story/the-thaw-of-the-third-pole-chinas-glaciers-in-retreat.)

TABLE 9.4
Regional Summary Statistics for Nonpolar Glaciers.

RGI Region*	Number	Area (km²)	Volume (10³ km³)
Alaska	27,108	86,677	18.98 ± 4.92
Western Canada and the United States	18,862	14,629	1.06 ± 0.27
Arctic Canada North	4,549	104,920	28.33 ± 7.35
Arctic Canada South	7,422	40,860	8.61 ± 2.23
Iceland	567	11,052	3.77 ± 0.98
Svalbard	1,615	33,932	7.47 ± 1.94
Scandinavia	3,417	2,947	0.30 ± 0.08
Russian Arctic	1,069	51,551	14.64 ± 3.80
North Asia	5,144	2,399	0.14 ± 0.04
Central Europe	3,927	2,091	0.13 ± 0.03
Caucasus Middle East	1,887	1,305	0.06 ± 0.02
Central Asia	54,429	49,295	3.27 ± 0.85
South Asia West	27,986	33,561	2.87 ± 0.74
South Asia East	13,119	14,734	0.88 ± 0.23
Low latitudes	2,940	2,341	0.10 ± 0.03
Southern Andes	15,908	29,368	5.34 ± 1.39
Antarctic periphery	2,751	132,771	46.47 ± 12.06
New Zealand	3,537	1,161	0.07 ± 0.02
Totals	196,237	615,594	96.02

*Randolph Glacier Inventory.

Data from Farinotti, D., Huss, M., Fürst, J.J., Landmann, J., Machguth, H., Maussion, F., Pandit, A., 2019. A consensus estimate for the ice thickness distribution of all glaciers on Earth. Nature Geoscience 12 (3), 168–173. https://doi.org/10.1038/s41561-019-0300-3.

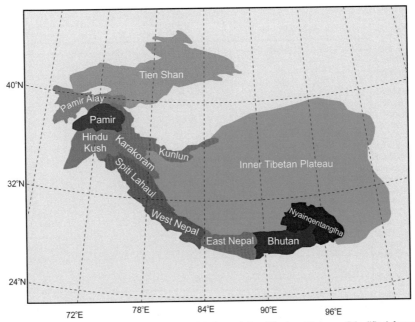

FIG. 9.20 Map of the High Mountain regions of Asia and the Tibetan Plateau. (Modified from Brun, F., Berthier, E., Wagnon, P., Kääb, A., Treichler, D., 2017. A spatially resolved estimate of high Mountain Asia glacier mass balances from 2000 to 2016. Nature Geoscience 10, 668–674. https://doi.org/10.1038/ NGEO2999.)

TABLE 9.5
High Mountain Asia Regional Mass Balances From ICESat Estimates.

Region	Glacier Area (km^2)	Mass Balance Change (m w.e.)
Bhutan	2,291	-0.76 ± 0.20
East Nepal	4,776	-0.31 ± 0.14
Hindu Kush	5,147	-0.42 ± 0.18
Inner TP	13,102	-0.06 ± 0.06
Karakoram	17,734	-0.09 ± 0.12
Kunlun	9,912	$+0.18 \pm 0.14$
Nyainqentanglha	6,378	-1.14 ± 0.58
Pamir Alay	1,915	-0.59 ± 0.27
Pamir	7,167	-0.41 ± 0.24
Spiti Lahaul	7,960	-0.42 ± 0.26
Tien Shan	10,802	-0.37 ± 0.31
West Nepal	4,806	-0.37 ± 0.15
Total	91,990	-0.34 ± 0.06

Data from Brun, F., Berthier, E., Wagnon, P., Kääb, A., Treichler, D., 2017. A spatially resolved estimate of high Mountain Asia glacier mass balances from 2000 to 2016. Nature Geoscience 10, 668–674. https://doi.org/10. 1038/NGEO2999.

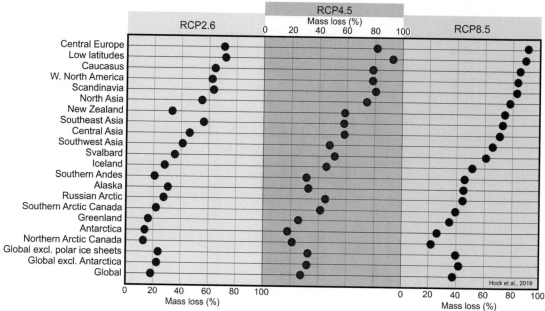

FIG. 9.21 Projected mass losses by 2100 in percent of the glacier mass in the year 2015 for 19 RGI regions from six glacier models using three RCP emission scenarios. Dots mark the arithmetic mean of the various model results. Regional results are sorted by the glacier models' mean mass loss according to the RCP8.5 scenario. (Modified from Hock, R., Bliss, A., Marzeion, B., Giesen, R., Hirabayashi, Y., Huss, M., Radic, V., Slangen, A., 2019. GlacierMIP-A model intercomparison of global-scale glacier mass-balance models and projections. Journal of Glaciology 65 (251), 453–467. https://doi.org/10.1017/jog.2019.22.)

Nyainqentanglha (E. Himalaya) to ~1.4 ± 0.8 Gt per year (~0.14 ± 0.08m w.e. per year) in the Kunlun region of NW China (Table 9.5). The assessment of glacial margin retreat rates by Li et al. (2019) provides a different measure of the health and potential longevity of glaciers, but they likewise concluded that the Kunlun region had by far the lowest rate of glacial retreat of High Mountain Asia.

Hock et al. (2019) estimated percent glacial mass loss for Asian highlands (Fig. 9.21), not differentiating the Asia High Mountain regions, based on modeling of regional climate changes under three IPCC scenarios. It was, therefore, necessary to determine the average of the three Asian highland values (Southwest Asia, Southeast Asia, and Central Asia). For RCP2.6, the average estimate of glacial mass loss by the end of this century is 47%. For RCP4.5, the average estimate of glacial mass loss by the end of this century is 58%. For RCP8.5, their average estimate of glacial mass loss by the end of this century is 71%.

Haeberli et al. (2020) also projected changes in mass balance for the High Asian Mountains during the 21st century (Fig. 9.22, 3). Their projections for changes in

mass balance by the year 2100, which are quite close to those of Hock et al. (2019), range from 44% loss under RCP2.6 conditions to 66% loss under RCP8.5 conditions.

New Zealand (Southern Alps)

Li et al. (2019) estimated that the glaciers of the Southern Alps experienced a shrinkage rate of 0.69% per year between 1980 and 2015, or a total of 31.5% loss of ice mass. Carrivick et al. (2020) determined volume changes for 400 mountain glaciers in the Southern Alps of New Zealand. Their assessment periods included the interval from preindustrial LIA to 1978, 1978 to 2009, and 2009 to 2019. Their study found that at least 60 ± 12 km³ (between 41% and 62%) of the LIA total ice volume has been lost. Furthermore, the rate of mass loss has accelerated greatly in the past decade, nearly doubling from −0.4 m w.e. per year during 1600−1978 AD to −0.7 m w.e. per year today.

The total ice-covered area mapped by Carrivick et al. (2020) for the LIA was 1492 ± 104 km². Assuming the 440 km² they were unable to map has remained the

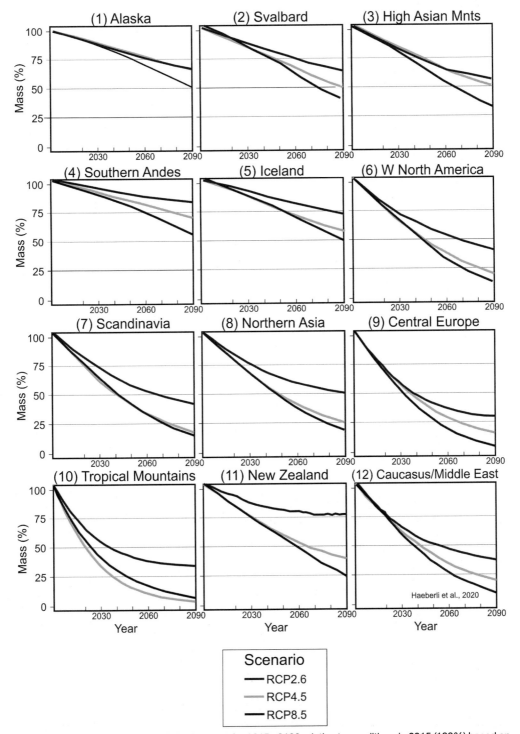

FIG. 9.22 Projected evolution of glacier mass for 2015–2100 relative to conditions in 2015 (100%) based on three RCP emission scenarios. Thick lines show the averages of 46–88 model projections based on four to six glacier models for the same RCP, and the shading marks ±1 standard deviation (not shown for RCP4.5 for better readability). (Modified from Haeberli, W., Huggel, C., Paul, F., Zemp, M., 2020. The response of glaciers to climate change: observations and impacts. In: Elias, S. (Ed.), Reference Module in Earth Systems and Environmental Sciences. Elsevier. https://doi.org/10.1016/B978-0-12-818234-5.00011-0.)

same since the LIA because it is mostly comprised of ice above the regional ELA (i.e., taking the most conservative approach), then combining that with the area that we have mapped produces a total glaciated LIA area of at least 1932 ± 135 km^2. Given that the 1978 Southern Alps glacial inventory yielded an area of 1463 km^2, Carrivick et al. (2020) reckoned a loss of at least 24% in glacier ice areal extent since the LIA.

The remaining ice in the Southern Alps is contained in small (<1 km^2) rapid-response glaciers that are increasingly restricted to high elevations close to the ELA. This ice is fast becoming restricted to shaded areas immediately beneath steep head walls, as is the case for glaciers in the European Alps. The few surviving valley floor glacier ablation tongues are stagnating beneath increasingly thick debris covers. They are rapidly losing mass because of their low elevation and the action of adjacent proglacial lakes that are enhancing ablation rates and causing tabular calving.

In the authors' view, the accelerating rate of ice loss reflects regional-specific climate conditions and suggests that peak glacial meltwater production is imminent if not already passed, which has profound implications for water resources and riverine habitats.

Hock et al. (2019) estimated percent glacial mass loss for the Southern Alps through the 21st century, based on modeling of regional climate changes under three IPCC scenarios (Fig. 9.21). For RCP2.6, the estimate of glacial mass loss by the end of this century is 33%. For RCP4.5, the estimate of glacial mass loss by the end of this century is 59%. For RCP8.5, their estimate of glacial mass loss by the end of this century is 74%.

Haeberli et al. (2020) also projected changes in mass balance for the Southern Alps during the 21st century (Fig. 9.22, 11). Their projections for changes in mass balance by the year 2100 range from 20% loss under RCP2.6 conditions to 75% loss under RCP8.5 conditions.

Europe

The Alps of Europe have by far the best-studied glaciers in the world. That being said, the Alps are rather topographically and climatologically complex, stretching from southeast France to central Austria and northern Italy. The WGMS provided information on the mass balance of three alpine glaciers in its annual report (2020): the Vernagtferner Glacier in the Austrian Alps, the Caresèr Glacier in the Italian Alps, and Silvretta Glacier in the Swiss Alps.

Concerning the Vernagtferner Glacier, the WGMS (2020) reported that the year 2015/16 showed a slightly less negative mass balance (−0.78 m w.e.) compared

with the mean mass balance value since 2000, which was of −0.82 m w.e. per year. The following year (2016/17) saw the second most negative mass balance (−1.34 m w.e.) since the beginning of observations. The only year with a more negative mass balance (−0.21 m w.e.) was 2002/03.

The report on Caresèr Glacier stated that the long-term (1981−2001) average value of the annual mass balance declined by −1.2 m w.e. per year. Between 2002 and 2017, the loss rate increased by 50% to −1.8 m w.e. per year. It seems highly likely that this glacier will disappear in the next few years, as within the past 15 years, the glacier has separated into several ice patches, and most of these are rapidly melting.

Since 1917 when records began to be kept, only 33 years have shown positive mass balances at the Silvretta Glacier. The remainder (69 years) have had negative mass balances. A consistent trend in mass balance loss began in 1981, with a record of −2.5 m w.e. in 2001/02. Since then, most years have experienced mass balance losses ranging from 0.75 to 1.5 m w.e. per year, capped by the year 2016/17, with a mass loss of −1.5 m w.e. Since then, the mass balance losses have not been so severe.

Davaze et al. (2020) quantified a spatially resolved estimate of individual annual mass balance time series for 239 glaciers in the European Alps, using both geodetic and snowline altitude estimates for the period from 2000 to 2016. The study showed that all glaciated regions of the Alps are experiencing mass balance losses, especially in the Austrian Alps and the eastern Bernese regions. When taken together, the glaciers of the European Alps experienced a mean annual mass loss of −0.74 ± 0.20 m w.e. per year from 2000 to 2016.

As shown in Fig. 9.23 (top), the study by Davanze et al. (2020) showed that glaciers on steep slopes and those situated at higher elevations experienced less mass loss than glaciers on shallow slopes and/or situated at lower elevations. The results from the last year of the study (2016) are quite startling. Glaciers on slopes greater than 23.8° had cumulative mass balance losses of about −8 m w.e., while glaciers on slopes less than 17° had cumulative mass balance losses that were 75% greater—about −14 m w.e. The study also revealed that seasonal weather patterns also play a role in glacier mass balance in the Alps. During years when mass balance losses are low, the main atmospheric flow is westerly in winter, advecting moist air from the Atlantic. The added snowfall helps maintain more positive mass balances. Conversely, high loss years tend to be associated

Mass balance changes in the Alps

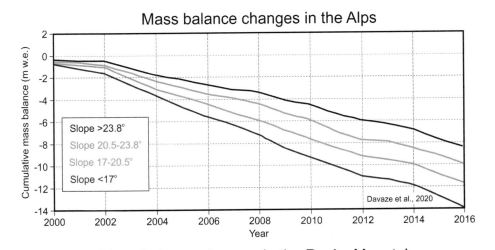

Davaze et al., 2020

Mass balance changes in the Rocky Mountains

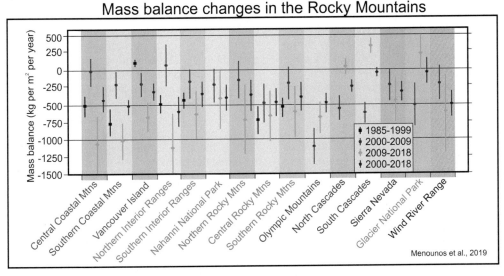

Menounos et al., 2019

Mass balance changes in the Andes

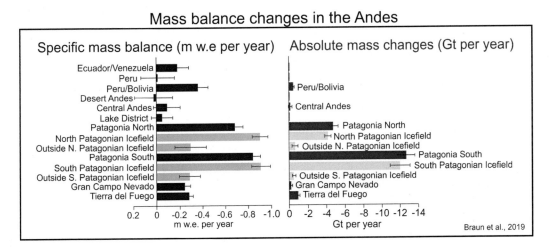

Braun et al., 2019

with the easterly flow, which delivers less winter precipitation to the region.

Hock et al. (2019) estimated percent glacial mass loss for the European Alps through the 21st century, based on modeling of regional climate changes under three IPCC scenarios (Fig. 9.21). For RCP2.6, the estimate of glacial mass loss by the end of this century is 72%. For RCP4.5, the estimate of glacial mass loss by the end of this century is 81%. For RCP8.5, their estimate of glacial mass loss by the end of this century is 90%. These figures once again emphasize the vulnerability of the alpine glaciers of Europe. If even the best-case scenario (RCP2.6) leads to a 72% reduction in alpine glaciers by the end of the century, their odds of survival much beyond the beginning of the 22nd century seem very small. The WGMS (2020) estimate of the current state of European alpine glaciers indicates that they have already lost an average of about one-third of their cumulative mass. Haeberli et al. (2020) also projected changes in mass balance for the Alps during the 21st century (Fig. 9.22, 9). Their projections range from 70% loss under RCP2.6 conditions to 95% loss under RCP8.5 conditions. These agree well with the Hock et al. (2019) estimates.

North America

Western North American (WNA) glaciers outside of Alaska cover 14,384 km^2 of mountainous terrain. These include glaciers in the coastal ranges of British Columbia, the Cascade Mountains of Washington and Oregon, the Sierra Nevada of California and Nevada, and the Rocky Mountains.

The Rocky Mountains stretch 4800 km from northern British Columbia, in western Canada, to New Mexico in the Southwestern United States. The Rockies are distinct from the Cascade Range and the Sierra Nevada, both of which lie farther west. The Rocky Mountains contain the highest peaks in central North America. The range's highest peak is Mount Elbert located in Colorado at 4401 m above sea level. Mount Robson in British Columbia, at 3954 m, is the highest peak in the Canadian Rockies.

There are currently 148 active glaciers in the Rocky Mountains. Of these, 16 occur in Colorado, 44 are found in Wyoming, 57 are in Montana, and 31 are in the Canadian Rockies of Alberta and British Columbia. But these numbers are somewhat misleading. In terms of the size of these glaciers, most of this glacier coverage (88%) is in the Northern Rockies of British Columbia and Alberta. Just 7% of glacier coverage is in the conterminous United States, and 5% is in the Yukon and Northwest Territories (Menounos et al., 2019). No comprehensive analysis of recent mass change exists for these glaciers, but now nearly all of them are shrinking.

Western North America

The WGMS (2020) plotted cumulative mass changes in WNA as the most extreme in the world, with a value of about 38 m w.e. The only other region approaching this level of glacial ice loss is Central Europe, at 35 m w.e. This level of current ice mass loss is borne out in the predictions for WNA glacier mass changes for the rest of this century. Hock et al. (2019) estimated the average percent glacial mass loss for RCP2.6 at 64%. For RCP4.5, the average estimate of glacial mass loss by the end of this century is 78%. For RCP8.5, their average estimate of glacial mass loss by the end of this century is 84% (Fig. 9.21). Haeberli et al. (2020) also projected changes in mass balance for the glaciers of WNA during the 21st century (Fig. 9.22, 6). Their projections range from 58% loss under RCP2.6 conditions to 86% loss under RCP8.5 conditions. These agree well with the Hock et al. (2019) estimates.

Menounos et al. (2019) generated over 15,000 DEMs from satellite images to assess the change in glacial masses for WNA over the period 2000–18. These glaciers lost a total of 117 ± 42 gigatons (Gt) of mass. They documented a fourfold increase in mass loss rates

FIG. 9.23 Graphs showing recent mass balance changes in mountain region glaciers. Top: mass balance changes in the Alps. Middle: mass balance changes in the Rocky Mountains. Bottom: mass balance changes in the Andes. (Top: After Davaze, L., Rabatel, A., Dufour, A., Hugonnet, R., Arnaud, Y., 2020. Region-wide annual glacier surface mass balance for the European Alps from 2000 to 2016. Frontiers in Earth Science, 8. https://doi.org/10.3389/feart.2020.00149; Middle: After Menounos, B., Hugonnet, R., Shean, D., Gardner, A., Howat, I., Berthier, E., Pelto, B., Tennant, C., Shea, J., Noh, M.J., Brun, F., Dehecq, A., 2019. Heterogeneous changes in western North American glaciers linked to decadal variability in zonal wind strength. Geophysical Research Letters 46 (1), 200–209. https://doi.org/10.1029/2018GL080942; Bottom: After Braun, M.H., Malz, P., Sommer, C., Farías-Barahona, D., Sauter, T., Casassa, G., Soruco, A., Skvarca, P., Seehaus, T.C., 2019. Constraining glacier elevation and mass changes in South America. Nature Climate Change 9 (2), 130–136. https://doi.org/10.1038/s41558-018-0375-7.)

TABLE 9.6
Mass Balance Changes in Western North American Glaciers, 2000–18.

Region	Area (km^2)	Mass Balance (kg per m^2 per year)			Mass Budget (Gt per year)
		Early[a]	Late[b]	Full[c]	
Central Coast	1,580	-42 ± 221	$-1,067 \pm 418$	-424 ± 163	0.669 ± 0.258
Southern Coast	7,180	-215 ± 190	$-1,027 \pm 258$	-517 ± 140	-3.709 ± 1.006
Northern Interior	253	75 ± 298	$-1,143 \pm 510$	-608 ± 222	-0.154 ± 0.056
Southern Interior	1,946	-175 ± 232	-647 ± 352	-353 ± 185	-0.686 ± 0.360
Nahanni	649	-220 ± 250	-419 ± 444	-407 ± 192	-0.264 ± 0.124
Northern Rockies	415	-148 ± 271	-724 ± 506	-362 ± 233	-0.150 ± 0.097
Central Rockies	422	-483 ± 284	-671 ± 370	-474 ± 202	-0.200 ± 0.086
Southern Rockies	1,350	-200 ± 240	-614 ± 376	-394 ± 205	-0.533 ± 0.277
Olympics	30	$-1,113 \pm 259$	-696 ± 235	-474 ± 144	-0.014 ± 0.004
North Cascades	250	-567 ± 184	46 ± 106	-245 ± 87	-0.061 ± 0.022
South Cascades	153	-632 ± 147	346 ± 112	-46 ± 61	-0.007 ± 0.009
Sierra Nevada	11	-234 ± 231	-448 ± 326	-318 ± 141	-0.004 ± 0.002
Glacier National Park	29	-522 ± 310	235 ± 268	-41 ± 158	-0.001 ± 0.005
Wind River	60	-202 ± 249	-652 ± 571	-503 ± 187	-0.030 ± 0.011
Total (Western North America)	14,341	-203 ± 214	-858 ± 320	-452 ± 162	-6.49 ± 2.32

[a] Records from 2000 to 2009.
[b] Records from 2009 to 18.
[c] Records from 2000 to 18.
Data from Menounos, B., Hugonnet, R., Shean, D., Gardner, A., Howat, I., Berthier, E., Pelto, B., Tennant, C., Shea, J., Noh, M.J., Brun, F., Dehecq, A., 2019. Heterogeneous changes in western North American glaciers linked to decadal variability in zonal wind strength. Geophysical Research Letters 46 (1), 200–209. https://doi.org/10.1029/2018GL080942.

between 2000–2009 (-2.9 ± 3.1 Gt per year) and 2009–18 (-12.3 ± 4.6 Gt per year), attributed to a shift in regional meteorological conditions driven by the location and strength of the upper-level zonal wind. The study by Menounos et al. (2019) showed that during the early interval (2000–09), mass balance change was less negative for glaciers in southern British Columbia (BC), whereas glaciers further south in the Cascade Mountains experienced high rates of mass loss. This pattern reversed itself from 2009 to 18 when glacier mass loss from the southern Coast Mountains of BC increased by a factor of 4.8, while glaciers in the South Cascades showed only a slight mass gain, and glaciers in the North Cascades showed no detectable mass change. Changes in precipitation regime seem to account for much of these trends. During the period 2000–07, the Mount Rainier region of the North Cascades experienced drier than average conditions; an interval of increased precipitation near Mount Rainier began after 2011. Simultaneously, there was a decrease in precipitation in the central Coast Mountains

of BC after 2012. Rates of mass loss for glaciers in the Canadian Rockies slowed from 2000 to 2009 and then increased from 2009 to 2018. The southern Coast Mountains of BC contain nearly half of the total ice cover of Western North America, and the rate of mass loss over the last 9 years was -7.4 ± 1.9 Gt per year, a loss of ice that is about 20% greater than the period 1985–99. Menounos et al. (2019) project that up to 90% of current glacier mass in the Canadian Rockies and Interior Ranges will be gone by the end of the 21st century (Clarke et al., 2015).

Rocky Mountains

A study by Castellazzi et al. (2019) quantified the meltwater input fluxes into the watersheds draining the Canadian Rocky Mountains (CRM). Their glacier mass balance model estimates a total ice mass change of ~ 43 Gt for the period 2002–15, corresponding to an average of -3.1 Gt per year. Model output indicates that the glacier cover of the CRM has experienced mass loss during 11 of the 14 years during the period of study. Glaciers across the entire study area have thinned at an average of 0.86 m w.e. per year, whereas the glacier cover west and east of the continental divide has thinned by 0.96 m w.e. per year and 0.67 m w.e. per year, respectively. Estimates of mass losses from specific mountain regions within the Canadian Rockies range from −0.67 m w.e. per year in the eastern ranges of Alberta to −1.14 m w.e. per year in the Peyto region near the continental divide in southeastern British Columbia. Ice mass changes across the entire study area for individual years range from a mass gain of 1 Gt in 2010 to a loss of 9.3 Gt in 2003 (Fig. 9.23, middle). However, the authors recorded a good deal of oscillation in glacier mass balance in the Canadian Rockies during the 21st century, even though the general trend is for lost ice mass.

As summarized by Menounos et al., 2019), there was considerable variation observed in glacial mass changes in the Rocky Mountains. From 2000 to 2009, the Central Rockies lost 478 kg per m² per year, while Northern and Southern Rockies lost only 146 and 193 kg per m² per year, respectively. During the most recent of the study intervals, 2009−18, all three regions lost >600 kg per m² per year. For the Northern Rockies, this represents almost a 500% increase in mass loss. Because the Central Rockies experienced a relatively large increase in mass loss from 2000 to 2008, the change in mass loss in the subsequent interval represents only a 40% increase. However, the Southern Rockies experienced more than a 300% increase in mass loss in 2009−18, compared with 2000−08. Interestingly, the differences in mass loss in the three Rocky Mountain regions tended to average out over the long term. Considering the full study interval of 2000−18, all three regions lost an average of about 385 kg per m² per year (−357, −400, and −395 kg per m² per year, respectively).

Glacier National Park (GNP) in northern Montana is at the heart of the Northern Rockies, yet its history of glacial mass changes is remarkably different from the Northern Rockies as a whole. The interval of 2000−09 saw a mass loss of −500 kg per m² per year in GNP glaciers, by far the greatest rate of loss recorded from the Northern Rockies. Even more remarkable was a net gain of ice mass in the park during 2009−18. The glaciers in GNP added 230 kg per m² per year during that interval. The Cascade Mountains of the Pacific Northwest region are the only other region in which the glaciers added ice volume during this interval. The long-term trend for ice mass change in GNP was nearly neutral, with a value of −37 kg per m² per year from 2000 to 18.

Sierra Nevada Mountains

The Sierra Nevada Mountains of California and Nevada are well south of the Cascades, Olympic, and coastal ranges of BC. They are also well west of the Rockies. Accordingly, the recent record of glacial ice mass changes is likewise different from these other regions. The interval 2000−09 saw a mass change of −250 kg per m² per year in the Sierras. Following the regional trend for western North America, the glacial ice of the Sierras lost 437 kg per m² per year during 2009−18. The long-term trend (2000−18) for this region was a loss of 322 kg per m² per year.

The Wind River Range

The Wind River Range, situated southeast of Grand Teton and Yellowstone National Parks, forms part of the Central Rockies. Unlike the rest of the region's glaciers, however, the interval 2000−09 saw only a decrease of about 235 kg per m² per year—about half of the combined average value for the Central Rockies during this period. During the subsequent interval in the Wind Rivers (2009−18), the glaciers lost an average of 604 kg per m² per year, a 250% increase in loss rate. The long-term average ice mass flux for the Wind River Range (2000−18) was about −500 kg per m² per year, a 20% greater loss than the average for the entire Central Rockies.

Menounos et al. (2019) predict that mountain glaciers in western North America will continue to lose mass throughout this century, even under moderate emission scenarios. These changes will reduce or eliminate the thermal buffering capacity of glaciers in the alpine and adjacent mountain ecosystems. The decrease in meltwater runoff will have negative effects on downstream ecosystems and water resources. If the last 18 years provide a suitable analog for the next 30−50 years, future glacier change will be modulated by decadal-scale climate variability.

South America

Braun et al. (2019) used synthetic aperture radar interferometry over the years 2000−2011/2015 to compute continent-wide, glacier-specific elevation, and mass changes for 85% of the glacierized area of South America (Table 9.7). They calculated the combined South American glacial mass loss rate as 3.06 ± 1.24 Gt per year. The largest contributions to this ice mass loss come from the Patagonian Ice Fields. Here the losses represent 83% mass loss for the entire Andes. Overall, the authors found that the glaciers in the southern Andes are losing ice much more rapidly than the glaciers in the central and northern Andes (see Table 9.7). This is especially true for the southern Patagonian region, where the glacial mass change rates exceed 12 Gt per year (Fig. 9.23, bottom). In the northern Patagonian region, glacial mass change rates are around 4 Gt per year. Elsewhere in the Andes, mass

TABLE 9.7
Glacial Ice Mass Changes, Andes Mountains.

Region	Study Interval	Glaciated Area (km^2)	Mass Change Rate (Gt per year)
Ecuador/Venezuela (Inner Tropics)	2000–2014	187	−0.3 ± 0.02
Peru-Bolivia (Outer Tropics)	2000–2013	1,280	−0.47 ± 0.11
Central Andes	2000–2011	1,660	−0.15 ± 0.20
Lake District	2000–2011	478	−0.02 ± 0.05
Patagonia-North	2000–2012	6,977	−4.77 ± 0.48
Northern Patagonian Icefield	2000–2012	4,653	−4.19 ± 0.32
Patagonia-South	2000–2012	15,101	−12.64 ± 0.98
Southern Patagonian Icefield	2000–2012	13,231	−12.02 ± 1.07
Gran Campo Nevado/Isla Riesco	2000–2012	1,254	−0.31 ± 0.06
Tierra del Fuego	2000–2011	3,545	−1.02 ± 0.11
Sum/area weighted average		31,751	−19.43 ± 0.60

Data From Braun, M.H., Malz, P., Sommer, C., Farías-Barahona, D., Sauter, T., Casassa, G., Soruco, A., Skvarca, P., Seehaus, T.C., 2019. Constraining glacier elevation and mass changes in South America. Nature Climate Change 9 (2), 130–136. https://doi.org/10.1038/s41558-018-0375-7.

change rates are mostly less than 1 Gt per year (Braun et al., 2019).

As discussed earlier, the surface mass balance of a glacier depends on the balance between accumulation and ablation or melting. Various factors affect melt, such as temperature and the radiation balance. Braun et al. (2019) observed that regional climatic changes in South America are quite variable. They noted that in the past two decades, surface temperatures increased significantly in the high Andes, especially in the tropical latitudes and in the Desert Andes. This observation agrees with previous studies.

Throughout the tropical latitudes, there is a significant increase in water vapor in the Amazon basin close to the Andes (more than 5% per decade), with only a modest rise in temperatures. The enhanced moisture explains the observed mass gain in higher elevations glaciers in the northern, wet outer tropics of Peru. Regional glaciers at lower elevations are subject to rising surface temperatures, leading to higher ablation rates and retreat at the glacier termini. A similar pattern exists for the cold, dry outer tropical zone of the Andes straddling the Peru–Bolivia border. Here, there is a small but significant increase in vertically integrated water vapor.

Hock et al. (2019) estimated percent glacial mass loss for mountain glaciers in the tropical latitudes, including the Andes and the few African glaciers, through the 21st century, based on modeling of zonal climate changes under three IPCC scenarios. The results for the low latitudes show the most extreme losses of any region in the world (Fig. 9.21). For RCP2.6, the estimate of glacial mass loss by the end of this century is 70%. For RCP4.5, the estimate of glacial mass loss by the end of this century is 94%. For RCP8.5, their estimate of glacial mass loss by the end of this century is 90%. Haeberli et al. (2020) also projected changes in mass balance for mountain glaciers in the tropics during the 21st century (Fig. 9.22, 10). Their projections range from 66% loss under RCP2.6 conditions to 97% loss under RCP8.5 conditions. These estimates agree well with the Hock et al. (2019) estimates.

For the Tierra del Fuego and Patagonia regions, the authors found no significant trend in water vapor conditions. However, surface temperatures in these regions are rising substantially, driving enhanced glacial melting. As discussed earlier, the greatest mass losses of glacial ice in the Andes are taking place in the NPI and SPI regions. The areas outside of these large ice fields contain smaller glaciers that are losing ice more slowly. This supports the hypothesis that the mass loss of the Patagonian Ice Fields is mainly driven by the dynamic action of the tidewater and lake calving glaciers.

The predictions of percent glacial mass loss for the mountains of Patagonia (Hock et al., 2019) through the 21st century are as follows. For RCP2.6, the estimate of glacial mass loss by the end of this century is 20%. For RCP4.5, the estimate of glacial mass loss by the end of this century is 31%. For RCP8.5, their estimate of glacial mass loss by the end of this century is 44% (Fig. 9.21). Haeberli et al. (2020) projected changes in mass balance for the glaciers of the southern Andes during the 21st century (Fig. 9.22, 4). Their projections range from 16% loss under RCP2.6 conditions to 44% loss under RCP8.5 conditions. Again, the two sets of estimates are in close agreement.

Africa

Barr and Chandler (2013) provide an overview of the current state of the African glaciers and trends in their decline according to the latest field observations and satellite imagery (Fig. 9.24). The East African glaciers that once covered several hundred square kilometers during the Pleistocene Epoch are now vulnerable. At the end of the 1990s, these glaciers had a total area of about 10.7 km², but they are now melting at an unprecedented pace. Between 1906 and 2006, they lost about 82% of their area. This massive reduction in ice atop Mt. Kenya, Mt. Kilimanjaro, and the Ruwenzori range has attracted global attention. Although Africa's share of the GHG emissions that contribute to global warming is relatively small, its ecosystems, including glaciated landscapes, are extremely vulnerable to the impacts of a changing climate. If these precious ice caps continue to recede, as they have since 1906, some experts project that they will entirely disappear within the next few decades.

Kilimanjaro

There are only three African mountains that retain glaciers today. These are Mount Kenya (5199 m above sea level), the Ruwenzori Mountains (5109 m asl), and Kilimanjaro, the best known of the three, and Africa's tallest mountain (5892 m asl). Three major ice bodies are found on the summit area of Kilimanjaro today: Furwängler Glacier and the Northern and Eastern ice fields. These are remnants of a former ice cap that covered the Kilimanjaro plateau at the end of the 19th century. Present-day climatological conditions are not favorable for maintaining these glaciers (Fig. 9.25). All three are suffering negative mass balances.

Stadelmann et al. (2020) concur that the glaciers on Kilimanjaro are unique indicators for climatic change in the tropical midtroposphere of Africa and emphasize that their disappearance seems imminent. They presented thickness maps for the Northern Ice Field (NIF) and Kersten Glacier with mean values of 26.6 and 9.3 m, respectively, in 2011. In the absence of direct measurements on Kersten Glacier, multitemporal satellite information was exploited to infer past thickness values in areas that have become ice-free and therefore allow glacier-specific calibration.

Bohleber et al. (2017) used ground-penetrating radar (GPR) data to investigate bed topography, ice thickness, and internal stratigraphy at Kilimanjaro's largest remaining ice body, the NIF. The GPR profiles reveal an ice thickness ranging from 6.1 ± 0.5 to 53.5 ± 1.0 m across the ice field. Ice thickness in the western third and all along the periphery of the NIF

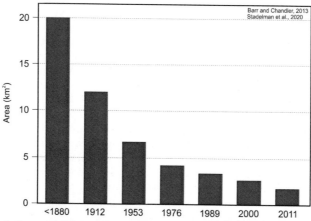

FIG. 9.24 Changes in the area of Mount Kilimanjaro's ice cover (in km²) since before 1880. (Data from Barr, J., Chandler, A., 2013. Africa without ice and snow. Environmental Development 5, 146–155. https://doi.org/10.1016/j.envdev.2012.10.003 and Stadelmann, C., Jakob Fürst, J., Mölg, T., Braun, M., 2020. Brief communication: glacier thickness reconstruction on Mt. Kilimanjaro. The Cryosphere 14 (10), 3399–3406. https://doi.org/10.5194/tc-14-3399-2020.)

FIG. 9.25 Landsat (top and middle) and Earth Observing-1 satellite (bottom) images of Mount Kilimanjaro in taken in 1973, 2000, and 2012, showing the decline of ice cover. (Images courtesy of NASA, in the public domain.)

is < 10 m. Three domes of ice in the center and eastern thirds are 40–50 m thick. They also discovered that the top 10 m of the ice field is suffused with glacial meltwater. Combining these data with a very high-

resolution digital elevation model, the authors extrapolated ice thickness, yielding an estimate of the total ice volume remaining at NIF's southern portion as $12 \times 10^6 \, \text{m}^3$.

Ruwenzori mountains

The Ruwenzori massif at 5109 m is popularly called the "Mountains of the Moon." It is a 50-km-wide block of mountains that rises between two faults in the Earth's crust. It straddles the Uganda-Democratic Republic of Congo border and has a total area of 3000 km². The range is heavily dissected into valleys and individual mountains that run about 100 km northwestward from the Equator. The range's central portion contains 25 peaks over 4500 m, the highest of which is the Margherita peak on Mount Stanley, which exceeds 5000 m. In recent times, the Ruwenzori glaciers have persisted on only three of the range's peaks: Mounts Stanley, Speke, and Baker. In 1906, 43 named glaciers were distributed over six mountains in the Ruwenzori Range, with a total area of 7.5 km². At that time, this represented about half the total glaciated area in Africa. By 2005, less than half of this ice volume remained, and it was reduced to only three mountains. Taylor et al. (2006) reported on the recent decline of these Ruwenzori glaciers. They declined from an area of $2.01 \pm 0.56 \, \text{km}^2$ in 1987 to an area of $0.96 \pm 0.34 \, \text{km}^2$ in 2003. The authors suggested that the spatially uniform loss of glacial cover at lower elevations, combined with regional meteorological data, indicates that increased air temperature is the main driver of this decline in ice cover. Air temperatures in these mountains increased by +0.5°C per decade between 1976 and 2006 without significant changes in annual precipitation. Taylor et al. (2006) predicted that glaciers in the Ruwenzori Mountains would disappear by 2026. As it happens, the glaciers on Mount Speke and Mount Baker disappeared by 2012. The Mount Stanley glacier is rapidly melting. The terminus of this glacier Stanley retreated upslope 60 m between 2012 and 2020 (Thymann, 2020).

Mount Kenya

Mount Kenya straddles the equator, 193 km northeast of Nairobi and about 480 km from the Kenyan coast. The mountain is Africa's second-highest, at 5199 m. It is the remnant of a large volcano, and its diameter is roughly 100 km. Until recently, its ice cap was composed of 10 glaciers that occupied the valleys of this very dissected mountain. Since the last complete glacier mapping of Mt. Kenya in 2004, strong glacier retreat and glacier disintegration have been reported. Prinz et al. (2018) compiled a glacier inventory of Mt.

Kenya to document the most recent changes. They derived glacier area and mass changes from orthophotos and DEMs extracted from satellite images. During 2004–16, the total glacier area on Mt. Kenya decreased by $121 \times 10^3 \, m^2$ (44%). The largest glacier (Lewis) lost 46% of its area and 57% of its volume during the same period. The mass loss of Lewis Glacier has accelerated since 2010 due to glacier disintegration, which led to the emergence of a rock outcrop that splits the glacier in two. If the current retreat rates prevail, Mt. Kenya's glaciers will be extinct before 2030 (Prinz et al., 2018).

Glacial melting on Mt. Kenya is likely to be due to both warming and drying of conditions at the peak, such that the glaciers are melting faster and at the same time receiving less snow. The drying of Mt. Kenya is a signal found across East Africa, both in the mountains like Mt Kenya and Kilimanjaro and in the lowlands. Precipitation amounts have become smaller and more variable; the wet seasons have become shorter. The drying is caused by changes in SST patterns over the Indian Ocean that control the moisture transport toward East Africa. This change in sea surface pattern was found to be a consequence of global warming. The warming of Mt Kenya means that more precipitation falls as rain rather than snow. The lack of fresh snow robs the glacier of mass input and causes a decrease in albedo, which causes a higher absorption of solar radiation and thus increased melt.

DRIVERS OF CURRENT CHANGES

As we have seen in this summary of the health of glaciers in various mountain ranges around the world, every region is unique and there is not one single driver of glacial volume changes. As they say in the electoral world, "All politics are local." The corollary in the realm of mountain glaciers is, "All mass balance changes are local." The complexities of local conditions come from the interactions of many variables, including differences in climatology, slope, aspect, the elevation of mountainsides, and latitude. Nevertheless, there are certain recurring themes in this narrative that merit discussion. One of these is global warming.

Impacts of Global Warming

Global warming is probably the single most important driver of glacial melting throughout the mountain ranges of the world. It is clearly more important in higher latitude regions than in the tropics. This is because global warming is being amplified in the high latitudes, as discussed in the introduction of this book. The losses in glacier mass in all the major mountain regions of the world have been at least partially attributed to rising temperatures, with special emphasis on this forcing factor on the glaciers of the Southern Alps, the ranges of Western North America, the temperate regions of the Andes, and the East African mountains. Concerning the last region, warming temperatures, both in the mountains themselves and in the nearby Indian Ocean (the prominent source of moisture), have caused the balance between rain and snow in highland precipitation to tilt toward rain.

Changing Precipitation Patterns

Much of the changes in precipitation that are affecting mountain glaciers are indirectly due to global warming. Increased temperatures have caused shifts in atmospheric circulation, increased ocean temperatures, and caused a predominance of mountain rainfall rather than snowfall. As discussed before, precipitation plays a major role in building and maintaining glaciers. No matter how cold it is, if there is no new snow, new ice will not form. From the history of Himalayan glaciations, we have seen how increased Westerly winter precipitation contributed to regional glacial advances during the LIA. Indeed, ice core evidence indicates that monsoon precipitation seems to have been the main driver of regional glacial expansion and contraction since the start of the 20th century.

The study of glaciers in the European Alps by Davanze et al. (2020) showed that during years when mass balance losses are low, the main atmospheric flow is westerly in winter, advecting moist air from the Atlantic. The added snowfall helps maintain more positive mass balances. Conversely, high loss years tend to be associated with an easterly flow, which delivers less winter precipitation to the Alps. Carrivick et al. (2020) suggest that peak glacial meltwater production in the Southern Alps may already have occurred or will occur soon.

In the study of mountain glaciers in western North America, Menounos et al. (2019) concluded that the spatial distribution of glacial mass change within given regions is linked to regional changes in atmospheric circulation that affect accumulation and ablation. Specifically, research has pointed to the importance of zonal wind in controlling glacier mass balance. Zonal winds flow parallel to lines of latitude (i.e., east and west). In the Menounos et al. (2019) study, anomalies in the zonal wind are tied with temperature and precipitation—two key elements in the control glacier mass balance. Orographically enhanced precipitation (i.e., moisture that rises upslope along mountain fronts, cools at it ascends, and falls as precipitation) in WNA is also favored when strong zonal flow delivers moist air masses originating over the Pacific Ocean. Conversely, regions of weak zonal wind have supported fewer and smaller mountain glaciers during the past 18 years.

Black Carbon and Dust

Hu et al. (2020) propound that one reason the large glaciers in the Third Pole region are thinning and retreating is due to the increased deposition of black carbon (BC) and mineral dust (MD). BC has generally been considered a more important contributing factor to glacial and ice sheet melting than MD. However, compared with the glaciers of the Arctic and the European Alps, the glaciers of the Third Pole receive a heavy burden of MD, sourced from large surrounding deserts, and widely distributed arid areas in the western and northern parts of the region. Once incorporated into glacial snow and ice, MD grains, most of which are much larger than BC (soot) particles, are more difficult for wind or water to dislodge from glacial surfaces during the melting process. MD therefore likely plays a more important role in the northern and western Third Pole due to the heavier dust storms that occur in that area. BC is derived from the incomplete combustion of fossil fuels. BC particle size ranges from 175 to 700 nm. A nanometer is 1×10^{-6} mm. In contrast to this, the particle sizes of mineral dust range from about 0.25 to 10 µm. Since 1 µm is equal to 0.001 mm, even the largest BC particle (700 nm = 0.0007 mm) is only slightly larger than the smallest MD grain (0.25 µm = 0.00025 mm).

Hu et al. (2020) concluded that the contribution of MD to glacier melting is quite persistent due to the low scavenging efficiency of MD during the melting process, so BC is a major factor in albedo reduction when the glacier surface appears as fresh snow, while MD plays an important role comparable with that of BC in snow/ice melting when the glacier surface appears as aged snow or bare ice in the Third Pole. Under the influence of global warming, fresh snow on glacier surfaces can quickly turn into aged snow or bare ice due to the strong melting process during the monsoon period. Because of this phenomenon, as well as the heavy MD burden, the radiative forcing caused by both MD and BC in the Third Pole glacier region must be considered in future studies. Hu et al. (2020) contend that when a glacier surface consists of aged snow and bare ice, then the relative contribution of MD to glacier melting is greater than or equal to that of BC. The importance of MD to glacier melting must therefore be emphasized in the water tower of Asia.

IMPACTS OF MOUNTAIN GLACIAL MELTING ON SOCIETY

Many regions have already begun to suffer from the loss of mountain glacial ice and the meltwater it provides.

Because the mass balance of glaciers around the world is declining, it is just a matter of time until mountain glaciers disappear. Their only chance of rebuilding would be if global climates cooled dramatically—a scenario that seems highly unlikely in the foreseeable future. Mountain glaciers not only supply water to the rivers below, but they also form an integral part of Alpine ecosystems. Their loss will greatly affect those ecosystems, exacerbating rising temperatures and robbing alpine soils of precious moisture. They act as local centers of refrigeration—cooling the adjacent ice-free regions.

Glacial Lakes

Glacial lakes play a large role in the modern activity of mountain glaciers, affecting glacier evolution, interrupting the flux of meltwater and sediments downvalley, and controlling long-term landscape evolution. These lakes, also called Proglacial lakes, are fed by meltwater from adjacent glaciers. They serve as another active agent in speeding up the process of glacial melting. The regions where this phenomenon is particularly important to consider are the mountains of Patagonia and the Southern Alps of New Zealand. For instance, Carrivick et al. (2020) described a major recession of ice termini in the Southern Alps between 2009 and 2019, especially where ice is flowing and calving into proglacial lakes. Sugiyama et al. (2016) noted that because of proglacial lake formation following the retreat of glacial margins, the number of freshwater calving glaciers is increasing around the world.

Shugar et al. (2020) assessed the state of glacial lakes throughout the world in a recent publication (Table 9.8). Based on their research, the volume of the world's glacial lakes grew by 48% from 1990 to 2018, and these lakes now hold 156.5 km^3 of water. Glacial lakes are an important source of freshwater for many of the world's poorest people, particularly in the mountains of Asia and South America. But the lakes also present a growing threat from outburst floods that can tear down villages, washing away roads, destroying pipelines, and other infrastructure. A glacial lake outburst flood (GLOF) is a type of lake flood that occurs when the dam containing a glacial lake fails. The dam may be formed of glacial ice or rocks and sediments in the glacier's terminal moraine. Failure can happen due to erosion, a buildup of water pressure, a snow or rock avalanche, an earthquake, or massive displacement of water in a glacial lake when a large portion of glacial ice collapses into it.

Satellite images show the number of glacial lakes rose by 53% between 1990 and 2018, expanding the amount of the Earth the lakes cover by about 51%. According to Shugar et al. (2020), there are now 14,394 glacial lakes spread over nearly 9000 km^2 of the Earth's

TABLE 9.8
Volume of Glacial Meltwater Lakes (km^3).

Region	Year	Estimate	1 Standard Deviation	Net Increase in Volume
Greenland	1990	38.1	30.5–54.6	12%
	2015	42.66	34.9–58.2	
High Arctic Canada	1990	1.38	1.3–1.4	33%
	2015	1.84	1.7–2.1	
Low Arctic Canada	1990	20.15	12.3–37.3	20%
	2015	24.22	16.4–40.3	
Russian Arctic	1990	4.54	2.3–9.5	56%
	2015	7.06	4.7–12.4	
Alaska	1990	10.95	7.7–17.8	95%
	2015	21.4	15–35.3	
Scandinavia	1990	2.38	1.9–3.4	232%
	2015	5.53	4.6–7.5	
Global Total	1990	105.7	91–141.6	48%
	2000	125.2	106–162	
	2005	136.9	116.5–181.5	
	2010	144	126–189.9	
	2015	156.5	135–207.5	

Data from Shugar, D.H., Burr, A., Haritashya, U.K., Kargel, J.S., Watson, C.S., Kennedy, M.C., Bevington, A.R., Betts, R.A., Harrison, S., Strattman, K., 2020. Rapid worldwide growth of glacial lakes since 1990. Nature Climate Change 10 (10), 939–945. https://doi.org/10.1038/s41558-020-0855-4.

surface. The fastest-growing lakes are in Scandinavia, Iceland, and Russia. These lakes more than doubled in area over the study period. Glacial lakes grew more slowly in other regions, such as Patagonia and Alaska, but many of the lakes in these regions are very large. For instance, three Patagonian glacial lakes exceed 1000 km^2 in area. Alaskan lakes have nearly doubled in volume to 21.4 km^3 since 1990. Lake volumes have grown most rapidly in the high latitudes, due to Arctic amplification of global warming.

Glacial lake outburst floods

Besides the rising vulnerabilities of human populations in some glaciated mountain ranges, the infrastructure for tourism, commerce, and energy security is increasingly exposed to glacial lake outburst floods (GLOFs). Frequent reassessment of the risks posed by glacier lakes is thus necessary. For example, consider the effects of GLOFs on tourism, hydroelectric power development, and highway infrastructure. An example of a potential threat to tourism concerns the Everest trekking and mountain climbing approach from Nepal and the communities that serve it. This trail is exposed to multiple GLOF hazards, including one from Imja Lake, which has recently undergone engineered GLOF hazard

mitigation. Hydroelectric power development may also be vulnerable to outburst floods. In the Himalaya-Karakoram region, many hydroelectric power plants exist in valleys with glacial lakes that have recently experienced GLOFs. In some cases, the hydroelectric plants have been destroyed. In other countries, such as Peru, Bhutan, Switzerland, and Austria, hydroelectric power has been developed in concert with engineered reductions of GLOF hazards. On a more positive note, recent work has demonstrated that deglaciated basins may be important storage basins for hydroelectric power development.

GLOF hazards also threaten highways that cross glaciated ranges. A prime example of this is the Karakoram Highway connecting China and Pakistan; this corridor carries billions of dollars in goods annually and has a regional security aspect. A study of the geomorphological hazards associated with this highway was prepared by Derbyshire et al. (2001). Their study showed that since its completion in 1979, the Karakoram Highway has been subject to damage and disruption by rockfall, sliding of rock and debris, debris flow, mudflow, dry powder flow, flash flooding by water and torrent gravels, basement undermining by abstraction, subsidence, and frost heaving. The road

TABLE 9.9
Estimates of Sea Level Equivalent by 2100 for Selected Regions.

Region	RCP2.6 (mm)	RCP4.5 (mm)	RCP8.5 (mm)
Alaska	12.9	16.2	21.7
Arctic Canada	14.2	17.3	23.2
Greenland Periphery	7.7	9.4	13
Svalbard	4.3	6.6	10.4
Russian Arctic	6.2	8.4	12.1
High Mountin Asia	8.7	13.8	17.6
Southern Andes	1.5	2.7	5
Antarctica	5.1	7.3	11.1
Global	65.7	88	121.7

Data from Huss, M., Hock, R., 2015. A new model for global glacier change and sea-level rise. Frontiers in Earth Science 3, 54. https://doi.org/10.3389/feart.2015.00054.

surface is regularly damaged by rockfall impact, floods, and frost shattering. The highway was also built near several glaciers, including Ghulkin Glacier in northern Pakistan. In 2008, GOFs from this glacier caused local property damage and undermined the Karakoram Highway in the upper Hunza Valley. According to Richardson and Quincy (2009), glacial meltwater builds up behind the terminal moraine of the glacier. Water eventually finds its way past the moraine, switching its point of exit when meltwater channels become choked with boulders because of glaciofluvial undercutting of steep moraine slopes.

Deglaciation is far advanced in places such as the Cordillera Blanca region of Peru, where the total lake volume is small, but the hazards and risks are exceptionally high. According to Shugar et al. (2020), this seeming contradiction is because most glacial margins have retreated upslope to cirques, where rocks and ice can fall off steep slopes directly into the lakes. As deglaciation and glacial lake formation proceed, and as development and exposure to hazards rise, increases in disasters are expected. The authors expect the global trend of glacial lake growth to continue, perhaps accelerating in a warming world, as glacier melting and margin retreat proceed.

Sea Level Rise

Sea level change through increased melt of glaciers and ice sheets is certainly the most far-reaching effect of ice melt on Earth. Huss and Hock (2015) developed a model (GloGEM) for calculating the 21st-century response of all 200,000 glaciers on Earth, excluding the Greenland and Antarctic ice sheets. Their model employed 14 GCMs and tested three of the IPCC emission scenarios. They used an approach that allows glaciers to retreat and advance while also adjusting the surface elevation across the entire glacier. They were able to account for the effect of grounded ice below sea level when converting net mass loss to sea level equivalent. For all the glaciers of the world, the authors predicted three sets of modeled sea level contributions by the year 2100. The results varied substantially by region (Table 9.9). For the best-case scenario, RCP2.6, they predicted a sea level contribution of 79 ± 24 mm for all regions combined. For a middle-case scenario, RCP4.5, they predicted a sea level contribution of 108 ± 28 mm. For the worst-case scenario, RCP8.5, they predicted a sea level contribution of 157 ± 31 mm. The corresponding changes in glacial ice volume for all regions combined are −25% ± 7% for RCP2.6, −33% ± 8% for RCP4.5, and −48% ± 9% for RCP8.5.

Water Supplied by Glaciers: Case Studies

In many parts of the world—including the western United States, South America, China, and India—glaciers are frozen reservoirs that provide a reliable water supply each summer to hundreds of millions of people and the natural ecosystems on which they depend. The following case studies exemplify the current and projected future consequences of mountain glacier melting. Across the Tibetan Plateau and in the Himalayan and Karakoram ranges, the glaciers number in the thousands, and there are hundreds of millions of people who rely on them along rivers like the Indus in Pakistan, the Ganges and Brahmaputra rivers of India, the Yellow and Yangtze in China, and the Mekong in Southeast Asia. Eventually, these rivers will be affected by glacial retreat, but the timing of these events will vary. The Indus River is more vulnerable to glacial extinction because it is more dependent on glacial melt than the Ganges, which receives much of its water from monsoon moisture. Glacial meltwater contributions to these and other rivers is likely to increase until about the 2050s, after which many of the regional glaciers may disappear (NSIDC, 2018).

Case study 1: Qilian Mountains, East Asia

The glaciers of High Mountain Asia are important contributors to streamflow in one of the most densely populated areas of the world. The region has been called "The Water Tower of Asia" (Zhaofu et al., 2020) because meltwater from its many glaciers plays such a central role in regional hydrology. In fact, this region is the source of the 10 major river systems that provide irrigation, power, and drinking water for over 1.3 billion people in Asia—nearly 20% of the world's population (Third Pole, 2020). For instance, the glaciers of the Qilian Mountains provide water sources that are currently used in hydroelectric generation, agricultural irrigation, and the supply of drinking water for both humans and farm animals in the Gansu Corridor of Northwest China. Further downstream, near Dunhuang, once a major junction on the ancient Silk Road, water flowing out of the mountains has formed a lake in the desert for the first time in 300 years (Pollard, 2020)

Across the Qilian Mountain region, glacial meltwater is pooling in lakes, causing devastating floods. Ironically, when water is needed most for irrigation later in the summer, the glacial meltwater dries up. The melting in the mountains is projected to peak by 2030, after which snowmelt will sharply decrease as regional glaciers disappear, according to the China Academy of Sciences (Pollard, 2020). An assessment of glacier mass changes in these mountains by Cao et al. (2019) showed that from 1972 to 2016, glacier coverage decreased by 40% (41.1 km^2), leaving only 62.1 km^2 remaining in 2016. The changes in Qilian glacial mass are part of glacial melt trends across the Tibetan plateau, including the glaciers that contribute to the headwaters of the upper Yellow Yangtze and other great Asian rivers.

Case study 2: Tien Shan mountains

There is ample evidence for glacier wastage in the Tien Shan Mountains of Central Asia. In the past two decades, glaciological research here has focused mainly on documenting changes in the glacier area as repeated

satellite imagery became widely available. However, from the perspective of changes in global sea level and in regional water resources, changes in glacier volume and mass are the most important. Shahgedanova et al. (2020) analyzed changes in glacier volume and mass to assess the impacts of projected climate and glacier change on river discharge in five glacierized catchments in northern Tien Shan, Kazakhstan. They used a conceptual hydrological model HBV-ETH. They also employed the PRECIS regional climate model under four different GCM scenarios (HadGEM2.6, HadGEM8.5, A1B using HadCM3Q0 and ECHAM5) to develop climate projections. The HadGEM2.6 scenario follows the conditions set out in the IPCC formulation of RCP2.6 (best-case scenario). The HadGEM8.5 scenario follows the conditions set out in the IPCC formulation of RCP8.5 (worst-case scenario).

All the climate scenarios predicted by the various models show statistically significant warming in the 21st century. Regional glaciers in the northern Tien Shan are projected to retreat rapidly until the 2050s and then stabilize afterward except under the HadGEM8.5 scenario, in which glacial margin retreat continues. The glaciers are projected to lose 38% −50% of their volume and 34%−39% of their area. Total river discharge in July−August is projected to decline in catchments with little glacial ice cover (2%−4%) by 20%−37%. This decline may have significant impacts on water availability for summer crop irrigation that dominates water consumption in these catchments. In catchments with medium glacial cover (10%−12%, such as the Kishi Almaty), a reduction in summer flow is projected for the less aggressive scenarios while under the strongest warming, more intensive glacier and snowmelt will sustain water supply for at least a few decades. In catchments with higher levels of ice cover (16% and over), no significant changes in summer discharge are expected, and spring discharge is projected to increase. In catchments with high levels of glacial ice cover (16% and over, such as the Ulken Almaty, and Talgar), summer flow is not expected to change significantly, but spring flow is predicted to increase under the aggressive scenarios. The Ulken Almaty River supplies water for municipal use by more than 2 million people in Almaty city. The municipal water supply is regulated by a reservoir. In the Talgar catchment, the projected increase in spring flow may require the development of flood prevention measures.

Case study 3: The Southern Alps

The remaining ice in the Southern Alps is contained in small (<1 km^2) glaciers, increasingly restricted to high elevations. This ice is fast becoming restricted to shaded areas immediately beneath steep head walls, as is the case for glaciers in the European Alps. Carrivick et al. (2020) observed that the accelerating rate of ice loss reflects regional-specific climate conditions and that peak glacial meltwater production is imminent if not already passed, which has profound implications for water resources and riverine habitats.

Chinn (2001) made predictions concerning water resource variations in the Southern Alps under climate change scenarios. His analysis showed that a climate warming of 1.5 and 3°C will generate increased flows from shrinking glaciers. Most of the remaining glaciers in New Zealand contain too little ice to contribute significantly to regional water resources. However, two drainages are exceptions to this. The Waitaki is a large, braided river that drains the Mackenzie Basin of the South Island of New Zealand. It runs 110 km southeast to enter the Pacific Ocean on the east coast. Its discharge is 356 m^3 per second. Climate warming as defined earlier would add 4−8 m^3 of water per second to the discharge of the Waitaki. The Clutha River is the longest river on the South Island, flowing south-southeast 338 km from Lake Wanaka in the Southern Alps to the Pacific Ocean. It is also New Zealand's swiftest river, discharging 614 m^3 per second. Chinn (2001) reckoned that the melting of the glaciers that feed the Clutha would add from 0.5 to 1 m^3 per second to the flow of the Clutha River.

Most of the Southern Alps glaciers have contributed little or nothing to the flow of rivers that supply water for agriculture or human consumption on the South Island pub New Zealand. However, glacier lakes are a vital part of electrical generation in the country. When Pleistocene glaciers retreated, they left basins that are now filled by lakes. The levels of most glacial lakes in the upper parts of the Waitaki and Clutha rivers are controlled for electricity generation. Hydroelectric reservoirs are common on the South Island, the largest of which is Lake Benmore, on the Waitaki River.

Case study 4: Glacier-fed water supply in the Andes

Glaciers do more than feed Andean rivers and lakes, they also serve as critical water savings banks that can be tapped for withdrawals during droughts. The glaciers and snowpack of the Andes provide slow, consistent meltwater that acts as a buffering mechanism during dry summer months and drought periods for many cities. As described by Cullen (2018), here are four cities in the Andes and their approach to the impending water crisis when the mountaintop glaciers disappear.

La Paz, Bolivia

La Paz is a metropolitan area of 2.3 million people. Perched at 3640 m above sea level (asl), it is the highest capital city in the world. La Paz relies on glaciers for about 15% of its water, but, during dry months, this number rises to 61%. During the driest month of a drought year, 85% of the city's water comes from glacial meltwater. The situation is expected to become critical as higher temperatures dry out local farms, lakes, and ecosystems, and droughts become more frequent and intense. As we have seen in this chapter, mountain glaciers in the Andes are shrinking. In fact, regional glaciers have lost 30% to 50% of their ice in the past 30 years, and many small glaciers are likely to disappear by 2030.

Santiago, Chile

Santiago's reliance on glacial melt as a buffer during hard times is not as well studied as that of La Paz, but some estimates suggest the city relies on glaciers for up to two-thirds of its water during the driest months of a drought year. The city is home to well over 5 million people. The Nature Conservancy performed a study that predicted that by 2070, Santiago's primary water source, the Maipo River, may face a 40% reduction in water availability due to glacier retreat and decreased precipitation due to climate change. The Maipo is fed by more than 1000 glaciers that cover almost 400 km^2. The meltwater trickles down through high-altitude wetlands into municipal catchments, groundwater reserves, and agricultural irrigation systems. Like near all of Chile's glaciers, all but the largest of these source glaciers are disappearing.

Quito, Ecuador

Quito is the second-highest capital city in the world, sitting at 2850 m asl. Zambrano-Barragán et al. (2011) described the response of Quito to the ongoing scarcity of water. The municipality of Quito, with a population approaching 2 million, adopted Quito's Climate Change Strategy (QCCS) in 2009. In coordination with key stakeholders, the city is implementing a series of adaptation and mitigation measures in key sectors. Actions fall under four strategic axes: (1) information generation and management; (2) use of clean technologies and good practices for adaptation and mitigation; (3) communication, education, and citizen participation; and (4) institutional strengthening and capacity building. Given the importance of water resources for the city and its surroundings, measures to face climate change in the water provision and risk management sectors are at the core of the strategy.

Quito is one of the few Latin American cities that have a climate change strategy. Together with the use of the city's ecological footprint as a planning tool, this puts the city at the vanguard of Andean country responses to climate change.

Lima, Peru

Lima is by the far the largest city in the Andean region of South America, with about 12.1 million people. Amid climate change and population growth concerns, Lima's water utility, SEDAPAL, recently invested $2.7 billion in a major plan that features green infrastructure, water recycling, and special considerations for the city's estimated 1.5 million underserved residents. A World Bank and SEDAPAL study on the implementation of the plan notes that two-thirds of the glaciers feeding the Rimac River's headwaters have disappeared, decreasing the river's glacier contributed volume by 90% in the past 40 years. A Nature Conservancy–led water fund, AQUAFONDO, is collaborating with SEDA-PAL to preserve upstream watersheds to help counteract the declining flow of glacial meltwater. The city now depends primarily on rainfall, which may become more uncertain as climate change progresses.

CONCLUSIONS

There are many parallels between the Arctic and alpine regions of the world—the terrestrial part of the cryosphere. Both were essentially too remote and hazardous to visit by all but the hardiest of people until recently— the outstanding exception being the people who lived in the Andean highlands and apparently spent considerable time on mountaintops since time immemorial. Both regions were poorly studied by scientists or natural historians until the last two to three centuries. Snow and ice have dominated both the Arctic and high alpine landscapes for many thousands of years, although both regions became more ice-bound during the LIA than they had been since the end of the last ice age. Permafrost soils dominate in both regions. These are now starting to melt, releasing water and GHGs.

Those are commonalities that kept human influence at a minimum in the terrestrial cryosphere, essentially until the Industrial Revolution. Indeed, one could argue that the notable increase in human activities in the Arctic and alpine regions was one of the hallmarks of the Anthropocene (see Elias, 2018). Unfortunately, another commonality is that the Arctic and the Alpine are now suffering under the effects of global warming.

Even though this warming comes from outside the cryosphere, it not only affects the cold regions, but it is also amplified there. Both the high-latitude and high-altitude regions are experiencing greater-than-average temperature increases. Ice is melting in response to this warming. The melting of Arctic ice comes in the form of thinning glaciers and ice sheets. These losses of ice are not greatly affecting the human populations of the Arctic. There is little reliance on glacial meltwater for human consumption, and there is no agriculture in the Arctic, except in greenhouses. Of course, the other part of the Arctic where ice is melting is the ocean. The loss of sea ice is having greater effects on both humans and other Arctic life, as discussed in Chapter 1.

As we have seen in this chapter, the loss of mountain glaciers is having significant effects on human populations in many regions of the world, but especially in Asia and South America. The Highlands of Central Asia have been called the water tower of this region. The height of that water tower is slowly being reduced, which is already affecting the downstream water supplies of many rivers. Likewise, in the Andes, people have come to rely on glacial meltwater to slake their thirst and irrigate their crops, especially during drought episodes. Some efforts are being made to compensate for the eventual loss of this resource as glaciers melt away, notably in Quito, Ecuador. But because of widespread poverty in this region, outside agencies and organizations are having to step in to help such projects succeed. Vital elements of water conservation plans for Andean cities include education of the public, improvements to water distribution infrastructure (e.g., fixing leaky pipes and improving reservoirs), and risk assessment planning.

Perhaps the most troubling aspect of the situation is the seeming inevitability of the loss of glacial ice. Even if all GHG emissions were to stop today, humankind has set in motion a climate warming steamroller that will keep going for decades or centuries to come. It seems that the Victorians were smart enough to invent the use of fossil fuels to run industries and transportation systems, but not sufficiently prescient to foresee the environmental disaster they were creating. But the post–World War II generations have not, for the most part, been any wiser. We have only refined and expanded the various technologies that pump GHGs into the atmosphere. Here again, we do not appear to be wise enough to put an end to the global warming problem anytime soon. If the 22nd century becomes a world without glaciers, planet Earth will be the poorer for it.

REFERENCES

American Chemical Society, 2020. It's Water Vapor, Not the CO_2. https://www.acs.org/content/acs/en/climatescience/climatesciencenarratives.

Barr, J., Chandler, A., 2013. Africa without ice and snow. Environmental Development 5, 146–155. https://doi.org/10.1016/j.envdev.2012.10.003.

Bohleber, P., Sold, L., Hardy, D.R., Schwikowski, M., Klenk, P., Fischer, A., Sirguey, P., Cullen, N.J., Potocki, M., Hoffmann, H., Mayewski, P., 2017. Ground-penetrating radar reveals ice thickness and undisturbed englacial layers at Kilimanjaro's Northern Ice Field. The Cryosphere 11 (1), 469–482. https://doi.org/10.5194/tc-11-469-2017.

Braumann, S.M., Schaefer, J.M., Neuhuber, S.M., Reitner, J.M., Lüthgens, C., Fiebig, M., 2020. Holocene glacier change in the Silvretta Massif (Austrian Alps) constrained by a new 10Be chronology, historical records and modern observations. Quaternary Science Reviews 245, 106493. https://doi.org/10.1016/j.quascirev.2020.106493.

Braun, M.H., Malz, P., Sommer, C., Farías-Barahona, D., Sauter, T., Casassa, G., Soruco, A., Skvarca, P., Seehaus, T.C., 2019. Constraining glacier elevation and mass changes in South America. Nature Climate Change 9 (2), 130–136. https://doi.org/10.1038/s41558-018-0375-7.

Brun, F., Berthier, E., Wagnon, P., Kääb, A., Treichler, D., 2017. A spatially resolved estimate of High Mountain Asia glacier mass balances from 2000 to 2016. Nature Geoscience 10, 668–674. https://doi.org/10.1038/NGEO2999.

Cao, B., Pan, B., Wen, Z., Guan, W., Li, K., 2019. Changes in glacier mass in the Lenglongling Mountains from 1972 to 2016 based on remote sensing data and modeling. Journal of Hydrology 578, 124010. https://doi.org/10.1016/j.jhydrol.2019.124010.

Carrivick, J.L., Tweed, F.S., 2013. Proglacial Lakes: character, behaviour and geological importance. Quaternary Science Reviews 78, 34–52. https://doi.org/10.1016/j.quascirev.2013.07.028.

Carrivick, J.L., James, W.H.M., Grimes, M., Sutherland, J.L., Lorrey, A.M., 2020. Ice thickness and volume changes across the Southern Alps, New Zealand, from the little ice age to present. Scientific Reports 10 (1), 13392. https://doi.org/10.1038/s41598-020-70276-8.

Castellazzi, P., Burgess, D., Rivera, A., Huang, J., Longuevergne, L., Demuth, M.N., 2019. Glacial melt and potential impacts on water resources in the Canadian rocky mountains. Water Resources Research 55 (12), 10191–10217. https://doi.org/10.1029/2018WR024295.

Chinn, T.J., 2001. Distribution of the glacial water resources of New Zealand. Journal of Hydrology New Zealand 40 (2), 139–187.

Clarke, G.K.C., Jarosch, A.H., Anslow, F.S., Radić, V., Menounos, B., 2015. Projected deglaciation of western Canada in the twenty-first century. Nature Geoscience 8 (5), 372–377. https://doi.org/10.1038/ngeo2407.

Cullen, K., 2018. Four Andean Cities Adapting to Glacier Retreat to Preserve Water Security. https://thecityfix.com/blog/4-andean-cities-adapting-glacier-retreat-preserve-water-security-kate-cullen/.

Davaze, L., Rabatel, A., Dufour, A., Hugonnet, R., Arnaud, Y., 2020. Region-wide annual glacier surface mass balance for the European Alps from 2000 to 2016. Frontiers in Earth Science 8, 149. https://doi.org/10.3389/feart.2020.00149.

Davies, B.J., Darvill, C.M., Lovell, H., Bendle, J.M., Dowdeswell, J.A., Fabel, D., García, J.L., Geiger, A., Glasser, N.F., Gheorghiu, D.M., Harrison, S., Hein, A.S., Kaplan, M.R., Martin, J.R.V., Mendelova, M., Palmer, A., Pelto, M., Rodés, Á., Sagredo, E.A., et al., 2020. The evolution of the Patagonian Ice Sheet from 35 ka to the present day (PATICE). Earth-Science Reviews 204, 103152. https://doi.org/10.1016/j.earscirev.2020.103152.

Derbyshire, E., Fort, M., Owen, L.A., 2001. Geomorphological hazards along the Karakoram highway: Khunjerab pass to the Gilgit river, Northernmost Pakistan. Erdkunde 55 (1), 49–71. https://doi.org/10.3112/erdkunde.2001.01.04.

Elias, S.A., 2018. Finding a "golden spike" to mark the anthropocene. In: DellaSala, D.A., Goldstein, M. (Eds.), Encyclopedia of the Anthropocene, vols. 1–5. Elsevier, Oxford, pp. 19–28.

Farinotti, D., Huss, M., Fürst, J.J., Landmann, J., Machguth, H., Maussion, F., Pandit, A., 2019. A consensus estimate for the ice thickness distribution of all glaciers on Earth. Nature Geoscience 12 (3), 168–173. https://doi.org/10.1038/s41561-019-0300-3.

Gardelle, J., Berthier, E., Arnaud, Y., Kääb, A., 2013. Region-wide glacier mass balances over the Pamir-Karakoram-Himalaya during 1999-2011. The Cryosphere 7, 1263–1286. https://doi.org/10.5194/tc-7-1263-2013.

Glasser, N.F., Harrison, S., Jansson, K.N., Anderson, K., Cowley, A., 2011. global sea-level contribution from the Patagonian icefields since the little ice age maximum. Nature Geoscience 4 (5), 303–307. https://doi.org/10.1038/ngeo1122.

Haeberli, W., Huggel, C., Paul, F., Zemp, M., 2020. The response of glaciers to climate change: observations and impacts. In: Elias, S.A. (Ed.), Reference Module in Earth Systems and Environmental Sciences. Elsevier. https://doi.org/10.1016/B978-0-12-818234-5.00011-0.

Herreid, S., Pellicciotti, F., 2020. The state of rock debris covering Earth's glaciers. Nature Geoscience 13 (9), 621–627. https://doi.org/10.1038/s41561-020-0615-0.

Hock, R., Bliss, A., Marzeion, B.E.N., Giesen, R.H., Hirabayashi, Y., Huss, M., Radic, V., Slangen, A.B.A., 2019. GlacierMIP-A model intercomparison of global-scale glacier mass-balance models and projections. Journal of Glaciology 65 (251), 453–467. https://doi.org/10.1017/jog.2019.22.

Hu, Z., Kang, S., Li, X., Li, C., Sillanpää, M., 2020. Relative contribution of mineral dust versus black carbon to Third Pole glacier melting. Atmospheric Environment 223, 117288. https://doi.org/10.1016/j.atmosenv.2020.117288.

Huss, M., Hock, R., 2015. A new model for global glacier change and sea-level rise. Frontiers in Earth Science 3, 54. https://doi.org/10.3389/feart.2015.00054.

Kaspari, S., Mayewski, P., Kang, S., Sneed, S., Hou, S., Hooke, R., Kreutz, K., Introne, D., Handley, M.,

Maasch, K., Qin, D., Ren, J., 2007. Reduction in northward incursions of the South Asian monsoon since ~1400 AD inferred from a Mt. Everest ice core. Geophysical Research Letters 34 (16), L16701. https://doi.org/10.1029/2007GL030440.

Kattel, D.B., Yao, T., Yang, K., Tian, L., Yang, G., Joswiak, D., 2013. Temperature lapse rate in complex mountain terrain on the southern slope of the central Himalayas. Theoretical and Applied Climatology 113 (3–4), 671–682. https://doi.org/10.1007/s00704-012-0816-6.

Kellerhals, T., Brutsch, S., Sigl, M., Knusel, S., Gaggeler, W.H., et al., 2010. Ammonium concentration in ice cores: a new proxy for regional temperature reconstruction? Journal of Geophysical Research 115, D16123. https://doi.org/10.1029/2009jd012603.

Körner, C., 2017. Alpine ecosystems. In: Reference Module in Life Sciences. Amsterdam, pp. 1–12. https://doi.org/10.1016/B978-0-12-809633-8.02180-41.

Körner, C., Paulsen, J., Spehn, E.M., 2011. A definition of mountains and their bioclimatic belts for global comparisons of biodiversity data. Alpine Botany 121 (2), 73–78. https://doi.org/10.1007/s00035-011-0094-4.

Larsen, D.J., Miller, G.H., Geirsdóttir, A., 2013. Asynchronous little ice age glacier fluctuations in Iceland and European Alps linked to shifts in subpolar North Atlantic circulation. Earth and Planetary Science Letters 380, 52–59. https://doi.org/10.1016/j.epsl.2013.07.028.

Li, Y.J., Ding, Y.J., Shangguan, D.H., Wang, R.J., 2019. Regional differences in global glacier retreat from 1980 to 2015. Advances in Climate Change Research 10 (4), 203–213. https://doi.org/10.1016/j.accre.2020.03.003.

Lindsey, R., 2020. Climate Change: Glacier Mass Balance. https://www.climate.gov/news-features/understanding-climate/climate-change-glacier-mass-balance.

Litaor, M.I., Williams, M., Seastedt, T.R., 2008. Topographic controls on snow distribution, soil moisture, and species diversity of herbaceous alpine Vegetation, Netwot Ridge, Colorado. Journal of Geophysical Research: Biogeosciences 113 (2), G02008. https://doi.org/10.1029/2007JG000419.

Ljungqvist, F.C., 2010. A new reconstruction of temperature variability in the extra-tropical northern hemisphere during the last two millennia. Geografiska Annaler, Series A: Physical Geography 92 (3), 339–351. https://doi.org/10.1111/j.1468-0459.2010.00399.x.

Lorrey, A., Fauchereau, N., Stanton, C., Chappell, P., Phipps, S., et al., 2014. The Little Ice Age climate of New Zealand reconstructed from Southern Alps cirque glaciers: a synoptic type approach. Climate Dynamics 42, 3039–3060. https://doi.org/10.1007/s00382-013-1876-8.

Menounos, B., Hugonnet, R., Shean, D., Gardner, A., Howat, I., Berthier, E., Pelto, B., Tennant, C., Shea, J., Noh, M.J., Brun, F., Dehecq, A., 2019. Heterogeneous changes in western North American glaciers linked to decadal variability in zonal wind strength. Geophysical Research Letters 46 (1), 200–209. https://doi.org/10.1029/2018GL080942.

Miles, K.E., Hubbard, B., Irvine-Fynn, T.D.L., Miles, E.S., Quincey, D.J., Rowan, A.V., 2020. Hydrology of debris-covered glaciers in high mountain Asia. Earth-Science Reviews 207, 103212. https://doi.org/10.1016/j.earscirev.2020.103212.

Miles, M.W., Andresen, C.S., Dylmer, C.V., 2020. Evidence for extreme export of Arctic sea ice leading the abrupt onset of the Little Ice Age. Science Advances 6 (38), eaba4320. https://doi.org/10.1126/sciadv.aba4320.

Miller, G.H., Geirsdóttir, A., Zhong, Y., Larsen, D.J., Otto-Bliesner, B.L., Holland, M.M., Bailey, D.A., Refsnider, K.A., Lehman, S.J., Southon, J.R., Anderson, C., Björnsson, H., Thordarson, T., 2012. Abrupt onset of the Little Ice Age triggered by volcanism and sustained by sea-ice/ocean feedbacks. Geophysical Research Letters 39 (2), L02708. https://doi.org/10.1029/2011GL050168.

NSIDC, 2018. Impact of High Asian Glaciers and Snowpack on Water Resources. Research Project: Contribution to High Asia. https://nsidc.org/charis/project-summary/.

NSIDC, 2019. State of the Cryosphere: Mountain Glaciers. https://nsidc.org/cryosphere/sotc/glacier_balance.html.

Nussbaumer, S.U., Steinhilber, F., Trachsel, M., Breitenmoser, P., Beer, J., Blass, A., Grosjean, M., Hafner, A., Holzhauser, H., Wanner, H., Zumbühl, H.J., 2011. Alpine climate during the Holocene: a comparison between records of glaciers, lake sediments and solar activity. Journal of Quaternary Science 26 (7), 703–713. https://doi.org/10.1002/jqs.1495.

Oosthoek, K.J., 2015. Little Ice Age. Environmental History Resources. https://www.eh-resources.org/little-ice-age/.

Pfiffner, O.A., 2021. Chapter 2. The geology of Switzerland. In: Reynard, E. (Ed.), World Geomorphological Landscapes. Springer, Switzerland. https://doi.org/10.1007/978-3-030-43203-4_2.

Pollard, M.Q., 2020. The Thaw of the Third Pole: China's Glaciers in Retreat. Reuters. https://widerimage.reuters.com/story/the-thaw-of-the-third-pole-chinas-glaciers-in-retreat.

Prinz, R., Heller, A., Ladner, M., Nicholson, L.I., Kaser, G., 2018. Mapping the loss of Mt. Kenya's glaciers: an example of the challenges of satellite monitoring of very small glaciers. Geosciences 8 (5), 174. https://doi.org/10.3390/geosciences8050174.

Protin, M., Schimmelpfennig, I., Mugnier, J.L., Ravanel, L., Le Roy, M., Deline, P., Favier, V., Buoncristiani, J.F., Aumaître, G., Bourlès, D.L., Keddadouche, K., 2019. Climatic reconstruction for the younger Dryas/early holocene transition and the little ice age based on paleo-extents of Argentière Glacier (French Alps). Quaternary Science Reviews 221, 105863. https://doi.org/10.1016/j.quascirev.2019.105863.

Richardson, S., Quincey, D.J., 2009. Glacier outburst floods from Ghulkin Glacier, upper Hunza valley. European Geophysical Union Meeting Abstracts 11.

Rowan, A.V., 2017. The 'Little Ice Age' in the Himalaya: a review of glacier advance driven by Northern Hemisphere temperature change. The Holocene 27 (2), 292–308. https://doi.org/10.1177/0959683616658530.

Ruegg, P., 2019. Team Calculates Volume of 215,000 Glaciers. Futurity. https://www.futurity.org/glacier-volume-1989722-2/.

Shahgedanova, M., Afzal, M., Hagg, W., Kapitsa, V., Kasatkin, N., Mayr, E., Rybak, O., Saidaliyeva, Z.,

Severskiy, I., Usmanova, Z., Wade, A., Yaitskaya, N., Zhumabayev, D., 2020. Emptying water towers? Impacts of future climate and glacier change on river discharge in the northern Tien Shan, Central Asia. Water (Switzerland) 12 (3), 627. https://doi.org/10.3390/w12030627.

Shugar, D.H., Burr, A., Haritashya, U.K., Kargel, J.S., Watson, C.S., Kennedy, M.C., Bevington, A.R., Betts, R.A., Harrison, S., Strattman, K., 2020. Rapid worldwide growth of glacial lakes since 1990. Nature Climate Change 10 (10), 939–945. https://doi.org/10.1038/s41558-020-0855-4.

Shukla, A.D., Sharma, S., Rana, N., Bisht, P., Juyal, N., 2020. Optical chronology and climatic implication of glacial advances from the southern Ladakh Range, NW Himalaya, India. Palaeogeography, Palaeoclimatology, Palaeoecology 539, 109505. https://doi.org/10.1016/j.palaeo.2019.109505.

Stadelmann, C., Jakob Fürst, J., Mölg, T., Braun, M., 2020. Brief communication: glacier thickness reconstruction on Mt. Kilimanjaro. The Cryosphere 14 (10), 3399–3406. https://doi.org/10.5194/tc-14-3399-2020.

Sugiyama, S., Minowa, M., Sakakibara, D., Skvarca, P., Sawagaki, T., Ohashi, Y., Naito, N., Chikita, K., 2016. Thermal structure of proglacial lakes in Patagonia. Journal of Geophysical Research: Earth Surface 121 (12), 2270–2286. https://doi.org/10.1002/2016JF004084.

Taylor, R.G., Mileham, L., Tindimugaya, C., Majugu, A., Muwanga, A., Nakileza, B., 2006. Recent glacial recession in the Rwenzori Mountains of East Africa due to rising air temperature. Geophysical Research Letters 33 (10), L10402. https://doi.org/10.1029/2006GL025962.

Third Pole, 2020. What is The Third Pole?. https://www.thethirdpole.net/about/.

Thompson, L.G., Mosley-Thompson, E., Davis, M.E., Lin, P.N., Henderson, K.A., Cole-Dai, J., Bolzan, J.F., Liu, K.B., 1995. Late glacial stage and holocene tropical ice core records from Huascarán, Peru. Science 269 (5220), 46–50. https://doi.org/10.1126/science.269.5220.46.

Thompson, L.G., Yao, T., Mosley-Thompson, E., Davis, M.E., Henderson, K.A., Lin, P.N., 2000. A high-resolution millennial record of the South Asian Monsoon from Himalayan ice cores. Science 289 (5486), 1916–1919. https://doi.org/10.1126/science.289.5486.1916.

Thompson, L.G., Mosley-Thompson, E., Brecher, H., et al., 2006. Abrupt tropical climate change: past and present. Proceedings of the National Academy of Sciences of the United States of America 103, 10536–10543. https://doi.org/10.1073/pnas.0603900103.

Thymann, K., 2020. Africa's First Mountains to Lose Their Glaciers. https://www.thestoryinstitute.com/rwenzori-mountains.

US Geological Survey, 2020a. How Old Is Glacier Ice? https://www.usgs.gov/faqs/how-old-glacier-ice?.

US Geological Survey, 2020b. Glaciers and Climate Project: Mass Balance Methods - Measuring Glacier Change. https://www2.usgs.gov/landresources/lcs/glacierstudies/massbal.asp.

Wayand, N.E., Marsh, C.B., Shea, J.M., Pomeroy, J.W., 2018. Globally scalable alpine snow metrics. Remote Sensing of Environment 213, 61–72. https://doi.org/10.1016/j.rse.2018.05.012.

World Glacier Monitoring Service, 2020. Global glacier change bulletin, 2016–2017. In: Zemp, M., Gärtner-Roer, I., Nussbaumer, S.U., Bannwart, J., Rastner, P., et al. (Eds.), Global Glacier Change Bulletin, vol. 3. World Glacier Monitoring Service. ISC(WDS)/IUGG(IACS)/UNEP/UNESCO/WMO.

Zambrano-Barragán, C., Zevallos, O., Villacís, M., Enríquez, D., 2011. Quito's climate change strategy: a response to climate change in the metropolitan district of Quito, Ecuador. In: Otto-Zimmermann, K. (Ed.), Local Sustainability. Springer. https://doi.org/10.1007/978-94-007-0785-6_51.

Zhaofu, H.Z., Shichang, K.S., Xiaofei, L.X., Chaoliu, L.C., Silanpää, M., 2020. Relative contribution of mineral dust versus black carbon to Third Pole glacier melting. Atmospheric Environment 223, 117288. https://doi.org/10.1016/j.atmosenv.2020.117288.

Arctic Lands

This section includes four chapters concerning Arctic lands: the Greenland Ice Sheet, Changes in Terrestrial Environments, Impacts of Global Change, and Impacts of Oil and Mineral Extraction.

The Arctic region comprises about 30 million km^2 of land and sea north of the Arctic Circle. The northern polar latitudes are quite different from their southern counterpart, where the continent of Antarctica dominates the polar region. The Arctic regions are almost an equal mix of land and sea, with little land extending beyond 80°N. The Arctic Ocean covers 15.6 million km^2, a little over half of the Arctic. This mix of sea and land makes Arctic climate considerably different from Antarctic climate, which is extremely continental in nature. Until quite recently, most of the Arctic Ocean was covered with sea ice for much of the year, with some of the ice being several meters thick. This kept the Arctic much colder than it would be without the sea ice cover—a fact that is becoming ever more important as we are losing that sea ice cover in the north.

The Greenland Ice Sheet (GrIS) is the last vestige of the ice sheets that covered the northern halves of North America and Europe during late Pleistocene glaciations. The GrIS is thus the only remaining ice archive of long-term climate change in the Northern Hemisphere, with ancient ice layers extending back at least 125,000 years to the last interglacial period. Amplified Arctic warming is causing the melting of the GrIS. The margins of the ice sheet have changed little, but the ice sheet is getting thinner, a fact recently established by airplane and satellite measurements. A vital research question is whether the melting of the GrIS has reached the point of no return. A number of scientists believe that this tipping point has already occurred, and it is only a matter of time before the ice sheet disappears. If the GrIS melts, global sea level will rise by about 6 m.

'Changes in Terrestrial Environments' covers a wide range of issues affecting Arctic landscapes. It is convenient to divide the biome into High Arctic and Low Arctic ecosystems. The High Arctic, also known as polar desert, occurs in the far north. The Low Arctic is slightly warmer (especially in summers) and much more extensive in area. Human interference with Arctic landscapes is mainly restricted to the Low Arctic. There are two principal categories of threats to Arctic terrestrial environments and biota. The principal threat is global warming. Its effects are being felt throughout the Arctic, in every type of environment. The intensified, rapid changes in Arctic thermal regimes exert stresses on regional fauna and flora—far greater climate changes than those of lower latitudes. The second principal threat comes from direct human interference on the landscape. This includes the development of industrial sites for the extraction of resources, a topic covered in depth in Chapter 13, and permafrost degradation due to the building of uninsulated or poorly insulated structures on frozen soils, a topic covered in depth in Chapter 14.

The term "global change" refers to the broad suite of human-caused changes to the biological world, the physical world, and the socioeconomic changes that are affecting Earth System functioning at the global scale. Human disruption of Arctic landscapes includes direct impacts, such as mining, nuclear testing, and petroleum extraction. These impacts are restricted to specific regions, but there are other indirect impacts that come from the south and affect the entire Arctic. These are air and water pollution, mostly from sources in the temperate zone. Toxic elements and compounds have drifted north and penetrated deep into the Arctic. Global climate change is forcing changes in the timing of the life history, population dynamics, and demography of northern biota. These encompass complex, interacting environmental changes.

The impacts of oil and mineral extraction used to be limited to very small areas in the Arctic, but they are expanding. This chapter revisits the conflict between preserving the Arctic and developing its resources for financial gain, national power and prestige, and geopolitical place marking. Since Renaissance times, outsiders have gone to the Arctic mainly to exploit its natural resources. These days, most northern countries have sought to balance the conservation of nature with resource extraction. The main exception is Russia, which continues to exploit mineral and petroleum resources throughout its Arctic regions with little or no effort to protect the environment.

CHAPTER 10

Greenland Ice Sheet

INTRODUCTION

An ice sheet is a mass of glacial land ice extending more than 50,000 square km. The two ice sheets on Earth today cover most of Greenland and Antarctica. During the last ice age, ice sheets also covered much of North America and Scandinavia. Together, the Antarctic and Greenland ice sheets (GIS) contain more than 99% of the freshwater ice on Earth. The Antarctic ice sheet covers almost 14 million square km, roughly the area of the contiguous United States and Mexico combined. The Antarctic ice sheet contains 30 million cubic km of ice. The GIS covers about 1.7 million square km, covering most of the island of Greenland.

Ice Sheet as Paleo-Archive

Ice sheets build up over thousands of years, from accumuted layers of snow amassed one by one each winter. As discussed below, chemical signals in the layers of compressed snow that eventually turn to ice allow scientists to reconstruct the history of global climate change from these frozen archives. The GIS contains a record of approximately the last 100,000 years. The Antarctic Ice Sheet record extends back more than a million years.

A Climate Change Indicator

Hvidberg et al. (2013) provide a concise summary of the dynamics of the GIS. Because it straddles the boundaries of the North Atlantic and Arctic Oceans, the GIS is in a unique position to register regional climate change. The GIS is not a static sheet of ice perched on Greenland bedrock and going nowhere. Rather, it is in a state of constant interaction with the atmosphere and neighboring ocean waters. As discussed in Chapter 3, in recent decades, it has become apparent to oceanographers and climatologists that the North Atlantic plays a key role in global climate change and that changes in North Atlantic thermohaline circulation determine the extent to which warm water from the lower latitudes enters the high northern latitudes. As the ocean currents stream along the Coast of Greenland, precipitation falls on the ice sheet as snow. The snow eventually turns to ice, being compressed by its own weight, and the ice

contains evidence concerning the state of the North Atlantic at the time the snow fell.

Snow accumulation rates in the central areas of the GIS have been sufficiently high through at least the last glacial cycle to produce consistent layers of ice so that annual ice layers may be resolved well back into the last glacial period. In one sense, ice is more of a deformable liquid than it is a solid. It flows under the force of gravity from the center of Greenland toward the coasts, slowing down as it encounters rough bedrock services and speeding up where the bedrock is smooth. All of this slowing down, speeding up, and spreading of the ice causes the annual layers observed in ice cores to thin and stretch since they were originally deposited as surface snow. The layers of ice provide a treasure trove of information, much of it resolved at the annual level, but the dynamics of the ice sheet must be understood before any useful paleoclimatic interpretations can be made.

Snow turns to ice

As layer upon layer of snow is deposited in the central areas of an ice sheet, the snow compresses under its own weight and gradually transforms into ice. "Firn" is the term used to describe the material in the intermediate stages of this transformation. The length of time required for snow to transform into ice depends largely on the temperature of the environment. In the interior of the GIS, temperatures are always well below freezing, so there is no melting, and the firn-ice transition depth is about 100 m. Below the firn-ice transition, the ice density remains practically constant at 917 kg per m^3.

The ice flows

In an idealized, steady-state section of the GIS from west to east in the cross section (Fig. 10.1), ice flows from the central areas toward the margins. The ice sheet accumulates snow in the center (the accumulation zone), and it loses ice mass at the margins (the ablation zone). If the ice sheet is in a steady state, the mass balance of the ice sheet is neutral, and the shape remains constant over time. If more snow accumulates than ice lost at the margins, then the ice sheet has a positive

Threats to the Arctic. https://doi.org/10.1016/B978-0-12-821555-5.00009-7

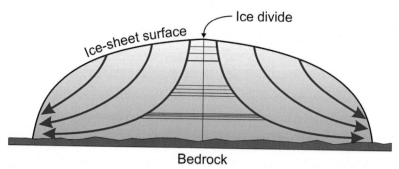

FIG. 10.1 Ice sheet flow cartoon. (After Hvidberg, C.S., Svensson, A., Buchardt, S.L., 2013. Dynamics of the Greenland ice sheet. In: Encyclopedia of Quaternary Science, second ed. Elsevier Inc., pp. 439–447. https://doi.org/10.1016/B978-0-444-53643-3.00327-7.)

mass balance and starts to expand and thicken. If the mass of the ice lost at the margins exceeds that of the snow accumulating at the center, then the ice sheet has a negative mass balance and it starts to thin and retreat along its edges.

One clarification needs to be made at this point: glaciers and ice sheets do not retreat. Only their margins may retreat from a previous position. This is because glacial ice is constantly flowing downhill, carrying rocks and other debris how long will it like a conveyor belt. Unless a glacier or ice sheet melts into oblivion, the ice will continue to flow downhill.

Equilibrium-line altitude (ELA) is the elevation on a glacier or ice sheet where accumulation and ablation balance on an annual basis. If net snow accumulation exceeds surface melting on a body of ice, then the ELA elevation moves down. Conversely, if surface melting exceeds snow accumulation, then the ELA elevation moves upslope. Thus, the ELA usefully integrates the combined effect of surface melting and net snow accumulation. ELA can be monitored in optical satellite imagery just prior to the first winter snow.

Ideally, a snowflake landing on the surface near the center of the ice sheet will gradually sink down, turn to ice, and be carried with the flow to the margin. Eventually, it emerges at the surface near the margin, where it is lost by surface melt and runoff.

Ablation is not just a slow, continual process at the margins of the GIS, but it also occurs as icebergs calve from a coastal glacier and drift out to sea. This process is aided by basal melting from beneath floating glacier tongues, or as basal melting and draining beneath the grounded ice sheet. Icebergs are becoming far more common along the coast of Greenland as the GIS is thinning and the ice is flowing more rapidly to the coast. During September of 2020, an iceberg measuring 119 km^2—more than twice the size of Manhattan Island—broke off a glacier at Nioghalvfjerdsfjorden at 79°N in northeast Greenland. According to a recent study published by the National Oceanic and Atmospheric Administration (Lindsey, 2020), ice loss from the GIS increased sevenfold from 34 billion tons a year from 1992 to 2001 to 247 billion tons a year from 2012 to 2016.

By the end of the summer season, all the fresh snow from the previous winter has been removed in the ablation zone; then toward the center surface of the ice sheet, the ice gets progressively younger.

The Importance of Ice Sheets

Ice sheets contain enormous quantities of frozen water. If the GIS melted, scientists estimate that sea level would rise about 6 m. This level of sea level rise would displace about 280 million people.

If the Antarctic ice sheet melted, sea level would rise by about 60 m. The GIS and Antarctic ice sheet also influence regional and global weather and climate. Large high-altitude plateaus on the ice caps alter storm tracks and create cold downslope winds close to the ice surface. The thousands of individual (often annual) layers of ice covering Greenland and Antarctica contain unique records of Earth's climate history (NSIDC, 2007).

As discussed earlier, far from being a static ice body, the GIS is a highly dynamic system of ice streams, ice divides, meltwater channels, and glacial outlets to the sea. The layers of ice represent many different time intervals of the Late Pleistocene and Holocene. Until the second half of the 20th century, very little was known about the thickness, age, or movements of the GIS.

Early Research on the Greenland Ice Sheet

The Cold War of the 1950 and 1960s brought the US Air Force to northwest Greenland where they built the Thule Air Force Base and a second, secret base called Camp Century that was built beneath the surface of the ice sheet. In 1959, the US Army Corps of Engineers built the subterranean city under the guise of conducting polar research—and scientists there did drill the first ice core ever used to study climate. But deep inside the frozen tunnels, the corps also explored the feasibility of Project Iceworm, a plan to store and launch hundreds of ballistic missiles from inside the ice. The military ultimately rejected the project, and the corps abandoned Camp Century in 1967.

The First Polar Ice Core

Led by US glaciologist Chester Langway, the US Army Cold Regions Research and Engineering Laboratory (CRREL) succeeded in drilling a core through the entire GIS at Camp Century in 1966 (Fig. 10.2). The core was 1390 m long and 12 cm in diameter and was analyzed by Langway and by Danish geochemist Willi Dansgaard, who was interested in extracting the oxygen isotopes ^{18}O and ^{16}O from the ice samples. The authors concluded that long-term variations in the isotopic composition of the ice reflect a sequence of climate changes during the past nearly 100,000 years. Their subsequent publication (Dansgaard et al., 1969) represented groundbreaking research in paleoclimatology that paved the way for all subsequent ice coring projects in Greenland, Antarctica, and elsewhere.

And so it came to be that the need for the US military to have a cover story (i.e., ice research) to hide their true intentions to build a secret nuclear missile base beneath the ice in northwest Greenland ended up launching a whole new line of research that revolutionized paleoclimatology. Ironically, the missile launching project was canceled because the ice sheet sitting over the tunnels was moving so quickly that the tunnels themselves were being inundated by the flowing ice.

There is one final twist to this story. When the Camp Century ice drill finally reached the bottom of the ice, the scientists continued drilling through about four additional meters of mud and rock. For decades, this bottom-most layer of ice and rock from the core was lost in the bottom of a freezer in Denmark. In 2019 it was rediscovered—in some cookie jars (Paul, 2019). Sediment samples from beneath the Greenland ice sheet are extremely rare, so when these lost samples reappeared, they generated a lot of enthusiasm with Earth scientists (Voosen, 2019). Although this research is still in progress, the samples have produced plant

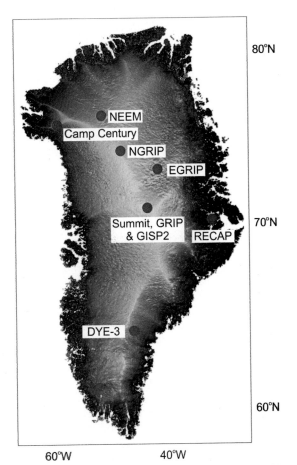

FIG. 10.2 Digital elevation model of Greenland (NASA, 2021), showing locations of ice core drilling sites discussed in the text. (From NASA, 2021. Digital Elevation Map of Greenland. https://icesat4.gsfc.nasa.gov/cryo_data/ant_grn_drainage_systems.php.)

fossils indicating that the camp century region was ice free as recently as 400,000 years ago (Brown, 2020).

This vignette of GIS research serves to demonstrate a variety of aspects of GIS research. It involves large international cooperative research programs, large amounts of money to do the research (in this case, supplied by the US military), and a healthy dose of curiosity-driven science. There is also some serendipity involved. The technology to extract drill cores from more than a kilometer of ice did not exist until the 1950s, and in fact, improvements in ice coring are still being made today. Willi Dansgaard happened to be in Greenland and looking for ancient snow samples when the Camp Century core was being drilled and was invited to take a look at the site.

HISTORY OF THE GREENLAND ICE SHEET

Studying the GIS is very costly, very difficult, and often dangerous to the researcher. Temperatures are substantially colder on the ice sheet than elsewhere in Greenland. The coldest region of the ice sheet has a mean annual temperature of $-31°C$, on the north–central part of the north dome. The warmest temperatures occur at the crest of the south dome, at about $-20°C$. Because of the chilling effects of the ice, temperatures remain well below freezing even at the height of summer. Except for the ability to store ice cores in a natural deep freeze, the very cold temperatures on the GIS make living and working there difficult. The ice sheet is far from being a single flat plateau. In fact, the ice ranges in thickness from 2000 to 3000 m. Working at high elevation sites becomes more difficult because the air is so thin. The GIS is enormous. With an area of $1,710,000 km^2$, the ice sheet is larger than the combined areas of Spain, France, the United Kingdom, Ireland, Belgium, the Netherlands, Denmark, and Sweden.

Because of the frigid temperatures, nearly all field work in Greenland takes place during the short summer season. Field research projects near the center of the GIS are enormously expensive, with daily costs of tens of thousands of dollars. The North Greenland Eemian Ice Drilling (NEEM) project cost $8.8 million. The Second Greenland Ice Sheet Project (GISP2) cost approximately $2 million per year for 5 years.

How Is the Greenland Ice Sheet Studied?

Because of all the difficulties discussed earlier, depending on the research topic, it is not always practical or even possible to conduct fieldwork on the GIS. However, within the past 30–40 years, remote sensing and satellite technology have combined to make possible various kinds of studies of the GIS possible. The following summaries exemplify these techniques and the kinds of results they are achieving.

Satellite surveys

Satellites are being used to measure the changing mass of the ice sheets in a number of ways (Briggs, 2016). First, they measure changes in the height of the surface of the ice sheet through time. This technique is called altimetry, and it allows glaciologists to estimate changes in ice volume. Satellite altimetry readings of the Earth's surface are made with both radar and laser measurements (mainly using CryoSat, Envisat, ERA, and ICESat satellites), which measure changes in the height of the surface over repeat surveys that are interpolated over the surface area of interest to estimate a volume change, which is converted into a mass change. Second, by measuring the change in the gravitational pull of the ice sheet, scientists can get a direct estimate of the change in the mass of the ice with a technique called gravimetry. For instance, the Gravity Recovery and Climate Experiment (GRACE) yielded a series of satellite gravimetry observations on the GIS from 2002 to 2017. Thirdly, by measuring the change in the speed at which the ice is flowing, and combining this with data about the snowfall and melting rates of the ice sheets, scientists are able to produce a budget of the inputs and outputs of the ice sheets. This is called the mass budget, or input-output method.

Mapping Greenland's bedrock

Greenland's bedrock topography exerts a strong influence on ice flow, grounding line migration, glacial calving dynamics, and subglacial drainage. Also, the bathymetry of coastal fjords controls the degree of penetration of warm Atlantic water (AW). This water rapidly melts and undercuts Greenland's tidewater glaciers. Morlighem et al., 2017 presented a compilation of Greenland bed topography that assimilates seafloor bathymetry and ice thickness data through a mass conservation approach. Their map (Fig. 4.4 in Chapter 4) reveals that if the GIS were to melt completely, global sea level would rise by 7.42 ± 0.05 m. Furthermore, it explains the recent calving front response of numerous outlet glaciers and reveals new pathways by which AW can access glaciers with marine-based basins, thereby highlighting sectors of Greenland that are most vulnerable to future oceanic forcing.

Role of fjords in Greenland ice sheet ice loss

Not all fjords allow the entrance of AW. For example, sills in some fjords block AW from reaching glacier calving fronts. AW is typically found at depths below 200–300 m below sea level. Thus, shallow fjords also block AW from reaching glacier fronts. Likewise, some glacier fronts are grounded on the ocean bed, but at depths above the AW level. When AW reaches the terminus of a glacier its calving front is exposed to strong ocean-induced melting. This melting may be enhanced by subglacial discharge, leading to glacier undercutting, enhanced calving, ice front retreat, flow acceleration, and glacier thinning. It is therefore critical to determine the locations that are currently exposed to AW and that may be exposed to AW in the future, that is, how far these glaciers need to retreat before the margin reaches higher ground (<200–300 m depth) or terminate on land (bed > 0 m).

Narrow and deep fjords have important implications for the current and future state of the GIS as they can provide pathways for AW to interact with glacier termini. To investigate the regions that are in contact with the ocean, the authors delineated locations that are continuously below sea level from the continental shelf to the ice sheet bed. As glaciers around Greenland retreat, these regions will remain in contact with the ocean. Scientists also determine the regions that are continuously below a depth of 200 and 300 m, respectively, and are currently connected to the ocean below these depths. Glaciers that retreat within these regions will potentially remain in contact with warm AW as they do so. Since submarine bed channels are widespread and extend far inland, these glaciers will remain vulnerable to ocean warming as they retreat for hundreds of km.

Satellite ice flow measurements

The ice volume estimates produced by many different laboratories around the world have all been slightly different. As a community effort to reconcile satellite measurements of ice sheet mass balance, a group of scientists formed IMBIE: the ice sheet mass balance intercomparison exercise, in 2011. IMBIE is a collaboration between scientists supported by the European Space Agency (ESA) and the National Aeronautics and Space Administration (NASA) and contributes to assessment reports of the Intergovernmental Panel on Climate Change (IPCC).

The first satellite acquiring data with which to make ice flow measurements was launched in the 1970s, followed by altimeters in the 1990s and a gravimeter in the 2000s. Data collected from more than 20 satellites have now been used by research groups around the world with a common aim of trying to develop the most accurate way of measuring the mass change of the ice sheets. Ice flow velocity estimates have been published by Joughin et al. (2018). They developed ice velocity maps produced by the Greenland Ice Mapping Project (GIMP) using Landsat 8 and Copernicus Sentinel 1A/B data. They examined populations of glaciers in northwest and southwest Greenland to produce a record of speedup since 2000. Collectively, these glaciers continue to speed up, but there are regional differences in the timing of periods of peak speedup.

Another form of satellite remote sensing is done with temperature radiometers, such as the European Union's Copernicus Sentinel-3's Sea and Land Surface Temperature Radiometer (SLSTR). These sensors measure the amount of energy radiating from Earth's surface.

Aerial surveys

Airborne radar has been used very extensively across the GIS for several decades. In the early 1980s, researchers started showing interest in using radar to measure ice thickness and map subice rock. This ice and bedrock mapping continues today with NASA's Operation Ice-Bridge airborne mission. The primary objective was to record ice sheet surface elevation changes from NASA's Ice, Cloud and Land Elevation Satellite, or ICESat, which stopped functioning in 2009. IceBridge also gathered data on other aspects of polar ice from snow on top to the bedrock below. Researchers used that funding to build their first radar depth sounder, which started flying aboard NASA aircraft in 1993. Over the years, the Center for Remote Sensing of Ice Sheets at the University of Kansas (CReSIS) has built a number of instruments—each more advanced that the last—leading to the radar IceBridge relies on today.

Low-frequency radar is able to pass through the ice to some degree. The radio waves are transmitted by downward-pointing antennas mounted beneath the aircraft in a rapid series of pulses (NASA, 2013). This multiple array of antennas allows researchers to survey a larger area and record several signals at once to get a clearer picture. Radar pulses travel down to the surface, through the ice to the bedrock below, and back up through the ice to the aircraft, where they are routed to the instrument's receiver and recorded on solid-state drives (Fig. 10.3, top). Each survey flight yields as much as 2 terabytes of data.

One of the biggest obstacles faced when building an ice-penetrating instrument is the nature of radar. Radar works by sending out radio waves and timing how long it takes for them to return. Radio waves travel through air virtually unimpeded, but materials such as rock and water act almost as mirrors. Radar can penetrate ice, but only on certain radio frequencies.

The application of various kinds of remote sensing has made it possible for scientists to study many properties of the GIS in real time without ever having to leave their home institutions. Nevertheless, some research must be done in the field, despite the difficulties and dangers of being on an ice sheet. These days, the GIS is becoming even more dangerous to traverse on foot because of the enormous quantity of meltwater coming off the surface and cutting through the ice toward the base, forming deep crevasses, tunnels, and caverns under the ice (Fig. 10.4). Sadly, fieldwork in Greenland during the summer of 2020 cost the life of one of the leading glaciologists of our time. Konrad Steffen, director of the Swiss Federal Institute for Forest, Snow and Landscape Research, was reported missing on

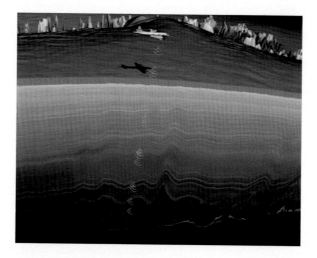

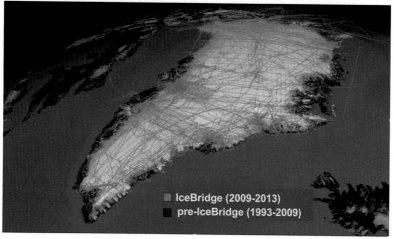

IceBridge (2009-2013)
pre-IceBridge (1993-2009)

FIG. 10.3 Above: NASA-sponsored research aircraft, transmitting radar pulses through the ice. Below: Map showing the radar altimetry tracks across Greenland, mostly as part of Project IceBridge, 2009–13. (From NASA, Starr, C. Radar Altimetry of GIS. https://www.nasa.gov/feature/goddard/land-facing-southwest-greenland-ice-sheet-movement-decreasing.)

August 8, 2020, and is believed to have fallen into a crevasse. The accident occurred a mere 100 m from the camp where Steffen was staying. His colleagues on the expedition reported that a snow bridge over a crevasse had collapsed under him. Steffen, a highly experienced glaciologist, had been doing fieldwork in Greenland since 1990.

Ice Coring

As described by Brook (2013a), ice coring began in the 1950 and 1960s through the pioneering efforts of scientists from several nations and is now conducted by a variety of national and international research teams, in the Arctic, Antarctic, and glaciated mountain regions of the world.

The snow that accumulates on the polar plateaus of Antarctica and Greenland is sampled by drilling vertical boreholes through the ice. Collecting these cores is a specialized engineering challenge. The drilling equipment used for ice coring was developed in the late 1950s to early 1960s and has continuously improved since that time. Ice coring locations are chosen to extract the most reliable paleoclimate information. The best drilling sites are generally on or near ice divides, where ice deposited at or near the same elevation as the current surface elevation can be obtained, and the

NASA, 2015; photo by Andrew Sole, University of Sheffield

FIG. 10.4 Meltwater stream eroding surficial snow and ice on the GIS. Courtesy of NASA, 2015. Photo by Andrew Sole, University of Sheffield. In public domain. (Data from NASA, 2015. https://svs.gsfc.nasa.gov/4249.)

distortion of the age versus depth curve due to thinning can be reliably modeled (Fig. 10.1). Ice cores of various lengths are collected for different purposes.

Deep ice cores require larger equipment and fluid-filled boreholes to avoid the collapse of the hole. Drilling fluids are typically petroleum-derived fluids such as refined kerosene, but other organic liquids have also been used. The physical characteristics of the fluid, particularly its freezing point and viscosity are critical parameters and can make choosing appropriate fluids a challenge. The deepest cores (>3,000m) in Greenland and Antarctica are collected during multiyear campaigns from scientific camps established near the drill rig. Perhaps the most unique aspect of ice cores is that they trap the ancient atmosphere when surface snowfall compacts to ice. This happens at the base of the firn, or unconsolidated snow layer, typically 50–100m below the surface in polar regions, and shallower at high accumulation sites with warmer temperatures.

Dating ice layers

Annual ice layers are not discernible in all ice cores, and the layers thin with age, making them difficult to detect in older sections of deep ice cores. In many cases, other techniques are needed to establish the ages of deep ice cores. The radiometric dating of ice had not been very successful until quite recently. Air bubbles trapped in ancient ice are now being dated by the Ar isotope method. This chronometric method makes use of the fact that ^{36}Ar and ^{38}Ar have essentially been constant throughout recent geologic time, while ^{40}Ar has been slowly increasing in the atmosphere because of the radioactive decay of ^{40}K. This allows the measurement of the isotope ratio $\delta^{40}Ar/^{38}Ar$ to be measured in ice cores to as old as 800ka. Over this interval, this ratio has risen at a rate of 0.066 ± 0.007 per million years. Yau et al. (2016a) dated trapped air bubbles from basal ice samples by this method.

Another common strategy for dating ice samples is to apply equations concerning the physics of ice flow and thinning to estimate the depth-age relationship for an ice core. Such estimates can be checked or constrained by assuming the ages of key climate transitions (e.g., glacial-interglacial transitions) recorded in ice core isotope (or other) records can be correlated to chronologies of ocean sediment records. The deepest Antarctic records have been dated in this way.

What is in the Ice?

Polar ice sheets contain more than frozen water. As shown in Table 10.1, teams of researchers analyze a wide range of materials trapped in the ice. The process begins with a visual inspection of the newly recovered core segment (visual stratigraphy) to begin the lengthy process of counting annual layers. For portions of an ice sheet that fall outside the zone of perpetual freezing conditions, it is also necessary to identify melt layers from summer seasons.

Scanning methods

In addition to visual inspection, X-ray and CT scans of ice core segments produce digitized radiographic images that provide enhanced layer counts, indicate levels of ice porosity, and identify inclusions such as dust and tephra. A recent addition to the ice core scanning toolkit is laser ultrasound, used to examine the mechanical property variations in ice cores (Mikesell et al., 2017).

Chemical stratigraphy

Chemical stratigraphy (glaciochemistry) is used to determine the chemical composition of ice layers from major ion species, including Na^+, Mg^{2+}, Ca^{2+}, K^+, NH^{4+}, Cl^-, NO^{3-}, and SO_4^{2-}. These ions represent about 95% of the soluble composition of the atmosphere, and their traces in the ice yield vital clues about the sources of continental dust, sea salt content, and biogenic processes. A closely associated line of research concerns the electrical conductivity of ice samples. In this procedure, an electric current is passed through the ice sample. Pure ice is a very poor conductor of electricity, while ice containing salts, various other chemical species, and other impurities is a far better conductor. Conductivity is used to measure these chemical variations in the ice. It provides measurements of ice pH

and is used to guide researchers to sections of a core with the potential to yield high-resolution chemical analyses.

Dust analysis

Even though Greenland is thousands of km away for many sources of glacial dust, the GIS has served as a repository for layer upon layer of these fine mineral particulates. Dust particles trapped in ice reveal intervals of increased aridity and windiness; chemical analyses have detected dust sources as far away as East Asia and North Africa. Recent dust fallout on the GIS is seasonally dominated by particles originating in the Taklamakan Desert, Xinjiang Province, Northwest China (Bory et al., 2002).

Tephrochronology

Tephra is volcanic ash. Many layers of Icelandic tephra are trapped in GIS ice layers due to Greenland's proximity to Iceland and its dozens of volcanoes that have been active in the Late Pleistocene. Some tephra layers are sufficiently thick to be seen by the naked eye, while others are cryptic—only visible as small groups of chards under the microscope (Davies, 2015). In either case, Icelandic tephras can often be traced to their source volcanoes and even to specific eruption events. Tephra has therefore been extremely useful to correlate synchronous layers between ice cores and chronostratigraphic tie points linking ice core records with the same tephra layers in marine and terrestrial deposits.

The best example of this phenomenon for the GIS comes from the discovery of the Vedde Ash, now considered the most important volcanic event marker layer for the correlation of Late Quaternary environmental archives in Europe and the North Atlantic (Lane et al., 2012). First defined from its type site localities near Ålesund, Western Norway, the Vedde Ash has now been traced across much of northern and central Europe, into northwest Russia, within North Atlantic marine sediments, and into the Greenland ice cores. The Vedde Ash is thought to derive from an eruption of the Katla volcano in Iceland (Fig. 10.5) that occurred 12.1 ka. Visible and cryptotephra deposits of the Vedde Ash have been found in numerous stratified sites with robust chronologies, allowing tephrochronologists to constrain the age of the eruption and to map its dispersal across the North Atlantic region.

Sea salt studies

Sea salt becomes incorporated in parts of the GIS closest to the coast at the time of deposition, but traces of sea salts occur throughout the ice sheet, transported by

TABLE 10.1
Ice Core Analytical Methods.

Research Group	Specific Proxy	Nature of Data	Paleoenvironmental application(s)	References
Ice stratigraphy	Visual stratigraphy	Annual layer counts	Establish chronology of visible layers; layer thickness based on snowfall history; past precipitation rates	Schwander (2013), Dunbar et al. (2017), Guillou et al. (2019)
		Melt layer observations	Track warm intervals with surficial melting	Brook (2013a), Fujita et al. (2020)
	X-ray and CT scans	Digitized radiographic images	Enhanced layer counts, ice porosity, inclusions (dust, tephra)	Brook (2013a), Freitag et al. (2013).
Ice chemistry	Chemical stratigraphy (glaciochemistry)	Chemical composition of ice layers from major ion species	Traces the input of chemical species from continental dust, sea salt, and biogenic processes	Kreutz and Koffman (2013), Simonsen et al. (2019)
	Electrical conductivity	Electric current passed through ice samples	Measures ice pH; high-resolution ice chemistry	Taylor (2013), Dreschhoff et al. (2020)
Particulates	Dust	Dust particles trapped in ice	Dusty intervals show increased aridity and windiness; chemical analysis may detect dust sources	McConnell (2013), Simonsen et al. (2019)
	Tephra	Volcanic ash trapped in ice	Especially in Greenland because of proximity to Icelandic volcanoes. Tephra used to correlate between ice cores, marine, and terrestrial deposits	Davies (2015), Guillou et al. (2019)
	Sea salt	Salt crystals trapped in ice	Salts blown off sea ice surfaces reflect sea ice coverage and proximity to core site	Kreutz and Koffman (2013), Yau et al. (2016)
Stable isotopes	18O/16O and D/H from trapped air bubbles in ice	Isotopic ratios	δ18O and δD of snow directly related to surface temperature of ice sheet	Landais et al. (2018), Holme et al. (2019)
Atmospheric CO_2 concentrations	CO_2 content of trapped air bubbles	Carbon dioxide concentration in sample bubbles	Reveals CO_2 content of global atmosphere at time of air bubble entrapment (~10 m below the surface of the ice)	Stauffer (2013), Petit and Raynaud (2020)
Atmospheric methane concentrations	CH_4 content of trapped air bubbles	Methane concentration in sample bubbles	Reveals methane content of global atmosphere at time of air bubble entrapment (~10 m below the surface of the ice)	Chappellaz (2013), Lee et al. (2020)
Aerosols and microparticles	Chemical species, isotopes from soluble and insoluble microparticles	Chemical composition of particles	Provenance of airborne dust, volcanic tephras, aerosols	McConnell (2013), Seki et al. (2015)
Borehole temperatures	Temperatures measured down the length of borehole	Thermistor readings at bottom of core being drilled	Long-term averages of past surface temperatures	Cuffey (2013), MacGregor et al. (2016)
Biological materials	Ancient DNA, pollen, spores, diatom frustules, plant fibers	Evidence of local (interglacial) and long-distance (windblown) organisms	Ancient DNA from Greenland reveals a coniferous forest at DYE-3 during an interglacial	Priscu et al. (2013), Willerslev et al. (2007)

FIG. 10.5 Katla volcano, Iceland: the source of the Vedde Ash. Image courtesy of Earth Chronicles News. (From Katla Volcano, Iceland, 2020. https://earth-chronicles.com/natural-catastrophe/iceland-woke-up-katla-volcano.html.)

winds. As discussed by Kreutz and Koffman (2013), sea salt aerosol reaching coastal and inland plateau sites was initially thought to come from bubbles bursting over open ocean water. The role of sea ice in this process was not clear, because ions associated with sea salt typically peak in GIS deposits during winter and are also elevated during glacial periods. It would thus appear that increased sea ice extent during winter and glacial periods should equate with a decrease in the amount of sea salt aerosol in GIS ice.

One explanation for this seeming paradox is that increased strength of ocean storms and enhanced transport of sea salt aerosols inland during winter and glacial periods more than offset the greater distance the aerosols must travel. Another possibility is that newly formed sea ice is the source of the sea salt aerosols. As the sea ice forms, it excludes salt molecules, creating a highly saline brine that becomes a very effective source of sea salt aerosols in winter.

Oxygen and hydrogen isotopes

As reviewed by Davies (2020), the most important proxies for paleoclimate reconstruction based on ice sheets are the ratios of stable isotopes from water: the $^{18}O/^{16}O$ ratio and the $^{2}H/^{1}H$ (deuterium/hydrogen) ratio. Snow precipitation over Greenland (and elsewhere) is dominated by $H_2^{16}O$ molecules (99.7%). But rare stable isotopes are present in minute quantities, such as $H_2^{18}O$ (0.2%) and $D^{16}O$ (0.03%). Past precipitation can be used to reconstruct past temperatures. δD and $\delta^{18}O$ are related to surface temperature in the middle and high latitudes. The relationship is consistent and linear over Greenland.

As Greenland snowpack is slowly converted to ice, tiny pockets between compressed snow crystals trap air bubbles that contain these stable isotopes. The air bubbles are extracted from ice core samples by melting, crushing, or grating the ice in a vacuum. The stable isotopes are measured with a mass spectrometer. Measuring changing ratios of δD and $\delta^{18}O$ through time in layers through an ice core provides a detailed record of temperature change, going back thousands of years. These ratios are always expressed in parts per thousand (‰).

More than 99% of all water molecules on Earth consist of two ^1H and one ^{16}O atom. The much rarer forms contain either the heavier ^{18}O isotope, denoted $H_2^{18}O$, or with one ordinary hydrogen atom replaced by a deuterium atom, denoted $HD^{16}O$, with the "D" standing for deuterium. Both the $H_2^{18}O$ and $HD^{16}O$ molecules are heavier than the standard $H_2^{16}O$ molecule, and because of their increased weight, the heavier water molecules evaporate more slowly from the surface of a water source (such as the North Atlantic) and drop out of the sky as precipitation more readily than the lighter $H_2^{16}O$ molecules. Thus, ocean waters tend to become enriched in $H_2^{18}O$ and $HD^{16}O$, while the snow that falls on Greenland is depleted in these heavier isotope water molecules. This difference between the $\delta^{18}O$ and δD values in North Atlantic seawater and polar glacial ice becomes greater during glacial intervals, as the snow that falls on the ice sheet remains there for many thousands of years, piling up increasing amounts of isotopically "light" water. The concentration of $H_2^{18}O$ in natural waters is only about 2‰. The amount of $H_2^{18}O$ in Arctic precipitation is depleted by about 40‰. The differences between the $H_2^{18}O$ content in precipitation are therefore very small, and a mass spectrometer is used to measure the relative proportions of $H_2^{18}O$ and $H_2^{16}O$ in ice air bubble samples.

Deuterium excess. As summarized in the document by the University of Copenhagen Center for Ice and Climate (UCCIC) (2020a), $\delta^{18}O$ and δD values indicate past temperatures near the ice-coring site. But these isotopes also reflect the climatic conditions in the areas from which their moisture sources originated. The δD values from ice core samples are about eight times the $\delta^{18}O$ values because the mass difference between deuterium and normal hydrogen is eight times larger than the mass difference between ^{18}O and ^{16}O. However, the standardized difference between the two isotopic records, reflected by the equation:

$\delta D - (8 \times \delta^{18}O)$ (called the deuterium excess) is variable and also carries information about the climate conditions of the moisture sources. The small differences between the two records shed light on moisture source temperatures—in this case, sea surface temperatures (SSTs).

We know from various terrestrial and marine proxy data sources that the end of the last glaciation, about 11.7 ka, saw abrupt, large-scale climate changes. All the parameters measured from the GIS ice core samples changed from glacial to interglacial values within a century, but how rapidly did the switch between glacial and interglacial climate modes take place? The deuterium excess measurements clearly indicate that this wholesale change in Greenland and adjacent North Atlantic climate regime was accomplished within 1—3 years (UCCIC, 2020b).

The deuterium excess record indicates that the temperature of the moisture source areas dropped at a time when the glacial was ending and the world was warming. This only makes sense if the source areas for Greenland precipitation changed location very abruptly at the end of the glaciation. This shift is so well marked in GIS ice cores that it has been used to define the formal ending of the last glaciation and the beginning of the Holocene. In the NGRIP core, the commencement of the Holocene is now defined from ice core measurements at a depth corresponding to 11,703 years before AD 2000.

Carbon dioxide and methane in ice cores

As reviewed by Petit and Raynaud (2020), in 1980, a method was found to determine the carbon dioxide content of ancient air trapped in polar ice, providing direct evidence that CO_2 is coupled to climate and affects global temperatures. Another milestone was reached in 1987 when a dry analyses technique for fossil gas extraction yielded a history of CO_2 changes throughout an ice core more than 2 km deep from the Vostok site in Antarctica. This provided a 160,000-year record of temperature and CO_2, demonstrating their relationship over a complete glacial cycle. It was found that CO_2 levels ranged from 290 ppm during the warm period to 190 ppm in the cold period. Subsequent research from Antarctica has demonstrated the climate-CO_2 correlation over the past 800,000 years.

Ice cores from Greenland preserve the history of atmospheric methane and have been used to infer past changes in methane sources. Natural sources of methane are dominated by microbially produced CH_4 emitted from wetlands in processes largely controlled by temperature and precipitation. The ice core methane record, therefore, provides important indications of the history of terrestrial hydroclimate (Lee et al., 2020).

The atmospheric concentration of methane is essentially uniform throughout the world. In fact, glaciologists wanting to make detailed comparisons of ice core data from different sites use methane levels to synchronize the records, putting the records on a common timescale. This is particularly important when attempting to synchronize ice core data from Greenland and Antarctica because there are very few reference horizons of, e.g., volcanic origin that can be found in both hemispheres and used for synchronization. But because the global atmosphere is relatively well mixed, the air bubbles trapped in ice cores from Antarctica and Greenland have roughly the same concentration of methane at any given time. Methane concentrations have changed abruptly and repeatedly in the past. The records of these changes can therefore be used to synchronize the ice cores (UCCIC, 2020c).

Although the GIS record terminates at or before the last (Eemian) interglacial period, Antarctic ice core CH_4 records extend back 2 million years (Yan et al., 2019). Different analytical techniques have been applied for extracting the gas from the ice and quantifying the CH_4 content. The typical reproducibility obtained on replicate measurements of neighboring samples is 10 parts per billion by volume (ppbv), which represents 1%—3% of the measured mixing ratio.

Changes in atmospheric CH_4 match Greenland climate fluctuations throughout the last glaciation. Warm Greenland interstadials are often referred to as Dansgaard-Oeschger events. Each of the Dansgaard-Oeschger events recorded in Greenland ice has a counterpart in atmospheric methane levels, although with a variable relationship in terms of amplitude. The CH_4 variations cover a range from 50 to 200 ppbv. Each Dansgaard-Oeschger event has been investigated for both CH_4 and temperature changes interpreted from the $\delta^{18}O$ and D/H records. The peaks and valleys of both signals are in phase with each other (Chappellaz, 2013)

AGE OF THE GREENLAND ICE SHEET

To get a better sense of the relative stability of the GIS in the face of climate change, the only practical method is to study the history of the ice sheet back through times of warmer-than-modern and colder-than-modern temperatures. In conjunction with this, it is vital to know how long the current ice sheet has been in existence. This aspect of the GIS is very different from the ice sheets covering Antarctica. The glaciation of Antarctica began about 45.5 million years ago and increased further about 34 million years ago. A sharp decrease in atmospheric CO_2 levels apparently triggered

Antarctic glaciation. Despite many changes in global temperature regimes, the Antarctic continent has remained glaciated to the present day.

We now know from multiple ice coring sites in Greenland that the current ice sheet is at least as old as the beginning of the last glaciation, roughly 115,000 years ago. The evidence from specific ice cores sheds additional light on this topic.

Evidence from the NEEM Ice Core

A 2540-m-long ice core was drilled at the North Greenland Eemian Ice Drilling (NEEM) site (77.45°N, 51.06°W) (Fig. 10.2). The top 1419 m of ice derives from the Holocene (i.e., the past 11,000 yr), and together with the glacial ice below, it can be matched to GIS ice core timescales down to 108,000 years before present (108 kyr BP). Below this, the ice is disturbed and folded. All indications are that this deeper ice predates the Last Interglacial (LIG), the Eemian (130–115 kyr BP). This pre-Eemian ice lies just above bedrock, and its low $\delta^{18}O$ ice values suggest that the ice layers are most probably from the glacial period before the Eemian (NEEM Community Members, 2017).

The NEEM ice core apparently recovered the entire Eemian interglacial period intact. The reconstructed Eemian sequence (128.5–114 kyr BP) shows a regular occurrence of melt features from 127 to 118.3 kyr BP, an indication of warmer temperatures at the study site than those of the last millennium. The isotope anomaly at 126 kyr BP implies that surface temperatures were

about 7.5°C warmer than today. This warming apparently caused substantial melting the GIS during the LIG. From 128 to 122 kyr BP, the surface elevation is estimated to have decreased by an average of 764 cm per year and remained at this level until 117–114 kyr BP, long after the commencement of the subsequent glacial interval.

Evidence from the GRIP and Dye-3 Ice Cores

As discussed earlier, Yau et al. (2016a) examined isotopic records from the GRIP and Dye-3 ice core samples (Fig. 10.2) to determine whether the GIS survived previous interglacials known to be warmer (~130 ka) or longer (~430 ka) than the present interglacial (Fig. 10.6). They presented geochemical analyses of the basal ice from the Dye-3 and GRIP cores that help characterize and date the ice. They dated trapped air bubbles from basal ice samples by measuring the $^{40}Ar/^{38}Ar$ ratios and the ^{17}O anomaly of O_2 in the trapped air. The resulting dates must be considered minimum ages because of possible sample contamination from bedrock radiogenic ^{40}Ar. The base of the GRIP ice core was found to be less than or equal to 970 ± 140 ka. The base of the Dye-3 ice core was found to be less than or equal to 400 ± 170 ka. Taken together, these basal ages confirm that the GIS did not completely melt in Southern Greenland during the LIG, nor did it completely melt at the Summit Ice Core site (Fig. 10.2) during the unusually long interglacial ~430 kyr before present.

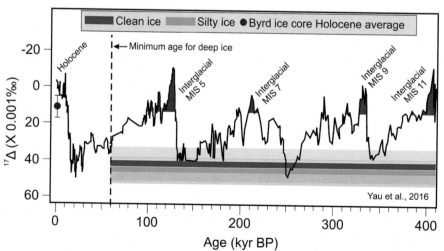

FIG. 10.6 Record of the ^{17}O anomaly of O_2 in trapped air bubbles from the DYE-3 core, showing potential correlations with interglacial warming peaks back through MIS 11. After Yau et al., 2016a. (Modified from Yau, A., Bendera, M., Blunier, T., Jouzel, J. 2016b. Setting a chronology for the basal ice at Dye-3 and GRIP: Implications for the long-term stability of the Greenland ice sheet. Earth and Planetary Science Letters 451, 1–9. doi:10.1016/j.epsl.2016.06.0530012-821X.)

About 5 million years ago (Ma), Northern Hemisphere ice sheets began to form as a result of a general climatic cooling. The GIS is thought to have initiated in the mountains along the east coast of Greenland, gradually advancing inland to the central plains (Hvidberg et al., 2013).

A recent estimate of the basal age of the GIS is 2.4 million years ago (NSIDC, 2007).

The Greenland Ice SheetTipping Point

A great deal of discussion in recent years has been devoted to the topic of "tipping points" in the climate system, particularly concerning the potential melting of the GIS. The concept of a tipping point focuses on a rapid shift between one relatively steady state of a system and another. For instance, when paddling a canoe, the upright position of the boat represents the first steady state and a capsized position represents a second steady state. If the occupants of the canoe lean too far out over the water, the boat reaches a tipping point beyond which capsizing is inevitable. This is a form of nonlinear behavior and as such is very difficult to incorporate in climate models which are generally designed to reflect more gradual changes. Climate scientists are attempting to determine whether there is a tipping point in the melting process of the GIS, beyond which a complete loss of the ice is inevitable. Just as with the other research questions that can only be addressed through the lens of deep history, this tipping point question is also best answered with geologic evidence.

Thomas et al. (2020) focused on the LIG (129−116 ka) to examine the question of tipping points for the GIS. This interglacial provides a useful analog for future change, as it was the warmest interglacial of the past 800,000 years and was also the most recent period during which global temperatures were comparable with low-end 21st-century projections (up to 2°C warmer, with temperature increase amplified over polar latitudes). Substantial environmental changes occurred during this time. The authors synthesized the nature and timing of potential high-latitude tipping elements during the LIG, including sea ice, the extent of the boreal forest, permafrost, ocean circulation, and ice sheets/sea level (Fig. 10.7). They also reviewed the

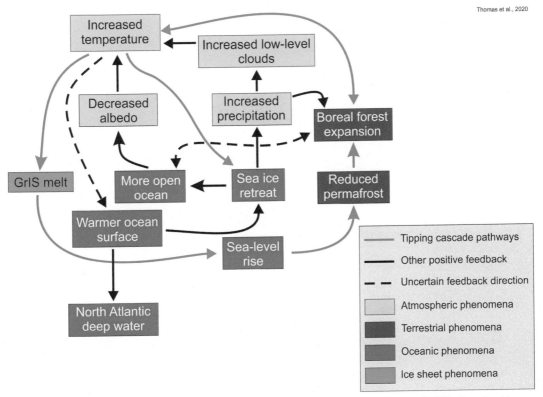

FIG. 10.7 GIS climate feedback flowchart, based on concepts by Thomas et al. (2020). GIS, Greenland ice sheet.

thresholds and feedbacks that likely operated at that time, including substantial ice mass loss from Greenland and Antarctica. Based on evidence from the sites throughout northern high latitudes, the LIG saw reduced summer sea-ice extent, poleward extension of boreal forests, and reduced areas of permafrost. Despite dating uncertainties, they report that tipping elements in the high latitudes all experienced rapid, abrupt onset (within 1—2 millennia of each other), followed by more gradual recovery to prior conditions (multimillennia). The authors identified important feedback loops between tipping elements, amplifying polar and global change during the LIG (Fig. 10.8). High levels of sensitivity and interconnections between polar tipping elements suggest the GIS is likely to reach tipping thresholds in the future.

Another piece of the LIG puzzle may be dropping into place. As reported by Guarino et al. (2020), stronger summertime insolation at high northern latitudes during the Eemian interglacial caused Arctic land summer temperatures to rise 4—5 °C higher than in the pre-industrial era. Previous climate model simulations failed to capture these elevated temperatures, possibly because they were unable to correctly capture LIG sea-ice changes. However, the latest version of the UK

Hadley Center climate model (HadGEM3) simulates a more accurate Arctic LIG climate, including elevated temperatures. Improved model physics result in a reconstruction of the complete loss of summer Arctic sea ice during the LIG. This ice-free Arctic Ocean provides a compelling solution to the long-standing puzzle of what drove LIG Arctic warmth. The same mechanism supports a fast retreat of future Arctic summer sea ice as part of amplified Arctic warming.

Another important point made by Thomas et al. (2020) is that GIS tipping points do not act in isolation. Rather, they are part of a global system of feedback loops. Ocean-atmosphere interactions mean that the tipping of one system into a different state could trigger the collapse of an interconnected system. This has been termed a "tipping cascade." As discussed in Chapter 13 and elsewhere in this book, Arctic amplification works through feedbacks between sea ice, ocean circulation, permafrost, and ice sheets to warm the Arctic more rapidly and to a greater extent than in the lower latitudes. In other words, changes in the net radiation balance (such as might be caused by an increase in greenhouse gas concentration) tend to produce a larger change in climatic conditions near the poles than the global average.

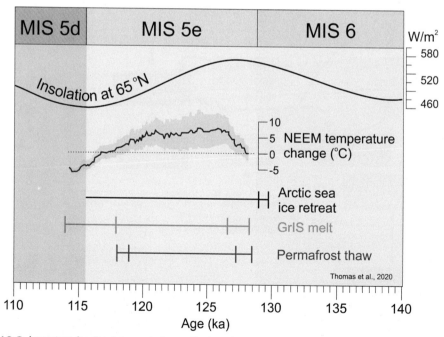

FIG. 10.8 Important feedback loops between tipping elements, amplifying polar and global change during the LIG. (Modified from Thomas, Z., Jones, R., Turney, C., Golledge, N., Fogwill, C., 2020. Tipping elements and amplified polar warming during the last interglacial. Quaternary Science Reviews 233, 106222. https://doi.org/10.1016/j.quascirev.2020.106222.)

Many elements within the Earth's system are increasingly recognized to display this nonlinear behavior, and the term "tipping point" is equivalent to terms such as positive feedbacks, hysteresis, irreversible transitions, multiple equilibria, bifurcations, and abrupt change. As might be expected from a system that shifts abruptly from one steady state to another, the process is likely not directly reversible, a behavior known as hysteresis. In such cases, the magnitude of force required to move in one direction is different from the force needed to move in the opposite direction. Alluding back to the metaphor, getting the canoe right-side-up again takes a great deal more effort than capsizing it.

Steffen et al. (2018) suggest that future destabilizing temperature thresholds for Earth's climate system that constitute tipping elements range from 1 to 5°C warming, but these thresholds are based on global averages and do not appear to take into account polar amplification. These tipping elements responded more-or-less simultaneously at the start of the LIG, but because of the difficulty of accurately dating events this far back in the past, it is not yet possible to pinpoint the timing of any leads and lags within intervals of 1,00−2000 yr.

The Greenland Ice Sheet Layer Cake Revealed

MacGregor et al., 2015 discuss the results of ice-penetrating radar surveys of the GIS, from which they have observed numerous widespread internal reflections of layers in the ice. The oldest reflections, dating to the Eemian period (MIS 5e), are found mostly in the northern part of the ice sheet. In outlet glacier zones, the radar reflections usually do not conform to the bed topography. Dated reflections are used to generate a gridded age volume for most of the ice sheet and also to determine the depths of key climate transitions that have not been directly observed. MacGregor et al., 2015 report that their radio-stratigraphy research provides a new constraint on the dynamics and history of the GIS.

MacGregor et al., 2015 examined the 479,595 km of 150 and 195 MHz ice-penetrating radar data collected over the GIS using several airborne platforms between 1993 and 2013. These data include 512 transects of varying lengths (Fig. 10.9). Ice-penetrating radar only works at low frequencies, between 120 and 240 MHz. Higher-frequency radar signals do not identify the ice layers.

The dated reflections of ice layers charted by MacGregor et al., 2015 yielded some surprising results. As shown, the portions of the ice sheet in North Central Greenland are at least 90,000 years old (areas marked in red in the figure). The ice sheet regions of northwest Greenland date from the last glaciation, between 74 and 26 ka. But nearly all of southern Greenland and the northwest coastal regions are covered in ice that is less than 10,000 years old—Holocene ice. As shown in Fig. 10.9, ice dating from the last glaciation reaches its maximum thickness near the center of Greenland and then tapers almost to nothing in the south. In contrast to this, Holocene age ice is thickest just south of the center on Greenland and remains thick to the

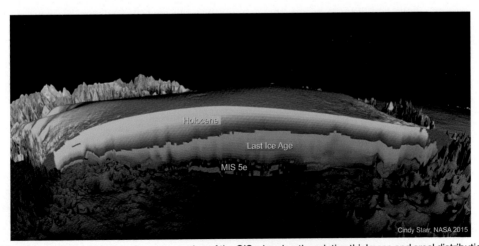

FIG. 10.9 Artist's rendering of a cross section of the GIS, showing the relative thickness and areal distribution of ice deposited during the Holocene, the last glaciation, and the LIG period (MIS 5e). Illustration by Cindy Starr, NASA, 2015. In the public domain. GIS, Greenland ice sheet. (From Starr, C., NASA, 2013. https://svs.gsfc.nasa.gov/4249.)

southern edge of the ice sheet. Remnants of ice predating the last glaciation remain in small pockets at the bottom on the ice sheet toward the center of Greenland.

HISTORY OF ARCTIC/SUBARCTIC CLIMATE FROM GREENLAND ICE SHEET EVIDENCE

Before ice accumulated on Greenland during the LIG period, there is geologic evidence for climate change and ice movement. The GIS is drained by ice streams that have advanced repeatedly to the edge of the continental shelf over millions of years, depositing glacially eroded sediments onto the continental margins. The geological component of these glacial outlets, known as trough-mouth fans, are characterized by km-thick sediment accumulations in front of shelf crossing troughs that mark the main ice-stream drainage route. The modern distribution of marine-terminating outlet glaciers on glaciated margins is dwarfed by the sizes attained by ice streams during glacial maxima, most recently 22−18 ka. Knutz et al. (2019) used a grid of seismic reflection data and borehole stratigraphic information to analyze the anatomy and spatial evolution of two paleo-ice streams that drained into Baffin Bay on the northwest margin of Greenland. They documented 11 major phases of shelf-edge ice advance and subsequent transgression since the first ice sheet expansion 3.3−2.6 million years ago. Each major advance and retreat presumably represents one complete glacial cycle. The glacial outlet system appears to have developed in four stages, each driven by tectonic and climatic changes. The authors infer that an abrupt change in ice flow conditions occurred during the mid-Pleistocene transition, about 1 million years ago. This large-scale change in Quaternary climate systems saw glacial-interglacial cycles switch from a more-or-less regular 41,000-year cycle to a more erratic 100,000-year cycle. Many mechanisms have been proposed to explain this transition, including declining atmospheric CO_2 levels and regolith removal in the past million years (Willeit et al., 2019). Whatever the cause of the mid-Pleistocene transition, Knutz et al. (2019) found that at that time, ice movement across the shelf margin in Greenland changed from widespread to a more focused flow (ice streams), establishing the basis for present-day glacial troughs.

Compared with the Antarctic ice sheet record, the Greenland record is relatively short. As discussed earlier, reliably dateable ice layers can only be traced back to the LIG period. Nonetheless, the Greenland records offer detailed evidence of climate change throughout the last glacial cycle and into the Holocene. In the following review, we will start with the LIG and work our way toward the top of the record.

Summary of Ice Core Records

Ice cores covering depths of 100−200 m have been taken at several locations on the GIS. These cores document ice deposition ranging from a few centuries to thousands of years, depending on core depth and local accumulation rates. Because multiple cores are available for comparison, it is possible to stack the records and thereby reduce the glaciological noise in the $\delta^{18}O$ data. Comparisons of results from cores taken at various locations make it possible to filter out local climatic variations affecting only parts of the ice sheet. The investigation of $\delta^{18}O$ records from the upper part of ice cores is complicated by water vapor diffusion in the pores of snow and firn in the top 60−70m of the ice sheet. This diffusion tends to smooth the high-frequency oscillations of $\delta^{18}O$ (Vinther and Johnsen, 2013).

The close relationship between $\delta^{18}O$ and site temperature makes $\delta^{18}O$ a valuable proxy for both the Greenland climate and Northern Hemisphere atmospheric circulation patterns. To exploit the potential of Greenland $\delta^{18}O$ records fully, it is essential to consider seasonal oscillations in the $\delta^{18}O$ records. In the vast areas of the GIS where annual accumulation exceeds 20 cm of ice equivalent, seasonal oscillations of $\delta^{18}O$ can be retrieved through most of the Holocene. In areas of lower snow accumulation, the seasonal $\delta^{18}O$ cycles are obliterated by the diffusion in the upper snow and firn layers.

During the past 50 years, a total of seven deep ice cores have been drilled all the way through the GIS. All of these records contain ice from the entire Holocene, the last glaciation, and the Eemian. Measurements of $\delta^{18}O$ have been carried out on the full length of all seven ice core records. The seven $\delta^{18}O$ records are very similar to each other over the past 40 ka. Below 40 ka, the Dye-3 record is disturbed by ice flow, as is the Renland core below 60 ka.

The Last Interglacial

The reconstruction of Greenland climate based on ice core evidence begins with the LIG interval. In Europe, they call this the Eemian interglacial, after a site in the Netherlands. On a more global basis, this interglacial is called Marine Isotope Stage (MIS) 5e.

A review of the GIS ice core $\delta^{18}O$ record is provided by Vinther and Johnsen (2013). By using the $\delta^{18}O$ values observed in Eemian ice, it is possible to further quantify the changes in the size of the GIS during the Eemian. Comparing present-day $\delta^{18}O$ values to those observed in the Eemian ice, it is clear that the Eemian had 2.5–3.5‰ higher $\delta^{18}O$ values than presently observed all across Greenland (Table 10.2) except for at the southern Dye-3 site. At Dye-3, the Eemian $\delta^{18}O$ was 4.7‰ higher than recent values. The elevated Eemian $\delta^{18}O$ values at the Dye-3 site indicate that the southern part of the GIS was lowered by some 400 m relative to the rest of the ice sheet (1.5‰ extra change at Dye-3 added to a 1‰ change due to ice flow—induced extra lowering of the Eemian $\delta^{18}O$). Given this information, it is obvious that most (if not all) of the 2.5–3.5‰ Eemian increase in $\delta^{18}O$ should be attributed to temperature change, not to elevation change. Hence, it is probably only the southern part of the Greenland ice sheet that experienced significant decreases in thickness during the Eemian. In fact, it could be argued that a slight increase in the thickness of the northern part of the GIS is supported by the $\delta^{18}O$ data, as NGRIP and Camp Century Eemian $\delta^{18}O$ seem to have increased a bit less than the Renland $\delta^{18}O$. The contribution to the global Eemian sea level rise from a thinning of the GIS is therefore believed to be modest, most likely less than 1m. The results from the LIG NEEM record are anomalous, as discussed in the following.

Estimates of changes in the thickness and surface temperature of the GIS during the LIG period were derived from the NEEM ice core records, northwest Greenland. The NEEM ice core record extends back to 128 ka, just prior to the LIG. The initial LIG temperature estimate was made using the average Holocene $\delta^{18}O$ ice-temperature relationship established from other central Greenland ice cores. $\delta^{18}O$ for ice from 126 ka was estimated to be 3.6‰ above the local preindustrial level, which translated into local surface air temperature warming of 7.5 ± 1.8°C. After accounting for upstream effects and for Greenland ice sheet elevation change based on the air content, this led to an estimate of 8 ± 4°C warming at the NEEM deposition site at 126 ka (NEEM community members, 2013).

The "NEEM paradox"

This remarkable warming event has been called "the NEEM paradox," because this level of warming should have brought about a great deal more melting of the ice sheet than has been reconstructed for Greenland as a whole. In fact, climate modelers predict that an 8°C warming of Greenland climate should be sufficient to melt the GIS completely. Does the GIS have more melting resilience than that predicted by most ice sheet models (ISMs)? Landais et al. (2016) attempted to address this question through an independent assessment of LIG Greenland surface warming using ice core air isotopic composition and relationships between accumulation rate and temperature. This more recent assessment matched the 5.2 ± 2.3°C warmer-than-modern temperature reconstruction obtained from the isotopic signatures of Eemian ice samples at the NGRIP site (Fig. 10.2) by the same method. Climate simulations performed with present-day ice sheet topography lead in general to a warming event smaller than reconstructed, but sensitivity tests show that larger amplitudes (up to 5°C) are produced in response to prescribed changes in sea ice extent and ice sheet topography.

Atmospheric simulations can explain up to 5°C annual mean warming with respect to the preindustrial period when accounting for a reduced Greenland ice

TABLE 10.2
Average Values and Differences of $\delta^{18}O$ During Three Late Quaternary Intervals (All Values Are Given in ‰).

Drill site	Elevation (m)	Latitude (°N)	Modern (M) $\delta^{18}O$	Stadial (S) $\delta^{18}O$	Eemian (E) $\delta^{18}O$	M-S	E-P
Camp Century[a]	2000	77.17	−29.5	−41.9	−27	−12.4	2.5
Dye-3[a]	2480	65.11	−27.6	−36.1	−22.9	−8.5	4.7
GISP2[a]	3200	72.35	−35	−42.2	−32.3	−7.2	2.7
GRIP[a]	3200	72.35	−35.1	−41.9	−31.6	−6.6	3.5
NEEM[b]	2450	77.45	−33	−42.5	−32.1	−9.5	0.9
NGRIP[a]	2917	75.1	−35.4	−44.5	−32.3	−9.1	3.1
Renland[a]	2340	71.18	−27.3	−32.1	−23.8	−5	3.5

[a] Vinther and Johnsen (2013).
[b] Landais et al. (2016).

sheet and a retreat in sea ice cover in the Nordic Seas. Recently, new information on climatic controls on NEEM δ^{18}O ice has become available from modern water isotope monitoring and multidecadal trends from shallow ice cores. The new modern data reveal a surprisingly large present-day response of NEEM δ^{18}O ice to temperature. If relationships established from the intraseasonal to the multidecadal scale remain valid for earlier warm periods such as the LIG, it also implies that the initially reconstructed temperature change based on NEEM δ^{18}O was overestimated. This is a cautionary tale in that the NEEM site locality was originally chosen on glaciological criteria (i.e., to maximize the chance of obtaining the longest possible core). No one could have guessed that the study site might yield anomalous oxygen isotopic results because of a more-or-less local meteorological phenomenon.

The Past 100,000 Years

MacGregor et al., 2015 developed an age model for the GIS based on radio-stratigraphy and age structure (Fig. 10.10). Apparently, the oldest ice (Eemian or older?) is in the north, and the youngest ice is in the south, with the regions in between mostly covered by ice from the last glaciation.

The world's climate system changed rather abruptly from warmer-than-modern conditions to substantially colder-than-modern, starting about 119 ka, and then the planetary climate went through about 25 glacial stadial-interstadial oscillations, as recorded in the GIS isotopic records shifts in δ^{18}O, leading to the warm Greenland interstadials that are often referred to as Dansgaard-Oeschger events (NOAA, 2013). The cause of D-O events remains under debate. During the last glaciation, large ice sheets fringed the North Atlantic. At certain times, these ice sheets released large amounts of freshwater into the North Atlantic. Heinrich events are an extreme example of this when the Laurentide ice sheet disgorged excessively large amounts of freshwater into the Labrador Sea in the form of icebergs (NOAA, 2013). Scientists have hypothesized that these enormous freshwater inputs sufficiently reduced ocean salinity to slow North Atlantic Deepwater Formation and hence curtail thermohaline circulation. Since thermohaline circulation plays an important role in transporting heat northward, a slowdown would cause the North Atlantic to cool. Later, as freshwater input decreased, ocean salinity and deepwater formation increased, and climate conditions recovered.

Evidence for changes in deepwater formation supports the freshwater forcing hypothesis. Measurements from deep-sea sediments in the North Atlantic indicate

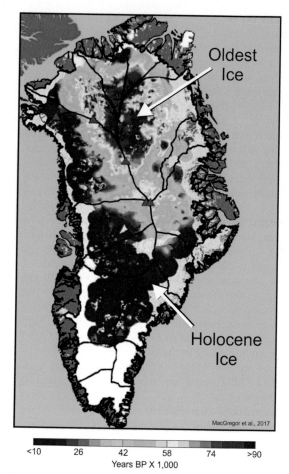

FIG. 10.10 Map of Greenland, showing estimated ages of ice in the GIS, ranging from greater than 90 ka in the north-central region of the ice sheet to less than 10 ka in the southern and western regions. (Modified from MacGregor, J., Fahnestock, M., Catania, G., Paden, J., Gogineni, S., 2015. Radiostratigraphy and age structure of the Greenland ice sheet. Journal of Geophysical Research: Earth Surfaces 1, 120, 220–241. https://doi:10.1002/2014JF003215229.)

a reduction of deepwater formation during Heinrich events. Evidence for freshwater forcing and reduced deepwater formation during D-O events is more ambiguous. The initial trigger for freshwater releases remains to be identified.

D-O events are characterized by mild climate (interstadial), lasting from a few centuries up to tens of thousands of years, which are interrupted by periods of full glacial conditions. High-resolution ice core records show that D-O events begin rapidly, typically within 50 years, while they end more gradually, over centuries.

These temperature changes are associated with reorganizations of atmospheric circulation that may have occurred within periods of one to three years (Botta et al., 2019).

RECAP Dust Record

Stable isotopes from air bubbles trapped in the ice are not the only source of paleoclimate information about the GIS. Records of past dust deposition have also been reconstructed from central Greenland ice cores. During the last glaciation, the ice core dust concentration was 10−100 times greater than in the Holocene due to enhanced continental aridity, increased wind strength, lower snow accumulation, and longer atmospheric particle lifetime. The dust record of the RECAP ice core was obtained from the Renland ice cap in the Scoresby Sund region of central East Greenland (Simonsen et al., 2019). The RECAP ice core (71.30°N, 26.72°W, 2315 m asl) was drilled in June 2015. The core reaches 584m to bedrock and contains a complete climate record back to 120 ka. The authors used the RECAP large dust particle record (larger than 8 μm) as an indicator of the presence of local dust sources. They presented new isotope geochemistry data constraining the likely sources of dust found on the Renland ice cap.

The large particle size of the RECAP Holocene dust limits its atmospheric residence time to less than a day; thus they considered only Greenland and Icelandic sources of the RECAP interglacial dust. High concentrations of large dust particles during the Eemian suggest ice sheet margins were located further inland than at present, supporting various lines of evidence indicating a smaller GIS during the Eemian. At the glacial onset, the ice sheet margin advanced to cover local dust sources from 113.4 to 111.0 ± 0.4 ka, with a small flux of large particles to Renland ice cap through the glacial. The RECAP dust record shows that dust sources in Kong Christian X Land became exposed from 12.1 ± 0.1 to 9.0 ± 0.1 ka, consistent with relative sea level estimates and previous measurements of ice sheet retreat. This dust record provides new constraints on the location of the GIS margin through the last glacial cycle, with potential implications for millennial-scale reconstructions of the GIS response to climate forcing.

Changes from stadial climatic conditions to the 10−15°C warmer interstadial climate are believed to have happened within a decade or two. The glacial climate is therefore characterized by a degree of instability that cannot be found during the Holocene.

A comparison between glacial $\delta^{18}O$ values observed in different Greenland ice cores shows important differences in the size of the $\delta^{18}O$ increase associated with the shift from stadial to Holocene climatic conditions. The Camp Century core shows the largest $\delta^{18}O$ changes while the changes at Renland are the most modest (see Table 10.2). It is speculated that the large changes in Camp Century (and NGRIP) $\delta^{18}O$ are related to those sites having a strongly depleted northwesterly source of precipitation during glacial stadials. Such a source could be due to a split in the polar jet stream, with branches both north and south of the Laurentide ice sheet. Precipitation arriving from the northwest would not be able to reach the Renland ice cap as it is situated east of the main ice sheet and therefore is shielded from any such precipitation. Hence, the relatively modest decrease in Renland $\delta^{18}O$ values during stadial events can be understood in terms of the Renland ice cap not receiving precipitation from the depleted northern branch of a split jet stream. Recent atmospheric modeling efforts suggesting a split jet stream during the glacial therefore enjoy significant support from observations of stadial $\delta^{18}O$ in Greenland ice cores.

Changes gradual and abrupt

Botta et al. (2019) propose the term "gradual climate change" for a change brought about by direct linear forcing, such as variation in solar insolation caused by fluctuations in Earth's orbit, and the term "abrupt climate change" for when the climate system crosses a tipping point and switches to a new state. Moreover, they proposed the use of "rapid climate change" for "a large-scale change in the climate system that takes place over a few decades or less, persists (or is anticipated to persist) for at least a few decades, and causes substantial disruptions in human and natural systems," as defined by the IPCC in 2013. Abrupt climate change is usually rapid, but not vice versa, as rapid climate change can also be the response of a rapid linear forcing. Following this terminology, the Late Pleistocene stadial-interstadial transitions can be considered abrupt events.

GREENLAND ICE SHEET LINKS WITH GLOBAL CLIMATE HISTORY

It may seem counterintuitive that a big block of ice sitting astride the North Atlantic and the Arctic Ocean could have an important role to play in global climate and oceanic circulation, but as the following section explains, the GIS is a major player in global environmental systems. This has to do with teleconnections. In atmospheric science, teleconnection refers to climate anomalies being related to each other at large distances. One example of this phenomenon is the Arctic

Oscillation (AO), a large-scale mode of climate variability, also referred to as the Northern Hemisphere annular mode. The AO is a climate pattern characterized by winds circulating counterclockwise around the Arctic at around 55°N latitude. When the AO is in its positive phase, a ring of strong winds circulating around the North Pole acts to confine colder air across polar regions. This belt of winds becomes weaker and more distorted in the negative phase of the AO, which allows easier southward penetration of colder, Arctic air masses and increased storminess into the midlatitudes.

Bipolar See-Saw

There are close connections between temperature variations in Antarctica and the Arctic, and the term "see-saw" has been used to describe an apparent inverse relation between peaks and troughs in temperature variations between the two hemispheres. The term "see-saw" is used for two distinct antiphase events. Seip et al. (2018) added a third phenomenon that shows a see-saw-like pattern and that also shows a particular lead-lag relation with CO_2. A multimillennial antiphase pattern was identified by Siddall et al. (2006). The antiphase pattern lasted from between 60 ka and 25 ka and showed cycle lengths of approximately 1.5–2 kyr. In their own study, Seidel et al. (2008) found a see-saw-like pattern in the smoothed time series between approximately 60 ka and 40 ka. The cycle length is approximately 20 kyr. Furthermore, there is a see-saw pattern, i.e., an antiphase between CO_2 and Greenland temperature. When the Greenland temperature shows a peak, CO_2 shows a trough, similar to the findings from Antarctica. There are several explanations for the see-saw phenomenon. The Atlantic meridional overturning has a role in most explanations, but Yao et al. (2017) attribute at least some of the forcing to the effects of multiple ocean surface temperatures. The effect of ocean warming in addition to the role of greenhouse gases (GHGs) may explain some of the changes between leading and lagging roles for CO_2 and temperature.

There are several periods in which carbon dioxide led temperature, even if the main pattern consists of temperature leading CO_2. The CO_2-leading periods may be associated with temperature anomaly events in the Southern and Northern Hemisphere. Some of the best evidence for this comes from short-term pauses in the glaciation-deglaciation sequence. For instance, during the recent deglaciation period, CO_2 led temperature during the Bølling-Allerød interstadial warming, which, together with the following Younger Dryas cold period, represents a short-term temperature oscillation at the end of the last deglaciation.

COMPARISONS WITH MARINE RECORDS

As we have seen, rapid, large-scale changes in climate have been documented from stable isotopes in ice cores over the last glacial period and subsequent deglaciation, highlighting the nonlinear character of Arctic climate and underscoring the probability of current and future rapid climate shifts in response to anthropogenic GHG forcing. Large-scale, rapid changes have likewise been observed in marine sediment records since the science of paleoceanography began in the mid-20th century. The pace and amplitude of environmental change during glacial-interglacial cycles in the northern high latitudes remain to be fully explained. Can the two sets of paleoenvironmental reconstructions be compared, side by side? Unfortunately, this is a bit like comparing apples to oranges. Ideally, ocean sediments would be deposited in clearly visible, undisturbed annual layers on the seafloor, just as layers of ice are laid down in an ice sheet or glacier. Sadly, this is not the case. Marine sediment sequences can be analyzed, and different species of marine fossils (e.g., foraminifera, diatoms, and coccolithophores) can tell us about the water bodies these organisms occupied. That line of research started in the 1960s. As in the Greenland ice core records, the oxygen isotope ratios of deep-sea sediments can be extracted from fossil shells in these deposits.

In the early 1960s, Nick Shackleton developed a mass spectrometer that could analyze the oxygen isotope ratios ($^{18}O/^{16}O$) in small numbers of foraminifera shells, thus revolutionizing the field of paleoceanography. While research continued to drive new methods, proxies, and statistical analyses to the paleoceanography toolkit, the arrival of paleoclimate reconstructions based on ice core data from both polar regions took place near the end of the 20th century. At the start of the 21st century, paleoclimate scientists were excited by the possibility of correlating climate change events on land and sea. But a major barrier stood in the way of progress.

Marine Versus Ice Sheet Chronologies

As discussed by Waelbroeck et al. (2019), a major impediment to our understanding of the interactions between past ocean circulation and climate changes has been the difficulty of accurately dating sediments and/or fossil remains from marine cores. Not only does this chronological uncertainty undermine the basic science of paleoceanography, but it also has greatly limited our ability to compare terrestrial records (e.g., the Greenland ice core records) with marine records. These researchers set out to find a way to establish firm chronologies for marine sediment sequences. They studied a set of 92 marine sediment cores taken from the Atlantic Ocean. They established age-depth models for these cores that are consistent with the Greenland GICC05 ice core chronology and computed the associated dating uncertainties, using a new deposition modeling technique. They focused on the past 40 ka because it is the time span covered by radiocarbon dating and the sole period for which it is possible to establish calendar age timescales for marine cores with a precision approaching that of ice core or speleothem records. Comparisons with a reliable standard were vital to this procedure. They chose the Greenland NGRIP ice core chronology because it is generally considered the best-dated continuous continental paleoclimatic archive over the past 50–75 ka. The NGRIP Greenland Ice Core Chronology 2005 (GICC05) calendar age scale was established by annual layer counting with estimated uncertainties of 50 years at 11,000 calendar years before 1950. The estimated age uncertainly of ice deposited during the 11,000–30,000 yr BP interval ranges from 100 to 450 years; the estimated age uncertainly of ice deposited during the 30,000–40,000 yr BP interval ranges from 450 to 800 years.

The reservoir effect

Radiocarbon dating of marine records is made difficult because of a difference between the $^{14}C/^{12}C$ ratio (expressed as $\Delta^{14}C$, in ‰) of ocean surface water and that of the contemporaneous atmosphere. New ^{14}C atoms are constantly being produced in the atmosphere as nitrogen atoms are bombarded by cosmic rays. However, ^{14}C is removed from ocean waters by radioactive decay in the water column, advection, and mixing with older waters. This difference in $\Delta^{14}C$ is termed the "reservoir age" of the surface waters. Previous studies have revealed that surface reservoir ages have not remained constant over time at high latitudes of the North Atlantic and Southern Ocean (i.e., poleward of ~38°N and of ~40°S) due to changes in the location and vigor of deepwater formation. Radiocarbon dates of a terrestrial and marine organism of equivalent age have a difference of about 400 radiocarbon years. Terrestrial organisms like trees primarily get ^{14}C from atmospheric CO_2, but marine organisms do not. Samples from marine organisms such as shells, whales, and seals appear much older.

In high-latitude regions, it is necessary to use an alternative dating strategy in lieu of ^{14}C dating of marine organisms. Waelbroeck et al. (2019) adopted a strategy that has been widely applied and has been adopted by the INTIMATE (Integration of Ice core, Marine and Terrestrial records of the North Atlantic) group in situations where surface reservoir ages cannot be assessed. This strategy consists of synchronizing the SST signal recorded in fossils or sediments taken from marine cores with the air temperature signal recorded in polar ice cores. This dating approach is based on the observed thermal equilibrium between the ocean's surface water and overlying air. Previous studies have demonstrated that changes in air and SSTs were synchronous across the last deglaciation.

Using the GICC05 and AICC2012 age scales as alignment targets for high-latitude SST records of the Northern and Southern Hemispheres, respectively, it is thus possible to directly compare marine records from both hemispheres on a common time frame. These new data sets should enable paleoclimate scientists to examine relative phases between Atlantic records (e.g., planktonic and benthic oxygen and carbon isotopes, Pa/Th). Researchers will be also able to use the spatial and temporal changes recorded in Atlantic sediments to constrain paleoclimate model simulations.

When SST reconstructions based on full census count data were not available, they used the percentage of the polar foraminifera species *Neogloboquadrina pachyderma* (left coiling) as a proxy for SST. This approach has been described and validated in a number of studies (see Fig. 10.11B).

Waelbroeck et al. (2019) also dated five cores taken from sites located north of 62°N by aligning their magnetic susceptibility (MS) records to the NGRIP ice δ18O signal. MS tie points and their associated uncertainties were defined using the same method described for the alignment of SST signals to ice core records. The MS records of four of these five cores have been previously shown to be in phase with the Greenland air temperature signal.

One example of their alignment procedure in fixing ages for events in the North Atlantic marine record is as follows. They set a first alignment target based on the observation that the cooling marking the beginning of Heinrich Stadial 1 in three independently dated North Atlantic cores is synchronous with the sharp increase in dust flux recorded in the Greenland ice cores and dated at 17.48 ± 0.21 ka on the GICC05 age scale. This cooling event coincided with an increase in dust transport from Asia to Greenland, as observed during other Greenland stadial intervals (Fig. 10.12A and B).

Comparisons with the Antarctic ice sheet

Brook (2013b) summarized correlations between the Greenland and Antarctic ice core records and compared

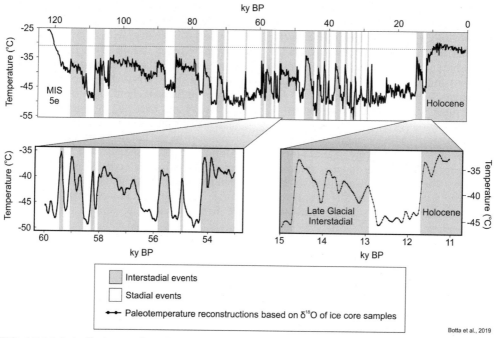

FIG. 10.11 Late Quaternary Greenland temperatures. Top: temperatures reconstructed for the Greenland NGRIP site over the last 120 ka. *Pink shaded* periods highlight the Holocene and the D-O warming events (interstadials). *Blue shaded* periods indicate Heinrich stadials. *White periods* indicate other stadial events. Bottom left: detail of the 60–53 ka year interval showing extreme climate variability. Bottom right: the 15–11 ka Late glacial oscillation with indication of the Bølling-Allerød warmings, Younger Dryas cooling, and the Holocene. (Modified from Botta, F., Dahl-Jensen, D., Rahbek, C., Svensson, A., Nogués-Bravo, D., 2019. Abrupt change in climate and biotic systems. Current Biology 29, R1045–R1054. https://doi.org/10.1016/j.cub.2019.08.066.)

$\delta^{18}O$ records from the Vostok core in Antarctica with the GISP2 core from Greenland. The two sets of records match remarkably well. As discussed earlier, the first ice coring attempts came at Camp Century in northwest Greenland. The ice recovered from this core suffered from poor time resolution, uncertain chronology, and poorly understood ice flow dynamics. Nevertheless, it demonstrated the potential of ice cores for climate research, thus enabling future projects to proceed.

The rhythm of climatic variability recorded in the Antarctic ice cores is very similar to that recorded in the Greenland ice cores. A difficulty in comparison is that the Greenland ice cores tend to have higher resolution (annual layers) than the Antarctic ice cores, which diffuse more rapidly with depth.

Correlation methods
The ice core records from Greenland and Antarctica have been linked using the records of atmospheric trace gases trapped in air bubbles in the ice (Fig. 10.14, below), such as CH_4. Atmospheric methane levels generally equilibrate rapidly throughout the world, so changes in atmospheric methane content should register the same time period in both polar ice sheets. There is a good match in both ice sheets between 10 and 52 ka. Beyond this age range, dating uncertainties are too large to allow accurate matching. Ice core records can also be linked using levels of cosmogenic isotopes such as beryllium-10 and volcanic eruptions (Davies, 2020).

Dunbar et al. (2017) reported the first identification of a New Zealand tephra (volcanic ash) in Antarctic ice. The Oruanui supereruption from Taupo volcano near the center of the North Island has been dated at 25,580 ± 258 cal yr BP, providing a key time marker for the Last Glacial Maximum (LGM) in Antarctica. The Oruanui supereruption is the second-largest known eruption globally in the past 100 ka.

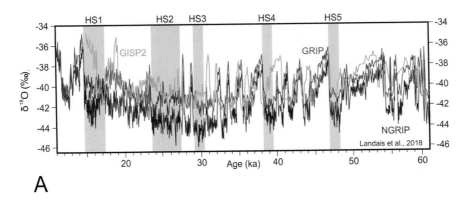

A

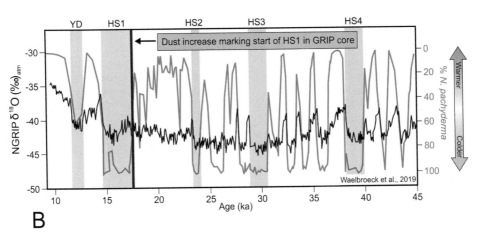

B

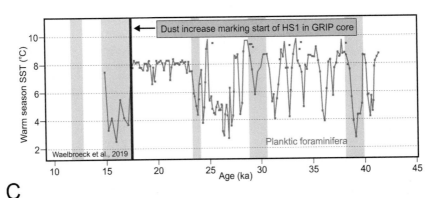

C

FIG. 10.12 **(A)** Comparison of the GISP, GRIP, and NGRIP δ18O records over the last 60 ka, highlighting Heinrich Events 1–5; **(B)** Comparison of NGRIP δ18O record with paleotemperature reconstruction based on percent composition of *Neoglobigerina pachyderma* from North Atlantic core sites MD99-2281 and MD04-284559; **(C)** planktic foraminifer-based warm-season SSTs from core MD99-2281. ((A) Modified from Landais, A., Capron, E., Masson-Delmotte, V., Toucanne, S., Rhodes, R., et al., 2018. Ice core evidence for decoupling between midlatitude atmospheric water cycle and Greenland temperature during the last deglaciation. Climates of the Past 14, 1405–1415. https://doi.org/10.5194/cp-14-1405-2018. (B) and (C) Modified from Waelbroeck, C., Lougheed, B. C., Vazquez Riveiros, N., Missiaen, L., Pedro, J., Dokken, T., Hajdas, I., Wacker, L., Abbott, P., Dumoulin, J.P., Thil, F., Eynaud, F., Rossignol, L., Fersi, W., Albuquerque, A.L., Arz, H., Austin, W.E.N., Came, R., Carlson, A.E., Ziegler, M., 2019. Consistently dated Atlantic sediment cores over the last 40 thousand years. Scientific Data, 6(1), 165. https://doi.org/10.1038/s41597-019-0173-8.)

Volcanologists reckon it produced about 1100 km³ of fine, wind-transported tephra over a period of weeks to months.

One of the first interpolar comparisons was made between the GISP2 core from Greenland (Figs. 10.2 and 10.13) and Byrd core from Antarctica. A range of quantitative analyses has been applied to the two data sets to facilitate correlations, with varying success (Yau et al., 2016b). Steig and Alley (2002) used lag-correlation analysis of filtered data to examine the relationships between the records. They concluded that the data are consistent with either a Southern Hemisphere lead of 1000–1600 yr or a 400–800 yr Southern Hemisphere lag and emphasized that neither is a complete description of the data. Schmittner et al. (2003) also used correlation analysis of filtered data. They argued that temperature changes in Greenland were in advance of changes of the opposite sign in Antarctica. Their analysis also revealed a high correlation for Antarctica leading Greenland by 1300 yr, but they emphasized that change in the Southern Hemisphere did not require an abrupt climate trigger. Stocker and Johnsen (2003) developed the concept of the "thermal bipolar see-saw" model discussed before, which allowed them to predict the Byrd isotope curve fairly accurately from the Greenland (GRIP in this case) record by incorporating a heat reservoir (the Southern Ocean), which delays and integrates the impact of a Northern Hemisphere climate changes on the Southern Hemisphere. They argued that almost all D-O events between 25 and 65 ka have a recognizable counterpart in the Byrd record. In summary, there is consensus that the largest of the D-O events were preceded by Antarctic warming and that the Antarctic warming terminated at about the time of abrupt warming in Greenland. Cooling then followed in both locations. The Byrd and Vostok records appear to show similar patterns for shorter events.

A more recent comparison between the two polar records has been published by Davies (2020). During the period 10,000 to 52,000 years ago (during an ice-age climate), the stable isotope records of the Antarctic and Greenland ice cores are out of phase with each other (the bipolar see-saw). Cooling episodes in Greenland coincide with warming episodes in Antarctica and vice versa. This is a driving mechanism that transfers energy alternately between the poles, driven by ocean circulation changes, such as changes in the course and speed of the Gulf Stream.

PALEOBIOLOGY OF LATE PLEISTOCENE GREENLAND

Sixty years ago, Willi Dansgaard and colleagues discovered several abrupt climate change events in Greenland during the last glacial period. As we have seen, since then, multiple ice cores retrieved from the GIS have verified the existence of 25 abrupt climate changes: Dansgaard-Oeschger warming events. These events are characterized by a rapid 10–15°C warming over a few decades followed by a stable period of centuries or millennia before a gradual return to full glacial conditions. Similar warming events have been identified in other geochemical and fossil archives across the Northern Hemisphere. These findings triggered a wide interest in the impact of abrupt climate change on regional biota, but ambiguous definitions have constrained our ability to assign biotic responses to the different types of climate change.

Botta et al. (2019) offer coherent definitions for different types of climatic change, including "abrupt climate change," and a summary of past abrupt climate change events. Biotic responses to abrupt climate change range from the genetic to ecosystem level and demonstrate that abrupt climatic and ecological

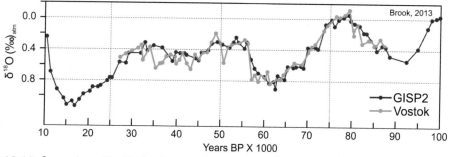

FIG. 10.13 Comparison of the Vostok (Antarctica) and GISP2 (Greenland) δ18O records over the past 100 ka, after Brook (2013b). (Modified from Brook, E., 2013c. Correlations between Greenland and Antarctica. In: Elias, S., Mock, C. (Eds.), Encyclopedia of Quaternary Science, second ed. Elsevier, pp. 1258–1266.)

Drilling at the EastGRIP ice core site, Greenland,
Photo by Helle Astrid Kjær, Creative Commons Attribution
4.0 International license.

Sections of the GISP-2 ice core.
A - from 53-54 m;
B - from 1836-1837 m;
C - from 3050-3051 m.
Photographs courtesy U.S. National
Ice Core Laboratory

Air bubbles trapped in a thin section of ice core.
Photo courtesy of Oregon State University,
licensed under CC BY-SA 2.0.

FIG. 10.14 Upper left: drilling at the EastGRIP ice core site, Greenland, photo by Helle and van Ommen, 2020, Creative Commons Attribution 4.0 International license; lower left: thin section of a polar ice core, showing trapped air bubbles. Photo by Tas van Ommen, Courtesy of Australian Antarctic Division; Right: three sections of the GISP-2 ice core: (A) 53—54 m; (B) 1836—1837 m (note obvious layers); (C) 3050—3051 m (note inclusion of mineral sediments). Ice core photos courtesy of US National Ice Core Laboratory. (From van Ommen, T., Kjær, A., 2020. Ice core researchers from AWI drilling at the EastGRIP ice core site, Greenland. https://commons.wikimedia.org/wiki/File:Ice_core_researchers_from_AWI_drilling_at_the_EastGRIP_ice_core_site,_Greenland_2.jpg.)

changes have been instrumental in shaping biodiversity. This much can be seen in the fossil record of Greenland, which must be considered "Ground Zero" for biotic responses to glaciation. However, identifying causal relationships between past climate change and biological responses remains difficult. Botta et al. (2019) point out the need to formalize and unify the definition of "abrupt change" across disciplines and further investigate past abrupt climate change periods to better anticipate and ameliorate future impacts on biodiversity from global change.

Geologists and biologists are investigating past climatic changes not only to determine what happened to the climate, the oceans, the land, and the cryosphere but also to gain an understanding of how the biota responded and to forecast biodiversity impacts from the current global warming. Biological responses to past events of climate change include migrations, ecological community turnovers, reorganization of geographical ranges, changes in population sizes, and extinctions. It is useful to have an unambiguous terminology for the topic, by providing definitions of abrupt, rapid, and fast climate change as well as abrupt ecological change.

In Quaternary science, the terms "rapid," "fast," and "abrupt" climate change are widely applied and often considered as synonyms. According to the US National Research Council, an abrupt climate change "occurs when the climate system is forced to cross some threshold." Despite many formal definitions, there has been a consensus that abrupt climate change involves a switch into a *new state following a tipping point*.

Botta et al. (2019) proposed the term "gradual climate change" for a change caused by *direct linear forcing*, such as variation in solar insolation caused by fluctuations in Earth's orbit. Moreover, they proposed the use of "rapid climate change" for "a large-scale change in the climate system that *takes place over a few decades or less*, persists (or is anticipated to persist) for at least a few decades and causes substantial disruptions in natural systems," as defined by the IPCC in 2013. Following this terminology, the Late Pleistocene stadial-interstadial transitions can be considered abrupt events (Fig. 10.11). Moreover, although the current climate change is being described as rapid, it is not clear whether it will lead to an abrupt climate event during this century, although that likelihood will increase within the coming centuries.

Biotic Responses

The ability of climate change to bring about an abrupt ecological change has as much to do with relevant biological thresholds as with the inherent properties of climate change, including regime shifts triggered by linear (nonabrupt) climate forcing and tipping points. Marine records provide a unique window into the pace and mode of biotic responses to abrupt climate change. There is solid evidence for a significant turnover of marine species following abrupt climate change during the last 20,000 years. During warm interstadial intervals, thermophilic marine species that were already adapted to those conditions were dominant and reappeared consistently at the beginning of subsequent warm climate episodes. However, the exact species composition of ecological communities differed from one warming event to the other. These changes in ecological communities in response to climate warming occurred within a few decades. Benthic foraminifera assemblages shifted from oxic to anoxic species within less than 40 years, and they also experienced regional increases in species diversity within a few decades to a few centuries.

Recent high-resolution studies on fossil pollen records show significant responses of plant communities to abrupt climate change but also to gradual climate change. It should be noted that the timing of biotic responses to environmental change, whether it is at the species level or the community level, is never synchronous. Species come and go from a given region based on their own specific requirements, so biological communities are always just temporary associations of plants and animals, never lasting more than a few centuries or millennia. They may appear to us to be extremely durable and long-lived, but that is only because human life is so short.

In some Late Pleistocene situations, ecological communities were able to maintain viable populations under changing climatic conditions, reaching dynamic equilibrium within 100 years after an abrupt climate change. This suggests some degree of resilience. However, even these "equilibrium" states were maintained through community turnover, with the invasion of species more suited to the new climate conditions, and the loss of less well-adapted species. It must also be reinforced here that the Arctic biota is particularly vulnerable to environmental change because the flora and fauna are so well adapted to a very narrow set of environmental conditions. There are virtually no substitute species that can come in from other regions and take advantage of changing environments. The Arctic has small food webs and so it has less ecological flexibility than other regions except perhaps for extreme deserts.

Interglacial forests at DYE-3

As discussed earlier, we know almost nothing of the preglacial biotic history of the Arctic and sub-Arctic regions that were repeatedly glaciated in the Late Pleistocene. These regions comprise about 10% of Earth's terrestrial surface, but in many subsequently glaciated regions, the existing landscape and evidence of its biota were obliterated by repeated glacial erosion, leaving little or no evidence of these ancient ecosystems. However, the field of ancient DNA research is providing new evidence that would otherwise be unobtainable. Willerslev et al. (2007) extracted ancient DNA and amino acids from sediments buried beneath the basal sections of two deep ice cores in Greenland, allowing them to produce a fairly detailed reconstruction of an ancient ecosystem that occupied southern Greenland sometime before the last glaciation. In fact, they demonstrated that the mountains of southern Greenland were once inhabited by a diverse coniferous woodland or forest.

The study samples were taken from the silt-rich sediments at the base of the Dye-3 core from south-central Greenland and from the long GRIP core from the summit of the GIS. The silty ice yielded only a few pollen grains and no macrofossils. However, the Dye-3 silty ice samples showed low levels of amino acid racemization, indicating good organic matter preservation.

The identifiable DNA sequences from the Dye-3 core samples reveal a community very different from that of modern Greenland. The taxa identified include trees such as alder (*Alnus*), spruce (*Picea*), pine (*Pinus)*, and members of the yew family (Taxaceae). These are indicative of a northern boreal forest ecosystem—not Arctic tundra. The other groups identified, including Asteraceae (composites), Fabaceae (legumes), and Poaceae (grasses), are mainly herbaceous plants and are represented by many species found today in northern regions but not restricted to boreal forest habitats. The presence of these herb-dominated families suggests an open forest, allowing sun-loving plants to thrive. Additionally, Willerslev et al. (2007) identified the DNA of taxa that are common in the Arctic and/or Boreal regions but lacked a 100% sequence identity between independent laboratories. These include yarrow (*Achillea*), birch (*Betula*), chickweed (*Cerastium*), fescue (*Festuca*), rush (*Luzula*), plantain (*Plantago*), bluegrass (*Poa*), saxifrage (*Saxifraga*), snowberry (*Symphoricarpos*), and aspen (*Populus*). Although not independently authenticated at the sequence level, the presence of these taxa adds further support to the conclusion of a northern boreal forest ecosystem at the Dye-3 site.

It remains uncertain whether the overlying clean ice of Dye-3 was deposited immediately after the lower silty section or much later. Dating evidence suggests the latter. Willerslev et al. (2007) applied four dating techniques to the silt-rich samples: $^{10}Be/^{36}Cl$ isotope ratios, single grain luminescence measurements, amino acid racemization coupled with modeling of the basal ice temperature histories of GRIP and Dye-3, and maximum likelihood estimates for the branch length of the invertebrate COI sequences. All four dating methods give overlapping dates for the silty ice between 450Ka and 800Ka, suggesting the Dye-3 silty ice and its forest community far predate the LIG. However, given the uncertainties in the dating procedures, the investigators could not rule out the possibility of an Eemian age for the Dye-3 basal ice.

Clarifying the pace of ancient changes

Radiocarbon dating remains the principal method of age determination for terrestrial events within the past 50 ka. However, the statistical uncertainty of radiocarbon dates increases with age, typically to several hundred years during the last glacial period or even more toward the limit of the dating technique. Especially in cold regions, low accumulation rates give many marine and terrestrial archives low temporal resolution, reducing our ability to assess the timing of biological responses. These problems may seem somewhat insurmountable, but instead of relying on paleoclimatic reconstructions from individual sites, Botta et al. (2019) recommend the development of spatially explicit high-resolution paleoclimatic models to provide deeper and more meaningful insights into the impacts of abrupt climate change on biological diversity.

HOW IS THE MODERN GREENLAND ICE SHEET STUDIED?

Since the end of the World War II, many glaciologists, climatologists, Quaternary scientists, field biologists, and others have spent their summers trekking around the edges or over the surfaces of the GIS. As shown in Table 10.3, a wide variety of scientific expeditions have explored various aspects of the GIS and adjacent

TABLE 10.3
International Research Projects in Greenland.

Project	Sponsor	Research Period
Tundra ecosystems of western Greenland	Fulbright Arctic Initiative	2016–18
East Greenland Ice Core Project (EGRIP)	Center of Ice and Climate at the Niels Bohr Institute and US NSF	2015-? (2020 field season canceled)
Greenland ice core research (24 projects)	Earth Institute, Columbia University, US	2014–20
Greenland Ecosystem Monitoring Strategy and Working Programme	Aarhus University, Denmark	2011–15
Glacier-Ocean Interactions	Greenland Climate Research Center, Nuuk	2011–20
Greenland Sea Ice	Greenland Climate Research Center, Nuuk	2008–20
Greenland Analogue Project (glacial processes)	Sweden, Finland, Canada	2008 and 2013
North Greenland Eemian Ice Drilling (NEEM)	Center of Ice and Climate at the Niels Bohr Institute and US NSF	2007–12
International Tundra Experiment (ITEX)	Norway, Canada, Iceland, Sweden, US, UK	2003–09
Study of Environmental Arctic Change Project (SEARCH)	NSF, NASA (US)	2001–02
Greenland Climate Network (GC-Net) GIS automated weather stations	NSF and NASA (US)	1999–2020
Ice Flow in the North East Greenland Ice Stream	NASA (US)	1997–99
Ecological research, Zackenberg Research Station	Aarhus University, Denmark	1995–2020
Studies of Earthquakes and Rapid Motions of Ice (SERMI)	Columbia University, NSF (the United States)	1993–2010
Man and the Biosphere (Vegetation mapping)	The United States	1992
Greenland Ice Sheet Project Two (GISP2)	NSF (the United States)	1988–93
Greenland Ice Core Project (GRIP)	Belgium, Denmark, France, Germany, Iceland, Italy, Switzerland, the United Kingdom, the European Union	1989–95
Greenland Environmental Observatory at the Summit site (GEOSummit)	NOAA (the United States)	1989–2020
International Arctic Systems for Observing the Atmosphere (IASOA)	Commission for Scientific Investigations in Greenland	1989–2020
Renland Ice Core	Swedish Natural Science Research Council, Nordic Council of Ministers	1988
DYE-3 Ice Core	The United States	1971–81
Greenland Ice Sheet Project (GISP)	Denmark, Switzerland, the United States	1971–81
Panarctic Flora Project	The United States, Russia, Norway	1981–2005
Vegetation survey, Tasiilaq, Southeast Greenland	German Research Foundation, Ministry of Education of the Czech Republic	1968–81
Camp Century Ice Core	US Army Corps of Engineers	1966
Greenland Botanical Survey	Botanical Museum, University of Copenhagen	1962–2020

unglaciated regions of Greenland. Authority over scientific research in Greenland, like other aspects of governance, has been evolving since the 1950s. In 1953, a new Danish Constitution incorporated Greenland into Denmark for the first time. The island gained representation in the Danish Parliament and was recognized as a Danish province. In 1979, the Danish government granted Greenland home rule, with Denmark retaining control of foreign relations, defense, currency matters, and the legal system. In 2008, Greenland's citizens approved a self-government referendum that gave them control of law enforcement, the coast guard, and the legal system. The act also gives Greenlanders control over foreign relations and international trade.

Field Studies

Since the 1960s, field studies in Greenland have been dominated by ice coring projects, supported by glaciological studies to determine the most suitable sites for drilling. Of the 26 research projects listed in Table 10.3, 14 are either directly or indirectly associated with ice drilling to obtain long cores. Eight of the studies concern the description and ecology of tundra vegetation types in Greenland, and four studies have examined climate change issues. The financial support for these projects has been provided by a host of organizations and governments, including Aarhus University, Denmark; the Botanical Museum of the University of Copenhagen; the Center of Ice and Climate at the Niels Bohr Institute, Denmark; the Commission for Scientific Investigations in Greenland; the Earth Institute of Columbia University, the United States; the Fulbright Arctic Initiative, and the Greenland Climate Research Centre, Nuuk.

Major research grants for Greenlandic field studies have been provided by the research councils of Belgium, the Czech Republic, Denmark, the European Union, France, Germany, Iceland, Italy, Norway, Russia, Sweden, Switzerland, the United Kingdom, and the United States.

Studies on the Greenland ice sheet

Because of logistic difficulties, most of the studies of the GIS since the 1990s have made use of remote sensing, with data generated either from aircraft or satellites. Some of the earliest field studies of the GIS were performed in northwest Greenland by scientists brought in by the US Air Force to consult on the construction of a secret base that was built beneath the surface of the ice. These early studies surprised the engineers when it became apparent that the ice sitting over the

tunnels was moving so rapidly that it was ruining the secret base.

Determining the thickness of the GIS was accomplished in a few select locations by the drilling of ice cores down to bedrock. Of course, these few coring sites represent only a tiny fraction of the ice sheet. Since the 1990s, scientists have relied on various remote sensing methods to estimate the thickness of the ice sheet across the width and breadth of Greenland.

Ice core drilling is an arduous task. Depending on the length of the ice core to be retrieved, this process may take anywhere from a few days during one summer to several weeks over the course of many summers. For instance, the first field season of the East Greenland Ice Core Project (EGPRIP) was during the summer of 2015; the project was expected to drill to the base of the ice sheet by 2020. However, the 2020 field had to be canceled because of the COVID-19 pandemic (Fig. 10.14).

Greenland ice sheet climate data

Until the development of automated weather stations, it was nearly impossible to collect sufficient climate data from most of the GIS. However, such data are vital to our understanding of current conditions, especially in the current environment of rapid climate change in the Arctic. Changes in climate for any region must be compared with a set of baseline data for that region—ideally a database of 30 years of climate data. Other than a few coastal stations, Greenland is lacking in these long-term climate data sets.

Another vital purpose for GIS climate data is its application to ice core studies. To interpret the paleoclimate signal from an ice core, one must have knowledge of the modern climate of the study area. But all of the ice coring locations in Greenland are in remote regions, often hundreds or thousands of km away from the nearest weather station. The ice coring research teams have attempted to gather as much climate data as possible from their study sites, but even in the best case, these data represent only a few years of climate.

In response to this problem, NASA and the NSF of the United States funded the Greenland Climate Network. GC-Net was established in 1999 and research teams positioned 18 automatic weather stations ranging in location from about 62°N to 78°N to collect climate information on the GIS. Hourly averaged data are transmitted via a satellite link throughout the year. The system is powered by batteries that are recharged by solar panels.

Studies on the margins of the Greenland ice sheet

Carrivick and Tweed (2019) reviewed the evidence for glacier outburst floods in Greenland, looking particularly for evidence of megafloods. Floods with a peak discharge of at least 1 million cubic meters per second are defined as megafloods. These are the largest glacier outburst floods, based on their volume and peak discharge. Some glacier lake outburst floods in Greenland may fall into this category, given the enormous lake volumes drained. The authors identified 14 potential megaflood sites, all but one in southern and southwest Greenland (Fig. 10.15). There are very few reported impacts of glacier outburst floods in Greenland. Ice dam failure has caused frequent flooding in Greenland compared with the volcanically triggered floods in Iceland, and this, combined with the proximity of the Greenland glacier lakes to the coast, means that most proglacial channels in Greenland are flood-hardened and most landscape impact is likely to be

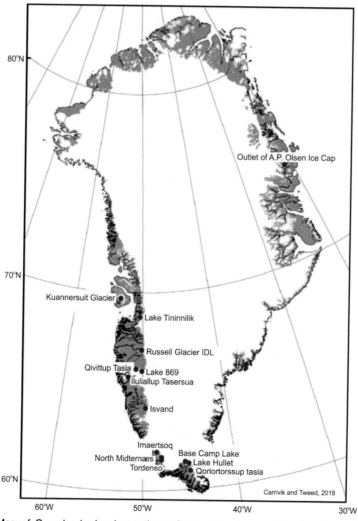

FIG. 10.15 Map of Greenland, showing outburst flood sites discussed by Carrivick and Tweed (2019). (Modified from Carrivick, J., Tweed, F., 2019. A review of glacier outburst floods in Iceland and Greenland with a megafloods perspective. Earth-Science Reviews 196, 102876. https://doi.org/10.1016/j.earscirev.2019.102876.)

offshore in estuaries and fjords. Future floods in Greenland will probably be caused by extreme weather and rapid ice melt.

Coastal glaciers and Greenland ice sheet ice loss. Mouginot et al. (2019) used surveys of thickness, surface elevation, velocity, and surface mass balance (SMB) of 260 glaciers situated along the periphery of the GIS to reconstruct the recent mass balance of the GIS since 1972. They calculated mass discharge into the ocean for 107 glaciers and indirectly for 110 glaciers using velocity-scaled reference fluxes. The decadal mass balance switched from a net gain of 47 ± 21 Gt per year in 1972−80 to net losses thereafter (Fig. 10.16). For reference, 1 Gt is one billion tons of ice and corresponds to 1 cubic km of water. The accelerating loss of mass seen in 2000−10 slowed down in 2010−18 due to a series of cold summers, illustrating the difficulty of extrapolating short records into longer-term trends.

About two-thirds of the mass loss is due to glacier dynamics, with the remainder controlled by SMB. Even in years of high winter accumulation, enhanced glacier discharge in summer has remained sufficiently high above equilibrium to maintain an annual mass loss every year since 1998.

The study showed that the net annual accumulations and losses of the ice sheet were nearly balanced from 1972 to 80. However, the analysis of the combined records reveals that the GIS has been consistently losing ice mass since the 1980s. The largest mass losses have come from the northwest, southeast, and central-west regions. These are controlled by tidewater glaciers. Several glaciers in North Greenland have played a more important role in the total mass loss than has been previously recognized.

Geothermal heat flux

Because of the thick mantle of the GIS, the bedrock topography of Greenland remains somewhat enigmatic. East and South Greenland are flanked by mountains reaching more than 2000 m in elevation. The presence of Tertiary basalts on mountain tops suggests a young, rapid uplift well away from tectonic plate boundaries, but the fastest Greenland orogeny (recent vertical movements) lies south of the highest mountains. Ice melting causes postglacial adjustments (isostatic rebound), but there is a poor fit between isostatic models and uplift rates in Greenland.

The thermal structure of the Greenland lithosphere is likewise poorly known. Seismic- and paleomagnetic-based models have yielded inconsistent results.

However, the current thinning of the GIS is taking place both at the surface and at the base of the ice sheet. It is therefore important to develop an understanding of Greenland's geothermal heat flux (GHF) radiating from bedrock up through the base of the ice sheet. Knowledge of GHF and lithosphere temperature anomalies is crucial for understanding the lithosphere dynamics of the island and estimating basal ice temperature, an important parameter in ice sheet melting. However, conventional heat flux measurements are nearly absent in Greenland.

The existing knowledge concerning the structure of the lithosphere is based mainly on observations made along the narrow ice-free zone along the coasts. Artemieva (2019) described a new method she used to calculate upper mantle temperature anomalies, lithosphere thickness, and GHF in Greenland, based on seismic data, topography, and ice thickness. More specifically, the study models the temperature heterogeneity of the lithosphere based on thermal isostasy analysis. Thermal isostasy is a geodynamic process in which regional variations in lithospheric temperature cause changes in elevation. Elevation changes result from variations in rock density in response to thermal expansion.

The results of this study are quite startling. As shown in Fig. 10.17, the highest levels of GHF occur in east-central Greenland, where values exceed 100 milliwatts per m^2 (hereafter mW/m^2). The North Central region of Greenland has GHF levels in and around 70 mW/m^2. The northwest and southeast regions show GHF values ranging from 40 to 60 mW/m^2. Finally, the region with the least GHF is in the south-central and southwest parts of Greenland, averaging about 35−55 mW/m^2.

Lithosphere thermal thickness varies from 200 to 270 km in the cratonic SW Greenland with the maximum thickness in the kimberlite province of West Greenland. A craton is a large stable block of the Earth's crust forming the nucleus of a continent. Most of the Proterozoic lithosphere is 120−190 km thick with a gradual decrease in thickness from SW Greenland toward the Summit station in Central Greenland (Fig. 10.2). The study identified an anomalous belt with a 100- to 120-km-thick lithosphere and increased GHF that extends from the northwest coast toward East Greenland. This belt may have resulted from the passage of the Iceland hot spot.

During the Late Cretaceous (ca 80 million yr ago), the North American continental plate carried Greenland north, where it glided over a relatively stationary hot spot—the same spot that later formed Iceland after

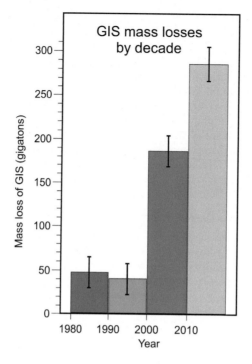

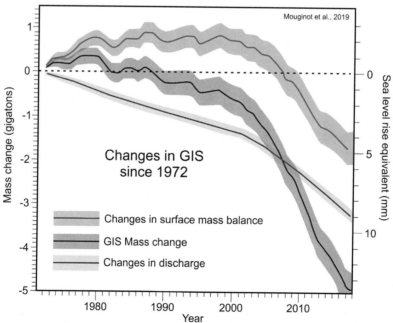

FIG. 10.16 Above: Bar graph illustrating mass losses of ice from the GIS by decade, since the 1980s. Below: Changes in surface mass balance *(blue line)*, GIS mass change *(purple line)*, and changes in discharge *(red line)* since 1972. Shaded areas around lines show the 95% confidence intervals of the data. GIS, Greenland ice sheet. (Modified from Mouginot, J., Rignot, E., Bjørk, A. A., van den Broeke, M., Millan, R., Morlighem, M., Noël, B., Scheuchl, B., Wood, M., 2019. Forty-six years of Greenland ice sheet mass balance from 1972 to 2018. Proceedings of the National Academy of Sciences of the United States of America, 116(19), 9239—9244. https://doi.org/10.1073/pnas.1904242116.)

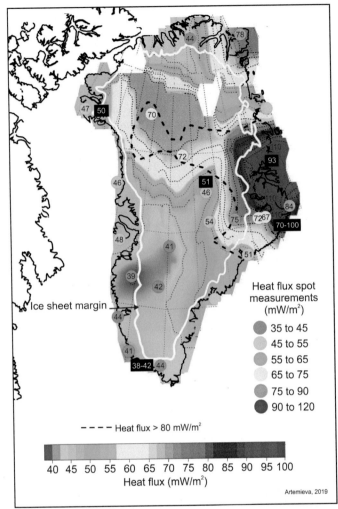

FIG. 10.17 Map of Greenland showing geothermal heat flux (mW/m²) by region and by spot measurements. Black boxes show heat flux measured in boreholes. (Modified from Artemieva, I., 2019. Lithosphere thermal thickness and geothermal heat flux in Greenland from a new thermal isostasy method. Earth Science Reviews 188, 469–481. https://doi.org/10.1016/j.earscirev.2018.10.015.)

Greenland had moved on. The hot spot punched out a new landmass from the crust beneath the sea. The track of this scar through Greenland still retains a measurable heat signature (Martos et al., 2018).

The region of east-central Greenland has high mantle temperatures and a strong GHF anomaly (>100 mW/ m²) centered in the Fjordland region. This anomaly cannot be explained solely on the basis of conductive heat transfer. In fact, Artemieva (2019) concluded that it requires advective heat transfer such as is found above active magma chambers. The high GHF of this region may significantly increase basal ice melting. An

extension of this anomaly ca. 500 km inland (GHF values from 70 to 95 mW/m²) may contribute substantially to basal ice melting, further expanding the ice drainage stream system in central Greenland. Basal melting releases liquid water that acts as a lubricant for glacial movement. Under such conditions, the ice sheet may expel ice more rapidly to coastal outlets through faster ice flow.

Effects of Black Carbon

Aerosol black carbon (soot) (BC) is released during the incomplete combustion of fossil fuels, biofuels, and biomass burning (Sharma et al., 2013). The largest BC emission sources that affect the Arctic are agricultural burning, wildfires, and on-road diesel vehicles, followed by residential burning, off-road diesel, and industrial combustion, including gas flaring. Most of the atmospheric BC deposited in the Arctic is the result of long-range transport from Russia, Europe, North America, and East Asia. BC affects the radiation balance in the Arctic by absorbing solar radiation when suspended in the atmosphere, by altering cloud properties, and, when deposited on snow and ice, by darkening the surface and enhancing the absorption of solar radiation and melt rates. A large proportion of the Arctic climate response to an increase in surface temperature is due to the snow/albedo effect.

Thomas et al. (2017) reported on BC concentrations from 22 snow pits sampled in northwest Greenland in April 2014, documenting a widespread BC aerosol deposition event, which was dated to have accumulated in the pits from two snowstorms between July 27 and August 2, 2013. Using modeling and remote sensing tools, they were able to link this deposition event to forest fires burning in Canada during summer 2013.

Massling et al. (2015) reported their measurements of BC in aerosols at the Villum Research Station (VRS) in northern Greenland. Their results showed a seasonal variation in BC concentrations with a ground-level maximum in winter and spring. BC particles, especially long-lived particles that become chemically altered on their way to their ultimate destination, affect the radiation budget by scattering and absorbing insolation. They form part of the "Arctic haze." In haze situations, solar energy can be reemitted via long-wave radiation. This results in a heating effect below and within the polluted aerosol layer.

Remote Sensing

As discussed earlier, many aspects of research on the GIS are now being performed by remote sensing rather than in the field. Some of the remote sensing is done with signals transmitted from airplanes flying over the surface of the ice. Most of it is being done using satellite data. While satellites are expensive to build and launch, they provide unequaled information about remote regions such as Greenland and much of the rest of the Arctic.

Satellite data

Satellites are being used in a wide variety of applications to study the GIS. Among the satellite missions launched since the 1990s, several have collected data for Arctic research. NASA's twin Gravity Recovery and Climate Experiment (GRACE) satellites, launched in 2003, were followed by GRACE Follow-On (GRACE-FO) satellites launched in 2018. These satellites measured the mass balance of the GIS every 30 days by detecting minute differences in gravity that are indicative of different thicknesses of ice. While most of the planet's mass does not move much in 30 days, its water and ice do, causing Earth's gravity to shift. By tracking these changes, GRACE and GRACE-FO can identify how much ice sheets and glaciers are shrinking.

The surface of the GIS has been mapped by conventional radar altimetry since the launch of the ERS-1 satellite, which was followed by ERS-2, Envisat, and currently CryoSat-2. The European Space Agency's two European Remote Sensing (ERS) satellites, ERS-1 and ERS-2, were launched into the same orbit in 1991 and 1995, respectively. Their payloads included a synthetic aperture imaging radar, radar altimeter, and instruments to measure ocean surface temperature and wind fields. ERS-2 added an additional sensor for atmospheric ozone monitoring. The two satellites acquired a combined data set extending over two decades. The ERS-1 mission ended on March 10, 2000, and ERS-2 was retired on September 05, 2011.

CryoSat-2 is a European Space Agency environmental research satellite that was launched in April 2010. It provides scientists with data about the polar ice caps and tracks changes in the thickness of the ice with a resolution of about 1.3 cm. CryoSat-2 is the first satellite with the capability to measure changes in the coastal regions of the ice sheet with radar altimetry. Its novel SAR Interferometric (SARIn) mode provides improved measurement over regions with steep slopes.

In 2003, NASA placed its Geoscience Laser Altimeter System (GLAS) onboard their ICESat spacecraft. ICESat's primary objectives were to determine the mass balance of the polar ice sheets and their contributions to global sea level change and to obtain essential data for prediction of future changes in ice volume and sea level. From 2003 to 2009, the ICESat mission provided multiyear elevation data needed to determine ice sheet mass balance as well as cloud property information, especially for stratospheric clouds common over polar areas. It also provided topography and vegetation data around the globe, in addition to the polar-specific coverage over the GIS and Antarctic ice sheet. The Ice, Cloud, and land Elevation Satellite-2 (ICESat-2) was a second-generation laser altimeter ICESat mission, operating since 2018. It measures the height of a changing Earth—one laser pulse at a time, 10,000 laser pulses a second. ICESat-2 carries a laser altimeter that detects individual photons, allowing scientists to measure the elevation of ice sheets, sea ice, forests, and more in unprecedented detail (Fig. 10.18).

MODIS (moderate-resolution imaging spectroradiometer) is a key instrument aboard NASA's Terra and Aqua satellites. Terra's orbit around the Earth is timed so that it passes from north to south across the equator in the morning, while Aqua passes south to north over the equator in the afternoon. Terra MODIS and Aqua MODIS are viewing the entire Earth's surface every 1−2 days, acquiring data in 36 spectral bands, or groups of wavelengths (see MODIS Technical Specifications).

One of the MODIS tasks is to monitor the level of albedo coming from the GIS. The ESA launched their Sentinel-1a satellite in 2014, followed by Sentinel 1b in 2016. These two sets of satellites use synthetic aperture radar (SAR) to map the Earth's surface. Greenland researchers are using Sentinel data to document the speed and direction of GIS movements. A new set of data are available every 12 days.

Thermal imaging of the planet is being carried out by the Copernicus Sentinel-3 satellite, launched by the ESA in 2016. The satellite is carrying a Sea and Land Surface Temperature Radiometer (SLSTR). The SLSTR has nine

European Space Agency's CryoSat 2 satellite conducting an ice survey from orbit.
Figure courtesy of the European Space Agency

ICESat-2 satellite. Figure courtesy of NASA

FIG. 10.18 Above: Artist's impression of the European Space Agency's CryoSat 2 satellite conducting an ice survey from orbit. Figure courtesy of the European Space Agency under Creative Commons Attribution-ShareAlike 3.0 IGO (CC BY-SA 3.0 IGO) License. Below: Artist's impression of the ICESat-2 satellite. Figure courtesy of NASA, in the public domain. (From European Space Agency, NASA, 2020. European Space Agency's CryoSat 2 Satellite and NASA's ICESat-2 Satellite. https://www.esa.int/Applications/Observing_the_Earth/CryoSat.)

bands covering the visible, shortwave infrared, and thermal infrared areas of the spectrum.

In summary, satellites are now providing daily or weekly updates on the thickness and movements of the GIS, its mass balance, the discharge of water and ice from its margins, the surface temperature of the ice, and its albedo. Now let us examine some of the results of these studies.

Recent temperatures

As mentioned earlier, there have been marked increases in air and ocean temperatures and reductions in summer cloud cover around Greenland in the past few decades. These changes have produced increases in surface runoff, supraglacial lake formation and drainage, iceberg calving, glacier terminus retreat, submarine melting, and ice flow, leading to widespread changes in the surface elevation, particularly near the margin of the ice sheet. The warming trend extends throughout the Arctic and made international headlines in the summer of 2020. That summer was marked by record high temperatures in several regions of the Arctic, as well as extreme wildfires in Siberia and a significant loss of sea ice. Observations from space offer a unique opportunity to understand the changes occurring in this remote region. Images captured by the Copernicus Sentinel-3 mission show the extent of smoke from the fires in the Chukotka region, the most north-easterly region of Russia, in June 2020. Wildfire smoke releases a wide range of pollutants including carbon monoxide, nitrogen oxides, and solid aerosol particles. In June alone, the Arctic wildfires were reported to have emitted the equivalent of 56 megatons of carbon dioxide, as well as significant amounts of carbon monoxide and particulate matter. These wildfires affect radiation, clouds, and climate on a regional, and global, scale.

The Arctic heat wave is also contributing to the thawing of permafrost. Near the surface, Arctic permafrost soils contain large quantities of organic carbon and the remains of ancient plants prevented from decomposing by subfreezing temperatures. As discussed further in Chapter 12, once these organic materials thaw, they decompose, releasing CO_2 and methane into the atmosphere. Deeper permafrost layers are dominated by mineral grains of sand, silt, and clay. The permanently frozen ground, just below the surface, covers around a quarter of the land in the Northern Hemisphere.

According to the Copernicus Climate Change Service, July 2020 was the third warmest July on record for the globe, with temperatures 0.5°C above the 1981–2010 average. In addition, the Northern Hemisphere saw its hottest July since records began—surpassing the previous record set in 2019. The Arctic has not escaped the heat. On June 20, the Russian town of Verkhoyansk, which lies above the Arctic Circle, recorded a staggering 38°C (European Space Agency, 2020). Extreme air temperatures were also recorded in northern Canada and on Svalbard. On August 11, Nunavut's Eureka Station, located in the Canadian Arctic at 80°N latitude, recorded a high of 21.9°C (Fig. 10.19). This has been reported as the highest temperature ever recorded so far north.

The temperature regime of the GIS during the summer of 2020 was more complicated. Temperatures from across Greenland were near average over the June 21 to August 1, 2020 period. Higher than average temperatures (up to 3.5°C warmer) were restricted to the far northern edge of the ice sheet. Temperatures 1.5–2°C lower than average occurred in the Scoresby Sund region of east-central Greenland (NSIDC, 2020).

Changes in albedo

Albedo, the proportion of the incident light or radiation that is reflected by a surface, is now being measured across the GIS by satellites. The MODIS satellite has recorded GIS albedo since 2000. Snow-covered surfaces on the ice sheet reflect 80% of insolation back to space, while snow-free surfaces have lower albedo. Bare ice has an albedo of about 40%, and even lower albedo if summer meltwater pools are standing on the surface. After a record low level in 2012, Greenland albedo was relatively high from 2013 to 2018 (Tedesco et al., 2018).

As shown in Fig. 10.20, albedo measurements of the GIS in late summer 2020 reveal that the northernmost regions exhibited high albedo ratings. Field observations verified that this region was largely snow-covered. In contrast to this, the southern sector of the ice sheet was apparently almost free of snow, and the albedo of this region was far lower (NSIDC, 2020). The lower panel of the figure compares the percent melting of the ice sheet surface in 2020 and long term average (1981–2010). The summer of 2020 saw surface melting at or above 30% in June and July. The long-term average melting peaks below 20% in July, so the summer melting in 2020 was one-third greater than average.

Greenland ice sheet elevation

The thinning of the GIS correlates with decreasing elevation of the ice surface through time. Sandberg Sørensen et al. (2018) developed a record of 25 years of surface

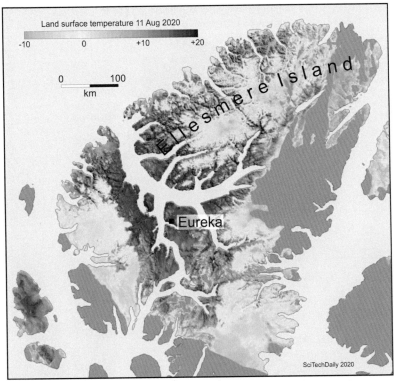

FIG. 10.19 Map of Ellesmere Island, Canada, showing regional extent of high surface temperatures on August 11, 2020, as recorded at the Eureka Weather Station. (From European Space Agency, 2020. Surface Temperatures on August 11, 2020, as Recorded at the Eureka Weather Station. https://www.esa.int/ESA_Multimedia/Images/2020/08/Extreme_temperatures_in_Eureka.)

elevation changes of the GIS (1992–2016), derived from satellite radar altimetry. The ice sheet has been mapped by this method since the launch of ERS-1 in 1991, which was followed by ERS-2, Envisat, and currently CryoSat-2 (Fig. 10.19). The recently launched Sentinel-3A is providing a continuation of the radar altimetry time series. Since 2010, CryoSat-2 has begun measuring elevational changes in the coastal regions of the ice sheet with radar altimetry, with its novel SAR Interferometric (SARIn) mode, which provides improved measurement over regions with steep slopes.

The 25 years of elevation changes were evaluated as 5-year running means. A clear acceleration in thinning is evident from the elevation declines observed since 2003.

The early 1990s show only modest elevation changes, with no clear spatial pattern of thinning or thickening. Starting in 2000, a clear pattern of ice sheet thinning appeared that dominated a large part of the ice sheet margin. The data highlight individual outlet glaciers because of their rapid thinning rates. Over time, these rapid ice flows have propagated further inland onto the main ice sheet. The past decade (2010–20) has been characterized by large climate variability in Greenland, and heat penetration of the ice sheet is highly climate dependent. This is especially evident in the occurrence of strong melt events where the formation of ice lenses creates strong reflective surfaces for the radar altimeters.

Mass Balance and Thickness of the Greenland Ice Sheet

The mass balance of a glacier or ice sheet is the net annual balance between the mass gained by snow deposition, and the loss of mass by melting (either at the glacier surface or at the base) and calving (production of icebergs). A negative mass balance means that an ice sheet is losing mass. When this happens to grounded glaciers and ice sheets, this mass loss directly

Albedo anomalies, Aug 31–Sep 3 2020
compared with same period 2000-2009

NASA 2020

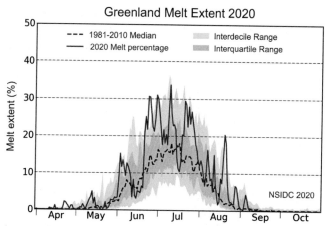

NSIDC 2020

FIG. 10.20 Albedo anomalies, August 31—September 3, 2020 compared with the same period 2000—09. Regions in *blue* or *purple* register higher than average albedo, indicating substantial snow cover. Regions in *red* or *yellow* registered lower than average albedo, indicating little or no snow cover. (From NASA, 2020. Greenland Albedo Anomalies, Aug 31—Sep 3, 2020. https://www.carbonbrief.org/guest-post-how-the-greenland-ice-sheet-fared-in-2020.)

contributes to sea level rise. This is one of the reasons why so much research is currently being focused on the mass balance have ice sheets. To reckon the total mass balance of an ice sheet or glacier, we must factor in ice lost from the calving of icebergs and from ocean melting along the coasts.

Of course, ice can melt both at the surface of an ice sheet and at its base, but we have far less information about what happens 2 or 3 km below the surface. Therefore, most mass balance estimates are just based on SMB. Although increases in GIS ice flow and surface melting have been largely driven by oceanic and atmospheric warming, the degree and trajectory of the current loss of ice mass remain uncertain. What is certain has been a steady increase in ice melting phenomena, such as increased surface runoff, supraglacial lake formation, and drainage, iceberg calving, glacier terminus retreat, submarine melting, and ice flow. All of these have contributed to the widespread thinning of the GIS, as detected by satellite measurements of changes in the surface elevation, particularly near the margin of the ice sheet.

The Ice Sheet Mass Balance Inter-Comparison Exercise ((IMBIE team 2019)), composed of more than 75 researchers, compared and combined 26 different satellite measurements of changes in the GIS volume, flow, and gravitational potential. They produced a revised estimate of the recent history of changes in GIS mass balance. The GIS was close to being stable in the 1990s, with summer losses in mass balance more-or-less equaled by winter increases. However, since the beginning of the 21st century, annual losses have substantially exceeded gains, peaking at a loss of 345 ± 66 gigatons per year in 2011. One gigaton (Gt) equals 1×10^{12} kg and corresponds to 1 cubic km of water. In all, Greenland lost 3902 ± 342 Gt of ice between 1992 and 2018, causing the mean sea level to rise by almost 11 mm. In 2019, there were greater mass balance losses on the western side of Greenland, and some mass balance gains in the east. Velicogna et al. (2020) report, based on GRACE-FO data, an exceptional summer loss of 600 Gt in 2019 following two cold summers (Fig. 10.21).

When losses of ice from iceberg calving and from ocean melting at its edge are included, the total mass loss of the GIS was 152 Gt for 2020 (Mottram, 2020). The ice has been lost throughout two main processes. Using three regional climate models, the IMBIE (2019) analysis showed that the reduced SMB has caused the loss of 1964 ± 565 gigatons $(1964 \times 10^{12}$ kg) (50.3%) through increased meltwater runoff. The remaining 1938 ± 541 Gt (49.7%) of ice loss was due to increased imbalances in glacial dynamics, which rose from 46 ± 37 Gt per year in the 1990s to 87 ± 25 Gt per year since then.

The principal mechanism of glacier discharge in Greenland is iceberg calving. Discharge rose from 46 ± 37 Gt per year in the 1990s to 87 ± 25 Gt per year since then. Between 2013 and 2017, the total rate of ice loss slowed to an average of 220 ± 30 Gt per year, as atmospheric circulation favored cooler conditions and as ocean temperatures fell at one of the more important glacial outlets, the terminus of Jakobshavn Isbræ. Cumulative ice losses from Greenland have been close to the IPCC's predicted rates for their high-end climate warming scenario, which forecasts an additional 50—120 mm of global sea level rise by 2100 when compared with their central estimate.

Although ice loss has dominated the recent history of the GIS, there is considerable variability in this process. During the 2000s, just four glaciers were responsible for half of the total ice loss due to increased discharge, whereas many others contribute today. Likewise, some neighboring ice streams have been observed to speed up over this period, while others have slowed down, suggesting diverse reasons for the changes that have taken place. Uncounted billions of liters of meltwater have sunk from the surface to the interior of the ice sheet, some of them reaching bedrock. The shifting patterns of meltwater drainage channels, some of which are presumably the size of large terrestrial rivers, are bound to be one of the factors causing changes in the glacial dynamics at the surface.

Another independent assessment of GIS mass balance was published by King et al. (2020). This team examined dynamic ice loss from the GIS by integrating historical velocity data from the LANDSAT 4 and 5 satellites with the AeroDEM digital elevation model. They also appended the extended time series (1985—2018), with annually resolved synthetic-aperture radar velocity mosaics between 1990 and 1999. They used this series to assess decadal trends in ice discharge from the GIS (Fig. 10.22), and where possible, resolve the seasonal amplitude through time. They reported that, beginning around the year 2000, ice discharged through outlet glaciers started to outpace annual snowfall, causing a negative mass balance. They consider that this change represents a tipping point that is irreversible in the near future. According to their assessment, annual mass loss will become the new norm for the ice sheet in the near future.

Mass balance trend of the Greenland Ice Sheet: 2002 to 2019 as measured by the GRACE/GRACE-FO satellites.

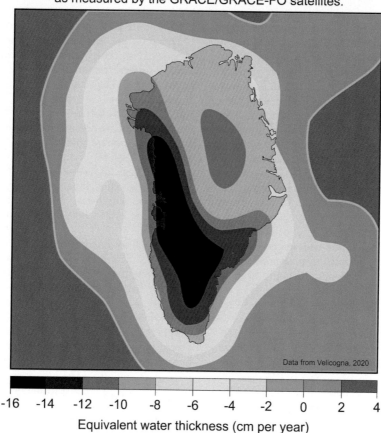

Data from Velicogna, 2020

-16   -14   -12   -10   -8   -6   -4   -2   0   2   4

Equivalent water thickness (cm per year)

FIG. 10.21 Mass balance trend of the Greenland Ice Sheet: 2002 to 2019, as measured by the GRACE/ GRACE-FO satellites. (Data from Velicogna, I., Mohajerani, Y., Felix Landerer, G., Mouginot, J., et al., 2020. Continuity of ice sheet mass loss in Greenland and Antarctica from the GRACE and GRACE follow-on missions. Geophysical Research Letters, 47, e2020GL087291. https://doi.org/10.1029/2020GL087291. Figure 2 on page 5.)

King et al. (2020) suggest that long-term thinning of the GIS throughout the 20th century led up to a mass retreat event in the early 2000s. This thinning was likely due to warming SSTs. The result was a "step-increase" in the rate of discharge through outlet glaciers. Before 2000, King et al. (2020) estimate that 420 Gt of ice were discharged annually. Since 2000, the annual discharge rate has increased to 480 Gt.

The ice loss estimates of King et al. (2020) are far greater than the estimates made by Hanna et al. (2020). According to several estimates cited by Hanna et al. (2020), the GIS lost 257 ± 15 Gt per year of mass during 2003—15 (Box et al., 2018), 262 ± 21 Gt per year during 2007—11 (Andersen et al., 2015), and 269 ± 51 Gt per year during 2011—14 (McMillan et al., 2016). According to Mouginot et al. (2019), the GIS lost 286 ± 20 Gt per year during 2010—18. Some of the differences between these numbers can be attributed to methods that consider either just the contiguous ice sheet or also including disconnected peripheral glaciers and ice caps, the latter being the case for GRACE-based estimates. However, GIS mass loss approximately quadrupled during 2002/3—2012/13 (Bevis et al., 2019) (Fig. 10.23). The GIS sea level contribution over 1992—2017 was approximately one and a half times the sea level contribution of Antarctica (Box et al., 2018).

Greenland mass loss is mainly driven by atmospheric warming, and, based on ice-core-derived melt

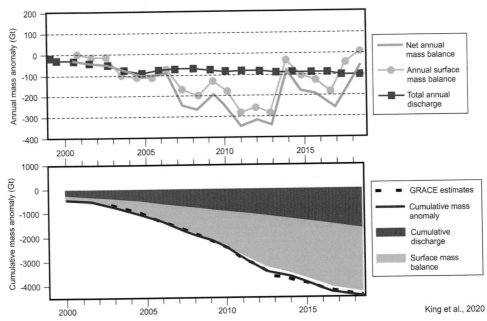

FIG. 10.22 Above: 2000—18 annual mass anomaly (Gt) of the GIS, expressed as net annual mass balance *(green line)*; annual surface mass balance *(blue line)*, and total annual discharge *(red line)*. Below: Cumulative mass anomaly (Gt) of the GIS *(purple line)*, Gravity Recovery and Climate Experiment (GRACE) estimates *(dashed line)*, cumulative discharge *(area in brown)*, and surface mass balance *(area in blue)*. GIS, Greenland ice sheet. (Modified from King, M., Howat, I., Candela, S., Noh, M., Jeong, S., et al., 2020. Dynamic ice loss from the Greenland ice sheet driven by sustained glacier retreat. Nature Communications: Earth and Environment 1(1). https://doi.org/10.1038/s43247-020-0001-2.)

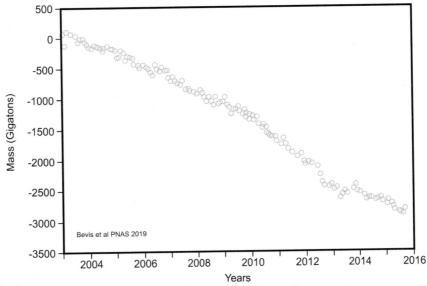

FIG. 10.23 Changes in mass (gigatons) of the GIS from 2003 to 15. GIS, Greenland ice sheet. (Modified from Bevis, M., Harig, C., Khan, S.A., Brown, A., Simons, F.J., Willis, M., Fettweis, X., Van Den Broeke, M.R., Madsen, F.B., Kendrick, E., Caccamise, D.J., Van Dam, T., Knudsen, P., Nylen, T., 2019. Accelerating changes in ice mass within Greenland, and the ice sheet's sensitivity to atmospheric forcing. Proceedings of the National Academy of Sciences of the United States of America, 116(6), 1934—1939. https://doi.org/10.1073/pnas.1806562116.)

information and regional model simulations, surface meltwater runoff increased by ~50% since the 1990s, a level unprecedented in the past 7000 years (Trusel et al., 2018). Enderlin et al. (2014) found an increasingly important role of runoff on total mass annual losses during their 2000—12 study period and concluded that SMB changes were the main driver of long-term (decadal or longer) mass loss. Just five marginal glacier near-termini regions, covering <1% of the GIS by area, were responsible for 12% of the net ice loss (McMillan et al., 2016), highlighting the potentially important role and sensitivity of ice dynamics.

Glacial discharge into the ocean

Mouginot et al. (2019) used surveys of thickness, surface elevation, velocity, and SMB of 260 Greenland glaciers from 1972 to 2018 to reconstruct the recent mass balance of the GIS (Fig. 10.16). They calculated mass discharge (D) into the ocean for 107 glaciers (85% of D) and indirectly for 110 glaciers (15%) using velocity-scaled reference fluxes (Fig. 10.24). Unlike the IMBIE team (2019), Mouginot et al. (2019) report that decadal mass balance switched from a net gain of 47 ± 21 Gt per year in 1972—80 to net losses thereafter, placing the start of negative SMB some 20 years earlier than the IMBIE reconstruction. However, as shown in Fig. 10.16, the pace of ice loss calculated by Mouginot et al. (2019) quickened considerably in the 21st

century. The accelerating loss of mass seen in 2000—10 slowed down in 2010—18 due to a series of cold summers, which illustrates the difficulty of extrapolating short records into longer-term trends. Cumulated since 1972, the largest contributions to global sea level rise are from northwest (4.4 ± 0.2 mm), southeast (3.0 ± 0.3 mm), and central west (2.0 ± 0.2 mm) Greenland, with a total 13.7 ± 1.1 mm for the ice sheet. The mass loss is controlled at 66 ± 8% by glacier dynamics (9.1 mm) and 34 ± 8% by SMB (4.6 mm). Even in years of high winter accumulation, enhanced glacier discharge in summer has remained sufficiently high above equilibrium to maintain an annual mass loss every year since 1998.

The study by Mouginot et al. (2019) concluded that the greatest losses from iceberg calving are in drainages that terminate in tidewater glaciers. Several glaciers in northern Greenland have played a more important role in the total mass loss than has been previously recognized. Future mass changes in the northern part of the GIS are expected to become some of the greatest contributors to sea level rise, because of the large reserve of ice above sea level and the potential for manifold increases in ice discharge.

Mankoff et al. (2020) focused their research efforts on estimating GIS discharge from 1986 through March 2020 (Fig. 10.24). Their data include all discharging ice

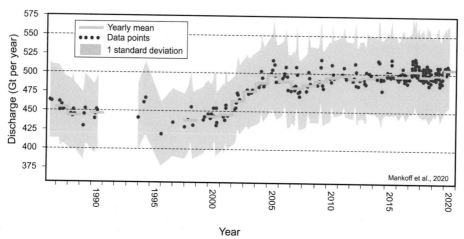

FIG. 10.24 GIS ice discharge (Gt per year) from 1986 to 2019, shown as discrete data points *(red dots)*, yearly mean *(yellow line)*, and one standard deviation of the data points *(blue area)*. Note that annual discharge until 2000 was generally at or below 450 Gt per year; by 2005, it climbed to 500 Gt per year and above. GIS, Greenland ice sheet. (Modified from Mankoff, K., Solgaard, A., Colgan, W., Ahlstrøm, A., Khan, S., & Fausto, R., 2020. Greenland ice sheet solid ice discharge from 1986 through March 2020. Earth System Science Data 12, 1367—1383. https://doi.org/10.5194/essd-12-1367-2020.)

that flows faster than 100 m per year. In addition to using annual time-varying ice thickness, their time series used velocity maps generated from satellite observations. The quality and quantity of those observations were quite sparse, to begin with, but toward the end of the project, the satellites provided near-complete spatial coverage, with ice flow velocity updates every 12 days.

According to their reconstruction, the average ice discharge from the studied outlet glaciers from 2010 through 2019 was about 487 ± 49 Gt per year—a level remarkably close to the estimate by King et al. (2020) (480 Gt per year). The 10% uncertainty stems primarily from the uncertain ice bed location (ice thickness). The discharge was relatively steady from 1986 to 2000; then it increased sharply from 2000 to 2005 and then has returned to a steadier pace in the past 15 years. However, regional and glacier variability is more pronounced, with recent decreases at most major glaciers and in all but one region offset by increases in the northwest region through 2017 and in the southeast from 2017 through March 2020.

Mankoff et al. (2020) noted that an update of the NSIDC 0646 ice velocity data has increased coverage and discharge data from the 1980s by 25—40 Gt per year, due to higher velocity estimates than the previous product (Mouginot et al., 2018a, b). This change demonstrates that refinements to velocity models may produce quite different results. The top individual contributing glaciers remain dynamic: Jakobshavn Glacier and Helheim Glacier, which has been the top Greenlandic contributor to sea level rise since 2018.

Climate drivers of changing surface mass balance. Bevis et al. (2019) examined the potential climate drivers behind recent oscillations in GIS SMB. They noted that the total mass of the GIS decreased at an accelerating rate from 2003 to mid-2013 and then stopped for more than a year. Satellite data from the GRACE and global positioning system (GPS) were used to examine the spatial patterns of mass losses. It was found that years of strongest net losses coincided with the negative phase of the North Atlantic Oscillation (NAO) (see Chapter 4). This phase of the NAO enhances summertime warming and insolation in the Arctic while reducing winter snowfall, especially in west Greenland. The combined effects of these changes have an adverse impact on the SMB of the GIS, causing reduced snowpack formation in winter and increased melting in the summer, as supported by a regional climate model MAR. The spatial pattern of accelerating mass losses appears to be closely linked with the pattern of NAO-driven atmospheric forcing.

Subglacial lakes. Few subglacial lakes have been identified from beneath the GIS despite extensive documentation of such lakes in Antarctica, where the periodic release of water can impact ice flow. Bowling et al. (2019) made an ice sheet–wide survey of Greenland subglacial lakes, identifying 54 candidates from airborne radio-echo sounding, and 2 additional lakes from ice surface elevation changes (Fig. 10.25). These water bodies range from 0.2 to 5.9 km in length and are mostly situated well away from ice divides, beneath relatively slow-moving ice. Based on their results and previous observations, the authors suggest three zones of subglacial lake formation in Greenland: (1) stable lakes in the northern and eastern regions above the equilibrium line altitude (ELA) but away from the interior; (2) hydrologically active lakes near the ELA that are recharged by surface meltwater; and (3) small, seasonally active lakes below the ELA, which form over winter and drain during the melt season. These observations provide important constraints on the basal thermal regime of the ice sheet and help refine our understanding of the subglacial hydrological system.

Interestingly, Bowling et al. (2019) did not identify many subglacial lakes beneath the fast-flowing warm based regions of the ice sheet, although they suggest that significant additional water storage may be taking place there. Long-term basal meltwater storage beneath the region above the ELA could be activated in the future as the ablation area migrates inland. The resulting increased input of meltwater to the bed at higher elevations could open new subglacial drainage pathways through enhanced sliding and potentially connect this dormant storage to the ice sheet margin.

Surface melting

In 2020, the level of ice sheet melting through October was well above the 1981 to 2010 average but below the levels of many previous summers of the past decade (NSIDC, 2020). While the northeast and southwest areas of the ice sheet had significantly more melt than average, the extent of melting over the southeast and northwest coasts was lower than average (Fig. 10.26). The 2020 surface runoff was 23.1 million km², lower than in several recent years.

The season was marked by a series of moderate spikes in melt extent, primarily in the southeast and northwest. The peak one-day melt extent for 2020 to date was July 10, when 551,000 km², or 34% of the ice sheet surface, melted. The maximum number of

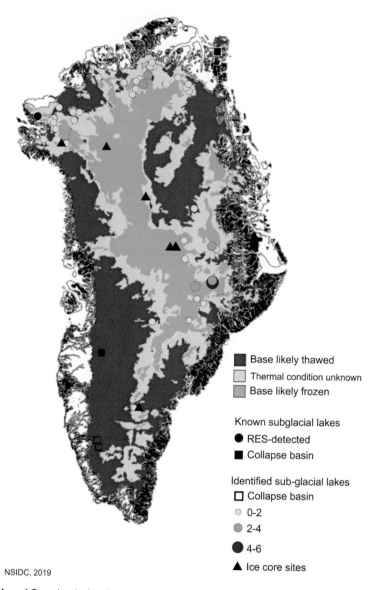

Base likely thawed

Thermal condition unknown

Base likely frozen

Known subglacial lakes
● RES-detected
■ Collapse basin

Identified sub-glacial lakes
□ Collapse basin
◌ 0-2
◉ 2-4
● 4-6
▲ Ice core sites

NSIDC, 2019

FIG. 10.25 Map of Greenland, showing the extent of regions with thawed base *(red)*, frozen base *(blue)*, or freeze/thaw condition unknown *(white)*. Known subglacial lakes *(black dots)*; identified subglacial lakes *(yellow, green, and purple dots)*. (Modified from Bowling, J., Livingstone, S., Sol, A., Chu, W., 2019. Distribution and dynamics of Greenland subglacial lakes. Nature Communications 10, 2810. https://doi.org/10.1038/s41467-019-10821-w.)

melt days occurred in two areas, along the southwest edge and northeast corner of the ice sheet, at around 65 out of 123 days from April 1 to August 1, 2020. The 2020 total melt extent was greater than about 75 percent of past years in the satellite record after late June.

Temperatures in the Greenland region were near average over the June 21 to August 1, 2020 period, with higher than average temperatures of up to 3.5°C limited to the far northern flank of the ice sheet, and lower than average temperatures of 1.5−2°C near the Scoresby Sund area.

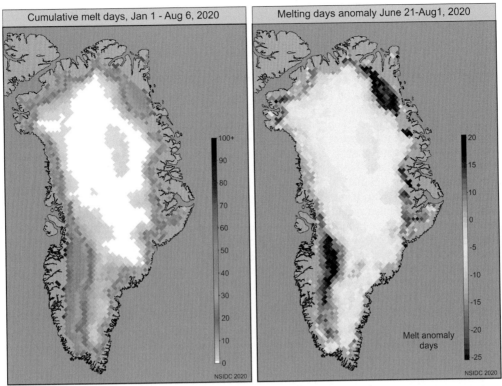

FIG. 10.26 Left: GIS cumulative melt days, January 1—August 6, 2020; Right: GIS melting day anomaly June 21—August 1, 2020. Note that the strongest melting anomalies fall within the northeast and southwest regions. Data from NSIDC, 2020, in the public domain. GIS, Greenland ice sheet. (From NASA, 2020. GIS Cumulative Melt Days, Jan 1—Aug 6, 2020; GIS melting day anomaly June 21—Aug 1, 2020. http://nsidc.org/greenland-today/.)

Wind patterns over the ice sheet favored a westward to southwestward flow, bringing air off the far northern Atlantic Ocean (Norwegian Sea) across the ice sheet. The variation in circulation this year is nearly the opposite of that of Summer 2019, which resulted in the melting and snowfall being less extreme this year compared with last year. Net mass balance was below average all along the western side of the ice sheet in 2020, but near average to slightly above average along most of the eastern coastal area.

Numerical Modeling

During the past 50 years, the speed of operation of supercomputers has been growing exponentially. In 1964, the first Cray supercomputer, the CDC 6600, reached a top speed of 3 million floating-point operations per second (megaFLOPS). In 1997, the ASCI Red supercomputer broke the 1 teraFLOPS (1 trillion FLOPS, or 1×10^{12}) barrier, making it 1 million times faster than the CDC 6600. More recently, the Chinese developed the Sunway TaihuLight supercomputer in 2016. This computer performs at 93 petaFLOPS (93 quadrillion [93×10^{18}] floating-point operations per second), making it 93 million times faster than the Asci Red. Finally, in 2020, the Japanese Fugaku supercomputer reached 415.53 PFLOPS, more than four times faster than the Sunway TaihuLight.

As computing power has increased through time, numerical models have improved in two significant ways. First, the more recent models include far more variables than older models, for instance, a recent model of the atmosphere might include dozens or hundreds of layers from the ground to the stratosphere. Second, the spatial resolution of numerical models has increased dramatically. Climate models from the 1990s had very large grid cells, making their results too indefinite. Recent climate models have reduced the grid cell size, considerably making their predictions far more precise, if not more accurate. Early climate models typically had about 10 vertical layers. Current models often have about 30 layers (UCAR, 2020). Typical model resolution in the mid-1990s had grid cells of about 200 by 300 km at midlatitudes. The T85 resolution, typical for current models, has a grid size of about 100 by 150 km across at midlatitudes. The highest resolution examples, T170 and T340, represent future directions in modeling as more powerful supercomputers come online. Such models will be able to produce projections for regional climates that will be much finer than those currently being generated. However, increased resolution (either spatial or temporal) comes with a price. As a rule of thumb, doubling the resolution of a model requires about 10 times the computing power, or that the model will take 10 times longer to run on the same computer.

Climate models are typically run with time steps of about 30 min. A climate model run that covers a century of events may involve 1,753,152 time steps—the number of half-hours in a century. All model parameters (temperature, wind speed, humidity, etc.) are recalculated at each of the thousands to millions of grid points in the model and at each of those time steps (UCAR, 2020). This is why only supercomputers are capable of running such models.

No model, however complex, will exactly match reality. Modelers attempt to refine their models to gain an understanding of how natural systems work. In the case of the GIS, models have been devised to examine the interactions of the atmosphere, cryosphere, ocean, unglaciated land, and the bedrock beneath the ice sheet. As we have seen in this chapter, this is an enormously complicated system with a very high number of variables not all of which work uniformly throughout Greenland. In fact, one of the great complexities of the GIS system concerns local environments: everything from the temperature of seawater in fjords to the albedo from mountain surfaces.

Mass balance

Advances in ice sheet modeling are expected through experiments such as the Ice Sheet Model Intercomparison Project for CMIP6 (ISMIP6), which will deliver process-based projections from stand-alone ISMs forced by output from coupled atmosphere-ocean GCMs in time for AR6 in 2022. These efforts will improve predictions of the ice dynamical response, particularly in Antarctica

where the spread among AR5 scenarios is large, through advanced representations of ice-ocean interactions that extrapolate GCM ocean forcing into ice shelf cavities. Modeling of surface processes is also improved by using RCMs to increase the spatial resolution of atmospheric GCM forcing and capture SMB variations found in steep topography at ice sheet margins.

Ice flow speed and patterns

The GIS exhibits large spatial variations in surface velocity, with a few fast-flowing outlets draining most of the interior. It is therefore critical to capture the complex flow pattern of GIS in models used for future sea level projections. Recent developments in ISMs such as efficient parallel computation, better representation of flow equations, detailed basal topography, and the inclusion of subglacial hydrology have contributed to greatly improve the representation of this spatially varying flow. In addition to these advances, inversion for basal friction using surface velocities has proved to be a powerful tool, and models now appear to be able to capture most of the complex flow patterns of the ice sheet. Inversions are useful to capture present-day velocity, but they mask information that is needed to evolve these conditions in time. Therefore, we cannot fully rely on inversions for future projections, as basal conditions may evolve because of changing climate and in turn influence ice dynamics.

The Northeast Greenland Ice Stream (NEGIS) drains more than 10% of the GIS and has recently undergone significant dynamic change, displaying high velocities all the way to the ice divide (Fig. 10.27). Despite its large impact on the GIS mass balance, NEGIS is not accurately represented in ISMs without inverting for basal friction. At present, NEGIS is reproduced in ISMs by inferring basal conditions using observed surface velocities. This approach helps estimate conditions at the base of the ice sheet but cannot be used to estimate the evolution of basal drag through time, so it lacks the predictive power to model the future evolution of the ice sheet under climate warming scenarios. NEGIS is suggested to be initiated by a GHF anomaly close to the ice divide, left behind by the movement of Greenland over the Icelandic plume. However, the heat flux underneath the ice sheet is largely unknown, except for a few direct measurements from deep ice core drill sites. Using the Ice Sheet System Model (ISSM), with ice dynamics coupled to a subglacial hydrology model, Smith-Johnsen et al. (2020) investigated the possibility of initiating NEGIS by

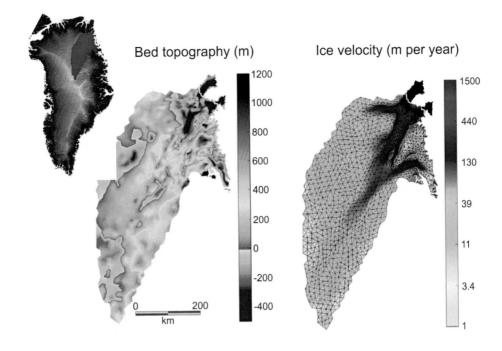

Smith-Johnsen et al., 2020

FIG. 10.27 Map of northeast Greenland Ice Stream. Left: Bed surface topography from BedMachine (Morlighem et al., 2017) interpolated onto the model mesh of Smith-Johnsen et al. (2020). Right: Interferometric synthetic aperture radar (InSAR)—derived surface velocities (m per year). Creative Commons Attribution 4.0 License. (Modified from Smith-Johnsen, S., de Fleurian, B., Schlegel, N., Seroussi, H., Nisancioglu, K, 2020. Exceptionally high heat flux needed to sustain the Northeast Greenland ice stream. The Cryosphere 14, 841–854. https://doi.org/10.5194/tc-14-841-2020.)

inserting heat flux anomalies with various locations and intensities. In their model experiment, a minimum heat flux value of 970 mW per m^2, located close to the East Greenland Ice Core Project (EGRIP) (Fig. 10.14, left; Fig. 10.2), is required locally to reproduce the observed NEGIS velocities, yielding basal melt rates consistent with previous estimates. The value cannot be attributed to GHF alone, and the authors suggest hydrothermal circulation as a potential explanation for the high local heat flux. By including high heat flux and the effect of water on sliding, they were able to successfully reproduce the main characteristics of NEGIS in an ISM without using data assimilation (Fig. 10.2).

Greenland ice sheet-ocean interactions
Bindschadler et al. (2013) were the first to develop a process-based case study modeling the response of the GIS to predict future climate changes. Their models indicated that Greenland is more sensitive than Antarctica to likely atmospheric changes in temperature and precipitation, while Antarctica is more sensitive to increased ice shelf basal melting. They ran a modeling experiment based on the IPCC RCP8.5 (worst-case) scenario. These model runs indicate additional contributions to sea level of 22.3 cm from Greenland in the coming century. A 200-year model run projected sea level rise contributions of 53.2 cm for the GIS. The estimates for the 21st century are in stark contrast to the most recent IPCC forecast of sea level rise in the next 100 years. Under the worst-case scenario (RCP8.5), the latest IPCC estimate (Oppenheimer et al., 2019) is that sea level will rise between 84 and 110 cm above its current level—less than half the sea level rise predicted in the Bindschadler et al.'s (2013) model.

As discussed by Morlighem et al. (2019), the margins of many glaciers along the northwest coast of Greenland have been retreating at an accelerated pace during the past 20 years, sometimes dramatically (Fig. 10.28). It has been suggested that the retreat of these glacier fronts is initiated by the presence of warm and salty subsurface AW in the fjords where the glaciers terminate. This warmer water is typically found 200–300 m below the surface. Surface runoff has also been increasing over the past decades, enhancing subglacial water discharge at the base of calving fronts. This freshwater flux boosts ocean circulation in the fjord, which in turn further increases the melting rate at the surface as well as the rate of undercutting at the calving face of marine-terminating glaciers. While it is expected that both surface runoff and ocean heat content will continue to increase over the next century, their effect on ice dynamics and ice discharge into the ocean remains unclear. Although they are in close proximity to each other, individual outlet glaciers along the coast respond differently to frontal forcing. It has been proposed that this heterogeneity in glacier behavior may be due to differences in bed topography and fjord bathymetry. For instance, AW may be blocked from access to glacial calving fronts by the presence of sills in the fjord. It has also been suggested that the termini of many glaciers are currently resting on pronounced ridges, or in regions of lateral constrictions. Under these circumstances, the glaciers' calving fronts become stabilized, preventing warm water from dislodging them from their current position.

Calving-front dynamics exert important controls on Greenland's ice mass balance. For example, the ice front retreat of marine-terminating glaciers may lead to a loss in resistive stress, ultimately resulting in glacier acceleration and thinning. Over the past decade, it has been suggested that such retreats may be triggered by warm and salty AW, which is typically found at a depth below 200–300 m. An increase in subglacial water discharge at glacier ice fronts due to enhanced surface runoff may also be responsible for an intensification of undercutting and calving. An increase in ocean thermal forcing or subglacial discharge, therefore, has the potential to destabilize marine-terminating glaciers along the coast of Greenland. It remains unclear which glaciers are currently stable but whose margins may retreat in the future and how far inland and how fast they will retreat. Morlighem et al. (2019) sought to quantify the sensitivity and vulnerability of these marine-terminating glaciers to the effects of ocean forcing and subglacial discharge using the ISSM. They relied on a parameterization of undercutting based on ocean thermal forcing and subglacial discharge and use ocean

temperature and salinity from the high-resolution ECCO2 model (Estimating the Circulation and Climate of the Ocean, Phase II). They simulated conditions at the fjord mouth to constrain the ocean thermal forcing. They found that some glaciers, such as Dietrichson Gletscher and Alison Glacier, are sensitive to small increases in ocean thermal forcing. Others, such as Illullip Sermia and Cornell Gletscher, are remarkably stable, even in an ocean warming scenario. Under the most intense experiment, they predicted that the margin of Hayes Gletscher will retreat more than 50 km inland by 2100, becoming lodged in a deep trough. Furthermore, its velocity is predicted to triple within 23 years. The model confirms that ice-ocean interactions can trigger extensive and rapid glacier retreat, but the topography of the bed beneath the ice controls the rate and magnitude of the retreat. This study demonstrated that accounting for ice front dynamics can lead to predictions of significantly more ice loss than models that employ a fixed calving front. Under current oceanic and atmospheric conditions, the authors predicted that this northwest sector of the Greenland ice sheet will contribute more than 1 cm to sea level rise and up to 3 cm by 2100 under the most extreme scenario.

Greenland ice sheet-atmosphere-ocean interactions

Hanna et al. (2020) summed up the current situation concerning GIS ISM-climate model coupling. They note that fully coupled simulations based on state-of-the-art atmosphere-ocean general circulation models (AOGCMs) and ISMs are an emerging field of research. If successful, this line of research will help improve our understanding of processes and feedbacks due to climate-ice sheet coupling in consistent modeling frameworks. However, coupling AOGCMs with ISMs is very difficult because these two kinds of models are run at different resolutions. Another hindrance to progress in this field is the enormous computational expense of running global climate models. The ISMIP6 project, begun in 2017, is leading and supporting current coupled modeling efforts.

For the GIS, coupled models have been applied to investigate several outstanding questions regarding ice-climate interaction, particularly on multicentury and multimillennia timescales. Some examples of the topics already addressed include the impacts of meltwater on ocean circulation, regional impact of ice sheet area change, effect of albedo and cloud change on future SMB, and elevation-SMB feedback. Ongoing work aims to include more interaction processes, such as the effects of ocean warming on ice sheet stability.

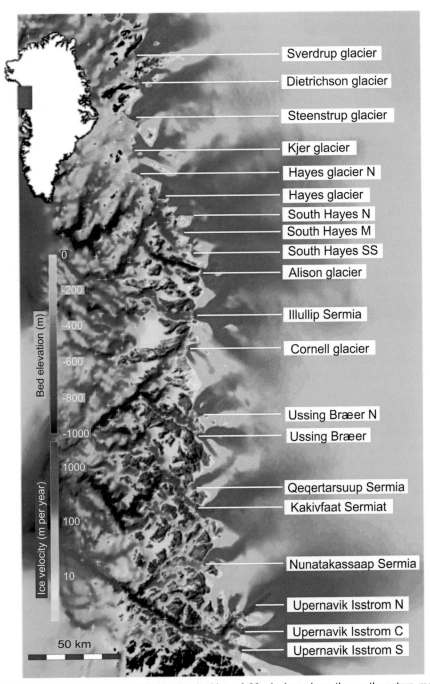

FIG. 10.28 Map showing the location and velocities of 20 glaciers along the northwestern margin of Greenland. Creative Commons Attribution 4.0 License. (Modified from Morlighem, M., Wood, M., Seroussi, H., Choi, Y., Rignot, E., 2019. Modeling the response of northwest Greenland to enhanced ocean thermal forcing and subglacial discharge. The Cryosphere, 13, 723–734. https://doi.org/10.5194/tc-13-723-2019.)

Despite considerable advances with RCMs and SMB models, there are significant remaining differences in absolute values between GIS SMB simulations for the past few decades. However, these are expected to be somewhat reconciled through a new SMB Model Intercomparison Project which is standardizing model comparisons and evaluation using in situ and satellite data. The results of this exercise should help to improve the models as well as highlight the more reliable models. This exercise may help to resolve significant disagreements between model reconstructions of GIS SMB, and especially accumulation, for the past 50–150 years.

PREDICTED FUTURE CHANGES

An article by Overland et al. (2019) synthesizes the latest observational trends and projections for the future of the Arctic, including Greenland. The Arctic region is already operating outside of previous experience, because of high temperatures and large sea ice deficits. Modeled changes of the Arctic cryosphere demonstrate that even limiting global temperature increases (near 2°C) will cause substantial changes to Arctic environments by the middle of this century. There will be less snow and sea ice, melted permafrost, altered ecosystems, and a projected annual mean Arctic temperature increase of +4°C.

Second, even under ambitious emission reduction scenarios, high-latitude glaciers and ice sheets including Greenland will continue to melt, leading to progressive global sea level rise throughout the century. Third, future Arctic changes may drive environmental changes in the lower latitudes through the release of GHGs from melting permafrost soils. The enhanced warming brought on by the increase in atmospheric GHGs may cause shifts in ocean and atmospheric circulation that affect regions far to the south. Amplified Arctic warming will likely raise atmospheric and ocean temperatures to levels not seen for millions of years. These elevated temperature regimes may prevent the achievement of a stabilized global climate. Arctic feedback loops driven by rising temperatures include decreases in albedo, heat storage shifts from loss of glaciers, loss of sea ice and snow cover, increased carbon releases from permafrost, shifts in clouds, and increases in water vapor, and atmospheric and ocean transport changes. The magnitude of Arctic environmental impacts and their potential to affect the larger global climate system is also open to question.

These trends cannot be predicted with certainty. For example, the rates of change for ongoing Arctic feedbacks in the cryosphere (the frozen water portion of the Earth system) are mostly positive but are incompletely understood. It appears extremely likely that these trends will strengthen in the coming decades. Hence there is a pressing need for more precise, detailed modeling of these phenomena. Continued scientific research is required to better underpin both mitigation and adaptation planning.

The threshold for summer air temperatures that would trigger irreversible wastage of the GIS was previously estimated to be an annual global temperature increase of 2–5°C. However, there is now an updated estimate based on a higher resolution simulation that explicitly incorporates albedo and elevation feedbacks. This revised estimate indicates that a rise in Greenland air temperatures of only 0.8–3.2°C would be sufficient to completely melt the GIS. Given this new estimate, it is likely that the GIS will suffer an irreversible loss under the RCP 4.5 scenario.

Given the short-term nature of the observational record, it is difficult to assess the historical importance of the current mass-loss trend. Unlike records of GHG concentrations and global temperature, there are no comparably long records for rates of GIS mass change. Briner et al. (2020) have predicted unprecedented mass loss from the GIS this century, by placing contemporary and future rates of GIS mass loss within the context of the natural variability over the past 12,000 years. They forced a high-resolution ISM with a group of climate histories constrained by ice core data. Their simulations focused on southwestern Greenland, the mass change of which is dominated by SMB. The results agree with an independent chronology of the history of the GIS margin. The largest preindustrial rates of mass loss (up to 6000 Gt per century) occurred in the early Holocene and were similar to the modern (2000–18) rate of around 6100 Gt per century. Simulations of future mass loss from the southwestern GIS, based on RCP scenarios corresponding to low (RCP2.6) and high (RCP8.5) GHG concentration trajectories, predict mass losses of between 8800 and 35,900 Gt during the 21st century. These rates of GIS mass loss exceed the maximum rates over the past 12,000 years. Because rates of mass loss from the southwestern GIS scale linearly with the GIS as a whole, Briner et al. (2020) predict that the rate of mass loss from the GIS will exceed Holocene rates this century.

Global Links

How will changes in the Arctic affect the rest of the world? The Arctic is linked to the global climate system through north-south heat and water exchanges, atmosphere and ocean circulation, river discharge, and the global carbon cycle. There is growing evidence that the Arctic cryosphere has the potential to affect humans outside the Arctic through sea level rise and influence on atmospheric circulation. Other physical global impacts include potential increases of carbon dioxide and methane releases from previously frozen ground and effects on ocean thermohaline circulation.

Dong et al. (2019) analyzed the strength and direction of the flow of the Gulf Stream in the North Atlantic from data collected between 1993 and 2016. During that 23-year interval, the course of the Gulf Stream shifted southward in ocean regions east of 65°W, accompanied by a weakening of the current, largely attributed to an increase in sea surface height (SSH) to the north of the current. Both ocean warming and the increase in mass contribute to the positive SSH, though ocean warming plays a larger role. Dong et al. (2019) did not discuss the possible effects of changes in and around the Arctic Ocean, such as the loss of sea ice, increased inputs of meltwater from the GIS, or the increase in both surface and deeper seawater. From an Arctic perspective, these are some of the "usual suspects" to be considered.

Shifts in Arctic sea ice and snow cover and increased surface temperatures are warming the lower atmosphere in the Arctic, which decreases air density and north-south horizontal pressure gradients and thus influences wind patterns and the jet stream. There is evidence for regional Arctic/midlatitude weather connections from Barents-Kara sea ice loss and cold air outbreaks into eastern Asia. Although there have been extensive new sea-ice-free areas in all years of the past decade, latitude and longitudinal phasing of the tropospheric jet stream pattern have not been conducive for North American midlatitude weather linkages in most years.

CONCLUSIONS

Even with the most up-to-date technology, unraveling the mysteries of the GIS has been and will continue to be herculean tasks. There are many different forcing factors that control the direction and speed of ice flow within the ice sheet. Even after decades of mapping ice sheet stratigraphy and bedrock topography, we are still in possession of only part of the story. Research tools, especially those onboard polar-orbiting satellites, are becoming more refined and transmitting more and better data concerning Greenland than anyone would have thought possible in the 20th century. We cannot afford to wait for even better equipment to come along, or for more sophisticated models to be written. The GIS is melting. In the short term, there is little we can do about it but to keep track of how much water is returning to the ocean.

In spite of contradicting estimates and theories about the Arctic in general and the GIS in particular, some facts have emerged in the past few decades of Arctic research. For instance, it is clear that global change is having amplified effects on the Arctic, such as sea ice loss, increased temperatures, ice sheet and glacier melting, and permafrost degradation. These changes will continue. There are multiple linkage mechanisms between these and other regional variables. Some will produce episodic changes; others will generate sustained unidirectional changes.

Has the GIS reached a tipping point? There are a number of scientists who believe that this has already happened—that the ice sheet cannot be saved. It is more difficult to predict when the ice sheet will disappear. One glaciological research group has recently modeled three different scenarios for the GIS in the coming centuries. To estimate the impact of parametric uncertainties for each of the three emissions scenarios, Aschwanden et al. (2019) performed a 500-member ensemble of simulations in which they varied 11 key model parameters governing atmospheric forcing, surface processes, submarine melt, calving, and ice dynamics. They also identified a control simulation for each scenario, which they used for interscenario comparisons.

They concluded that the GIS will look significantly different in a thousand years. Depending on the emission scenario, the GIS will have lost the following percentages of its ice:

Best-case scenario (RCP 2.6): 8%—25% ice loss, contributing 0.59—1.88 m to global mean sea level (gmsl).

Middle ground scenario (RCP 4.5): 26%—57% ice loss, contributing 1.86—4.17 m to gmsl.

Worst-case scenario (RCP 8.5): 72%—100% ice loss; contributing 5.23—7.28 m to gmsl.

The authors recognized that their model still has imperfections. In particular, they noted that better quantification of feedbacks that interlink and amplify mass loss processes will help clarify the extent to which these feedbacks make changes in the GIS irreversible. Because of these feedbacks, delaying mitigation of GHG emissions is likely to increase sea level rise, even with future significant reductions of emissions.

Once again, the fate of the GIS, like the fate of sea ice, Arctic tundra ecosystems, and global sea level, appears to be in our hands. In spite of all the global climate conferences (the Kyoto Protocol of 1998, the Bali Roadmap of 2007, the Copenhagen Agreement of 2009, the Cancun Agreements of 2010, the Durban Platform of 2011, and the Paris Agreement of 2015), we seem, for the most part, to be ambling down the "worst-case scenario" path with our hands in our pockets.

Greenland, like much of the Arctic, cannot be "rescued" by teams of environmental scientists or engineers rushing north in environmental SWAT teams. Nearly all the environmental hazards faced by the GIS are external. In fact, they are global. To bring down the surface temperature of the GIS, Mr. Jones of Indianapolis, Indiana, along with billions of other human inhabitants of the planet, will have to reduce his carbon emissions. Direct action is required to alleviate the less obvious but hugely important problem of global warming.

REFERENCES

Andersen, M.L., Stenseng, L., Skourup, H., Colgan, W., Khan, S.A., et al., 2015. Basin-scale partitioning of Greenland ice sheet mass balance components (2007–2011). Earth and Planetary Science Letters 409, 89–95. https://doi.org/10.1016/j.epsl.2014.10.015.

Artemieva, I.M., 2019. Lithosphere thermal thickness and geothermal heat flux in Greenland from a new thermal isostasy method. Earth-Science Reviews 188, 469–481. https://doi.org/10.1016/j.earscirev.2018.10.015.

Aschwanden, A., Fahnestock, M.A., Truffer, M., Brinkerhoff, D.J., Hock, R., Khroulev, C., Mottram, R., Abbas Khan, S., 2019. Contribution of the Greenland ice sheet to sea level over the next millennium. Science Advances 5 (6), eaav9396. https://doi.org/10.1126/sciadv.aav9396.

Bevis, M., Harig, C., Khan, S.A., Brown, A., Simons, F.J., Willis, M., Fettweis, X., Van Den Broeke, M.R., Madsen, F.B., Kendrick, E., Caccamise, D.J., Van Dam, T., Knudsen, P., Nylen, T., 2019. Accelerating changes in ice mass within Greenland, and the ice sheet's sensitivity to atmospheric forcing. Proceedings of the National Academy of Sciences of the United States of America 116 (6), 1934–1939. https://doi.org/10.1073/pnas.1806562116.

Bindschadler, R.A., Nowicki, S., Abe-OUCHI, A., Aschwanden, A., Choi, H., Fastook, J., Granzow, G., Greve, R., Gutowski, G., Herzfeld, U., Jackson, C., Johnson, J., Khroulev, C., Levermann, A., Lipscomb, W.H., Martin, M.A., Morlighem, M., Parizek, B.R., Pollard, D., Wang, W.L., 2013. Ice-sheet model sensitivities to environmental forcing and their use in projecting future sea level (the SeaRISE project). Journal of Glaciology 59 (214), 195–224. https://doi.org/10.3189/2013JoG12J125.

Bory, A.J.M., Biscaye, P.E., Svensson, A., Grousset, F.E., 2002. Seasonal variability in the origin of recent atmospheric mineral dust at NorthGRIP, Greenland. Earth and Planetary Science Letters 196 (3–4), 123–134. https://doi.org/10.1016/S0012-821X(01)00609-4.

Botta, F., Dahl-Jensen, D., Rahbek, C., Svensson, A., Nogués-Bravo, D., 2019. Abrupt change in climate and biotic systems. Current Biology 29 (19), R1045–R1054. https://doi.org/10.1016/j.cub.2019.08.066.

Bowling, J.S., Livingstone, S.J., Sol, A.J., Chu, W., 2019. Distribution and dynamics of Greenland subglacial lakes. Nature Communications 10, 2810. https://doi.org/10.1038/s41467-019-10821-w.

Box, J.E., Colgan, W.T., Wouters, B., Burgess, D.O., O'Neel, S., Thomson, L.I., Mernild, S.H., 2018. Global sea-level contribution from Arctic land ice: 1971–2017. Environmental Research Letters 13 (12), 125012. https://doi.org/10.1088/1748-9326/aaf2ed.

Briggs, K., 2016. Measuring the Changing Mass of the Greenland and Antarctic Ice Sheets from Space. Centre for Polar Observation and Modelling. https://cpom.org.uk/measuring-the-changing-mass-of-the-greenland-and-antarctic-ice-sheets-from-space/.

Briner, J.P., Cuzzone, J.K., Badgeley, J.A., Young, N.E., Steig, E.J., et al., 2020. Rate of mass loss from the Greenland ice sheet will exceed Holocene values this century. Nature 586, 70–78. https://doi.org/10.1038/s41586-020-2742-6.

Brook, E.J., 2013a. Ice core methods: Overview. In: Elias, S.A., Mock, C. (Eds.), Encyclopedia of Quaternary Science, second ed. Elsevier Inc., pp. 1145–1156.

Brook, E.J., 2013b. Correlations between Greenland and Antarctica. In: Elias, S.A., Mock, C. (Eds.), Encyclopedia of Quaternary Science, second ed. Elsevier Inc., pp. 410–415. https://doi.org/10.1016/B978-0-444-53643-3.00322-8

Brook, E.J., 2013c. Correlations between Greenland and Antarctica. In: Elias, S.A., Mock, C. (Eds.), Encyclopedia of Quaternary Science, second ed. Elsevier, pp. 1258–1266.

Brown, J.E., 2020. UVM Today: Secrets under the ice. https://www.uvm.edu/uvmnews/news/secrets-under-ice.

Carrivick, J.L., Tweed, F.S., 2019. A review of glacier outburst floods in Iceland and Greenland with a megafloods perspective. Earth-Science Reviews 196, 102876. https://doi.org/10.1016/j.earscirev.2019.102876.

Chappellaz, J., 2013. Ice core methods: methane studies. In: Elias, S.A., Mock, C. (Eds.), Encyclopedia of Quaternary Science. Elsevier, Amsterdam, pp. 1199−1207.

Cuffey, K.M., 2013. Ice core methods: borehole temperature records. In: Elias, S.A., Mock, C. (Eds.), Encyclopedia of Quaternary Science. Elsevier, Amsterdam, pp. 1167−1173.

Dansgaard, W., Johnsen, S.J., Møller, J., Langway Jr., C.C., 1969. One thousand centuries of climatic record from camp century on the Greenland ice sheet. Science 166, 377−380. https://doi.org/10.1126/science.166.3903.377.

Davies, B., 2020. Are the records from the Antarctic ice cores at odds with the records from the Greenland ice cores? AntarcticGlaciers. http://www.antarcticglaciers.org/question/records-antarctic-ice-cores-odds-records-greenland-ice-cores.

Davies, S.M., 2015. Cryptotephras: the revolution in correlation and precision dating. Journal of Quaternary Science 30 (2), 114−130. https://doi.org/10.1002/jqs.2766.

Dong, S., Baringer, M.O., Goni, G.J., 2019. Slow down of the Gulf stream during 1993−2016. Scientific Reports 9 (1), 6672. https://doi.org/10.1038/s41598-019-42820-8.

Dreschhoff, G., Jungner, H., Laird, C.M., 2020. Deuterium−hydrogen ratios, electrical conductivity and nitrate for high-resolution dating of polar ice cores. Tellus B: Chemical and Physical Meteorology 72 (1), 1−6. https://doi.org/10.1080/16000889.2020.1746576.

Dunbar, N.W., Iverson, N.A., Van Eaton, A.R., Sigl, M., Alloway, B.V., Kurbatov, A.V., Mastin, L.G., McConnell, J.R., Wilson, C.J.N., 2017. New Zealand super-eruption provides time marker for the last glacial maximum in Antarctica. Scientific Reports 7 (1), 12238. https://doi.org/10.1038/s41598-017-11758-0.

Enderlin, E.M., Howat, I.M., Jeong, S., Noh, M.J., van Angelen, J.H., et al., 2014. An improved mass budget for the Greenland ice sheet. Geophysics Research Letters 41, 866−872. https://doi.org/10.1002/2013GL059010.

European Space Agency, 2020a. Monitoring the Arctic Heatwave: Alarmingly High Temperatures, Extreme Wildfires and a Significant Loss of Sea Ice. https://scitechdaily.com/monitoring-the-arctic-heatwave-alarmingly-high-temperatures-extreme-wildfires-and-a-significant-loss-of-sea-ice.

European Space Agency, 2020b. Surface Temperatures on August 11, 2020, as Recorded at the Eureka Weather Station. https://www.esa.int/ESA_Multimedia/Images/2020/08/Extreme_temperatures_in_Eureka.

European Space Agency, NASA, 2020. European Space Agency's CryoSat 2 Satellite and NASA's ICESat-2 Satellite. https://www.esa.int/Applications/Observing_the_Earth/CryoSat.

Freitag, J., Kipfstuhl, S., Laepple, T., 2013. Core-scale radioscopic imaging: a new method reveals density-calcium link in Antarctic firn. Journal of Glaciology 59 (218), 1009−1014. https://doi.org/10.3189/2013JoG13J028.

Fujita, K., Matoba, Iizuka, Y., Takeuchi, N., Aoki, T., 2020. Physically based summer temperature reconstruction from ice layers in ice cores. Climates of the Past. https://doi.org/10.5194/cp-2019-97.

Guarino, M.V., Sime, L.C., Schröeder, D., Malmierca-Vallet, I., Rosenblum, E., Ringer, M., Ridley, J., Feltham, D., Bitz, C., Steig, E.J., Wolff, E., Stroeve, J., Sellar, A., 2020. Sea-ice-free Arctic during the last interglacial supports fast future loss. Nature Climate Change 10, 928−932. https://doi.org/10.1038/s41558-020-0865-2.

Guillou, H., Scao, V., Nomade, S., Van Vliet-Lanoë, B., Liorzou, C., Guðmundsson, Á., 2019. 40Ar/39Ar dating of the Thorsmork ignimbrite and Icelandic sub-glacial rhyolites. Quaternary Science Reviews 209, 52−62.

Hanna, E., Pattyn, F., Navarro, F., Favier, V., Goelzer, H., van den Broeke, M.R., Vizcaino, M., Whitehouse, P.L., Ritz, C., Bulthuis, K., Smith, B., 2020. Mass balance of the ice sheets and glaciers − progress since AR5 and challenges. Earth-Science Reviews 201. https://doi.org/10.1016/j.earscirev.2019.102976.

Helle, A., van Ommen, T., 2020. Ice Core Researchers from AWI Drilling at the EastGRIP Ice Core Site, Greenland. https://commons.wikimedia.org/wiki/File:Ice_core_researchers_from_AWI_drilling_at_the_EastGRIP_ice_core_site,_Greenland_2.jpg.

Holme, C., Gkinis, V., Lanzky, M., Morris, V., Olesen, M., et al., 2019. Varying regional δ18O−temperature relationship in high-resolution water isotopes from east Greenland. Climates of the Past 15, 893−912. https://doi.org/10.5194/cp-15-893-2019.

Hvidberg, C.S., Svensson, A., Buchardt, S.L., 2013. Dynamics of the Greenland ice sheet. In: Elias, S.A., Mock, C. (Eds.), Encyclopedia of Quaternary Science, second ed. Elsevier Inc., pp. 439−447. https://doi.org/10.1016/B978-0-444-53643-3.00327-7.

IMBIE Team, 2019. Mass balance of the Greenland ice sheet from 1992 to 2018. Nature 579, 233−238. https://doi.org/10.1038/s41586-019-1855-2.

Joughin, I., Smith, B.E., Howat, I., 2018. Greenland ice mapping project: ice flow velocity variation at sub-monthly to decadal timescales. Cryosphere 12 (7), 2211−2227. https://doi.org/10.5194/tc-12-2211-2018.

Landais, A., Capron, E., Masson-Delmotte, V., Toucanne, S., Rhodes, R., et al., 2018. Ice core evidence for decoupling between midlatitude atmospheric water cycle and Greenland temperature during the last deglaciation. Climates of the Past 14, 1405−1415. https://doi.org/10.5194/cp-14-1405-2018.

Katla, 2020. Katla Volcano, Iceland. https://earth-chronicles.com/natural-catastrophe/iceland-woke-up-katla-volcano.html.

King, M.D., Howat, I.M., Candela, S.G., Noh, M.J., Jeong, S., et al., 2020. Dynamic ice loss from the Greenland ice sheet driven by sustained glacier retreat. Nature

Communications: Earth and Environment 1 (1), 1–7. https://doi.org/10.1038/s43247-020-0001-2.

Knutz, P.C., Newton, A.M.W., Hopper, J.R., Huuse, M., Gregersen, U., Sheldon, E., Dybkjær, K., 2019. Eleven phases of Greenland ice sheet shelf-edge advance over the past 2.7 million years. Nature Geoscience 12 (5), 361–368. https://doi.org/10.1038/s41561-019-0340-8.

Kreutz, K.J., Koffman, B.G., 2013. Glaciochemistry. In: Elias, S.A., Mock, C. (Eds.), Encyclopedia of Quaternary Science, second ed. Elsevier Inc, pp. 326–333. https://doi.org/10.1016/B978-0-444-53643-3.00312-5.

Landais, A., Masson-Delmotte, V., Capron, E., Langebroek, P.M., Bakker, P., Stone, E.J., Merz, N., Raible, C.C., Fischer, H., Orsi, A., Prié, F., Vinther, B., Dahl-Jensen, D., 2016. How warm was Greenland during the last interglacial period? Climate of the Past 12 (9), 1933–1948. https://doi.org/10.5194/cp-12-1933-2016.

Lane, C.S., Blockley, S.P.E., Mangerud, J., Smith, V.C., Lohne, Ø.S., et al., 2012. Was the 12.1ka Icelandic Vedde ash one of a kind? Quaternary Science Reviews 33, 87–99. https://doi.org/10.1016/j.quascirev.2011.11.011.

Lee, J.E., Edwards, J.S., Schmitt, J., Fischer, H., Bock, M., Brook, E.J., 2020. Excess methane in Greenland ice cores associated with high dust concentrations. Geochimica et Cosmochimica Acta 270, 409–430. https://doi.org/10.1016/j.gca.2019.11.020.

Lindsey, R., 2020. NOAA Climate Change: Global Sea Level. https://www.climate.gov/news-features/understanding-climate/climate-change-global-sea-level.

MacGregor, J.A., Fahnestock, M.A., Catania, G.A., Paden, J.D., Prasad Gogineni, S., Young, S.K., Rybarski, S.C., Mabrey, A.N., Wagman, B.M., Morlighem, M., 2015. Radiostratigraphy and age structure of the Greenland ice sheet. Journal of Geophysical Research: Earth Surface 120 (2), 212–241. https://doi.org/10.1002/2014JF003215.

MacGregor, J.A., Fahnestock, M.A., Catania, G.A., Aschwanden, A., Clow, G.D., et al., 2016. A synthesis of the basal thermal state of the Greenland Ice Sheet. Journal of Geophysical Research Earth Surface 121, 1328–1350. https://doi.org/10.1002/2015JF003803.

Mankoff, K.D., Solgaard, A., Colgan, W., Ahlstrøm, A.P., Abbas Khan, S., Fausto, R.S., 2020. Greenland ice sheet solid ice discharge from 1986 through March 2020. Earth System Science Data 12 (2), 1367–1383. https://doi.org/10.5194/essd-12-1367-2020.

Martos, Y.M., Jordan, T.A., Catalán, M., Jordan, T.M., Bamber, J.L., Vaughan, D.G., 2018. Geothermal heat flux reveals the Iceland hotspot track underneath Greenland. Geophysical Research Letters 45 (16), 8214–8222. https://doi.org/10.1029/2018GL078289.

Massling, A., Nielsen, I.E., Kristensen, D., Christensen, J.H., Sorensen, L.L., et al., 2015. Atmospheric black carbon and sulfate concentrations in Northeast Greenland. Atmospheric Chemistry and Physics 15, 9681–9692. https://doi.org/10.5194/acp-15-9681-2015.

McConnell, J.R., 2013. Ice core methods: microparticle and trace element studies. In: Elias, S.A., Mock, C. (Eds.), Encyclopedia of Quaternary Science, second ed. Elsevier Inc, pp. 1207–1212.

McMillan, M., Leeson, A., Shepherd, A., Briggs, K., Armitage, T.W.K., Hogg, A., Kuipers Munneke, P., van den Broeke, M., Noël, B., van de Berg, W.J., Ligtenberg, S., Horwath, M., Groh, A., Muir, A., Gilbert, L., 2016. A high-resolution record of Greenland mass balance. Geophysical Research Letters 43 (13), 7002–7010. https://doi.org/10.1002/2016GL069666.

Mikesell, T.D., van Wijk, K., Otheim, L.T., Marshall, H.P., Kurbatov, A., 2017. Laser ultrasound observations of mechanical property variations in ice cores. Geosciences 7 (3), 47. https://doi.org/10.3390/geosciences7030047.

Morlighem, M., Williams, N.C., Rignot, E., An, L., Arndt, J.E., et al., 2017. BedMachine v3: complete bed topography and ocean bathymetry mapping of Greenland from multibeam echo sounding combined with mass conservation. Geophysical Research Letters 44, 11051–11061. https://doi.org/10.1002/2017gl074954.

Morlighem, M., Wood, M., Seroussi, H., Choi, Y., Rignot, E., 2019. Modeling the response of northwest Greenland to enhanced ocean thermal forcing and subglacial discharge. Cryosphere 13 (2), 723–734. https://doi.org/10.5194/tc-13-723-2019.

Mottram, R., 2020. Guest Post: How the Greenland Ice Sheet Fared in 2020. https://www.carbonbrief.org/guest-post-how-the-greenland-ice-sheet-fared-in-2020.

Mouginot, J., Bjørk, A.A., Millan, R., Scheuchl, B., Rignot, E., 2018a. Insights on the surge behavior of Storstrømmen and L. Bistrup Bræ, Northeast Greenland, over the last century. Geophysical Research Letters 45 (20), 11–205. https://doi.org/10.1029/2018GL079052.

Mouginot, J., Rignot, E., Millan, R., Wood, M., Scheuchl, 2018b. Annual Ice Velocity of the Greenland Ice Sheet (1972–1990), V5 Dataset. https://doi.org/10.7280/D1MM37.

Mouginot, J., Rignot, E., Bjørk, A.A., van den Broeke, M., Millan, R., Morlighem, M., Noël, B., Scheuchl, B., Wood, M., 2019. Forty-six years of Greenland ice sheet mass balance from 1972 to 2018. Proceedings of the National Academy of Sciences of the United States of America 116 (19), 9239–9244. https://doi.org/10.1073/pnas.1904242116.

NASA, 2013. Airborne Radar Looking through Thick Ice during NASA Polar Campaigns. https://www.nasa.gov/content/goddard/.

NASA, 2015. Land-facing, Southwest Greenland Ice Sheet movement decreasing. https://www.nasa.gov/feature/goddard/land-facing-southwest-greenland-ice-sheet-movement-decreasing.

NASA, 2015. https://svs.gsfc.nasa.gov/4249.

NASA, 2020a. GIS Cumulative Melt Days, Jan 1–Aug 6, 2020; GIS Melting Day Anomaly June 21–Aug1, 2020. http://nsidc.org/greenland-today/.

NASA, 2020b. Greenland Albedo Anomalies, Aug 31–Sep 3 2020. https://www.carbonbrief.org/guest-post-how-the-greenland-ice-sheet-fared-in-2020.

NEEM Community Members, 2017. Interglacial reconstructed from a Greenland folded ice core. Nature 493, 489–494.

NOAA, 2013. Heinrich and Dansgaard–Oeschger Events. https://www.ncdc.noaa.gov/abrupt-climate-change/.

NSIDC, 2007. Quick Facts on Ice Sheets. https://nsidc.org/cryosphere/quickfacts/icesheets.html.

NSIDC, 2020. Greenland Ice Sheet Today, Aug 12, 2020. https://nsidc.org/greenland-today/2020/08/greenlands-2020-summer-melting-a-new-normal/.

Oppenheimer, M., Glavovic, B.C., Hinkel, J., van de Wal, R., Magnan, A.K., 2019. Sea level rise and implications for low-lying islands, coasts and communities. In: Pörtner, H.O., Roberts, D.C., Masson-Delmotte, V., Zhai, P., Tignor, M. (Eds.), IPCC Special Report on the Ocean and Cryosphere in a Changing Climate. IPCC.

Overland, J., Dunlea, E., Box, J.E., Corell, R., Forsius, M., Kattsov, V., Olsen, M.S., Pawlak, J., Reiersen, L.O., Wang, M., 2019. The urgency of Arctic change. Polar Science 21, 6–13. https://doi.org/10.1016/j.polar.2018.11.008.

Petit, J.R., Raynaud, D., 2020. Forty years of ice-core records of CO_2. Nature 579 (7800), 505–506. https://doi.org/10.1038/d41586-020-00809-8.

Priscu, J.C., Christner, B.C., Foreman, C.M., Royston-Bisho, G., 2013. Ice core methods: biological material. In: Elias, S.A., Mock, C. (Eds.), Encyclopedia of Quaternary Science, second ed. Elsevier Inc, pp. 1156–1167.

Sandberg Sørensen, L., Simonsen, S.B., Forsberg, R., Khvorostovsky, K., Meister, R., Engdahl, M.E., 2018. 25 years of elevation changes of the Greenland ice sheet from ERS, Envisat, and CryoSat-2 radar altimetry. Earth and Planetary Science Letters 495, 234–241. https://doi.org/10.1016/j.epsl.2018.05.015.

Schmittner, A., Saenko, O.A., Weaver, A.J., 2003. Coupling of the hemispheres in observations and simulations of glacial climate change. Quaternary Science Reviews 22 (5–7), 659–671. https://doi.org/10.1016/S0277-3791(02)00184-1.

Schwander, J., 2013. Ice core methods: chronologies. In: Elias, S.A., Mock, C. (Eds.), Encyclopedia of Quaternary Science, second ed. Elsevier Inc, pp. 1173–1181.

Seidel, D.J., Fu, Q., Randel, W.J., Reichler, T.J., 2008. Widening of the tropical belt in a changing climate. Nature Geoscience 1, 21–24. https://doi.org/10.1038/ngeo.2007.38.

Seip, K.L., Grøn, Ø., Wang, H., 2018. Carbon dioxide precedes temperature change during short-term pauses in multimillennial palaeoclimate records. Palaeogeography, Palaeoclimatology, Palaeoecology 506, 101–111. https://doi.org/10.1016/j.palaeo.2018.06.021.

Seki, O., Kawamura, K., Bendle, J., Izawa, Y., Suzuki, I., et al., 2015. Carbonaceous aerosol tracers in ice-cores record multi-decadal climate oscillations. Scientific Reports 5, 14450. https://doi.org/10.1038/srep14450.

Sharma, S., Ishizawa, M., Chan, D., Lavoué, D., Andrews, E., Eleftheriadis, K., Maksyutov, S., 2013. 16-year simulation of arctic black carbon: transport, source contribution, and sensitivity analysis on deposition. Journal of Geophysical Research Atmospheres 118 (2), 943–964. https://doi.org/10.1029/2012JD017774.

Siddall, M., Stocker, T.F., Blunier, T., Spahni, R., McManus, J.F., Bard, E., 2006. Using a maximum simplicity paleoclimate model to simulate millennial variability during the last four glacial periods. Quaternary Science Reviews 25 (23–24), 3185–3197. https://doi.org/10.1016/j.quascirev.2005.12.014.

Simonsen, M.F., Baccolo, G., Blunier, T., Borunda, A., Delmonte, B., Frei, R., Goldstein, S., Grinsted, A., Kjær, H.A., Sowers, T., Svensson, A., Vinther, B., Vladimirova, D., Winckler, G., Winstrup, M., Vallelonga, P., 2019. East Greenland ice core dust record reveals timing of Greenland ice sheet advance and retreat. Nature Communications 10 (1), 4494. https://doi.org/10.1038/s41467-019-12546-2.

Smith-Johnsen, S., De Fleurian, B., Schlegel, N., Seroussi, H., Nisancioglu, K., 2020. Exceptionally high heat flux needed to sustain the Northeast Greenland ice stream. Cryosphere 14 (3), 841–854. https://doi.org/10.5194/tc-14-841-2020.

Starr, C., NASA, 2013. Greenland Ice Sheet stratigraphy. https://svs.gsfc.nasa.gov/4249.

Stauffer, B., 2013. Ice core methods: CO_2 studies. In: Elias, S.A., Mock, C. (Eds.), Encyclopedia of Quaternary Science, second ed. Elsevier Inc, pp. 1181–1189.

Steffen, W., Rockström, J., Richardson, K., Lenton, T.M., Folke, C., et al., 2018. Trajectories of the Earth system in the anthropocene. Proceedings of the National Academy of Sciences United States of America 115, 8252–8259. https://doi.org/10.1073/PNAS.1810141115.

Steig, E.J., Alley, R.B., 2002. Phase relationships between Antarctic and Greenland climate records. Annals of Glaciology 35, 451–456. https://doi.org/10.3189/172756402781817211.

Stocker, T.F., Johnsen, S.J., 2003. A minimum thermodynamic model for the bipolar seesaw. Paleoceanography 18 (4). https://doi.org/10.1029/2003PA000920.

Taylor, K., 2013. Ice core methods: conductivity studies. In: Elias, S.A., Mock, C. (Eds.), Encyclopedia of Quaternary Science, second ed. Elsevier Inc, pp. 1189–1192.

Tedesco, M., Box, J.E., Cappelen, J., Fausto, R.S., Fettweis, X., et al., 2018. Greenland Ice Sheet. Arctic Report Card: Update for 2018. NOAA. https://arctic.noaa.gov/Report-Card/Report-Card-2018/ArtMID/7878/ArticleID/781/Greenland-Ice-Sheet.

Thomas, J.L., Polashenski, C.M., Soja, A.J., Marelle, L., Casey, K.A., Choi, H.D., Raut, J.C., Wiedinmyer, C., Emmons, L.K., Fast, J.D., Pelon, J., Law, K.S., Flanner, M.G., Dibb, J.E., 2017. Quantifying black carbon deposition over the Greenland ice sheet from forest fires in Canada. Geophysical Research Letters 44 (15), 7965–7974. https://doi.org/10.1002/2017GL073701.

Thomas, Z.A., Jones, R.T., Turney, C.S.M., Golledge, N., Fogwill, C., Bradshaw, C.J.A., Menviel, L., McKay, N.P., Bird, M., Palmer, J., Kershaw, P., Wilmshurst, J., Muscheler, R., 2020. Tipping elements and amplified polar warming during the last interglacial. Quaternary Science Reviews 233, 106222. https://doi.org/10.1016/j.quascirev.2020.106222.

Trusel, L.D., Das, S.B., Osman, M.B., Evans, M.J., Smith, B.E., Fettweis, X., McConnell, J.R., Noël, B.P.Y., van den Broeke, M.R., 2018. Nonlinear rise in Greenland runoff in response to post-industrial Arctic warming. Nature 564 (7734), 104−108. https://doi.org/10.1038/s41586-018-0752-4.

University Corporation for Atmospheric Research (UCAR), 2020. Climate Modeling. https://scied.ucar.edu/longcontent/climate-modeling.

University of Copenhagen Center for Ice and Climate, 2020a. Isotopes Reveal Changing Moisture Sources at the End of the Glacial. http://www.iceandclimate.nbi.ku.dk/research/past_atmos/past_temperature_moisture/isotopes_reveal/.

University of Copenhagen Center for Ice and Climate, 2020b. Isotopes and the Delta Notation. http://www.iceandclimate.nbi.ku.dk/research/past_atmos/past_temperature_moisture/isotopes_delta_notation/.

University of Copenhagen Center for Ice and Climate, 2020c. Synchronizing Ice Cores via the Global CH_4 Record. http://www.iceandclimate.nbi.ku.dk/research/strat_dating/synch_ice_core_rec/synch_ch4/.

van Ommen, T., Kjær, A.H., 2020. Ice Core Researchers from AWI Drilling at the EastGRIP Ice Core Site, Greenland. https://commons.wikimedia.org/wiki/File:Ice_core_researchers_from_AWI_drilling_at_the_EastGRIP_ice_core_site,_Greenland_2.jpg.

Velicogna, I., Mohajerani, Y., Geruo, A., Landerer, F., Mouginot, J., Noel, B., Rignot, E., Sutterley, T., van den Broeke, M., van Wessem, M., Wiese, D., 2020. Continuity of ice sheet mass loss in Greenland and Antarctica from the GRACE and GRACE follow-on missions. Geophysical Research Letters 47 (8), e2020GL087291. https://doi.org/10.1029/2020GL087291. Figure 2 on page 5.

Vinther, B.M., Johnsen, S.J., 2013. Greenland stable isotopes. In: Elias, S.A., Mock, C. (Eds.), Encyclopedia of Quaternary Science, second ed. Elsevier Inc, pp. 403−409. https://doi.org/10.1016/B978-0-444-53643-3.00323-X.

Voosen, P., 2019. Ancient soil from secret Greenland base suggests Earth could lose a lot of ice. Science: Earth News. https://doi.org/10.1126/science.aba0351.

Voosen, P., 2019. Ancient soil from secret Greenland base suggests Earth could lose a lot of ice. Science 369, 1043−1044. https://doi.org/10.1126/science.aba0351.

Waelbroeck, C., Lougheed, B.C., Vazquez Riveiros, N., Missiaen, L., Pedro, J., Dokken, T., Hajdas, I., Wacker, L., Abbott, P., Dumoulin, J.P., Thil, F., Eynaud, F., Rossignol, L., Fersi, W., Albuquerque, A.L., Arz, H.,

Austin, W.E.N., Came, R., Carlson, A.E., Ziegler, M., 2019. Consistently dated Atlantic sediment cores over the last 40 thousand years. Scientific Data 6 (1), 165. https://doi.org/10.1038/s41597-019-0173-8.

Willeit, M., Ganopolski, A., Calov, R., Brovkin, V., et al., 2019. Mid-pleistocene transition in glacial cycles explained by declining CO_2 and regolith removal. Science Advances 5, eaav7337. https://doi.org/10.1126/sciadv.aav7337.

Willerslev, E., Cappellini, E., Boomsma, W., Nielsen, R., Hebsgaard, M.B., et al., 2007. Ancient biomolecules from deep ice cores reveal a forested southern Greenland. Science 317 (5834), 111−114. https://doi.org/10.1126/science.1141758.

Yao, S.L., Luo, J.J., Huang, G., Wang, P., 2017. Distinct global warming rates tied to multiple ocean surface temperature changes. Nature Climate Change 7 (7), 486−491. https://doi.org/10.1038/nclimate3304.

Yan, Y., Bender, M.L., Brook, E.J., Clifford, H.M., Kemeny, P.C., Kurbatov, A.V., Mackay, S., Mayewski, P.A., Ng, J., Severinghaus, J.P., Higgins, J.A., 2019. Two-million-year-old snapshots of atmospheric gases from Antarctic ice. Nature 574 (7780), 663−666. https://doi.org/10.1038/s41586-019-1692-3.

Yau, A.M., Bender, M.L., Robinson, A., Brook, E.J., 2016a. Reconstructing the last interglacial at summit, Greenland: insights from GISP2. Proceedings of the National Academy of Sciences of the United States of America 113 (35), 9710−9715. https://doi.org/10.1073/pnas.1524766113.

Yau, A.M., Bender, M.L., Blunier, T., Jouzel, J., 2016b. Setting a chronology for the basal ice at Dye-3 and GRIP: implications for the long-term stability of the Greenland ice sheet. Earth and Planetary Science Letters 451, 1−9. https://doi.org/10.1016/j.epsl.2016.06.0530012-821X.

FURTHER READING

Bjørk, A.A., Kruse, L.M., Michaelsen, P.B., 2015. Brief communication: getting Greenland's glaciers right − a new data set of all official Greenlandic glacier names. The Cryosphere 9, 2215−2218. https://doi.org/10.5194/tc-9-2215-2015.

Gisela, D., Högne, J., Claude, M.L., 2020. Deuterium−hydrogen ratios, electrical conductivity and nitrate for high-resolution dating of polar ice cores. Tellus B: Chemical and Physical Meteorology 1−6. https://doi.org/10.1080/16000889.2020.1746576.

Khan, S.A., 2017. Greenland Ice Sheet Surface Elevation Change. GEUS Data Center. https://doi.org/10.22008/promice/data/DTU/.

Khan, S.A., Sasgen, I., Bevis, M., Van Dam, T., Bamber, J.L., Willis, M., Kjær, K.H., Wouters, B., Helm, V., Csatho, B., Fleming, K., Bjørk, A.A., Aschwanden, A., Knudsen, P., Munneke, P.K., 2016. Geodetic measurements reveal similarities between post−last glacial maximum and present-day mass loss from the Greenland ice sheet. Science Advances 2 (9), e1600931. https://doi.org/10.1126/sciadv.1600931.

Millan, R., Mouginot, J., Rabatel, A., Jeong, S., Cusicanqui, D., Derkacheva, A., Chekki, M., 2019. Mapping surface flow velocity of glaciers at regional scale using a multiple sensors approach. Remote Sensing 11 (21), 2498. https://doi.org/10.3390/rs11212498.

Mouginot, J., Rignot, E., 2019. Glacier Catchments/Basins for the Greenland Ice Sheet. https://doi.org/10.7280/D1WT11.

NOAA/NESDIS/NCEI Ocean Climate Laboratory, 2019. Global ocean heat and salt content. Global Ocean Heat and Salt Content. https://www.nodc.noaa.gov/OC5/3M_HEAT_CONTENT/.

Will, S., Johan, R., Katherine, R., Carl, F., et al., 2018. Trajectories of the Earth system in the anthropocene. Proceedings of the National Academy of Sciences 115, 8252–8259. https://doi.org/10.1073/pnas.1810141115.

CHAPTER 11

Changes in Terrestrial Environments

INTRODUCTION

The Arctic tundra is thought by many as a rather monotonous landscape with a small number of herbaceous and moss species clinging to life in an inhospitable climate. In fact, this biome is as diverse in vegetation communities as the boreal forest or temperate grassland biomes to the south. The simplest definition of Arctic tundra is the land beyond the northern limit of trees. Tundra covers about 360,000 km^2 in Alaska, 2,480,000 km^2 in Canada, 2,167,000 km^2 in Greenland and Iceland, and 2,560,000 km^2 in Russia.

In recent years, a distinction has been drawn that separates Alpine environments from Arctic tundra (Körner et al., 2011). At high latitudes (>65−70°N), alpine vegetation merges with Arctic tundra. Despite a number of common taxa and the overwhelming influence of low averages temperatures, the Arctic tundra ecosystem is very different from the alpine zone in terms of climate, land surface structure, and vegetation. Hence, many recommend that alpine vegetation not be referred to as "alpine tundra." Most Arctic tundra grows at elevations near sea level. Here the atmosphere is far denser than the atmosphere of high mountains. The greenhouse effect is far stronger at sea level, diminishing the difference between daytime and nighttime temperatures. Low elevations in the Arctic receive far less UV radiation than alpine regions. An important feature of alpine life is its geographic isolation. Mountain tops, with their high biological diversity, represent habitat islands or archipelagos surrounded by lowlands, where most alpine organisms cannot survive. Therefore, in this chapter, the word "tundra" refers only to Arctic vegetation, not Alpine.

Although finer divisions of the Arctic tundra biome have been made by ecologists, in the context of this chapter, it is convenient to divide the biome into High Arctic and Low Arctic ecosystems (Fig. 11.1). The High Arctic, also known as polar desert, occurs in the far north, such as the Canadian Arctic Archipelago, northern Greenland, and the islands north of Siberia. Here, the vegetation cover is generally sparse, and the floral diversity is quite low (Fig. 11.2). High Arctic climates are extremely cold, with mean July temperatures

(TMAX) from 1 to 4°C and mean January temperatures (TMIN) in the −20 to −40°C range. Mean annual temperatures are well below freezing, from about −6 to −15°C. Mean annual precipitation (MAP) is extremely low, ranging from about 60 to 400 mm, with almost all falling as snow.

The Low Arctic ecosystem is much more floristically diverse and extensive in area. The vegetation ranges from tall shrubs to prostrate herbs, lichens, and mosses, depending on latitude, slope, aspect, and available moisture (Fig. 11.2). Mean July temperatures in the Low Arctic range from 4 to 12°C; mean January temperatures range from about −15°C near the coasts of Arctic Scandinavia to −50°C in the coldest parts of interior north-eastern Siberia. Mean annual temperatures are below freezing, ranging from about −1 to −15°C. Mean annual precipitation is low, ranging from about 120 to 550 mm.

Przybylak (2016) presented temperature profiles for many Arctic meteorological stations, as summarized in Fig. 11.3. These sites are divided into three climatic types: (1) sites with continental climate (solid lines in Fig. 11.3); (2) sites with a maritime climate (dashed lines in Fig. 11.3; and (3) sites with coastal climate (dotted lines in Fig. 11.3). There are three well-defined types of annual temperatures cycles in the Arctic: maritime, coastal, and continental. Jan Mayen has a maritime type climate. In terms of the thermal continentality of Arctic climate, Jan Mayen has a continentality below 20%. This type of Arctic climate exhibits a very small annual range of temperatures, slightly exceeding 10°C. The mean temperatures of the summer months (June−August) vary only between 4 and 5°C. Similarly, the mean monthly winter temperatures (December to March) vary only between −5°C and −6°C.

The Maly Karmakuly and Egedesminde sites have a coastal type climate, characterized by a continentality index of about 40%−50%. This type is transitional between the maritime and continental types. The annual range of temperatures is about 20°C. Winter air temperatures are markedly lower than the maritime type and also significantly higher than in the continental type.

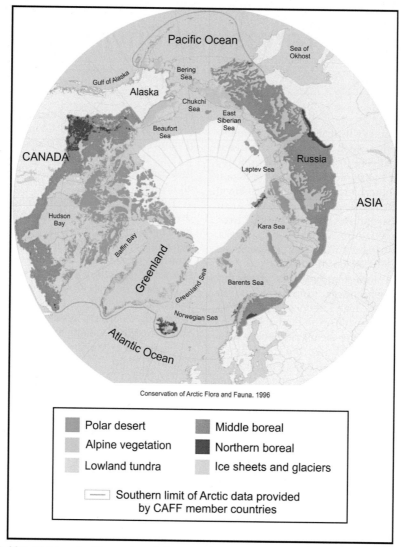

FIG. 11.1 Map of the major biomes (vegetation zones) of the Arctic. (Modified from Arctic Portal Library, 2011. CAFF Map No.33 - Major Biomes (Vegetation Zones) of the Arctic. (2011). Retrieved August 2, 2020, from http://library.arcticportal.org/1364/.)

Summer air temperatures are similar or higher than those of the maritime type.

The rest of the stations shown in Fig. 11.3 have a continental type climate, with a continentality above 60%, even though most of them are situated on or near seacoasts. The continental effect at these sites is greatly enhanced by proximity to sea ice that blocks the warmth of the ocean from reaching land for much or all the year. Continental climate is characterized by a high annual temperature range, about 40°C. It also has the lowest winter temperatures, with monthly mean values between −30°C and −35°C. Summer temperatures here are relatively high, especially in southern parts of the Arctic, where they reach values near 10°C. In the High Arctic, however, they reach 1−3°C. At the North Pole, the mean temperature of the warmest month (July) remains slightly below freezing (−0.5°C). Colder still is the climate at the center of the Greenland Ice Sheet (Eismitte Station), where mean monthly temperatures range from about −42°C in February to about −12°C in July. Continental climate dominates almost 80% of the Arctic landscapes.

FIG. 11.2 Top: High Arctic Tundra photo: Alexandra Fiord, Ellesmere Island. Note open-top environmental chamber used in research by the International Tundra Experiment. Photo by Cassandra Elphinstone, Creative Commons Attribution-Share Alike 4.0 International license. Bottom: Low Arctic Tundra photo: Arctic coastal plain, Alaska. Photo by Donald A. Walker, Circumpolar Arctic Vegetation project. Creative Commons Attribution-NonCommercial-ShareAlike License. (From Elphinstone, C., 2020. High Arctic Tundra photo: Alexandra Fiord, Ellesmere Island. Note Open-Top Environmental Chamber Used in Research by the International Tundra Experiment. https://en.wikipedia.org/wiki/International_Tundra_Experiment.)

Because the mean annual air temperatures of all these tundra regions are below freezing, the entire region, both Low and High Arctic, has permanently frozen soils, or permafrost. Permafrost exists where the ground remains frozen for at least 2 years in a row. The upper 0.4—4 m of these soils generally thaw in summer, creating an active zone in which plants can take root, and draw moisture and nutrients. Below

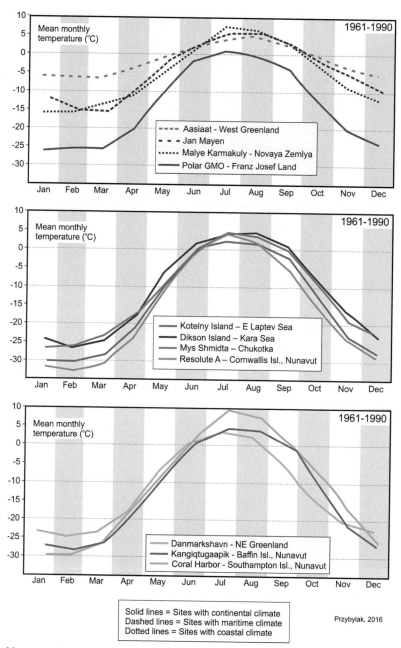

FIG. 11.3 Mean monthly temperature profiles for a selection of sites with meteorological stations in the Arctic. (1) sites with continental climate (solid lines); (2) sites with a maritime climate (dashed lines); and (3) sites with coastal climate (dotted lines). (Modified from Przybylak, R., 2016. The Climate of the Arctic, Second ed. Springer. Multiple pages.)

this active layer, the soil remains frozen, often to depths of hundreds of meters. The coldest parts of Siberia have permafrost depths exceeding 1500 m.

Tundra plants are highly adapted to survive harsh conditions: very short growing season, low summer temperatures, brutally cold winters, and lack of moisture. Many tundra plants are perennials. This life strategy allows them time to slowly accumulate sufficient metabolic energy to set seeds, a process that may take several growing seasons to accomplish. Likewise, many tundra insects live multiple years, unlike their counterparts in more temperate regions. Again, this multi-year life strategy allows them to slowly accumulate enough energy to reproduce. There are very few species of year-round resident terrestrial vertebrates in the Arctic. Most birds found on the Arctic coastal plain are only summer visitors, taking advantage of the rapid growth of plants in the continuous daylight, and the lack of predators they face in lower latitudes, to quickly raise a brood of chicks in the short, intense Arctic summer. The few resident birds, such as ptarmigan, burrow under the snow in winter, thus avoiding the deathly cold air temperatures at the surface. These birds build up as much fat as possible in the summer and survive the winter by a combination of burning fat reserves and foraging on dwarf willow and birch shrubs beneath the snow. Grizzly bears hibernate through the Arctic winter, while polar bears go out on the sea ice to hunt seals through the winter months and then hibernate during the summer.

TERRAIN FEATURES

Although the Arctic is commonly thought to be ice-covered, less than 40% of land surfaces are buried by permanent ice. The remainder is ice-free, largely due to aridity. Arctic terrains are extremely variable, ranging from steep mountain slopes to flat coastal plains. Large parts of north-eastern Siberia, Alaska, and the Yukon remained unglaciated during the Pleistocene, with the Siberian and Alaskan regions connected by a land bridge during low sea level episodes, forming the ancient ecosystem called Beringia. Landscapes that were part of Beringia tend to have better-developed soils than the regions that were glaciated during the Late Pleistocene. The latter regions had their ancient soils stripped away by erosion from slowly moving ice sheets and glaciers. The following summaries are essentially snapshots of localities from regions across the Arctic, starting in Alaska and moving west through Russia, Scandinavia, Greenland, and Canada. The sites include both the Low and High Arctic ecosystems.

Arctic Coastal Plain of Alaska

The Arctic Coastal Plain of Alaska (ACPA) comprises lowland tundra with a land area of 49,870 km^2, with freshwater lakes and ponds covering an additional 10,160 km^2. The ecoregion is mainly a smooth plain rising very gradually from the Arctic Ocean to the foothills of the Brooks Range, 180 m above sea level. Locally, permafrost-related features mark the terrain surface. Pingos (ice-cored hills) rise 6−70 m above the surrounding area, and other ice-related features, such as extensive networks of ice-wedge polygons, oriented lakes (ranging from a few meters to 15 km in length), peat ridges, and frost boils, are common.

Continuous permafrost underlies the entire North Slope of Alaska. The permafrost table is at or near the ground surface, with an active layer of less than 0.50 m (except beneath the larger rivers, where thawing may be deeper). The year-round frozen ground dictates that the ACPA is very poorly drained. Thaw lakes cover 20%−50% of the land surfaces (Fig. 11.4). In many areas, lakes are rectangular and oriented north-northwest. This orientation is related to the effects of predominant winds on the permafrost shorelines of thaw lakes. Thaw lakes expand approximately 1 m/year in places and range from less than 1−7 m in depth. Streams originate in the highlands of ecoregions to the south. Streams west of the Colville River tend to be sluggish and meandering. The climate, measured at Utqiaġvik (formerly known as Barrow), is quite frigid and arid, given its coastal location (Fig. 11.5). TMAX occurs in July, with a long-term average of 4.9°C. TMIN falls in February, at −25.7°C. MAP at Utqiaġvik falls just above the hyper-arid level, at 114 mm/year (Alaska Climate Research Center, 2020).

Northeastern Siberia: Chukotka Peninsula

The Chukotka Peninsula is the easternmost region of Asia, bordered on the north by the Chukchi and East Siberian Seas, and on the east by the Bering Sea. The ecoregion stretches 700 km from the mouth of the Lena River in the northwest to the eastern tip of the Chukchi Peninsula in the east. The Chukchi Peninsula projects eastward to the Bering Strait between Siberia and the Alaska Peninsula. Ecologically, Chukotka comprises three distinct areas: the northern Arctic desert, the central tundra, and the taiga in the south. About half of Chukotka is above the Arctic Circle. The region is very mountainous, containing the Chukotsky Mountains and the Anadyr Highlands. Chukotka's rivers spring from its northern and central mountains. The terrain is mostly treeless Arctic plains on alluvial deposits and widespread groundwater saturation. Regional climates

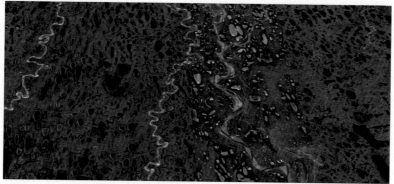

Landsat photo of ACPA thaw lakes

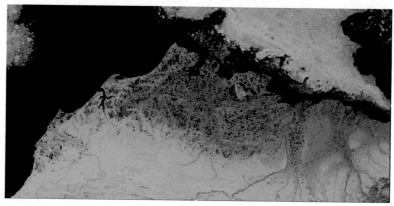

Infrared photo of western ACPA

FIG. 11.4 2007 infrared photo of the western ACPA, showing the density of thaw lakes. Image courtesy of Jeff Schmaltz, MODIS Rapid Response Team, Goddard Space Flight Center. *ACPA*, Arctic Coastal Plain of Alaska. (From NASA Earth Observatory. North Slope of Alaska. (2007). Retrieved from https://earthobservatory.nasa.gov/images/7863/north-slope-of-alaska.)

are characterized by long, cold winters and very short summers. MAP on the north coast is in the arid category, at 184 mm/year. Regional TMIN averages −26.9°C, and TMAX is 8.7°C.

New Siberian Islands

The New Siberian Islands form an archipelago in the Russian High Arctic. The islands lie north of the East Siberian mainland, situated between the Laptev and East Siberian Seas.

The islands encompass a land area of about 29,000 km² and consist of Kotelny, Faddeyevsky, and Bunge Land islands and two smaller islets: Zheleznyakov and Matar. The New Siberian Islands are low-lying, essentially forming the northern perimeter of the East Siberian Lowland. Their highest point is located on Bennett Island, with an elevation of 426 m. The

climate is Arctic and severe. Snow cover is present for 9 months of the year. TMIN ranges from −28 to −31°C. TMAX on the coast is 2.9°C. The climate is also arid, with MAP at or below 132 mm.

Northwestern Siberia: The Taimyr Peninsula

The Taimyr Peninsula is the northernmost part of Eurasia, lying in north-central Siberia. The peninsula is bounded on the west by the Kara Sea and the Gulf of Yenisei, and on the east by the Laptev Sea. The peninsula encompasses 400,000 km². The center of the peninsula is dominated by the Byrranga Mountains, reaching elevations of 500–1150 m. The rest of the peninsula consists of tundra-covered lowlands. As on the ACPA, the lowlands are dominated by poorly drained landscapes dotted with thaw lakes. Cape Chelyuskin climate (77°44′N) is bitterly cold, with a

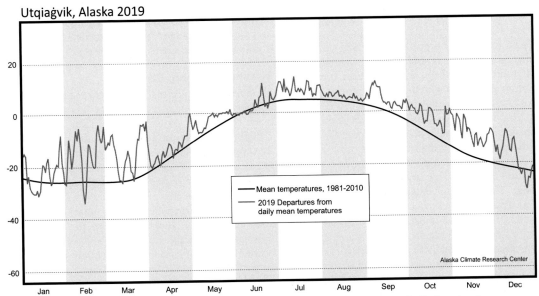

FIG. 11.5 The 2019 temperature record for Utqiaġvik (formerly known as Barrow), Alaska, showing mean temperatures from 1981 to 2010, historic daily low and high temperatures, and departures above and below daily mean temperatures. Note that 90% of days were warmer than the long-term average. (Data from the Alaska Climate Research Center, 2020. Alaska Statewide Climate Summary. Retrieved from http://akclimate. org/sites/Default/Files/202005_May_summary.pdf.)

TMIN of −28.2°C and TMAX of 1.4°C. Despite its coastal location, the MAP of Cape Chelyuskin falls in the arid category, at 205 mm.

Arctic Scandinavia

This northernmost region of Scandinavia, exemplified by the Finnmark region of Norway, encompasses terrains ranging from barren coastal areas facing the Barents Sea to sheltered fjord areas and river valleys with gullies and tree vegetation. About half of the county is above the tree line, and large parts of the other half are covered with small Downy birch. The interior parts of the region lie on the Finnmarksvidda plateau, with elevations from 300 to 400 m, and numerous lakes and river valleys. This plateau is one of the coldest regions in Western Europe. TMIN is −17.1°C, and TMAX is 13.1°C. The mean annual precipitation is semiarid, at 366 mm.

Greenland

Greenland is located between the Arctic Ocean and the North Atlantic Ocean, northeast of Canada, and northwest of Iceland. Greenland is the largest island in the world. The island has 44,087 km of coastline. The land

area of Greenland is 2,166,086 km², of which 410,449 km² is ice-free and 1,755,637 km² is ice-covered. The climate is Arctic to Subarctic, with cool summers and cold winters. The ice-free terrain is dominated by a narrow, mountainous, barren, rocky coast. The highest elevation is the summit of Gunnbjørn Fjeld, the highest point in the Arctic (3694 m). There are few long-term climate records outside of the southern coastal regions. One interesting record comes from the Eismitte Meteorological Station at the center of the Greenland ice sheet (70°53′N). Here, the TMAX is −12.2°C, and the TMIN is −47.2°C. MAP at Eismitte is almost hyperarid at 109 mm. The Danmarkshavn Meteorological Station is on the northeast coast of Greenland, at latitude: 76°45′N. Here, TMIN is −24.3°C in February, and TMAX is 3.7°C in July. MAP at the site is a mere 139 mm, despite its coastal location.

Greenland has 310 known species of vascular plants, with species diversity decreasing northward. The High Arctic tundra of northern Greenland is, overall, poorly vegetated. Well-developed High Arctic plant communities are concentrated on the east coast of Greenland, gradually becoming more sparse from south to north. On Peary Land, northernmost Greenland, vegetation covers only about 5% of the terrain. Annual precipitation is low at 25–200 mm, all falling as snow. Much of this is blown into drifts, uncovering most land surfaces. Snow drifts provide the only source of freshwater during the brief summer, and the sparse vegetation communities occur in association with them. These are generally sedge fens and cottongrass, heaths of Arctic bell heather (*Cassiope tetragona*), and snowbed vegetation of mosses. Large expanses have virtually no vegetation at all, especially at the interior and at higher altitudes.

Jameson Land is located on the central east coast at the transition between Low and High Arctic tundra ecoregions. This is the largest lowland area in Greenland. From sea level to 400 m, the vegetation is dominated by dwarf scrub of *Betula nana, Cassiope tetragona*, and *Vaccinium uliginosum*. Moist protected sites have 75% plant cover, generally a heath rich in mosses.

Low Arctic tundra covers the ice-free coasts of southern Greenland. The ground is covered in stunted vegetation yet teeming with wildflowers and wild berries in the lowland areas during summer. The Low Arctic tundra lies below 75°N latitude at Melville Bay on the west coast and 70°N at Scoresby Sund on the east coast. Vegetation within the Low Arctic ecoregion varies according to altitude, distance from the exposed coastline, and available moisture. Some regions closest to the Greenland ice sheet are quite arid, such as along the west coast, which exhibits a dwarf-scrub heath and steppe-like vegetation. At the head of fjords where the climate is warmer sub-Arctic, sheltered areas support scrub and low forests composed of green alder (*Alnus crispa*), downy birch (*Betula pubescens*), and Greenland mountain ash (*Sorbus groenlandica*).

As elsewhere in the Arctic, Greenland climates are rapidly warming (Fowler et al., 2019). A recent study by Saros et al. (2019) quantified environmental responses to recent climate changes in West Greenland using long-term monitoring. Based on more than 40 years of climate data, they found that after 1994, mean June temperatures increased by 2.2°C and mean winter precipitation doubled from 21 to 40 mm. Since 2006, mean July air temperatures rose an additional 1.1°C. Nonlinear environmental responses occurred simultaneously or shortly after these abrupt climate shifts. These included increasing discharge rates from the Greenland ice sheet, increasing dust deposition, advancing dates of plant phenology, earlier ice-out dates in lakes, and greater diversity of algal functional traits. Modeled timing of ice melt off of regional lakes shifted 6 days earlier than historic values in 1993. The average date of the emergence of at least 50% of regional plant species came 13 days earlier in 2009

than it had before. However, the plant phenology record represents community-averaged green-up dates, and the initial changes in phenology dates are probably dominated by responses of early-season species.

Ellesmere Island, Canada

Ellesmere Island forms part of the Canadian Arctic Archipelago. Ellesmere Island lies less than 800 km from the North Pole and less than 27 km from Greenland. Its landscapes are composed mainly of ice-capped mountains, and the coastline is incised by fjords and bays. The lowland landscapes are typically composed of gravel riverbeds, marshy ground soaked from melting glaciers, and dry clay cracking from the brittle and dry atmosphere of this polar desert. Its sole lake, Lake Hazen, is the largest in the Arctic. At its northern coast, Cape Columbia (83°7′N) is the most northerly point of land in Canada. Ellesmere comprises an area of 196,235 km^2; the total length of the island is 830 km, making it Canada's third-largest island. The Arctic Cordillera mountain system covers much of Ellesmere Island, making it the most mountainous island in the Arctic Archipelago. Ellesmere Island is bounded on the east by Nares Strait, which separates the island from Greenland. On the west side are Eureka and Nansen Sounds, separating Ellesmere from Axel Heiberg Island. The southern coast lies on Jones Sound and Cardigan Strait, separating Ellesmere from Devon Island.

Large portions of Ellesmere Island are covered with glaciers and ice, covering a total of 76,900 km^2, or 39% of the island. These include the Manson and Sydkap Ice Fields in the south, Prince of Wales Ice Field and the Agassiz Ice Cap along the central-east side of the island, and the Northern Ellesmere Ice Fields. The climate is bitterly cold and dry, a true polar desert. At the Eureka Meteorological Station (79°59′N), the TMAX is 6.1; TMIN is −36.5, and MAP is in the hyperarid category at 79 mm.

Ellesmere falls within the harshest, coldest ecoregion of the Arctic. Across the island, TMAX ranges between 0 and 3°C. Less than 5% of the landscape has a cover of vascular plants, and vascular plant growth is very low to the ground, barely exceeding the height of mosses. Woody plants are absent, and less than 50 plant species grow here.

Cambridge Bay, Nunavut, Canada

Cambridge Bay, now called by its original name, Ikaluktutiak, is situated on the southeast coast of Victoria island, at 69°06′N latitude. Victoria Island is a large island near the southern margin of the Canadian Arctic Archipelago. The island is about 515 km long and 270–600 km wide, with a land area of 217,291 km^2.

Victoria Island is an island of peninsulas, having a heavily indented coastline with many inlets. In the east, pointing northward is the Storkerson Peninsula, which ends with the Goldsmith Channel, the body of water separating Victoria from Stefansson Island. The Storkerson Peninsula is separated from the island's north-central areas by Hadley Bay, a major inlet. Another broad peninsula is found in the north, Prince Albert Peninsula. This ends at the Prince of Wales Strait. The terrain rises from a deeply indented coast to about 655 m in the northwest. The TMIN at Ikaluktutiak is $-32.5°C$ in February; the TMAX there is $8.9°C$ in July. MAP at the meteorological station is a meager 142 mm, placing it in the arid zone.

Victoria Island is fully within the High Arctic ecozone, as is most of the Canadian Arctic Archipelago. Victoria Island spans two bioclimate subzones, reflecting the climatic gradient across the island. The more northerly region has a TMAX of $5–7°C$. Shrubs here are less than 15 cm tall, and floral diversity is relatively low, with 75–150 species found in local floras. The more southerly region has a TMAX of $7–9°C$. Vascular plants in this warmer region cover 50%–80% of the landscape, with herbaceous and dwarf shrub layers 10–40 cm tall, and 125–250 species are found in local floras.

Herschel Island

Herschel island lies just off the north coast of the Yukon in the Beaufort Sea. It is the most northwesterly point of the territory, situated close to the Alaska-Yukon border. The island is small, with an area of only 116 km^2. It is approximately 15 km by 8 km between shorelines, with a rolling tundra terrain that ranges in height from sea level to 182 m. The island is subject to very high rates of coastal erosion due to the ice-rich nature of the underlying permafrost, and its soil heaves and rolls down its hillsides from the effects of frost creep and solifluction. From November to early June, Herschel Island is locked in ice. Herschel Island's climate is characterized by long, cold winters followed by short, but intense, summers. Strong steady winds are prevalent throughout the year. July is the warmest month, with TMAX of $7.4°C$. TMIN is $-27°$. MAP on the island ranges from 161 to 254 mm, in the arid range.

The island's soils are composed of glacial and marine deposits, underlain by ice-rich permafrost. Most of the island is composed of level to gently sloping stable uplands. Prominent geomorphic features include numerous retrogressive thaw slumps, most of which were activated by coastal erosion of ice-rich permafrost. Herschel Island is situated in the Yukon Coastal Plain ecoregion. The vegetation of this ecoregion is described as Arctic tundra, with continuous ground cover and no trees present. There are over 200 species of plants on Herschel Island, which occur in a diversity of habitats. One major plant community consists primarily of tussocks of tussock cottongrass (*Eriophorum vaginatum*) with a varying cover of the potentially canopy-forming willow species *Salix pulchra*.

Terrestrial Climates

As noted often in this book, Arctic temperatures are confidently predicted by the IPCC (2014) to rise dramatically in the coming decades, especially under the RCP8.5 (worst-case) scenario. Indeed, temperatures are already rising throughout the northern high latitudes. The average annual land surface air temperature north of 60° N for October 2018–August 2019 was the second warmest since 1900 (NOAA, 2019). The warming air temperatures are driving changes in the Arctic environment that affect ecosystems and communities on a regional and global scale. North American Arctic snow cover in May 2019 was the fifth-lowest in 53 years of record. June snow cover was the third-lowest. One of the most news-worthy of the current warming trends has taken place in Northeast Siberia (NASA Earth Observatory, 2020). After several months of warm weather, the Siberian town of Verkhoyansk reported a daytime temperature of $38°C$ ($100.4°F$) on June 20, 2020. This represents the northernmost temperature reading above $100°F$ ever observed and the highest temperature on record in the Arctic. Fig. 11.6 shows the extent of Arctic warming during the summer of 2020. The map is based on data from the moderate-resolution imaging spectroradiometer (MODIS) on NASA's Aqua satellite. Siberian TMAX in 2020 ranged from 1 to $>8°C$, with the greatest degree of warming over northern Siberia, extending to western Alaska (NASA Earth Observatory, 2020). The temperature anomaly map depicts land surface temperatures (LSTs), not air temperatures. LSTs may be significantly hotter or cooler than air temperatures.

The persistent high-pressure atmospheric pattern that brought the extreme heat has exacerbated wildfires, prompting dozens to burn in the region's forest and shrub ecosystems. Some of those ecosystems grow on top of carbon-rich layers of peat and permafrost. Fig. 11.7 shows smoke streaming from several active wildfires in Russia's Sakha region, drifting north of the Arctic Circle in northeastern Siberia. As discussed in the following, much of Earth's terrestrial carbon is stored in the northern high latitudes. The forests in the Sakha region are dominated by Dahurian larch

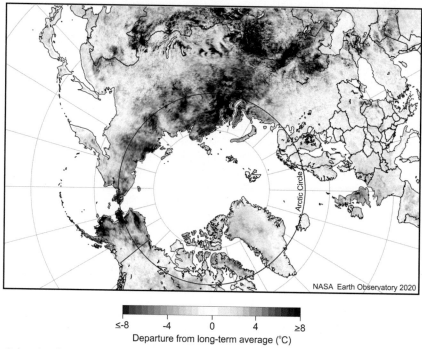

≤-8   -4   0   4   ≥8
Departure from long-term average (°C)

FIG. 11.6 Land surface temperature anomalies, March 19—June 20, 2020, in comparison with average spring temperatures 2003—18. (Image from NASA Earth Observatory, 2020. Heat and Fire Scorches Siberia. Retrieved from https://earthobservatory.nasa.gov/images/146879.)

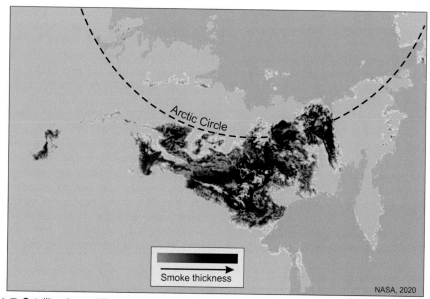

FIG. 11.7 Satellite photo of Eastern Siberia, showing the extent of smoke from regional fires, summer, 2020. (Image courtesy of the European Centre for Medium-Range Weather Forecasts. (n.d.). Retrieved from https://atmosphere.copernicus.eu/another-active-year-arctic-wildfires.)

(*Larix gmelinii*), a deciduous conifer that sheds its needles each winter. The extremely cold temperatures of regional winters limit the decomposition of needles. Over the centuries, this buried fuel builds up. When fires are initiated, this fuel facilitates massive conflagrations that can burn millions of hectares, as happened in the summer of 2020.

The Intergovernmental Panel on Climate Change (IPCC, 2013) Fifth Assessment Report (AR5) developed a series of possible future warming scenarios from Coupled Model Intercomparison Project Phase 5 (CMIP5) of the World Climate Research Program. These scenarios are referred to as representative concentration pathways (RCPs). The Fifth Assessment by the IPCC (2013) states that the Arctic will continue to warm more rapidly than the global mean and that the mean warming over land will be larger than over the ocean and larger than global average warming. In a recent study, Overland et al., (2020) used three sets of RCPs to predict potential future warming trends in the Arctic in the coming decades. Their primary interest was in RCP4.5, a "middle of the road" scenario in which aggressive but not implausible mitigation leads to temperature rises somewhat above the global mitigation aim of $+2°C$ by the end of the 21st century. This level of warming above the historic long-term average (1900−50) corresponds to greenhouse gas (GHG) concentrations stabilizing near 540 ppm in the second half of the century and globally averaged air temperatures in 2100 of $+2.4 \pm 0.5$ C. A more rigorous mitigation of atmospheric GHG levels, a low emission scenario, is termed RCP2.6. This ambitious scenario requires a cessation of anthropogenic GHG emissions over the next few decades and assumes negative emissions in the second half of the 21st century. RCP2.6 produces global warming of $\sim 1.6°C$ for 2046−65 and stabilizes more of the Arctic cryosphere than the RCP4.5 scenario. At the other end of the spectrum of future scenarios is a high-end, "business as usual" emission scenario, RCP8.5. This scenario assumes that no measures will be taken to lower GHG emissions.

The best-case scenario (RCP2.6) prediction for mean annual air temperatures in the Arctic is a warming of $+3.3 \pm 0.9°C$ for the years 2050 to 2100 but increases by 3−4°C from the Barents Sea to the North Pole (IPCC 2013). Under RCP2.6, mean annual precipitation (MAP) increases by 0%−10% in the Low Arctic and by 10%−20% in the High Arctic. Under RCP8.5, average surface temperatures increase by 5−7°C on Greenland and by 7−9°C through the rest of the Arctic. MAP increases by 30%−40% in the Low Arctic and by 40%−50% in the High Arctic.

The RCP4.5 prediction for mean annual Arctic air temperatures is $+4.0 \pm 0.9°C$ for 2050 and $+5.2 \pm 1.5°C$ for 2100. The worst-case scenario (RCP8.5) for mean annual Arctic temperatures is $+4.8 \pm 1.1°C$ for 2050 and $+9.5 \pm 1.9°C$ for 2100. Both RCP4.5 and RCP8.5 temperature predictions would mean extreme transformations of Arctic landscapes and biota by the end of the 21st century. The most severe warming under all three RCPs is predicted to occur during Arctic winters. Under RCP2.6, winter temperatures will warm $+4.7 \pm 1.1°C$ by 2100. Under RCP4.5, winter temperatures will warm $+7.1 \pm 2.1°C$ by 2100, and under RCP8.5, winters will warm by a remarkable $+13.3 \pm 2.2°C$ by 2100 (Fig. 11.8).

In a report about the remarkably warm temperatures in Siberia (Hersbach et al., 2020), European scientists examined historical temperature data in their global ERA5 reanalysis, finding that temperatures have been unusually warm in the region since January 2020. Since the ERA5 data begins in 1979, the European team also looked to GISTEMP, a NASA temperature record with data back to 1880. They could not find any other examples in either data set of such an intense heat wave in this part of Siberia persisting for such an extended period.

Changing snow conditions

Global climate models predict that terrestrial northern high-latitude snow conditions will change substantially over the 21st century. Results from a Community Climate System Model simulation of 20th- and 21st-century climate (Lawrence and Slater, 2009) show increased winter snowfall (10%−40%), shallower snow depths, and a shortened snow season (14 fewer days in spring; 20 fewer days in autumn). Lawrence and Slater (2009) were able to isolate the effects of changes in snowfall, snow depth, and snow season length on soil temperatures. In their model runs, they found that increased snowfall is effectively a soil warming agent, accounting for 10%−30% of total soil warming at 1 m depth. The resultant warming of the Arctic soils was found to contribute between 10% and 30% of total soil warming during the 21st century. The thicker snow cover was also modeled to counter the snowpack shallowing influence of warmer winters and shorter snow seasons. A shortened snow season enhances soil warming due to increased solar absorption. Snow depth and snow season length trends tend to be positively correlated, but they exert opposing effects on soil temperature. A shortening of the snow season enhances soil warming due to lengthening of the annual soil heating period. Consequently, on the

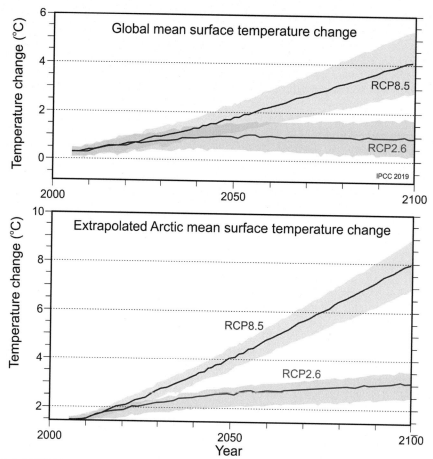

FIG. 11.8 IPCC (2014) predictions of surface temperature departures from the mean of 1986–2005 values. Top: Forecast changes in global surface temperature departures under the RCP2.6 (best-case) scenario under the RCP8.5 (worst-case) scenario; bottom: Arctic mean surface temperature changes from 2006 to 2100 relative to 1986–2005, as determined by multimodel simulations. Modified by the author from a global surface temperature forecast from the IPCC, 2014 Synthesis Report on Climate Change. (From IPCC, 2014. Synthesis Report. Contribution of Working Groups I, II and III to the Fifth Assessment Report of the Intergovernmental Panel on Climate Change. IPCC.)

century timescale, the net change in snow state can either amplify or mitigate soil warming. Snow state changes explain less than 25% of total soil temperature change by 2100.

Freshwater pond temperatures

Alaska's Arctic Coastal Plain is one of the wettest regions of the Arctic. Approximately 30% of Alaska's Arctic Coastal Plain consists of fresh surface water, with much of this total contributed by mosaics of small tundra ponds. A prominent pond type occupies low-centered polygons that form in drained lake basins because of thermokarst activity. The Arctic Coastal Plain

is a landscape low in relief and underlain with continuous permafrost that restricts groundwater flow to levels near the surface. In some regions, tundra ponds are becoming desiccated, as evaporation increases in response to a combination of increasing water temperatures and vapor pressure deficits, with consequent impacts on the number, size, distribution, and limnology of these ponds. Tundra thaw ponds in ice-rich permafrost are a dynamic feature of the land surface on the Arctic Coastal Plain: ponds often increase in size through thermal erosion and coalescence before draining and reforming, in a cyclical pattern occurring over centuries.

McEwen and Butler (2018) examined temperature dynamics across a 42-year period in a low-centered tundra polygon pond on the Arctic Coastal Plain of northern Alaska to assess potential changes in thermal dynamics for ponds of this type. Using water temperature data from a pond near Barrow (now Utqiaġvik), Alaska, studied intensively during 1971–73, and again in 2007–12, they developed an empirical model coupling historical air temperatures to measured pond temperatures for four summers. Then they used the model to predict summer pond temperatures from 1974 to 2008, for which direct aquatic temperature records do not exist. Average pond temperatures during the growing season (May 1 through October 31) increased by 0.5°C per decade or 2.2°C during the 42 years. Interestingly, the average date of spring thaw for the pond remained June 2 throughout the 42-year study interval. However, average pond temperature during the first 30 days of the growing season increased from 1971 to 2012, suggesting that ponds have become warmer in early spring. The average date of pond sediment freeze-up during the 42 years shifted later by 15 days, from September 28 in 1971 to October 13 in 2012. These changes correspond to a growing season that has increased in length by 14 days, from 118 days in 1971 to 132 days in 2012. Contemporary temperature measurements in other shallow tundra ponds in northern Alaska indicate that tundra ponds on Alaska's Arctic Coastal Plain have undergone a significant change in thermal dynamics over the past four decades.

Assessing such long-term thermal dynamics is crucial to our understanding of biological changes in these Arctic ponds. Initial snow loss from the surface of pond ice and the surrounding tundra lowers the albedo of both aquatic and terrestrial habitats, allowing increased transmission and absorption of solar energy by dark sediments and early meltwater. Heat transfer via inflowing terrestrial meltwater further degrades the remaining snow and ice in tundra pond basins. Exposed patches of organic sediments and stained water have a strong capacity to absorb additional heat, promoting rapid thaw of residual bedfast ice and subsequent warming of both water and sediments. Spring thaw initiates a period of intense biological activity in these ponds. The date of thaw and subsequent increases in water temperatures both influence the seasonality of ecological processes ranging from algal production rates and plant growth to zooplankton population dynamics and the timing of insect emergence 8 days earlier in 2009–13 than in the 1970s.

BLACK CARBON DEPOSITION

Black carbon (BC) is the soot-like by-product of wildfires and fossil fuel consumption, able to be carried long distances via atmospheric transport. BC comes from the burning of fossil fuels, like coal and diesel, and forest fires, and cookstoves (NASA, 2020). The majority reaching the Arctic comes from North America and Eurasia. BC warms the Earth in two ways. Airborne particles absorb sunlight and generate heat in the atmosphere. This can affect cloud formation and rain patterns. When the carbon covers snow and ice, the Sun's radiation is absorbed instead of being reflected to the atmosphere. This again generates heat and speeds up melting (NOAA, 2020). Because these black particles absorb more heat than white snow, the study of BC concentrations in glaciers and polar ice sheets is important for predicting future melt rates. Scientists have previously studied BC in areas with obvious nearby sources (such as a coal mine in Svalbard, Norway) but less is known about its complex interactions in snow-covered areas further removed from human impact. The exact sources of BC are often difficult to pinpoint in remote areas; molecular analysis of BC can be used, along with analysis of wind patterns. For instance, Khan et al. (2017) showed that Greenland ice sheet has recently received deposits of BC from wildfires burning thousands of km away in the Canadian Arctic. They found that the range of dissolved black carbon (DBC) concentrations indicates that significant amounts of DBC persist in both pristine and human-impacted snow and glacial meltwater. Wildfires are predicted to increase due to climate change, and the darkening of the surface of the Greenland ice sheet (known as albedo cannibalism) is already increasing meltwater. Lengthening of the summer melt season as a result of global warming may result in longer periods between snowfalls, lengthening the exposure of BC on snow and ice surfaces (Fig. 11.9), which could further exacerbate surface albedo reduction in the cryosphere.

Biomass burning and fossil fuel combustion are both sources of BC that are transported by winds to the cryosphere. Organic matter from these sources can be stored in glaciers for millennia and mobilized during glacial melting. In addition to ice sheet and glacial melting on land, recent studies suggest that BC is also contributing to the acceleration of sea ice melting in the Arctic. Loss of this ice would lead to more rapid warming and possibly irreversible climate change.

BC undoubtedly contributes to Arctic climate warming, yet source attributions have so far been inaccurate

J. Stoeve, NSIDC 2020

FIG. 11.9 Black carbon deposits on pack ice in the Bering Sea. (Photograph by J. Stroeve, NSIDC, 2018. Arctic Sea Ice News and Analysis. Retrieved from http://nsidc.org/arcticseaicenews/2018/08/.)

due to poor observational constraints and uncertainties in emission inventories. Year-round, isotope-constrained observations by Winiger et al. (2019) have revealed strong seasonal variations in BC sources with a consistent and synchronous pattern at all Arctic sites (Fig. 11.10). These sources were dominated by emissions from fossil fuel combustion in the winter and by biomass burning in the summer. The annual mean source of BC to the Circum-Arctic was $39 \pm 10\%$ from biomass burning. Comparison of transport model predictions with the observations showed good agreement for BC concentrations, with larger discrepancies for fossil/biomass burning sources. The consistency in seasonal source contributions of BC throughout the Arctic provides a strong justification for targeted emission reductions to limit the impact of BC on climate warming in the Arctic and beyond. Observation-based Arctic BC studies are scarce and have rarely lasted more than a year, especially with regard to data on source diagnostic dual-isotopic composition ($\delta^{13}C$ and $\Delta^{14}C$). The study by Winiger et al. (2019) yielded year-round $\delta^{13}C/\Delta^{14}C$-based source apportionment of elemental carbon (EC) from the Arctic sites Alert (Canadian High Arctic; $n = 9$), Zeppelin (Svalbard; $n = 11$), and Utqiaġvik (formerly known as Barrow) (north Alaska; $n = 10$), covering about 3 years. The $^{14}C/^{12}C$ isotope ratio of an EC sample allows determination of the biomass burning fraction (f_{bb}; containing contemporary ^{14}C) relative to the fossil fuel combustion fraction (f_{fossil}; devoid in ^{14}C). The $^{13}C/^{12}C$ ratio helps to further distinguish between

various types of fossil fuel sources (e.g., natural gas, coal, or oil). Last, these observations of atmospheric BC are compared with results from an atmospheric transport model, which includes both anthropogenic and natural-fire BC emissions, and has shown great potential to accurately simulate observational data. The observational data show that both fossil and biomass emissions contribute substantially to the levels of BC in the Arctic, but with opposite seasonal trends.

One inescapable fact about soot generation and dispersal is that nearly 2 billion people worldwide use open fire stoves to do their cooking and/or home heating. These stoves are a major source of BC. Not only is soot the second leading cause of climate change, but it also leads to 1.8 million premature deaths each year from respiratory and heart ailments (NOAA, 2020).

ARCTIC AMPLIFICATION

Since the mid-20th century, average global temperatures have warmed about 0.6°C, but temperatures have increased about twice as fast in the Arctic as in the midlatitudes, a phenomenon known as "Arctic amplification" (NASA Earth Observatory, 2009) (Fig. 11.11).

An example of this dramatic level of warming was documented by satellite observations (NASA, 2020) for 2018–19 (Fig. 11.12, above). From October through December 2018, temperatures in the interior of Alaska wear as much as 6°C warmer than average (Fig. 11.12, center). By January 2019, the high-temperature zone

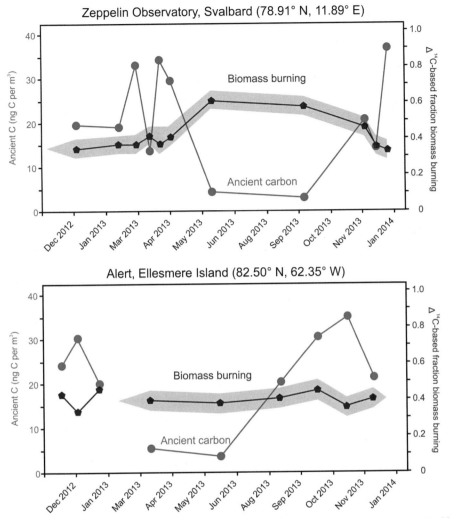

FIG. 11.10 Seasonal changes in sources of black carbon (BC) in the Arctic, sampled from November 2012 to January 2014. Above: BC sources analyzed from samples taken at the Zeppelin Observatory, Svalbard (78.91°N, 11.89°E). Below: BC sources analyzed from samples taken at the Alert Station, Ellesmere Island (82.50°N, 62.35°W). (Modified from Winiger, P., Barrett, T.E., Sheesley, R.J., Huang, L., Sharma, S., Barrie, L.A.,Gustafsson, Ö., 2019. Source apportionment of circum-Arctic atmospheric black carbon from isotopes and modeling. Science Advances 5 (2). https://doi.org/10.1126/sciadv.aau8052.)

had shifted North into Arctic Alaska, raising temperatures as much as 6°C above average. The surface temperature trend for the Arctic became consistently higher than the global average about the beginning of the 21st century (shaded area in Fig. 11.12, below).

Rises in surface and lower troposphere air temperatures through the 21st century are projected to be especially pronounced over the Arctic Ocean during the cold season. This Arctic amplification is largely driven by the loss of sea ice cover, allowing for strong heat transfers from the ocean to the atmosphere (Jones et al., 2014; Serreze et al., 2009). As the coldest region of the Northern Hemisphere, the Arctic influences climate patterns well south of the Arctic Circle (Cohen et al., 2014). Arctic amplification was formerly minimized by abundant sea ice cover over the Arctic Ocean and adjacent northern seas. This helped keep the Arctic atmosphere cold. The whiteness of the ice, often enhanced by a layer of snow, reflects most of the insolation to space.

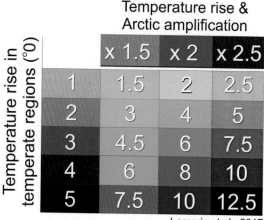

FIG. 11.11 Predicted Arctic amplification values compared with temperature increases in the midlatitudes. (After Lameris, T.K., Scholten, I., Bauer, S., Cobben, M.M.P., Ens, B.J., Nolet, B.A., 2017. Potential for an Arctic-breeding migratory bird to adjust spring migration phenology to Arctic amplification. Global Change Biology 23 (10), 4058–4067. https://doi.org/10.1111/gcb.13684.)

The open ocean reflects only 6% of the insolation and absorbs the rest, while sea ice reflects 50%–70% of the incoming energy. Snow has an even higher albedo than sea ice; thick sea ice covered with snow reflects up to 90% of the insolation (NSIDC, 2020). Arctic amplification occurs in all seasons but is the strongest in autumn and winter.

The ice pack also physically insulates the Arctic atmosphere from the underlying Arctic Ocean. With the rapid loss of sea ice cover in recent years, more dark open water is exposed, readily absorbing insolation in summer. As discussed in Chapter 1, this contributes to a positive feedback loop in which the newly warmed ocean waters cause even more melting of the sea ice. The reduction of sea ice strips the insulation separating the seawater from the atmosphere, so that heat from the ocean escapes, warming the atmosphere in the autumn and winter. As shown in Fig. 11.13, the most pronounced heating effects occur in the Arctic autumn, delaying the annual formation of sea ice as winter approaches.

The open Arctic Ocean also warms the land. At a minimum, Arctic amplification increases regional temperatures 1.5 higher than the temperatures in the temperate latitudes. At a maximum, Arctic amplification increases regional temperatures 2.5 higher than the temperatures in the temperate latitudes, so that a 4°C warming in the midlatitudes becomes a 10°C warming in the Arctic (Fig. 11.11).

Though sea ice loss is the greatest contributor to Arctic amplification, Arctic snow cover in spring and summer has decreased at an even greater rate than sea ice (Cohen et al., 2014). For instance, June snow cover has decreased at nearly double the rate of September sea ice. The decrease in spring snow cover has contributed to both the rise in warm season surface temperatures over the Northern Hemisphere and the decrease in summer Arctic sea ice. The combined rapid loss of sea ice and snow cover in the spring and summer has played a role in amplifying Arctic warming.

CHANGES IN LENGTH OF THE GROWING SEASON

Several processes are thought to contribute to Arctic amplification, including local radiative effects from increased GHG forcing, changes in the snow- and ice-albedo feedback induced by a diminishing cryosphere, aerosol concentration changes and deposits of BC on snow and ice surfaces, changes in Arctic cloud cover and water vapor content, and an increase in emission of long-wave radiation to space. In addition to these local drivers of Arctic amplification, Arctic temperature change is also sensitive to variations in the poleward transport of heat and moisture from lower latitudes (Cohen et al., 2014).

Historically, the Arctic growing season has ranged from 50 to 60 days. The Arctic is well known for its very short growing season, but there is increasing evidence that the growing season has been starting earlier and ending later since the late 20th century and the early 21st century. As Arctic temperatures warm, the growing season is expanding. One recent study helps to shed light on this phenomenon.

A different study from the same region as the tundra pond experiment was carried out by Arndt et al. (2019). They aimed to investigate the widely reported phenomenon of Arctic greening. Satellite imagery produces images in which Arctic Greening is quantified through a normalized difference vegetation index (NDVI). Arctic NDVI has largely increased in recent decades, based on low spatial resolution satellite imagery of the pan-Arctic region. The study by Arndt et al. (2019) focused on one well-studied area of the Arctic Coastal Plain of Alaska using high spatial resolution (4m) multispectral satellite imagery to analyze the greening trend near Utqiaġvik over 14 years from 2002 to 2016. They found that tundra vegetation has been greening despite the long-term stability of local vegetation community composition. Their study concluded that the greening is linked to the number of thawing degree days (i.e., the number of days in the year when soil

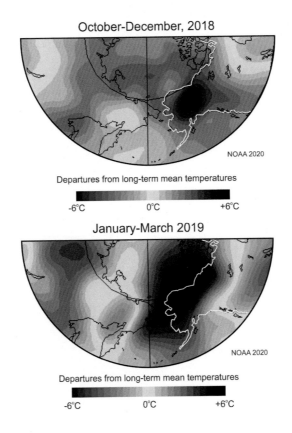

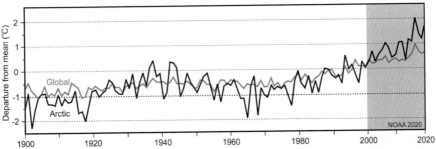

FIG. 11.12 Departures from long-term mean temperatures in the Arctic. Above: Temperature anomalies from October–December 2018. Middle: Temperature anomalies from January–March 2019. Below: Comparison of global versus Arctic temperature anomalies since 1900. Gray area indicates Arctic temperatures consistently warmer than the global average in the 21st century. (All data and images courtesy of NASA. (n.d.). Retrieved from https://arctic.noaa.gov/Report-Card/Report-Card-2019/ArtMID/7916/ArticleID/835/Surface-Air-Temperature.)

temperatures are above freezing), which increased by an average of about 1.8 days per summer over the 14-year study period. This suggests that the growing season has increased by 25 days in this Arctic ecosystem and that greening is occurring due to a lengthening growing season that stimulates plant productivity without any significant change in vegetation communities. Their research also concluded that vegetation communities in wetter locations greened about twice as fast as those found in drier conditions.

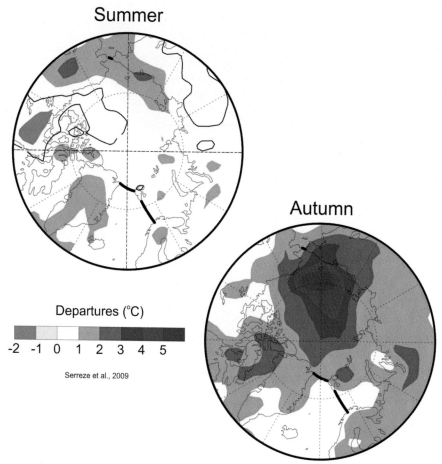

FIG. 11.13 Maps of the Arctic Ocean, showing NCEP/NCAR modeled anomalies in surface air temperature for 2003–07, relative to 1979–2007. Means for summer (above), defined as June–August and (b) autumn (below), defined as September–November. (Figures after Serreze, M.C., Barrett, A.P., Stroeve, J.C., Kindig, D.N., Holland, M.M., 2009. The emergence of surface-based Arctic amplification. The Cryosphere 3 (1), 11–19. https://doi.org/10.5194/tc-3-11-2009.)

As we have seen, there is compelling evidence from experiments and observations that climate warming prolongs the growing season in the Arctic. Until now, the timing of the start, peak, and end of the growing season, which are used to model influences of vegetation on biogeochemical cycles, were mostly quantified using aboveground phenological data (Blume-Werry et al., 2016). Yet over 80% of the plant biomass in Arctic regions is below ground, and the timing of root growth affects biogeochemical processes by influencing plant water and nutrient uptake, soil carbon input, and microbial activity. Blume-Werry et al. (2016) measured the timing of above- and belowground productivity in three plant communities along an Arctic elevation gradient over two growing seasons. They found that belowground production peaked later in the season and was more uniform through the growing season than aboveground productivity. Most importantly, the growing season lasted about 50% longer below the ground. Their results strongly suggest that traditional aboveground estimates of phenology in the Arctic, including remotely sensed information, do not present the whole picture of plant productivity, intensity, or duration. Studies that include root phenology enhance our knowledge of carbon and nutrient cycling in Arctic soils, which will aid in the development of terrestrial biosphere models, and shed new light on how Arctic ecosystems will respond to climate warming.

Seasonality of Photosynthetic Activity

Xu et al. (2013) examined the recent history of Arctic vegetation photosynthetic activity, using Normalized Difference Vegetation Index (NDVI) data from satellite imagery to serve as a proxy for vegetation photosynthetic activity. Vegetation photosynthetic activity in the North depends on the seasonal cycle of temperature and not on the difference between annual maximum and minimum temperatures.

Most of the Arctic has gained more than 3 days of growing season per decade since the 1980s. Global temperature is increasing, especially over northern lands (>50°N), owing to positive feedbacks. As this increase is most pronounced in winter, temperature seasonality (S_T)—conventionally defined as the difference between summer and winter temperatures—is diminishing over time, a phenomenon that is analogous to its equatorward decline at an annual scale. Trends in the timing of these thresholds and cumulative temperatures above them may alter vegetation productivity, or modify vegetation seasonality (S_v), over time.

Empirical evidence suggests that in addition to the direct effects of warming, several other factors influence the relationship between temperature seasonality (S_t) and vegetation seasonality (S_v). These include (1) warming-induced disturbances such as summer droughts, midwinter thaws, increased frequency of fires, and outbreaks of pests; (2) the shrinking and draining of lakes from thawing permafrost, desiccation of ponds, and the colonization of the growing banks by vegetation; (3) the interacting effects of temperature and precipitation; (4) climate feedbacks that enhance wintertime snow amount on land asymmetrically between Eurasia and North America; (5) feedbacks from declining snow cover extent on land leading to longer growing seasons and promoting vegetation compositional/structural changes; (6) enhanced nitrogen mineralization in warmer soils insulated by increased shrub cover and so on; (7) anthropogenic influences (pollution from metal smelters, herding practices of grazing herbivores, and so on); and (8) changes in wild herbivore populations. These factors could have contributed to an amplification of the ratio between S_t and S_v in the Arctic.

PHENOLOGY

Phenology is the term used to describe cyclic and seasonal natural phenomena, especially concerning climate and plant and animal life. Van Gestel et al. (2019) noted that the warming of the Arctic has brought about changes in the timing of the seasonal cycle of tundra plant growth and development. Advanced early-season plant growth has been observed across the Arctic, and this has been attributed to earlier snowmelt and warmer air temperatures. Warming resulted in earlier initiation of reproductive events (flowering and setting seeds) by 5–10 days. Soil warming may also be an important driver of the timing of plant leaf-out and senescence (die-back).

As can readily be appreciated, the increases in Arctic temperatures, expansion of the length of the growing season, and other global warming phenomena are disrupting the timing of seasonal events in the lives of Arctic animals and plants. If each organism lived independently of all others, then changes in an organism's life cycle might or might not be important. However, this is not the case. All organisms live in ecosystems in which they have relationships with other taxa. If, on the other hand, changes in phenology were the same for all the different taxa (i.e., all flowering plants in a plant community set seed on August 12th, whereas before they all set seed on August 18th), then perhaps there would be no serious disruption to ecosystem functioning. Unfortunately, this is also not the case. Each organism has its inherent life cycle, and research has shown that the phenology of each species is different from others.

Migratory Bird Phenology

As discussed earlier, Arctic amplification is causing the earlier onset of spring in the Arctic than in temperate regions. Lameris et al. (2017) used a dynamic state variable model to test the potential reproductive success of long-distance migrant barnacle goose (*Branta leucopsis*) in advancing their migratory schedules under climate warming scenarios, which include Arctic amplification. Their study focused on Arctic Scandinavia and adjacent regions of northwest Russia. As shown in Fig. 11.14, the window for nesting expands from 23 to 78 days under an amplified Arctic warming of 8°C above the long-term average.

PREDATION ON MIGRATORY BIRDS

Birds are often subjected to strong food web interactions, such as predation, which have pervasive impacts on population dynamics and extinction risk. Eggs and nestlings are life stages in which birds are particularly vulnerable to predation. Hence, factors determining nest predation have been the targets of many studies. Yet, there are few studies available about how climate change may affect nest predation. Arctic birds build their nests on the ground, often in landscapes with sparse vegetation cover. Therefore, their nests can be expected to be particularly vulnerable to predation because they are often clearly exposed (that is, visible because of little cover) and easily accessible to any predator species that are present. A critical concern is whether the ongoing warming influences nest predation risk. A new study has shown that the nest

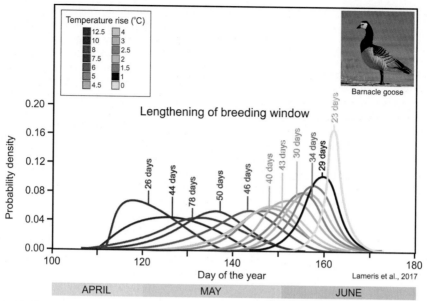

FIG. 11.14 Chart showing the modeled timing of the breeding window for the barnacle goose (*Branta leucopsis*) at the breeding site at Kolokokova Bay, NW Russia (68°35′N, 52°20′E). Note that the timing of the opening of the window advances with increasing temperatures. (After Lameris, T.K., Scholten, I., Bauer, S., Cobben, M.M.P., Ens, B.J., Nolet, B.A., 2017. Potential for an Arctic-breeding migratory bird to adjust spring migration phenology to Arctic amplification. Global Change Biology 23 (10), 4058–4067. https://doi.org/10. 1111/gcb.13684.)

predation of Arctic shorebirds has increased steeply, providing evidence for the ecological mechanisms that may be involved. The most fundamental response of tundra ecosystems to climate warming is an increase in plant biomass: the tundra is greening. Although the increased vegetation cover could provide a lower exposure of bird nests to predators, food web theory predicts that increased primary productivity in the tundra will render species at intermediate trophic levels (such as many ground-nesting birds) more suppressed by predation. In particular, generalist consumers (omnivores such as corvids and foxes) that feed on a variety of food items from several trophic levels, including bird nests, are expected to become more abundant as primary productivity increases. Although this expectation is derived from general food web theory, consumers in a tundra ecosystem may be particularly sensitive to a warming-induced increase in primary productivity because primary productivity is initially low and temperature-limited in cold regions.

Ims et al. (2019) tested a prediction from food web theory that increased primary productivity (greening of tundra) in the warming Arctic leads to a higher risk of nest predation in tundra ecosystems. By exploiting landscape-scale spatial heterogeneity in areas of primary productivity across alpine tundra ecotones and supplied

with experimental nests in sub-Arctic Scandinavia, they found that predation risk indeed increased with primary productivity. The productivity-predation risk relationship was independent of the simultaneous effects of rodent (lemmings and voles) population dynamics and vegetation cover at nest sites. Their study contributes to an improved understanding of how climate change may affect Arctic ecosystems and threaten endemic biodiversity through a trophic cascade. Trophic cascades are powerful interactions that help shape the nature of entire ecosystems. They occur when a trophic level in a food web is suppressed. For example, a top-down cascade will occur if predators are effective enough in predation to reduce the abundance or alter the behavior of their prey, thereby releasing the next lower trophic level from predation (or herbivory if the intermediate trophic level is a herbivore).

CASE STUDY: THE BARNACLE GOOSE IN NW RUSSIA

Heavy-bodied birds such as geese need to stop along their northern migration to feed, to have enough metabolic energy to reach their Arctic breeding grounds. The model by Lameris et al. (2017) predicts that the barnacle goose is likely to experience substantially reduced reproductive success with increasing Arctic amplification. These birds, and likely many other species, rely on biological cues in their temperate winter habitats to guide them concerning the timing of their departure for the Arctic in the spring. As reproductive success is largely determined by the timing of spring arrival, changes in spring arrival may have considerable effects on individual fitness and eventually on population dynamics. However, under Arctic amplification conditions, the temperate regions warm later than the Arctic and thus the environmental cues in the temperate region will be out of sync with the arrival of spring in the high latitudes. The timing of springtime biological events in the Arctic is advancing as the climate warms and the growing season lengthens. The optimal time window for bird and mammal reproduction is also advancing considerably. Migrants such as birds need time to prepare for their migration, and they must time their journey based on cues in their wintering grounds that may be thousands of km south of the Arctic Circle.

Plant Phenology

The result of the rapid environmental changes taking place in the Arctic has been an alteration of the phenology of many species, creating potential disconnects between organisms that rely on each other for survival. Species phenology in the Arctic is often tightly linked to snowmelt date, although different species and trophic levels can differ widely in their responses. Whether increased Arctic temperatures will cause future snowmelt dates to become earlier or later, the timing of key events in the life cycles of plants and animals, their phenology, is likely to be significantly affected (Gillespie et al., 2016). Air temperatures also affect plants and insects differently, and a key issue in phenology research is in determining whether differential responses to climate factors of insects and plants will be sufficient to disrupt their interactions. If closely associated species such as insect pollinators and flowering plants alter their phenologies in differing ways, the overlap between their populations (their phenological synchrony) may become disrupted, with potentially deleterious effects to one or both groups. Arctic plant-pollinator communities are less diverse than those in temperate regions, and this together with the short season length is likely to contribute to instability among the interactions.

Competition from aperiodic species

It has been suggested that the end of plant growth periods during autumn is triggered by external cues, such as day length, light quality, or temperature. These concepts lead to the hypothesis that earlier or later snowmelt dates

will lengthen or shorten the duration of these periods, respectively, and thereby affect plant performance. Semenchuk et al. (2016) tested whether snowmelt date controls phenology and phenological period duration in High Arctic Svalbard using a melt timing gradient from natural and experimentally altered snow depths. They investigated the response of early- and late-season life cycle events (phenophases) from both vegetative and reproductive phenological periods of eight common High Arctic plant species: *Alopecurus magellanicus, Bistorta vivipara, Cassiope tetragona, Dryas octopetala, Luzula arcuata, Pedicularis hirsuta, Salix polaris,* and *Stellaria crassipes*. They found that all phenophases follow snowmelt patterns, irrespective of timing of occurrence, vegetative, or reproductive nature. Three of four phenological period durations based on these phenophases were fixed for most species, defining the studied species as periodic. The authors concluded that (1) species more-or-less locked into a life cycle schedule might not be able to adapt to changing growing-season durations, as opposed to aperiodic species; (2) while changing snowmelt patterns might change phenology timing, its effects may be modulated by increasing growing-season temperatures accelerating and shortening some periods. This may apply specifically for periodic species and give them a further disadvantage compared with aperiodic species. In conclusion, periodic species are likely to be limited in their ability to adapt to changing snowmelt dates and may be outcompeted by some aperiodic taxa invading Svalbard, such as *Rumex longifolius, Ranunculus acris, Ranunculus repens, Epilobium montanum, Deschampsia cespitosa,* and *Poa pratensis*.

Tundra Productivity and Arctic Greening

Plants capture and store solar energy through photosynthesis. They demonstrate net primary productivity (NPP), which is how much carbon dioxide vegetation takes in during photosynthesis minus how much carbon dioxide the plants release during respiration (metabolizing sugars and starches for energy). Because of the harsh climates and brief growing season for terrestrial vegetation, the biological productivity of Arctic landscapes is quite low. In low Arctic tundra environments, NPP averages about 600–1000 g/m^2. In the High Arctic tundra, NPP averages only 150 g/m^2. For comparison, the average NPP for grasslands and savannahs is 500–700 g/m^2, the average for the boreal forest is 800 g/m^2, and at the high end, the average NPP for tropical rainforest is 2000 g/m^2.

Several elements contribute to Arctic plant productivity. All plants require moisture, soil nutrients, minerals, and above freezing temperatures to develop and mature. Moisture is an important factor that exerts control over plant NPP through its effects on photosynthesis. The extraction of minerals and nutrients from Arctic soils can effectively only take place during the growing season when there is a thawed active layer at the top permafrost table. As we have seen, recent changes in Arctic climate have increased air and soil temperatures and caused the growing season to expand by days to weeks, depending on location. These climate changes have driven changes in NPP, but not uniformly across whole regions, and not in a straightforward one-to-one relationship between temperatures in NPP. A number of recent studies have shed light on changes in primary in the Arctic in response to global change.

Greening of the Arctic

Arctic inhabitants and visitors alike have noticed a greening-up of the Arctic in recent years. Bare ground is becoming vegetated, and where plants once grew, they are now growing taller. This may be the biggest biological signal of climate change anywhere on the planet. Arctic North America is getting greener, and this has been measured through a NASA study that provides the most detailed look yet at plant life across Alaska and Canada. In a changing climate, almost a third of the land cover, much of it Arctic tundra, is taking on the appearance of landscapes found in warmer ecosystems. The NASA Earth Observatory (Ramsayer, 2017) study involved the analysis of 87,000 images taken from Landsat satellites, converting the images into NDVI data that reflects the amount of healthy vegetation on the ground. The study found that western Alaska, northern Quebec, and other regions became greener between 1984 and 2012. The new Landsat study further supports previous work that has shown changing vegetation in Arctic North America.

Increased productivity (greening) has been shown to be a fundamental response of the tundra ecosystem to global warming that is likely to have cascading effects by changing trophic interactions in the food web. By empirically substantiating the prediction from food web theory that Arctic greening leads to increased predation pressures on vulnerable prey species, the study by Ims et al. (2019) contributes to an improved understanding of how climate change may affect Arctic ecosystems through a trophic cascade. Determining such changed interactions in tundra food webs may also help implement biodiversity conservation under climate change. Although the ongoing greening of the Arctic may be impossible to counteract through local environmental management, actions taken to halt the increase of generalist predators may nevertheless be a management option to preserve Arctic birds in a warming climate.

CHANGES IN PLANT COMMUNITIES

The impact of local factors is particularly important in Arctic ecosystems, which are constrained by the predominance of abiotic factors such as low temperature, permafrost, and limited moisture availability that can limit biological activity. There have been a number of changes to Arctic plant communities in response to global warming. Some of the changes have been predicted on the basis of biotic response to past warming events, such as occurred during the Pleistocene and early Holocene. Probably foremost among these is a general northward advance of treeline into landscapes previously occupied by Arctic tundra. While some regional tree lines have shifted, the more typical scenario has been the northward spread shrubby plants into herbaceous tundra.

Changes in Shrub Growth

The recent warming of the Arctic has been accompanied by increasing vegetation biomass that is mainly attributed to the expansion of tall, deciduous shrubs into the tundra ecosystems. This "shrubification" is expected to alter important attributes of the ecosystem such as carbon sequestration, snow cover, productivity, fire regimes, and hydrology.

Recent evidence indicates that an expansion of canopy-forming shrubs is currently taking place at sites on the North Slope of Alaska, on the coast of the Northwest Territories, in Northern Quebec, and northern Russia. In Arctic Alaska, a canopy cover of alder (*Alnus viridis* subsp. *crispa*) shrubs has increased by 14%–20% on average within the past 40 years, with increases up to 80% in some areas (Myers-Smith et al., 2011). In addition, studies of population structures of shrub and tree species indicate an advance of shrubs up slopes in alpine tundra ecosystems in sub-Arctic Sweden and Arctic sites in Norway. These reports reflect widespread changes in shrub cover in the absence of localized disturbances. Growing season temperatures are warming in Alaska and western Canada, and on Herschel Island, mean annual temperatures have increased over the past few decades. Willows (*Salix* spp.) are well adapted to invading ecosystems when conditions change. Willow shrub growth is most sensitive to temperatures in the early growing season period of the year. The study by Myers-Smith et al. (2011) concluded that willow species are increasing in canopy cover and height on Herschel Island.

Expansion of tall shrubs in Arctic North America

As discussed by Duchesne et al. (2018), Arctic shrub expansion has been linked to warming using repeat aerial photography, observational and experimental studies, shrub ring analysis, and regional "greening" trends observed in remote sensing imagery. Though rising temperatures generally correlate well with shrub growth, other local factors such as herbivory, soil moisture, or disturbance sometimes override temperature effects. In reciprocal action, shrub expansion will likely alter the regional climate and ecology of the Arctic. An increase in shrubs would change the snow distribution pattern, affecting many regional ecological and hydrological processes. Shrub proliferation will also raise surface temperatures through a decrease in albedo. Darker surfaces absorb more insulation than the lighter-colored herbaceous vegetation.

Duchesne et al. (2018) presented an evaluation of changes in tall shrub cover on the North Slope of Alaska between 2000 and 2010, using NASA satellite imagery. In this context, shrubs taller than 0.5 m are considered "tall shrubs." At a spatial resolution of 250 m, estimated tall shrub cover on the North Slope ranged from 0% to 21% of land surfaces, although the vast majority of sites had less than 2% tall shrub cover. High shrub cover values were identified along floodplains, creeks, and sloped terrain. Time-series analyses indicated that tall shrubs expanded across the Alaskan North Slope from 2000 to 2010. All told, 176,524 km² of the North Slope of Alaska experienced a change in tall shrub fractional cover. This represents about 72% of the North Slope region. Tall shrubs may have expanded throughout a larger area, but there is insufficient precision in the MISR-based estimates to make an indisputable determination. Nevertheless, 94% of the locations that exhibited a robust change showed a positive trend toward an increase in shrub cover. The highest rates of tall shrub expansion were in the southwestern portion of the study area.

A recent ecological assessment of the Arctic by Hirawake et al. (2021) suggests that the northern forest-tundra boundary is more likely to be controlled by nonlinear (tipping-point) processes than by macroclimate because tundra vegetation shifts depend on local factors, such as soil nutrients, moisture levels, and biological interactions, as well as regional factors such as atmospheric temperature and surface albedo. Hirawake et al. (2021) investigated vascular plant composition and abiotic environments in the tundra-boreal forest ecotone on the East shore of Hudson Bay during the summer of 2016 to clarify local factors promoting shrub expansion. The investigation revealed a positive colonizing relationship between shrub species, suggesting that such local biotic interaction could provide positive feedback to shrub expansion.

Shrub expansion effects on tundra plant communities

Another unforeseen change in Arctic vegetation has simply been that in many places the plants are growing taller and they used to. The composition of plant communities has not necessarily changed, but the resident plant species are increasing in stature. This is especially true in the High Arctic. Some of the research highlights of these phenomena are discussed in the following.

Shrub expansion has most likely been driven by climate change, as shrub vegetation has been shown to be highly sensitive to changes in temperature. Shrub annual growth and growth rings are positively correlated with climate and can represent year-to-year variation in temperature, where a general increase in shrub growth is expected as a major response to global warming (Boscutti et al., 2018). In addition, the age of shrubs or their ramets have been shown to reflect environmental conditions. Other shrub traits, such as shoot length, leaf number, abundance, and biomass, have commonly been found to be sensitive indicators of environmental change and ecosystem function. Changes in plant growth resulting from climate warming can cause considerable modifications in vegetation traits, which can in turn influence species composition, ecosystem functions, and thus ecosystem services, such as regulation of nutrient cycles, gas exchanges, or biomass stock. One major effect of climate-related vegetation alteration concerns changes in biodiversity. In particular, changes in the abundance and height of shrubs could lead to tangible shifts in both the structure and species composition of a plant community.

Elevational shifts in shrubs

Boscutti et al. (2018) analyzed the growth of bilberry (*Vaccinium myrtillus* L.) and its associated plant communities along an elevation gradient of ca. 600 vertical meters in the eastern European Alps. They assessed the ramet age, ring width, and shoot length of *V. myrtillus*, and the shrub cover and plant diversity of the community. At higher elevation, ramets of *V. myrtillus* were younger, with shorter shoots and narrower growth rings. Shoot length was positively related to shrub cover, but shrub cover did not show a direct relationship with elevation. Increased shrub cover reduced species richness in alpine plant communities, as well as affecting species composition (beta diversity), but these variables were not influenced by elevation. Their findings suggest that changes in plant diversity are driven directly by shrub cover and only indirectly by climate, as represented in this study by changes in elevation.

Shrub-induced changes in tundra productivity

Recently observed shifts in Arctic tundra shrub cover have uncertain impacts on 21st-century net ecosystem carbon exchanges. Using the parameters established by the IPCC for the RCP8.5 (worst-case) scenario, Mekonnen et al. (2018) applied an ecosystem model to examine the effects of tundra plant dynamics on ecosystem carbon balances from 1980 to 2100 in the North America Arctic. Tundra productivity was modeled to increase from enhanced carbon fixation and N mineralization under recent and future climates. Averaged across the region, by the year 2100, the model-predicted increases in the relative dominance of woody plants, raising ecosystem annual NPP by 244 g/cm^2. This level of NPP offset simultaneously increases in plant respiration (139 g/cm^2), resulting in an increasing net carbon sink over the 21st century. However, by 2100, seasonal changes in carbon uptake and increased ecosystem respiration outside of the growing season are predicted to result in larger carbon losses during autumn and winter (140 g/cm^2). Soil temperatures are predicted to increase more slowly than air temperatures ($\sim 0.6°C$ for every $1°C$ increase in air temperature over the 21st century). This slower soil warming leads to greater increases in CO_2 fixation versus soil respiration rates, which will also contribute to the tundra remaining a carbon sink through 2100.

Changes in Herbaceous Plant Size

The s-Tundra Working Group of the German Center for Integrative Biodiversity Research explored biome-wide relationships between temperature, moisture, and seven key plant functional traits both across space and over three decades of warming at 117 Arctic tundra locations (Bjorkman et al., 2018). They found strong spatial relationships between temperature and community height, specific leaf area (SLA), and leaf dry matter content (LDMC). Both height and SLA increased with summer temperature, but the temperature-trait relationship for SLA was much stronger at wetter sites than at drier sites. LDMC was negatively related to temperature, and more strongly so at wetter than drier sites. Community woodiness decreased with temperature, but the ratio of evergreen to deciduous woody species increased with temperature, particularly at drier sites. These spatial temperature-trait relationships indicate that long-term climate warming should cause pronounced shifts toward communities of taller plants with more resource-acquisitive leaves (high SLA and low LDMC), particularly where soil moisture is high. Both leaf area and leaf nitrogen content decreased with warmer temperatures in dry sites but increased with warmer temperatures in wet sites. Thus, future warming-driven changes in traits and associated ecosystem functions (for example, decomposability) will probably depend on soil moisture conditions at a site. Furthermore, future changes in water availability (for example, because of changes in precipitation, snowmelt timing, permafrost, and hydrology) could cause substantial shifts in these traits and their associated functions, irrespective of warming. Throughout the 27 years of monitoring, they found that plant height increased rapidly at nearly every survey site. Looking to the future, the team concluded that the consequences of future warming will most clearly be tied to rapid changes in plant height.

The Response of Peatlands to Warming

Peat deposits have accumulated during the Holocene across the Arctic. Peat is particularly abundant in the world's largest wetland—the West Siberian Lowland—which is largely underlain by vulnerable discontinuous permafrost and projected to experience a further decrease in permafrost extent. We have only a poor understanding of how Arctic peatlands will respond to future climate change (Zhang et al., 2018).

As discussed by Andresen et al. (2018), wetlands represent a significant portion of the Arctic landscape. These wetlands support aquatic vegetation on saturated soils and shallow ponds perched above the shallow permafrost layer. Arctic wetlands are oases for productivity in the otherwise arid environments of the North. The timing, extent, and magnitude of plant primary productivity play a fundamental role in Arctic hydrology, carbon fluxes, and energy balance. There have been measurable increases in wetland plant biomass and cover in tundra ponds over recent decades. These changes are associated with increased nutrient availability and a longer thaw season. Unfortunately, the lengthening growing season and elevated summer temperatures have accelerated anaerobic bacterial decomposition of ancient organic matter from the bottoms of these ponds, and this has resulted in a significant rise of methane emissions to the atmosphere (Andresen et al., 2017). With the projected warming of the Arctic over the next century, changes in plant biomass and phenology in Arctic wetlands may either contribute to or mitigate warming.

A study by Andresen et al. (2018) explored the influences of climate and nutrients on the productivity of Arctic wetlands. They employed a Green Excess Index (GEI) derived from Red, Green, and Blue digital image brightness values from digital repeat photography (aka phenocams) tracking the interannual variability in seasonal greening and aboveground biomass for two dominant aquatic emergent graminoids on the Arctic Coastal Plain of northern Alaska: water sedge (*Carex aquatilis*) and pendant grass (*Arctophila fulva*). They analyzed images from 4 years of photographs and found that there were strong differences in the timing and intensity of greenness among species, as reflected in both seasonal and interannual greening trends. The number of thawing degree-days (TDD, days above $0°C$) was a good predictor of GEI in both aquatic plant species.

Based on long-term climate records and TDD thresholds, the date of the commencement of wetland greening has shifted 16 days earlier over the past 70 years.

Changes in Animal Communities
Polar bears and sea ice loss

Polar bears (*Ursus maritimus*) require sea ice for hunting seals. The predicted loss of sea ice in the coming decades threatens to drive polar bears to extinction. The lack of sea-ice cover shortens the polar bears' seal hunting season and makes dispersal to and from land over ice-covered seas more difficult. Wildlife biologists estimate that the current population of polar bears in the Arctic stands at about 25,000. In some areas, the bears remain on the sea ice year-round, but in others, the melting in spring and summer forces them to come ashore. Polar bears can fast for months. While fasting, bears move as little as possible to conserve energy. But sea-ice loss and population declines create new problems, such as having to expend more energy searching for a mate, that could further affect survival. Even under more modest warming projections, in which GHG emissions peak by 2040 and then begin to decline, many regional populations would still be extirpated.

The southernmost polar bear populations, such as those who spend the winter on the Beaufort Sea shelf north of the coast of Alaska, are particularly vulnerable because of the complete loss of coastal sea ice for increasing parts of the year. The lack of ice forces bears ashore each summer where they draw on body energy reserves for survival and lactation due to the absence of high-energy foods on land. Essentially, even if they find some food such as bird's eggs or small mammals, they will lose weight the entire time they are on land and can only replenish their fat stores once they are back at sea hunting seals. Prolonged ice absence from productive continental shelf waters now forces increasingly long fasts for land-bound bears. Lengthening fasts have already lowered body condition, reproduction, survival, and abundance in some near-shore populations, and similar trends are expected throughout the Arctic as ice loss continues (Molnár et al., 2020). No one knows how long bears can fast before substantial declines in lactation (and therefore cub recruitment) and adult survival occur. Because of resource differences, such fasting thresholds may vary somewhat between different subpopulations, affecting the rate of population declines (Fig. 11.15). The evidence shows that the subpopulations of the southern Beaufort Sea off northeastern Alaska, and in western Hudson Bay in Canada, have already been severely affected by the loss of sea ice.

Molnár et al. (2020) used numerical modeling to predict the nature, timing, and order of future population declines by estimating the threshold numbers of days that polar bears can fast before cub recruitment and/or adult survival decline rapidly (Fig. 11.15). Their model suggests that bear cub recruitment and survival thresholds may already have been exceeded in some subpopulations. They also suggest that if the current trends in GHG emissions persist, the accompanying amplified warming will cause a steep decline in reproduction and survival rates, jeopardizing all but a few High Arctic subpopulations by 2100. Moderate emissions mitigation would prolong the persistence of the species but is unlikely to prevent some regional extinction within this century.

Modeling efforts by Molnár et al. (2020) produced the following predictions for the polar bear subpopulation that spend their summers on the ACPA. One of these subpopulations hunts seals on the pack ice of the Southern Beaufort Sea. The other hunts on the pack ice of the Chukchi Sea. The model was designed to predict cub recruitment, and adult male and adult female survival. Cub recruitment refers to the process of raising cubs to the point at which they are independent.

Under the RCP4.5 (medium-case) scenario for the Southern Beaufort Sea subpopulation, the model predicts that disturbance of cub recruitment will be "Very Likely" by 2025. By 2050, disturbance of adult male survival will be "Likely," and by 2060, disturbance of adult female survival will also be "Likely." Under the RCP8.5 (worst-case) scenario, disturbance of cub recruitment will "Very Likely" occur by 2030. Disturbance of adult male survival will be "Likely" by 2035, and disturbance of adult female survival will be "Likely" by 2050.

For the Chukchi Sea subpopulation, under the RCP4.5 scenario, the model predicts that disturbance of cub recruitment will be "Very Likely" by 2050. By 2055, disturbance of adult male survival will be "Likely," and by 2065, disturbance of adult female survival will also be "Likely." Under the RCP8.5 (worst-case) scenario, disturbance of cub recruitment will "Very Likely" occur by 2060, and it will become "Inevitable" by 2090. Disturbance of adult male survival will be "Very Likely" by 2075, becoming "Inevitable" by 2100. Disturbance of adult female survival will be "Likely" by 2045, becoming "Very Likely" by 2090.

Expansion of grizzly bears into the High Arctic

Ursus arctos, called the grizzly bear in western North America and the brown bear in Eurasia and coastal Alaska, is well adapted to living on Arctic tundra. Although the brown bear has previously been spotted as far north as 74 °N in Siberia, the range of North American grizzlies has traditionally been considered to end on the mainland. Although grizzly bear and

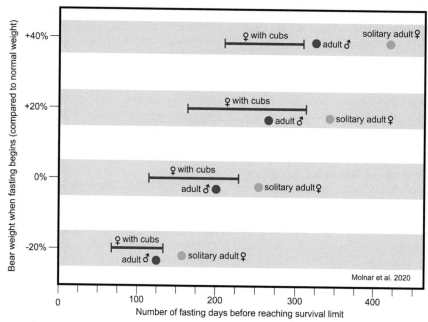

FIG. 11.15 The number of fasting days before solitary adult male and female polar bears and female polar bears with cubs reach their survival limit, based on their weight at the start of fasting. (Data from Molnár, P.K., Bitz, C.M., Holland, M.M., Kay, J.E., Penk, S.R., Amstrup, S.C., 2020. Fasting season length sets temporal limits for global polar bear persistence. Nature Climate Change 10 (8), 732–738. https://doi.org/10.1038/s41558-020-0818-9.)

polar bear ranges overlap along the northern coasts of North America, Asia, and Europe, polar bears are generally found offshore hunting seals, separated from grizzly bears during their mating season. However, in the past 20 years, it has become increasingly common to observe grizzly bears north of their traditional range, sometimes 500 km or more from the mainland coast, well into the range of polar bears in the Canadian Arctic Archipelago. Grizzlies have been seen on the more southerly of the islands in the Canadian Arctic Archipelago, ranging from Hudson Bay in the east to the Beaufort Sea in the west. The most northerly sighting thus far has been off the northwest coast of Banks Island approximately 75°N (Doupé et al., 2007).

Polar Bear Hybridization with Grizzly Bears

The recent northward advance of brown bears onto the Canadian Arctic Archipelago has brought this species into more regular contact with polar bears, with some interesting consequences. In 2006, a hybrid bear was found near Sachs Harbour, Northwest Territories. Over the past decade, brown bears have been mating with polar bears in this region. Using genetic evidence, Pongracz et al. (2017) were able to trace eight hybrid individuals to a single female polar bear who mated with two grizzly bears in recent years. The primary conservation concern raised by this hybridizing is that the polar bear population could become subsumed by northward-invading grizzly bears, causing the loss of the distinctive polar bear genotype. However, polar bears and grizzly bears have different mating habits. Thus, there are behavioral barriers that should prevent polar bears from being genetically swamped by grizzly bears, in whose presence polar bears have long maintained an independent evolutionary course. Pongracz et al. (2017) observed that grizzly bear range expansion into the High Arctic is driven by males that disperse farther distances from their birthplace home ranges than females do.

The impact of sea-ice loss on Pacific walrus

The Pacific walrus is a subspecies of walrus (*Odobenus rosmarus*) found in the Bering, Chukchi, Laptev, and East Siberian Seas. As reviewed by Post et al. (2013), the loss of sea-ice cover forces walrus to haul out onto land to rest and to bear pups, increasing overcrowding of walrus populations on beaches, which facilitates disease transmission within populations (Fig. 11.16). The presence of sea ice benefits walrus in many ways. It

FIG. 11.16 Flowchart illustrating terrestrial and marine ecosystem damage if Arctic sea-ice cover is lost. (Concepts from Post, E., Bhatt, U.S., Bitz, C.M., Brodie, J.F., Fulton, T.L., Hebblewhite, M., Kerby, J., Kutz, S.J., Stirling, I., Walker, D.A., 2013. Ecological consequences of sea-ice decline. Science 341 (6145), 519–524. https://doi.org/10.1126/science.1235225.)

provides a platform for walrus to hunt for mollusks on the seafloor (their main source of food). It also provides a safe place for walrus pupping (Fig. 11.17). The walrus pups are safer on floating ice near shore than if they are forced to share overcrowded beaches with adults and other pups.

Throughout the Arctic, the loss of sea ice is limiting the amount of space available for walruses to congregate. Floating summer sea ice is also receding further north to the point where the water is too deep for the animals to dive and feed. This forces them to desert the ice and seek refuge ashore (World Wildlife Fund, 2020).

Researchers from the US Geological Survey first observed large haul-outs along the coast of the Chukchi Sea off Alaska's Point Lay in 2007, when summer Arctic sea ice reached its second-lowest minimum extent in recorded history (Laustsen, 2016). As discussed in Chapter 1, the extent of summer sea ice has continued to decline in Arctic waters, and the number of walruses coming ashore has grown considerably. In 2014, around 35,000 walruses hauled out along a small stretch of beach at Point Lay, northwest Alaska. These massive haul-outs are extremely dangerous for walruses. The crowded animals are easily spooked; any unusual

sound or scent may cause a deadly stampede. In their rush to the ocean, the large adults, weighing up to 700 kg, often trample other walruses, especially young calves, which are susceptible to injuries and death. In 2015, stampedes at a haul-out near Cape Schmidt, Russia, caused more than 500 walrus deaths.

When on land, the animals are far away from their food sources. Satellite tagging by US Geological Survey wildlife biologists documented walruses making long treks to their feeding grounds. In 2011, USGS scientists discovered that half of the animals they had tagged made round trips of about 400 km (250 miles) between Point Lay and the southern part of Hanna Shoal in the Chukchi Sea. While this is a great distance for an adult walrus to travel, it would be even more difficult for a calf accompanying its mother. Any rebound in walrus numbers is also greatly hampered because walruses have the lowest reproductive rate of any pinniped species, and mothers invest considerable care in their young (Groc, 2017).

Pacific walrus numbers reached record-lows in the early 1960s but rebounded in the 1980s following significant conservation efforts. Their population size reached 200,000 in the 1990s but is once again in decline. The current population is estimated at

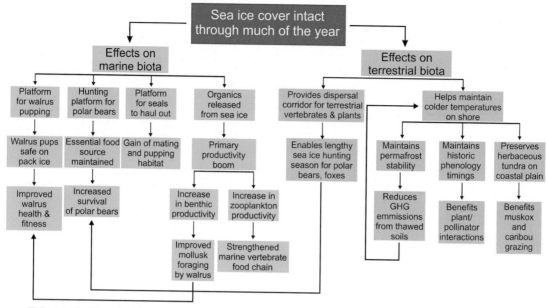

FIG. 11.17 Flowchart illustrating terrestrial and marine ecosystem benefits of sea-ice cover in the Arctic. (Concepts from Post, E., Bhatt, U.S., Bitz, C.M., Brodie, J.F., Fulton, T.L., Hebblewhite, M., Kerby, J., Kutz, S.J., Stirling, I., Walker, D.A., 2013. Ecological consequences of sea-ice decline. Science 341 (6145), 519–524. https://doi.org/10.1126/science.1235225.)

129,000 individuals. The IUCN now lists this species as "Vulnerable."

The conservation status of the Atlantic subspecies of walrus (*Odobenus rosmarus* ssp. *divergens*) was last assessed by the IUCN in 2014. At that time, the population estimate was 100,000 individuals, and the conservation status for this subspecies was "Data deficient." The most recent population estimates of Atlantic walruses show a serious decline. There is now an estimated total of 22,500 individuals including populations of 6000 in Norway and Russia, 12,000 in Canada, and 4500 in Greenland (Seal Conservation Society, 2020). The Atlantic walrus once ranged south to Sable Island, Nova Scotia, and as late as the 18th century was found in large numbers in the Greater Gulf of St. Lawrence region, sometimes in colonies of up to 7000 to 8000 individuals. The Atlantic subspecies of walrus was nearly eradicated by commercial harvest. Their populations were devastated in the 19th and early 20th centuries, mainly by European sealers and whalers. Overhunting led to the near-extirpation of the Atlantic subspecies. The Atlantic walrus has not been able to recover from these historic losses and is still well below its preexploitation level of several hundred thousand. It is currently believed that there are eight subpopulations of Atlantic walrus, scientists concluding in 1995 that four of these

were stable or increasing, two to be declining and two to have unknown status.

Declines in caribou herds

The migratory tundra caribou (*Rangifer tarandus*) in North America undergoes synchronized, cyclic, or quasi-cyclic fluctuations of population size spanning periods ranging from 40 to 90 years. The ecological processes behind these fluctuations remain largely unknown. They may be related to climatic oscillations, density-dependent interactions with forage plants, or interactions with predators. The recent warming of the Arctic has been accompanied by increasing vegetation biomass mainly attributed to the expansion of tall, deciduous shrubs into the tundra ecosystems. This "shrubification" is expected to alter important attributes of the ecosystem such as carbon sequestration, snow cover, productivity, fire regimes, and hydrology. Fauchald et al. (2017) demonstrated that the loss of sea-ice cover in the Arctic Ocean is the principal driver of climate-induced changes in the regions of Arctic tundra that constitute caribou summer ranges and that this is the most significant environmental change affecting regional caribou-plant dynamics. They found no evidence of impacts of caribou abundance on regional vegetation. Rather, they found that the ongoing climate warming has boosted plant biomass on caribou summer pastures. One would suppose that this increase in plant productivity would benefit caribou herds, but this is not the case. In fact, there has been a decline in caribou populations. This result suggests that the greening of the coastal tundra regions has led to a deterioration of caribou pasture quality. The shrub expansion in Arctic North America involves plant species with strong antibrowsing defenses. Fauchald et al. (2017) concluded that this phenomenon might signal a climate-driven shift in caribou-plant interactions. The previous system was characterized by low plant biomass modulated by cyclic caribou populations. The newly developing system is characterized by the dominance of nonedible shrubs with strong antibrowsing defenses. Thus, both the quality and quantity of caribou pasture are declining, contributing to a reduction in the size of migratory caribou herds.

Red fox versus Arctic fox

In ecology, the competitive exclusion principle specifies that species with similar habitat requirements (i.e., complete competitors) cannot both live in the same territory. One seeming exception to this rule is that of the red fox (*Vulpes vulpes*) and the Arctic fox (*Vulpes lagopus*) in the Arctic. Here, both fox species occupy the same habitat. They are both medium-sized predators with largely overlapping ecological niches, suggesting the likelihood of strong competitive interactions. They both largely rely on microtine rodents for food and use similar strategies to subsist in the harsh Arctic environment, such as food caching and foraging on the sea ice for alternative foods such as seal carcasses in winter (Gallant et al., 2012).

Red foxes have some competitive advantages. They can weigh twice as much as Arctic foxes and can be up to 70% longer. They also tend to grow larger at higher latitudes. The red fox is thus physically dominant and can kill the smaller Arctic fox. Red foxes are currently using preexisting Arctic fox dens, sometimes evicting Arctic foxes in the process. On the other hand, the Arctic fox is highly adapted to Arctic conditions. It has better insulation and has a lower critical temperature than the red fox. It is better adapted to travel on snow, with a lower foot load (body mass divided by foot surface area) than the red fox. The Arctic fox uses less metabolic energy at rest than the red fox, and the Arctic fox is also able to decrease its basal metabolic rate in winter. Thus, the larger red fox has higher metabolic energy requirements and needs more food and a larger territory to sustain itself compared with the Arctic fox. Coexistence between the two species thus likely depends on a harsh environment or low ecosystem productivity to prevent the dominant red fox from occupying all of the landscape and completely excluding the Arctic fox, through interference competition for food and dens.

The red fox expansion has coincided with Arctic climate warming, and climate change has generally been considered to be the ultimate determinant of the outcome of the competition between the two species. Is there a direct cause-and-effect relationship between climate warming and red fox expansion into the Arctic? As we have seen from other examples, biological interactions are rarely as simple as this. Climate change hypotheses notwithstanding, wildlife fieldwork over long periods of time have raised serious questions about the validity of the climate change theory. Gallant et al. (2012) conducted aerial and ground fox den surveys on Herschel Island and the coastal mainland of the northern Yukon to investigate the relative abundance of red and Arctic foxes. They compared the results of their 2008—10 research with previous regional field studies from the past four decades. The results of this and the other surveys all indicate little change in the relative abundance of the two species. North Yukon fox dens are mostly occupied by Arctic foxes, with active red fox dens occurring sympatrically, i.e., they exist in the same geographic area and thus frequently encounter one another. While vegetation changes have been reported, there is no indication that secondary productivity (i.e., increases in microtine rodent populations) has increased. Their study shows that in the western North American Arctic, the competitive balance between red and Arctic foxes has changed little in the past 40 years. Such results clearly challenge the hypotheses linking climate to red fox expansion and that the negative effects of climate warming may be overriding the positive effects of milder temperatures and longer growing seasons.

The negative effects of climate warming on microtine rodents (mainly voles and lemmings) may be considerable. These include reduced snow cover duration, which decreases thermal buffering of beneath-snow habitats and limits predation cover and increased frequency of winter thaw events that limit rodent access to food through ice cover. Even with potentially positive effects of warming on microtines (i.e., enhanced NPP), their unreliability as a food source for foxes may remain significant. For example, in this primarily top-down food web, any increase in microtine numbers due to increased summer primary production would likely draw increased hunting pressure from many other regional predators. These include raptors such as the snowy owl (*Nyctea scandiaca*), rough-legged hawk

(*Buteo lagopus*), peregrine falcon (*Falco peregrinus*), gyr-falcon (*Falco rusticolus*), and merlin (*Falco columbarius*), as well as mammalian predators including the grizzly bear (*Ursus arctos*), wolf (*Canis* lupus), Canada lynx (*Lynx canadensis*), wolverine (*Gulo gulo*), and least weasel (*Mustela nivalis*). This increased predation pressure leaves both species of foxes no better off than during years of low microtine numbers.

Despite the high densities of microtines that appear periodically on Herschel Island, red foxes have not excluded Arctic foxes from the island, nor have they excluded Arctic foxes from the coastal plain. On the mainland, the reason may be the limited distribution and abundance of alternative prey, especially Arctic ground squirrels (*Urocitellus parryii*), which are much larger than microtines and help sustain red foxes during summers with low microtine numbers. Ground squirrels have not successfully colonized Herschel Island and are absent over large areas on the coastal plain, probably because of poor burrowing opportunities. Gallant et al. (2012) noted that red fox dens on the northern Yukon mainland coincided with areas of active ground squirrel. The current red fox distribution appears to be associated with two areas of higher food abundance: one linked to ground squirrels and the other to a fluctuating lemming population. When compared with the wider Arctic fox distribution, this further indicates that food is a limiting factor that has a bigger impact on the larger red fox. This favors the coexistence of red and Arctic foxes in northern Yukon.

The competition between the red fox and Arctic fox in northern Scandinavia has been quite different from that found in North America. Arctic foxes have been fully protected in Scandinavia since the drastic decline from overhunting in the early 19th century (Selås and Vik, 2007). Despite this protection, their numbers have failed rebound since that time. This lack of recovery has been thought to result from (1) increased interspecific competition with red foxes that have colonized the Arctic fox's range, or (2) changes in prey dynamics due to increased variation in climatic conditions. Hamel et al. (2013) used a large-scale field study combined with extensive regional red fox removals in northern Norway to test the validity of these two hypotheses. In their study, Arctic foxes were never observed at sites where red foxes live, even in localities where only a few red fox observations had been made. The historical record makes it clear that Arctic foxes do not recolonize areas in northern Norway unless the red fox has been removed. The odds of Arctic fox recolonization increased when lemming abundance was very high, but the presence of red foxes remained the most

limiting factor for Arctic fox recolonization. The results obtained by Hamel et al. (2013) lend support to both hypotheses but highlight the presence of red foxes as the most limiting factor for Arctic fox recolonization in this region. The red foxes of Arctic Scandinavia clearly dominate over the Arctic foxes, so here the competitive exclusion principle appears to be in effect.

Global change impacts on tundra-breeding birds

Arctic avian diversity declines with latitude and is low compared with temperate and tropical regions. Only 2% of global bird species are known to be regular Arctic breeders, and fewer still are Arctic specialists. Although diversity is low overall, groups such as the geese (Anserini) and shorebirds (Scolopacidae, Charadriidae) achieve their highest diversity at Arctic latitudes, dominating the bird communities in many locations. Moreover, nearly all Arctic-breeding bird species migrate to warmer regions during the nonbreeding season, connecting the Arctic to all regions of the planet (Kulbelka et al., 2018).

Effects of rapid climate change are already measurably affecting tundra ecosystems, and this includes bird communities. For instance, there may be mismatches between the timing of the hatch of tundra birds' chicks and the emergence of their arthropod prey. After the breeding season, Arctic-breeding birds disperse south through boreal to tropical habitats, some traveling as far as the Antarctic. Throughout their migrations, they face many anthropogenic pressures, from modification and loss of habitats to sport and subsistence hunting in flyways around the world.

Tundra-breeding birds face diverse large conservation challenges, from accelerated rates of Arctic climate change to threats associated with highly migratory life histories. A recent study by Kubelka et al. (2018) considered two groups of birds that spend summers in the Arctic: wading birds and waterfowl. Wading birds are water birds, especially species with long legs, that habitually wade in shallow water in search of food. Fish, aquatic and terrestrial invertebrates, amphibians, and crustaceans are their common foods. Waterfowl are ducks, geese, or other large aquatic birds. Ducks mainly forage for food while swimming and do not wade. Geese are grazing birds that eat a variety of different items. They eat roots, shoots, stems, seeds, and leaves of grass and grain, bulbs, and berries.

According to Smith et al. (2020), over half of all circumpolar Arctic wading bird taxa (waders) are declining, amounting to 51% of 91 taxa with known trends. At the same time, almost half of all waterfowl are increasing (49% of 61 taxa). These opposite trends have brought changes in avian community composition in some regions. Widespread, and in some cases accelerating, changes underscore the urgent conservation needs faced by many Arctic terrestrial bird species.

Predation on shorebird nests

As discussed earlier, the study by Ims et al. (2019) showed that nest predation of Arctic shorebirds has increased steeply in recent years, and the authors investigated the underlying causes for this, in terms of the ecological mechanisms involved.

In simplest terms, the reason for the increased predation on Arctic birds during a time of increased regional NPP is as follows. Increased productivity increases the quantity and quality of herbaceous vegetation on tundra landscapes. In regions occupied by voles and lemmings, these changes in vegetation can lead to population booms in these rodents. This, in turn, causes the population sizes of generalist predators to increase. However, vole and lemming populations in the Arctic are known to oscillate between boom and bust conditions, even when food is readily available. In Scandinavia, lemming populations have been seen to remain relatively stable for long periods between outbreaks, while vole populations maintain more-or-less regular population cycles with a period of 4−5 years. However, these patterns may soon change (Ims et al., 2011). Increased climatic warming and the concurrent dampening of vole population cycles may decrease the frequency, amplitude, and geographic range of lemming outbreaks in tundra ecosystems. But thus far, when the rodent populations crash, the enlarged predator populations must seek prey elsewhere, and that is when predation on bird nests becomes extremely high.

Increased NPP (greening) has already been demonstrated as a fundamental tundra ecosystem response to global warming. The greening effect is likely to have cascading effects in Arctic ecosystems through the alteration of trophic interactions in food webs. A determination of the nature of these changes in tundra food webs will help implement biodiversity conservation under climate change. Although the ongoing greening of the Arctic may be impossible to counteract through local environmental management, actions taken to halt the increase of generalist predators may nevertheless be a management option to preserve Arctic birds in a warming climate. Indeed, such actions are currently implemented in northern Fennoscandia to safeguard the critically endangered population of lesser white-fronted goose (*Anser erythropus*).

Permafrost Melting

As discussed by Wild et al. (2019), the destabilization of permafrost and peat deposits in the warming Arctic involves a range of processes that act on different temporal and spatial scales. Rising temperatures cause a gradual deepening of the seasonally thawed active layer at the surface of permafrost soils and a northward recession of the permafrost boundary at the southern margin of the permafrost zone. The combination of rising temperatures and increasing precipitation further stimulates abrupt surface collapse and degradation of deeper organic carbon deposits. Ice-rich permafrost deposits are particularly vulnerable to collapse (thermokarst), including organic-rich deposits that accumulated in the Holocene as well as Pleistocene deposits that are still widespread, especially across north-eastern Siberia.

Thermokarst is the most widespread form of abrupt permafrost thaw. Thermokarst is caused by the melting of bodies of ground ice beneath the surface, triggering land surface collapse. Water pooling in collapsed areas leads to the formation of taliks. These are areas of thawed ground surrounded by permafrost. They are frequently found beneath expanding lakes, and they accelerate permafrost thaw far faster and more deeply than the effects of increasing air temperatures alone. Remote sensing and field observations reveal that localized abrupt thaw features, including thermokarst lakes, thermoerosional gullies, thaw slumps, and peat-plateau collapse scars, are extensive across the Low Arctic in regions that have ice-rich permafrost. Two decades of observations have shown that thermokarst lakes and other abrupt thaw features are hot spots of soil carbon conversion to atmospheric methane (Anthony et al., 2018, Fig. 11.18).

The extent of permafrost melting

Most permafrost in the Northern Hemisphere occurs between latitudes of 60°N and 68°N. North of 67°N, permafrost declines sharply as land gives way to the Arctic Ocean. Permafrost regions currently occupy approximately 22.8 million km^2 of the Northern Hemisphere, constituting about 24% of exposed land surfaces (NSIDC, 2017). According to the model predictions of the IPCC (2019) (Fig. 11.19), under the best-case scenario (RCP2.6), the Northern Hemisphere's permafrost area will decrease by 4.5 million km^2 by 2100. Under RCP4.5, the Northern Hemisphere permafrost area will decrease by more than 6.5 million km^2 by 2100. Under the worst-case scenario (RCP8.5), the Northern Hemisphere area will decrease by more than 10.5 million km^2 by 2100, representing a loss of

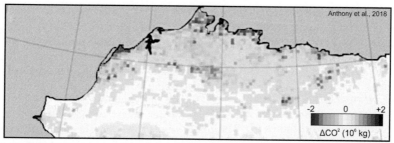

FIG. 11.18 Carbon flux implications in northwestern Alaska for 1999–2014 determined with Landsat satellite trend analysis: the net change in carbon flux over the 15-year observation period. (Image after Anthony, K., Schneider von Deimling, T., Nitze, I., Frolking, S., Emond, A., Daanen, R., Anthony, P., Lindgren, P., Jones, B., & Grosse, G., 2018. 21st-century modeled permafrost carbon emissions accelerated by abrupt thaw beneath lakes. Nature Communications 9 (1). https://doi.org/10.1038/s41467-018-05738-9.)

almost half of the Northern Hemisphere permafrost area.

As illustrated in Turestsky et al. (2019), major regions of the Arctic currently experience rapidly melting permafrost (marked in shades of red in Fig. 11.20). This abrupt permafrost thawing is expected to occur in less than 20% of frozen land in the coming decades. Other Arctic regions are experiencing more gradual permafrost melting. In climatic terms, the estimated rate of permafrost loss may be approximately 4.0 million km^2 per 1 degree warming, and permafrost landscapes are likely to get wetter or drier in the future depending on microtopographic features (Zhang et al., 2018).

The collapse of the Batagay mega slump in Siberia

One of the most dramatic permafrost events in modern times has been the collapse of an enormous permafrost region in northern Siberia (Stone, 2020). Known to locals as the "gateway to the underworld," Batagay is the largest thaw slump on the planet. Once just a gully on a slope logged in the 1960s, the scar has expanded year by year, as the permafrost thaws and meltwater carries off the sediment. Now more than 900 m wide, it demonstrates the vulnerability of Arctic permafrost to thermal degradation, even in one of the coldest regions of the Northern Hemisphere. Global warming is causing permafrost degradation all across Siberia. Bursts of pent-up methane gas in thawing permafrost have pockmarked Russia's Yamal and Gydan peninsulas with holes tens of meters across.

Dates from ice and soil gathered at Batagay show that it holds the oldest exposed permafrost in Eurasia, spanning the past 650,000 years. The ice wedges from the

Lower Ice Complex are likely the oldest (>0.5 Ma) ever analyzed isotopically and also point to a very cold winter climate during formation (Opel et al., 2020).

Role of the Arctic in the global carbon cycle

Terrestrial Arctic ecosystems play a key role in the global carbon (C) cycle. Much of Earth's terrestrial carbon is stored in the northern high latitudes. Until the past few years, Arctic permafrost has served as an enormous carbon sink for the planet. All that is changing as the Arctic warms. Thawing permafrost throughout the Arctic could be releasing an estimated 300–600 million tons of net carbon per year to the atmosphere (Richter-Menge et al., 2019). In numerical terms, the amount of organic carbon stored in the top 3 m of soils in the northern permafrost region accounts for 1035 ± 150 Pg (Pg), which is about 50% of the global soil organic carbon pool (1 Pg = 1 × 10^{18} gr) (Cannone et al., 2019).

Carbon accumulation in Arctic soils and peatlands

As illustrated in Fig. 11.21, estimates of the carbon content in the upper 2 m of Arctic soil range from less than 20 kg/m^2 of surface area to as much as 260 kg/m^2 (NOAA, 2019). The response of Arctic peatlands to recent warming remains uncertain even though future warming may result in major changes in C accumulation in these high-latitude peatlands (Zhang et al., 2018). This is partly because warming increases the growing season length and therefore plant productivity, while at the same time, plant physiology and decay rates of plant litter are affected by changes in soil moisture conditions.

While ancient carbon has formed a sink in the North, the decomposition of organic remains releases

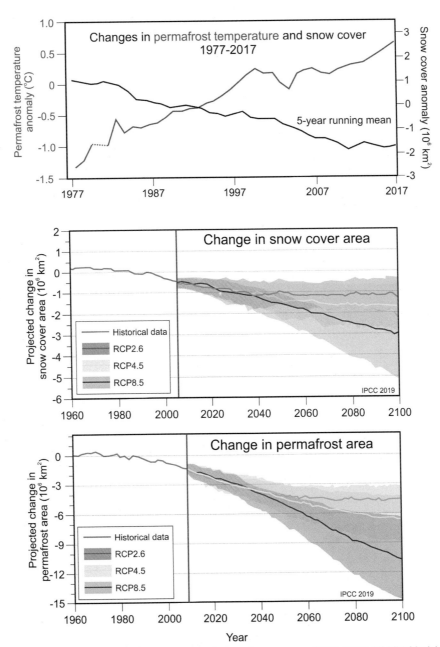

FIG. 11.19 Top: Changes in permafrost temperature and snow cover, 1977–2017. Middle: Modeled future changes in snow cover area through the year 2100. Bottom: Modeled future changes in the northern permafrost area through the year 2100. Modified from IPCC, 2019. (Modified from Pörtner, H., Roberts, D., Masson-Delmotte, P., Zhai, P., & Tignor, M., 2019. IPCC, 2019. IPCC Special Report on the Ocean and Cryosphere in a Changing Climate. IPCC.)

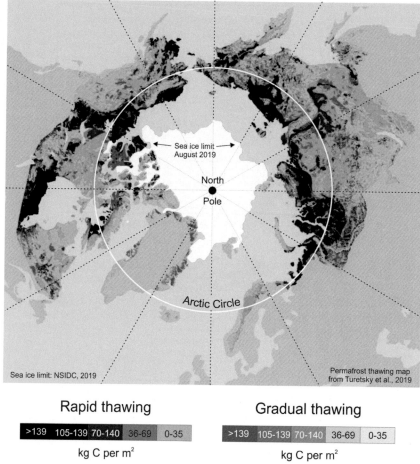

Rapid thawing   Gradual thawing

>139	105-139	70-140	36-69	0-35

kg C per m²

>139	105-139	70-140	36-69	0-35

kg C per m²

FIG. 11.20 Map of the current extent of Arctic permafrost, showing rapidly thawing regions (shades of red on the map) and more slowly thawing regions (shades of blue). (Figure modified from Turetsky, M.R., Abbott, B.W., Jones, M.C., Walter Anthony, K., Olefeldt, D., Schuur, E.A.G., Sannel, A.B.K., 2019a. Permafrost collapse is accelerating carbon release. Nature 569 (7754), 32–34. https://doi.org/10.1038/d41586-019-01313-4. with the addition of 2019 Arctic sea ice minimum from NSIDC (2019).)

C to the atmosphere, mostly as carbon dioxide and methane. Also, active biological processes draw C into plant and animal tissues. As occurs to a greater or lesser extent in every ecosystem, some amount of C is stored, while another amount of C is released to the atmosphere. Therefore, the carbon balance of any region represents the difference between sources and syncs of carbon in any given year. A major study in northern Siberia, based on observations at 100 field sites, determined that northern permafrost released on average about 600 million tons more carbon than that absorbed by vegetation each year from 2003 to 2017 (Opel et al., 2020).

Release of Arctic carbon to the atmosphere
As discussed by Turetsky et al. (2019b), soils in the permafrost region hold twice as much carbon as the atmosphere. One assumption built into current models of GHG release and climate is that permafrost thaws gradually from the surface downward. Deeper layers of organic matter are exposed over decades or even centuries, and some models are beginning to track these slow changes. But these models overlook an even more troubling problem. Frozen soil does not just trap carbon, but it physically holds the landscape together. Across the Arctic, as permafrost is melting, the ground surface is suddenly collapsing as pockets of ice within

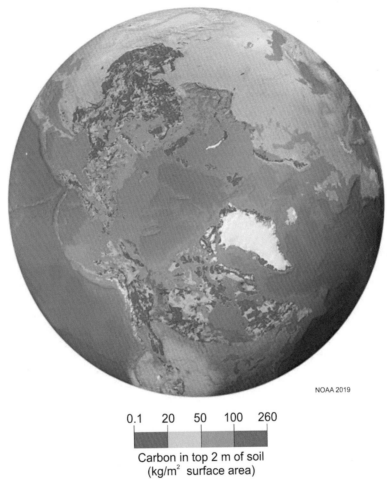

NOAA 2019

0.1   20   50   100   260

Carbon in top 2 m of soil
(kg/m^2 surface area)

FIG. 11.21 Map of current Arctic permafrost, showing the carbon content of permafrost soils in kg/m^2. (From NOAA Arctic Program, 2019. Arctic Report Card: Update for 2019. Retrieved from https://arctic.noaa.gov/Report-Card/Report-Card-2019.)

it melt. This means that instead of a few centimeters of soil thawing each year, several meters of soil can become destabilized within days or weeks. The land can sink and be inundated by swelling lakes and wetlands.

Predictions based on numerical modeling suggest that slow and steady thawing will release around 200 billion metric tons (200 Pg) of carbon over the next 300 years under a business-as-usual (RCP8.5) warming scenario. But Turetsky et al. (2019b) caution that this could be a vast underestimate. Around 20% of frozen lands have features that increase the likelihood of abrupt thawing, such as large quantities of ice in the ground or on unstable slopes. Here, permafrost thaws quickly and erratically, triggering landslides and rapid

erosion. Lakes that have existed for generations can disappear, or their waters can be diverted. Worse than this, the most unstable regions also tend to be the most carbon-rich. For example, 1 million km^2 of Siberia, Canada, and Alaska contain pockets of Yedoma, thick deposits of permafrost from the last ice age. These deposits are often 90% ice, making them extremely vulnerable to warming. Moreover, because of the glacial dust and grasslands that were folded in when the deposits formed, Yedoma contains 130 billion metric tons of organic carbon, the equivalent of more than a decade of global human GHG emissions (Turetsky et al., (2019b)).

As we have seen, permafrost thaw and the subsequent mobilization of carbon (C) stored in previously frozen soil organic matter (SOM) have the potential to exert strong positive feedback to climate. As the northern permafrost region experiences as much as a doubling of the rate of warming as the rest of the Earth, the vast amount of C in permafrost soils is vulnerable to thaw, decomposition, and release as atmospheric GHGs (Natali et al., 2015). Diagnostic and predictive estimates of high-latitude terrestrial C fluxes vary widely among different models depending on how permafrost dynamics and the seasonally thawed "active layer" above it, are represented. Hayes et al. (2014) used a process-based model simulation experiment to assess the net effect of active layer dynamics on this "permafrost carbon feedback" in recent decades, from 1970 to 2006, over the circumpolar domain of continuous and discontinuous permafrost. Over this period, the model estimates a mean increase of 6.8 cm in active layer thickness across the domain, which exposes a total of 11.6 Pg C of thawed SOM to decomposition. According to their experiment, mobilization of this previously frozen C results in an estimated cumulative net source of 3.7 Pg C to the atmosphere since 1970, all directly tied to active layer dynamics. This estimated net C transfer to the atmosphere from permafrost thaw represents a significant factor in the overall ecosystem carbon budget of the entire Arctic and a substantial additional contribution on top of the combined fossil fuel emissions from the eight Arctic nations over this period.

Bringing together a wide range of recent observations, Schuur et al. (2015) proposed that GHG emissions from warming permafrost are likely to occur at a magnitude similar to other historically important biosphere carbon sources but that the emissions from permafrost will be only a fraction of current fossil fuel emissions. At the proposed rates, the observed and projected emissions of CH_4 and CO_2 from thawing permafrost are unlikely to cause abrupt climate change during the next few years. Instead, permafrost carbon emissions are going to contribute to GHG warming of the planet for several centuries, continuing to add to the amplified warming of the Arctic and driving the pace of climate change to higher rates than those expected on the basis of human activities alone.

Schuur et al. (2015) also observed that the first studies bringing widespread attention to permafrost carbon estimated the amount of organic carbon stored in northern permafrost soils at almost 1700 billion tons (170 Pg). Those estimates were based on carbon stored at depths greater than 1 m in permafrost, below the traditional zone of soil carbon accounting. Deeper carbon measurements were initially rare, and it was not even possible to quantify the uncertainty for the permafrost carbon pool size estimate. However, important new analyses are reporting large quantities of carbon preserved in deeper layers of permafrost at many previously unsampled locations, and a substantial portion of this deep permafrost carbon is susceptible to future thaw. The permafrost carbon pool is now thought to comprise organic carbon in the top 3 m of the soil. Carbon is stored in deposits deeper than 3 m, including those Pleistocene sediments in the Yedoma region, an area of deep sediment deposits that cover unglaciated parts of Siberia and Alaska, as well as carbon stored in frozen terrestrial sediments that now lie on shallow ocean shelves in the Arctic.

Model scenarios show the potential release of permafrost carbon in the range 37–174 Pg C by 2100, given the IPCC worst-case scenario trajectory of RCP8.5. The average of these wide-ranging model estimates is 92 Pg of carbon. Most models estimate that 59% of total permafrost carbon emissions will occur after 2100. While such estimates are undoubtedly uncertain, they demonstrate the gathering momentum of climate change brought about by the release of permafrost carbon, causing a cascading release of GHGs as microbes slowly decompose newly thawed permafrost carbon.

C released by erosion

Peat deposits have accumulated across the Arctic since the end of the last glaciation. Peat is particularly abundant in the world's largest wetland, the West Siberian Lowland, which is mainly underlain by vulnerable discontinuous permafrost and projected to experience a further decrease in permafrost extent. Rising temperatures are expected not only to accelerate the degradation of frozen peat carbon deposits to CO_2 or CH_4 in situ (i.e., at their thawing location, but also to trigger a more catastrophic release of carbon into streams and rivers. As discussed by Wild et al. (2019), part of this fluvial carbon eventually reaches the Arctic Ocean. Once this terrestrial carbon enters seawater, thousands of km away from point of thaw, a significant portion decomposes, giving rise to CO_2 efflux (the flowing out of a substance or particle to the atmosphere), leading to severe ocean acidification. Although this phenomenon is known to be occurring, it will be exceedingly difficult to incorporate this transported carbon into global carbon budget models. This is because large spatial heterogeneity in organic carbon thaw and remobilization is expected on local, drainage basin, and subcontinental scales, so a large number of site-specific observations from multiple regions are needed to meet this upscaling challenge.

Wild et al. (2019) used a decade-long, high-temporal resolution record of ^{14}C in dissolved and particulate organic carbon (DOC and POC, respectively) to disentangle the sources of particulate organic carbon (PP-C) being released in a series of river drainage basins across Siberia, including the Ob, Yenisey, Lena, and Kolyma rivers. They found that DOC was dominated by recent organic carbon, and POC was dominated by PP-C ($63 \pm 10\%$) and represents the best method to detect spatial and temporal dynamics of PP-C release. They found that there are distinct seasonal patterns in the release of carbon from frozen deposits. It appears that DOC primarily stems from gradual leaching of surface soils, while POC reflects the abrupt collapse of deeper deposits. Higher dissolved PP-C export by the Ob and Yenisey rivers is derived from soils in the discontinuous permafrost zone that facilitates leaching. Higher particulate PP-C export by Lena and Kolyma derives mainly from the thermokarst-induced collapse of Pleistocene deposits. This process can be a result of lateral erosion causing ground ice to be exposed or lateral degradation where warm surface water penetrates the adjacent shore, causing patches of thawed ground (taliks) to form and ground ice to thaw.

Release of greenhouse gases from frozen soils

The two GHGs emitted from Arctic ecosystems are carbon dioxide and methane. CO_2 is emitted from the aerobic decomposition of organic matter (i.e., bacterial decomposition in the presence of oxygen). CH_4 is emitted from the anaerobic decomposition of organic matter (i.e., bacterial decomposition in the absence of oxygen). Based on the analysis of their ages, it has become apparent that methane and carbon dioxide being released into the atmosphere from melting permafrost come from different sources of terrestrial carbon. Anthony et al. (2018) used radiocarbon dating of CH_4 and CO_2 samples taken from numerous Arctic thermokarst lakes. They reported that the methane currently being released from Arctic lake bottoms has a median age of about 17,500 yr BP, indicating that the methane is coming from organic deposits that formed during the last glaciation. In contrast to this, the ^{14}C age of CO_2 currently being emitted from permafrost is almost modern, ranging in age from 567 to 700 yr BP.

An increase in the volume of newly thawed sublake sediments through the expansion of existing thermokarst lakes and the formation of new ones is likely to yield disproportionately large releases of methane to the atmosphere during this century. Using the conditions prescribed by the RCP8.5 scenario, Anthony et al. (2018) estimate a 27 Tg per year increase in methane emissions from newly formed lakes by midcentury. This level of methane emissions represents a threefold rise in global anthropogenic CH_4 emissions, compared with those recorded from 2003 to 2012 (10 Tg/year).

A bubble of methane gas, swelling beneath Siberia's melting permafrost for an indeterminate period, burst open during the summer of 2020, forming a 50-m-deep crater (Casella, 2020). Researchers who went to the site found chunks of ice and rock thrown hundreds of meters from the funnel-shaped hole. Although a highly unusual geological event, this giant hole is at least the 17th such "funnel" discovered to date in the region and the largest of its kind in recent years. Air near the bottom of the crater contained unusually high concentrations of methane—almost 10%—in tests conducted at the site on 16 July, says Andrei Plekhanov, an archaeologist at the Scientific Centre of Arctic Studies in Salekhard, Russia. Plekhanov and his team believe that the rapid development and sudden collapse of this funnel is linked to the abnormally hot Yamal summers of 2012 and 2013, which were warmer than usual by an average of about 5°C. Analysis of regional photographs as old as the 1970s revealed that these Siberian funnels have existed for at least the past 50 years, but have recently expanded, adding strength to the argument that melting permafrost is the most likely driver of these collapses. Another study from 2017 identified 7000 gas pockets under the Yamal Peninsula, in the immediate vicinity of the newly discovered funnel. A study of a funnel that appeared on the Yamal Peninsula in 2014 also concluded that methane released from melting permafrost was the cause of the funnel's formation (Moskvitch, 2014).

The amplified warming of the Arctic is causing greater thawing of permafrost than has been previously observed. Based on numerical modeling, Hayes et al. (2014) estimated that the warmth-enhanced decomposition from newly exposed frozen SOM accounts for the release of both CO_2 (4.0 Pg C) and CH_4 (0.03 Pg C) but is partially compensated by CO_2 uptake (0.3 Pg C) associated with enhanced net primary production of vegetation. Increases in ecosystem respiration will very likely result from warmer soils and thawing permafrost. However, this uptake of CO_2 by Arctic vegetation may be partially offset by enhanced plant productivity through greater availability of nutrients and extension of the growing season. Also, in areas dominated by snow, ice, and permafrost, the importance of the transition seasons (late spring and early autumn) has been identified. During these transitional periods, net ecosystem exchange may show large fluctuations (Cannone et al., 2019). New regional and winter season

measurements of ecosystem carbon dioxide flux independently indicate that permafrost region ecosystems
are releasing net carbon (potentially 0.3 to 0.6 Pg C/
year) to the atmosphere (Schuur, 2019). These observations indicate that the GHG feedback to accelerated
climate change may already be underway.

In summary, the balance between uptake and release
of CO_2 from Arctic tundra is full of complexities. There
are a series of interactions between abiotic and biotic
factors that play important roles in the CO_2 flux from
herbaceous tundra. The abiotic factors include thaw
depth, ground surface temperature, and soil moisture.
The biotic factors include phenology and leaf area index
(LAI). The latter concerns the amount of foliage in the
plant canopy and serves as a proxy for plant productivity. LAI is one of the leading indicators of net primary
production, water and nutrient use, and carbon balance
(Bjorkman et al., 2018).

Changes in soil microbial communities

In recent years, efforts have been made to accurately
assess global methane (CH_4) budgets to understand the
dynamics of the sinks (CH_4 oxidation) and sources
(CH_4 production) of this important GHG. Singh
(2013) discusses changes in the concentration of atmospheric methane, the most abundant organic trace gas
in the atmosphere, and the role it plays in climate change.
As Arctic soils thaw, the methane molecules trapped inside are partially digested by methanotroph bacteria,
thus lowering the amount of methane gas that escapes
to the atmosphere. It has been estimated that approximately 30 Tg (1 Tg $= 10^{12}$ g) of atmospheric methane
per year is oxidized by aerobic methanotrophic bacteria
in upland soils, accounting for about 6% of the global atmospheric CH_4 sink. Since the methane cycle is affected
by feedback mechanisms from ongoing global warming,
the scientific community is now focusing on estimating
potential climatic feedbacks from a rapidly changing
climate.

As discussed by Singh (2013), various environmental factors exert influences on methane oxidation
by methanotrophic bacteria in the soil. Microbial processes have a central role in the global fluxes of the
key biogenic GHGs, CO_2, CH_4, and N_2O (nitrous
oxide) and are likely to respond rapidly to climate
change. Global climate change will affect the activity
of microorganisms that break down methane (methanotrophic bacteria). These soil bacteria play a crucial
role in the global carbon cycle. It has been found that
soil nitrogen levels increase with temperature; soil
nitrogen suppresses methane oxidation rates by methanotrophs. Under global warming conditions, increasing

soil temperatures lead to greater N availability in the
soil, as well as lowering soil moisture. Over time, soil
inorganic nitrogen suppresses the activity and diversity
of soil methanotrophs. Thus, changes in such environmental factors as temperature, soil moisture, and soil
nitrogen status are bound to influence rates of methane
oxidation by methanotrophs.

However, in laboratory-controlled soil column experiments, increased temperatures had no appreciable
effect on methanotrophic communities and activities
(Singh, 2013). Another factor that could affect the
methane oxidation activity of methanotrophs is soil
moisture content. Soil moisture is the vehicle that facilitates the transport of atmospheric methane into the
mineral soil, along with soil porosity.

How will the microorganisms in permafrost respond
to global warming? Though it is well known that microorganisms can survive under permafrost conditions
over geologically significant time and that they are
well adapted to respond to permafrost thaw, questions
remain regarding the similarities in the microbial functional response at different permafrost locations. It is
well known that a wide variety of microorganisms live
in Arctic permafrost. These include bacteria that function as aerobic and anaerobic heterotrophs, methano-
gens, iron reducers, sulfate reducers, and nitrogen
fixers (Messan et al., 2020).

Bacteriologists and soil scientists have also come to
realize that microbial community structure and functional response undergo changes when frozen soils
thaw. Specifically, community structure changes with
increased temperature or thaw stage. However, a study
of permafrost collected in interior Alaska led to the
conclusion that different microbial community
responses to thawing may have similar functional outcomes. As discussed earlier, microbial activity and the
production and emission of GHGs such as carbon dioxide, methane, and nitrogen oxide have been shown to
increase following permafrost thaw.

Permafrost microbial communities

Microbial growth or metabolic activity have been reported in permafrost bacteria living at temperatures
down to $-10°C$ (Rivkina et al., 2000). Significant
numbers of viable bacteria (10^2 to 10^8 cells/gram of
sediment) are known to be present in Arctic permafrost
as old as 3 million years. The ratio of aerobic to anaerobic bacteria in permanently frozen soils and sediments
apparently varies according to regional geological history, so that the metabolic pathways of these bacteria
harken back to the time their initial deposition. For
instance, permafrost sediments of alluvial, lake, and

marine origin that formed under anoxic conditions contain high proportions of anaerobic bacteria.

Rivkina et al. (2000) performed laboratory experiments on the viability and metabolic activity of bacteria found in samples of Siberian permafrost. Bacterial activity was measured in a laboratory at temperatures between 5 and −22°C. Metabolic activity followed a sigmoidal pattern similar to other biological growth curves. At all temperatures, the growth phase was followed by a stationary phase within 200−350 days. The minimum bacteria doubling times ranged from 1 day (5°C) to about 160 days (−22°C).

Although the physical structure of permafrost makes metabolic activity possible, it has often been assumed that microorganisms in permafrost are in a state of stasis, or suspended animation. However, permafrost soils are known to contain liquid water. The most biologically important aspect of this water is that it makes possible the mass transfer of molecules through the otherwise almost impenetrable medium of frozen materials. Mass exchange of molecules is greatest in sites with low ice content and least in sites with high ice content.

Messan et al. (2020) analyzed microbial metabolites originating from Alaskan permafrost. They found that permafrost thaw induced a shift in microbial metabolic processes. The thawed permafrost metabolites from different locations were highly similar. Samples of intact permafrost yielded several metabolites with antagonistic properties, illustrating the competitive survival strategy required for these microbes to survive in a frozen state. Interestingly, the study showed that the concentration of these antagonistic metabolites decreased with warmer temperature, indicating a shift in microbial ecological strategies in thawed permafrost. These findings illustrate the impact of change in temperature and spatial variability as permafrost undergoes thaw. This represents knowledge that will become crucial for predicting permafrost biogeochemical dynamics as Arctic landscapes continue to warm.

RELEASE OF NUTRIENTS FROM FROZEN SOILS (PONDS, LAKES, RIVERS)

The combination of increased nutrient inputs and warmer temperatures is a key factor enhancing NPP in Arctic wetlands (Andresen et al., 2018). On the Arctic coastal plain of Alaska, these authors measured the increase in nutrients that are being released into ponds, lakes, and streams from adjacent patches of thawing permafrost. Permafrost has been thawing in the study area, with the thaw depth of ponds in Utqiaġvik having

increased by about 11 cm over the past 60 years. The total number of thawing degree-days has also increased by 13 days over the past 40 years, which has likely enhanced permafrost thaw, thus contributing to the observed rise in pond nutrients over recent decades. Thawing degree-days are defined as the cumulative number of degree-days when air temperatures are above 0°C.

The role of permafrost thaw on changes in Arctic water chemistry remains poorly understood. As the study by Andresen et al. (2018) indicates, permafrost degradation may well lead to substantial increases in nutrient availability to tundra ponds and lakes. Likewise, a portion of the nutrient and carbon enriched waters liberated by melting permafrost eventually discharge to the ocean where the nutrients fertilize marine phytoplankton and the carbon eventually sinks to the ocean floor, decomposes, and adds to the flux of CO_2 and methane entering the atmosphere.

Reyes and Lougheed (2015) designed an experiment to examine rapid nutrient release from permafrost and active layer soils under different warming scenarios, including changes in redox (oxidation-reduction reaction) environments and microbial activity. They found that permafrost soils tended to release more nutrients and at a higher rate than active layer soils, especially under warmer conditions. This suggests that the active layer may be depleted in nutrients and that the water column may be resupplied with these ions from thawing permafrost. In their experiments, they found that the release of micronutrients (Ca, Fe, Mg, S, Si) from incubated permafrost soils was greatly facilitated by increased temperatures. However, temperature was not necessarily the primary driver of macronutrient (N, P) release from permafrost. Other physicochemical parameters are influenced by temperature. These include oxidation-reduction potential, microbial activity, dissolved organic carbon, and iron. All of these were shown to play a role in nutrient release from permafrost. These experiments demonstrated that nutrient release via the thaw of permafrost soils is an important contributor to essential nutrients in Arctic aquatic ecosystems. In field studies, the stimulation of aquatic plant growth was detected within 1 day of warming. Further warming of the Arctic is likely to result in substantial changes to nutrient availability and cycling in these dominant habitats.

The biogeochemical response of permafrost to warming is certainly not simple and straightforward. Temperature changes are unlikely to be the only driver of nutrient release from soils. As the study by Rivkina et al. (2000) demonstrates, permafrost soils often host very active microbial communities that may enhance decomposition and nutrient release, even at below-freezing temperatures.

CONCLUSIONS

There are two principal categories of threats to Arctic terrestrial environments and biota. The principal threat comes from global warming. This is an over-arching hazard, as its effects are being felt throughout the Arctic, in terrestrial, freshwater, and marine environments. The current intensified, rapid changes in thermal regime affect a wide range of environmental parameters, exerting various stresses on regional fauna and flora. Because of Arctic amplification, the biota of all Arctic landscapes is currently facing far greater climate changes than the lower latitudes. The second principal threat comes from direct human interference on the landscape. This includes the development of industrial sites for the extraction of resources, a topic covered in depth in Chapter 15, and permafrost degradation due to the building of uninsulated or poorly insulated structures on frozen soils, a topic covered in depth in Chapter 16.

There are so many take-home messages in this chapter that they are quite difficult to summarize in a few sentences. One aspect that is an environmental threat to Arctic lands is simply the interconnectedness of the ecosystems, in both the Low Arctic and High Arctic and between the two regions. For instance, sea ice, although a marine phenomenon, has enormous effects on the climate of the land. In fact, as we have seen, it is the principal driver of the Arctic amplification phenomenon which is felt hundreds and thousands of kilometers inland from the Arctic Ocean. Loss of sea ice cover means the loss of the principal habitat of both marine mammals in terrestrial mammals that make their living on the sea ice in winter. Beyond that, the regional warming brought on by the loss of sea-ice cover is affecting trophic relationships of plants and animals on land, such as the increased threat to the eggs and chicks of shorebirds.

The amplified temperatures are causing permafrost to melt and its southern edges to retreat northward. The melted permafrost is contributing to GHG warming of the atmosphere through the release of CO_2 and methane released from ancient organic deposits in the frozen soils, as these are broken down by bacteria at or near the surface.

The interconnectedness of Arctic ecosystems has remained invisible to most of humanity through the centuries because of the remote aspects and hostile

climates of the Arctic. Sadly, it is only now, when we have done so much harm to Arctic environments, that we are finally realizing the nature and fragility of the North. A final irony in this situation is that many people, if they ponder the question at all, consider the Arctic flora and fauna to be extremely tough and durable. How else could life endure in such an extreme environment? However, the truth is that their ability to withstand such trials as extremely long, bitterly cold winters and extremely short, cool summers with little precipitation have made them extremely vulnerable to conditions that seem to us more amenable to life, such as warmer temperatures and longer growing seasons. Polar bears and walruses need sea ice, not warm beaches. Arctic vegetation thrives on top of frozen soils, not on midlatitude prairies.

In other words, the very adaptations that allow polar biota to survive in the Arctic prevent them from adapting to any other kind of environment. If the Arctic biota had tens or hundreds of thousands of years to adapt to climate change, then their descendants might succeed surprisingly well. The pace and amplitude of the current environmental changes make this impossible. Adaptation is a very slow process taking thousands of generations to reach an endpoint. Climate forecasters tell us that the new warm Arctic will be arriving in very short order and, in some cases, is already here.

REFERENCES

Alaska Climate Research Center, Alaska Statewide Climate Summary, 2020. Atmospheric and Oceanographic Sciences Library. Springer. http://akclimate.org/sites/Default/Files/202005_May_summary.pdf.

Andresen, C.G., Lara, M.J., Tweedie, C.E., Lougheed, V.L., 2017. Rising plant-mediated methane emissions from arctic wetlands. Global Change Biology 23 (3), 1128–1139. https://doi.org/10.1111/gcb.13469.

Andresen, C.G., Tweedie, C.E., Lougheed, V.L., 2018. Climate and nutrient effects on Arctic wetland plant phenology observed from phenocams. Remote Sensing of Environment 205, 46–55. https://doi.org/10.1016/j.rse.2017.11.013.

Anthony, K., Schneider von Deimling, T., Nitze, I., Frolking, S., Emond, A., Daanen, R., Anthony, P., Lindgren, P., Jones, B., Grosse, G., 2018. 21st-century modeled permafrost carbon emissions accelerated by abrupt thaw beneath lakes. Nature Communications 9 (1). https://doi.org/10.1038/s41467-018-05738-9.

Arctic Portal Library, 2011. CAFF Map No.33 - Major Biomes (Vegetation Zones) of the Arctic. (2011). http://library.arcticportal.org/1364/.

Arndt, K.A., Santos, M.J., Ustin, S., Davidson, S.J., Stow, D., Oechel, W.C., Tran, T.T.P., Graybill, B., Zona, D., 2019. Arctic greening associated with lengthening growing seasons in Northern Alaska. Environmental Research Letters 14 (12). https://doi.org/10.1088/1748-9326/ab5e26.

Bjorkman, A.D., Myers-Smith, I.H., Elmendorf, S.C., Normand, S., Rüger, N., et al., 2018. Plant functional trait change across a warming tundra biome. Nature 562, 57–84. https://doi.org/10.1038/s41586-018-0563-7.

Blume-Werry, G., Wilson, S.D., Kreyling, J., Milbau, A., 2016. The hidden season: growing season is 50% longer below than above ground along an arctic elevation gradient. New Phytologist 209 (3), 978–986. https://doi.org/10.1111/nph.13655.

Boscutti, F., Casolo, V., Beraldo, P., Braidot, E., Zancani, M., Rixen, C., 2018. Shrub growth and plant diversity along an elevation gradient: evidence of indirect effects of climate on alpine ecosystems. PloS One 13 (4), e0196653. https://doi.org/10.1371/journal.pone.0196653.

Cannone, N., Ponti, S., Christiansen, H.H., Christensen, T.R., Pirk, N., Guglielmin, M., 2019. Effects of active layer seasonal dynamics and plant phenology on CO_2 land-atmosphere fluxes at polygonal tundra in the high Arctic, Svalbard. Catena 174, 142–153. https://doi.org/10.1016/j.catena.2018.11.013.

Cassella, C., 2020. Giant Gaping Void Emerges in Siberia, the Latest in a Dramatic Ongoing Phenomenon. https://www.sciencealert.com/another-giant-gaping-crater-was-suddenly-found-in-siberia-the-largest-in-recent-years.

Cohen, J., Screen, J.A., Furtado, J.C., Barlow, M., Whittleston, D., et al., 2014. Recent Arctic amplification and extreme mid-latitude weather. Nature Geoscience 7, 627–637. https://doi.org/10.1038/NGEO2234.

Doupé, J.P., England, J.H., Furze, M., Paetkau, D., 2007. Most northerly observation of a grizzly bear (Ursus arctos) in Canada: photographic and DNA evidence from Melville Island, Northwest territories. Arctic 60 (3), 271–276.

Duchesne, R.R., Chopping, M.J., Tape, K.D., Wang, Z., Schaaf, C.L.B., 2018. Changes in tall shrub abundance on the North Slope of Alaska, 2000–2010. Remote Sensing of Environment 219, 221–232. https://doi.org/10.1016/j.rse.2018.10.009.

Fauchald, P., Park, T., Tømmervik, H., Myneni, R., Hausner, V.H., 2017. Arctic greening from warming promotes declines in caribou populations. Science Advances 3 (4), e1601365. https://doi.org/10.1126/sciadv.1601365.

Fowler, R.A., Barry, C.D., Northington, R.M., Osburn, C.L., Pla-Rabes, S., Mernild, S.H., Whiteford, E.J., Grace Andrews, M., Kerby, J.T., Post, E., 2019. Arctic climate shifts drive rapid ecosystem responses across the West Greenland landscape. Environmental Research Letters 14 (7). https://doi.org/10.1088/1748-9326/ab2928.

Gallant, D., Slough, B.G., Reid, D.G., Berteaux, D., 2012. Arctic fox versus red fox in the warming Arctic: four decades of den surveys in North Yukon. Polar Biology 35 (9), 1421–1431. https://doi.org/10.1007/s00300-012-1181-8.

Gillespie, M.A.K., Baggesen, N., Cooper, E.J., 2016. High Arctic flowering phenology and plant-pollinator interactions in response to delayed snow melt and simulated warming. Environmental Research Letters 11 (11), 115006. https://doi.org/10.1088/1748-9326/11/11/115006.

Groc, I., 2017. Walrus Habitat on the Edge. World Wildlife Magazine.

Hamel, S., Killengreen, S.T., Henden, J.A., Yoccoz, N.G., Ims, R.A., 2013. Disentangling the importance of interspecific competition, Food availability, and habitat in species occupancy: recolonization of the endangered Fennoscandian Arctic fox. Biological Conservation 160, 114–120. https://doi.org/10.1016/j.biocon.2013.01.011.

Hayes, D.J., Kicklighter, D.W., McGuire, A.D., Chen, M., Zhuang, Q., Yuan, F., Melillo, J.M., Wullschleger, S.D., 2014. The impacts of recent permafrost thaw on land-atmosphere greenhouse gas exchange. Environmental Research Letters 9 (4), 045005. https://doi.org/10.1088/1748-9326/9/4/045005.

Ims, R.A., Yoccoz, N.G., Killengreen, S.T., 2011. Determinants of lemming outbreaks. Proceedings of the National Academy of Sciences 108, 1970–1974. https://doi.org/10.1073/pnas.1012714108.

Hersbach, H., Bell, B., Berrisford, P., Hirahara, S., Horányi, A., et al., 2020. The ERA5 global reanalysis. Quarterly Journal of the Royal Meteorological Society 149, 1999–2049. https://doi.org/10.1002/qj.3803.

Hirawake, T., Uchida, M., Abe, H., Alabia, I.D., Hoshino, T., et al., 2021. Response of Arctic biodiversity and ecosystem to environmental changes: findings from the ArCS project. Polar Science 27, 100533. https://doi.org/10.1016/j.polar.2020.100533 (in press).

Ims, R.A., Henden, J.A., Strømeng, M.A., Thingnes, A.V., Garmo, M.J., Jepsen, J.U., 2019. Arctic greening and bird nest predation risk across tundra ecotones. Nature Climate Change 9 (8), 607–610. https://doi.org/10.1038/s41558-019-0514-9.

IPCC, 2013. Climate Change 2013. In: Stocker, T.F., Qin, D., Plattner, D.K., Tignor, M., Allen, S.K., et al. (Eds.), The Physical Science Basis. Contribution of Working Group I to the Fifth Assessment Report of the Intergovernmental Panel on Climate Change. Cambridge University Press, Cambridge, United Kingdom and New York, NY, USA. ISBN 978-1-107-05799-1.

IPCC, 2014. Synthesis Report. Contribution of Working Groups I, II and III to the Fifth Assessment Report of the Intergovernmental Panel on Climate Change. (2014). IPCC.

IPCC, 2019. IPCC Special Report on the Ocean and Cryosphere in a Changing Climate. Cambridge University Press, Cambridge, United Kingdom and New York, NY, USA, p. 1190.

Jones, J., Screen, J.A., Furtado, J.C., Barlow, M., Whittleston, D., Coumou, D., Francis, J., Dethloff, K., Entekhabi, D., Overland, J., 2014. Recent Arctic amplification and extreme mid-latitude weather. Nature Geoscience 7 (9), 627–637. https://doi.org/10.1038/ngeo2234.

Khan, A.L., Wagner, S., Jaffe, R., Xian, P., Williams, M., Armstrong, R., McKnight, D., 2017. Dissolved black carbon in the global cryosphere: concentrations and chemical signatures. Geophysical Research Letters 44 (12), 6226–6234. https://doi.org/10.1002/2017GL073485.

Körner, C., Paulsen, J., Spehn, E.M., 2011. A definition of mountains and their bioclimatic belts for global comparisons of biodiversity data. Alpine Botany 121 (2), 73–78. https://doi.org/10.1007/s00035-011-0094-4.

Kubelka, V., Šálek, M., Tomkovich, P., Végvári, Z., Freckleton, R.P., Székely, T., 2018. Global pattern of nest predation is disrupted by climate change in shorebirds. Science 362 (6415), 680–683. https://doi.org/10.1126/science.aat8695.

Lameris, T.K., Scholten, I., Bauer, S., Cobben, M.M.P., Ens, B.J., Nolet, B.A., 2017. Potential for an Arctic-breeding migratory bird to adjust spring migration phenology to Arctic amplification. Global Change Biology 23 (10), 4058–4067. https://doi.org/10.1111/gcb.13684.

Laustsen, P.C., 2016. Walrus Sea-Ice Habitats Melting Away. https://www.usgs.gov/center-news/walrus-sea-ice-habitats-melting-away.

Lawrence, D.M., Slater, A.G., 2009. The contribution of snow condition trends to future ground climate. Climate Dynamics 34, 969–981. https://doi.org/10.1007/s00382-009-0537-4.

McEwen, D.C., Butler, M.G., 2018. Growing-season temperature change across four decades in an arctic tundra pond. Arctic 71 (3), 281–291. https://doi.org/10.14430/arctic4730.

Mekonnen, Z.A., Riley, W.J., Grant, R.F., 2018. 21st century tundra shrubification could enhance net carbon uptake of North America Arctic tundra under an RCP8.5 climate trajectory. Environmental Research Letters 13 (5), 054029. https://doi.org/10.1088/1748-9326/aabf28.

Messan, K.S., Jones, R.M., Doherty, S.J., Foley, K., Douglas, T.A., Barbato, R.A., 2020. The role of changing temperature in microbial metabolic processes during permafrost thaw. PloS One 15 (4), e0232169. https://doi.org/10.1371/journal.pone.0232169.

Molnár, P.K., Bitz, C.M., Holland, M.M., Kay, J.E., Penk, S.R., Amstrup, S.C., 2020. Fasting season length sets temporal limits for global polar bear persistence. Nature Climate Change 10 (8), 732–738. https://doi.org/10.1038/s41558-020-0818-9.

Moskvitch, K., 2014. Mysterious Siberian crater attributed to methane. Nature News. http://doi:10.1038/nature.2014.15649.

Myers-Smith, I.H., Hik, D.S., Kennedy, C., Cooley, D., Johnstone, J.F., Kenney, A.J., Krebs, C.J., 2011. Expansion of canopy-forming willows over the twentieth century on Herschel Island, Yukon territory, Canada. Ambio 40 (6), 610–623. https://doi.org/10.1007/s13280-011-0168-y.

NASA Earth Observatory, 2009. Arctic Amplification. https://earthobservatory.nasa.gov/.

NASA Earth Observatory, 2020. Heat and Fire Scorches Siberia. https://earthobservatory.nasa.gov/images/146879.

Natali, S.M., Olefeldt, D., Romanovsky, V.E., Schaefer, K., Turetsky, M.R., Treat, C.C., Vonk, J.E., Koven, C.D., Kuhry, P., Lawrence, D.M., 2015. Climate change and the permafrost carbon feedback. Nature 520 (7546), 171–179. https://doi.org/10.1038/nature14338.

NOAA, 2020. Ocean Today: Black Carbon. https://oceantoday.noaa.gov/blackcarbon/.

NOAA Arctic Program, 2019. Arctic Report Card: Update for 2019. https://arctic.noaa.gov/Report-Card/Report-Card-2019.

NSIDC, 2017. State of the Cryosphere: Permafrost and Frozen Ground. https://nsidc.org/cryosphere/sotc/permafrost.htm.

NSIDC, 2018. Arctic Sea Ice News and Analysis. http://nsidc.org/arcticseaicenews/2018/08/.

NSIDC, 2019. Arctic Sea Ice at Minimum Extent for 2019. https://nsidc.org/news/newsroom/arctic-sea-ice-minimum-extent-2019.

NSIDC, 2020a. Ocean Today: Black Carbon. https://oceantoday.noaa.gov/blackcarbon/.

NSIDC, 2020b. All about Sea Ice. Thermodynamics: Albedo. In: https://nsidc.org/cryosphere/seaice/processes/albedo.html.

Opel, T., Wetterich, S., Meyer, H., Murton, J., 2020. Ground-Ice Stable-Isotope Paleoclimatology at the Batagay Megaslump. EGU General Assembly EGU2020-3748, pp. 2020–3748. https://doi.org/10.5194/egusphere-egu2020-3748.

Overland, J.E., Hanna, E., Hanssen-Bauer, I., Kim, S.-J., Walsh, J.E., 2020. Arctic Report Card: Surface Air Temperature Update for 2019. NOAA. https://arctic.noaa.gov/Report-Card/Report-Card-2019.

Pongracz, J.D., Paetkau, D., Branigan, M., Richardson, E., 2017. Recent hybridization between a polar bear and grizzly bears in the Canadian Arctic. Arctic 70 (2), 151–160. https://doi.org/10.14430/arctic4643.

Post, E., Bhatt, U.S., Bitz, C.M., Brodie, J.F., Fulton, T.L., Hebblewhite, M., Kerby, J., Kutz, S.J., Stirling, I., Walker, D.A., 2013. Ecological consequences of sea-ice decline. Science 341 (6145), 519–524. https://doi.org/10.1126/science.1235225.

Przybylak, R., 2016. The Climate of the Arctic, second ed. Springer.

Ramsayer, K., 2017. NASA Studies Details of a Greening Arctic. https://www.nasa.gov/feature/goddard/2016/nasa-studies-details-of-a-greening-arctic.

Reyes, F.R., Lougheed, V.L., 2015. Rapid nutrient release from permafrost thaw in arctic aquatic ecosystems. Arctic Antarctic and Alpine Research 47 (1), 35–48. https://doi.org/10.1657/AAAR0013-099.

Richter-Menge, J.M., Druckenmiller, M.L., Jeffries, M., 2019. Arctic Report Card 2019. https://www.arctic.noaa.gov/Report-Card.

Rivkina, E.M., Friedmann, E.I., McKay, C.P., Gilichinsky, D.A., 2000. Metabolic activity of permafrost bacteria below the freezing point. Applied and Environmental Microbiology 66 (8), 3230–3233. https://doi.org/10.1128/AEM.66.8.3230-3233.2000.

Saros, J.E., Anderson, N.J., Juggins, S., McGowan, S., Yde, J.C., et al., 2019. Arctic climate shifts drive rapid ecosystem responses across the West Greenland landscape. Environmental Research Letters 14, 074027. https://doi.org/10.1088/1748-9326/ab2928.

Schuur, T., 2019. NOAA Arctic Program, Arctic Essays: Permafrost and the Global Carbon Cycle. https://arctic.noaa.gov/Report-Card/Report-Card-2019.

Schuur, E.A.G., McGuire, A.D., Schädel, C., Grosse, G., Harden, J.W., et al., 2015. Climate change and the permafrost carbon feedback. Nature 520, 171–179. https://doi.org/10.1038/nature14338.

Seal Conservation Society, 2020. Walrus (*Odobenus rosmarus*). https://www.pinnipeds.org/seal-information/species-information-pages/walrus#.

Selås, V., Vik, J.O., 2007. The arctic fox *Alopex lagopus* in Fennoscandia: a victim of human-induced changes in interspecific competition and predation? Biodiversity and Conservation 16 (12), 3575–3583. https://doi.org/10.1007/s10531-006-9118-6.

Semenchuk, P.R., Gillespie, M.A.K., Rumpf, S.B., Baggesen, N., Elberling, B., Cooper, E.J., 2016. High Arctic plant phenology is determined by snowmelt patterns but duration of phenological periods is fixed: an example of periodicity. Environmental Research Letters 11 (12), 125006. https://doi.org/10.1088/1748-9326/11/12/125006.

Serreze, M.C., Barrett, A.P., Stroeve, J.C., Kindig, D.N., Holland, M.M., 2009. The emergence of surface-based Arctic amplification. The Cryosphere 3 (1), 11–19. https://doi.org/10.5194/tc-3-11-2009.

Singh, J.S., 2013. Anticipated effects of climate change on methanotrophic methane oxidation. Climate Change and Environmental Sustainability 1, 20–24. https://doi.org/10.5958/j.2320-6411.1.1.003.

Smith, P.A., McKinnon, L., Meltofte, H., Lanctot, R.B., Fox, A.D., et al., 2020. Status and trends of tundra birds across the circumpolar Arctic. Ambio 49, 732–748. https://doi.org/10.1007/s13280-019-01308-5.

Stone, R., 2020. Siberia's "gateway to the underworld" hit by heat wave. Science (New York, N.Y.) 369 (6504), 612–613. https://doi.org/10.1126/science.369.6504.612.

Turetsky, M.R., Abbott, B.W., Jones, M.C., Walter Anthony, K., Olefeldt, D., Schuur, E.A.G., Koven, C., McGuire, A.D., Grosse, G., Kuhry, P., Hugelius, G., Lawrence, D.M., Gibson, C., Sannel, A.B.K., 2019a. Permafrost collapse is accelerating carbon release. Nature 569 (7754), 32–34. https://doi.org/10.1038/d41586-019-01313-4.

Turetsky, M.R., Abbott, B.W., Jones, M.C., Anthony, K.W., Olefeldt, D., et al., 2019b. Permafrost Carbon Pools: How Much Permafrost Carbon Is Available to Release into the Atmosphere? https://arctic.noaa.gov/Report-Card/Report-Card-2019/.

van Gestel, N., Natali, S., Andriuzzi, W., Chapin III, F.S., Ludwig, S., et al., 2019. Long-term warming research in high-latitude ecosystems: responses from polar ecosystems and implications for future climate. In: Mohan, J.E. (Ed.), Ecosystem Consequences of Soil Warming. Elsevier, Amsterdam, pp. 441–487. https://doi.org/10.1016/B978-0-12-813493-1.00016-8.

Wild, B., Andersson, A., Bröder, L., Vonk, J., Hugelius, G., McClelland, J.W., Song, W., Raymond, P.A., Gustafsson, Ö., 2019. Rivers across the Siberian Arctic unearth the patterns of carbon release from thawing permafrost. Proceedings of the National Academy of Sciences of the United States of America 116 (21), 10280–10285. https://doi.org/10.1073/pnas.1811797116.

Winiger, P., Barrett, T.E., Sheesley, R.J., Huang, L., Sharma, S., Barrie, L.A., Yttri, K.E., Evangeliou, N., Eckhardt, S., Stohl, A., Klimont, Z., Heyes, C., Semiletov, I.P., Dudarev, O.V., Charkin, A., Shakhova, N., Holmstrand, H., Andersson, A., Gustafsson, Ö., 2019. Source apportionment of circum-Arctic atmospheric black carbon from isotopes and modeling. Science Advances 5 (2), eaau8052. https://doi.org/10.1126/sciadv.aau8052.

World Wildlife Fund, 2020. Climate Change Puts the Pacific Walrus Population on Thin Ice. https://www.worldwildlife.org/stories/climate-change-puts-the-pacific-walrus-population-on-thin-ice.

Xu, L., Myneni, R.B., Chapin III, F.S., Callahan, T.V., Pinzon, P.E., et al., 2013. Temperature and vegetation seasonality diminishment over northern lands. Nature Climate Change 3 (6), 581–586. https://doi.org/10.1038/nclimate1836.

Zhang, H., Gallego-Sala, A.V., Amesbury, M.J., Charman, D.J., Piilo, S.R., Väliranta, M.M., 2018. Inconsistent response of arctic permafrost peatland carbon accumulation to warm climate phases. Global Biogeochemical Cycles 32 (10), 1605–1620. https://doi.org/10.1029/2018GB005980.

FURTHER READING

Elphinstone, C., 2020. High Arctic Tundra Photo: Alexandra Fiord, Ellesmere Island. Note Open-Top Environmental Chamber Used in Research by the International Tundra Experiment. https://en.wikipedia.org/wiki/International_Tundra_Experiment.

NASA Earth Observatory, North Slope of Alaska, 2007. Atmospheric and Oceanographic Sciences Library. https://earthobservatory.nasa.gov/images/7863/north-slope-of-alaska.

Ozinga, W.A., Penuelas, J., Poorter, H., Poschlod, P., Reich, P.B., Sandel, B., Schamp, B., Sheremetev, S., Weiher, E., Ordoñez, J.C., 2018. Plant functional trait change across a warming tundra biome. Nature 562 (7725), 57–62. https://doi.org/10.1038/s41586-018-0563-7.

Wall, D.H., Natali, S., Andriuzzi, W., Chapin, F.S., Ludwig, S., Moore, J.C., Pressler, Y., Salmon, V., Schuur, T., Simpson, R., 2019. Long-term warming research in high-latitude ecosystems: responses from polar ecosystems and implications for future climate. In: Ecosystem Consequences of Soil Warming: Microbes, Vegetation, Fauna and Soil Biogeochemistry. Elsevier, pp. 441–487. https://doi.org/10.1016/B978-0-12-813493-1.00016-8.

NOAA, 2019. Arctic Report Card - 2019 . https://arctic.noaa.gov/Report-Card/Report-Card-2019/ArtMID/7916/ArticleID/835/Surface-Air-Temperature.

CHAPTER 12

Impacts of Global Change

INTRODUCTION

To many people, the term global change refers to climate change, in particular, climate warming. However, this is but one aspect of global change. According to Steflen et al. (2004), "global change' refers to the broad suite of human-caused changes to the biological world, the physical world (e.g., climate change, ice sheet change, sea level change, etc.), and the socioeconomic changes that are affecting Earth System functioning at the global scale. Global change encompasses the transformations of global environments that have occurred over the past few centuries, including major changes in land use and land cover, the growth of cities, the growth of international travel and commerce, and increases in the pollution of the air, water, and land. Global change also includes human interference in natural cycles, such as the damming of rivers and streams, alterations to the nitrogen and carbon cycles, disruption of marine and terrestrial food chains, and reduction of biological diversity, sometimes leading to species extinctions. Steflen et al. (2007) also emphasized the importance of linkages and interactions between these various changes, which also form part of global change and are just as important as the individual changes themselves. Global changes often do not occur as linear progressions, but rather as nonlinear phenomena. This makes their future impacts more difficult to predict.

This chapter discusses a broad range of global changes that affect the Arctic and adjacent regions. As with all aspects of human disruption, the impacts on the Arctic are often more severe than elsewhere, because of the fragility of Arctic environments and ecosystems. Except for the whaling industry that devastated Arctic whale populations until the early 20th century, human impacts on the Arctic were relatively small until the end of World War II. Likewise, the lack of scientific research in the Arctic until after World War II meant that scientific knowledge of human impacts on the Arctic was negligible. Starting in the 1950s, human impacts on the Arctic have risen year by year, accompanied by increases in Arctic scientific research. What follows is a brief description of some aspects of global change in the Arctic.

ARCTIC TERRESTRIAL IMPACTS

Human disruption of Arctic landscapes has not been as extensive as it is in lower latitudes. This is mostly due to the climate, with a growing season that is too short and too cold to grow crops. This had to be learned the hard way. When British fur traders established outposts in sub-Arctic Canada on behalf of the Hudson's Bay Company, they attempted to grow such crops as potatoes, but with little success. This lack of agriculture is in strong contrast to the more temperate parts of the world where 40% of all landscapes are farmland.

Human disruption of Arctic landscapes includes direct impacts, such as mining, nuclear testing, and petroleum extraction. All of these impacts are restricted to specific regions, but there are other indirect impacts that come from the south and affect the entire Arctic. These are air and water pollution, mostly from sources in the temperate zone. Toxic elements and compounds have drifted North and penetrated deep into the Arctic.

As described by Law and Stohl (2007), pilots flying over Arctic North America in the 1950s observed widespread haze every winter and early spring. By the 1970s, scientists realized that the haze was air pollution transported from the middle latitudes. Arctic haze continues to damage the air quality of Arctic villages and towns. It contains, among other things, acidic sulfate compounds. These can come to Earth with precipitation, or they can be directly deposited on the ground, leading to increased acidity of soil and water. Long-range transport of pollution to the Arctic also carries other toxic substances, such as mercury, heavy metals, and persistent organic pollutants (Fig. 12.1). Their effects on ecosystems and human health are discussed in the following.

Heavy Metals in Snow, Ice, Soils

As discussed by Barbante et al. (2017), studies of ice core samples from the Arctic that date from the past two centuries have demonstrated human impacts on the Arctic. In most of the world, levels of heavy metals and trace elements in the environment are controlled by local geochemistry. However, in the Arctic, natural levels are swamped by anthropogenic emissions

Threats to the Arctic. https://doi.org/10.1016/B978-0-12-821555-5.00012-7

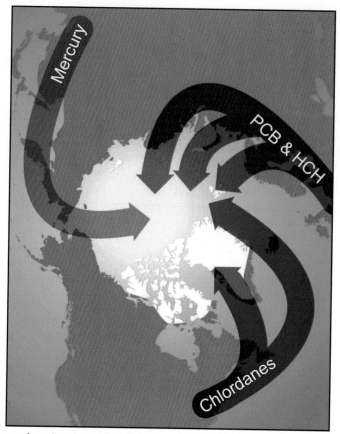

FIG. 12.1 Sources of persistent organic pollutants and mercury that make their way to the Arctic Ocean.

transported from lower latitudes. Some metallic elements, in small quantities, are essential for life. These include copper (Cu), iron (Fe), and zinc (Zn). However, these can accumulate in the environment at toxic levels. The heavy metals, including cadmium (Cd), mercury (Hg), and lead (Pb), are considered toxic at any level (Fig. 12.2). Some elements from the Platinum Group, including iridium (Ir), osmium (Os), palladium (Pd), platinum (Pt), and rhodium (Rh), while not necessarily toxic, have biological effects when they occur in concentrations much higher than natural (Barbante et al., 2001). Heavy metals and trace elements are dispersed globally and enter various biogeochemical cycles after deposition. Emissions of toxic elements from the midlatitudes are often more reactive with biological compounds because of their chemical form (speciation) or particle size. Speciation controls the solubility of the compounds, altering their absorption and facilitating their entry into the food web and other geochemical cycles. The fallout of these various elements from the

atmosphere is an important tracer of the transport processes that carry them to the Arctic. They are markers for human activity and are preserved in glacial ice and snow.

Lead contamination

The first traces of exotic metals recovered from Arctic ice samples came from the Greeks and Romans, who began mining and metal smelting in the Bronze Age. This early metallurgy left deposits of lead, copper, and other trace metals in Greenland ice. The level of atmospheric lead fallout from Roman times was not exceeded until lead additives were put in gasoline in the 1920s. The extensive use of fossil fuels during the 20th century also caused increases in the number of other elements emitted into the atmosphere, mainly because of the use of coal for power generation. Regulatory limitations on the use of lead additives in gasoline from the 1970s onward with the advent of the Clean Air Act and their prohibition in the 1980s in the United States and the

1990s in Europe caused atmospheric lead concentration to decline rapidly.

Perryman et al. (2020) examined soil samples from sites throughout Alaska, looking for traces of heavy metals, including lead. Arctic communities are particularly susceptible to threats to water security from climate change, and the potential liberation of metals through permafrost thaw may additionally threaten water quality in the Arctic. The study by Perryman et al. (2020) found high concentrations of heavy metals in soils across Alaska, suggesting that permafrost thaw and associated changes in the environment may present an unaccounted-for pathway of heavy metal exposure. Their study emphasizes the need for a better understanding of how permafrost thaw alters the mobility and cycling of heavy metals.

Mercury contamination

Mercury (Hg) is a naturally occurring element that bonds with organic matter and, when converted to methylmercury, is a potent neurotoxicant. Mercury has no local sources in the Arctic. Its presence there is mainly due to long-range air transport from Asia, Russia, North America, and Europe. It occurs in the atmosphere in the vapor phase (Hg^0) or as particulate or ionic mercury (Hg^{2+}). Mercury has well-known toxic effects and bioaccumulates through the food chain. The mercury biogeochemical cycle in the Arctic is associated with springtime atmospheric chemistry and is dominated by photochemical reactions (Barbante et al., 2017).

Schaefer et al. (2020) discussed that evidence from soil measurements indicates that permafrost regions store an estimated 1656 ± 962 gigagrams (Gg) of mercury in the top 3 meters of soil, of which 793 ± 461 Gg of mercury is frozen in permafrost (1 Gg equals one million kg). Observations indicate accelerated permafrost thaw over the past 30—40 years. Model projections estimate a 30%—99% reduction in the extent of permafrost in the Northern Hemisphere by 2100. When permafrost thaws, microbial decay of the stored organic matter will resume and release mercury, but how much, where, and when remain unclear.

St Pierre et al. (2018) documented high concentrations of methyl mercury downstream of retrogressive thaw slumps in the Western Canadian Arctic. Retrogressive thaw slumps are thermokarst features created by the rapid thaw of ice-rich permafrost and can mobilize vast quantities of sediments and solutes that are transported away from the slump site as they enter nearby streams (Fig. 12.3). They found that fluvial concentrations of

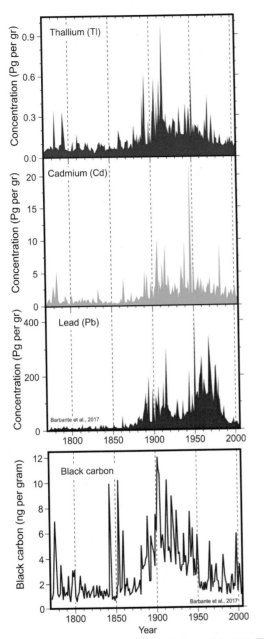

FIG. 12.2 Annually averaged concentrations of thallium (Tl), cadmium (Cd), lead (Pb), and black carbon (BC) in central Greenland precipitation from 1772 through to 2003 from ice core samples from the D4 and Summit sites in Greenland. (Modified from Barbante, C., Spolaor, A., Cairns, W., Boutron, C., 2017. Man's footprint on the Arctic environment as revealed by analysis of ice and snow. Earth-Science Reviews 168, 218—231. https://doi.org/10.1016/j.earscirev.2017.02.010.)

FIG. 12.3 Large retrogressive thaw slump, northwest Alaska, triggered by lateral erosion of the Selawik River. (Photo by Kenji Yoshikawa, University of Alaska Fairbanks, in the public domain.)

total mercury (THg) and methylmercury (MeHg) downstream of thaw slumps on the Peel Plateau (Northwest Territories, Canada) were up to two orders of magnitude higher than upstream, reaching concentrations of 1270 ng per liter and 7 ng per liter, respectively, the highest ever measured in uncontaminated sites in Canada.

St Pierre et al. (2018) also studied mean open water season yields of THg and MeHg. Downstream from the thaw slump, they tested stream water and found that the yield for THg was 610 mg/km^2 per day and the yield for MeHg was 2.61 mg/km^2 per day. These downstream mercury concentrations were up to an order of magnitude higher than those for the nearby large Yukon, Mackenzie, and Peel rivers. The retrogressive thaw slump released a tiny fraction of the mercury being stored in Arctic permafrost soils. The authors estimated that about 5% of the mercury stored for centuries or millennia in northern permafrost soils (88 Gg) is susceptible to release into modern-day mercury biogeochemical cycling from further climate changes and thermokarst formation.

Schaefer et al. (2020) used numerical models to forecast the potential release of mercury from permafrost soils during the next two centuries, based on low and high greenhouse gas emissions scenarios. By 2200, the RCP8.5 (worst-case) scenario shows annual permafrost mercury emissions to the atmosphere comparable with current global anthropogenic emissions. By 2100, mercury concentrations in the Yukon River are projected to increase by 14% for the RCP2.6 (best-case) scenario, but double for the RCP8.5 scenario. Fish mercury concentrations do not exceed the US Environmental Protection Agency guidelines for the RCP2.6 scenario by 2300, but for the RCVP8.5 scenario, fish in the Yukon River exceed EPA guidelines by 2050. Thus, the results of this study indicate minimal increases in mercury concentrations in water and fish for the low emissions scenario and high impacts for the high emissions scenario.

Arctic Natives at the apex of the food chain, including the Yúpik, Iñupiat, and Inuit peoples, are vulnerable to mercury poisoning, especially those who are heavily exposed to dietary mercury, because of bioaccumulation and biomagnification.

Volatile organic compounds

Volatile organic compounds (VOCs) are pervasive gaseous constituents in the atmosphere originating from anthropogenic and natural sources. As a key precursor of tropospheric ozone and aerosol formation, VOCs significantly affect air quality and climate change. Li et al. (2020) discussed how VOCs undergo a sequence of oxidative chemical processes and yield various oxidation products in the atmosphere. The impact of thawing permafrost on carbon dioxide (CO_2) and methane (CH_4) emissions has been extensively studied and evaluated. Li et al. (2020) examined VOC emissions from carbon-rich permafrost soils, because these emissions may cause important climate feedbacks as permafrost melts, especially in the Arctic, where anthropogenic sources are weak and aboveground vegetation is scarce. Upon permafrost thawing, organic vapors can be released directly from the trapped gases accumulated in the permafrost or via postthaw microbial breakdown of soil organic matter. Previous studies have shown that under aerobic conditions, the active layer microbiota would largely consume VOC emissions from thawing permafrost and prevent their entrance to the air. However, in water-saturated thawing soils, anaerobic conditions slow down the microbial VOC consumption and may thus enable the release of volatile compounds into the atmosphere. Under the scenario of RCP 8.5, 70.4 petagrams (Pg; 1×10^6 g) of permafrost soil organic carbon in water-saturated soils is projected to thaw by 2100 versus 172.4 Pg C in nonsaturated soils. The study by Li et al. (2020) showed that the emission of VOCs increases with temperature (Fig. 12.4). In total, 165 organic ions were identified as VOCs released from soil samples. Generally, the most abundant compounds showed both overlapping and differences between permafrost and the active layer. The cumulative fluxes of acetic acid ($C_2H_4O_2$) and acetone (C_3H_6O) were the highest for both permafrost and active layer. Other VOCs detected included acetaldehyde (C_2H_4O) and ethanol (C_2H_6O).

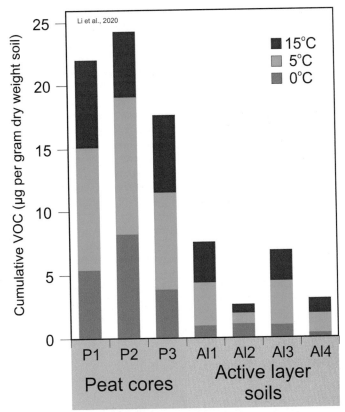

FIG. 12.4 Stacked cumulative volatile organic compound (VOC) fluxes at different temperatures for thawing permafrost and the active layer. P1, P2, and P3 are three peat core replicates from the same depth of frozen permafrost soils. AL1, AL2, AL3, and AL4 are four active layer replicates from the same active layer soil depth. (Modified from Li, H., Väliranta, M., Mäki, M., Kohl, L., Sannel, A., Pumpanen, J., Koskinen, M., Bäck, J., Bianchi, F., 2020. Overlooked organic vapor emissions from thawing Arctic permafrost. Environmental Research Letters 15 (10). https://doi.org/10.1088/1748-9326/abb62d.)

These results highlight the potential for substantial releases of VOC from thawing permafrost under climate warming, which introduces additional uncertainties to global climate modeling. Particularly, the first observation of sesquiterpene, diterpene, and other SVOC and IVOC emissions from thawing soils may play a more important role in atmospheric chemistry with the continued warming trend. In the Arctic, with sparse vegetation cover and much lower anthropogenic influences, such high volatile emissions from thawing soils might contribute significantly to regional and global climate feedbacks.

Radioactive Materials on Land and at Sea

Like heavy metals, almost all radioactive materials in the Arctic have been transported there from other regions. There are many sources of these radioactive nuclides that reach the Arctic through the air, the sea, and by direct human transportation. Radionuclides (spent nuclear fuel) have been released from radioactive waste storage facilities. One example is the radioactive element, technetium (^{99}Tc), which is present in Barents and North Atlantic seawater near the Norway coast, as a result of discharge of this radionuclide from the Sellafield nuclear power generation plant in northwest England, especially in the 1990s.

The Barents Sea is surrounded on three sides (west, south, and east) by industrial and military facilities, all of which are a potential source of radiation hazard. There are also sea and river ports at which ships transport nuclear materials and radioactive substances. As a result, the southern Barents Sea has been the most radioactive ocean region since the nuclear age began.

Yakovlev and Puchkov (2020) presented an assessment of current natural and anthropogenic radionuclide concentrations in sediment samples taken from the Barents Seafloor. Since the Cold War in the second half of the 20th century, the Arctic zone of Russia has been significantly affected by radioactive contamination. Between 1954 and 1990, the Soviet government exploded 87 nuclear weapons in the atmosphere above Novaya Zemlya, as well as 3 explosions carried out underwater and 42 explosions carried out underground. These tests left a legacy of long-lived radionuclides, including cesium-137 (^{137}Cs), plutonium-238 (^{238}Pu), and plutonium 239–240 ($^{239-240}$Pu) (Fig. 12.5). Several underground nuclear explosions were also carried out in Arctic Russia for geological purposes.

The naturally occurring radionuclides ^{226}Ra, ^{232}Th, and ^{40}K are important because of the proximity of Russian oil and gas fields; oil, gas, gas condensate, and associated formation water have been found to contain elevated levels of these radionuclides. Therefore, testing for background levels of these nuclides is important to establish a baseline for comparison with samples from the oil and gas fields.

Radioactive contamination in the Barents Sea

Yakovlev and Puchkov (2020) reported that the highest radionuclide activities in the Barents Sea region were found within the deep-water shelf of the sea. The current level of activity of the radionuclide ^{137}Cs is low and does not exceed 6.5 Becquerels (Bq) per kg. However, the authors expressed concern that Arctic warming may release a secondary source of radiation pollution to the Barents Sea. If the Novaya Zemlya ice sheet melts, it will release huge quantities of radioactive materials that were deposited there during Soviet nuclear weapons tests of the 1950s and 1960s.

Gwynn et al. (2012) reported on the results of Norwegian radiological monitoring of the Barents Sea in 2007, 2008, and 2009. Activity concentrations of the synthetic radionuclides ^{137}Cs, strontium-90 (^{90}Sr), ^{239}Pu, and ^{240}Pu, and Americium-241 (^{241}Am) in seawater were as much as an order of magnitude less than in those measured in previous decades. Unlike the study by Yakovlev and Puchkov (2020), the Norwegian study found that activity concentrations of ^{137}Cs in

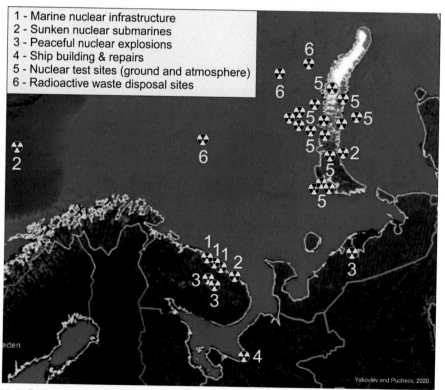

FIG. 12.5 Radioactive sites in the Barents Sea region, showing six different sources of contamination. (Modified from Yakovlev, E., Puchkov, A., 2020. Assessment of current natural and anthropogenic radionuclide activity concentrations in the bottom sediments from the Barents Sea. Marine Pollution Bulletin 160. https://doi.org/10.1016/j.marpolbul.2020.111571.)

surface sediments were low, with higher values in coastal sediments along the Norwegian mainland than from locations in the open ocean. Tests for radioactive materials in marine biota also showed levels of ^{137}Cs and ^{99}Tc that were as much as an order of magnitude less than in previous decades.

Nuclear power station accidents

Nuclear power station accidents have also released radioactive particles that have found their way to the Arctic. For example, the 2011 nuclear disaster at the Fukushima Daiichi Nuclear Power Plant resulted in the release of a radioactive cloud containing ^{129}I, ^{131}I, ^{134}Cs, and ^{137}Cs. The cloud released from Fukushima reached Svalbard within a few weeks, and the radionuclides were deposited with precipitation. SILAM, a global-to-mesoscale atmospheric dispersion model, showed that an even greater deposition may have occurred over Greenland.

The radioactive fall-out from the accident at the Chernobyl nuclear power plant in 1986 released $2-10$ Bq per m^2 in NW Norway, the Kola Peninsula, and northwestern Russia. In comparison with the Fukushima accident (Table 12.1), the Chernobyl accident released more than twice as much cesium-134 and almost six times as much cesium-137. It also released more than 70 times the amount of strontium-90 and almost 800 times the amount of plutonium-238.

TERRESTRIAL MAMMAL IMPACTS

Like other kinds of fieldwork in the Arctic, the study of terrestrial mammals has been quite limited both spatially and temporally. For one thing, almost all the fieldwork has been conducted during the summer, so our knowledge of their winter activities is exceptionally poor (Hutchison et al., 2020). However, except for a few hibernating species (e.g., the Arctic ground squirrel), most Arctic mammals remain active through the winter. As discussed by Hutchison et al. (2020), during winter, active generalist carnivores feed exclusively on resident herbivores. These include hares, voles, and lemmings, the latter two remaining active under snow or ice cover. Carnivores may also exploit marine resources (e.g., polar bears hunting seals from the pack ice), and above-snow carrion (e.g., wolves and foxes). The caching of food by such animals as grizzly bears, wolves, wolverines, and weasels is a strategy for conserving food resources, especially in landscapes of low productivity, such as the Arctic (Smith and Reichman, 1984). This activity is carried out by animals that bring food to their offspring.

Arctic warming is causing increased winter temperatures, ice loss on land and sea, and earlier spring snowmelt. Davidson et al. (2020) made use of the newly developed Arctic Animal Movement Archive (AAMA) to study various aspects of animal responses to these climate changes. The AAMA is a collection of more than 200 standardized terrestrial and marine animal

---

TABLE 12-1
Comparison of Amounts of Radionuclides Released by the Chernobyl and Fukushima Daiichi Accidents.

Radionuclide	Half-Life	PBQ RELEASED INTO THE ENVIRONMENT[A]	
		Chernobyl	Fukushima Daiichi
Xenon-133	5 days	6500	11,000
Iodine-131	8 days	~1760	160
Cesium-134	2 years	~47	18
Cesium-137	30 years	~85	15
Strontium-90	29 years	~10	0.14
Plutonium-238	88 years	1.5×10^{-2}	1.9×10^{-5}
Plutonium-239	24,100 years	1.3×10^{-2}	3.2×10^{-6}
Plutonium-240	6540 years	1.8×10^{-2}	3.2×10^{-6}

[a] 1 PBq = 1015 Bq.
Data from the Japanese Ministry of the Environment, 2019. Booklet to Provide Basic Information Regarding Health Effects of Radiation, vol. 254. Government of Japan.

tracking studies from 1991 to the present. These changes profoundly affect the environmental conditions that shape animal resources and behavior, including food availability, interspecific competition, predation, and increased human disturbances. The impacts of climate change on Arctic mammals include range shifts, phenological mismatches, changes in foraging behavior, and predator-prey dynamics.

Shifts in Migration Patterns

In Arctic Russia, the same species as caribou (*Rangifer tarandus*) is called reindeer. While caribou are wild animals, reindeer are domesticated, with herds managed by people who use reindeer for food and transportation. Cairns (2020) related the current situation with reindeer migration in the Taimyr Peninsula. In August 2020, the people of the village of Khatanga witnessed a herd of Taimyr reindeer migrating through their village weeks earlier than usual. Up to 200 young calves perished or were abandoned by the herd because they were not sufficiently strong to cross the broad Khatanga River. This passage was part of the reindeer's spring and fall migration. In autumn, Taimyr reindeer leave their calving and summer ranges in the Taimyr Peninsula to their winter ranges in the boreal forest. This migration involves trekking great distances and crossing many obstacles. However, in recent years, climate change has caused a shift in the timing of their migration to earlier in the year. Because of this, the migration is starting when young calves are too small to make the journey. The drowning of calves at the Khatanga River follows an earlier mass death from September 2019, when the Yenisei reindeer population of the Taimyr Peninsula completely disappeared. A Russian study also documented great population losses in the Tarey reindeer herd. A group of reindeer that had numbered 44,000 in 2017 was reduced to only a few thousand individuals by 2019. Cairns (2020) also reported that a combination of climate change and poaching caused a sudden decline in 2019, and the early migration in 2020 was also a result of climate change. During the past decade, the Taimyr reindeer population has been cut almost in half.

Changes in Seasonal Adaptations

Davidson et al. (2020) found that all terrestrial Arctic mammal species exhibited lower movement rates during winter relative to summer. As temperatures increased in summer, wolves and black bears slowed their movement rates, whereas moose increased their movement rates. In winter, only barren-ground caribou increased movement rates as temperature increased.

Snow impeded wolves, boreal caribou, and moose, whereas all species were generally insensitive to summer precipitation. These patterns may reflect asynchronous responses to climate change within and across trophic levels. Climate-driven variation in animal activity is likely to affect species interactions, altering energy expenditure, encounter rates, and foraging success with demographic implications for both predators and prey.

Gagnon et al. (2020) reported on changes in the demography of the Porcupine Caribou Herd (PCH) in the northwest Yukon and northeast Alaska. They assessed the influence of environmental change on the spring and fall body condition of caribou from the herd. Body condition is a key variable to assess animal fitness because it represents the energy reserves that an animal must daily and seasonal activities. Body condition is also tied to overwinter survival, age at first reproduction, and the likelihood of pregnancy. The Porcupine herd (named for the Porcupine River) is the only large North American herd that has not declined since the 2000s (Fig. 12.6). Using observations recorded between 2000 and 2010 by an indigenous community-based monitoring program, Gagnon et al. (2020) analyzed temporal trends in caribou condition and quantified the effects of weather and critical weather-dependent variables (insect harassment and vegetation growth), on caribou condition. Both spring and fall body condition in the PCH improved from 2000 to 2010, despite a continuous population increase of ca. 3.6% per year.

Effects of climate change on caribou/reindeer

Spring and fall caribou condition are influenced by weather on the winter and spring ranges, particularly snow conditions and spring temperatures. Both snow conditions and spring temperatures improved during this study period, contributing to the observed caribou population increase.

The results of the study by Gagnon et al. (2020) suggest that variations in snow depth, temperature, and melt date are essential drivers of spring and fall caribou condition in the PCH. Early calf survival and recruitment are closely tied to female caribou condition in late winter. Recruitment is the process by which new individuals are added to a population, whether by birth and maturation or by immigration. Unusually deep winter snows increase the probability of caribou being in poor condition in spring by as much as a factor of 10. This adverse winter weather likely exerts large impacts on the PCH, as has been shown for other northern ungulates.

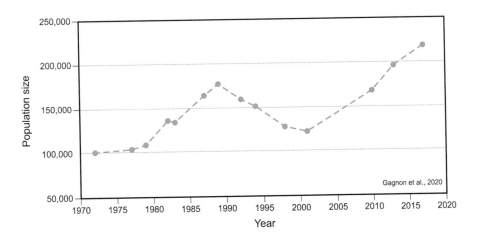

FIG. 12.6 (Above) Graph showing population trends in the Porcupine caribou herd since 1970. (Below) Photograph of the Western Arctic Caribou Herd in 2017. Photo by Mario Davalos, used with permission. (Above: From Gagnon, C., Hamel, S., Russell, D., Powell, T., Andre, J., Svoboda, M., Berteaux, D., 2020. Merging indigenous and scientific knowledge links climate with the growth of a large migratory caribou population. Journal of Applied Ecology 57 (9), 1644–55. https://doi.org/10.1111/1365-2664.13558.)

Gagnon et al. (2020) also observed that the negative effects of a long snow season with deep spring snow have carryover effects on the caribou condition in the subsequent fall. This supports previous studies indicating that PCH females are not able to compensate during the summer for poor spring condition. This, in turn, depresses fecundity rates in the fall. One climate variable that seems to be a good indicator of fall condition is the number of growing degree days (GDD) in May. When the number of May GDDs is large, it increases the probability that caribou will enjoy excellent fall condition.

This study indicates that a large migratory caribou population (Fig. 12.7, top) can grow and improve condition even though global caribou populations are in decline, and despite environmental perturbations brought on by climate warming. The study also demonstrates the folly of generalizations about the influence of climate on caribou populations. Finally, it testifies how data from indigenous community-based monitoring can remarkably improve our understanding of wildlife ecology.

Barten et al. (2001) compared habitat use, forage characteristics, and group size among female caribou

FIG. 12.7 Top: Arctic caribou herd crossing a river. Middle: Polar bear swimming in the Barents Sea. Bottom: Polar bear mother and cub near Hudson Bay, Canada. (Top: Courtesy of Corel Corp. Middle: Photo courtesy of Magnus Andersen, Norwegian Polar Institute. Bottom: Courtesy of Corel Corp.)

from the Mentasta caribou herd, Alaska over 2 years before, during, and after the birthing season to test hypotheses involving the acquisition of forage and risk of predation. This herd lives in the Wrangell-St Elias mountains of southeast Alaska. During peak

parturition, females with young sought out sites with fewer predators. These sites had less forage, and the forage was more variable in quality compared with sites used by females without young. However, the authors could not demonstrate a nutritional cost to maternal females from their analyses. Barten et al. (2001) suggested that increasing population density might intensify intraspecific competition among females for birth sites and thereby increase nutritional costs of using high-elevation areas with less forage but fewer predators.

The relative contribution of weather fluctuations to animal population dynamics seems to be particularly large in marginal habitats, such as the High Arctic, so Arctic tundra food webs tend to display relatively strong population synchrony both within and across species. Hansen et al. (2019) studied the relationship between reindeer population size and climate parameters on Svalbard. The Svalbard reindeer (*Rangifer tarandus platyrhynchus*) are wild, not domesticated. On the whole, these High Arctic islands have experienced warmer summers in recent years that have improved both reindeer carrying capacity and population size. Hansen et al. (2019) found that the main driver of reindeer population size on Svalbard is a different aspect of climate warming: the negative effects of winter rain-on-snow (ROS). These events melt the surface snow, and the rainwater and meltwater percolate to the permafrost soil surface where they refreeze, encasing the vegetation in ice. When these rainstorms occur in regions where reindeer populations are dense, large-scale die-offs can occur. Such ice-locked tundra has potentially ecosystem-wide consequences. For instance, icing events have been shown to cause over-winter body mass loss, reduced skeletal growth, and reduced survival and fecundity in both muskox and reindeer. These large herbivores starve because they cannot browse the ice-encased dried tundra vegetation beneath the snow.

By comparing one coastal and one "continental" reindeer population over four decades, the authors showed that locally contrasting reindeer abundance trends can be explained by spatial differences in climate change and responses to weather. The coastal population experienced a larger increase in ROS and a stronger density-dependent ROS effect on population growth rates than the continental population. In contrast, the continental population benefitted from stronger summer warming and showed the strongest positive response to summer temperatures. Accordingly, contrasting net effects of a recent climate regime shift—with increased ROS and harsher winters, yet higher summer temperatures and improved carrying capacity—led to negative and positive abundance trends in the coastal and continental population, respectively. The longer and warmer plant growth season of the interior region boosts ecological carrying capacity due to increased primary production and "Arctic greening."

Changes in caribou parturition timing

As discussed by Davidson et al. (2020), the timing of parturition (giving birth to young) is key to the demography of wildlife populations and can show an adaptive response to climate shifts. Demography is the science of populations; demographers study population dynamics through three main processes: birth, migration, and aging (including death). For many mammals, the period from late pregnancy through the weaning of offspring has the highest energetic demands and thus is timed to occur when vegetation productivity is highest. Caribou populations live in five different ecotypes across boreal and Arctic North America (Fig. 12.8). Davidson et al. (2020) found differences in parturition

timing and trends among the five populations. The boreal populations calved earliest, followed by the mountain populations. Barren-ground caribou calved later than the other populations, despite occupying a similar latitudinal range as the northern boreal caribou populations. Most importantly, barren-ground and northern woodland caribou (but not southern woodland caribou) have been trending toward earlier parturition [0.4—1.1 days/year].

Changes in Polar Bear Ecology

Kazlowsky (2020) reviewed ecological shifts in polar bears in response to climate change for the World Wildlife Fund, focusing on Svalbard bear populations. Here on the Barents Sea coast, polar bears are experiencing the fastest loss of sea ice recorded throughout the Arctic.

During the winter of 2019, temperatures in Svalbard climbed above freezing several times. The north coast, usually bound in pack-ice through the winter, was entirely ice-free for much of the season. In the 21st century, the "new normal" has been sea-ice extent declining year by year. Polar bears are adapted for life on the ice,

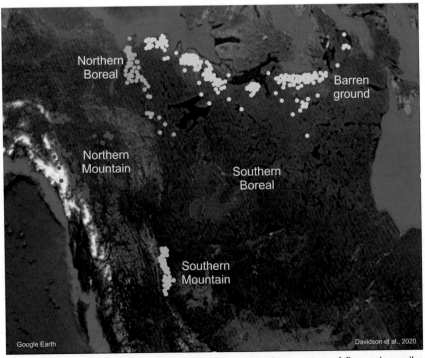

FIG. 12.8 Map of northern North America showing the approximate range of five major caribou herds. (Modified from Davidson, S., Bohrer, G., Gurarie, E., LaPoint, S., Mahoney, P., Boelman, N., Eitel, J., Prugh, L., Vierling, L., Jennewein, J., Grier, E., Couriot, O., Kelly, A., Meddens, A., Oliver, R., Kays, R., Wikelski, M., Aarvak, T., Ackerman, J., et al., 2020. Ecological insights from three decades of animal movement tracking across a changing Arctic. Science 370 (6517), 712–715. https://doi.org/10.1126/science.abb7080.)

hunting seals. Wildlife conservation officers on Svalbard have taken note of the bears' attempts to adapt to this rapidly deteriorating situation by finding new food sources on land and adjusting their winter activities, but it remains unclear whether they can adapt quickly enough to survive here. Wildlife biologists have been tracking Svalbard polar bears using satellite tags that reveal the bears' movements in real time. Their research provides some fascinating insights into this rapidly changing situation.

Polar bear survival strategies

Svalbard polar bears have developed two strategies to find food. One group, the pelagic bears, spend much of their lives on the sea ice. Local bears, however, spend life on the coast. Since sea ice around Svalbard began its decline 10—15 years ago, wildlife biologists have witnessed the bears quickly adopt new survival strategies. Pelagic bears wander the sea ice and mainly come to shore only to hibernate in their dens. These bears follow in the footsteps of thousands of previous generations that have hunted pinnipeds on the Arctic sea ice since time immemorial. They prey mainly on ringed or bearded seals that they stalk on the ice or catch by waiting beside breathing holes. This hunting technique remains successful, as long as there is sea ice to work from. The ever-growing problem for these bears is finding a way back to land.

Sea ice is systematically retreating from the Svalbard coast, or it fails to form at all. Polar bears used to be able to walk ashore from the sea ice to the land, but now they are becoming increasingly separated from their coastal denning grounds. In some years, some bears have failed to reach land on Svalbard. This has caused them either not to den or to seek new dens in the western Russian Arctic. Those that do reach land in time to build dens may emerge to an ice-free coast and limited hunting opportunities.

For coastal bears, those that prefer to spend their lives on Svalbard's coast, hibernating and finding a place to raise cubs is easy. Finding food on land, however, is often difficult. They attempt to hunt seals that they find resting on floating glacial ice in the many Svalbard fjords. The bears swim from the shore (Fig. 12.7, middle), get close to the seals, and then attempt to leap out of the water to catch the seals on floating ice. When floating ice is sparse, the coastal bears search for food at bird colonies or scavenge on the beach. While these food sources keep the bears alive day by day, they are not as rich in calories as seal meat and blubber. Also, the bears must walk substantial distances to find them, burning more calories on the way. Thus,

this strategy may not be sustainable for bears in the long term. It is also taking a toll on local bird populations. On some coasts in Svalbard, 90% of bird nests on the ground have been raided by bears. Some bears adopt a far more dangerous tactic by searching for birds' nests on high cliffs (Fig. 12.9). Polar bears are built for swimming, not rock climbing. Polar bears are intelligent and adaptable, but the pace of climate change on Svalbard may be too much for them.

Polar bear bird nest predation in Arctic Canada

As discussed earlier, polar bears have difficulty capturing prey in open water and come ashore during ice-free periods in regions with seasonal sea ice. They rely primarily on stored fat throughout the ice-free period. In Arctic Canada, they have been observed feeding on terrestrial foods including grasses, marine algae, berries, carrion, and occasionally caribou, fish, rodents, and birds. Despite numerous accounts of bears taking terrestrial foods, these behaviors have traditionally been considered opportunistic, and the food considered to be of limited energetic value. However, a report published by Smith et al. (2010) documented intense, systematic exploitation of bird's nests on islands in northern Hudson Bay. The authors presented observations of at least six bears consuming large numbers of snow goose (*Chen caerulescens*) eggs at two locations in the Low Arctic of eastern Canada in 2004 and 2006 (Coats and Southampton Islands, Nunavut). They also provided two records of polar bear consumption of the eggs and chicks of cliff-nesting thick-billed murres (*Uria lomvia*) in 2000 and 2003. In both localities, the bears came ashore before the hatch of nests, and the subsequent predation was catastrophic for the birds at a local scale. Bears have also been documented preying on eggs and chicks in a snow goose colony at Churchill, Manitoba. This has happened six times in the past 40 years, with four attacks occurring since 2000.

Here, as elsewhere, polar bears are being forced ashore by the progressively earlier breakup of sea ice. In 1985—91, polar bears came ashore on Coats Island from July 1 to July 10. In 2000—07, they came ashore as early as June 21. This 3-week differential represents a substantial loss of quality seal hunting time on the pack ice.

The observations made by Smith et al. (2010) suggest that polar bears are consuming substantial quantities of eggs in the Canadian Arctic and that bear predation on nests is severely affecting the reproductive success of snow geese. In fact, the evidence points to a complete breeding failure at the study site on

FIG. 12.9 Above: A colony of about 60,000 breeding pairs of Brünnich's guillemot (Uria lomvia) at Alkefjellet, Svalbard. (Photo by Andreas Weith, Creative Commons Attribution-Share Alike 4.0 Generic license.) Below: Polar bear mother and two cubs climbing up Guillemot Island (Ukkusiksalik National Park, Nunavut, Canada). (Photo by Ansgar Walk, Creative Commons Attribution-Share Alike 2.5 Generic license.)

Southampton Island and a nearly complete failure at the Coats Island site. These nest attacks represent more than random events. The extent of the nutritional benefits to individual bears remains unknown but is worthy of investigation.

Modeling polar bear reproductive fitness

Reimer et al. (2019) used numerical modeling to develop a better understanding of the optimal foraging responses of polar bears to changing Arctic environments and the fitness consequences of those responses. In this context, fitness here is defined as an individual's expected reproductive success, resulting from life history strategies that successfully balance competing factors. Environmental shifts caused by climate change may alter the relative success of various strategies, due to changes in their costs and benefits. Using data gathered from Beaufort Sea populations of polar bears, Reimer et al. (2019) studied these shifting optimal responses based on the analysis of multiple interacting factors, accounting for an individual's need to balance survival with reproduction, often over multiple years and reproductive attempts. Beaufort Sea polar bear females give birth to a litter of one to three cubs which remain with their mother until they are weaned (Fig. 12.7, bottom). Weaning typically occurs in the spring of their second year, so a female may successfully wean a litter every 3 years at most.

Female polar bears in different reproductive states vary in their choice of foraging habitat during the spring feeding period. Sea-ice habitat used by polar bears in the southern Beaufort Sea can be broadly grouped into two types: pack ice and land-fast ice. Pack ice and the floe edge are high-quality habitats for foraging polar bears. The floe edge is where the ice that is still attached to the land, having frozen over the winter months, meets the sea. The sea has abundant prey, including ringed seals (*Pusa hispida*) and bearded seals (*Erignathus barbatus*). Nearshore, fast ice provides lower quality foraging habitat. The main available prey here is ringed seal pups and, to a lesser extent, their mothers. Pack ice is the preferred foraging habitat for male bears of all ages and females without offspring. Female polar bears accompanied by dependent offspring (especially females with cubs of the year, COYs), however, are found more often in the fast ice. This foraging strategy, though offering lower quality food resources, may be an adaptation by mother bears to avoid the risks of (1) losing cubs to predation by adult males and (2) losing cubs to hypothermia from prolonged exposure to polar waters. Pack-ice hunting involves more swimming in the open ocean than fast-ice hunting. Polar bear cubs are not as well insulated against cold as adult bears.

Reimer et al. (2019) used their model to predict the optimal foraging decisions throughout a female polar bear's lifetime. They also calculated the energetic thresholds below which it is better for females to abandon reproductive attempts. In recent decades, the ice-free period in the southern Beaufort Sea has increased by approximately 10−20 days per decade. For polar bears, this results in a shorter feeding period during which they must attempt to acquire the necessary reserves to survive the longer summer fasting period. These changing ice conditions have already been linked with smaller bear body size, reduced recruitment, and bear population declines in the Beaufort Sea.

In light of these current trends of progressively earlier sea ice melting in spring, the authors adjusted their model by shortening the spring feeding period by up to 3 weeks, which led to predictions of riskier foraging behavior and higher reproductive thresholds. The model predicted that if the spring feeding period was reduced by 1 week, the expected fitness of a female polar bear declined by 15%. If the spring feeding period was reduced by 3 weeks, then the expected fitness declined by 68%. Furthermore, the modeling results suggest that a female with cubs may spend more time hunting on the pack ice as breakup occurs earlier, despite the increased difficulty of getting to shore for the summer denning period. There is some empirical evidence suggesting that such a shift in female hunting habitat choice may already be happening as the date of ice breakup falls ever earlier in the year.

PLANT COMMUNITY IMPACTS

As we have seen in the discussion about the impacts of global change on animal life in the Arctic, biological responses to environmental impacts are complex and often unexpected, such as winter rain showers in Svalbard causing starvation of reindeer. This section describes global change impacts on plant communities, highlighting once again that there is no uniform biological response to these changes, but rather a series of recently discovered complex patterns that the scientific community never guessed beforehand. As with animal life in the Arctic, it is foolish to make broad sweeping generalizations about plant responses to environmental change.

Is the Arctic Greening?

One of the generalizations that has dominated Arctic vegetation literature in recent years has been the concept of Arctic Greening—the idea that the Arctic tundra is becoming more lush and more dominated by shrubs than by herbaceous plants, in response to climatic warming. Like most generalizations in biology, there is a strong element of truth in the "Greening of the Arctic" concept. Across the Arctic, the rapid warming of surface air temperatures appears to be driving broadscale greening trends linked with increases in biological productivity. These changes have been registered by satellite remote sensing through the analysis of the Normalized Difference Vegetation Index (NDVI). An understanding of this index is quite important to the remainder of this section of the chapter.

As explained by the NASA Earth Observatory (2000), to determine the density of green on a patch of land, researchers who analyze satellite imagery must observe the distinct wavelengths of visible and near-infrared sunlight reflected by the plants. The pigment in plant leaves (chlorophyll) strongly absorbs visible light (wavelengths from 0.4 to 0.7 μm) for use in photosynthesis. On the other hand, leaf cell structure strongly reflects light in the near-infrared range (wavelengths from 0.7 to 1.1 μm). In any given pixel of satellite imagery data, the greater the ratio between the near-infrared and visible wavelengths, the denser the vegetation. Nearly all satellite vegetation indices employ this difference formula to quantify the density of plant growth on the Earth. The result of this formula is called the NDVI.

For the most part, NDVI interpretations of a "Greening Arctic" have been corroborated by independent analyses of repeat photography and vegetation surveys, both of which show increases in the cover and density of upright shrubs and graminoids at the expense of moss and lichen cover.

Potter and Alexander (2020) discussed changes in vegetation phenology and productivity in Alaska over the past two decades, based on information from the moderate-resolution imaging spectroradiometer (MODIS) satellite data sets. They examined Alaskan data from 2000 to 2018 and used phenology metrics derived from NDVI time series at 250 m resolution. This allowed them to track changes in the total integrated greenness cover, maximum annual NDVI, and the start of the season timing date over the past two decades. They found that trends in the timing of the start of the growing season showed significantly earlier spring vegetation greening (at more than 1 day per year) across the northeastern Brooks Range Mountains, as well as some regions in the boreal zone. Total integrated greenness cover and maximum annual NDVI increased significantly across the western Arctic Coastal Plain. Interestingly, the start of the growing season (SOST) was not uniformly earlier across Arctic Alaska and adjacent parts of Canada. The study indicated that many of the tundra regions of Alaska and northwest Canada experienced SOSTs 2–3 days later than the long-term average from 2000 to 2018. This includes parts of the Brooks Range, the northern Yukon, and the northern parts of the Northwest Territories. Potter and Alexander (2020) concluded that their analyses have identified a new database of localized study locations across Alaska where vegetation phenology has recently shifted notably and where land cover types and ecosystem processes could be changing rapidly.

A more extensive study of tundra vegetation response to climate that included 18 years of fieldwork on Qikiqtaruk Island (formerly known as Herschel Island) was reported by Myers-Smith et al. in 2019. The project represented a unique collaboration between government scientists, local inhabitants, park rangers, and academic researchers. This long-term vegetation monitoring program showed an increase in total plant cover at the study sites and a decrease in bare ground tundra. Their study revealed changes in tundra plant community composition that indicate a shift toward increased deciduous shrub cover, shrub expansion, and graminoid cover, while mosses and lichen species tend to be decreasing (Fig. 12.10). The authors noted that the same changes have been observed in warming experiments, suggesting that these vegetation changes are at least partly driven by rising temperatures. In addition to increased land coverage, shrub and graminoid plants were also found to be growing taller and reproducing more frequently. However, rates of shrub expansion varied greatly across sites, and some localities showed no evidence of change or growth stimulated by climate change. This is not surprising, given the heterogeneity of the study plots. These varied in a variety of site-level factors including soil moisture, topography, disturbance, herbivory, and interactions between plants.

The timing of leaf emergence (SOST) and flowering has advanced at some, but not all tundra ecosystem regions. Satellite-observed NDVI estimates of green-up in the tundra indicate advancement by 1.6–4.7 days per decade, corresponding with earlier snowmelt. However, some tundra sites have experienced decreased growing season length and delayed phenology as a result of higher snowfall, so summer temperatures are far from the only driver of Arctic vegetation change. Myers-Smith et al. (2019) pointed out that studies indicate a decrease in the satellite greening trend and reveal heterogeneity in satellite observations across different sensor platforms. Nonecological factors, including atmospheric change, drift in satellite sensors, or earlier snowmelt, may produce false greening signals. For example, the NDVI, the most common index of remotely sensed tundra greening, is also sensitive to other landscape-level parameters such as snow duration or standing water. Thus, there is current uncertainty whether the greening patterns observed by satellites actually indicate changes in vegetation patterns across the Low- and High-Arctic tundra biomes.

Lack of greening in interior regions

As reviewed by Andruko et al. (2020), the recent widespread expansion of deciduous shrubs across much of the Arctic has been largely attributed to climate warming. "Greening" has been observed in many Low-Arctic tundra regions of North America in the past 30 years, but especially in coastal and near-coastal regions such as northern Alaska, the northwest coast of Canada, and northern Quebec. These observations come from satellite-based studies. However, these studies show only patchy, small-scale vegetation greening in more inland tundra regions of Canada, such as the central and eastern Canadian Arctic, ranging from the central Northwest Territories to Nunavut. Two possible explanations for this disparity between coastal and interior "greening" trends were discussed by Andruko et al. (2020). Either vegetation change within the interior continental region has not occurred over the past 30

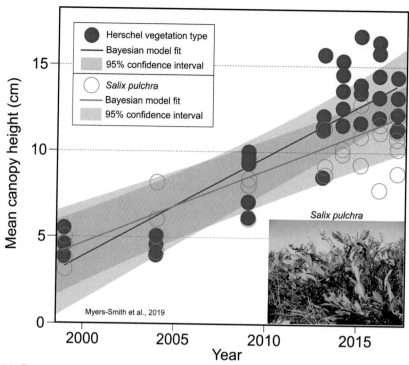

FIG. 12.10 Changes in plot canopy height from 1999 to 2017. Purple: Herschel vegetation type; green: tealeaf willow (*Salix pulchra*). (Modified from Myers-Smith, I., Grabowski, M., Thomas, H., Angers-Blondin, S., Daskalova, G., Bjorkman, A., Cunliffe, A., Assmann, J., Boyle, J., McLeod, E., McLeod, S., Joe, R., Lennie, P., Arey, D., Gordon, R., Eckert, C., 2019. Eighteen years of ecological monitoring reveals multiple lines of evidence for tundra vegetation change. Ecological Monographs 89 (2). https://doi.org/10.1002/ecm.1351.)

years, or it has not been adequately detected from satellite imagery because of the low spatial resolution (30 m pixels) of the satellite data available for that period. Furthermore, although remote sensing observations have been largely supported by various ground-truthing studies in near-coastal Arctic regions (e.g., Myers-Smith et al., 2019), there has been very little ground-truthing in the continental interior region.

Andruko et al. (2020) investigated growth rates of dwarf birch (*Betula glandulosa*) in Low-Arctic tundra in the continental interior of Canada. Percent birch cover was determined from 100 m² replicate plots. Individual shrub stature measurement data sets for five representative habitat types were compared between 2006 and 2016 and evaluated in relation to environmental characteristics. The study found that birch height, lateral dimensions, and patch groundcover all increased by 20%–25% between 2006 and 2016. The extent of these increases was similar among the habitat types.

However, unlike some other Arctic regions, there is very limited evidence of recent warming at this site.

Based on the absence of significant habitat-type growth rate differences, the lack of significant correlation between annual climate and stem secondary growth, and little evidence of climatic amelioration from regional meteorological datasets, the authors contended that climate change has not been the principal cause of the increases in birch growth and areal extent. Instead, they proposed that release from caribou impacts following the recent severe herd decline may explain the net shrub growth. These results suggest that the recent severe caribou herd declines may be at least as significant as climate warming in driving birch shrub expansion in the Canadian central Low Arctic.

Further east on the Ungava Peninsula, Morrissette-Boileau et al. (2018) assessed the dynamics and potential impacts of caribou grazing of shrub species in the Low-Arctic tundra area near Deception Bay, a region in the summer range of the Rivière-aux-Feuilles migratory caribou herd. They surveyed the abundance and stem mortality of all shrub species along systematically located transects in a 54 km² area. Their results showed that shrubs covered only 11.5% of the study area and most of these individuals were the dwarf shrub species, *B. glandulosa*. Stem mortality of established individuals was greater for diamondleaf willow (*Salix planifolia*) (45.6%) and grayleaf willow (*Salix glauca*) (46.5%) than for shrub birch (9.3%). Using tree-ring analysis, the authors established the age structure of the two dominant erect shrub species. They found that a high number of shrub birch and, to a lesser extent, of diamondleaf willow recruits (71.0 and 4.6 per 100 m^2, respectively) became established after 1999 when the Rivière-aux-Feuilles herd was abundant. Thus, their results do not support the hypothesis that large herbivores counteracted the positive effects of climate change on shrub recruitment. However, since the structure of the shrub recruits is prostrate rather than erect, as is usually observed with shrub expansion in the warming Arctic, it is likely that herbivory, and/or, the harsh climate, is constraining change in plant community structure. The development of erect shrub cover in the study region depends on the capacity of small recruits to overcome both abiotic factors, such as erosion of tall branches by snow crystals, and herbivory.

Northern Treeline Response to Warming

Across the Arctic, the rapid warming of surface air temperatures and trends in NDVI indicate widespread increases in productivity. These broad trends have been corroborated by analysis of repeat photography and vegetation surveys, many of which show increases in the cover and density of upright shrubs and graminoids, accompanied by decreases in moss and lichen cover. The individual responses of species to changing climate, taken together, result in reorganization of biological communities and shifts in biome boundaries. Therefore, the current global warming should result in shifts of major biomes across the globe. Such shifts, however, may be more varied than simple uniform poleward displacement.

As discussed by Travers-Smith and Lantz (2020), tree density in the sub-Arctic forest has increased in some regions, but the northern treeline in many regions has remained stable, exhibiting a slower climate response than other northern vegetation types. The forest-tundra ecotone (FTE) is an important bioclimatic zone with feedbacks from forest advances and corresponding tundra disappearance driving widespread ecological and climatic changes. Global vegetation models often assume equilibrium between climatic conditions and vegetation distribution. These models have generally forecasted northward shifts in vegetation in association with warming. However, several lines of evidence suggest that the range of some sub-Arctic plant species has not shifted in response to recent climate changes and are thus lagging behind the climate signal.

Treeline stability in northwest Canada

Travers-Smith and Lantz (2020) studied northern treeline dynamics of green alder and white spruce in the Tuktoyaktuk Coastal Plain of the Canadian Northwest Territories. They employed repeat photography of the region, based on high-resolution air photographs taken in the 1970s and in the 2000s to quantify changes in the distribution and abundance of alder and spruce near their northern limits. While they did find increases in alder and spruce stem density over time, they detected no northward advance of either species onto the tundra. In fact, they found no change in their range limits. The authors suggested that this treeline stagnation is caused by leading-edge disequilibrium (LED) in both tree species.

LED occurs when changes in the composition and distribution of plant communities lag behind changes in climate. This process results in relatively species-poor communities, given that they contain only a fraction of the potential species diversity that could exist in the new climate conditions. This lag in vegetation response may be due to slow vegetation growth, low seedling survival, limited and short-range seed dispersal, herbivory, and competition with established species. Variations in the timing of LED also differ according to plant functional type and terrain conditions. Understanding the duration and timing of lagged responses is therefore critical to accurately predict the timing and magnitude of near-term vegetation change.

In the case of the Tuktoyaktuk Plain region, LED was likely caused by low stand density and temperature limitation of reproduction at the northern treeline margin. Significantly greater change in species occupancy was found in a burned area, suggesting that the increased frequency of fire plays a significant role in the timing and magnitude of short-term vegetation change.

The authors concluded that (1) stem density has increased for both spruce and alder, but the location of the tree and shrub ecotones has not changed significantly since the 1970s; (2) vegetation change in the study area exhibits aspects of LED, which is not adequately represented in climate-based projections of near-term vegetation change; (3) disequilibrium in spruce and alder is likely driven by low stand density at the range margin as well as temperature limitations of vegetation growth and reproduction; (4) alder is likely less affected by LED because it exhibits a gradual decline in occupancy across the study area, is generally more abundant than spruce, and is less limited by seed dispersal; (5) burned areas exhibited the greatest change in spruce and alder occupancy, suggesting that fire and other disturbances will strongly influence the timing and magnitude of future vegetation shifts.

Assessment of global northern treeline

Rees et al. (2020) delved into the northern treeline problem, focusing on the circumarctic FTE. They assessed changes in FTE position in relation to climate history during the 20th century, using data from 151 sites across the circumarctic area and site-specific climate data. All sites were characterized by similar qualitative patterns of behavior, and about half of the sites showed northern advancement of treeline. However, rising temperatures did not appear to be the main driver of vegetation change. Rather, the main climate variable linked with changes in treeline communities was precipitation rather than temperature, both qualitatively and quantitatively. Precipitation regime is apparently also important during the nongrowing season as well as in the growing season.

During the 20th century, northern treeline advances occurred at different rates in the various regions. The smallest shifts were in Eastern Canada, at about 10 m per year. The largest shifts were in Western Eurasia, at about 100 m per year. Rees et al. (2020) noted that these rates of treeline movement were one to two orders of magnitude smaller than expected if vegetation distribution remained in equilibrium with climate. Given this resistance to climate change, it would seem that northern advances in treeline will not occur at the rates predicted for the 21st century by computer models, which are about 103−104 m per year. There is a lack of empirical evidence (i.e., ground-truthing) for swift forest movements, and the discrepancy between computer models and actual FTE response invalidates equilibrium model-based assumptions about vegetation response to climate change.

Evidence for plant migration lag has also been interpreted from vegetation changes in the boreal forest—not just at the northern treeline. For instance, Buchwal et al. (2020) noted that the divergence between tree growth and mean summer temperatures in the boreal zone generally began around 1960 and that the shrub growth divergence began later, near the end of the 20th century. They plotted a map of regions where shrub growth response has been positive (increasers in Fig. 12.11), where it has been negative (decreasers in Fig. 12.11), and where it has remained stable (neutral in Fig. 12.11). There are few obvious regional trends

in these changes in shrub growth across the Arctic. These authors hypothesized that the pattern of delayed response to climatic amelioration in the growth of woody vegetation has been spreading north as the magnitude of climate warming increases and the geography of growth-limiting factors, favoring precipitation over temperature, shifts northward. A thorough assessment of these phenomena at the tundra biome level will require a broader data set of shrub-ring chronologies covering both longer time spans and the most recent decade.

How fast can the northern treeline move? An estimate published in the IPCC AR5 suggests that trees can migrate up to about 1500 m/year, with a median of around 100 m/year. However, this is a global estimate that fails to consider differences between species or biomes. One way of estimating future potential rates of tree movement in the North is to examine rates of tree expansion at the end of the Last Ice Age, about 11,000 years ago. Initially, palynologists reconstructed tree migration rates at around 1 km/year, but such estimates failed to take into account the presence of forest tree refugia in regions such as northwest Russia and the southern Yukon territory of Canada (Edwards et al., 2014) and have been revised downward to around 250 m/year or as low as 100 m/year. Consequently, these results for latitudinal FTE movement would not contradict predictions of models that imposed maximum advance rates of the order of 100 m/year. If viable seed dispersal is the rate-limiting process, it is reasonable to suppose that while climate-driven altitudinal FTEs may be able to keep pace with climate change at a centennial scale and at optimal sites, latitudinal FTEs are unable to do so.

Rees et al. (2020) advise caution when assessing the implications of global change-related biotic and abiotic dynamics, including land-atmosphere feedbacks and carbon sequestration.

INSECT IMPACTS

The Arctic regions are generally inhospitable to insects and other invertebrates. The winters are very long and incredibly cold. Air temperatures that are lethal to insects persist for most months of the year. Summers are short, and the length of the growing season decreases with latitude. Insect and other arthropod diversity are extremely low here. As discussed by Høye (2020), Arctic summer air temperatures fall close to the lower thermal limits of arthropod species; thus, warming could remove temperature constraints on their movement and survival. It would seem, then, that

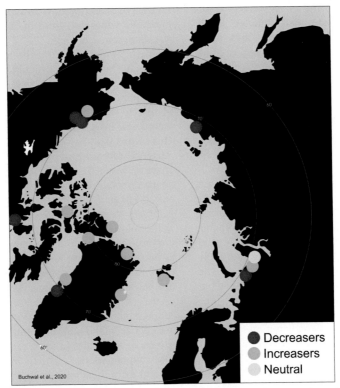

FIG. 12.11 Map of the Arctic, showing the locations of 23 shrub-ring chronology sites with an indication of sea-ice extent-shrub growth response type. (Modified from Buchwal, A., Sullivan, P., Macias-Fauria, M., Post, E., Myers-Smith, I., 2020. Divergence of Arctic shrub growth associated with sea ice decline. Proceedings of the National Academy of Sciences of the United States of America, In press. https://doi.org/10.1073/pnas. 2013311117.)

global warming should facilitate Arctic arthropod viability, reproduction, and range expansion and diversity. However, arthropod survival in cold regions has developed through a very long history of adaptations, and substantial climatic warming would not benefit most of these cold-adapted organisms. As Everatt et al. (2015) so succinctly put it, "invertebrates, more so than any other animal group, are at the whim of their environment."

Insect Diversity

Insect diversity decreases with latitude. This is because biological productivity likewise decreases toward the poles. One useful measure of the biological productivity of an ecosystem is its net primary productivity or NPP. NPP is the rate at which all the plants in an ecosystem produce useful chemical energy; it is equal to the difference between the rate at which the plants in an ecosystem produce useful chemical energy (GPP) and the rate at which they use some of that

energy during respiration. In Arctic tundra ecosystems, 87% of the estimated NPP is produced in the Low Arctic. Low-Arctic NPP averages about 75–125 gr/m^2 per year. High-Arctic NPP ranges from <1 to 25 g/m^2 per year (Gould et al., 2003). A survey of the insect life of all the High-Arctic regions of the world by Hodkinson (2013) found only 296 species. By comparison, the boreal forest regions of Canada have more than 32,000 insect species.

A basic concept in ecology is that low diversity renders an ecosystem much more vulnerable to disturbance. A paucity of species in a biological community decreases the system's resilience in the face of disruption, leading to population declines and local extinctions, and opening the door to the invasion of new species. Food chains may collapse, mutualistic species interactions are more likely to fall apart, and keystone species may be depleted by disease or parasitism. In other words, the system itself becomes more vulnerable to disturbances such as climate change. This lack of diversity in Arctic insects is one of its greatest weaknesses. As long as the environment remained stable, as it was for much of the past 10,000 years, these small Arctic terrestrial food chains, including a handful of insect species, remained relatively stable. But the pace and amplitude of modern environmental changes in the Arctic have the potential to unravel these fragile webs.

Insect Physiology and Behavior

Some northern arthropods manage to avoid the persistently cold macroclimatic conditions by seasonally occupying habitats with warmer, protected microclimates. Such microclimates can be quite different from the surrounding macroclimates measured at weather stations. Arctic microclimates vary across small spatial scales. Based on studies of the thermal environments of Arctic insects, spiders, and mites, we have gathered sufficient data to realize that the temperatures actually experienced by some arthropods in the field can be affected by fine-scale behavior on a daily to seasonal basis. For instance, because insects are poikilotherms (i.e., having a body temperature that varies with the temperature of their surroundings), many flying insects (flies, moths, butterflies, bees) spend time basking in sunshine to warm their flight muscles sufficiently to allow them to fly. Thus, they need open ground habitats to accomplish this basking behavior. Given that Arctic warming is facilitating the growth of tall shrubs in some regions, the increased shading from shrub expansion on the tundra may affect the activity of flying and surface-active arthropods by limiting their ability to behaviorally thermoregulate.

As insects move through life stages, their temperature sensitivity also varies. Likewise, Arctic spiders exhibit large temperature variation responses among age classes. Similarly, there are indications that heat hardening in Arctic insects can occur within hours suggesting that thermal performance may be a much more plastic trait than previously believed. Temperature thresholds for development and activity are commonly very low in Arctic species. At the cold end of the year, the development of many species of midges and mosquitoes has been observed taking place throughout the winter, even when the lakes remain ice-covered and the lake water temperature is close to freezing. Tundra species of shore bugs (Saldidae) and crane flies (Tipulidae) are active at 0.5°C (Danks, 1981). Such species have antifreeze in their bodies, as reviewed by Denlinger and Lee (2010).

Bowden et al. (2015) analyzed a long time series of data on body size variation in two High-Arctic butterfly species: the Arctic fritillary, *Boloria chariclea*, and the northern clouded yellow butterfly, *Colias hecla*. They measured wingspans from individuals collected annually between 1996 and 2013 from the Zackenberg Research Station, Greenland, and found that wing length in both species significantly decreased at a similar rate in response to warmer summers. These results might seem counterintuitive from the perspective of insects living south of the Arctic, but as we have seen, Arctic insects are geared toward low temperatures. This means that their physiology works best at temperatures near freezing. In particular, their whole system of enzymes works best at low temperatures. Because cold-adapted metabolism is facilitated by enzymes that work *only* at low temperatures, Arctic insects exposed to temperatures above about 15—20°C often die from metabolic failure. Specifically, the weak stability of these enzymes at even moderate temperatures causes them to become ineffective and even break down (Georlette et al., 2004). Cold-adapted insects produce enzymes that have a specific activity at low temperatures that is up to 10 times higher than the enzymes in warm-adapted insects. Therefore, exposure to warmer-than-expected Greenland summer temperatures by these two species of butterflies (and by extension, to many other cold-adapted species of Arctic insects) caused them to expend extra metabolic energy, just to stay alive. This, in turn, limited the amount of energy available to form their wings. As summarized by Bowden et al. (2015), even though longer, warmer summers may mean a longer feeding period for the caterpillars of these butterflies, the higher metabolic cost of obtaining food suggests that the larvae cannot compensate for the energy losses. While some species may be capable of increasing their feeding rate to offset increased metabolic costs associated with warming, this ability appears to be rare. Likewise, Barrio et al. (2016) studied respiration rates in the Arctic woolly-bear moth (*Gynaephora groenlandica*) at two field sites at different elevations. They found that respiration rates were significantly higher and growth rates significantly lower in the warmer temperatures encountered at lower elevation sites, adding support to the idea that cold-adapted insects expend more metabolic energy when exposed to warm temperatures. As body size (in this case, wing length) is strongly related to dispersal ability and fecundity, the results of Bowden et al. (2015) suggest that many Arctic insect species may face severe challenges in response to ongoing rapid climate change.

Life history traits

Life history traits such as the timing of reproduction (phenology), adult body size, and clutch size are sensitive to climatic variation. Because biological productivity is low in the Arctic, many arthropods live multiple years, becoming dormant each winter and active each summer, until they have gathered enough metabolic energy to allow them to reproduce—an aspect of life history that is energy-intensive. The proportion of species taking longer than 1 year to complete their life cycle increases with latitude (Danks, 1981). Relatively few High-Arctic species complete their life cycles in a single year. The duration of the life cycle also depends on habitat, since relatively rapid development is possible in warm microhabitats, such as inside plants or in warm shallow ponds. If Arctic temperatures rise sufficiently, a strong disconnect between this multiyear life cycle and climate may threaten these species' existence, especially if warmer climates facilitate the invasion of warm-adapted species that may be able to outcompete the native (cold-adapted) fauna.

Insect phenology

The phenological shifts of Arctic insects are likely to be substantial but remain poorly studied, particularly at the species level. Most phenological studies of Arctic arthropods are based on data from Zackenberg in North-East Greenland. Here, arthropods have shown strong phenological shifts and clear responses to climate change. Timing of snowmelt is a good predictor of interannual and spatial variation in phenology. For instance, the emergence of an Arctic butterfly in spring may track the timing of snowmelt and is synchronous among males and females. This contrasts with the widespread butterfly habit of protandry (males emerging before females) at lower latitudes. Early flying Arctic butterflies may be out of synch with the timing of regional flower blooming. Without the life-sustaining nectar these flowers provide, the early-arriving butterflies will not be able to complete their life cycle, as the development of eggs in the females requires energy acquired through nectar consumption. Because of climate change, these insects may end up "coming too early to the party."

A study of Arctic specimens of the Napaea fritillary butterfly *Boloria napaea* by Ehl et al. (2019) determined that the females generally invest more time in gaining and keeping resources (and hence in egg production), while males spend more time (and hence resources) in searching for females to optimize their reproductive success. Males spent more than two-thirds of the observed time flying and only rarely rested, perhaps because the reproductive season is so short there.

In their Arctic study area, the authors determined that the flight period of *B. napaea* lasted only 26 days, from July 20 to August 15 in 2016. This was considerably shorter than the flight period of *B. napaea* populations in the Alps, which are about 68 days. In the Arctic, the flight period starts later and ends sooner. However, Ehl et al. (2019) noted that the onset of the Arctic flight period differs considerably from year to year. Researchers working at the same site (the Abisko research station in northern Sweden) have recorded sightings of this species as early as late June. This variability underscores the strong link between the timing of adult emergence and prevailing weather conditions. There is clearly weather-related interannual variation in the phenology of the species. More work is needed to assess variation in phenology among populations and species and to understand the extent of phenological adaptation of Arctic arthropods to climate change.

Population Dynamics, Community Composition, and Diversity

Climate-driven changes in arthropod abundances and distributions can affect food webs and nutrient cycling. Arthropods play an essential role in ecosystem processes such as pollination, aquatic-terrestrial linkages, nutrient cycling, predator-prey, and plant-herbivore interactions. For instance, increasing summer temperatures could enhance insect herbivory. Changes in herbivory could have unexpected downstream effects such as those found for plant-emitted VOCs. Many Arctic plants rely on insects for pollination. However, as we have seen with the life history of the Napaea fritillary butterfly, the unpredictable nature of Arctic climate means that insect activity varies substantially with weather conditions, and Arctic plant pollination could suffer from changes in pollinator phenology driven by changeable weather conditions at the start of the growing season. In contrast, Høye (2020) pointed out that mosquitos serve as disease vectors. Many of the diseases that have been documented in the lower latitudes are not currently found in Arctic mosquitos. Future changes in environmental conditions and the introduction of new mosquito species or arboviruses could bring about novel species interactions that have consequences for polar wildlife and even humans. This kind of scenario has been played out in other ecosystems, with devastating effects.

ARCTIC MARINE IMPACTS

Global change is triggering worldwide reorganizations of marine life. Marine organisms are being exposed to changes in ocean temperature, both at the surface and at the depth. The water they live in is becoming more acidic, more polluted, and more clogged with plastic, both at the surface and on the seafloor. Overfishing has led to the collapse of many fish stocks, driving reorganization of food webs as some species become marginalized or lost to an ecosystem while others gain importance. As with many ecological changes, the changes in ocean systems start at the bottom and work their way up. When the phytoplankton and zooplankton of the ocean are affected by global change, disturbances in their kinds and numbers in a given region will have immediate effects on the macro plankton, the larger invertebrates, the fish, and the marine mammals.

As discussed by Alabia et al. (2020), in recent decades, the Arctic Ocean and adjacent seas have undergone unprecedented warming, accompanied by marine ecosystem changes. These changes include substantial food web restructuring: some species are thriving, while others are dying out, causing trophic cascades. For example, since 2015, the Bering and Chukchi seas have seen August mean sea surface temperatures increase by 2−8°C. The rapid decline in sea-ice cover across the Pacific Arctic, combined with ecosystemwide changes of the past few years, indicates that a potential transformation is under way. Are these changes temporary, or do they signal the onset of a new state for this ecosystem? In the Pacific sector of the Arctic, for instance, prolonged alternating episodes of warming and cooling have significantly reshuffled spatial distributions and species abundances of marine benthic communities. We are already seeing the expansion of temperate species into Arctic waters. Consequently, novel interactions between the incoming and resident Arctic taxa are bound to occur, with important consequences for food webs and the ecosystem.

Plankton

As reviewed by Hirawake et al. (2021), the Arctic marine ecosystem needs sea ice to sustain itself. The ice serves as a platform for marine life and plays a crucial role in controlling physical and chemical processes that affect biological activity. Therefore, sea-ice reduction, accompanied by ocean temperature increases, is altering marine environments, starting with changes in primary production, almost all of which is carried out by phytoplankton.

Sea ice ties with phytoplankton

Sea-ice formation and melt play important roles in Arctic marine biogeochemical cycles. Micronutrients and trace metals needed by phytoplankton, such as iron, become entrained in ice that forms along shelf sediments and are then transported northward through the accumulation in sea ice (Evans and Nishioka, 2018). Sea-ice formation, chemical reduction, and brine injection are the processes driving first the accumulation and then the release of iron and other trace metals from the ice into the Arctic Ocean and adjacent seas (Evans and Nishioka, 2019). The reduction of sea-ice cover will inevitably decrease the concentration of micronutrients and trace metals in Arctic seawater, limiting phytoplankton photosynthesis, growth, and reproduction.

Hirawake et al. (2021) also reported that resting stage cells of diatoms in Arctic ocean sediments are thought to be the seed population for the phytoplankton bloom that takes place in spring. However, this phenomenon is poorly studied in Arctic waters. High densities of viable resting stage cells representing typical Arctic diatom species have been taken from sediment samples from the Chukchi Sea. Cells kept in seawater survived for 6 months, so these could be the seeds for ice algae and phytoplankton blooms. Apparently, there are (poorly known) cycles of the density and type of resting cells in these waters. The number of germinating cells sampled in sediments from the northern Bering Sea in 2018 was 10−100 times the number identified in 2017. Hirawake et al. (2021) reported that phytoplankton species were dominant in the 2017 samples, while ice algae species were dominant in 2018. These results suggest that phytoplankton community composition and abundance of resting stage cells are greatly affected by sea-ice conditions.

Beaufort Sea study

Ice algal biomass is at a minimum during the polar night, but even under thick snow and sea-ice cover microalgae begin to grow in the bottom layer of ice as daylight gradually increases during spring. The ice algal growth season typically runs from March until the onset of snow and ice melt. Hence, sea-ice algae increase until the sea ice melts to the point of exposing the shade-adapted algae to intolerable light intensities, causing their release into the water column during the thaw period. Once released, sea-ice algae may rapidly sink toward the seafloor, or they may play a role in seeding the pelagic bloom.

A study of sea-ice algae in the Beaufort Sea by Nadaï et al. (2021) found that the export of the ice-obligate algae *Nitzschia frigida* at least one to 3 weeks earlier in 2016 than during other years sampled indicated an early onset of snowmelt. Snowmelt and sea-ice breakup occurred at the two southernmost sites over the Mackenzie shelf break, and there was a completely ice-free

Beaufort Sea during September 2016. The early release of ice algae accompanied early peaks in diatom fluxes and higher diatom and phytoplankton carbon fluxes during spring and summer. Arctic warming is causing earlier snowmelt and breakup of sea ice in the Beaufort Sea, and these events directly affect the timing and magnitude of microalgal export and its contribution to particulate organic carbon flux in the Arctic Ocean. The ongoing sea-ice reduction in the Arctic Ocean with global warming may increase phytoplankton carbon fluxes to the seafloor and potential carbon sequestration at depth.

Rise of toxic algae

Toxic algae were also found by Hirawake et al. (2021) in waters of the Chukchi and Bering Sea shelves. Under warming conditions, these algae could germinate on the surface of sediments and cause poisoning of marine life as well as humans. Vegetative cells of *Alexandrium tamarense* were found on the shelves in summer. *A. tamarense* is a species of dinoflagellate that produces a neurotoxin known to cause the human illness clinically known as paralytic shellfish poisoning (PSP). Multiple species of phytoplankton are known to produce this kind of toxin (saxitoxin), including at least 10 other species of *Alexandrium*. High densities of this toxic genus were found in Bering Shelf Water, mainly in warmer waters. Densities far exceeded the levels causing PSP. Dinoflagellate density was found to be positively correlated with water temperature. It is clear that increases in water temperature could enhance toxic algal blooms, which, in turn, bring severe negative effects, from plankton feeders up through higher marine predators to humans.

Benthos

Benthic ecosystems in the Arctic Ocean and adjacent seas are likewise undergoing reorganization because of global change. As with so many aspects of Arctic environments, much of the change to the benthos is linked with changes in sea-ice cover. For instance, Hirawake et al. (2021) discuss how marine ecosystems on the Bering and Chukchi Sea shelves are changing in response to strong pelagic-benthic coupling. This, in turn, is caused by elevated primary production of large diatoms (ice algae and phytoplankton) and low zooplankton grazing pressure. Both processes are considerably affected by sea-ice extent and the seasonal timing of sea-ice retreat. The authors clearly identified this process in their analysis of satellite and long-term monitoring data of benthic biomass. They compared

optically retrieved phytoplankton size with benthic macrofaunal biomass and found a large northward shift in biomass as well as a strong positive correlation between phytoplankton size and sediment chlorophyll-*a* concentration, signaling food availability for the benthic community.

Of course, the changes in the benthic communities are not happening in a vacuum. Other aspects of Arctic Ocean ecology are being affected by global change. Zooplankton advection is one of the potential mechanisms that could also change pelagic-benthic coupling. The benthic flora may soon suffer increased predation from more warm-adapted zooplankton, small fish, and crustaceans that are on the verge of becoming established north of Bering Strait. For instance, Landeira et al. (2018) reported the most northward expansion yet known of the planktonic larvae of the tanner crab *Chionoecetes bairdi* to the Chukchi Sea as early as 1992. In an earlier study, Landeira et al. (2017) reported that in 2007 and 2008, *C. bairdi* larvae were only distributed in the southern Bering Sea and their presence had a strong correlation with that of their adult populations. Populations transported from the Bering Sea or the Pacific Ocean into the Arctic Ocean would have difficulty surviving and reproducing under the current climate conditions. However, they may become viable under future ocean warming. Warm water temperature has been shown to have a clear positive effect on the growth of the dominant zooplankton species (*Calanus glacialis*) in the area. Landeira et al. (2018) further concluded that water temperature that is a few degrees higher than the current level on the Chukchi Sea shelf would lead to eventual recruitment of *C. bairdi* postlarvae.

Marine Fish

Changes in water temperature are the primary driver in the range shifts of marine fish species, but some of their range shifts may also be attributed to nonclimatic factors that are related to species-specific sensitivity and exposure to climate regimes. As discussed by Hirawake et al. (2021), distribution shifts of major marine fish and invertebrate taxa collected from bottom trawl surveys along the Bering Sea shelf were compared with local shifts in the climate regime. Numerical model outputs showed that observed and modeled rates of distribution shifts in marine taxa were poorly correlated with the rate of climate change. Distribution shifts of many marine fish taxa appear to be lagging behind changes in ocean temperatures, which in turn are driven by climate regime shifts. This disconnect suggests that these fish may end up outside their preferred climate envelopes. This could eventually lead to the loss of regional species in a rapidly warming ocean.

Changes in marine fish biogeography

Biogeographical transition zones (BTZs) are biodiversity-rich regions straddling distinct biogeographic zones where the physical environment and ecological factors promote the fusion of biological communities through the combination of species living near their physiological tolerance thresholds. One part of the global ocean where such BTZs exist is at the interface between boreal and polar waters (Alabia et al., 2020). Because of their unusual species richness, BTZs are often regarded as hot spots of biodiversity, making them more resilient to short-term climate shifts that arise from cyclic oceanographic fluctuations. On the other hand, the inhabitants of BTZs remain vulnerable to climate changes that are sufficient to push the ocean conditions well beyond species' tolerance levels. Furthermore, because BTZs represent the leading or trailing edge of the distribution range for many of these species, such disruptions are likely to facilitate the rapid thermophilization of these communities as warm-adapted species increasingly dominate over cold-adapted species. Thermophilization describes a shift in biological communities toward more warm-adapted species, driven by climate warming.

Alabia et al. (2020) investigated the mechanisms and consequences of climate-driven thermophilization, using numerical model projections through the 21st century (2026–2100) to study marine biodiversity in the Eastern Bering and Chukchi Seas, a BTZ representing the boreal-to-Arctic boundary. Overall, the model runs projected changes in species distributions resulting in poleward increases in species richness and functional redundancy in the last quarter of this century (2076–2100). Functional redundancy may keep an ecosystem functioning despite the loss of one or a few species. Some species perform similar roles in communities and ecosystems and may therefore fill the role of lost species with little impact on ecosystem processes (Rosenfeld, 2002). This has generally been lacking in the Arctic Ocean and adjacent seas, because of species diversity kept low by the Arctic temperature regime. Future poleward shifts of boreal species in response to warming and sea-ice changes are projected to alter the composition of Arctic marine communities as larger, longer-lived, and more predatory taxa expand their ranges northward. Drawing from the existing evidence of other Arctic regions, these changes are anticipated

to decrease the stability of Arctic marine ecosystems while increasing their vulnerability.

The study by Alabia et al. (2020) uncovered significant differences in species distributions and biodiversity patterns in the Pacific sector of the Arctic during the 21st century. These differences are even now being driven by changes in thermal and sea-ice regimes in Pacific Arctic waters. Biodiversity patterns are projected to change, with the biota of polar waters shifting from Arctic- to boreal-dominated communities. These changes will have profound impacts on the ecosystem structure and functioning.

In the Arctic marine systems of the present day, there are well-defined differences in biodiversity between the fish faunas of the Arctic Ocean and subarctic waters, with the latter group showing higher taxonomic richness. This assessment concurs with the reported contemporary patterns of regional fish faunas: in regions with warmer, ice-free, biologically productive shelves, such as the Eastern Bering Sea, there is higher fish species abundance. As the regional climatic warming continues to reduce sea-ice cover and to increase ocean temperatures, it is expected that the eastern coast of the South Chukchi Sea will see increasing fish species richness in comparison with the rest of the Arctic basin. This species enrichment will result from climate-driven northern movement of boreal fish species to coastal waters of the Chukchi Sea.

As discussed by Alabia et al. (2020), the strong environmental gradients that currently exist between the sub-Arctic and the Arctic Ocean preclude the survival and colonization of more warm-adapted boreal species in the harsh, perpetually cold polar waters. Unlike the boreal fish fauna, polar fish species have evolved physiological and behavioral adaptations that enable them to cope and survive in extremely cold conditions. However, under future warming scenarios, the advantages of cold-adapted over warm-adapted fish taxa will lessen. This prediction agrees with the results of current Arctic-wide studies that show fish community homogenization as the Arctic loses its extreme habitats and becomes more suitable for generalist species. The changes in biodiversity and species composition posited for the Pacific sector of the Arctic are already apparent in the Atlantic Arctic; both regions are experiencing ecosystem-wide changes in fish faunas.

Freshwater Lake Biota

Primicerio et al. (2007) assessed the impacts of climate change on Arctic lakes, particularly the effects on phenology and community dynamics. Climate change generates variation in ice and snow cover phenology, thickness, and texture that triggers a broad spectrum of physical, chemical, and biotic responses in Arctic lakes. Changes in community structure and dynamics caused by these environmental responses are symptomatic of the ecological impacts of climate change. Range expansion and changes in phenology consistent with global warming scenarios are being documented for an increasing number of taxa.

Arctic lakes are characterized by an extended period of ice cover, cold water, and low productivity. Arctic warming will result in a prolonged ice-free season, stronger thermal stratification, and enhanced nutrient resuspension. These changes are expected to increase lake productivity. Further productivity increases may come from indirect effects of warming on external inputs of nutrients and organic carbon. These environmental changes may also shift the primary source of lake productivity from benthic to pelagic organisms. Increased temperature and production will facilitate successful invasion and introduction of species adapted to warmer and more productive waters.

Climate change is projected to be the main driver of biodiversity changes in Arctic freshwaters throughout the 21st century. Many endemic cold stenothermal species populating these regions are even now forced into deep, cool waters of lakes to survive. Warming waters will very likely eliminate these cold stenotherms. The loss of even a few endemic species will have a strong impact on regional diversity, considering the current low diversity of Arctic lakes. However, the expected increase in productivity will be followed by increased species richness.

Lake trout range expansion

Campana et al. (2020) studied Arctic freshwater fish productivity and colonization in response to climate warming. They used geospatial analysis of the population dynamics of one of the most abundant north temperate freshwater fish species (lake trout) to forecast changes in demography, productivity, and potential colonization success in response to the climate warming scenarios proposed by the IPCC. They used lake morphometry data to assess the lake trout habitat suitability of almost a half-million lakes in the Canadian Arctic. Due to anticipated climate change, lake trout (*Salvelinus namaycush*) productivity in existing habitats is projected to increase by 20% by 2050. Although many ecosystems are likely to be negatively affected by climate warming, the phenotypic plasticity of fish

will allow a rapid relaxation of the current environmental constraints on growth in the far north, as well as enhanced colonization of bodies of water in which there are few potential competitors.

The assessment by Campana et al. (2020) was that 93% of all Canadian Arctic lakes larger in area than 10 ha already have viable lake trout populations. Nearly all the lakes that currently lack trout were predicted to lie north of 69°N, mainly between 69 and 74°N. The authors predicted that the greatest effect of warming on Canadian Arctic lake trout will be their northward colonization of High-Arctic lakes, rather than density changes in lakes already occupied.

For mammals, birds, and marine fishes, colonization of new habitats as the environment changes is largely a continuous extension of range, but this is not the case for freshwater fish. They often colonize new lakes that have no apparent physical connectivity. The study by Campana et al. (2020) identified 30,832 lakes in the High Arctic that will likely become habitable for lake trout by 2050. Some of these lakes currently lack fish altogether, whereas others contain resident or anadromous Arctic charr (*Salvelinus alpinus*). The research team observed some Arctic lakes that are currently home to only a single adult fish. This suggests that colonization of new lakes can occur through passive dispersal of eggs by birds, or as free embryos transported along dry-bed streams under flood conditions. Climate models predict up to a 50% increase in precipitation in the Canadian Arctic in coming years, so flood conditions conducive to increased connectivity between lakes should increase in frequency. Although the time frame for successful colonization by lake trout is unknown, experiments with other fish species demonstrate that lakes may become colonized shortly after they become habitable. Strontium-calcium ratios in fish otoliths (calcareous accretions that form part of the inner ear of fishes) indicate that some lake trout are capable of short periods at sea. Given this capability, lake trout may be able to colonize lakes on islands north of the Arctic mainland.

The team's findings indicate that the relatively simple, species-poor Arctic freshwater ecosystem would respond most to abiotic influences, primarily temperature. Poikilothermic fish have greater phenotypic plasticity than homeotherms, fostering demographic traits such as juvenile growth and sexual maturation that can respond rapidly to temperature change. In contrast to this, Arctic warming has had little effect on the productivity of terrestrial vertebrates, other than through changes in habitat availability. Although climate is a limiting factor on the range boundaries of most freshwater and terrestrial organisms, the response of Arctic freshwater fish to climate change is quite unlike that of regional birds and mammals.

Sub-Arctic lake impacts

At the southern end of the Arctic, global change is altering lake ecology in different ways. The sub-Arctic landscape is scattered with lakes that provide abundant and well-defined ecosystems for food web diversity studies. Increasing temperature and productivity have been shown to shift sub-Arctic lake communities toward more numerous, diverse, smaller-bodied, warmer-water-adapted taxa which are more reliant on pelagic energy sources. However, we do not know how increasing temperature and productivity affect energy transfer efficiency and thus biomass distribution across different trophic levels. Aquatic food webs in many Arctic areas are based on the production and transfer of lipids from primary producers to top consumers, where seasonal storage of lipids is important for the survival and reproduction of long-lived organisms at higher trophic levels.

Keva et al. (2021) conducted a study in 20 sub-Arctic lakes in Finnish Lapland, spanning climatic and chemical gradients, to test how temperature and productivity jointly affect the structure, biomass, and community fatty acid content. They found that lake communities are shifting from cold to warmer water adapted taxa with increasing temperature and productivity. The trophic condition of the lakes has changed from ultraoligotrophic (extremely poor in nutrients) to eutrophic (nutrient-rich) conditions. In response, the phytoplankton flora has changed from diatom-dominated communities to cyanobacteria dominance. Copepods contributed 50%—80% of the total zooplankton community biomass along this gradient and calanoids had the highest biomass percentage in every lake type. In general, the increasing temperatures and productivity have caused lake communities to shift toward the dominance of warmer, murky-water-adapted taxa, with a general increase in the biomass of primary producers, and secondary and tertiary consumers. Primary invertebrate consumers did not show equally clear trends. This process altered various trophic pyramid structures toward an hourglass shape in the warmest and most productive lakes.

Climate warming in Arctic regions is likely to move aquatic ecosystems toward a predominance of generalist species in the community structure. These organisms can tolerate a wider range of environmental conditions than typical Arctic inhabitants and will gain advantages in development. This indicates that the full recovery of Arctic ecosystems in a warming climate may not be possible.

GLOBAL CHANGE EFFECTS ON ARCTIC BIRDS

The interconnectedness of ecosystems and climate is amply demonstrated in Arctic birds. Hirawake et al. (2021) noted that the seasonal migration of seabirds may be closely linked to zooplankton distribution and abundance in Arctic seas. In particular, the short-tailed shearwater *Ardenna tenuirostris* (Fig. 12.12, top right), an abundant marine top predator in the Pacific, was also seen in the Bering Sea in spring and summer and the Chukchi Sea in autumn. *A. tenuirostris* distributions have been found to be concentrated in the waters where large-sized krill were more abundant, suggesting that the changes in the zooplankton advection and size are crucial determinants of shearwaters' distribution. Sea ice in the northern Bering Sea exhibited an anomalously early retreat in the winter of 2017/2018. Crested and least auklets (*Aethia cristatella* and *Aethia pusilla*; Fig.12.12, lower left and right, respectively) are also zooplankton feeders. The populations on St. Lawrence Island failed to fledge their chicks during the summer of 2018.

Shifting Migration Patterns

Chmura et al. (2020) reviewed studies of migration patterns of birds that spend summer in the Arctic. Ample evidence demonstrates that the environmental conditions experienced by these birds on their wintering grounds and/or during the spring migration to the Arctic play a role in the timing and success of their breeding season. The authors evaluated whether breeding and prebasic molt also have carryover effects on autumn migration. They monitored nests of Gambel's white-crowned sparrows (*Zonotrichia leucophrys*) (Fig. 12.12, top left) that breed in Low-Arctic Alaska over 3 years and tracked the birds' autumn migratory departure. Gambel's white-crowned sparrows are long-distance migrants that experience both rapid fluctuations in environmental conditions and an extremely short reproductive window on their breeding grounds. They found that reproductive timing and weather parameters contributed to variation in an autumn departure from the breeding site—not molt timing. Birds that ended looking after their chicks late in the summer departed from the breeding grounds later than other

Gambel's white-crowned sparrow   Short-tailed shearwater

Curlew sandpiper   Red-necked stint

Crested auklet   Least auklet

FIG. 12.12 Top left: Gambel's white-crowned sparrow photo by Wolfgang Wander, Creative Commons Attribution-Share Alike 3.0 Unported license. Top right: Short-tailed shearwater photo by J.J. Harrison, Creative Commons Attribution-Share Alike 3.0 Unported license. Middle left: Curlew sandpiper photo by J.J. Harrison, Creative Commons Attribution-Share Alike 3.0 Unported license. Middle right: Red-necked stint photo by J.J. Harrison, Creative Commons Attribution-Share Alike 3.0 Unported license. Lower left: Crested auklet photo by F. Deines, US Fish and Wildlife Service, in the public domain. Lower right: Least auklet photo by Greg Thompson, Alaska Maritime National Wildlife Refuge, CC BY 2.0.

birds. The authors also observed that, on average, birds departed 2.5 h after sunset and shifted the hour of departure as sunset advanced over the migration season. This study, in conjunction with observations of migration from earlier studies, raises the possibility that global climate change may delay autumn migratory departure in breeding populations of Gambel's white-crowned sparrows in Alaska.

Lisovski et al. (2020) note that many migratory birds are declining worldwide. Long-distance migrants seem especially vulnerable to rapid global change, yet the rate of decline across populations and species varies greatly within flyways. They tested the hypothesis that differences in migration strategy (notably stopover-site use) may cause these variations in resilience to global change. Migratory birds rely on a range of widely separated sites distributed along the routes connecting their breeding and nonbreeding locations. Such sites are considered to have disproportionate importance in terms of area and site use. Both models and empirical data suggest that the quantity and quality of such stopover sites can limit population abundance and affect individual survival. The authors compared the migration strategies of two very closely related shorebird species, the curlew sandpiper (*Calidris ferruginea*) (Fig. 12.12, middle left) and the red-necked stint (*Calidris ruficollis*) (Fig. 12.12, middle right), migrating from the same wintering grounds in Australia to similar breeding sites in the high Russian Arctic. These two species have quite different levels of resilience to different population trajectories in recent times.

Using geolocator tags, Lisovski et al. (2020) tracked the migrations of the two species. They found that curlew sandpipers use fewer stopover areas along the flyway compared with red-necked stints. During northward migration, curlew sandpipers have a higher dependency on fewer sites, in terms of both the percentage of individuals visiting key stopover sites and the relative time spent at those sites. While curlew sandpipers mainly stopover in the Yellow Sea region, which has recently experienced a sharp decline in suitable habitat. In contrast to this, red-necked stints make use of additional sites and spread their relative time en-route across sites more evenly. These different migration strategies may explain why curlew sandpiper populations that use the East Asian-Australasian Flyway are declining rapidly (5.5%–9.5% per year) while the numbers of red-necked stints remain relatively stable (−3.1%–0%). The use of a larger number of sites for rest and feeding during the spring migration appears to confer an advantage to the red-necked stints. Conversely, the curlew sandpiper's reliance on just a few sites along its northern migration, combined with the decline in habitat quality of those sites, is conferring a disadvantage to the species.

Changes in Arctic Bird Phenology

Colella et al. (2020) discussed the phenology of Arctic birds. When phenology directly influences individual fitness, it can affect the genetic variability and adaptive capacity of populations. Climatic warming may trigger earlier migration and breeding in birds, disproportionally affecting species with environmentally constrained or short breeding seasons, common in the Arctic. Shifts in migratory phenology have been documented for numerous migratory Arctic birds over the past half-century. Along the central Arctic coast of Alaska, 16 species now arrive an average of 12 days earlier in spring. Research indicates that earlier egg-laying in sea-ice-obligate birds, such as the black guillemot (*Cepphus grylle mandtii*), is based on their phenotypic plasticity, suggesting that this species may have a limited ability to respond to environmental change beyond existing plasticity. The phenotype of an organism is its physical appearance, as distinguished from its genetic makeup. Phenotypic plasticity is the ability of one genotype of a species to produce more than one phenotype when exposed to different environments. Changing phenologies are expected to continue driving mismatches in the timing of breeding, migration of Arctic birds, and the availability of critical prey species, particularly arthropods, birds, and fish.

INVASIVE SPECIES

Exotic organisms enter new biological communities by several means. The process of human-caused introduction is different from biological colonization, in which species spread to new areas through natural means such as storms that transport flying organisms to new regions, or the transport of organisms from a mainland region to an island by rafting on floating debris.

Many human-caused introductions are accidental. For instance, the seeds of lowland plants may be transported to an alpine region in the tire treads of an automobile without the car owner's knowledge, or an insect or spider "hitch-hikes" from one continent to another inside an unsuspecting airline passenger's luggage. Other introductions are deliberate, such as the transportation of exotic game-fish species to streams or lakes by people interested in fishing for them. Whether accidental or deliberate, species introductions are brought about by human action and therefore constitute an element of global change. In the middle and low latitudes, accidental introductions account for only about 10% of invasive species. In the Arctic, nearly all introductions are accidental. The process that leads to an organism becoming an invasive species, Introduction-Escape-Naturalization-Invasion, may take up to a century to complete. Those species that do become established and spread beyond the place of introduction are considered "naturalized."

Invasive species lack predators, pathogens, and diseases in their new environment, so there is little to keep their numbers in check. Invasive plants produce copious amounts of seed with high viability. They have highly successful dispersal mechanisms. Exotic plants are often attractive to wildlife. They thrive on disturbance and are very opportunistic. They are fast growing and reproduce rapidly. They are habitat generalists, so they do not have specific or narrow ecological requirements. Some invasive plants demonstrate allelopathy, the production and dispersal of chemicals that kill or inhibit the growth of other plants nearby. Some have longer photosynthetic periods. They are the first to leaf out in the spring and last to drop leaves in autumn. Some alter soil and habitat conditions where they grow to better suit their own survival and expansion.

Terrestrial Plants

The Arctic is one of only a few areas of the world where ecosystems remain minimally affected by nonnative species. This near-pristine condition concerning introduced plant species has been maintained by limited large-scale human disturbance, low human population size, light traffic volumes, harsh climatic conditions, and short growing seasons. However, climate change and increasing human activity are affecting large parts of the Arctic, possibly diminishing many of these constraints. In other words, milder climates, longer growing seasons, and increasing levels of anthropogenic disturbance may force a shift in the balance between native and nonnative Arctic flora.

Pathways of invasive species generally reflect the movements of people, which are closely tied to commerce and trade. The volume and rate of globally traded goods have increased dramatically in recent decades, facilitating the transport of nonnative species. As discussed by Wasowicz et al. (2020), the Arctic is no exception.

Nonnative plant species may arrive in a new region by one of six pathways: intentional release, escape from confinement, transport contaminant, transport-stowaway, corridor, or unaided. Some groups of species, such as shrubs and trees, have been almost entirely intentionally released.

Shipping is one of the most important pathways of introduced species in the Arctic. As discussed in Chapter 8, shipping in Arctic waters has been steadily growing in recent decades. Many marine species from other parts of the world have been accidentally transported and released into Arctic waters from the ballast of cargo ships, or they have hitch-hiked on the hulls of ships, and then they have been released into Arctic waters as hulls are cleaned. The footwear of travelers is also a significant pathway of viable nonnative seed to high latitudes. For example, the average visitor to Svalbard transports approximately four seeds on their hiking boots, with 40% of visitors transporting at least one species.

Wasowicz et al. (2020) developed a comprehensive list of nonnative vascular plants found in the Arctic. They used this list to explore the geographic distribution of these plants among 23 subregions of the Arctic, to analyze the extent of their naturalization and invasion, and to examine the pathways of their introductions. Of the 341 nonnative taxa, 188 are naturalized in at least one of the 23 regions. A small number of taxa (11) are considered invasive (Table 12.2); these plants are known from just three regions in relatively close proximity to human population centers. In several Arctic regions, the authors found no naturalized nonnative taxa. Most Arctic regions have a low number of naturalized taxa. Analyses of the nonnative vascular plant flora identified two main biogeographic clusters within the Arctic: American and Asiatic.

Based on the assessment of Wasowicz et al. (2020), the proportion of nonnative taxa in the Arctic flora is 8.6%. The 341 nonnative taxa recorded for the Arctic belong to 39 families and 180 genera. The greatest number of nonnative plant taxa in the Arctic belongs to the grass family (Poaceae: 51 taxa), the sunflower family (Asteraceae: 48 taxa), and the cabbage family (Brassicaceae: 45 taxa). The genera richest in Arctic nonnative taxa are dock (*Rumex*: 12 taxa), grass (*Poa*: 8 taxa), buttercup (*Ranunculus*: 7 taxa), clover (*Trifolium*: 7 taxa), and vetch (*Vicia*: 7 taxa). Goosefoot (*Chenopodium album*) is the most widespread nonnative taxon in the Arctic (recorded in 13 of the 23 regions), followed by chickweed (*Stellaria media*: 11 regions), and wild buckwheat (*Fallopia convolvulus*: 11 regions). Most of the other nonnative taxa have very limited distributions in the Arctic.

Four of the six pathways have contributed significantly to the introduction of nonnative plants in the

TABLE 12.2
Invasive Nonnative Plant Taxa Recorded in the Arctic.

Species	Common Name	Family	Regions[a]	Origin	Life Form[b]
Anthriscus sylvestris	Cow parsley	Apiaceae	NI	Europe, Asia	hc
Bromus inermis	Smooth brome	Poaceae	AN,AW	Europe, Asia	hc
Caragana arborescens	Siberian pea tree	Fabaceae	AW	Asia	Ph
Cirsium arvense	Creeping thistle	Asteraceae	AN	Europe, Asia	Gn
Hordeum jubatum	Foxtail barley	Poaceae	AN,AW	Asia, N America	hc
Leucanthemum vulgare	Oxeye daisy	Asteraceae	AN,AW	Europe, Asia	hc
Linaria vulgaris	Common toadflax	Plantaginaceae	AN	Europe, Asia	hc
Lupinus nootkatensis	Blue lupine	Fabaceae	NI	N America	hc
Melilotus albus	Honey clover	Fabaceae	AN	Europe, Asia	hc
Prunus padus	European bird cherry	Rosaceae	AN	Europe, Asia	Ph
Vicia cracca	Tufted vetch	Fabaceae	AN,AW	Europe, Asia	hc

[a] *NI*, North Iceland; *AN*, North Alaska-Yukon; *AW*, Western Alaska.
[b] hc, chamaephyte; Gn, nonbulbous geophyte; Ph, phanerophyte.
Data from Wasowicz, P., Sennikov, A.N., Westergaard, K.B., Spellman, K., Carlson, M., Gillespie, L.J., Saarela, J.M., Seefeldt, S.S., Bennett, B., Bay, C., Ickert-Bond, S., Väre, H., 2020. Non-native vascular flora of the Arctic: taxonomic richness, distribution and pathways. Ambio 49 (3), 693–703. https://doi.org/10.1007/s13280-019-01296-6.

Arctic, though the proportion of species varies greatly among the different pathways. Escape from confinement is responsible for the introduction of 48% of invasive vascular plant taxa. Transport-stowaway was the second-most active pathway for invasive taxa (37% of all introductions) and most active pathway for naturalized taxa (contributing to the importation of 19% of naturalized taxa). Unaided spread and spread through corridors do not play a significant role in the Arctic. The identification of these pathways is important in developing biosecurity measures at local and regional scales. It may also help in developing strict international biosecurity measures that do not yet exist in the Arctic.

Plant invasion in the Arctic is currently limited to a local scale and that there are no universally successful invaders in many Arctic regions. Regions with a long history of human settlement and relatively high population density are among the most affected by nonnative plant species. These include northern Iceland, Jan Mayen, northern Fennoscandia, Svalbard, and the Kanin-Pechora region of Russia.

Marine Species

The Arctic marine biome is connected to the lower latitudes by the North Pacific and the North Atlantic. By flowing northward through the European Arctic Corridor (the main Arctic gateway where 80% of in- and outflow takes place), North Atlantic Waters (NAWs) transport ocean heat, nutrients, and planktonic organisms to the Arctic Ocean. NAWs enter the Arctic Ocean through two main branches that split around 70°N. One branch flows northward toward Fram Strait, while the second one turns eastward into the Barents Sea. The NAW inflow largely controls the physical and sea-ice conditions of the region. The recent warming of the European Arctic Corridor associated with NAW inflow has been suspected to trigger poleward intrusions of temperate phytoplankton and species from higher trophic levels. By carrying biomass and nutrients produced elsewhere, bioadvection has recently been proposed as an "essential mechanism" for ecosystem dynamics in the Arctic Ocean. In this context, bioadvection may be defined as biologically driven transport of a substance or organism by bulk motion of seawater.

Oziel et al. (2020) used satellite-derived altimetry data to verify substantial increases in the velocity of the surface waters of the North Atlantic current over the past 24 years. They also substantiated evidence that the North Atlantic current has been transporting temperate-zone phytoplankton northward to Arctic waters. In particular, they determined that a well-known temperate species, *Emiliania huxleyi*, has been transported northward into the Barents Sea. *E. huxleyi* is a coccolithophore and has been used previously as a tracer for temperate waters. Their study also showed that bioadvection is a major mechanism responsible for the recent poleward intrusions of southern species like *E. huxleyi*. Previously, it was thought that water temperature was the main driver of such warm-adapted species intrusions.

E. huxleyi is usually associated with the temperate surface waters of the NAW in summer. It typically forms part of the postspring algal bloom, characterized by low nutrients, low light, and strong stratification. Unlike neritic (shallow water) diatom species, *E. huxleyi* does not form winter resting spores, so it is generally considered only a summer visitor in the Barents Sea. Several factors combine to keep *E. huxleyi* out of the Barents Sea year-round. Some of these factors are considered bottom-up controls, such as winter darkness, cold temperatures, and intense vertical mixing. Other factors are considered top-down controls, such as predation by zooplankton, small fish, and crustaceans, as well as viral infections that cause lysis—the splitting open of an organism's cells. Together, these factors prevent *E. huxleyi* from year-to-year survival in the coldest, north-easternmost parts of the Barents Sea (Oziel et al., 2020). The coastal regions and fjords of the Norwegian Sea are suspected to be the source of *E. huxleyi* for the European Arctic Corridor, because of their high abundance there (i.e., 115×10^6 cells per L).

E. huxleyi is expanding poleward and doubling its areal extent in the Barents Sea, providing evidence for the ongoing "Atlantification" of the Arctic Ocean. Both the Arctic and Atlantic oceans have distinct ecological signatures, and the latter is clearly "invading" the former.

Will *E. huxleyi* become established in Arctic waters year-round? This coccolithophore faces several obstacles in its northward path. First, it is advected with the NAW surface currents, and must survive its voyage between marine ecosystems. It will have to avoid being eaten by zooplankton and other small predators, as well as subduction under the polar mixed layer in Fram Strait, which would send it south again. Once in the Barents Sea, it may find more favorable blooming conditions in summer but will have to adapt to the cold, dark winter conditions. The fate of *E. huxleyi* in the Barents Sea is of major importance as it determines the potential "seeding effect" of the species in the Arctic regions. *E. huxleyi* already survives the low light, low nutrient, oligotrophic, and highly stratified conditions of the NAW in summer, but its expansion, growth, and blooming in the Arctic Ocean at some point will be limited by the bottom-up and top-down constraints discussed earlier.

Prevention of marine invasive species

As discussed by Hewitt and Campbell (2007), many introductions of exotic marine invertebrate species into Arctic waters have occurred accidentally, associated with commercial shipping, cruise ships, recreational boating, and other marine vessels. Shipping has been widely recognized as the most active vector of marine invasions and has been the focus of intense activity at the International Maritime Organization (IMO). Recently, after 13 years of negotiation within the IMO Marine Environment Protection Committee (MEPC), the International Convention on the Control and Management of Ship's Ballast Water and Sediments has been adopted. One of the primary functions of the Marine Environment Protection Committee is the prevention and control of marine pollution from ships. This MEPC Convention provided a comprehensive suite of obligations to parties once the Convention had entered into force, with several guidelines supporting implementation. At the core of the Convention is the recognition that current technological solutions are limited. Consequently, the current best practice of the exchange of coastal water (and organisms) for oceanic water (and organisms) outside of specific depth and distance from shore limits will be progressively replaced with a discharge standard that applies in any location. A subsequent resolution in 2018 (MEPC.306(73)) (International Maritime Organization, 2018) amends the guidelines for ballast water management, as follows: The Ballast Water Management Convention requires that discharge of ballast water shall only be conducted through ballast water management and that each ship shall have on board and implement a ballast water management plan approved by the Administration, taking into account Guidelines developed by the IMO, to be found in the MEPC.127(53) management plan, "Guidelines for ballast water management and development of ballast water" (MEPC 2020) (Table 12.3).

Presently, ballast water is managed to reduce invasion threats using a practice known as ballast water exchange or saltwater flushing (International Maritime Organization, 2004). In theory, this practice should reduce the abundance and richness of species contaminating ballast water by either purging the water (releasing the organisms into a lethal habitat) or killing the organisms through osmotic shock. In practice, ballast water exchange can effectively reduce invasion risk between freshwater ecosystems using a marine (saline) exchange en route. However, efficacy is less apparent when shipping connects marine ecosystems.

TABLE 12.3

List of Guidelines for the Implementation of the International Convention for the Control and Management of Ships' Ballast Water and Sediments.

Guidelines for sediment reception facilities (G1)
Guidelines for ballast water sampling (G2)
Guidelines for ballast water management equivalent compliance (G3)
Guidelines for ballast water management and development of ballast water management plans (G4)
Guidelines for ballast water reception facilities (G5)
Guidelines for ballast water exchange (G6) Revokes
Guidelines for risk assessment under regulation A-4 of the BWM Convention (G7)
Guidelines for approval of ballast water management systems (G8)
Procedure for approval of ballast water management systems that make use of active substances (G9)
Guidelines for approval and oversight of prototype ballast water treatment technology programmes (G10)
Guidelines for ballast water exchange design and construction standards (G11)
Guidelines on design and construction to facilitate sediment control on ships (G12)
Guidelines for additional measures regarding ballast water management including emergency situations (G13)
Guidelines on designation of areas for ballast water exchange (G14)
Ballast water management systems installed on ships on or after October 28, 2020 should be approved taking into account the 2016 guidelines (G8)

Data from Marine Environment Protection Committee (MEPC), 2020. Marine Environment Protection Committee (MEPC) 2020. Index of IMO Resolutions: Guidelines for Approval of Ballast Water Management Systems. https://www.imo.org/en/KnowledgeCentre/IndexofIMOResolutions/Pages/MEPC.aspx.

Requirements to install ballast water treatment systems in ships to limit (or even eliminate) nonindigenous species transfer should be realized in coming years under the International Convention for the Control and Management of Ships' Ballast Water and Sediments (International Maritime Organization, 2004). However, technological and logistical hurdles are expected to delay the immediate impact of this requirement, and until such time, some level of regional species introduction threat from this source will likely remain. Marine biological invasion threats to the Arctic are poorly understood. While the number of documented established marine nonindigenous species, including invasive species, is low in the region, detection effort is also substantially lower compared with other global regions. Potentially rapid changes in climate for the coming century in combination with the pronounced effect of changes in the Arctic region are expected to promote the establishment of nonindigenous species. Increasing surface temperatures and changing salinity levels are forecast for Arctic waters. These changes will likely reduce environmental barriers currently preventing the colonization of more temperate species. This applies to species that may be introduced through human agency, but also to lower-latitude species able to expand their ranges into Arctic waters. Recent efforts quantifying the vulnerability of Arctic ecosystems to ship-mediated marine species introduction and invasion indicate that some level of threat exists presently and is set to increase as climate change progresses; however, conclusions have been drawn largely in the absence of biological samples.

Case Study: Introduced Marine Invertebrates in Svalbard

Ware et al. (2016) evaluated risks associated with nonindigenous propagule loads discharged with ships' ballast water while docked in Svalbard, as a case study for the wider Arctic. They sampled and identified transferred propagules (invertebrate and fish eggs) using traditional and DNA barcoding techniques. Then they assessed the suitability of the Svalbard coast for nonindigenous species under contemporary and future climate scenarios using ecophysiological models based on critical temperature and salinity reproductive thresholds (Figs. 12.13 and 12.14).

(2050 and 2100) global warming scenarios. Invasion hot spots were in Hudson Bay, Northern Grand Banks/Labrador, Chukchi/Eastern Bering seas, and Barents/White seas, suggesting that these regions could be more vulnerable to invasions. Globally, both benthic and planktonic organisms showed a future poleward shift in suitable habitat. Results from this study may help prioritize management efforts in the face of climate change in the Arctic marine ecosystem.

Species introductions through ship hull fouling

Chan et al. (2015) examined the processes of marine species introductions in the Canadian Arctic. Ships' hull fouling and ballast water are leading vectors of nonnative marine species globally, yet few studies have examined their magnitude in the Arctic. To determine the relative importance of these vectors in Canada's Arctic, Chan et al. (2015) collected hull and ballast water samples from 13 and 32 vessels, respectively, at Churchill, Manitoba. They compared total abundance and richness of invertebrates transported on hulls versus those in ballast water and found that hull fouling contributed to greater abundance and diversity richness of exotic species than ballast water. Thus, the likelihood of a high-risk introduction is greater for the former than the latter vector. They discovered viable, widespread exotic barnacle species in hull samples. This further emphasizes the prominence of hull fouling over ballast water as a vector for invasive species transport.

These results contradict a common assumption that organisms transported on ships' hulls have lower survivorship than those in ballast water because fouling taxa are directly exposed to the ambient environment and can experience extreme wave turbulence and fluctuations in water temperature and salinity during transoceanic voyages. Conversely, organisms in ballast tanks are relatively protected from the external environment, though they may suffer high mortality within tanks owing to starvation, predation, light and oxygen limitation, or toxicity associated with antifouling applications. Nevertheless, certain vessels were found to have their hulls fouled with abundant "hitch-hikers," suggesting that existing antifouling practices are not sufficient to manage hull fouling. Although the IMO has introduced Guidelines for the Control and Management of Ships' Biofouling to Minimize the Transfer of Invasive Species (International Maritime Organization, 2011), hull maintenance remains a voluntary practice in Canada and is generally only carried out by vessel owners to reduce hydrodynamic drag and fuel consumption.

Svalbard coastal waters appear vulnerable to the introduction of several invertebrate species. These are well-known invasive species, including the bay barnacle *Amphibalanus improvisus* and the green crab *Carcinus maenas*. These species have caused wide-ranging impacts elsewhere including fouling, parasite introduction, reduction of indigenous diversity and abundance, and trophic cascades.

Ships discharging ballast water in Svalbard were found to carry high densities of zooplankton (mean 1522 ± 335 individuals per m³), predominately comprised of indigenous species. However, ballast water exchange did not prevent nonindigenous species introduction. Nonindigenous coastal species were present in all except one of 16 ballast water samples (mean 144 ± 67 individuals per m³), despite five of the eight ships exchanging ballast water en route. Of a total of 73 taxa, 36 species were identified, including 23 nonindigenous species. Of those 23, sufficient data permitted evaluation of the colonization potential for eight widely known species of invaders. Modeled suitability indicated that the coast of Svalbard is presently unsuitable for seven of the eight species. However, under the conditions predicted by the IPCC RCP8.5 climate scenario, modeled suitability will favor colonization for six species by the year 2100.

Future invasion threats will be greatest in situations in which ships transfer organisms from one marine ecosystem to another. For this type of shipping route, ballast water treatment technologies may be required to prevent invasive species from entering Arctic waters.

Invasion hot spots

Goldsmit et al. (2020) attempted to predict places where invasive marine taxa might become established—so-called "invasion hot spots" (Table 12.4). As discussed earlier, the risk of marine invasions in the Arctic is expected to increase with climate warming, greater shipping activity, and resource exploitation. Cold-tolerant planktonic and benthic marine aquatic invasive species (AIS) were scored to produce a list of the top species with the highest relative likelihood of invasion and impact since these ecological groups include the dominant known highest risk marine invasive species. Planktonic and benthic AIS with the greatest potential for invasion and impact in the Canadian Arctic were identified, and the 23 riskiest species were modeled to predict their potential spatial distributions at pan-Arctic and global scales (Fig. 12.15; Table 12.5). Modeling was conducted under present environmental conditions and two intermediate future

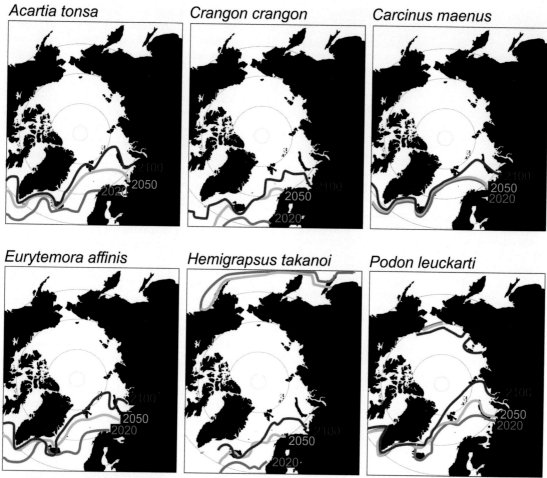

FIG. 12.13 Projected ecophysiological thresholds for six assessed nonindigenous species. Thresholds were based on the number of days required at critical minimum temperature and salinity values for successful reproduction. Thresholds were projected into the future based on ocean climates forecast under the representative concentration pathway (RCP) 8.5 emissions scenario. (Data from Ware, C., Berge, J., Jelmert, A., Olsen, S.M., Pellissier, L., Wisz, M., Kriticos, D., Semenov, G., Kwaśniewski, S., Alsos, I.G., 2016. Biological introduction risks from shipping in a warming Arctic. Journal of Applied Ecology 53 (2), 340–349. https://doi.org/10.1111/1365-2664.12566.)

Some of these AIS have the ability to tolerate current environmental conditions in Hudson Bay and may have adverse effects on resident communities if successfully established.

The king crab in the Barents Sea

Sundet and Hoel (2016) discussed the deliberate introduction of a crustacean species into the Barents Sea in the 1960s. The red king crab (*Paralithodes camtschaticus*) was taken from North Pacific waters of the Russian Far East and released into Kola Bay in the Barents Sea waters of the Soviet Union on several occasions during the 1960s and once during the 1970s. The main purpose of the endeavor was to enhance the food supply in north-western Russia and to increase the economic output of regional fisheries. The king crab introduction was not reported to neighboring countries. By 1977, it had invaded most of the Kola Peninsula's coastal waters, crossed the Norwegian-Russian border, and become abundant in small inlets close to the border. It was not until 1992, however, that the crab came to the attention of Norwegian management and research institutions, because of the problems it caused in local gillnet fisheries.

Hans Hillewaert

Crangon crangon [Brown shrimp]

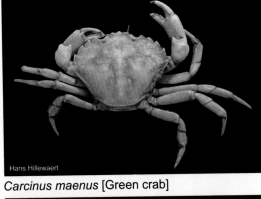

Hans Hillewaert

Carcinus maenus [Green crab]

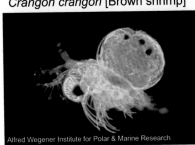

Alfred Wegener Institute for Polar & Marine Research

Podon leuckarti [Marine water flea]

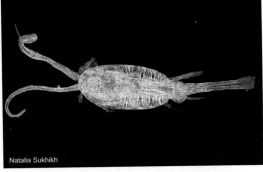

Natalia Sukhikh

Eurytemora affinis [Copepod]

Centre for Biodiversity Genomics

Hans Hillewaert

Hemigrapsus takanoi [Asian shore crab]

FIG. 12.14 Photos of species shown in Fig. 12.13. Photos of *Carcinus maenus*, *Crangon crangon*, and *Hemigrapsus takanoi* by Hans Hillewaert, Creative Commons Attribution=ShareAlike 4.0 International License. Photo of *Eurytemora affinis* courtesy of the Norwegian Institute for Water Research. Photo of *Podon leuckarti* courtesy of the Alfred Wegener Institute for Polar and Marine Research, Creative Commons Attribution 3.0.

TABLE 12.4

Predicted Hot Spots of Aquatic Invasive Species Richness by Ecological Group Under Present and Future (2050 and 2100) Conditions.

Ecological Group	INVASIVE SPECIES HOT SPOTS		
	Present-day	2050	2100
Zoobenthos	Bering Sea, **Hudson Bay coast, Canadian Maritime coast,** S. Greenland coast, E. coast of Iceland, **Barents Sea coast,** Barents Sea	Bering Sea, **Chukchi Sea coast,** Beaufort Sea coast, E. Hudson Bay and **Hudson Bay coast, E. coast of mainland Canada,** S. Greenland, E. coast of Iceland, Barents Sea, **Barents Sea coast,** Kara Sea, **Kara Sea coast**	Bering Sea, Chukchi Sea, Beaufort Sea coast, Hudson Bay and **Hudson Bay coast, E. coast of mainland Canada,** E. Canadian Arctic Archipelago, S. Greenland, E. coast of Iceland, Barents Sea, **Barents Sea coast,** Kara Sea, **Kara Sea coast,** East Siberian Sea coast
Phytobenthos	Bering Sea coasts, Chuchi Sea coasts, Hudson Bay coasts, Canadian Maritime coast, S. Greenland coast, Barents Sea coast	Bering Sea coasts, Chuchi Sea coasts, Hudson Bay coasts, Canadian Maritime coast, S. Greenland coast, Barents Sea coast	Bering Sea coasts, Chuchi Sea coasts, Hudson Bay coasts, Canadian Maritime coast, S. Greenland coast, Barents Sea coast
Zooplankton	Bering Sea, S. Hudson Bay coast, Canadian Maritime coast, E. coast of Iceland, Barents Sea coast, Kara Sea coast, S. Svalbard coast	Bering Sea, Chukchi Sea, S. Hudson Bay, Canadian Maritime coast, E. coast of Iceland, Barents Sea coast, Barents Sea, Kara Sea coast; S. Svalbard coast	Bering Sea, S. Hudson Bay coast, Canadian Maritime coast, E. coast of Iceland, Barents Sea coast, Barents Sea, Kara Sea coast; S. Svalbard coast
Phtyoplankton	Bering Sea coasts, Hudson Bay coasts, **Canadian Maritime coast,** S. Greenland coast, **E. coast of Iceland, Barents Sea coast,** Kara Sea coast; S. Svalbard coast	Bering Sea coasts, **Hudson Bay coasts, Canadian Maritime coast,** S. Greenland coast, **E. coast of Iceland, Barents Sea coast,** Kara Sea coast; S. Svalbard coast	Bering Sea coasts, **Hudson Bay coasts, Canadian Maritime coast,** S. Greenland coast, **E. coast of Iceland, Barents Sea coast,** Kara Sea coast; Svalbard coasts

Data from Goldsmit, J., McKindsey, C.W., Schlegel, R.W., Stewart, D.B., Archambault, P., Howland, K.L., 2020. What and where? Predicting invasion hotspots in the Arctic marine realm. Global Change Biology 26 (9), 4752–4771. https://doi.org/10.1111/gcb.15159.

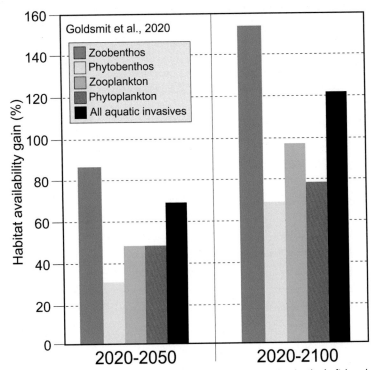

FIG. 12.15 Bar chart showing increases in marine invasive species in the Arctic. Left-hand side: Projected increases between 2020 and 2050. Right-hand side: Projected increases between 2020 and 2100. (Data from Goldsmit, J., McKindsey, C., Schlegel, R., Stewart, D., Archambault, P., Howland, K., 2020. What and where? Predicting invasion hotspots in the Arctic marine realm. Global Change Biology 26 (9), 4752–4771. https://doi.org/10.1111/gcb.15159.)

Since its introduction, the red king crab population has expanded to most areas in the southern Barents Sea, from Kolguyev Island in the east to the Tromsø district in the west. The first objective of the existing management regime is to limit further expansion of the population. It has therefore been important to monitor the development of the stock west of 26°E. Since 2011, annual surveys of the stock have been carried out by sampling at fixed stations in the area using traps.

Sundet and Hoel (2016) reported that there is no indication of a significant northward expansion of the red king crab from the Norwegian coast, and such a migration seems unlikely because the prevailing ocean currents along the northern Norwegian coast flow east. It is, therefore, unlikely that larvae could be transported by currents to Svalbard from existing hatching areas.

Large numbers of king crab on the fishing grounds of eastern Finnmark (North Norway) have interfered with traditional gillnet and longline fisheries,

prompting regional fishermen to demand compensation for lost income. Difficult trade-offs were posed by the dual management objectives established by the Norwegian government, which included (1) preventing the geographical expansion of the crab and (2) exploiting the resource to provide income to coastal communities. The government proposed an open-access fishery west of 26°E to prevent further west- and southward expansion of the crab population and a regular commercial fishery east of that longitude. This management regime commands the consideration of other countries, owing to its handling of the dilemmas inherent in the management of introduced species. Balancing commercial exploitation of the red king crab in one region and eradicating it in another seems to constitute a pragmatic approach to the problem, providing significant benefits to the local communities in the north. The present dual-approach management regime has been in effect for 10 years and appears to be working as intended.

TABLE 12.5
List of Potential Invasive Species for Arctic Waters.

Taxonomic group	Species	Common name	Ecological Group
Crustacea	*Amphibalanus eburneus*	Ivory barnacle	Zoobenthos
Crustacea	*Carcinus maenas*	Green crab	Zoobenthos
Crustacea	*Chionoecetes opilio*	Snow crab	Zoobenthos
Crustacea	*Paralithodes camtschaticus*	Red king crab	Zoobenthos
Tunicata	*Botrylloides violaceus*	Violet tunicate	Zoobenthos
Tunicata	*Botryllus schlosseri*	Golden star tunicate	Zoobenthos
Tunicata	*Molgula manhattensis*	Sea grape	Zoobenthos
Tunicata	*Ciona intestinalis*	Vase tunicate	Zoobenthos
Mollusca	*Mya arenaria*	Soft shell clam	Zoobenthos
Mollusca	*Littorina littorea*	Common periwinkle	Zoobenthos
Bryozoa	*Membranipora membranacea*	Coffin box bryozoan	Zoobenthos
Chlorophyta	*Codium fragile*	Dead man's fingers	Phytobenthos
Rhodophyta	*Dumontia contorta*	Dumont's tubular weed	Phytobenthos
Phaeophycea	*Sargassum muticum*	Japanese wireweed	Phytobenthos
Phaeophycea	*Undaria pinnatifida*	Wakame	Phytobenthos
Copepoda	*Acartia tonsa*	No common name	Zooplankton
Cnidaria	*Aurelia limbata*	Brown-branded moon jelly	Zooplankton
Ctenophora	*Mnemiopsis leidyi*	Warty comb jelly	Zooplankton
Dinoflagellata	*Alexandrium tamarense*	No common name	Phytoplankton
Dinoflagellata	*Dinophysis caudata*	No common name	Phytoplankton
Dinoflagellata	*Dinophysis dens*	No common name	Phytoplankton
Dinoflagellata	*Gonyaulax polygramma*	No common name	Phytoplankton
Dinoflagellata	*Kryptoperidinium triquetrum*	No common name	Phytoplankton

Data from Goldsmit, J., McKindsey, C.W., Schlegel, R.W., Stewart, D.B., Archambault, P., Howland, K.L., 2020. What and where? Predicting invasion hotspots in the Arctic marine realm. Global Change Biology 26 (9), 4752–4771. https://doi.org/10.1111/gcb.15159.

Since Sundet and Hoel (2016) wrote their article in 2016, there have been some interesting and unexpected developments. According to Staalesen (2019), the king crab is expanding not only southward along the coast of Norway, but also toward the north. In a few years, the crab is expected to be found near the coast of Bear Island in the Barents Sea, and by 2030, it is expected to reach Svalbard. Sundet and Hoel reported that sea currents have kept the king crab mainly in the central and eastern Barents Sea. However, the ocean currents are changing in this region. According to Carsten Hvingel, research leader at the Norwegian Institute of Marine Research, sea currents west of the Island of Senja in Norway will accelerate the northward spread of king crabs. The crab can live in deep waters but moves toward shallow areas for spawning. It will likely reproduce near the shores of Bear Island and Svalbard, but not elsewhere in the Barents Sea. However, more recent data have revealed that the king crab's range now extends north to 71.3°N in the Barents Sea and east into the Pechora Sea as far as 55.6°E (Protozorkevich and van der Meeren, 2020). The biggest part of the regional king crab population is still found on the Russian side of the Barents Sea. But it could soon become more dominant on the Norwegian side, where there are no catch regulations west of the 26th parallel.

Ecological costs of king crab invasion

Oug et al. (2018) assessed the effects of king crab predation on the species composition, ecological function, and sediment quality of the Barents Sea floor in three fjord areas in the Varanger region with low, moderate, and very high crab abundances. Compared with data from 1994, most benthic invertebrate species were markedly reduced in abundance, in particular nonmoving burrowing and tube-dwelling polychaetes, bivalves, and echinoderms. The study by Oug et al. (2018) highlights some of the ecological costs of the crab invasion. Following the crab invasion, there was a relative reduction of suspension and surface deposit-feeding species, an increase in mobile and predatory organisms, and an increase in taxa with planktotrophic larval development. The composition of the benthic invertebrate community changed from tube-building, deep deposit-feeding, and fairly large size organisms in localities with small crab populations. With extremely large crab populations, the benthic invertebrate community changed to free-living, shallow burrowing, and rather small-sized organisms. Benthic organisms that rework sediment (both downward and upward) were reduced in number, whereas surficial modifiers increased. These changes imply that benthic invertebrate diversity has been significantly diminished because of king crab predation and that because of this, biological sediment mixing was reduced, leading to a degraded sedimentary environment.

The king crab is not the only exotic crustacean in the Barents Sea. In 1996, Russian fishermen discovered the first snow crabs in the area. Snow crabs are slightly smaller than the king crab. Since then, it has expanded explosively across the region. Genetic studies show links to crabs on the Canadian east coast. An article by Kaiser (2018), a research scientist at the University of Southern Denmark reveals some interesting facets of the snow crab invasion of the Barents Sea. Until recently, snow crabs could only be found in Alaska, Pacific Russian, and Atlantic Canadian waters. But it is thought that increased marine traffic allowed the species to hitch a ride to the Barents Sea from elsewhere in the Arctic. For instance, ballast water may have carried the eggs or larvae of the crab across the Atlantic from northern Canada. A female snow crab produces from 16,000 to 160,000 eggs at a time. Eggs hatch from late spring to early summer to become minute pelagic larvae. Either crab eggs, larvae, or both became established in the Barents Sea, and the new territory has proved quite amenable—they are thriving. Climate change in the Barents Sea has extended the snow crab fishing season, as more of the Barents remains ice-free well into the autumn. At the same time, native snow crab populations in the western Atlantic have been declining, possibly due to climate changes and warming waters.

The crab is now found across most of the Barents Sea with the biggest numbers in the Russian part of the region (Fig. 12.16). It is expected to expand northward into the High Arctic. Snow crab was found for the first time near the shores of Svalbard (Spitsbergen) in 2017 (Protozorkevich and van der Meeren, 2020). According to Russian marine researchers, the snow crab is also found in large numbers east of Novaya Zemlya, in the Kara Sea.

Thermal tolerance of snow crab

Siikavuopio et al. (2017) investigated the effects of temperature on the survival, food intake, metabolic rate, and growth of captive snow crabs (*Chionoecetes opilio*). The crabs were held at three different temperatures, 3, 6, and 9°C. The results showed that temperature had a significant effect on snow crab survival. As might be expected for a truly cold-adaptive crustacean, the crab survival rate at 3°C (61%) was significantly higher than the rate at 6°C (33%) and 9°C (28%). Oxygen consumption rates (hence the metabolic rates) of crabs held at 6 and 9°C were significantly higher than at 3°C. In summary, the current study shows that the Barents Sea snow crab thrives within a narrower temperature range than the red king crab (*Paralithodes camtschaticus*).

Invasive Parasites

Kutz et al. (2015) reported on an exotic bacterium that has been killing muskoxen in the High Arctic. *Erysipelothrix rhusiopathiae*, a bacterium commonly known from domestic swine and poultry, was recently associated with multiple unusual mortality events in muskoxen (*Ovibos moschatus wardi*) on two islands in the Canadian Arctic Archipelago. The muskoxen is one of only two large herbivores in the Canadian Arctic. It provides important ecosystem services and a healthy source of food, a focus for cultural activities, and a source of income through tourism and commercial and sport hunting. The detection of *E. rhusiopathiae* as a significant cause of mortality in muskoxen has potential implications for wildlife conservation, food safety, and food security. During the summer of 2012, approximately 150 dead muskoxen of all age classes were observed on Banks Island in the Canadian Northwest Territories. Wildlife officials sampled four carcasses (one bull and two calves found dead and one bull that was found near-death and was euthanized) and submitted tissues

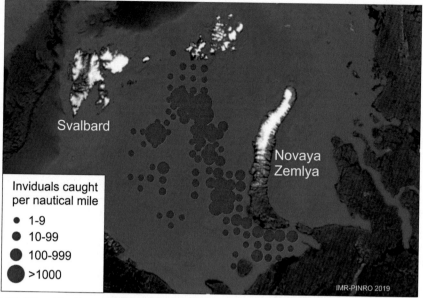

FIG. 12.16 Map of the Barents Sea, showing relative numbers of snow crab caught in 2019. (Modified from Protozorkevich, D., Meeren, v.d., 2019. Survey report from the joint Norwegian/ Russian ecosystem survey in the Barents Sea and adjacent waters. IMR/PINRO Joint Report Series 80.)

to the Canadian Wildlife Health Cooperative (CWHC) in Saskatoon. The animals were in good bodily condition and appeared to have died rapidly with the terminal thrashing of legs from the two bulls and one calf. Various internal organ tissues were cultured and tested positive for *E. rhusiopathiae*, as confirmed by RNA sequencing.

Kutz et al. (2015) noted that the pattern and severity of the recent mortality events are consistent with what would be expected when a new pathogen is introduced into a naïve host population. Full-genome sequencing of isolates from muskoxen across the Canadian High Arctic was initiated to further evaluate this hypothesis. There is a history of muskox die-offs on Banks Island and elsewhere. The species has suffered several population declines in historic times, leading to fragmentation and near extirpations. These declines have left them with exceptionally poor genetic diversity. This lack of genetic diversity likely diminishes their ability to overcome infectious diseases. Second, the Arctic environment is warming rapidly; melting sea ice and extremes in heat are some of the key characteristics (Landrum and Holland, 2020). High-temperature extremes have previously led to nonspecific pneumonia and mortalities of muskoxen in Norway and in captivity and may have contributed to at least some of the recent mortality events in Canada.

The outbreak of *E. rhusiopathiae* infection in muskoxen, combined with other unusual mortality events in this species, emphasizes the importance of understanding the mode of introduction of new pathogens into the Arctic, the resilience of their hosts, and the complex host-pathogen relationships. *E. rhusiopathiae* may have been accidentally transported to Arctic Canada by transport from the midlatitudes. Increased mining and other industrial activities, along with the rise in regional tourism in the Arctic, are creating more opportunities for invasive species to be transported here.

As discussed by Mavrot et al. (2020), between 2001 and 2014, the Banks Island muskox population declined from 69,000 to 14,000 individuals. On Victoria Island, the population dropped from 19,000 in 2001 to 11,000 in 2015. Data from the interval of 1992–2014 from the rest of the island indicate a decrease from 24,000 to 10,000 animals. In North American wildlife, sporadic large-scale outbreaks of infection from *E. rhusiopathiae* have previously been reported in American bison (*Bison bison*), moose (*Alces alces*), pronghorn antelope (*Antilocapra americana*), white-tailed deer (*Odocoileus virginianus*), and wolf (*Canis lupus*). Mavrot et al. (2020) carried forward the study of complex host-pathogen relationships between this bacterium and Arctic mammals.

To understand the epidemiology of this bacterium as it concerns Arctic wildlife, Mavrot et al. (2020) used an *E. rhusiopathiae* enzyme-linked immunosorbent assay (ELISA) to test 958 blood samples collected between 1976 and 2017 from muskoxen from seven regions in Alaska and the Canadian Arctic. ELISA is a laboratory technique that uses antibodies linked to enzymes to detect and measure the amount of a substance in a solution, such as serum. Enzyme-linked immunosorbent assays may be used to help diagnose disease.

A single genotype of *E. rhusiopathiae* was implicated as the cause of death during the muskoxen die-offs in the declining populations of Banks and Victoria Islands in the Northwest Territories and Nunavut, Canada. Subsequently, different genotypes were isolated from carcasses of Alaskan muskoxen, as well as woodland caribou (*Rangifer tarandus caribou*) and moose in Canada, during periods of unusually high mortality of these species. The data verified the link between large herbivore population declines and exposure to *E. rhusiopathiae*.

THE PROPOSED INTRODUCTION OF EARTHWORMS TO ARCTIC SOILS

In a recent publication, Blume-Werry et al. (2020) proposed the systematic introduction of earthworms (Lumbricidae) into the Arctic. Past glaciations eradicated earthworms from Arctic soils. Their slow natural dispersal rates (5−20 m per year) presumably have kept them from recolonizing Arctic soils. Despite this, the authors argue that the deliberate introduction of earthworms into formerly glaciated landscapes may lead to rapid colonization. Specifically, they propose the introduction of two taxa that represent geoengineering earthworms (*Lumbricus* sp. and *Aporrectodea* sp.). The logic of their argument is as follows: (1) Arctic plant growth is predominantly nitrogen (N) limited; (2) this limitation is generally attributed to slow soil microbial processes due to low temperatures (Adamczyk et al., 2020); (3) Arctic plant-soil N cycling is substantially constrained by the lack of larger detritivores; (4) earthworms can mineralize and physically translocate litter and soil organic matter. The authors posit that these new functions provided by earthworms will increase shrub and grass N concentration and increase the height or number of floral shoots while enhancing fine root production and vegetation greenness. Finally, the authors proposed that by boosting plant N and greening in the Arctic, the human introduction of earthworms may lead to substantial changes in the structure and function of Arctic ecosystems.

If such a large-scale introduction of a group of exotic organisms were to take place, one of several outcomes may be predicted. First, the introduced worms may not be able to tolerate the Arctic climate and permafrost soils, in which case they would rapidly die-out. Most of the northern sites where earthworms have thus far been introduced are in sub-Arctic regions. Second, the introduced earthworms could become established and then outcompete many of the native soil organisms, driving them to extinction. Third, and most importantly, the earthworms may become established, but their introduction would cause a series of unforeseen and unintended consequences for tundra ecosystems.

In fact, most of these issues have already been discussed by Wackett et al. (2018), who argue that some earthworm species (*Lumbricus rubellus*, *Lumbricus terrestris*, and *Aporrectodea* sp.) with great geomorphological impact (geoengineering species) should be considered potentially dangerous invasive species in the Fennoscandian Arctic birch forests, where they were first introduced by farmers and more recently through recreational fishing and gardening. They appear to outcompete previously established litter-dwelling earthworm species (i.e., *Dendrobaena octaedra*) that likely colonized the Fennoscandian landscape rapidly following deglaciation. The authors strongly suggest that geoengineering earthworms may pose a potent threat to some of the most remote and protected Arctic environments in northern Europe, based on the following arguments. First, these earthworm species are known to be drivers of dramatic environmental changes in other formerly glaciated environments, including negative effects on native plants and preferential selection for nonnative plants and graminoids, increased greenhouse gas emissions, and depletion of key nutrients in surface soils. Second, measured earthworm biomasses in Fennoscandian Arctic birch forests adjacent to anthropogenic source areas are comparable with biomasses in North American temperate and boreal forests where invasive earthworms are recognized as a potent threat. Third, it is known from previous studies that earthworms are almost impossible to eradicate once they have invaded a system. Finally, relatively pristine regions, such as the Fennoscandian tundra environment, must be protected from invasive species, however, well-intentioned their introduction on the landscape. This area is one of the few remaining northern European biomes where anthropogenic spreading of earthworms is likely still limited.

LIGHT POLLUTION IN THE POLAR NIGHT

One other change merits brief mention, here: the increase in light pollution in the Arctic. As discussed by Berge et al. (2020), until the 21st century, most scientists assumed that Arctic ecosystems enter a resting state during the winter darkness. More recent research has revealed that Arctic ecosystems remain highly active through the winter, with continuous biological interactions across all trophic levels and taxonomic groups. To the human eye, there is virtually no difference in illumination between night and day in the Arctic winter interval when the sun never rises. Nevertheless, many marine organisms stay active and can adjust their behavior to the daily cycle of background solar illumination.

Across the globe, nighttime artificial illumination is increasing annually by 6% on average. It is becoming one of the fastest-spreading environmental challenges of the Anthropocene. Berge et al. (2020) have written a report on the ways in which artificial light during the polar night is disrupting Arctic fish and zooplankton behavior down to depths as great as 200 m. It is too easy for those of us who live in the midlatitudes to forget that the Arctic polar night has been one of the last undisturbed and pristine dark environments on the planet. It could very well be that the moon and stars provide important cues to guide the distribution and behavior of Arctic organisms. We know almost nothing of this phenomenon. As human activities increase in the Arctic, the darkness of the polar night is being shattered by strong illumination from artificial light. Berge et al. (2020) showed that normal shipboard working lights disrupt fish and zooplankton behavior down to at least 200 m depth.

CONCLUSIONS

Global change threats to Arctic environments and biota are too many and diverse to be fully covered in this chapter. The aim here has been to point out the kinds of threats that face Arctic ecosystems, today and in the future. We know the history of some human disturbances from the minute traces they have left in the Greenland ice sheet. Perhaps surprisingly, the ice core records demonstrate that heavy metal contamination of the Arctic is far older than expected. It did not arise during the Industrial Revolution, but rather in Greek and Roman times, as products of metal smelting. Mercury pollution of the Arctic is a pervasive problem on land and sea. The toxicity of the element, combined with the ease with which it bonds with organic matter in the form of methylmercury, makes it a potent poison. One of the more frightening aspects of mercury pollution is the expected release of large quantities as permafrost melts, releasing its load of stored mercury to the environment for decades or centuries to come.

Radioactive isotopes of several species (cesium, plutonium, etc.) are drifting around Arctic seas, buried in ice sheets, residing in sediments on the Barents Sea-floor, and being held in a large number of nuclear waste containers, from metal drums to decommissioned Russian submarines. The level of radioactive contamination has been decreasing since the end of the Cold War, but some of these nuclides have half-lives of thousands of years.

Environmental changes are driving changes in migration patterns in reindeer in the Taimyr Peninsula, and causing population declines of reindeer in Svalbard. In Arctic North America, global change is forcing fluctuations in caribou life history, population dynamics, and demography. These environmental changes are not just straightforward climate warming. They represent complex changes in weather patterns and vegetation cover. For instance, spring and fall caribou condition has been shown to be influenced by winter and spring weather, particularly snow conditions and spring temperatures.

Any way you look at it, polar bears are in deep trouble. Their sea-ice hunting platform is steadily disappearing. This is the habitat they evolved to exploit. They got the name *Ursus maritimus* for a reason. Their body shape, fur, physiology, and behavior are all adapted for hunting seals from pack ice and paddling around in Arctic seas. It is highly doubtful that they can successfully switch to a terrestrial lifestyle. The survival strategies listed in this chapter are only stop-gap measures. These bears require up to 12,000 calories a day to keep healthy. A guillemot egg contains about 180 calories, so a bear would have to consume at least 66 eggs per day to survive on this diet.

During the 21st century, much has been written on the "Greening of the Arctic." This is a real phenomenon, to be sure, but it has not uniformly occurred across the whole Arctic, nor has it persisted as many thought it would. The latest Arctic Report Card (NOAA, 2020) notes that "Since 2016, tundra greenness trends have diverged strongly by continent, declining sharply in North America but remaining above the long-term average in Eurasia." The discrepancy between reports of greening based on the interpretation of satellite data and ground observations is apparently due in part to problems in the NDVI interpretation of older (grainier) satellite imagery. More recent studies (using satellite imagery with higher definition) indicate a decrease in the satellite greening trend and reveal heterogeneity in satellite observations across different sensor platforms. False greening signals may be interpreted from atmospheric changes, drift in satellite sensors, or earlier snowmelt. Thus, there is current uncertainty in whether the greening patterns observed by satellites truly indicate changes in vegetation patterns across the Low- and High-Arctic tundra biomes.

Not all Arctic regions are warming or "greening-up" to the same degree. As discussed earlier, climate change has apparently not been the principal factor driving birch shrub expansion in the Canadian central Low Arctic. Instead, shrubby plant release from caribou, the pressure of caribou browsing, following a severe herd decline, may explain recent shrub growth in this region.

These are just a few examples of the effects of global change on Arctic environments and biota. In writing this chapter, I was struck by a few key aspects of global change effects in the Arctic, starting with complexity and interconnectedness. As far as scientific research is concerned, you might say that we are on the right track when we discover that systems we thought were quite simple turn out to be quite complex. Nearly every good scientific investigation produces this result. Another theme of this chapter and book is the interconnectedness of the physical and biological elements of Arctic ecosystems.

This chapter also discusses many unexpected results, such as a warmth-driven change in ocean currents allowing king crab to invade more northerly sectors of the Bering Sea, or warmer temperatures causing metabolic stress in Arctic caterpillars and snow crabs.

Another recurring theme concerns unintended consequences of human actions, from cargo ships transporting invasive marine organisms to the Arctic to the as-yet-known consequences of "seeding" the Arctic with earthworms to boost biological productivity.

The bottom line is that there is still much that we do not know about the structure and function of Arctic ecosystems, both marine and terrestrial. Unfortunately, the negative effects of global change have become well established in the Arctic, whether or not science has come to grips with understanding these impacts.

REFERENCES

Adamczyk, M., Perez-Mon, C., Gunz, S., Frey, B., 2020. Strong shifts in microbial community structure are associated with increased litter input rather than temperature in high Arctic soils. Soil Biology and Biochemistry 151, 108054. https://doi.org/10.1016/j.soilbio.2020.108054.

Alabia, I.D., Molinos, J.G., Saitoh, S.I., Hirata, T., Hirawake, T., Mueter, F.J., 2020. Multiple facets of marine biodiversity in the Pacific Arctic under future climate. The Science of the Total Environment 744, 140913. https://doi.org/10.1016/j.scitotenv.2020.140913.

Andruko, R., Danby, R., Grogan, P., 2020. Recent growth and expansion of birch shrubs across a low arctic landscape in continental Canada: are these responses more a consequence of the severely declining caribou herd than of climate warming? Ecosystems 23 (7), 1362–1379. https://doi.org/10.1007/s10021-019-00474-7.

Barbante, C., Veysseyre, A., Ferrari, C., Van De Velde, K., Morel, C., Capodaglio, G., Cescon, P., Scarponi, G., Boutron, C., 2001. Greenland snow evidence of large scale atmospheric contamination for platinum, palladium, and rhodium. Environmental Science and Technology 35 (5), 835–839. https://doi.org/10.1021/es000146y.

Barbante, C., Spolaor, A., Cairns, W.R., Boutron, C., 2017. Man's footprint on the Arctic environment as revealed by analysis of ice and snow. Earth-Science Reviews 168, 218–231. https://doi.org/10.1016/j.earscirev.2017.02.010.

Barrio, I.C., Bueno, C.G., Hik, D.S., 2016. Warming the tundra: reciprocal responses of invertebrate herbivores and plants. Oikos 125 (1), 20–28. https://doi.org/10.1111/oik.02190.

Barten, N.L., Bowyer, R.T., Jenkins, K.J., 2001. Habitat use by female caribou: tradeoffs associated with parturition. Journal of Wildlife Management 65 (1), 77–92. https://doi.org/10.2307/3803279.

Berge, J., Geoffroy, M., Daase, M., Cottier, F., Priou, P., Cohen, J.H., Johnsen, G., McKee, D., Kostakis, I., Renaud, P.E., Vogedes, D., Anderson, P., Last, K.S., Gauthier, S., 2020. Artificial light during the polar night disrupts Arctic fish and zooplankton behaviour down to 200 m depth. Nature Communications Biology 3, 102. https://doi.org/10.1038/s42003-020-0807-6.

Blume-Werry, G., Krab, E.J., Olofsson, J., Sundqvist, M.K., Väisänen, M., Klaminder, J., 2020. Invasive earthworms unlock arctic plant nitrogen limitation. Nature Communications 11 (1), 1766. https://doi.org/10.1038/s41467-020-15568-3.

Bowden, J.J., Eskildsen, A., Hansen, R.R., Olsen, K., Kurle, C.M., Høye, T.T., 2015. High-arctic butterflies become smaller with rising temperatures. Biology Letters 11 (10), 20150574. https://doi.org/10.1098/rsbl.2015.0574.

Buchwal, A., Sullivan, P.F., Macias-Fauria, M., Post, E., Myers-Smith, I.H., 2020. Divergence of Arctic shrub growth associated with sea ice decline. Proceedings of the National Academy of Sciences of the United States of America 17 (52), 33334–33344. https://doi.org/10.1073/pnas.2013311117.

Cairns, P., 2020. Climate Change Continues to Devastate Taimyr Reindeer. World Wildlife Fund. https://arcticwwf.org/newsroom/stories/climate-change-continues-to-devastate-taimyr-reindeer-in-russia/.

Campana, S.E., Casselman, J.M., Jones, C.M., Black, G., Barker, O., Evans, M., Guzzo, M.M., Kilada, R., Muir, A.M., Perry, R., 2020. Arctic freshwater fish productivity and colonization increase with climate warming. Nature Climate Change 10 (5), 428–433. https://doi.org/10.1038/s41558-020-0744-x.

Chan, F.T., Macisaac, H.J., Bailey, S.A., 2015. Relative importance of vessel hull fouling and ballast water as transport vectors of nonindigenous species to the canadian arctic. Canadian Journal of Fisheries and Aquatic Sciences 72 (8), 1230—1242. https://doi.org/10.1139/cjfas-2014-0473.

Chmura, H.E., Krause, J.S., Pérez, J.H., Ramenofsky, M., Wingfield, J.C., 2020. Autumn migratory departure is influenced by reproductive timing and weather in an Arctic passerine. Journal of Ornithology 161 (3), 779—791. https://doi.org/10.1007/s10336-020-01754-z.

Colella, J.P., Talbot, S.L., Brochmann, C., Taylor, E.B., Hoberg, E.P., Cook, J.A., 2020. Conservation genomics in a changing Arctic. Trends in Ecology & Evolution 35, 149—161. https://doi.org/10.1016/j.tree.2019.09.008.

Danks, H.V., 1981. Arctic arthropods: a review of systematics and ecology with particular reference to the North American fauna. Entomological Society of Canada, Ottawa.

Davidson, S.C., Bohrer, G., Gurarie, E., LaPoint, S., Mahoney, P.J., Boelman, N.T., Eitel, J.U.H., Prugh, L.R., Vierling, L.A., Jennewein, J., Grier, E., Couriot, O., Kelly, A.P., Meddens, A.J.H., Oliver, R.Y., Kays, R., Wikelski, M., Aarvak, T., Ackerman, J.T., et al., 2020. Ecological insights from three decades of animal movement tracking across a changing Arctic. Science 370 (6517), 712—715. https://doi.org/10.1126/science.abb7080.

Denlinger, D.L., Lee, R.E., 2010. Low temperature biology of insects. Cambridge University Press, pp. 1—390. https://doi.org/10.1017/CBO9780511675997.

Edwards, M.E., Armbruster, W.S., Elias, S.A., 2014. Constraints on post-glacial boreal tree expansion out of far-northern refugia. Global Ecology and Biogeography 23 (11), 1198—1208. https://doi.org/10.1111/geb.12213.

Ehl, S., Holzhauer, S.I.J., Ryrholm, N., Schmitt, T., 2019. Phenology, mobility and behaviour of the arcto-alpine species Boloria napaea in its arctic habitat. Scientific Reports 9 (1), 3912. https://doi.org/10.1038/s41598-019-40508-7.

Evans, L.K., Nishioka, J., 2018. Quantitative analysis of Fe, Mn and Cd from sea ice and seawater in the Chukchi sea, Arctic ocean. Polar Science 17, 50—58. https://doi.org/10.1016/j.polar.2018.07.002.

Evans, L.K., Nishioka, J., 2019. Accumulation processes of trace metals into Arctic sea ice: distribution of Fe, Mn and Cd associated with ice structure. Marine Chemistry 209, 36—47. https://doi.org/10.1016/j.marchem.2018.11.011.

Everatt, M.J., Convey, P., Bale, J.S., Worland, M.R., Hayward, S.A.L., 2015. Responses of invertebrates to temperature and water stress: a polar perspective. Journal of Thermal Biology 54, 118—132. https://doi.org/10.1016/j.jtherbio.2014.05.004.

Gagnon, C.A., Hamel, S., Russell, D.E., Powell, T., Andre, J., Svoboda, M.Y., Berteaux, D., 2020. Merging indigenous and scientific knowledge links climate with the growth of a large migratory caribou population. Journal of Applied Ecology 57 (9), 1644—1655. https://doi.org/10.1111/1365-2664.13558.

Georlette, D., Blaise, V., Collins, T., et al., 2004. Some like it cold: Biocatalysis at low temperatures. FEMS Microbiology Reviews 28, 25—42. https://doi.org/10.1016/j.femsre.2003.07.003.

Goldsmit, J., McKindsey, C.W., Schlegel, R.W., Stewart, D.B., Archambault, P., Howland, K.L., 2020. What and where? Predicting invasion hotspots in the Arctic marine realm. Global Change Biology 26 (9), 4752—4771. https://doi.org/10.1111/gcb.15159.

Gould, W.A., Raynolds, M., Walker, D.A., 2003. Vegetation, plant biomass, and net primary productivity patterns in the Canadian Arctic. Journal of Geophysical Research: Atmospheres 108, D2, 8167. https://doi.org/10.1029/2001jd000948.

Gwynn, J.P., Heldal, H.E., Gäfvert, T., Blinova, O., Eriksson, M., Sværen, I., Brungot, A.L., Strålberg, E., Møller, B., Rudjord, A.L., 2012. Radiological status of the marine environment in the Barents Sea. Journal of Environmental Radioactivity 113, 155—162. https://doi.org/10.1016/j.jenvrad.2012.06.003.

Hansen, B.B., Pedersen, Å.Ø., Peeters, B., Le Moullec, M., Albon, S.D., Herfindal, I., Sæther, B.E., Grøtan, V., Aanes, R., 2019. Spatial heterogeneity in climate change effects decouples the long-term dynamics of wild reindeer populations in the high Arctic. Global Change Biology 25 (11), 3656—3668. https://doi.org/10.1111/gcb.14761.

Hewitt, C.L., Campbell, M.L., 2007. Mechanisms for the prevention of marine bioinvasions for better biosecurity. Marine Pollution Bulletin 55 (7—9), 395—401. https://doi.org/10.1016/j.marpolbul.2007.01.005.

Hirawake, T., Uchida, M., Abe, H., Alabia, I.D., Hoshino, T., et al., 2021. Response of Arctic biodiversity and ecosystem to environmental changes. Findings from the ArCS project, Polar Science, 27, 100533. https://doi.org/10.1016/j.polar.2020.100533.

Høye, T.T., 2020. Arthropods and climate change — arctic challenges and opportunities. Current Opinion in Insect Science 41, 40—45. https://doi.org/10.1016/j.cois.2020.06.002.

Hodkinson, I.D., 2013. Terrestrial and freshwater invertebrates. In: Meltofe, H. (Ed.), Arctic Biodiversity Assessment: Status and Trends in Arctic Biodiversity. Conservation of Arctic Flora and Fauna. Akureyri, Iceland, pp. 247—277.

Hutchison, C., Guichard, F., Legagneux, P., Gauthier, G., Bêty, J., 2020. Seasonal food webs with migrations: multi-season models reveal indirect species interactions in the Canadian Arctic tundra. Philosophical Transactions of the Royal Society A: Mathematical, Physical and Engineering Sciences 378, 20190354. https://doi.org/10.1098/rsta.2019.0354.

International Maritime Organization, 2004. International Convention for the Control and Management of Ships' Ballast Water and Sediments (BWM). BMW.2-Circ (Vol. 40). IMO. Resolution MEPC.207(62).

International Maritime Organization, 2011. Issue 62. Guidelines for the Control and Management of Ships' Biofouling to Minimize the Transfer of Invasive Species, vol. 207. IMO. Resolution MEPC.207(62).

International Maritime Organization, 2018. Amendments to the Guidelines for Ballast Water Management and Development of Ballast Water Management Plans (G4). Resolution MEPC (Vol. 306). IMO; Resolution MEPC.306(73).

Japanese Ministry of the Environment, 2019. Booklet to Provide Basic Information Regarding Health Effects of Radiation, vol. 254. Government of Japan.

Kaiser, B., 2018. How the saga of Barents Sea snow crab illustrates the complexity of climate change. Arctic Today. https://www.arctictoday.com/saga-barents-sea-snow-crab-illustrates-complexity-climate-change/.

Kazlowsky, S., 2020. A Tale of Two Bears in a Changing Arctic. World Wildlife Fund. https://arcticwwf.org/newsroom/stories/a-tale-of-two-bears-in-a-changing.

Keva, O., Taipale, S.J., Hayden, B., Thomas, S.M., Vesterinen, J., Kankaala, P., Kahilainen, K.K., 2021. Increasing temperature and productivity change biomass, trophic pyramids and community-level omega-3 fatty acid content in subarctic lake food webs. Global Change Biology 27 (2), 282–296. https://doi.org/10.1111/gcb.15387.

Kutz, S., Bollinger, T., Branigan, M., Checkley, S., Davison, T., Dumond, M., Elkin, B., Forde, T., Niptanatiak, A., Orsel, K., Hutchins, W., 2015. Erysipelothrix rhusiopathiae associated with recent widespread muskox mortalities in the Canadian Arctic. Canadian Veterinary Journal 56 (6), 560–563. http://www.canadianveterinarians.net/publications/cvj-current-issue.aspx#.Ubld5_nWW5I.

Landeira, J.M., Matsuno, K., Yamaguchi, A., Hirawake, T., Kikuchi, T., 2017. Abundance, development stage, and size of decapod larvae through the Bering and Chukchi Seas during summer. Polar Biology 40 (9), 1805–1819. https://doi.org/10.1007/s00300-017-2103-6.

Landeira, J.M., Matsuno, K., Tanaka, Y., Yamaguchi, A., 2018. First record of the larvae of tanner crab Chionoecetes bairdi in the Chukchi Sea: a future northward expansion in the Arctic? Polar Science 16, 86–89. https://doi.org/10.1016/j.polar.2018.02.002.

Landrum, L., Holland, M.M., 2020. Extremes become routine in an emerging new Arctic. Nature Climate Change 10 (12), 1108–1115. https://doi.org/10.1038/s41558-020-0892-z.

Law, K.S., Stohl, A., 2007. Arctic air pollution: origins and impacts. Science 315 (5818), 1537–1540. https://doi.org/10.1126/science.1137695.

Li, H., Väliranta, M., Mäki, M., Kohl, L., Sannel, A.B.K., Pumpanen, J., Koskinen, M., Bäck, J., Bianchi, F., 2020. Overlooked organic vapor emissions from thawing Arctic permafrost. Environmental Research Letters 15 (10), 104097. https://doi.org/10.1088/1748-9326/abb62d.

Lisovski, S., Gosbell, K., Minton, C., Klaassen, M., 2020. Migration strategy as an indicator of resilience to change in two shorebird species with contrasting population trajectories. Journal of Animal Ecology. https://doi.org/10.1111/1365-2656.13393.

Marine Environment Protection Committee (MEPC), 2020. Marine Environment Protection Committee (MEPC) 2020. Index of IMO Resolutions: Guidelines for Approval of Ballast Water Management Systems. https://www.imo.org/en/KnowledgeCentre/IndexofIMOResolutions/Pages/MEPC.aspx.

Mavrot, F., Orsel, K., Hutchins, W., Adams, L.G., Beckmen, K., Blake, J.E., Checkley, S.L., Davison, T., Francesco, J.D., Elkin, B., Leclerc, L.M., Schneider, A., Tomaselli, M., Kutz, S.J., 2020. Novel insights into serodiagnosis and epidemiology of Erysipelothrix rhusiopathiae, a newly recognized pathogen in muskoxen (Ovibos moschatus). PLoS One 15 (4), e0231724. https://doi.org/10.1371/journal.pone.0231724.

Morrissette-Boileau, C., Boudreau, S., Tremblay, J.P., Côté, S.D., 2018. Revisiting the role of migratory caribou in the control of shrub expansion in northern Nunavik (Québec, Canada). Polar Biology 41 (9), 1845–1853. https://doi.org/10.1007/s00300-018-2325-2.

Myers-Smith, I.H., Grabowski, M.M., Thomas, H.J.D., Angers-Blondin, S., Daskalova, G.N., Bjorkman, A.D., Cunliffe, A.M., Assmann, J.J., Boyle, J.S., McLeod, E., McLeod, S., Joe, R., Lennie, P., Arey, D., Gordon, R.R., Eckert, C.D., 2019. Eighteen years of ecological monitoring reveals multiple lines of evidence for tundra vegetation change. Ecological Monographs 89 (2), e01351. https://doi.org/10.1002/ecm.1351.

Nadaï, G., Nöthig, E.M., Fortier, L., Lalande, C., 2021. Early snowmelt and sea ice breakup enhance algal export in the Beaufort Sea. Progress in Oceanography 190, 102479. https://doi.org/10.1016/j.pocean.2020.102479.

NASA Earth Observatory, 2000. Remote Sensing: Normalized Difference Vegetation Index (NDVI). https://earthobservatory.nasa.gov/features/MeasuringVegetation/measuring_vegetation_2.php.

NOAA, 2020. Arctic Report Card: update for 2020. https://arctic.noaa.gov/report-card/report-card-2020.

Oug, E., Sundet, J.H., Cochrane, S.K.J., 2018. Structural and functional changes of soft-bottom ecosystems in northern fjords invaded by the red king crab (Paralithodes camtschaticus). Journal of Marine Systems 180, 255–264. https://doi.org/10.1016/j.jmarsys.2017.07.005.

Oziel, L., Baudena, A., Ardyna, M., Massicotte, P., Randelhoff, A., Sallée, J.B., Ingvaldsen, R.B., Devred, E., Babin, M., 2020. Faster Atlantic currents drive poleward expansion of temperate phytoplankton in the Arctic Ocean. Nature Communications 11 (1), 1705. https://doi.org/10.1038/s41467-020-15485-5.

Perryman, C.R., Wirsing, J., Bennett, K.A., Brennick, O., Perry, A.L., Williamson, N., Ernakovich, J.G., 2020. Heavy metals in the Arctic: distribution and enrichment of five metals in Alaskan soils. PLoS One 15 (6), e0233297. https://doi.org/10.1371/journal.pone.0233297.

Potter, C., Alexander, O., 2020. Changes in vegetation phenology and productivity in Alaska over the past two decades. Remote Sensing 12 (10), 1546. https://doi.org/10.3390/rs12101546.

Primicerio, R., Rossetti, G., Amundsen, P.A., Klemetsen, A., 2007. Impact of climate change on arctic and alpine lakes: effects on phenology and community dynamics. In: Ørbæk,

J.B., Kallenborn, R., Tombre, I., et al. (Eds), Arctic Alpine Ecosystems and People in a Changing Environment. Springer, Berlin Heidelberg, pp. 51–69. https://doi.org/10.1007/978-3-540-48514-8_4.

Protozorkevich, D., Meeren, v.d., 2019. Survey report from the joint Norwegian/Russian ecosystem survey in the Barents Sea and adjacent waters. In: IMR/PINRO Joint Report Series.

Protozorkevich, D., van der Meeren, G.I. (Eds.), 2020. Survey report from the joint Norwegian/ Russian ecosystem survey in the Barents Sea and adjacent waters August-October 2019. IMR/PINRO Joint Report Series, 1-2020, p. 93.

Rees, W.G., Hofgaard, A., Boudreau, S., Cairns, D.M., Harper, K., Mamet, S., Mathisen, I., Swirad, Z., Tutubalina, O., 2020. Is subarctic forest advance able to keep pace with climate change? Global Change Biology 26 (7), 3965–3977. https://doi.org/10.1111/gcb.15113.

Reimer, J.R., Mangel, M., Derocher, A.E., Lewis, M.A., 2019. Modeling optimal responses and fitness consequences in a changing Arctic. Global Change Biology 25 (10), 3450–3461. https://doi.org/10.1111/gcb.14681.

Rosenfeld, J.S., 2002. Functional redundancy in ecology and conservation. Oikos 98 (1), 156–162. https://doi.org/10.1034/j.1600-0706.2002.980116.x.

Schaefer, K., Elshorbany, Y., Jafarov, E., Schuster, P.F., Striegl, R.G., Wickland, K.P., Sunderland, E.M., 2020. Potential impacts of mercury released from thawing permafrost. Nature Communications 11, 4650. https://doi.org/10.1038/s41467-020-18398-5.

Siikavuopio, S.I., Whitaker, R.D., Sæther, B.S., James, P., Olsen, B.R., Thesslund, T., Hustad, A., Mortensen, A., 2017. First observations of temperature tolerances of adult male snow crab (Chionoecetes opilio) from the Barents Sea population and the effects on the fisheries strategy. Marine Biology Research 13 (7), 744–750. https://doi.org/10.1080/17451000.2017.1313989.

Smith, C.C., Reichman, O.J., 1984. The evolution of food caching by birds and mammals. Annual Review of Ecology and Systematics 15, 329–351. https://doi.org/10.1146/annurev.es.15.110184.001553.

Smith, P.A., Elliott, K.H., Gaston, A.J., Gilchrist, H.G., 2010. Has early ice clearance increased predation on breeding birds by polar bears? Polar Biology 33 (8), 1149–1153. https://doi.org/10.1007/s00300-010-0791-2.

St Pierre, K.A., Zolkos, S., Shakil, S., Tank, S.E., St Louis, V.L., Kokelj, S.V., 2018. Unprecedented increases in total and methyl mercury concentrations downstream of retrogressive thaw slumps in the western Canadian arctic. Environmental Science and Technology 52 (24), 14099–14109. https://doi.org/10.1021/acs.est.8b05348.

Staalesen, A., 2019. Invasive Arctic crab species in Norway are expanding to new shores. Arctic Today, 14 November 2019. https://www.arctictoday.com/invasive-arctic-crab-species-in-norway-are-expanding-to-new-shores/.

Steflen, W., Andreae, M.O., Bolin, B., Cox, P.M., Crutzen, P.J., et al., 2004. Abrupt changes: the Achilles' heel of the Earth System. Environment 46, 9–20. https://doi.org/10.1080/00139150409604375.

Steflen, W., Crutzen, P.J., McNeill, J.R., 2007. The Anthropocene: Are humans now overwhelming the great forces of Nature? Ambio 36, 614–621. https://doi.org/10.1579/0044-7447.

Sundet, J.H., Hoel, A.H., 2016. The Norwegian management of an introduced species: the Arctic red king crab fishery. Marine Policy 72, 278–284. https://doi.org/10.1016/j.marpol.2016.04.041.

Travers-Smith, H.Z., Lantz, T.C., 2020. Leading-edge disequilibrium in alder and spruce populations across the forest–tundra ecotone. Ecosphere 11 (7), e03118. https://doi.org/10.1002/ecs2.3118.

Wackett, A.A., Yoo, K., Olofsson, J., Klaminder, J., 2018. Human-mediated introduction of geoengineering earthworms in the Fennoscandian arctic. Biological Invasions 20 (6), 1377–1386. https://doi.org/10.1007/s10530-017-1642-7.

Ware, C., Berge, J., Jelmert, A., Olsen, S.M., Pellissier, L., Wisz, M., Kriticos, D., Semenov, G., Kwaśniewski, S., Alsos, I.G., 2016. Biological introduction risks from shipping in a warming Arctic. Journal of Applied Ecology 53 (2), 340–349. https://doi.org/10.1111/1365-2664.12566.

Wasowicz, P., Sennikov, A.N., Westergaard, K.B., Spellman, K., Carlson, M., Gillespie, L.J., Saarela, J.M., Seefeldt, S.S., Bennett, B., Bay, C., Ickert-Bond, S., Väre, H., 2020. Non-native vascular flora of the Arctic: taxonomic richness, distribution and pathways. Ambio 49 (3), 693–703. https://doi.org/10.1007/s13280-019-01296-6.

Yakovlev, E., Puchkov, A., 2020. Assessment of current natural and anthropogenic radionuclide activity concentrations in the bottom sediments from the Barents Sea. Marine Pollution Bulletin 160, 111571. https://doi.org/10.1016/j.marpolbul.2020.111571.

FURTHER READING

Baccolo, G., 2020. Did You Know the Surface of Melting Glaciers is One of the Most Radioactive Places on Earth? European Geosciences Union Blogs: Cryosphere Sciences. https://blogs.egu.eu/divisions/cr/2020/05/29/did-you-know-the-surface-of-melting-glaciers-is-one-of-the-most-radioactive-places-on-earth/.

Baccolo, G., Łokas, E., Gaca, P., Massabò, D., Ambrosini, R., Azzoni, R.S., Clason, C., Di Mauro, B., Franzetti, A., Nastasi, M., Prata, M., Prati, P., Previtali, E., Delmonte, B., Maggi, V., 2020. Cryoconite: an efficient accumulator of radioactive fallout in glacial environments. The Cryosphere 14 (2), 657–672. https://doi.org/10.5194/tc-14-657-2020.

Bellard, C., Cassey, P., Blackburn, T.M., 2016. Alien species as a driver of recent extinctions. Biology Letters 12 (3), 20150623. https://doi.org/10.1098/rsbl.2015.0623.

Mogrovejo-Arias, D.C., Brill, F.H.H., Wagner, D., 2020. Potentially pathogenic bacteria isolated from diverse habitats in Spitsbergen, Svalbard. Environmental Earth Sciences 79 (5). https://doi.org/10.1007/s12665-020-8853-4.

Oh, Y., Zhuang, Q., Liu, L., Welp, L.R., Lau, M.C.Y., Onstott, T.C., Medvigy, D., Bruhwiler, L., Dlugokencky, E.J., Hugelius, G., D'Imperio, L., Elberling, B., 2020. Reduced net methane emissions due to microbial methane oxidation in a warmer Arctic. Nature Climate Change 10 (4), 317–321. https://doi.org/10.1038/s41558-020-0734-z.

Zheng, J., Xu, X., Jia, G., Wu, W., 2020. Understanding the spring phenology of Arctic tundra using multiple satellite data products and ground observations. Science China Earth Sciences 63 (10), 1599–1612. https://doi.org/10.1007/s11430-019-9644-8.

CHAPTER 13

Impacts of Oil and Mineral Extraction

INTRODUCTION

Once again in this chapter, we revisit the tension between preserving the Arctic and developing its resources for financial gain, national power and prestige, and geopolitical place marking. As we will see, some governments are far more interested in the latter than in the former, for a variety of reasons. Other countries are striving to find a balance between conservation and development. The world is in a transitional phase between ways of thinking about the Arctic. For centuries, people from midlatitudes have exploited natural resources in the Arctic. After exhausting the resource, they have often moved on to another place, leaving behind the remains of their installations and accommodations. For centuries, outsiders considered the polar regions as supply areas for raw materials (Hacquebord and Avango, 2009). This old paradigm was established with some of the early visits of Europeans to the Arctic. While the Norse sought to develop a new homeland in southern Greenland, British and other explorers were sent to the Arctic strictly for financial gain, whether to find a less expensive route to the Orient via the Northwest Passage or to exploit the natural resources of the Arctic and adjacent territories, such as walrus tusks, narwhal teeth, mineral wealth, and so on. These explorers had no concern for the environment, other than finding the best ways to exploit it.

The new paradigm has been taking shape in the past 50-or-so years and gaining increased traction in the 21st century. Because of various books, research articles, movies, and television programs (especially the Arctic episodes of David Attenborough's BBC many nature series) revealing the natural beauty and fragility of the Arctic and its peoples, the people of the temperate and tropical zones started taking more notice of the Arctic and its problems.

In the past, public opinion showed little or no interest in the indigenous peoples of the Arctic, much less in the livelihoods they obtained from their environment. Much of this was due to sheer ignorance about Arctic natives. This problem is remarkably well exemplified by the abortive attempt by the US government to set off a hydrogen bomb in Alaska.

Project Chariot

In 1957, the US Atomic Energy Commission developed a project to use hydrogen bombs to create a new harbor in Arctic Alaska. Some government leaders thought it would be a good thing to develop peacetime uses of nuclear weapons. They chose Cape Thompson, a locality in northwestern Alaska. The story of this project is told in "The Firecrackers Boys" by O'Neil (1995). Without launching any inquiries, the AEC considered this region to be uninhabited, and it was well away from population centers, so the resulting radiation would not poison the population. As it turned out, quite a few Inupiat Eskimo families did live and hunt at Cape Thompson. They had no need of a new harbor—or any harbor, for that matter. They could launch their kayaks and umiaks (larger skin boats) from just about any shore, and they pulled these lightweight boats out of the water on their return to land. Also, in the 1950s, the waters along the cape were blocked by sea ice for 8—9 months of the year. Nevertheless, the project got the backing of the newly formed state government in Alaska. As soon as the plans were made public, opposition came from the tiny Inupiat Eskimo village of Point Hope, from the scientists under AEC contract to perform an environmental impact study and from a handful of conservationists.

The environmental studies commissioned by the AEC indicated that the mushroom clouds from the blasts would emit huge numbers of radioactive particles that would land on the native villages, regional lakes and streams, and the regional vegetation. Thus, the Inupiat people would get radiation poisoning from three different sources: first by direct contact with radioactive nuclides in their villages, second from their drinking water, and third from the caribou meat that formed the primary food source for these people. Radioactive vegetation, eaten by caribou, would end up in people's digestive systems. The people of Point Hope wrote to President Kennedy, petitioning him to end the project. The project was finally canceled in 1962, as the AEC faced increased public uneasiness over the environmental risk and potential danger to the Inupiat people. With that sobering tale, I will now launch into the two

Threats to the Arctic. https://doi.org/10.1016/B978-0-12-821555-5.00013-9

biggest resource exploitation threats to the Arctic in the 21st century.

Mining in the Arctic

Because the Arctic was so remote to Europeans for most of history, little or no mining took place there until the 20th century. There were just too many logistic headaches. Even if miners had managed to find valuable minerals, they would have had no way to transport them cheaply back to their home countries. Mineral-laden rocks are very heavy and bulky, and Europeans had no way to make smelters in the Arctic so that they could reduce ores and extract valuable base metals while leaving the slag behind. Even if this had been possible and precious metals could have been transported out of the Arctic by boat, pack ice stood in the way. The Little Ice Age brought pack ice to most Arctic waters, so this physical barrier also prevented people from attempting to exploit the Arctic's natural resources.

Metal use by inuit groups

Long before contact with Europeans, the native peoples of Greenland and Arctic Canada were making good use of iron and copper to make tools. As established by Buchwald (1992), the Inuit of northwest Greenland produced iron objects from minute explosion fragments of the Cape York meteor shower, an iron meteorite-strewn field stretching from Melville Bay to the Thule area. The oldest of these iron objects found thus far date from about AD 800. The iron tools fabricated by these people include knives, ulus, and harpoon blades. The tools clearly had considerable value, as the archaeological evidence shows that they were traded across Smith Sound and at least 2500 km into Canada as far as the northwest coast of Hudson Bay. It appears that the trade south along the Greenland coast was much more restricted, only one knife having been identified at Jakobshavn. A second form of iron was also used by prehistoric Inuit from western Greenland. Telluric (native) iron deposits occur in the Disko Bay area, and tools made from this source of iron constitute about 50% of all examined objects from regional archaeological sites. Meteoritic iron and telluric iron were cold hammered into objects of suitable size and shape. Sometimes a hole was drilled to accommodate a rivet. The tools were probably finished by grinding. There is no evidence that iron forging or heat treatment was carried out.

Throughout Arctic Canada, metal in one form or another appears to have been highly prized, long before European contact. Morrison (1987) attempted to set the record straight about Inuit use of copper to make tools in Arctic Canada. The extensive use of native copper is one of the most distinctive characteristics of historic Inuit groups living in the Coronation Gulf area. It had been previously thought that the copper artifacts found in regional archaeological sites in the central Canadian Arctic must have been made in historic times, either by Europeans or by Natives who copied European technology. However, Native use of copper for tools such as adze blades, arrowheads, and knife blades was noted by the earliest European visitors, so Native use of copper clearly predates European contact. Several copper sources were apparently used, including deposits in the Bathurst Inlet area, in the hills behind Prince Albert Sound, and, most importantly, in the lower Coppermine Valley. Native copper was widely available in what has become known as the Copper Inuit territory and seems to have been used extensively by most or all local groups. Prehistoric tools made almost exclusively of copper from this region include ulu blades, scraper blades, sewing needles, knife blades, grooving tool blades, adze blades, fishhooks, fish gaff prongs, ice chisels, and lance heads. Further research has shown that these copper tools date back to the Thule culture, circa AD 1300. Copper tools are suitable for most purposes. Copper's malleability and toughness were sufficient to ensure a high curation value. In other words, pieces of used cooper tools were reworked into new tools on a regular basis.

Coal mining on Svalbard

As reviewed by Hacquebord and Avango (2009), a coal mining industry developed on Spitsbergen at the beginning of the 20th century, an industry that is still in business today (Fig. 13.1). Svalbard is rich in coal recourses, and the main human activity during the first half of the 20th century was coal mining by Norway and Russia. Three phases can be distinguished in the development of this mining industry. During the initial phase, a great number of exploration and prospecting expeditions investigated the coastal zones of west Spitsbergen, searching first and foremost for coal, but also other minerals (notably gold). The earliest expeditions were organized by companies established by Norwegian ship's captains, but soon bona fide mineral prospectors entered the scene. These prospectors had previous experience in mineral exploration and mining. They were supported by investors who had both capital and political support from their respective governments, including Norway, Russia, the United States, and Great Britain. These companies laid claim to easily accessible coal seams coming to the surface close to the coast. Some prospectors worked out of temporary stations: their tent camps were used for only a few days. These camps left no other traces than rock piles resulting

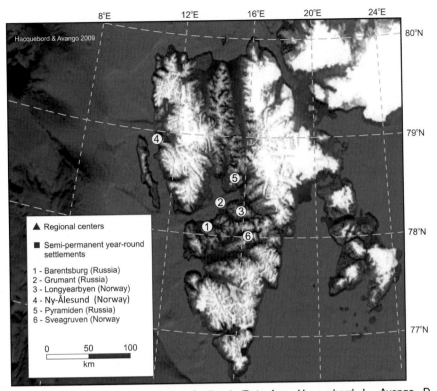

FIG. 13.1 Map of coal mining centers on Svalbard. (Data from Hacquebord, L., Avango, D., 2009. Settlements in an arctic resource frontier region. Arctic Anthropology 46(1—2), 25—39. https://doi.org/10. 1353/arc.0.0028.)

from geological investigations and thus had a very little environmental impact. Permanent seasonal camps were also established: prefabricated wooden buildings shipped from Europe. These camps were set up at sites with favorable conditions for small-scale mining: a short distance between the coal seam and the shore, reasonable harbor conditions (deep water, shelter from the winds), flat dry land for buildings, tools, and coal storage, as well as access to freshwater. One of these was the prospecting camp of Kulgrubekompagniet Isefjord on Spitsbergen, established in 1900.

By 1905, new companies established permanent year-round mining settlements where underground mining was carried out year-round. The first permanent year-round settlement was called "Advent city," established in 1905 by the British Spitsbergen Coal and Trading Company. This enterprise lasted only 3 years before labor unrest forced its shutdown. The second permanent mining settlement was Longyear city, established by the American Arctic Coal Co. in 1906 at the modern site of Longyearbyen, the regional capital of Svalbard. The second phase of the Spitsbergen coal

rush halted temporarily during World War I. By the end of the 1920s, only two nations were involved in the industry: Norway and Russia.

Norwegian-sponsored mining. During the 20th century, Norway operated mines at Longyearbyen, Lunckefjell, Sveagruvan, and Ny Ålesund (Figs. 13.1 and 13.2). The Ny-Ålesund site was the northernmost human settlement in the world. Of the Norwegian-operated mines, it was the first to close, following an accident in 1965. As reported by Williams (2019), following a downturn in coal prices, the Norwegian state-owned enterprise SNSK decided to suspend operations at the Lunckefjell and Sveagruvan mines in 2015. The latter mine had been the largest coal facility on Svalbard, with estimated reserves of over 30 million metric tons. At the time, the new colliery was still in the process of entering full commission after its official opening in February 2014. It had been expected to produce 2 million tons per year for at least 7 years. SNSK has now concluded the closure of the two sites permanently and has initiated a

Hayward Company coal stockpile, Advent Bay, 1914 Photographer unknown.

Abandoned coalmine, Longyearbyen, Svalbard   photo by Hylgeriak/Wikipedia,
Creative Commons Attribution-Share Alike 3.0 Unported, 2.5 Generic license.

FIG. 13.2 Photos of old coal works in Svalbard. Above: Hayward Company coal stockpile, Advent Bay, 1914, photographer unknown. Below: Abandoned coal mine, Longyearbyen, photo by ericpaints.it. (Courtesy ericpaints.it@rinthe, Abandoned coalmine, Longyearbyen, July 5, 2018, Svalbard.)

decommissioning project to clean up the areas following more than 100 years of industrial activity. Only the relatively small Mine No. 7 at Longyearbyen is being retained, to supply a local coal-fired power station and to export for metallurgical use in Europe.

Russian-sponsored mining

The Soviet Union operated mines at Grumant City, Pyramiden, and Barentsburg (after 1932). Barentsburg started as a Dutch mining town in the 1920s. In 1932, the Dutch sold their concession to the Soviet Union. Since then, the Russian state-owned Arktikugol Trust

has been operating on Svalbard, and the main economic activity in Barentsburg is coal mining by Arktikugol. Over its history, Arktikugol has mined more than 22 million metric tons of coal in Svalbard. By the beginning of the 21st century, played-out coal seams caused mining to become uneconomical at the Pyramiden and Grumant City mines. Closure of Pyramiden took place in 1998, and all activities were moved to Barentsburg. However, the Barentsburg infrastructure was dilapidated, subsidies by the Russian government were rapidly climbing, and more than 17% of man-hours were being spent on accident resolutions. A fire started in one of the coal seams in January 2006, which eventually was extinguished by Norwegian authorities. Current coal production at the Barentsburg mine is about 100,000 tons per year. The maintenance of this mine by the Russian government appears to be more politically motivated than economic, as it allows the Russian government to maintain a foothold on Svalbard.

Pollution From Svalbard Coal Mining

The environmental impact of coal mining on Svalbard has become increasingly intense over the past 100 years. The mining pollutants include metals/metalloids, persistent organic pollutants (POPs), and acidic wastewater. The metal contaminants include copper, cadmium, lead, arsenic, selenium, and mercury (Cu, Cd, Pb, As, Se, Hg). After these pollutants have entered local ecosystems, they tend to accumulate in terrestrial, aquatic, and marine organisms; their concentrations become biomagnified in higher trophic levels, increasing their toxicity. Until quite recently, studies concerning potential toxic element (PTE) contamination at the Ny-Ålesund site have focused on surface soil or marine sediments, with little research on the history and status of PTEs contamination over the past 100 years. Yang et al. (2020) collected a sediment profile at the site and tested the samples for the presence and concentrations of six typical PTEs (Cu, Pb, Cd, Hg, As, Se) in the sediments (Fig. 13.3). They assessed the historical pollution status in Ny-Ålesund using the pollution load index (PLI), geoaccumulation index, and enrichment factor (EF). The PLI geoaccumulation index (Igeo) and EF have been widely used to assess environmental pollution status. The PLI is the geometric mean of the concentration factor (CF) of each heavy metal in a soil or sediment sample. The CF is the ratio of the heavy metal concentration of the sediment to its geochemical background. The EF is the comparison between the heavy metal content in the sediments and its content in the geochemical background.

The results showed that the contents of PTEs increased rapidly during the past 100 years. Pb, Cd, and Hg showed a clear signal of enrichment and were the main polluters among the PTEs analyzed. These pollutants are still contaminating the local environment, even though their source—the coal mine—was closed more than 50 years ago.

Effects of Svalbard Mining Pollution on Wildlife

As mentioned earlier, mining at Ny-Ålesund ceased operations in 1963 following a major accident the previous year. The Norwegian authorities cleaned up the contamination from the incident in and around the village but did not attempt to decontaminate the mine area itself. Mercury contamination persists in the

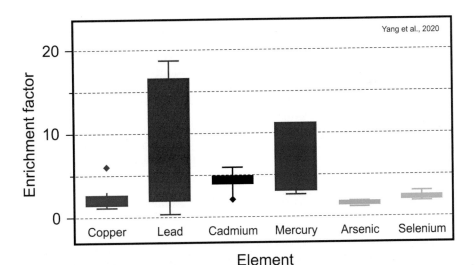

FIG. 13.3 Enrichment factors of heavy metals recovered from the upper unit of the sediment profile near the abandoned coal mine at Ny-Ålesund, Svalbard. (Modified from Yang, Z., Yuan, L., Xie, Z., Wang, J., Li, Z., Tu, L., Sun, L., 2020. Historical records and contamination assessment of potential toxic elements (PTEs) over the past 100 years in Ny-Ålesund, Svalbard. Environmental Pollution 266. https://doi.org/10.1016/j.envpol.2020.115205.)

vicinity, and Scheiber et al. (2018) studied the effects of this and other trace metals on the physiology and behavior of young barnacle geese (*Branta leucopsis*). The geese were from a breeding population in Kongsfjord, Svalbard. A group of goslings was taken from nests in unpolluted areas, to graze the vegetation in the area surrounding the abandoned coal mine. Here they were exposed to trace metals. A control group of goslings grazed on uncontaminated grounds. Both groups of geese were given three stress tests: group isolation, individual isolation, and on-back restraint. Stress-related behaviors in this species include vigilance such as watching for predators, escape actions such as movement or pecking, and group cohesion behaviors such as vocalizations and reestablishing spatial proximity. Such behaviors may be beneficial in the short term, but if they are persistent, they may waste metabolic energy or call the attention of a potential predator.

Pollutants can target various parts of the endocrine system in vertebrates, including the hypothalamic-pituitary-adrenal (HPA) stress axis. The HPA axis is an important hormonal response system to stress, triggering the release of stress hormones known as glucocorticoids and primarily cortisol. The actions of this hormone system are normally well regulated to ensure that once the body has responded to stressful events, the system can return to a normal state just as rapidly (Stephens and Wand, 2012). The effects of contaminants on physiological systems may therefore translate into behavioral changes and reduced fitness.

The goslings in the control group were calmer and excreted lower levels of corticosterone metabolites. The study showed that trace metal contamination from the mine decisively affected the stress physiology and behavior in this species of goose. In a complementary study, the authors demonstrated that goose grazing in the former mining area resulted in higher levels of mercury in their livers and concentration-related variations in dopamine (D2) receptors in the brains of barnacle goslings.

Pollution From Swedish Copper Mining

Fischer et al. (2020) evaluated the disproportionate water quality impacts from the century-old Nautanen copper mines in northern Sweden. The small-scale copper mines at Nautanen were only in operation for 6 years when they were abandoned in 1908. Research on mine waste degradation and heavy metal contaminant leaching from very old mines are lacking from the Arctic, in contrast to the relatively well-documented impacts of historic mines in the temperate regions. However, the Arctic is likely to be mined more extensively in the near future. The authors determined

that there were about 10,000 tons (5500 m³) of unconfined tailings and up to 10,000 tons (5000 m³) of metallurgical slag products in the Nautanen industrial area as of 2017. While in operation, the Nautanen mines produced about 68,000 tons of waste rock, most of which were dumped outside the mine entrances. Despite the small spatial scale of the Nautanen mining site, Fischer et al. (2020) found that the average concentrations of copper, zinc, and cadmium on-site (990, 280, and 1.0 µg/L, respectively) and downstream of the mining site (150, 50, and 0.2 µg/L, respectively) were generally considerably above local (7.2, 3.0, and 0.010 µg/L, respectively) and regional (1.6, 11, and 0.10 µg/L, respectively) background values (Fig. 13.4). In particular, downstream copper concentrations were consistently high throughout the surveyed 25-year period (1993−2017). The study shows that small, abandoned mining sites, which are numerous in the Arctic, could add disproportionately large amounts of metals to the surface water systems. Such effects need to be accounted for in assessments of total pollutant pressures in the relatively sensitive Arctic environment.

Copper and Nickel Mining on the Kola Peninsula

As reviewed by Hønneland and Jørgensen (2018), the Pechenganickel Combine operates four small mines and nickel and copper smelters in the northern parts of the Kola Peninsula, close to the Norwegian border (Fig. 13.5). Mineral production in the region began in the early 1940s when it was a part of Finland. At the end of World War II, the Petsamo (Pechenga) district was annexed by the Soviet Union, and mining and smelting operations were resumed at Nikel in 1946. The owner of the smelters, Norilsk Nikel, is the world's leading producer of nickel and palladium, and the company's sales constitute approximately US $3 billion annually. Pechenganickel sends its smelter output to the Severonickel Combine at Monchegorsk in the central Kola Peninsula. The combine has both smelting and refining facilities for processing nickel and copper. Production started in 1939, and since the late 1960s, the combine has processed ore from both Pechenganickel and Norilsk. Today, Severonickel operates Russia's largest capacity nickel refinery.

It has been shown that acidification and metal enrichment of waters are linked with sulfur dioxide and metal emissions from copper-nickel smelters, contaminating regional catchments. As discussed by Moiseenko et al. (2015), acidification and soil and water pollution in the Kola region of northwest Russia and the neighboring Fenno-Scandinavian countries (Finland, Sweden, and Denmark) have been caused by emissions from two large

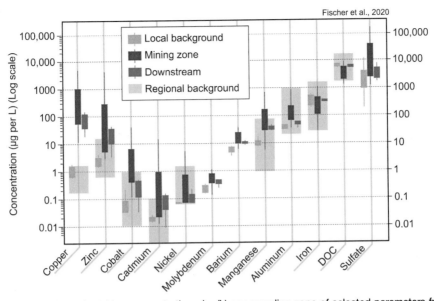

FIG. 13.4 Box plots of total log concentrations (µg/L) per sampling zone of selected parameters from the Nautanen copper mines site, measured in 2017 The plots show the results from the local background, from the mining site, and downstream from the mines. Vertical bars represent 1.5 times the interquartile length. Light green boxes indicate regional background ranges. (Modified from Fischer, S., Rosqvist, G., Chalov, S., Jarsjö, J., 2020. Disproportionate water quality impacts from the century-old nautanen copper mines, Northern Sweden. Sustainability (Switzerland) 12(4). https://doi.org/10.3390/su12041394.)

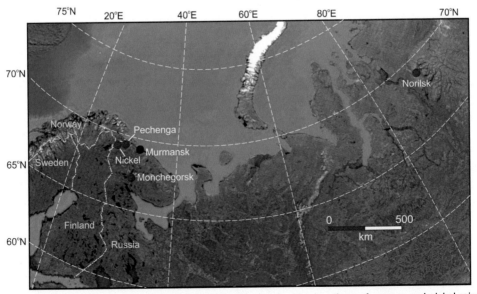

FIG. 13.5 Map of Scandinavia and northeast Siberia, showing the locations of copper and nickel mines discussed in the text.

copper-nickel smelting plants: Pechenganikel and Severonikel (Fig. 13.5). These production facilities are situated in the western part of the Kola region and have emitted SO_2 and heavy metals for more than 60 years. The emissions have resulted in high concentrations of anthropogenic SO_4 and heavy metals in surface waters, leading to acidification of streams and lakes (Norseth, 1994). Specifically, the regional surface waters are enriched in nickel, cadmium, arsenic, antimony, and selenium because of emissions from the copper-nickel smelters. Sulfur dioxide emissions produce acid rain, which causes water acidification and more intensive leaching of metals into the lake systems.

Between 1970 and 1990, annual emissions of dust from the smelters were as high as 64,000 tons per year, including up to 3000 tons per year of nickel and 2000 tons per year of copper during the 1980s. In addition to the metal pollution, SO_2 emissions comprised about 500,000 tons per year from 1987 through the 1990s. Sulfur deposition from 1980 to 1990 was more than 3 g per m^2 per year in the areas immediately downwind of the smelters and less than 0.3 g per m^2 per year further east on the Kola Peninsula.

To assess the effects of acid deposition, critical loads and their exceeds were calculated. The critical loads were exceeded by 56% in the lakes of the Kola region (Moiseenko et al., 2020). Water EFs by chemical element showed that the waters of the Kola region are enriched with Ni, Cd, As, Sb, and Se because of emissions from the copper-nickel smelters (Fig. 13.6). The high values of EF and the high levels of water acidity

are explained by the local and transboundary pollution impacts on the catchment of small lakes.

However, as reported by Moiseenko et al. (2015), the air and water quality of the Kola Peninsula is now starting to improve (Fig. 13.7). Reduced emissions from Cu-Ni smelting plants have led to improved water quality. During the 21st century, emissions of SO_2 have declined by 33%, emissions of copper have declined by 40%, and emissions of nickel have declined by 36%. Moiseenko et al. (2015) sampled the water quality of 75 lakes on the Kola Peninsula in 5-year intervals from 1990 to 2010. They tested for major anions and cations, dissolved organic carbon (DOC), and heavy metals. They found that the concentrations of Ni and Cu in regional lakes have decreased 5- to 10-fold over the past 20 years. Since 1990, the acid-neutralizing capacity (ANC) of the lakes has increased. This is due to the reduction of strong acids in water (sulfate, chloride). They found increased concentrations of DOC and nutrients in Kola lake waters over 20 years. This phenomenon has been explained by the authors through two mechanisms: a reduction in strong acid deposition and climate warming.

One final update on efforts to limit the pollution from the Kola Peninsula copper and nickel smelters was published by Nornickel (2019). On December 23, 2020, the Norilsk Nickel company announced that it is shutting down the smelting facilities at Nikel in Russia's Murmansk region. It is the company's oldest production facility. The shutdown forms part of Nornickel's program to reduce the environmental impact of their Kola Peninsula operations. With the closure of

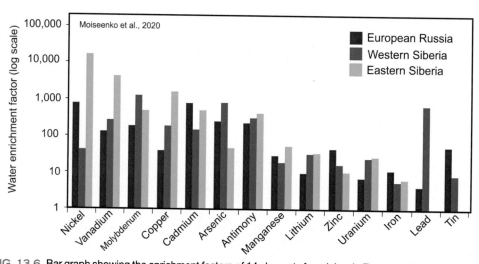

FIG. 13.6 Bar graph showing the enrichment factors of 14 elements from lakes in European Russia (*red bars*), Western Siberia (*blue bars*), and Eastern Siberia (*green bars*). (Modified from Moiseenko, T., Gashkina, N., Dinu, M., Kremleva, T., Khoroshavin, V., 2020. Water chemistry of arctic lakes under airborne contamination of watersheds. Water (Switzerland) 12(6). https://doi.org/10.3390/w12061659.)

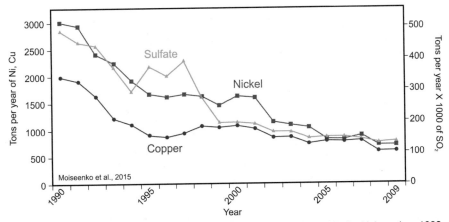

FIG. 13.7 Emissions of nickel, copper, and sulfate from the Ni-Cu smelters in the Kola region, 1990—2009. (Modified from Moiseenko, T., Dinu, M., Bazova, M., De Wit, H., 2015. Long-term changes in the water chemistry of arctic lakes as a response to reduction of air pollution: case study in the Kola, Russia. Water, Air, and Soil Pollution 226(4). https://doi.org/10.1007/s11270-015-2310-0.)

the Nikel facility, Norilsk Nickel claims that the hazardous atmospheric emissions at the Russian-Norwegian border will cease.

The Norilsk Mining Complex, Central Siberia

As reported by Kirdyanov et al. (2020), the northern regions of Siberia have become contaminated by large concentrations of anthropogenic air pollutants since the mid-20th century. Located about 300 km north of the Arctic Circle in central Siberia (69° 21′N, 88° 11′E), Norilsk is the world's northernmost city with over 100,000 inhabitants. Since the onset of Russia's heavy industrialization in the 1930s, pollution from the Norilsk mining complex has destroyed vast areas of pristine taiga and tundra habitats on the Taimyr Peninsula. Refined nickel production started at Norilsk in 1942, and mining activities have rapidly expanded since then. In 2018, the Norilsk complex released about 1,805,000 tons of pollutants, of which almost 98% was sulfur dioxide (SO_2). The surrounding region of ca. 24,000 km² is the most polluted in the world. Kirdyanov et al. (2020) combined evidence from dendroecology, biogeochemistry, and numerical modeling to assess the direct and indirect effects of this industrial pollution on the functioning and productivity of regional ecosystems at various spatial and temporal scales. They found that within this zone, nearly all trees are dead or dying, coincident with large-scale increases in atmospheric sulfur dioxide as well as copper, and nickel particulates. Norilsk's unrestricted emissions of highly noxious substances are not only extremely harmful to humans, but also flora and fauna.

Trofimov et al. (2020) likewise performed a study on air pollution from the Norilsk mining complex, with a focus on the transport of anthropogenic aerosols in cloud formations. They analyzed cloud responses to large-scale aerosol perturbations. The areal extent of the polluted cloud areas detected in MODIS satellite images extended to tens of thousands of km². Most of the clouds they analyzed originated from the aerosol pollution hot spot of the Norilsk factories (Fig. 13.8). Aerosols have been shown to exert an atmospheric cooling effect, but the extent of this effect remains poorly known, especially the changes in cloud properties. Their analysis of satellite data broke new ground in atmospheric chemistry, as they examined very large-scale polluted cloud areas. The clouds emanating from the Norilsk region show close ties between pollution aerosols and cloud moisture levels. The data from these large-scale polluted clouds indicate that the effect of aerosols on cloud droplet size has a stronger impact on Earth's climate than the effect caused by the cloud water changes per se.

Water pollution from the Norilsk site has likewise been extensive. Just like the nickel and copper smelters on the Kola Peninsula, the Norilsk complex has been the source of extensive water pollution from sulfur dioxide and toxic metals. Moiseenko et al. (2020) report that nickel, copper, cadmium, and other metal particulates are emitted from the Norilsk smelters and end up being deposited in regional watersheds. Prevailing winds can spread pollutant emissions from the Norilsk industrial complex as far west as 500—600 km, to the Gydan Peninsula.

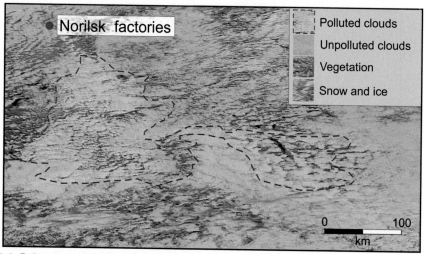

FIG. 13.8 Polluted clouds in the Norilsk region, as compiled from MODIS near-infrared composite satellite images. Image taken on 27 June 2000. (Modified from Trofimov, H., Bellouin, N., Toll, V., 2020. Large-scale industrial cloud perturbations confirm bidirectional cloud water responses to anthropogenic aerosols. Journal of Geophysical Research: Atmospheres, 125(14). https://doi.org/10.1029/2020JD032575.)

The most egregious water pollution catastrophe associated with the industrial complex took place during the summer of 2020 (Fig. 13.9). More than 150,000 barrels of diesel oil from the Norilsk mining complex were accidentally discharged into a nearby river in June of that year. This represents the largest oil spill ever recorded in the Arctic. Diesel oil is used as a backup fuel for the Norilsk-Taimyr Energy coal-fired heating and power plant. One of the fuel storage tanks failed due to corrosion of the bottom. The company had been ordered by the Russian regulatory agency for natural resources to remove rust from these tanks and to restore an anticorrosion coating, as well as to conduct inspections of the tank bottoms. These orders were to have been completed by 2015, but Norilsk-Taimyr Energy failed to comply. To make matters worse, the power plant operators did not report the incident for 2 days, while trying to contain the situation on their own. It is also thought that the storage tank sank because of melting permafrost, which weakened its supports. The spill contaminated an area of 350 km², according to a Russian state media report. Within 10 days of the spill, the oil had traveled by river about 20 km north of Norilsk. Initially, about 21,000 metric tons of diesel oil contaminated the nearby Ambarnaya River and surrounding subsoil. From the river, the oil flowed into Lake Pyasino, according to the governor of the Krasnoyarsk region (BBC News, 2020).

According to the Russian news agency TASS, 2020, cleanup efforts for the diesel spill were anticipated to be difficult as there are no roads in this region, and the rivers are too shallow for boats and barges. It was estimated that the immediate cost of emergency cleanup activities would be 10 billion rubles (US $146 million), with a total cleanup cost of 100 billion rubles (US $1.5 billion). The entire remediation project was estimated to take 5−10 years, with the Norilsk Nickel Company to pay the costs.

As of October 2020, the Norilsk Company reported that it had completed three main cleanup stages. They reported that the spill was fully localized and that most of the fuel and water mixture had been collected. The company pumped more than 25,000 m³ of this oil-water mixture through newly installed pipelines from temporary tanks to a storage facility for subsequent separation. The Russian Ministry of Industry and Trade has been supporting the cleanup operation since the first days of the accident (Nornickel, 2019).

Shishikin et al. (2014) studied the condition of the vertebrate fauna in the Norilsk industrial complex pollution impact zone. The research team carried out long-term faunal studies from this region. They found that proximity to the industrial site was actually linked with an increase in biomass of small herbivorous mammals, as well as birds of prey. On the face of it, this result appears counterintuitive, but the research team found a straightforward explanation. The emissions from the industrial complex poison the previously dominant moss-lichen vegetation cover of the region. The moribund moss-lichen cover has given way to various species of

Six days before the oil spill

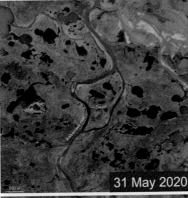

Two days after the oil spill

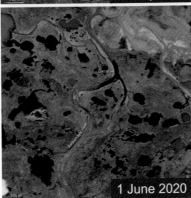

Three days after the oil spill

FIG. 13.9 Satellite imagery of the diesel fuel spill near the Norilsk nickel mine, Krasnoyarsk region, Siberia. Top: Photo of the Ambarnaya River before the spill. Middle: Photo of the Ambarnaya River 2 days after the spill. The red color in the water is the spilled diesel fuel. Bottom: Photo of the Ambarnaya River 3 days after the spill, showing the downstream spread of the diesel fuel. Copernicus Sentinel-2 Satellite imagery. Courtesy of the European Space Agency. (Data from the European Space Agency, 2020. Satellite Imagery of the Diesel Fuel Spill Near the Norilsk Nickel Mine. 2020. https://www.esa.int/ESA_Multimedia/Images/2020/06/Arctic_Circle_oil_spill.)

grasses and other herbs that provide more nutritional value for small herbivores, such as voles and hares. The internal effects of toxic substances notwithstanding these animals benefit at least in the short term. As the herbivore population thrives on the grass-rich diet, the raptors benefit from more prey animals.

The Verkhne-Munskoye Diamond Field

Novoselov et al. (2020) discussed the Verkhne-Munskoye diamond field, located on the Muna River, a tributary of the Lena River (Fig. 13.10). The diamond field is in the Olenek Evenk National District of Yakutia. In 2011, a section of the Olenek National District had its protection as a nature reserve withdrawn by a special Russian federal order to allow exploration and production of diamonds. The Verkhne-Munskoye deposit was commissioned in October 2018, yielding gem-quality diamonds that November. Fig. 13.10 shows the Diamond province of the Republic of Sakha (Yakutia) with the location of the considered diamond mining fields. The area of economic activity is 791 ha for open-pit mining of four diamond pipes. Overburden stripping works are being carried out using explosives. The Olenek Evenk National District otherwise belongs to a category of lands set aside for traditional uses by Native peoples.

Surveys of indigenous peoples in Yakutia show that they attribute the deterioration of regional water quality to diamond mining along the tributaries of the river. In particular, mining activities cause the previously frozen banks of the Anabar river to collapse, and hydraulic mining is causing river levels to drop. Development of alluvial diamond mining, for example, in the Ebelyakh river basin, a tributary of the Anabar, is causing substantial hydrological changes. The mining company is extracting diamonds from the river channels, even near Native settlements. This has led to the contamination of drinking water. The impacts of diamond mining in the Olenek Evenk National District will include a change in indigenous livelihoods, changes in the landscape, alteration and disruption of river channels, waste generation, as well as a decrease in the resource productivity of traditional lands. This, in turn, will negatively affect reindeer husbandry, hunting, fishing, and the gathering of wild plants for food and medicine.

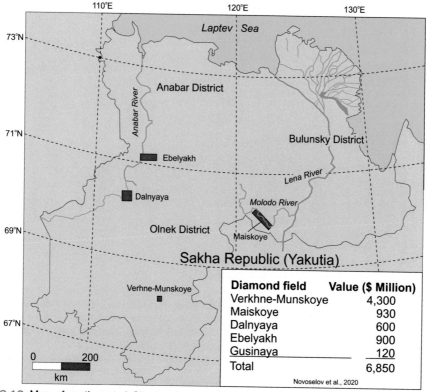

Diamond field	Value ($ Million)
Verkhne-Munskoye	4,300
Maiskoye	930
Dalnyaya	600
Ebelyakh	900
Gusinaya	120
Total	6,850

Novoselov et al., 2020

FIG. 13.10 Map of north-central Siberia, showing the locations of diamond mining localities (in *red*). (Modified from Novoselov, A., Potravny, I., Novoselova, I., Gassiy, V., 2020. Sustainable development of the arctic indigenous communities: the approach to projects optimization of mining company. Sustainability (Switzerland) 12(19), 1–18. https://doi.org/10.3390/su12197963.)

Nevertheless, the mining giant Alrosa company envisages the implementation of several further diamond mining projects in the region.

The Ekati Diamond Mine in Arctic Canada

St-Gelais et al. (2018) assessed diamond mine impacts on aquatic ecosystems in the Northwest Territories, Canada, which is the first Canadian diamond mine and has been in operation since 1998. The mine is estimated to contain 105.4 million carats of diamond reserves. Mining activities during the initial years of development were focused on six surface mines, which had produced 67.8 million carats as of January 2017. The expected life span of the Ekati mines runs to 2034, with a possible extension to 2042 (Mining Technology, 2020).

Unlike the diamond mining underway at Verkhne-Munskoye diamond field in Siberia, the Ekati venture benefitted from extensive, rigorous environmental impact studies, both before and during site mining activities. One of these was the study of aquatic ecosystems by St-Gelais et al. (2018). Their study was conducted in the Koala watershed, one of the main watersheds influenced by the Ekati Diamond Mine. As a component of the Ekati Diamond Mine Aquatic Effects Monitoring Program, phytoplankton and zooplankton communities along with multiple water quality parameters and physical variables in Koala watershed lakes were sampled annually over 19 years, including 2 years before mining operations began. They used these data to study how temporal changes in water quality affected downstream plankton communities. Diamond mining activities usually affect aquatic organisms through changes in water quality, as several concentrated compounds are released from a mine into the nearby environment. These releases may increase trace element concentrations, such as copper and aluminum, which have toxic effects on organisms. The study by St-Gelais et al. (2018) found that the diamond mining activities caused a shift in phytoplankton functional composition of both diatom and rotifer communities. They found little change in the composition of crustacean zooplankton. Thus, the research validated the efforts of the mining company to maintain regional water quality.

Black Angel Zinc Mine, Greenland

Thomassen (2003) described the operations and history of the Black Angel lead-zinc mine at Maarmorilik, West Greenland (Fig. 13.11). The site lies at 71°N on the west coast of Greenland. The mine was primarily developed to exploit zinc resources but also yielded ores of iron, lead, and silver. The first excavations in Maarmorilik took place in the 1930s, with operations continuing until 1945, and again from 1973 to 1990 when the mine was closed. During 17 years of operation, the mine produced 12 million metric tons of zinc ore. The total net earnings of the operation from 1973 to 1996 were 1154 million Danish Kroner ($US 190 million).

Because of the extremely remote location (the nearest permanent settlement is 775 km south of the mine), the mine operators simply discarded the mine tailings through a pipeline to the bottom of the nearby Affarlikassaa fjord. All told, the mining operations left behind about 8 million tons of tailings contaminated with heavy metals including iron, copper, arsenic, silver, zinc, cadmium, mercury, and lead. After a few years of mining, some of these elements, especially dissolved lead, zinc, and cadmium, were found in increasing amounts in the fjord water. Comprehensive monitoring of heavy metals in the marine biota showed that lead pollution in the coastal areas near Maarmorilik was the most serious problem. As a result, the authorities requested the mine operators (Greenex) to take steps to reduce the heavy metals in the tailings and terrestrial waste dumps. The tailings were discharged at a depth of 30 m through a pipeline, from where they settled on the bottom of the Affarlikassaa fjord at 50–60 m. Compliance with the government ordered cleanup from 1978 to 1985 saw the lead content of the tailings fall from 0.44% to 0.15%. The zinc content of the tailings fell from 1.10% to 0.23%, and the cadmium content fell from 57 to 14 ppm. Four waste dumps totaling 2–3 million tons had been disposed of outside the mine in its first 10 years of operation. The waste dump materials contain 0.1%–0.8% lead and 0.3%–2.3% zinc. One of these dumps of c. 400,000 tons reached into the tidal zone, and the highest lead and zinc values in marine biota (seaweed and mussels) occurred in this area. After mine was closed, this dump was removed. Part of the waste was placed on the bottom of the Affarlikassaa fjord and the other part in a concentrate storage area. Before abandonment, an environmental agreement was in place to ensure the ongoing monitoring of the fjords around Maarmorilik by the National Environmental Research Institute (NERI). Thus, since the closure, the environment around the site has been monitored annually by analyzing for lead and zinc in seawater, lichens, seaweed, and marine animals. As anticipated, these studies show a marked decline in heavy metal concentrations compared with earlier.

Hansson et al. (2020) performed a heavy metal pollution study more than two decades after the mine

FIG. 13.11 Photo of the Black Angel Mining town seen from the mine portal, summer 1990. Photo courtesy of the Government of Greenland. (Photo from Thomassen, B., 2003. The Black Angel Lead-Zinc Mine at Maarmorilik in West Greenland. Bureau of Minerals and Petroleum, Government of Greenland, vol. 12. Bureau of Minerals and Petroleum, Government of Greenland, p. 6.)

was closed, in which they compared metal concentrations accumulated in blood, liver, muscle, and otoliths from two Arctic predatory fish species, Greenland cod (*Gadus ogac*) and shorthorn sculpin (*Myoxocephalus scorpius*). The fish specimens were collected along a distance gradient near the former mine. They found that the sediment samples collected within 1 km of the former mine are still highly contaminated with both Pb and Zn classified as bad (Category IV—toxic effects following short-term exposure) to very bad (Category V—severe acute toxic effects), thereby posing a considerable risk to the surrounding environment. Within the same region, however, they found that sediment concentrations of cadmium and mercury were within safe levels.

The authors considered that metal accumulations in their fish samples from Maarmorilik, as well as the accumulations in their dietary prey, are more influenced by contaminants dissolved in the ambient seawater than those contained in the sediment. This is indicated by their metal concentration data from blood and otolith samples that showed the same trends as the concentrations found in seawater. Furthermore, even though all metals could be detected in blood, and Pb and Zn could be detected in otoliths, a significant decrease between stations was only observed for Pb, which was the only contaminant showing a clear gradual decrease in seawater. Other studies that used fish blood to evaluate lead exposure have found similar results. This finding may explain why the cod specimens they examined

had lower concentrations than sculpin in both blood and otoliths collected from the most contaminated area. Cod generally swim and feed closer to the top of the water column. Because of this, they are less prone to exposure (either directly or via their diet) to the highly contaminated sediments of the seafloor. Here, lead concentrations of >2000 µg per gram have been measured. Sculpin, on the other hand, swim and hunt their prey near the seafloor. They are thus more prone to exposure to contaminants contained in the sediment.

PETROLEUM EXTRACTION IN THE ARCTIC

The petroleum industry is currently the largest economic enterprise in the Arctic. Today, the Arctic produces about 10% of the world's oil and about 25% of its natural gas. Between 2016 and 2019, the banks of the world invested over $US16 billion in Arctic petroleum exploration and extraction (Kirsch et al., 2019).

Oil and gas are produced in four of the states that possess land or continental shelf territories in the Arctic. These are Russia, Alaska (US), Canada, and Norway. A total of 61 large oil and gas fields have been discovered inside the Arctic regions of these four countries (Jørgensen-Dahl, 2020). Of the 61 large fields, 42 are in Russia, most of which (35) are in the West Siberian Basin. Of the 18 large fields outside Russia, 6 are in Alaska, 11 are in Canada's Northwest Territories, and 1 is in Norway. Of these 18 fields, 15 have not yet gone into production. These include 11 in Canada, 2 in Russia, and 2 in Alaska. Russia is by far the world's leader in natural gas resources. Russian reserves comprise about 80% of the total reserves in the Arctic and 25% of the world's total reserves. Arctic petroleum assets represent an estimated 22% of all recoverable oil and gas resources in the world. More than 80% of these resources are marine, although relatively close to shore. Untapped petroleum resources lie between the shoreline and the 500 m contour of the continental shelves, within 200 nautical mile limits of the respective countries (Jørgensen-Dahl, 2020). Undeveloped natural gas is about three times more abundant than oil in the Arctic.

History of Arctic Petroleum Development

Drilling for oil and natural gas in the Arctic dates to the 1930s in Russia and the 1940s in Alaska. Spencer et al. (2011) outlined the history of Arctic petroleum exploration and extraction. The first exploratory drilling in the Arctic was government-sponsored and carried out for strategic purposes, first in the 1930s on the south shore of the Laptev Sea in what was then the USSR and then in what was then the US Territory of Alaska in 1944. The first significant petroleum discoveries in onshore Arctic basins were made from 1960 to 1970. These discoveries included the Eagle Plain, Sverdrup Basin and Mackenzie Delta regions of Canada, the Timan-Pechora region of West Siberia, and the North Slope of Alaska. The finds in West Siberia and Timan-Pechora extended existing oil fields northward, but the 1968 discovery of the Prudhoe Bay field in Alaska opened a major new petroleum province. Offshore drilling started in the Beaufort Sea in 1973, in West Greenland in 1976, and in the Barents Sea in 1980. Russian offshore drilling started in the Pechora Sea in 1982 and in the Kara Sea in 1987. The discoveries in the Beaufort, Pechora, and Kara seas extended proven petroleum provinces northward, while the Barents Sea petroleum provinces were new discoveries. These included the Norwegian sector in the southwest Barents Sea (1981) and the supergiant Shtokman gas field in the Russian sector (1988). There are now nine main Arctic petroleum provinces containing a total of 444 discoveries. These regions, combined, contain estimated recoverable resources of 14.3 trillion liters (3.8 trillion US gallons) of oil and 47,232 trillion liters (1669 trillion cubic feet) of gas. Four provinces dominate the oil and gas resources: West Siberia—South Kara, Arctic Alaska, Barents Sea East, and Timan-Pechora (Fig. 13.12).

Offshore boundary disputes

As discussed by Gulas et al. (2017), global interest in the exploitation of Arctic petroleum resources was driven during the late 20th and early 21st centuries by relatively high global energy prices and declining Arctic sea-ice cover. However, oil company interest in Arctic offshore drilling has always been tied very closely to global energy prices. Recent low oil prices combined with Western sanction policies against Russia's Arctic oil and gas industry, the new Conference of Parties 21 (COP21) low carbon economy climate agenda, and logistical and technical challenges of Arctic oil and gas operations have curbed interest in the Arctic.

The United Nations Convention on the Law of the Sea (UNCLOS), established in 1984, grants certain areas of the Arctic seabed to the five circumpolar nations (Canada, the United States, Russia, Norway, and Denmark). The Convention specifies exclusive economic zones (EEZs) that shall not extend beyond 200 nautical miles (nm) from the shore. Upon ratification of the UNCLOS, Arctic coastal countries were given 10 years to make scientifically proven claims to an extended continental shelf which, if validated, might provide exclusive rights to resources on or below the seabed of that extended shelf area. Norway, Russia, Canada, and Denmark launched projects to provide a basis for seabed claims on extended continental shelves beyond their EEZ. For example, the Lomonosov Ridge extending 2000 km across the Arctic Ocean from Russia to Canada is currently claimed by Russia as an extension of the Asian continental shelf, while both Canada and Denmark claim it is an extension of the North American continental shelf. Although Arctic coastal states have made different claims as to the outer delimitation of their continental shelves, sovereignty issues are regulated under UNCLOS rules. The Russian government, in cooperation with national oil and gas developers, has ambitious plans to develop oil and gas fields on the undisputed continental shelf regions to the north and east of Siberia (Fig. 13.13) (Kontorovich et al., 2019). The oil and gas complex forms a significant part of the budget of the Russian Federation, ranging from 35% to 52%. D.A. Medvedev, Chairman of the Government of the Russian Federation, recently noted that oil companies pay an average of two-thirds of their revenue to the budget.

Oil Spill Preparedness

For oil spills in Arctic regions, the breakdown of petroleum compounds from weathering will necessarily be

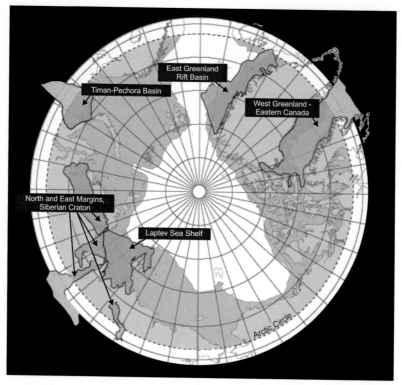

Spencer et al., 2011

FIG. 13.12 Discovered and yet-to-find petroleum resources of the Arctic, shown in green. (After Spencer, A., Embry, A., Gautier, D., Stoupakova, A., Sørensen, K., 2011. Chapter 1: an overview of the petroleum geology of the Arctic. Geological Society Memoirs 35, 1–15. https://doi.org/10.1144/M35.1.)

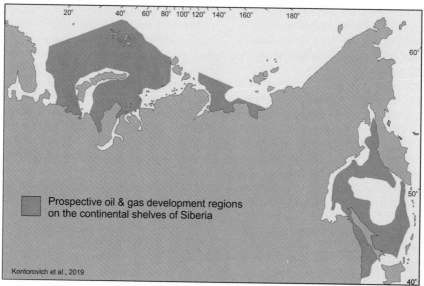

FIG. 13.13 Map of northern Asia, showing continental shelf regions of Arctic Russia with potential for oil and gas development. (Modified from Kontorovich, A., Burshtein, L., Livshits, V., Ryzhkova, S., 2019. Main directions of development of the oil complex of Russia in the first half of the twenty-first century. Herald of the Russian Academy of Sciences 89(6), 558–566. https://doi.org/10.1134/S101933161906008X.)

TABLE 13.1
Strengths and Weaknesses of the Major Oil Spill Remote Detection Methods.

Platform	Sensors Used	Strengths	Weaknesses
Airborne remote sensing	Side-looking airborne radar, synthetic aperture radar, infrared cameras and sensors	Airborne remote sensing most effective method for identifying presence of oil on water	Nonradar sensors blocked by darkness, cloud, fog, precipitation
Satellite radar systems	Synthetic aperture radar	Ability to document changing ice conditions in vicinity of spills; works in darkness, foul weather; can search very large areas	Not yet tested for major Arctic oil spills
Surface systems	Ground penetrating radar; hand-held infrared (IR) systems	Can detect presence of thin oil films under and within ice layers; low-cost, hand-held IR systems detect oil under certain conditions	GPR response may confound oil with other materials; successful detection requires careful interpretation
Integrated systems	SLAR, visual-light cameras, IR/UV scanners, GPS; data sent to geographic information system (GIS)	Can map location of images on electronic chart system. Aircraft platforms can access multiple sensors, feed data directly to GIS	Cannot search as large regions as satellite radar systems

Data from Potter, S., Buist, I., Trudel, K., 2012. Spill response in the arctic offshore. In: Scholz, D. (Ed.), The American Petroleum Institute and the Joint Industry Programme on Oil Spill Recovery in Ice. The American Petroleum Institute.

slower. A summary of Arctic studies and considerations regarding experimental spills and oil fates can be found in a report from the American Petroleum Institute (Potter et al., 2012) (Table 13.1).

With seasonal sea-ice present, mixing and breakup effects largely cease when the oil sits in the ice, and this affects weathering and degradation. The persistence of oil components for sea-ice spills is therefore likely to be greater for the Arctic than for open water spills in warmer climates. Gulas et al. (2017) have reviewed the state of preparedness for offshore oil and gas operations in the countries bordering the Arctic Ocean.

Specific risks to offshore oil and gas operations in the Arctic were assessed by Necci et al. (2019). They retrieved qualitative and quantitative data from the World Offshore Accident Database (WOAD) covering the period 1970–2013 to identify the riskiest operations of Arctic offshore activities in the harsh environmental conditions to better understand the associated risks (Fig. 13.14). Extreme operating conditions may disrupt the offshore infrastructure and cause major accidents, posing the threat of oil spills. Offshore operations in the Arctic regions need to manage difficulties that arise from extremely low temperatures. In low-temperature regimes, precipitation can be abundant and in the form of snow, freezing rain, sleet, or

ice pellets. Visibility can be very limited, because of fog, winter darkness, or snowstorms. In harsh environments, severe storms with high winds and rough seas occur throughout the year. Harsh environments encompass a variety of atmospheric and marine phenomena, such as strong winds, high waves, and low temperatures, icebergs, and icing, which, by themselves or combined, exert significant stresses on the offshore infrastructure, which can lead to incidents.

Canada

Offshore oil and gas exploration in the Beaufort Sea shelf regions of northwestern Canada includes an extensive and complex regulatory framework, overseen by both federal and territorial governments. Canada's regulatory frameworks for offshore operations are comprehensive. The government acknowledges that a major oil spill in the Canadian Arctic would be highly challenging to tackle, because of the region's distance from population centers and an inability by the few small Native villages to support a large influx of response personnel. Furthermore, the often-ice-covered waters of the Beaufort Sea would greatly complicate Canadian Coast Guard ship-based operations, although this situation is changing as lengthier open water seasons have been occurring in Canada's Arctic waters.

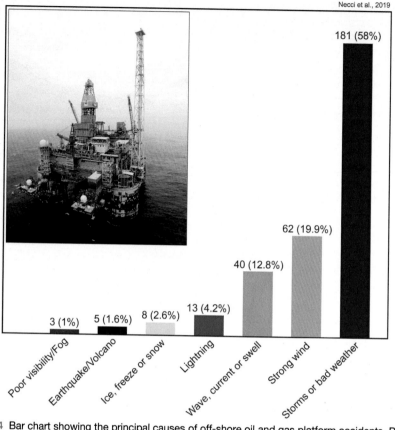

Necci et al., 2019

FIG. 13.14 Bar chart showing the principal causes of off-shore oil and gas platform accidents. Photo of oil rig. Courtesy of Wikipedia, Creative Commons Attribution 2.0 Generic license. (Modified from Necci, A., Tarantola, S., Vamanu, B., Krausmann, E., Ponte, L., 2019. Lessons learned from offshore oil and gas incidents in the Arctic and other ice-prone seas. Ocean Engineering 185, 12–26. https://doi.org/10.1016/j.oceaneng.2019.05.021.)

So, despite huge investments in oil and gas exploration in the Canadian Arctic, as yet there has been no significant commercial production.

The United States

Recently, the US Bureau of Ocean Energy and Management and the Bureau of Safety and Environmental Enforcement issued the proposed regulations to ensure safe and responsible exploratory drilling for offshore Alaska. These regulations ensure rigorous standards of equipment design and the conduct of exploration programs suitable for Arctic conditions. These regulations also certify that proponents should develop and implement an Oil Spill Response Plan designed for Arctic conditions and equipment, training, and personnel for oil spill response in the Arctic. However, the US Coast Guard (responsible for planning, preparedness, and supervision of oil spill response) only has three operational polar icebreakers, so the United States remains ill-prepared to respond to major spills in Arctic Alaskan waters.

Russian Federation

Oil and gas exploitation in the Russian Arctic has unfortunately been the cause of widespread pollution problems. Regulation of oil and gas in Russia is nominally administered by the Ministry of Natural Resources and Ecology, but multiple agencies have responsibility for the oil-rich regions of Siberia. Despite Russian federal legislation recognizing the need for accident response policies, there are no requirements for an oil spill emergency response plan. The federal policy entitled "Fundamentals of State Policy of the Russian Federation in the Arctic in the Period up to 2020 and

Beyond" mainly emphasizes Russia's intention to increase economic development in the Arctic, particularly oil and gas. The policy makes no specific mention of improvements to oil spill prevention methods or postspill cleanup operations. In the Western Russian Arctic, where oil exploitation developments continue, the regional emergency and rescue authority has but nine rescue vessels, many of which are typically out of commission at any given time. This lack of preparedness has brought strong criticism by environmental groups in Russia and abroad.

Norway

The decline in North Sea oil output has turned Norwegians' attention toward the development of Arctic resources. Norway is internationally acknowledged as a leader in pollution abatement and accident preparedness. Norwegian policy promotes an environmentally focused resource management approach between governments, petroleum corporations, and Arctic residents. The Petroleum Act provides legal governance of all Norwegian petroleum activities, including the Arctic. Norway's oil spill response equipment depots include 6 oil recovery vessels, 5 emergency towing vessels, surveillance aircraft, and 10 coast guard vessels with oil booms, skimmers, and pumping systems. These are managed by the Norwegian Coastal Administration and private companies.

Denmark

Denmark's exploitation and management of oil and gas resources inside the Arctic Circle are regulated by the Greenland Mineral Resources Act (MRA). The Mineral License and Safety Authority (MLSA) has published specific guidelines for oil and gas approval conditions that are based on Norwegian standards. Under the MLSA, a combination of emergency committees from Greenland and Denmark is jointly responsible for offshore accidents. Approvals for drilling licenses are issued under strict safety requirements in accordance with the MRA. Environmental impact assessments are completed by licensees for proposed oil exploitation.

To summarize, Norway, Denmark, and Greenland are the best prepared for oil spills in their territorial waters. Canada faces challenges because its offshore oil deposits are so far from population centers. Interestingly, Canadian government rules and regulations appear to be so constraining that no petroleum companies have attempted to drill for oil off the Arctic Canadian coast. The United States has a good set of regulations concerning oil spills, but preparedness for accidents is extremely underfunded, to the point that any offshore oil spills North of the Alaskan coast would likely be catastrophic because of lack of containment and remediation. Finally, the Russian government appears to be much more committed to reaping financial gains from the oil industry than it is to preparedness for oil spills.

As Gulas et al. (2017) have pointed out, any major oil spill in Arctic waters would likely have extensive, long-term environmental impacts. The logistics of response to an Arctic oil spill are difficult at best. Even if a rapid response team were to arrive quickly at an oil spill, few effective technologies currently exist to contain and clean up oil spills in Arctic sea ice, regardless of the season. The development of technologies and adequate response techniques, particularly for oil under ice and broken ice, should be of high priority.

An Ice-Free Northeast Passage and Natural Gas Exports

As discussed in Chapter 8, between the 1980s and 2010s, Arctic warming caused the retreat of coastal sea ice, expanding the average number of ice-free days nearshore along the Northeast Passage rose from 84 to nearly 150. A Chinese ship made the first nonstop voyage through the passage in 2015. Since then, vessels have made thousands of trips, including a record of 2700 journeys in 2019. On November 9, 2020, the passage iced over for the winter after remaining open for a record 112 days, but not before hosting another record. The ice-free season extended sufficiently long to allow an all-time monthly high of Russian exports of liquefied natural gas (LNG) through the passage to ports in Asia.

McDonnell (2020) examined the ramifications of an ice-free Northeast Passage for the cheap exportation of natural gas to Asian countries to the south. Natural gas is in increasing demand as worldwide electric generation moves away from coal-fired plants. That is especially true in Asia, and China is the biggest potential customer. The country's recently announced climate goals include a doubling in gas consumption by 2035.

Most of China's gas currently travels by pipeline from other Asian countries and southern Russia. But as discussed earlier, the Russian Arctic has the largest natural gas resources in the world, so if this region continues to become more accessible through the Northeast Passage, LNG would be relatively cheap to export to China, and Chinese gas purchases would serve as a significant boost to the Russian economy. LNG is natural gas that has been liquefied by cooling to a temperature of $-162°C$, for shipping and storage. The volume of LNG is about 600 times smaller than its volume in its gaseous state in a natural gas pipeline. Moreover, the opening of the Northeast Passage gives Russian natural gas producers more flexibility than operating pipelines. Shipping of LNG would allow

Russia to supply multiple customers in different ports—in particular, forging cheaper transit between vast gas reserves in western Russia and ports in China, presenting an opportunity to beat its competitors to Asian markets.

Russia's Yamal LNG Terminal opened in 2017 and nearly tripled the country's LNG export capacity. The terminal is near the vast natural gas reserves on the Yamal Peninsula. In a 2018 paper, German energy economists called an ice-free Northeast Passage a "game-changer" for Russia, because the Yamal Terminal would allow the country to directly challenge US ambitions in the Asian gas market. This prediction appears to have been accurate. In September 2020, the Yamal terminal shipped a record 700,000 tons of LNG to Asia.

In addition to the commencement of oil and LNG shipments to southerly ports in Asia, the Russians also are shipping oil to Norway. This oil transportation did not begin in earnest until 2002, when there was a dramatic increase in oil shipment as 5 million metric tons were delivered westward through the Barents Sea (Bambulyak et al., 2015). In 2004, some 12 million metric tons of petroleum cargoes were shipped to Norway. In 2010, petroleum flow peaked at 15 million metric tons. The Snøhvit gas field and Melkøya LNG in the Norwegian part of the Barents Sea came online in 2007. From 2010 to 2014, between 4 and 5 million metric tons of liquid gas cargoes were offloaded at Melkøya each year.

Role of Arctic Marine Microbes in Oil Spill Remediation

Vergeynst et al. (2018) considered the potential role of microorganisms in degrading and reducing the environmental impacts of oil spilled in Arctic waters, particularly in seawaters off Greenland. They pointed out that the self-cleaning capacity and biodegradation potential of natural Arctic microbial communities have yet to be discovered. There are several key elements in the microbial breakdown of crude oil. First, the presence of hydrocarbon-degrading bacteria is essential. This process also requires the availability of nutrients in the water. However, even if these requirements are met, the breakdown of ocean-borne crude oil in Arctic waters is not a straightforward proposition. For instance, the process is greatly influenced by oceanographic conditions, the low temperature of surface waters, presence or absence of sea ice, sunlight regime, the concentration of suspended sediment plumes, and presence or absence of phytoplankton blooms. The effects of several of these aspects of Arctic seawater are as yet poorly understood but have the potential to prevent or disrupt the microbial breakdown of

hydrocarbons. These include the oligotrophic (nutrient-poor) environment, poor microbial adaptation to hydrocarbon degradation, mixing of stratified water masses, and massive phytoplankton blooms and suspended sediment plumes merit to be topics of future investigation.

Oil Spills From Drilling Sites and Pipelines

The impact of the oil and gas complex on natural systems can be traced in a number of areas: direct extraction of resources (oil, gas, gas condensate, surface, and artesian waters, nonmetallic building materials, timber, etc.); use of resources (water resources, land resources, flora and fauna resources, fish resources); entry of pollution into the environment (emissions into the atmosphere, discharges into water bodies and onto the relief, injection of wastewater into underground horizons, waste, noise, and vibration, electromagnetic radiation, thermal and light pollution, etc.) and change of relief and landscape. Large sources of air pollution are oil and gas companies that flare associated gas. During the development of oil fields, about 10 billion m^3 of associated petroleum gas (APG) is produced, while the annual volume of flared gas is about 5 billion m^3.

Oil production in Arctic Eurasia focuses on western Siberia, where 68% of Russian gas and oil extraction take place. The biggest of these are in the Ob River basin, where 300 oil fields have pumped 65 billion barrels of oil since 1965, four times the quantity of oil pumped through the Trans-Alaska pipeline. The Russian government estimates that at least 5% of the oil recovered from western Siberian wells has been spilled on the tundra surface since the 1960s. This amounts to 3 billion barrels of oil spilled in the Ob River and adjacent basins. A report by Salmina (2010), a regional ecologist, states that there is no fish life in the Ob, Nadym, Pur, and Sob rivers because of pollution from oil production. The pollution is not limited just to the oil itself. Salts and heavy metals are in the brine that comes to the surface with oil from wells. These by-products are dumped on the land. The Russian oil industry tacitly acknowledged the level of water pollution in oil field regions when it installed water purification plants on-site, as the local water is unsafe to drink. The Yamal-Nenets Autonomous District, through which the Ob River flows, accounts for 90% of natural gas production in Russia, or 20% in the world. At the end of 2014, 234 hydrocarbon deposits were discovered in the region, including seven new fields added after geological exploration works during 2005—10 (Bambulyak et al., 2015). As shown in Fig. 13.15, more than 100 fields are in commercial production.

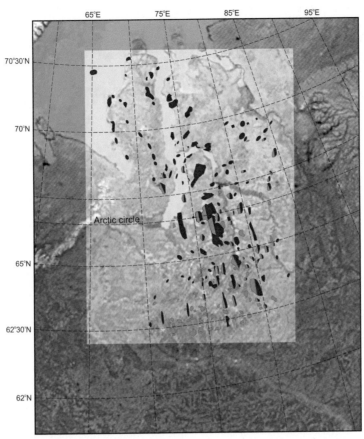

FIG. 13.15 Map of the oil fields (green) and gas fields (red) in the Yamal-Nenets Autonomous District in the north of Western Siberia. (Data from Bambulyak, A., Frantzen, B., Rautio, R., 2015. Oil Transport from Russian Part of the Barents Region 2015 Status Report.)

The Middle Ob River runs through the central part of Western Siberia in Khanty-Mansi Autonomous Okrug and brings its waters through the territory of Yamal-Nenets Autonomous Okrug to the Gulf of Ob of the Arctic Ocean. Rusak et al. (2019) took water samples and bottom sediment samples at 21 control points during 2003—08 and 26 control points during 2014—18 on a 180 km stretch of the river during various hydrological seasons. They compared the levels of pollutants in the Middle Ob collected in 2003—08 versus those collected in 2014—18. They tested water and river sediment samples for the concentration of salts, organic compounds, heavy metals, petroleum hydrocarbons, and biogenic ions (Table 13.2).

They identified the main pollutants, which are phenols, hydrocarbons, iron, manganese, chrome, copper, and phosphate ions, most of which come from waste products and spills from the petroleum industry. Unfortunately, the level of phenol pollution tripled by 2014—18.

The main industrial pollutants of regional rivers are oil hydrocarbons, phenols, and anionic surfactants. These come from oilfields and pipelines, because of normal operations but especially from accidents. The data obtained on the concentration of oil hydrocarbons shows a high level of oil pollution. Data samples of water collected in 2003—08 exceeded the maximum permissible concentration (MPC) of oil hydrocarbons up to three times, and subsequent water tests in 2014—28 revealed oil pollution levels that exceeded the MPC up to 16 times. Phenols occur in several petroleum fractions and crude oil. Phenol concentrations in water samples from 2003 to 08 were up to four times the MPC. Phenol levels were dramatically higher in samples from 2014 to 18, with concentrations as much as 27 times the MPC. Hydrocarbon concentrations exceeded the maximum permissible level established by the regional government (the Khanty-Mansi Autonomous Okrug) in river sediments (20 mg/kg)

TABLE 13.2
Chemical Composition of Water Samples Collected From the Middle Ob River.

| Compound/Element | MPC[b] (mg/dm³) | CONCENTRATIONS | | | |
| | | 2003–08 | | 2014–18 | |
		Average	Exceeds MPC	Average	Exceeds MPC
BOD[a]	3	2.47	Below MPC	3.51	1.17×
Ammonium	0.5	1.35	2.7×	0.41	Below MPC
Nitrites	0.02	0.11	5.5×	N/A	N/A
Nitrates	0.5	1.35	2.7×	0.41	Below MPC
Phosphates	0.2	0.47	2.35×	0.24	1.2
Oil hydrocarbons	0.05	0.048	Below MPC	0.09	1.8
Phenols	0.001	0.00068	Below MPC	0.0021	2.1×
Iron	0.1	2.8	28×	2.72	27×
Zinc	0.01	0.045	4.5×	0.0081	Below MPC
Manganese	0.01	0.0864	8.64	0.125	125×
Copper	0.001	0.005	5×	0.0024	2.4×

[a] Biochemical oxygen demand.
[b] Maximum permissible concentration.
Data from Rusak, S. N., Shornikova, E. A., Kurilenko, M. I., Homenushko, T. I., 2019. Transboundary aspects of river water pollution and water quality estimation of the Middle Ob river. IOP Conference Series: Earth and Environmental Science (p. 012018). https://doi.org/10.1088/1755-1315/400/1/012018.

and in surface water samples by factors of 2.7 and 4, respectively. This confirms the fact that hydrocarbon pollutants accumulate in river bottom sediments (Rusak et al., 2019).

Heavy Metal Pollution

Vinogradova and Kotova (2019) examined the sources of heavy metal (HM) pollution in Russia's northern seas. They found anthropogenic heavy metals (Pb, Cd, As, Zn, Ni, Cr, and Cu) on the surface of the four Russian Arctic seas and matched the HM concentrations with values obtained from the mouths of rivers entering the northern seas. They found that the concentrations of Pb, Cd, Zn, Ni, and Cu from river sources exceed their atmospheric fluxes by more than 10 times in the White Sea (source: Northern Dvina River), the Pechora Sea (source: Pechora River), the Kara Sea (source: Ob and Yenisei rivers), and the Laptev Sea (source: Lena River). These HMs come from the effluent of regional petroleum facilities.

WILDLIFE HAZARDS FROM PETROLEUM OPERATIONS

Arctic wildlife, both marine and terrestrial, face a number of hazards from the petroleum industry. These can

be straightforward threats, such as direct contamination of animals and plants from oil spills. But threats may also be more subtle, such as disturbances to marine mammals and fish from the shock waves and noise associated with piledriving to create offshore drilling platforms.

Oil Threats to Marine Organisms

According to Nevalainen et al. (2019), the types of oil that may be spilled in the ocean differ in their likely fate and their relative lethality. These factors, in turn, depend on several processes that may vary considerably in space and time. The authors described four oil groups that represent the potential fates and environmental consequences of the different oils that may be shipped in the Arctic. There are some general trends. Specifically, the lighter the oil, the better it mixes with and dissolves in water. In contrast, the heavier the oil, the more it coats the shoreline or sinks to the seafloor. Oil dispersed over a wide area can be highly toxic. The heavier, more solid oil causes physical harm to biota. Extra-heavy oils are assumed to sink to the seafloor after being spilled. Here they form thick, sticky layers that can remain on the seabed for a particularly long time. Heavy oils float as thick, sticky slicks that adhere to the shore and stick to sea ice; this grade of oil may also end up under the pack ice. Medium oils float as thin slicks. These may also adhere to shorelines and ice but to a lesser extent than heavy oils since medium oils disperse faster and evaporate more easily. Medium oil may also finish up under the sea ice. Light oils spread rapidly to very thin slicks that evaporate relatively quickly. They are not likely to adhere to either the shoreline or to ice, though they may end up under the ice.

Aune et al. (2018) examined the seasonal ecology of marine life in ice-covered Arctic seas, as a means of informed decision-making for oil spill responses. One fundamental ecological factor for the biota of Arctic seas is that upward of 53% of the primary production in Arctic water masses is transported toward the bottom. This is roughly twice the percentage of primary production that is transported to the bottom of the North Atlantic, suggesting that benthic processes play a much larger role in Arctic waters than pelagic processes. Furthermore, the extent of sea ice cover directly affects primary production rates. As such, interannual and long-term variation in ice and water mass conditions are environmental drivers of marine species distributions and ecosystem functioning.

Marine microbe response to oil spills

Boccadoro et al. (2018) examined marine microbial community responses and migration of petroleum compounds during a sea-ice oil spill experiment in Svalbard. They used experimental chambers filled with seawater, to which they added various dispersants. Hydrocarbon-degrading microbes were detected in ice cores, and the analysis of metabolically active bacterial populations in the different layers of sea-ice indicates significant population shifts following oil exposure, whether dispersant addition or oil burning was carried out or not. The presence of dispersant in the system was associated with the most pronounced and fastest population shifts out of all exposures, as well as lower bacterial diversity as measured by the Shannon index. Microorganisms were metabolically most active in the bottom layer of the sea ice and our data confirmed the predominance of *Oleispira* and *Colwellia aestuarii* in sea ice. Migration of polycyclic aromatics through the sea-ice layer was observed when dispersant was added to the oil, and the presence of oil-degrading organisms below the ice layer was consistent with biodegradation taking place. Given the thickness and concentration of the oil layer frozen into the ice, the bioavailability of the hydrocarbons was nevertheless limited. Consequently, much of the bulk of the oil remained intact.

The sea-ice oil exposure experiments conducted by Boccadoro et al. (2018) in Svalbard showed that much of the oil frozen into the ice in February remained largely unaltered through the rest of the winter. This study found that two species of bacteria: *Oleispira antarctica* and *Oleispira gap-e-97* serve as important microbes in the breakdown of hydrocarbon compounds in Arctic seawater. There is good evidence that these PAHs were being actively degraded by the bacteria. To test the hypothesis of active oil biodegradation in sea ice and seawater, short-term incubation experiments were performed. These experiments showed that the bacteria present in sea ice can degrade petroleum hydrocarbon components within weeks. Furthermore, the residues left from burning the oil can trigger an increase of sea-ice bacterial diversity. The addition of chemical dispersants clearly has the effect of reducing bacterial diversity. The study found that the bacterium *Colwellia rossensis*, first described from specimens in Antarctic waters, increased in numbers over time in the underice environment in the experimental chamber containing oil premixed with dispersant. This species was present in the control experimental chambers, but it was nearly absent from the chambers containing oil-exposed sea ice.

The addition of dispersant enhanced PAH migration through sea ice and stimulated bacterial response through increased bioavailability. However, it also lowered bacterial diversity. In theory, the greater the microbial diversity, the higher the potential for breaking down crude oil components, since each taxon has a characteristic substrate preference and optimum metabolic activity conditions. The dispersant used in this study certainly increased the availability of the oil compounds to be digested by the bacteria, but as Boccadoro et al. (2018) concluded, the question remains: how might the decline in biodiversity influence microbial biodegradation in the environment?

Arctic oil spills will likely break down more slowly than spills in warmer waters, and this process is further impeded by the presence of sea ice. The usual processes of mixing and breakup of oil patches on the water's surface largely cease when the oil sits in the ice, and this affects weathering and degradation. The persistence of oil components for sea-ice spills is therefore likely to be greater for the Arctic than elsewhere. Experimental studies of oil spills frozen into sea ice in Svalbard have concluded that the composition of the oil remains largely unaltered until the ice melts in the spring and the oil is released into the sea. The springtime release of oil into the water will certainly affect local food webs, but the process is bound to vary depending on ice drift, location, oil quality and quantity, weather conditions, and the presence of vulnerable species.

Studies on Arctic sea ice date back to 1966. Brakstad et al. (2008) performed a field study on oil biodegradation in sea ice in a Svalbard fjord in 2004, showing that oil spills cause a significant change in the ice microbial community. Moreover, biodiversity was found to decline with oil contamination. The study indicated the possibility of slow biodegradation of oil in contaminated ice but was not able to show this conclusively. It was inferred by Brakstad et al. (2008) that the concentrated nature of oil spills on sea ice results in very low bioavailability and that this likewise slows the pace of biodegradation.

Brine channels forming in sea ice cause variable salinity and channels in the ice that may mobilize the oil/bacteria system and enable biodegradation. The influence of the conventional oil spill mitigation measures such as skimming, the use of dispersants, and oil burning on surface waters. However, these remediation methods have not previously been tested on sea ice, so little is known about the effects of these measures on microbial activity or bacterial species diversity. However, some studies have demonstrated microbial ability to degrade chemically and physically dispersed oil in Arctic seawater. Oil spill response measures would normally be carried out when the sea is ice-free, either just before or immediately after the melting of the oil encasing sea ice. However, dispersant or burning techniques could be used up to the time of sea-ice formation in the fall, either as a means of dispersing or removing oil just before freezing or to enhance oil breakdown in winter and spring.

Investigation of the effects of such treatments on sea-ice biota is also important for oil spill contingency planning, as effects on the heterogeneity of the bacterial community, as well as biodegradation, could be enhanced or hindered. Understanding the migration of oil within sea ice and potentially the underlying water column and its degradation by naturally present microorganisms at Arctic temperatures represent the two main steps in establishing the fate of oil in sea ice. This will be one of the main scenarios for oil spill contingency planning in areas with ice.

Nevalainen et al. (2019) examined the vulnerability of Arctic biota to oil spills. In their assessments, vulnerability consists of two components: exposure potential and sensitivity (Table 13.3). Exposure potential

TABLE 13.3
Risk Assessment for Arctic Marine Biota Regarding Contact With Oil.

Taxonomic Group	Probability of Oil Exposure	Sensitivity to Oil Exposure
Polar bear	Low	High
Ice-dwelling seals	Low	Very low
Bottom-feeding marine mammals	Moderate	Very low
Other seals	Low	Very low
Toothed whales	Moderate	Very low
Baleen whales	Moderate	Very low
Omnivorous birds	Low	High
Diving fish-feeding birds	Low	High
Surface fish-feeding birds	Moderate	High
Bottom-feeding birds	Low	High
Pelagic fish	Moderate/high	Very low
Sea-ice associated fish	High	Very low
Bottom-feeding fish	Moderate	Very low
Foraging fish	High	Very low
Surface invertebrates	High	Very low
Benthic invertebrates	Moderate	Very low
Ice-associated invertebrates	High	Very low
Water column invertebrates	Moderate/high	Very low

From Nevalainen, M., Vanhatalo, J., Helle, I., 2019. Index-based approach for estimating vulnerability of Arctic biota to oil spills. Ecosphere 10(6). https://doi.org/10.1002/ecs2.2766.

describes the probability of an individual coming into contact with spilled oil. This depends on an organism's habitat use and behavior. Sensitivity describes two aspects of oil spill encounters: (1) the probability of death due to contact with oil, and (2) the probability of a population recovering from the dieback. Sensitivity takes into account both the immediate death from contact with oil, the long-term impacts of oiling, and a population's ability to replace the lost individuals through reproduction and migration. Overall exposure potential consists of 10 variables related to use of habitat, behavior, and offspring exposure potential, whereas overall sensitivity includes 13 variables related to physical and chemical sensitivity, recovery potential, and offspring sensitivity.

Oil threats to plankton

Phytoplankton and zooplankton typically form the base of a marine food web. Studies indicate that lipid-rich species such as krill (*Calanus*) may potentially bioaccumulate oil compounds in their tissues. However, little is known how these groups are affected by oil exposure in the long term, or if such effects propagate through the food chain, although the latter seems highly likely. The long-term impact of oil spills on plankton and the potential cascading effects on higher trophic levels remain to be assessed and quantified. Numerical modeling is one viable approach, as discussed in the following. Many marine species are most sensitive to oil toxicity and oil-related damage during their early life stages. During times of peak productivity, such as the rapid rise in herbivorous zooplankton following the spring algal bloom, large proportions of a given population may potentially be at risk.

As discussed by Aune et al. (2018), it is quite difficult for marine biologists to assess the response of zooplankton such as krill to oil spills. Biologists assume that the zooplankton move toward deeper water masses in response to exposure to oiled surface waters, especially in winter. Experimental exposure studies indicate that lipid-rich species such as *Calanus* bioaccumulate oil compounds. However, little is known about how these groups are affected by oil exposure in the long term, or if such effects propagate through the food chain. The long-term impact on plankton and the potential cascading effects on higher trophic levels would certainly depend on the size of the spill, and this could, for instance, be assessed and quantified using numerical modeling in a future model study. For instance, an oil spill in early spring would have a higher risk of affecting the common Arctic copepod *Calanus glacialis*, which moves toward the surface in February-March and stays

in these upper water layers until August-October. The sea-ice algae bloom begins around mid-March, with a peak just after the sea ice starts to break up. The main peak of the algal bloom occurs less than a month after sea ice breaks up. In *C. glacialis*, the egg production normally lasts for more than 2 months and peaks about the same time as the ice breaks up. Many plankton species complete their major life-history events before sea ice starts to form again in autumn, other species remain active and reproduce all year round.

The fish and invertebrate groups studied by Nevalainen et al. (2019) have similar, medium sensitivity to oil spills, but their exposure potential differs greatly depending on the type of oil. For bottom-feeding (demersal) fish and benthic invertebrates, the exposure potential is highest with extra-heavy oil, since it sinks to the seafloor. However, their exposure potential is significantly lower than that of other fish and invertebrates when exposed to other types of oil. For nondemersal fish and invertebrates, the exposure potential is highest with medium oil. This is because medium oil is likely to disperse down through the water column in greater amounts than other oil types. The differences between oil types are particularly high for fish and invertebrates associated with ice, as the exposure potential of these groups depends strongly on the fate of oil, that is, whether the oil ends up under ice or not.

Agersted et al. (2018) studied the effects of oil contamination on High-Arctic krill (copepod) species *Calanus hyperboreus*. They noted the lack of knowledge on bioaccumulation and toxic effects of oil components for High-Arctic organisms. Arctic marine organisms have unique biochemical and physiological adaptations that may alter their sensitivity to petroleum hydrocarbons. In addition to increased greater bioaccumulation potential due to higher lipid content, many taxa also have lower metabolic rates that contribute to slower uptake of contaminants and delayed toxicological effects, as well as physiological adaptations to extreme cold, such as the presence of blood antifreeze compounds (Bejarano et al., 2017). Therefore, when making environmental risk assessments of Arctic plankton response to oil spills, comparisons with temperate or tropical species may be less than helpful. Because of their higher lipid content, bioaccumulation of lipophilic contaminants (such as crude oil and other hydrocarbon compounds) is much more likely to take place in polar organisms. Crude oil contains several lipophilic components that show a high affinity for accumulation in lipid-rich tissues.

Copepods of the genus *Calanus* are the primary grazers during spring and summer in Arctic marine ecosystems. Krill form a major portion of the diets of numerous fish, seabirds, and baleen whales and have an important role in transferring lipid-based energy from primary production to these vertebrates. Arctic krill have a high lipid content (>50% of their dry weight) and are particularly rich in wax esters, a high energy storage lipid. Toxicity and tolerance level tests have been carried out by Agersted et al. (2018) on *Calanus* specimens, establishing tolerance and lethal concentration levels for some oil compounds and associated chemicals. One group of chemicals showing high toxicity are the polycyclic aromatic hydrocarbons (PAHs). These are a group of substances found in crude oil. Exposure of adult *Calanus* to the PAHs impacts their digestion and reproduction. Several studies have demonstrated that PAHs bioaccumulate in adult *Calanus*. However, specific knowledge concerning the bioaccumulation of different types of crude oil components is still scarce for *C. hyperboreus*. The results of the study by Agersted et al. (2018) show that even short-term oil exposure may result in long-term bioaccumulation and internal damage in these copepods. Because of the processes of elimination and purification of impurities of the contaminant oil components from *C. hyperboreus*, there is considerable risk that these toxic oil compounds will be transferred up the food web to pelagic fish, seabirds, and baleen whales.

Oil threats to fish eggs and larvae

Eggs of several marine fish species are spawned along the coasts of Norway and Russia and then carried northward by ocean currents into the Barents Sea. This includes the eggs and larvae of species such as the Northeast Arctic cod (*Boreogadus saida*) and the Northeast Arctic haddock (*Melanogrammus aeglefinus*) (Nahrgang et al., 2016). The eggs and larvae are largely retained in Atlantic water masses, far from the ice zone, whereas adult individuals may conduct summer feeding migrations further north, mainly in deeper waters where they are less likely to be exposed to oil in the case of an accidental spill. For many fish larvae, the pelagic zone is probably a less exposed and therefore safer habitat, with fewer predators and higher food availability; however, in the case of an oil spill, the larvae will more likely be exposed to oil.

Many Arctic fish species with free-living pelagic larvae are known to aggregate in surface waters and beneath the pack ice. This behavior increases their liability to exposure to crude oil that is likely to be trapped under sea ice if an accident were to occur in winter. Nahrgang et al. (2016) exposed the buoyant eggs of polar cod to realistic concentrations of a crude oil water-soluble fraction, mimicking the exposure of fish eggs under the ice to the water-soluble fraction of oil leaking from the surface to beneath the pack ice via brine channels (Fig. 13.16). The study showed that the proportion of viable, free-swimming larvae decreased significantly with exposure to oil. Specific oil-related injuries included increased incidence and severity of spine curvature, yolk sac alterations, and a reduction in spine length. These effects compromise the motility, feeding capacity, and predator avoidance of polar cod during the critical larval stage. The results suggest that the viability and fitness of polar cod early life stages are significantly reduced when exposed to aqueous hydrocarbons, even at extremely low concentrations.

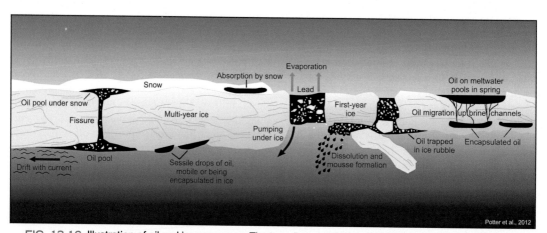

FIG. 13.16 Illustration of oil and ice processes. The term "mousse" represents the formation of water-in-oil emulsions. "Encapsulated" oil refers to bodies of oil that have been trapped inside sea ice formations. (Modified from Potter, S., Buist, I., Trudel, K., 2012. Spill response in the arctic offshore. In: The American Petroleum Institute and the Joint Industry Program on Oil Spill Recovery in Ice. The American Petroleum Institute, (p. 6).)

Oil threats to mollusks and echinoderms

Counihan (2018) examined the physiological effects of oil, dispersant, and dispersed oil on the bay mussel, *Mytilus trossulus*, in Arctic and sub-Arctic conditions. Bay mussels are ideal for coastal monitoring programs because they are sessile organisms that bioaccumulate pollutants through filter-feeding and are ubiquitous around Alaska and other northern cold water coastal environments. The effects of chemical dispersants on Arctic marine organisms are poorly known. These chemicals are used to mitigate the impact of an oil spill by accelerating natural degradations. However, the reduced water-repellent characteristics (hydrophobicity) of oil that has been treated with chemical dispersants likely increase the oil's bioavailability to marine organisms. Counihan (2018) tested the effects of non-dispersed oil, the dispersant Corexit 9500, and oil dispersed with different concentrations of Corexit 9500 on bay mussels in seawater. The tests ran for 3 weeks. Counihan found that most of the physiological responses occurred during the first 7 days of exposure, with mussels exhibiting significant cytochrome P450 activity and stress-response (heat shock) protein levels. Cytochrome P450 activity indicates the organism's efforts to detoxify and facilitate the excretion of foreign chemicals (McDonnell and Dang, 2013). Mussels exposed to nondispersed oil also experienced higher levels of mortality. After 21 days, mussels in all treatments exhibited evidence of genetic damage, tissue loss, and continued stress response.

Oil Threats to Sea Birds

The results of the study by Nevalainen et al. (2019) show that of all the Arctic marine biota, sea birds have the highest vulnerability to oil spills (Table 13.3). They do not have the highest exposure potential, but they are very sensitive to contact with oil, and therefore, their vulnerability is high. Bottom-feeding birds seem to have slightly lower exposure potential when compared with other bird groups, and omnivorous birds have somewhat lower sensitivity than other birds. The relative order of vulnerability of birds depends on an accident scenario, but omnivorous birds seem to always have lower vulnerability than other birds. They have a highly opportunistic diet and may therefore be less likely to get into physical contact with oil while foraging for food and to consume oiled food. Moreover, their breeding colonies are typically located on high cliffs, where the oil is unlikely to reach them. Bottom-feeding birds may spend less time in the water than other kinds of seabirds. For example, eiders may prefer to rest on land. On the other hand, marine birds are used to flying long distances to find food, so the oil in their habitat does not automatically harm them. It has also been suggested that oil on the water surface may prevent seabirds from locating schooling fish, which may further lower their exposure potential. In the study by Nevalainen et al. (2019), the difference between bird groups is explained mostly by differing use of habitat and the fate of their main prey after an oil spill.

As reviewed by Aune et al. (2018), not all sea birds are liable to encounter spilled oil on the sea surface. Some sea birds skim the surface of the water to catch fish or other prey. These make little contact with the water. Others float on the surface and then dive after prey. The divers spend more time on the sea surface and therefore are more susceptible to encounter an oil slick, making them more vulnerable to an oil spill. The authors' review of 45 oil spills from shipping accidents concluded that there is no correlation between the volume of oil spilled and numbers of injured and killed seabirds. Bird population size, density, and geographical distribution affect the likelihood of an oil spill encounter. These factors, in turn, are shaped by seasonal movements, life history traits, and food availability. Likewise, not all oil spills are the same. There are differences in the distribution of oil, such as the amount of oil on the water, the type of oil, air and sea surface temperature, wave height, wind velocity, and ocean currents.

Hazards for Migrating Seabirds

The Barents Sea is home to about 16−20 million birds in summer. The Lancaster Sound region of eastern Canada holds about 1.7 million seabirds, while the guillemot population is estimated to be about 7 million adult breeding birds (mostly Brünnich's guillemots, *Uria lomvia*) in the Eastern Canadian Arctic. Few colonial seabirds breed in the Beaufort Sea region.

As discussed by Lewis and Prince (2018), seasonal variations in populations are important in the Arctic because birds and marine mammals migrate. Fish and invertebrate populations can have very patchy distributions, but with less pronounced seasonal variations. Oil spills have a range of deleterious effects, perhaps most dramatically in the mortality of birds that encounter a floating slick, and this was a major driver in the initial development of oil spill dispersants. In certain seasons, there might be substantial populations of birds and cetaceans in the Arctic that would be at risk from a major spill.

An element of Net Environmental Benefit Analysis (NEBA) that is sometimes misunderstood is nevertheless quite pertinent for dispersant use. This is the aspect of "balancing" the trade-offs of the impacts in different resource categories. On the one hand, dispersing the oil into the water column will protect seabirds by removing the oil from the sea surface. On the other hand, the dispersed oil that moves down through the water column is acutely toxic to marine organisms until it dilutes to the level of a few parts per million (Lewis and Prince, 2018). This is sometimes misrepresented as a "zero-sum game" in which the "gains" of one participant (the birds) are offset by the "losses" of another participant (the organisms in the upper water column).

An assessment of the threats to little auks (*Alle alle*) wintering in the North Atlantic from marine pollution was done by Fort et al. (2013). They showed that most of the global population of little auks, most likely millions of birds, spend their summers in the Arctic and their winters off the Newfoundland coast. Unfortunately, their wintering area coincides with a high level of oil-related risks. With its strategic shipping lanes and several offshore oil production platforms, the waters off Newfoundland are some of the most vulnerable areas in the sub-Arctic. Until recently, the wintering locality of many Arctic species of seabirds remained poorly known. However, the latest studies involve bird-borne technology that allows tracking of individual bird's movements. This, combined with at-sea surveys indicate that several seabird species gather here in huge numbers during winter, including common guillemots (*Uria aalge*), Brünnich's guillemots (*Uria lomvia*), kittiwakes (*Rissa tridactyla*), and little auks. Chronic oil pollution from shipping is having significant impacts on these species, killing thousands of individuals each year. Little auks apparently follow just one annual migration route from the shores of Greenland to Newfoundland, based on data showing only one overwintering hotspot observed for all tracked birds from Greenland. Unfortunately, this relatively concentrated distribution may make the overwintering birds more susceptible to even a single point, small-scale pollution event. Little auks are thus among the species worst affected by oil spills in the North Atlantic (Fig. 13.17).

Crucially, seasonal overlaps between the marine industries (shipping and oil extraction) that are the source of oil pollution and seabird hot spots are bound to increase in the Arctic, as sea ice retreats and human activities expand northward. As discussed in Chapter 8, the reduction in sea-ice cover is even now increasing the number of days that existing Arctic shipping lanes are used, and new routes are being planned. Oil exploration, which has already expanded throughout the North Atlantic sector of the Arctic, will likely result in the building of many new oil platforms. Accordingly, it is more important to understand the large-scale distribution patterns of northern seabirds. Their hot spots were not chosen at random by these birds. Rather, these localities are biologically productive, with abundant food resources (mainly fish) and greater biodiversity. These hot spots need international protection before new human industries overwhelm them.

Oil Threats to Marine Mammals

For marine mammals to be affected by an oil spill, they must be at the oil spill location when the accident happens, or shortly thereafter. In other words, there must be an overlap in both time and space between the presence of the marine mammals and the spreading of the oil spill. In addition to the exposure level, the degree to which specific species are affected by exposure to oil also depends on their population status, local density within the affected area, and their geographical distribution outside of these areas. The distribution of marine mammals is generally driven by the distribution and abundance of their main prey but also depends on the timing and routes of seasonal migration between feeding and breeding grounds. Detailed knowledge of such processes is of crucial importance for the assessment of the ecological consequences in a NEBA process. Not much is known about how whales are affected by oil, but their feeding strategy will largely determine the level of risk of contamination from oil at the surface. Right whales, such as the North Atlantic right whale (*Eubalaena glacialis*) and the bowhead whale (*Balaena mysticetus*), are skim feeders. They often swim at the surface with their mouths open, filtering zooplankton from the water. This feeding pattern obviously makes them more vulnerable to surface oil. On the other hand, baleen whales, such as the humpback whale (*Megaptera novaeangliae*), feed both at the surface and at depth, probably making them moderately vulnerable to drifting oil.

The results of the study by Nevalainen et al. (2019) suggest that toothed whales typically face higher risks to oil exposure than baleen whales. Both whale groups have similar levels of sensitivity, but toothed whales have higher exposure potential especially related to medium or light oil during the breeding season (from spring to autumn). Explanatory factors include toothed whales' larger group sizes and their tendency to gather in estuaries, both of which make them more vulnerable to oil spills. The main differences between the whale groups arise from differing pod size and distributional patterns.

The effects of oil contamination on Arctic seals has received little scientific attention, but the general

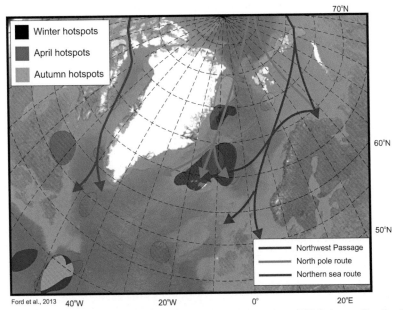

FIG. 13.17 Above: Photo of little auks on Svalbard, 2002. Photo by Michael Haferkamp, Creative Commons Attribution-Share Alike 3.0 Unported license. Below: Map showing the overlap between the nonbreeding distributions of tracked little auks and the development of future human activities in the Arctic, including oil and gas activities and shipping routes. Little auk hot spots are defined by kernel 50% density contours in autumn *(green regions)*, winter *(red regions)*, and spring *(purple kernels)*. (Modified from Fort, J., Moe, B., Strøm, H., Grémillet, D., Welcker, J., Schultner, J., Jerstad, K., Johansen, K., Phillips, R., Mosbech, A., 2013. Multicolony tracking reveals potential threats to little auks wintering in the North Atlantic from marine pollution and shrinking sea ice cover. Diversity and Distributions 19(10), 1322–1332. https://doi.org/10.1111/ddi.12105.)

understanding is that they are relatively unharmed by spilled oil. In the study by Nevalainen et al. (2019), the differences between ice seals and other seals were minor. Other seals have slightly higher exposure potential, and therefore, vulnerability. The level of sensitivity of both groups falls in the medium range. It also appears that the vulnerability of seals is only marginally higher during spring when they have their offspring. Even though offspring are undoubtedly more sensitive to oil contamination than adults during the time they have soft, fluffy hair, their exposure potential is low as they stay on ice or land, well away from the water. That being said, even if the individuals do not come into direct contact with oil, an oil spill during the seal breeding season would harm seal populations, for example, through changes in the food web.

Polar bears are typically considered to be highly vulnerable to oil contamination. In the study by Nevalainen et al. (2019), polar bears are second only to sea birds in their oil spill vulnerability. However, their results deemphasize oil risks to polar bears during the summer. For the time being, summer is the active season for Arctic shipping. This may be of practical significance for the protection of polar bears. However, the authors noted that there is significantly more uncertainty in the polar bears' habitat use (including use of ice) and diet during summer compared with other seasons.

Movement of Oil Pollutant Compounds up the Food Chain

Larsen et al. (2016) used a contaminant tracer in an ecological model to predict the spread of pollutants in an Arctic marine food web. They noted that immediate, negative impacts are to be expected from oil spills through the acute mortality of marine organisms, especially from heavy fuel oil (HFO). They suggested that marine diesel oil (MDO) is the preferred fuel for ships operating in Arctic waters. However, PAHs are toxic components in both types of fuel. These compounds are readily taken up by marine organisms (i.e., they are highly bioavailable) and can move up the food chain. A spill of MDO following a shipwreck could therefore have impacts beyond the spill site and long after the diesel has spread and evaporated.

At an accidental MDO oil spill site, the authors combined measurements of PAH concentrations in seawater and the tissues of blue mussels (*Mytilus edulis*). They also conducted experiments on king crab (*Paralithodes camtschaticus*) and blue mussels. From these tests, they obtained PAH concentration values for use as inputs into the numerical model. The model predicted that the pollution in the mussels would spread throughout the local marine food web, especially to the top predators of mussels. These include the king eider (*Somateria spectabilis*) and Atlantic walrus (*Odobenus rosmarus rosmarus*). Their model also predicted that PAH compounds would be passed from snow crabs (*Chionoecetes opilio*) to their predators: seals and toothed whales.

CONCLUSIONS

This chapter revisits the conflict between preserving the Arctic and developing its resources for financial gain, national power and prestige, and geopolitical place marking. Since Renaissance times, Europeans have gone to the Arctic to exploit its natural resources. After exhausting a given resource, they have moved on to other places and projects, leaving behind the remains of their industry. During most of the past 500 years, mining for minerals in the Arctic has been nearly impossible logistically, because of the harsh climate, impenetrable ice-choked seas, dangerously cold, long winters, and all-too-short summers. Underlying all these obstacles are the great distances between most of Europe and Arctic destinations. Until the later 19th century, explorers, adventurers, and entrepreneurs who set out in wooden ships to the Arctic knew that they might be making a one-way trip. The odds were small that they would ever return to Europe.

Following the Industrial Revolution, Arctic visitors could travel in relative safety in steam-driven iron ships. Once they arrived at their destination, they could make use of powered equipment to drill in the ground and process coal measures and mineral ores. The first of these enterprises was coal mining on Svalbard at the beginning of the 20th century. Presaging future Arctic mining developments, Svalbard coal mining enterprises were sponsored by companies in Russia, Norway, the Netherlands, and the United States.

World War II brought the Arctic to the attention of the European and North American nations. Airbases, weather stations, ships harbors, and other installations were built in Iceland, Greenland, and elsewhere. By that time, the world's navies and commercial shipping had switched energy sources from coal to oil. Domestic and industrial power and heating also shifted away from coal to oil and natural gas. The world's thirst for petroleum products has grown ever since. We are now burning well over 35 billion barrels of oil per year. That translates into 5.6 trillion liters (1.47 trillion gallons) of oil per year—more than triple the level of oil consumption in 1950.

Mineral production on the Kola Peninsula began in the early 1940s when the region was part of Finland. The USSR "annexed" the region in 1946 and carried on nickel and copper mining that continues today. The current output of these metals from the Kola Peninsula mines and smelters is worth US $3 billion per year. Meanwhile, the sister company, the Norilsk mining complex in central Siberia, has been mining and smelting various metals since the 1930s. The ore deposits of the Norilsk region are the largest nickel-copper-palladium deposits in the world.

In summary, it is fair to say that even in recent decades, European and North American countries have continued to exploit Arctic mineral and petroleum resources without paying due diligence to the environmental costs of these enterprises. Things are changing, and the world expects mining and oil companies to be better stewards of Arctic landscapes than they were in the past.

Probably the most troubling aspect of industrial development in the Arctic is the continued drilling for oil both on land and on the continental shelves. The literature review presented in this chapter demonstrates that neither the scientific community nor the oil companies have sufficient knowledge of the potential environmental impacts of Arctic oil spills, especially at sea. The chemistry of the petroleum compounds is complicated, and their effect on marine biota is only now beginning to be understood. Some countries, such as Norway and Denmark, have established workable oil spill remediation plans and have acquired the necessary ships, booms, chemicals, and other equipment. Canada is also reasonably well prepared for Arctic oil spills, insofar as this is possible in very remote Arctic regions. Russia makes a good case for oil spill readiness in its Arctic waters, but only on paper. The United States remains ill-prepared to respond to major oil spills in Arctic Alaskan waters.

Ready or not, a major oil spill in Arctic waters is an accident waiting to happen. Any such spill is bound to have extensive, long-term environmental impacts. Most of the Arctic remains remote from population centers, so a rapid response to an Arctic oil spill is essentially out of the question. When a response team does arrive, few technologies currently exist to effectively contain and clean up oil spills in Arctic sea ice, at any time of the year. Oil spill cleanup technologies and adequate response techniques must be developed as soon as possible, particularly for oil trapped under ice.

One of the major ironies of this situation is that the people of the world are looking for ways to reduce greenhouse gas emissions, so we can avoid the worst of global warming. In other words, now is not the time to be developing new petroleum resources in the Arctic or anywhere else. Instead, we should be shutting down petroleum drilling, transportation, and refining facilities while we find the best ways to switch to renewable energy.

REFERENCES

Agersted, M.D., Møller, E.F., Gustavson, K., 2018. Bioaccumulation of oil compounds in the high-Arctic copepod *Calanus hyperboreus*. Aquatic Toxicology 195, 8–14. https://doi.org/10.1016/j.aquatox.2017.12.001.

Aune, M., Aniceto, A.S., Biuw, M., Daase, M., Falk-Petersen, S., Leu, E., Ottesen, C.A.M., Sagerup, K., Camus, L., 2018. Seasonal ecology in ice-covered Arctic seas - considerations for spill response decision making. Marine Environmental Research 141, 275–288. https://doi.org/10.1016/j.marenvres.2018.09.004.

Bambulyak, A., Frantzen, B., Rautio, R., 2015. Oil Transport from Russian Part of the Barents Region 2015 Status Report. The Norwegian Barents Secretariat and Akvaplanniva, Norway.

BBC News, 2020. Russian Arctic Oil Spill Pollutes Big Lake Near Norilsk. BBC News. https://www.bbc.com/news/world-europe-52977740.

Bejarano, A.C., Gardiner, W.W., Barron, M.G., Word, J.Q., 2017. Relative sensitivity of Arctic species to physically and chemically dispersed oil determined from three hydrocarbon measures of aquatic toxicity. Marine Pollution Bulletin 122 (1–2), 316–322. https://doi.org/10.1016/j.marpolbul.2017.06.064.

Boccadoro, C., Krolicka, A., Receveur, J., Aeppli, C., Le Floch, S., 2018. Microbial community response and migration of petroleum compounds during a sea-ice oil spill experiment in Svalbard. Marine Environmental Research 142, 214–233. https://doi.org/10.1016/j.marenvres.2018.09.007.

Brakstad, O.G., Nonstad, I., Faksness, L.G., Brandvik, P.J., 2008. Responses of microbial communities in Arctic sea ice after contamination by crude petroleum oil. Microbial Ecology 55 (3), 540–552. https://doi.org/10.1007/s00248-007-9299-x.

Buchwald, V.F., 1992. On the use of iron by the Eskimos in Greenland. Materials Characterization 29 (2), 139–176. https://doi.org/10.1016/1044-5803(92)90112-U.

Counihan, K.L., 2018. The physiological effects of oil, dispersant and dispersed oil on the bay mussel, *Mytilus trossulus*, in Arctic/Subarctic conditions. Aquatic Toxicology 199, 220–231. https://doi.org/10.1016/j.aquatox.2018.04.002.

European Space Agency, 2020. https://www.esa.int/ESA_Multimedia/Images/2020/06/Arctic_Circle_oil_spill, 2020.

Fischer, S., Rosqvist, G., Chalov, S.R., Jarsjö, J., 2020. Disproportionate water quality impacts from the century-old nautanen copper mines, Northern Sweden. Sustainability (Switzerland) 12 (4), 1394. https://doi.org/10.3390/su12041394.

Fort, J., Moe, B., Strøm, H., Grémillet, D., Welcker, J., Schultner, J., Jerstad, K., Johansen, K.L., Phillips, R.A., Mosbech, A., 2013. Multicolony tracking reveals potential threats to little auks wintering in the North Atlantic from marine pollution and shrinking sea ice cover. Diversity and Distributions 19 (10), 1322–1332. https://doi.org/10.1111/ddi.12105.

Gulas, S., Downton, M., D'Souza, K., Hayden, K., Walker, T.R., 2017. Declining Arctic Ocean oil and gas developments: opportunities to improve governance and environmental pollution control. Marine Policy 75, 53–61. https://doi.org/10.1016/j.marpol.2016.10.014.

Hacquebord, L., Avango, D., 2009. Settlements in an Arctic resource frontier region. Arctic Anthropology 46 (1–2), 25–39. https://doi.org/10.1353/arc.0.0028.

Hansson, S.V., Desforges, J.P., van Beest, F.M., Bach, L., Halden, N.M., Sonne, C., Mosbech, A., Søndergaard, J., 2020. Bioaccumulation of mining derived metals in blood, liver, muscle and otoliths of two Arctic predatory fish species (Gadus ogac and *Myoxocephalus scorpius*). Environmental Research 183, 109194. https://doi.org/10.1016/j.envres.2020.109194.

Hønneland, G., Jørgensen, A.K., 2018. Chapter 6: Air pollution control. In: Hønneland, G., Jørgensen, A.K. (eds.), Implementing international environmental agreements in Russia. Manchester University Press, Manchester, UK, pp. 145–162. https://doi.org/10.7765/9781526137432.00012.

Jørgensen-Dahl, A., 2020. Oil and Gas Reserves in the Arctic. Arctis Knowledge Hub. http://www.arctis-search.com/Arctic+Oil+and+Gas.

Kirdyanov, A.V., Krusic, P.J., Shishov, V.V., Vaganov, E.A., Fertikov, A.I., Myglan, V.S., Barinov, V.V., Browse, J., Esper, J., Ilyin, V.A., Knorre, A.A., Korets, M.A., Kukarskikh, V.V., Mashukov, D.A., Onuchin, A.A., Piermattei, A., Pimenov, A.V., Prokushkin, A.S., Ryzhkova, V.A., et al., 2020. Ecological and conceptual consequences of Arctic pollution. Ecology Letters 23 (12), 1827–1837. https://doi.org/10.1111/ele.13611.

Kirsch, A., Disterhoft, J.O., Marr, G., Aitken, G., Ham, C., et al., 2019. Banking on Climate Change: Fossil Fuel Finance Report Card 2019. Rainforest Action Network, San Francisco, p. 110.

Kontorovich, A.E., Burshtein, L.M., Livshits, V.R., Ryzhkova, S.V., 2019. Main directions of development of the oil complex of Russia in the first half of the twenty-first century. Herald of the Russian Academy of Sciences 89 (6), 558–566. https://doi.org/10.1134/S1019331619006008X.

Larsen, L.H., Sagerup, K., Ramsvatn, S., 2016. The mussel path - using the contaminant tracer, ecotracer, in ecopath to model the spread of pollutants in an Arctic marine food web. Ecological Modelling 331, 77–85. https://doi.org/10.1016/j.ecolmodel.2015.10.011.

Lewis, A., Prince, R.C., 2018. Integrating dispersants in oil spill response in arctic and other icy environments. Environmental Science and Technology 52 (11), 6098–6112. https://doi.org/10.1021/acs.est.7b06463.

McDonnell, T., 2020. A Brutal New Climate Feedback Loop Is Brewing in the Arctic. Quartz. https://qz.com/1928866/how-the-northeast-passage-is-helping-russia-sell-more-natural-gas/.

McDonnell, A.M., Dang, C.H., 2013. Basic review of the Cytochrome P450 system. Journal of the Advanced Practitioner of Oncology 4, 263–268. https://doi.org/10.6004/jadpro.2013.4.4.7.

Mining Technology, 2020. The World's Top 10 Biggest Diamond Mines. https://www.mining-technology.com/features/feature-the-worlds-top-10-biggest-diamond-mines/.

Moiseenko, T.I., Dinu, M.I., Bazova, M.M., De Wit, H.A., 2015. Long-term changes in the water chemistry of arctic lakes as a response to reduction of air pollution: case study in the Kola, Russia. Water, Air, and Soil Pollution 226 (4), 98–110. https://doi.org/10.1007/s11270-015-2310-0.

Moiseenko, T.I., Gashkina, N.A., Dinu, M.I., Kremleva, T.A., Khoroshavin, V.Y., 2020. Water chemistry of arctic lakes under airborne contamination of watersheds. Water (Switzerland) 12 (6), 1659. https://doi.org/10.3390/w12061659.

Morrison, D.A., 1987. Thule and historic copper use in the copper inuit area. American Antiquity 52 (1), 3–12. https://doi.org/10.2307/281056.

Nahrgang, J., Dubourg, P., Frantzen, M., Storch, D., Dahlke, F., Meador, J.P., 2016. Early life stages of an arctic keystone species (*Boreogadus saida*) show high sensitivity to a water-soluble fraction of crude oil. Environmental Pollution 218, 605–614. https://doi.org/10.1016/j.envpol.2016.07.044.

Necci, A., Tarantola, S., Vamanu, B., Krausmann, E., Ponte, L., 2019. Lessons learned from offshore oil and gas incidents in the Arctic and other ice-prone seas. Ocean Engineering 185, 12–26. https://doi.org/10.1016/j.oceaneng.2019.05.021.

Nevalainen, M., Vanhatalo, J., Helle, I., 2019. Index-based approach for estimating vulnerability of Arctic biota to oil spills. Ecosphere 10 (6), e02766. https://doi.org/10.1002/ecs2.2766.

Nornickel, 2019. Clean-up Progress Update on the Accident, 28 October 2020. https://www.nornickel.com/news-and-media/press-releases-and-news/updates-on-the-clean-up-operation-following-diesel-spill-in-norilsk/.

Norseth, T., 1994. Environmental pollution around nickel smelters in the Kola Peninsula (Russia). The Science of the Total Environment 148 (2–3), 103–108. https://doi.org/10.1016/0048-9697(94)90389-1.

Novoselov, A., Potravny, I., Novoselova, I., Gassiy, V., 2020. Sustainable development of the arctic indigenous communities: the approach to projects optimization of mining company. Sustainability (Switzerland) 12 (19), 1–18. https://doi.org/10.3390/su12197963.

O'Neil, D., 1995. The Firecracker Boys. St. Martin's Press, New York. ISBN-10 : 0312134169.

Potter, S., Buist, I., Trudel, K., 2012. Spill response in the arctic offshore. In: Scholz, D. (Ed.), The American Petroleum Institute and the Joint Industry Programme on Oil Spill Recovery in Ice. The American Petroleum Institute.

Rusak, S.N., Shornikova, E.A., Kurilenko, M.I., Homenushko, T.I., 2019. Transboundary aspects of river water pollution and water quality estimation of the Middle Ob river. In: IOP Conference Series: Earth and Environmental Science, 400, p. 012018. https://doi.org/10.1088/1755-1315/400/1/012018.

Salmina, Y., 2010. River pollution in oil production areas in Siberia. Novosibirsk Region Social Committee for Water Protection, Novosibirsk, 4 pp.

Satellite Imagery of the Diesel Fuel Spill Near the Norilsk Nickel Mine, 2020. https://www.esa.int/ESA_Multimedia/Images/2020/06/Arctic_Circle_oil_spill.

Scheiber, I.B.R., Weiß, B.M., De Jong, M.E., Braun, A., Van Den Brink, N.W., Loonen, M.J.J.E., Millesi, E., Komdeur, J., 2018. Stress behaviour and physiology of developing Arctic barnacle goslings (*Branta leucopsis*) is affected by legacy trace contaminants. Proceedings of the Royal Society B: Biological Sciences 285 (1893), 20181866. https://doi.org/10.1098/rspb.2018.1866.

Shishikin, A.S., Oreshkov, D.N., Uglova, E.S., 2014. Condition of the fauna in the impact zone of the Norilsk industrial complex. Contemporary Problems of Ecology 7 (6), 723–731. https://doi.org/10.1134/S1995425514060134.

Spencer, A.M., Embry, A.F., Gautier, D.L., Stoupakova, A.V., Sørensen, K., 2011. Chapter 1: an overview of the petroleum geology of the Arctic. Geological Society Memoirs 35, 1–15. https://doi.org/10.1144/M35.1.

St-Gelais, N.F., Jokela, A., Beisner, B.E., 2018. Limited functional responses of plankton food webs in northern lakes following diamond mining. Canadian Journal of Fisheries and Aquatic Sciences 75 (1), 26–35. https://doi.org/10.1139/cjfas-2016-0418.

Stephens, M.A.C., Wand, G., 2012. Stress and the HPA axis: role of glucocorticoids in alcohol dependence. Alcohol Research: Current Reviews 34 (4), 468–483. http://llpubs.niaaa.nih.gov/publications/arcr344/468-483.pdf.

TASS, 2020. Norilsk Nickel to Pay Emergency Relief Costs, Says Putin. TASS. https://tass.com/emergencies/1164601.

Thomassen, B., 2003. The Black Angel Lead-Zinc Mine at Maarmorilik in West Greenland., vol. 12. Bureau of Minerals and Petroleum, Government of Greenland. ISSN 1602-818x.

Trofimov, H., Bellouin, N., Toll, V., 2020. Large-scale industrial cloud perturbations confirm bidirectional cloud water responses to anthropogenic aerosols. Journal of Geophysical Research: Atmospheres 125 (14), e2020JD032575. https://doi.org/10.1029/2020JD032575.

Vergeynst, L., Wegeberg, S., Aamand, J., Lassen, P., Gosewinkel, U., Fritt-Rasmussen, J., Gustavson, K., Mosbech, A., 2018. Biodegradation of marine oil spills in the Arctic with a Greenland perspective. The Science of the Total Environment 626, 1243–1258. https://doi.org/10.1016/j.scitotenv.2018.01.173.

Vinogradova, A.A., Kotova, E.I., 2019. Pollution of Russian northern seas with heavy metals: comparison of atmospheric flux and river flow. Izvestiya - Atmospheric and Oceanic Physics 55 (7), 695–704. https://doi.org/10.1134/S0001433819070119.

Williams, J., 2019. Norwegian Mine to Sell Longwall System. World Coal, 6 December 2019. https://www.worldcoal.com/mining/06122019/norwegian-mine-to-sell-longwall/.

Yang, Z., Yuan, L., Xie, Z., Wang, J., Li, Z., Tu, L., Sun, L., 2020. Historical records and contamination assessment of potential toxic elements (PTEs) over the past 100 years in Ny-Ålesund, Svalbard. Environmental Pollution 266, 115205. https://doi.org/10.1016/j.envpol.2020.115205.

Arctic People

The last section of the book includes three chapters: Impacts of Permafrost Degradation, Threats to Native Ways of Life, and Changing Political Landscape of the Arctic. These chapters all concern the human aspects of the Arctic. There are two groups of peoples involved. The first group are the Native peoples of the north. Native groups have been in the Arctic for thousands of years. Prior to the 20th century they traveled freely across Arctic land- and seascapes. They made no territorial claims, as this concept was not part of their worldview. The second group is composed of outsiders, that is, people from the south that come north for various reasons. These include school teachers, medical teams, mining and petroleum engineers, ships' crews (both civilian and military), soldiers, and air force personnel. Other than some trading post operators, fur trappers, and the occasional visit by whaling ships, the Natives pretty much had the Arctic to themselves until World War II. They left such a small imprint on the land that their traces are very hard to find. They fed themselves exclusively from regional resources—what are now called "country foods." But now the Native groups of the north have had a wide range of changes thrown at them from outside. One of these is permafrost degradation.

Permafrost is widely distributed in the Arctic. Upward of 10 million people live in northern regions underlain by permafrost. Those who came north to extract resources learned the hard way that permafrost must be treated with respect, or it will thaw with disastrous consequences. Keeping the ground frozen requires the proper installation and maintenance of buildings, factories, roads, railroads, ports, and airports. Not only that, but the situation is getting worse. Arctic permafrost conditions are rapidly changing because of climbing regional temperatures. 21st-century Arctic engineering is continuously being modified to prevent the thawing of permafrost soils. One pervasive problem seems to be that regional and national governments are highly reluctant to replace crumbling (or drowning) infrastructure with fully modern materials and methods. Instead, governments typically resort to funding much smaller (but generally ineffective) temporary fixes. Sadly, this tactic simply forestalls the inevitable failure of the patched-up infrastructure.

Until recently, Indigenous peoples who lived on the edge of the Arctic Ocean made effective use of sea ice. It has always been the Native's winter highway to reach hunting areas for marine mammals, fish, and sea birds, and to travel between villages. It was an aspect of life that Native peoples could rely on during winter, but now it has become unreliable or absent for much of the year. Sea ice is essentially indispensable to the Arctic Native way of life, but in recent years Arctic Native peoples have had climate change thrust upon them from the outside world. The climate has become unpredictable, and the land and sea are unfamiliar. As one Inuit hunter described their current situation, the Arctic environment has become "a friend acting strangely." Climate change is disrupting Native access to traditional foods from both land and sea. Native peoples across the Arctic who used to rely almost exclusively on hunting or fishing are now forced to eat manufactured food products loaded with sugars and other carbohydrates. Their reliance on marine resources for food means that their food security depends on clean ocean waters and a healthy marine environment. Arctic natives are now rightfully demanding a voice in the governance of their homelands and adjacent waterways, such as the waters of the Northwest Passage in Canada.

The Arctic region caught the attention of the two superpowers during the Cold War, as Russia and the United States faced off at the point where the two countries practically meet. Billions of dollars (and rubles) were spent on the installation of long-range radar sites, missile batteries, submarines capable of launching nuclear missiles, and other elements of armed force. In the 1990s, there was a drop in the Arctic's geopolitical and geostrategic relevance. This enabled various regional cooperative schemes to be established in the Arctic. However, the strategic importance of the North has risen again. When the sea ice started to melt in the 21st century, it changed everything. Suddenly there was a new, pressing interest in the Arctic by countries bordering the Arctic Ocean, as well as by China. Control of Arctic airspace and sea lanes matters a great deal, but there are numerous geopolitical complications.

CHAPTER 14

Impacts of Permafrost Degradation

INTRODUCTION

Permafrost is widely distributed in the Arctic, and according to Lee (2020), about 10 million people live in northern regions underlain by permafrost. Maintaining this population and the supporting infrastructure requires the installation and maintenance of buildings, factories, roads, railroads, ports, and airports. As we will see in this chapter, much of the existing Arctic infrastructure was built with little or no regard for the melting of the ground beneath. However, permafrost conditions are rapidly changing because of climbing regional temperatures. 21st-century Arctic engineering is continuously being modified to prevent the thawing of permafrost soils. However, human-induced thermokarst ("anthropokarst?") abounds in many Arctic countries, especially in Russia.

In Arctic Siberia, the first postwar buildings were made of wood. Many people emigrated from European Russia in the 1950s to take part in gold and diamond mining. The wooden structures they built tend to handle the permafrost better because they are smaller, lighter, and more flexible. However, from the 1950s onward, Soviet city planners designed big, inflexible apartment buildings made of concrete. The largest population center in Arctic Russia, Yakutsk (population about 328,000), is filled with six- and nine-story buildings sitting on stilts. These are piles driven at least 10 m down from the surface.

Essentially all houses, schools, and other structures in the Alaskan Arctic communities sit on stilts. This technology has been widely used in Alaska since the 1970s. There are many approaches to designing and building on permafrost. Almost all of them are passive methods to limit building heat from entering the ground below. The simplest passive approaches elevate a structure above the ground, allowing air to pass freely between the building's floor and the ground. The other passive system, the thermosiphon, uses refrigeration systems with special chemical fluids. These fluids change from liquid to gas at temperatures near 0°C. As the gas rises above the ground level, it releases heat as it converts back to a fluid. The cooled fluid sinks to the bottom of the system, where it again collects heat,

expands, and rises to release heat above the surface. The system has no moving parts and requires no external energy to operate. Thermopiles are piles containing thermosiphons (Fig. 14.1). The US Army Corps of Engineers developed the thermopile in the 1960s. These piles are used where a structural load must be carried. These passive cooling structures have the least impact on the landscape, leaving it in its natural condition. A cheaper (but less satisfactory) alternative is to build a structure on a thick pad of gravel.

Piles and thermopiles remain the most popular way to prevent thawing in Arctic communities. Simple stilts provide a certain level of protection for the permafrost as they create a separation between the heat source and the frozen ground, and they cast shadows on the ground beneath the building. They also limit the amount of ground insulating snow from accumulating on the surfaces beneath the building, leaving the ground exposed to extremely cold winter temperatures.

Arctic climate is tough on human-built structures. As of now, many of the structures found in northern villages from Alaska to Svalbard are prefabricated in temperate regions and shipped north by barge during the summer. The life span of buildings in the Arctic is often less than 50 years. This was not a major problem until recently, but climate warming is accelerating permafrost thawing. There are increasing numbers and kinds of societal challenges in Arctic communities that are related to permafrost degradation. In short, permafrost thawing is causing damages to human infrastructure. As discussed in Chapter 14, thawing permafrost may well have contributed to the oil spill at Norilsk. Company engineers believe that permafrost thaw weakened the support of an oil storage tank. Given the current rate of permafrost degradation, this kind of news story may soon become commonplace.

As reviewed by Lee (2020), the infrastructure required to maintain Arctic communities and their associated industries includes buildings, factories, roads, railroads, ports, and airports. Unfortunately, much of the current infrastructure, especially roads and airports, has been built directly over permafrost, and these frozen soils are rapidly changing in response to amplified Arctic

Threats to the Arctic. https://doi.org/10.1016/B978-0-12-821555-5.00014-0

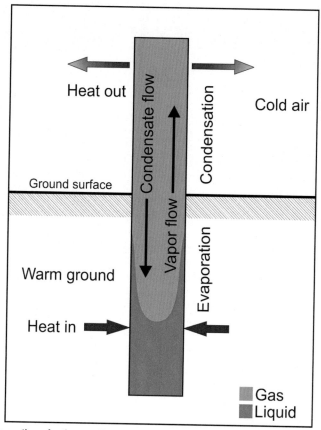

FIG. 14.1 Cross section of a thermopile, showing the mechanism for cooling permanently frozen soils.

warming. The design and construction of Arctic infrastructure must include in-depth consideration of the harsh climate as well as the protection of permafrost from melting. One of the problems with Arctic infrastructure is that its life span—measured in decades—is often far shorter than the timeframe of climate change—measured in centuries. Until recently, this mismatch was not a huge problem. But now the rapid pace of climate change is accelerating permafrost thawing at speeds not seen before in the Arctic. Recently, increasing numbers of news stories tell of Arctic community challenges and disasters related to permafrost degradation. Permafrost thawing is causing the destabilization of some northern communities because of its devastating effects on infrastructure. As discussed in the following, up to 80% of the buildings in Arctic Russian communities have cracked, sunk, or otherwise become deformed by the melting permafrost beneath them.

A recent modeling study (Schneider von Deimling et al., 2020) suggests that human infrastructure such as roads may be accelerating permafrost degradation. This is a highly significant result. Previously, we were only concerned about climate warming affecting permafrost degradation. But the results from this study suggest that the infrastructure itself is also adding to the problem, accelerating the degradation. Now the life spans of infrastructures are no longer free from the timescale of climate change.

THE NATURE OF PERMANENTLY FROZEN GROUND

As reviewed by (Harris, 2021), during both warming and cooling of Arctic soils, the initial change results in warmer or colder surface temperatures with a characteristic curvature near the surface (Fig. 14.2). These are very slow processes. For instance, in the warming of soil with 30% ice content by weight, the rate of raising or lowering of the permafrost table is 2 m per century.

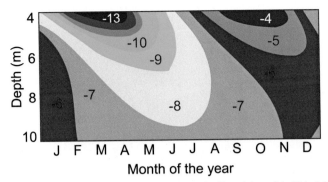

FIG. 14.2 Ground temperatures measured at 4–10 m depth in Shargin's well in Yakutsk during 1937. Note the gradual penetration of the warming (summer) and cooling (winter) waves into the ground, becoming weaker with depth. (Modified from Harris, A., 2021. Permafrost and related landforms. In: Alderton, D., Elias, S. (Eds.), Encyclopedia of Geology, second ed. Elsevier, pp. 385–411. https://doi.org/10.1016/b978-0-08-102908-4.00190-9.)

Permafrost ground increases with latitude. Permafrost types have been boiled down to three categories: (1) continuous permafrost, in which more than 70% of the land is underlain by permafrost, (2) discontinuous permafrost, which occurs in regions where permafrost occurs in some areas but not in others, and (3) sporadic permafrost, in which less than 30% of the land is underlain by permafrost. The exact definitions of the three kinds of permafrost regions vary among different scientists. Almost all landscapes exhibit some taliks under water bodies. Taliks are areas of thawed ground surrounded by permafrost (Fig. 14.3). In areas with greater depths of winter snow cover (>80 cm), the temperature of the snow at the ground surface at winter's end can be used as an indicator of the presence or absence of permafrost, as well as the probable thickness of the active layer (Fig. 14.4).

In areas of sporadic permafrost, some substrates develop permafrost while others at adjacent sites do not. One example is peat, which has enormous pore space if not compacted. The first autumn snows are melted by the warm ground surface, wetting the peat. As the air temperatures drop, the water changes to ice, which is a good conductor of heat. Additional water migrates to the cold surface layer from above or below, and heat readily escapes from the ground to the atmosphere during the long, cold winter nights. In the spring, the ice melts before the ground warms, and the water in the peat evaporates. The peat pore spaces become filled with air. Unlike water, air is a poor heat conductor, and so the underlying layers of peat remain cooler than the soils in the surrounding soils, resulting in a thin active layer beneath which permafrost forms and is maintained.

PERMAFROST AND THE GLOBAL CARBON CYCLE

Martens et al. (2020) discuss the ways in which past Arctic warming events may have caused large-scale permafrost thaw and carbon remobilization, thus affecting atmospheric CO_2 levels. Using carbon isotopes and biomarkers, the authors demonstrate that the three most recent warming events recorded in Greenland ice cores—(1) Dansgaard-Oeschger event 3 (~28 ka BP), (2) the Bølling-Allerød interstadial event (14.7–12.9 ka BP), and (3) early Holocene warming (starting at 11.7 ka BP) caused massive remobilization and carbon degradation from permafrost across northeast Siberia. This amplified the release of carbon trapped in permafrost soils by an order of magnitude, particularly during the last deglaciation. At that time (about 14 ka BP to the early Holocene), global sea levels rose in response to the melting of continental ice sheets. This caused rapid flooding of the land area thereafter known as the East Siberian Arctic Shelf.

As discussed in Chapter 12, Arctic permafrost currently holds a stock of organic carbon estimated to be 1300 Pg (Pg = 1×10^{15} g), about twice the inventory of atmospheric CO_2. Permafrost thaw is not a uniform process in either space or time. It includes gradual changes, such as deepening of the active layer and loss of permafrost extent, as well as rapid thaw by melting landscape collapse, or thermokarst. Rapid changes also include erosion of coastlines and riverbanks. These types of rapid erosion frequently occur during storm events. All of these are nonlinear phenomena that are difficult to program into climate models.

The last deglaciation, the time interval from the Last Glacial Maximum (26–20 ka BP) to the early Holocene

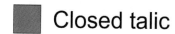 Closed talic

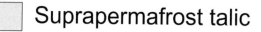 Suprapermafrost talic

Harris, 2020

FIG. 14.3 Block diagram of typical permafrost area showing the names and arrangements of the typical parts. Closed taliks can also form tubes parallel to the ground surface through which groundwater moves downslope. (Modified from Harris, S.A., 2021. Permafrost and Related Landforms, Elsevier BV, pp. 385–411. https://doi.org/10.1016/b978-0-08-102908-4.00190-9.)

(10−8 ka BP), was marked by rapid, large-scale climate change. This period witnessed a rise of atmospheric CO_2 level by 80 ppmv (parts per million by volume), the equivalent of 200 Pg of carbon. It is also suggested to have caused a broad reorganization of organic carbon pools in Arctic permafrost.

The results of the study by Martens et al. (2020) are sobering because they reinforce conclusions from modern studies from the Arctic Ocean and current modeling studies that simulate large injections of CO_2 into the atmosphere during deglaciation. As feared by many climatologists, Arctic warming by only a few degrees may enough to trigger rapid, large-scale permafrost thawing, creating a permafrost—climate change feedback loop (Fig. 14.5).

Abrupt Permafrost Thawing

As discussed by Turetsky et al. (2020), the current applications of Earth System models to the northern

permafrost regions have focused primarily on how warming increases the thickness and moisture content of the active layer. As discussed earlier, the thickening of the active layer will take place gradually across much of the permafrost zone. However, these authors emphasize that areas with excess ground ice are subject to thermokarst as part of permafrost degradation. In contrast to the slow, steady thickening of active layers, thermokarst formation is an abrupt process, and thus far, thermokarst has not been sufficiently embodied in coupled models. Abrupt thaw will probably occur in less than 20% of permafrost regions, but it could release as much as half the carbon held in frozen soils. Organic materials ranging from tiny particles to ancient shrub branches will suddenly be exposed to the atmosphere and sunlight through collapsing ground, rapid erosion, and landslides. As soon as these materials are exposed at the surface, bacterial decomposition will ensue, releasing greenhouse gases (GHGs) into the atmosphere. In

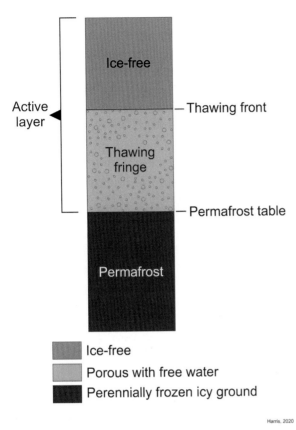

Active layer

Ice-free

— Thawing front

Thawing fringe

— Permafrost table

Permafrost

Ice-free

Porous with free water

Perennially frozen icy ground

Harris, 2020

FIG. 14.4 Diagram showing the upper layers of a permafrost profile in sediments. (Modified from Harris, A., 2021. Permafrost and related landforms. In: Alderton, D., Elias, S. (Eds.), Encyclopedia of Geology, second ed. Elsevier, pp. 385–411. https://doi.org/10.1016/b978-0-08-102908-4.00190-9.)

new location. If these materials end up in fluvial deposits, carbon sequestration will minimize GHG emissions. If they are carried by streams to the ocean, they will avoid decomposition for a long time.

In lowlands, the newly released water causes local inundation, starving the underlying sediments of oxygen. In this saturated condition, anaerobic decomposition takes place. This causes the release of methane into the atmosphere. In the 20 years after its release, methane is 84 times more potent as a GHG than carbon dioxide. Methane does not remain as long in the atmosphere as CO_2, but its potency as a GHG is remarkable.

PREDICTIONS OF FUTURE PERMAFROST RETREAT

Throughout much of the Arctic, the southern edge of the permafrost zone is already retreating northward, and even some High-Arctic regions are experiencing the symptoms of permafrost degradation, such as thickening of the active layer, and soil temperatures rising perilously close to 0°C, or even climbing above this level. This section highlights regional studies that shed light on the extent and nature of permafrost degradation, leading to predictions of future changes. For instance, Hassol (2004) projected changes in permafrost boundaries by the end of this century (Fig. 14.6). More recently, Hjort et al. (2020) predicted the southern limit of continuous permafrost for the interval 2041–60 (Fig. 14.7). (Spoiler alert—permafrost landscapes will be disappearing in all but the coldest regions by the beginning of the next century.)

As discussed by Turetsky et al. (2020), during the period 2000–2300, the area of abrupt thaw will increase threefold under climate warming projections associated with RCP8.5 (representative concentration pathway) GHG emissions. Studies clearly show that modern climate change is contributing to rapid increases in thaw rates. In 1900, it is estimated that abrupt thaw affected 905,000 km² (~5% of the entire permafrost region). With warming and associated thaw, the total area of abrupt thaw increased to 1.6 million km² by 2100, and numerical models predict that this level will increase to 2.5 million km² by 2300. Undisturbed permafrost vulnerable to abrupt thaw is projected to cover 1.5 million km² and 620,000 km² by 2100 and 2300, respectively. Increases in abrupt thaw area are driven by the initiation and expansion of newly formed active features that change through time into more mature features that stabilized, potentially allowing permafrost to form again.

upland areas, abrupt thaw causes the development of thaw slumps, gullies, and active layer detachments. In poorly drained areas, abrupt thaw creates collapse scar wetlands and thermokarst lakes. Regardless of landform types, an abrupt thaw liberates water previously frozen in the soil. The flow of recent meltwater runs downslope, adding to local permafrost soil erosion. The export of carbon from its source in a thaw slump or gully to regions downstream is responsible for more than 50% of carbon loss during permafrost degradation (Nature Geoscience Editorial Board, 2020). As the thawed carbon enters inland waters, it is subject to more complex physical, chemical, and biological processes as it travels to the ocean. Some of the recently liberated organic materials are simply reburied in a

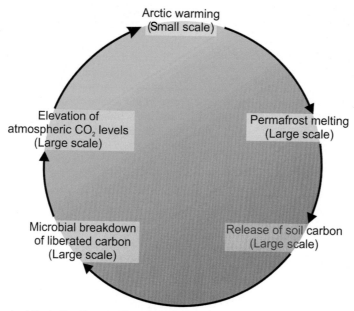

FIG. 14.5 Flowchart illustrating the amplification of Arctic warming through the release of CO_2 from carbon previously held in permafrost soils. (Concepts from Martens, J., Wild, B., Muschitiello, F., O'Regan, M., Jakobsson, M., Semiletov, I., Dudarev, O.V., Gustafsson, Ö., 2020. Remobilization of dormant carbon from Siberian-Arctic permafrost during three past warming events. Science Advances 6 (42). https://doi.org/10.1126/sciadv.abb6546.)

These simulations suggest net cumulative abrupt thaw carbon emissions on the order of 80 ± 19 Pg by 2300. The results also suggest that future abrupt thaw carbon losses will be equivalent to about 40% of the mean net emissions attributed to gradual thaw. Remarkably, most of the rapid carbon release stems from newly formed features that cover less than 5% of the total permafrost region. As of now, these abrupt thaw phenomena are not incorporated into any Earth System models, so they remain unresolved Earth System feedbacks to climate change. The results generated by Turetsky et al. (2020) suggest that abrupt thaw over the 21st century will lead to CO_2 feedback of 3.1 Pg of carbon per °C of global temperature increase and CH_4 feedback of 1180 teragrams (Tg) of carbon per °C global temperature increase under RCP8.5. One Tg of carbon equals 10^{12} g or 1 million metric tons. Forecasting to the year 2300, they estimate abrupt thaw feedbacks of 7.2 Pg of CO_2 per °C increase and 1970 Tg of CH_4 per °C increase. These estimates suggest that the CO_2 feedback from abrupt thaw is currently small but will become amplified beyond the 21st century. In contrast, the abrupt thaw CH_4 feedback is currently more substantial than the CO_2 feedback and is expected to vary less over time due to the balance between expanding thaw areas versus wetland and lake drying with continued warming.

The Western Russian Arctic

Vasiliev et al. (2020) present long-term data on permafrost degradation in the western regions of the Russian European Arctic and Northwestern Siberia, drawing on data compilations from the Thermal State of Permafrost and Circumpolar Active Layer Monitoring site networks. These sites are recording some of the highest rates of permafrost degradation.

Previous studies of permafrost in the Russian Arctic have already documented permafrost degradation manifested by warming temperatures, increasing annual active layer thickness, and declining permafrost extent. The long-term records presented by these authors from the western Russian Arctic show drastic permafrost system degradation from the mid-1970s to 2018 in response to rapid climate change. Regionally, mean annual air temperatures have increased 0.05–0.07°C per year, and precipitation has increased 1–3 mm per year (since the late 1990s as winter snow). The warming air temperatures and additional snowpack are driving permafrost degradation.

1980-1999

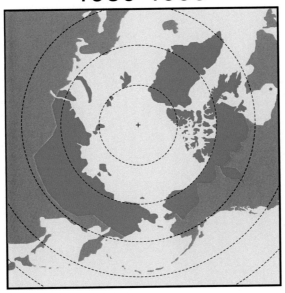

2080-2095 (RCP8.5)

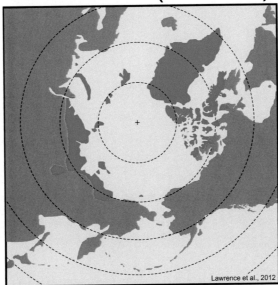

FIG. 14.6 Projected changes in the northern limits of permafrost by the year 2100. (Modified from Hassol, J., 2004. Arctic Climate Impact Assessment: Impacts of a Warming Arctic. University of Cambridge Press, 87.)

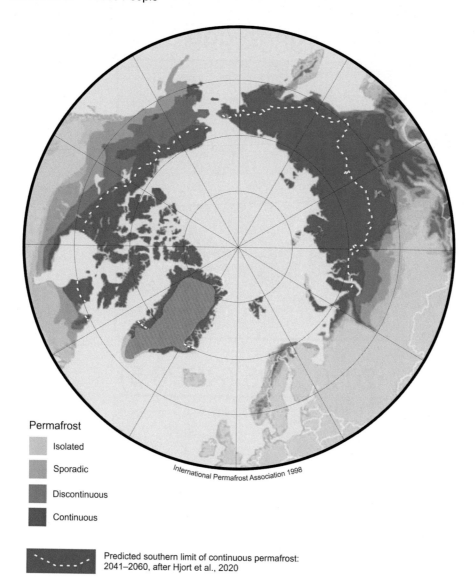

Permafrost

Isolated

Sporadic

Discontinuous

Continuous

International Permafrost Association 1998

- - - Predicted southern limit of continuous permafrost:
2041–2060, after Hjort et al., 2020

FIG. 14.7 Map of modern boundaries of isolated, sporadic, discontinuous, and continuous permafrost, overlain with the predicted southern limit of continuous permafrost: 2041–60. (Modified from Hjort, J., Karjalainen, O., Aalto, J., Westermann, S., Romanovsky, V., Nelson, F., Etzelmüller, B., Luoto, M., 2018. Degrading permafrost puts Arctic infrastructure at risk by mid-century. Nature Communications 9 (1). https://doi.org/10.1038/s41467-018-07557-4 3.)

While climatic factors explain much of the trends in permafrost extent and active layer thickening across large regions, the authors found that local vegetation and soil variability can significantly offset these trends. For example, permafrost temperatures in peatlands, mires, and bogs were found to respond to atmospheric variability to a smaller degree than well-drained landscapes with little organic material, such as sandy tundra

and blowouts. The variable response of different landscapes underlain by permafrost to changing climatic conditions, therefore, warrants further investigation to better inform large-scale models and economic development in permafrost regions. The plea for the inclusion of this variable in large-scale permafrost models is similar to the arguments put forward by Turtetsky et al. (2020) concerning the lack of consideration of rapid release of carbon in thermokarst areas in Earth System models. Finally, Vasiliev et al. (2020) report that mean annual near-surface soil temperatures throughout most of the region have increased above 0°C, indicating that vertical permafrost thaw is ongoing.

The Eastern Russian Arctic

Shifting to northeastern Siberia, Nitzbon et al. (2020) discuss evidence that ground ice distribution governs permafrost thaw pathways, thus shaping future landscape evolution under warming climates. Like Turetsky et al. (2020), these authors emphasize the importance of thermokarst activity resulting in characteristic landforms across ice-rich permafrost terrain. In the continuous permafrost zone of eastern Siberia, thermokarst is expressed in the transition from low-centered to high-centered ice-wedge polygons, or the formation of thaw lakes, thaw slumps, and gullies. These rapidly developing features cause changes at the landscape level. In contrast to the gradual thawing of permafrost in ice-poor terrain, thermokarst processes can cause severe permafrost degradation within a few years or decades and have thus been referred to as rapid or abrupt thaw. The future contribution of thermokarst processes to global-scale permafrost degradation is highly uncertain. Recent efforts using simple conceptual models allow first-order estimates and emphasize the global relevance of abrupt thaw in thermokarst terrain but are at the same time limited by strong assumptions on model parameters that mask the underlying physical processes.

Permafrost in the northeast Siberian Arctic lowlands is highly susceptible to thermokarst, as the landscapes' history led to an abundance of ice- and organic-rich permafrost deposits. The region contains large parts of the Yedoma domain. Yedoma deposits formed through a rise of the permafrost table during the deposition of additional sediment on the ground surface (syngenetic deposition). Yedoma is ice-rich silt that underlies hundreds of thousands of square km in central and northern Siberia (Murton et al., 2015). Because of its ice-rich sediments, Yedoma is more liable to thermokarst formation in warming conditions. Earth System models generally predict that the northeast Siberian Arctic lowlands are one of the most stable permafrost regions, with near-surface permafrost largely remaining thermally stable beyond 2100, even under the strong RCP8.5 warming scenario. However, these models fail to take into account thermokarst-inducing processes and associated feedbacks, making the model projections highly questionable. So here, as in the western Russian Arctic, existing Earth System models appear to

be inadequate to the task of accurate forecasting of landscape response to climate change.

Nitzbon et al. (2020) produced their own numerical model that considers thermokarst-inducing processes. Their model projections show substantial permafrost degradation, involving widespread landscape collapse in northeastern Siberia, especially under the IPCC RCP8.5 (worst-case) scenario, while under moderate warming (RCP4.5), thawing is moderated through the action of stabilizing feedback (Table 14.1). They estimated that by 2100, thaw-affected carbon could be up to three times the level projected by models that ignore thermokarst-inducing processes, and up to 12 times that level under RCP8.5.

WILDFIRE ON PERMAFROST LANDSCAPES

We have seen how the gradual rise in Arctic temperatures is causing permafrost to melt, either gradually or catastrophically, depending on the ice content of the frozen sediments. Now let us turn our attention to the most rapid possible warming of permafrost landscapes, namely, wildfires. Holloway et al. (2020) recently reported on the effects of wildfire in northern regions. Wildfires are becoming more common, more widespread, and more long-lived. This was demonstrated quite strongly in the fires of eastern Siberia during the summer of 2020.

Changes in the frequency and extent of wildfires are expected to lead to substantial and irreversible alterations to many permafrost landscapes under a warming climate. Holloway et al. (2020) reviewed the literature on this topic from the past decade. Tundra fires (Fig. 14.8) are rarer than forest fires, but even tundra fire frequency has increased over the past several decades and is expected to double by the year 2100. From 1951 to 2012, the average interval between fires in three Alaskan tundra regions was 13 years (Noatak basin), 18 years (Seward Peninsula), and 22 years (Southwestern Alaska) (Rocha et al., 2012) (Table 14.2). Within the past decade, however, there have been major regional fires almost every summer, presumably driven by the heat associated with global warming and amplified in the Arctic. The predicted increased frequency and magnitude of wildfires in combination with climate warming is expected to accelerate permafrost degradation and alter the postfire recovery of permafrost landscapes.

Impacts of Fire in Permafrost Regions

Recent studies have advanced our understanding of the effects of wildfire on surface and belowground

TABLE 14.1
Modeled Permafrost Degradation and Ground Saturation by 2100.

Soil Moisture	Lake Basins	Holocene Deposits	Yedoma Deposits
WELL-DRAINED, UNSATURATED DEPTH (CM)			
RCP4.5	61.1	58.3	112
RCP8.5	150	177	255
WELL-DRAINED, SATURATED DEPTH (CM)			
RCP4.5	90.1	90.2	131
RCP8.5	152	178	257
WATER-LOGGED, UNSATURATED DEPTH (CM)			
RCP4.5	0	0	0
RCP8.5	0	0	0
WATER-LOGGED, SATURATED DEPTH (CM)			
RCP4.5	113	142	402
RCP8.5	430	586	869

Data from Nitzbon, J., Westermann, S., Langer, M., Martin, L.C.P., Strauss, J., Laboor, S., Boike, J., 2020. Fast response of cold ice-rich permafrost in northeast Siberia to a warming climate. Nature Communications 11 (1). https://doi.org/10.1038/s41467-020-15725-8.

temperatures, and on changes in the thickness of the active layer. Studies of regions where permafrost is ice-rich have shed new light on rates of ground subsidence and the development of thermokarst. These thermal and geomorphic changes commence immediately after wildfires and alter the hydrology and biogeochemistry of permafrost landscapes, including the release of previously frozen carbon. In many locations, permafrost has been resilient, with key characteristics such as active layer thickness returning to prefire conditions within a few decades. Earth System modeling indicates that some tundra environments will remain resilient, while uplands and areas with thin organic layers and dry soils will experience rapid and irreversible permafrost degradation. Furthermore, as the Arctic warms, the permafrost that exists near its southern limit is losing

FIG. 14.8 Tundra wildfire burning in Bering Land Bridge National Park during the summer of 2019. Photo courtesy of the National Park Service, in the public domain. (From National Park Service (US), 2020. Wildfire in the Arctic. https://www.nps.gov/subjects/arctic/wildfire.htm.)

TABLE 14.2
Climate and Fire Statistics, Alaskan Tundra Regions 1951–2012.

	North Slope	Noatak	Seward Peninsula	Southwest
Average summer air temperature (°C)	8.6	10.5	9.9	11.2
Annual precipitation (cm)	16.6	21.9	19.6	45.7
Ecoregion area (km^2)	15,221	15,049	47,307	76,623
Total area burned, 1951–2012 (km^2)	1606	2155	6800	2469
Total number of fires, 1951–2012	40	156	174	173
Mean fire interval (years)	NA	13	18	22

Data from Rocha, A.V., Loranty, M.M., Higuera, P.E., MacK, M.C., Hu, F.S., Jones, B.M., Breen, A.L., Rastetter, E.B., Goetz, S.J., Shaver, G.R., 2012. The footprint of Alaskan tundra fires during the past half-century: Implications for surface properties and radiative forcing. Environmental Research Letters 7 (4). https://doi.org/10.1088/1748-9326/7/4/044039.

resiliency, because of a combination of warming and changes in vegetation cover. More work is needed to relate modeling to empirical studies.

In summer 2020, an estimated 20 million hectares of the Siberian landscape burned, an area larger than Greece (NASA, 2020). Around half of the fires in Arctic Russia in 2020 burned through areas with peaty soil, a large source of carbon. Warm temperatures, such as the record-breaking heat wave of June 2020, thawed and dried frozen peatlands, making them highly flammable. Peat fires often burn longer than forest fires— sometimes they remain burning just beneath the surface for multiple years. These fires release vast amounts of carbon into the atmosphere. A study by Nichols et al. (2019) examined the history of carbon accumulation in a bog in southern Kamchatka. Peatlands are an important part of the global carbon cycle, storing at least 550 Gt of carbon as partially decayed organic matter. Previous studies of northern peatlands have focused mainly on sphagnum-dominated, ombrotrophic bogs. These bogs depend on atmospheric moisture for their nutrients. However, sedge-dominated fens are also important carbon-storing environments, and these have received far less attention. This is yet another example of the existing numerical models of the impacts of global warming failing to consider the complexities of existing ecosystems. In fact, a suitable subtitle for this chapter might be, "What the modelers are getting wrong."

Ecological Benefits of Tundra Fires

It is easy to look at tundra fires as an instrument that delivers large volumes of GHGs into the atmosphere, and therefore a "bad thing." But such fires are a natural aspect of tundra ecosystems. Ecologists remind us of a variety of benefits of tundra fires (NPS, 2020). For

instance, fire plays a key role in the regulation of the permafrost table throughout the Arctic. Without fire, organic matter thickens, further insulating the underlying soils from solar warming. This chilling effect causes the permafrost table to rise, stifling ecosystem productivity. Vegetation communities become less diverse, and wildlife habitat decreases. Thus, fires act to rejuvenate Arctic soils. Fire removes insulating organic matter, facilitating the warming of the soil. Combustion and increased decomposition rates return nutrients to the soil from peaty layers. Under these conditions, the vegetation cover in recently burned areas experiences increased productivity and plant diversity. These vegetation changes reshape wildlife distribution and habitat use. For instance, patchy fires create a mosaic of habitats frequently used by snowshoe hares and ptarmigan. Small mammals, such as voles and lemmings, often thrive in recently burned areas, creating large colonies in the remaining duff, and feeding on newly sprouting vegetation. Caribou, on the other hand, tend to avoid recently burned areas because their favored foods (*Cladonia rangiferina* and other lichens) take far longer to return after a fire, as much as 100–150 years. Ultimately, these fire-related changes to wildlife habitat and animal distribution also affect subsistence users who rely on the availability of these animals.

THAW LAKE DRAINAGE ON ALASKAN TUNDRA

Water that melts on top of the permafrost in relatively flat topography collects in thaw lakes or thermokarst lakes. These result from the thawing of ice-rich permafrost or the melting of snow and ice, but the frozen ground underneath them prevents the water from draining into the soil. In cold regions, evaporation is slowed, so the meltwater that accumulates in these shallow lakes tends to remain all summer. The pooling of liquid water in these depressions causes further thawing of the underlying permafrost and along the lake edges, gradually extending the lake. These lakes are often oval-shaped, and over decades to centuries, thaw lakes migrate across the landscape, driven chiefly by the prevailing winds (Cyman, 2010). Wind pushes the water toward the leeward end of the lake, where it erodes the permafrost and extends the lake in that direction. Most of the thaw lakes along the Arctic Coastal Plain of Alaska are less than 2½ m deep. Because they are so shallow, generally they freeze to the bottom in winter.

Nitze et al. (2020) examined the recent history of thaw lake development in Arctic Alaska, specifically the Baldwin and northern Seward peninsulas. These areas are characterized by an abundance of thermokarst lakes that are highly dynamic and prone to lake drainage like many other regions at the southern margins of continuous permafrost. Using synthetic aperture radar and remote sensing data, they analyzed the recently widespread catastrophic drainage of thaw lakes. They compared regional weather data, climate model outputs, and lake ice growth simulations to test for potential drivers and future pathways of lake drainage in this region.

The winter of 2019-2020 (2.04 degrees Celsius above the twentieth-century average) was the warmest and wettest on record in Arctic Alaska. In the summer of 2018-2019, 192 lakes either completely or partially drained. This number exceeded the average thaw lake drainage rate by a factor of 10 and was twice the drainage rates of the previous record-holding years of 2005 and 2006. The primary drivers of the record-setting 2018 lake drainage event revolve around climate. There was abundant rain- and snowfall in the previous winter, combined with mean annual air temperatures close to 0°C. These exceptionally warm, wet conditions may have led to the destabilization of permafrost around the lake margins. Rapid snowmelt and great volumes of meltwater promoted the breaching of thaw lake margins.

The coming together of these unusual conditions caused the sudden drainage of regional thaw lakes. Remarkably, even some of the largest lakes of the study region were catastrophically drained—lakes that likely had been in existence for thousands of years. Nitze et al. (2020) hypothesized that permafrost destabilization and rapid thaw lake drainage will not just become more frequent, but under the ongoing changes in Arctic climate, temperature increases will cause massive thaw lake drainages, and these will become the norm in

coming years, developing into the dominant drivers of landscape change in Arctic Alaska and elsewhere. At nearby Kotzebue, the mean annual temperature in 2019 was just below 0°C (5.7°C warmer than the 20th century average). Assuming that the mean annual temperatures rise above the freezing point of water, the permafrost regime of this region will cease to exist.

GROUND-ICE CHARACTERISTICS IN WARM PERMAFROST, TIBETAN PLATEAU

The Qinghai-Tibet Plateau is one of the highest plateaus in the world, with an average elevation over 4000 m above sea level. Accordingly, this region is also the largest high-elevation permafrost area on Earth, covering about 1 million km². This situation is about to change. Lin et al. (2020) report that permafrost on the plateau is so warm (near 0°C) and ice-rich that it has become quite unstable and is likely to thaw permanently in the coming decades. Permafrost warming has been particularly notable along the Qinghai-Tibet Highway. Lin et al. (2020) worked to better understand the distribution and characteristics of near-surface ground ice. They sought to assess the negative impacts caused by permafrost degradation to prevent hazards to infrastructure. They also studied terrain responses to increased thermokarst processes.

Using borehole data, they examined ground ice conditions at eight sites on Beiluhe Basin, an area representative of the warm permafrost conditions where the mean annual ground temperature is currently just below freezing (above $-1.5°C$). The results of the study indicate that surface vegetation cover and topography affected moisture conditions within the active layer, which, in turn, controls ground ice content near the top of permafrost. Permafrost in Beiluhe Basin was found to be characteristically ice-rich, with ground ice concentrated in the upper 2—3 m of the permafrost zone. This warm, ice-rich permafrost is very sensitive to thermal disturbance. Responses to warming include subsidence, thermokarst lake initiation, and expansion, thaw slumping. All of the aforementioned result in hazards to infrastructure.

LONG-TERM PERMAFROST MONITORING SIBERIA

As reported by Vasiliev et al. (2020), seven sites distributed between the Kanin and Gydan Peninsulas in the western Russian Arctic were identified as representing dominant regional landscape forms. Long-term permafrost monitoring began at these sites in the mid-1970s.

The results of this monitoring program reveal three distinguishable stages of permafrost degradation along the latitudinal gradient through the western Russian Arctic.

The Three Stages of Permafrost Degradation

The initial stage of permafrost degradation is a progressive thawing, increasing down from the active layer. This first phase occurs while permafrost temperatures remain relatively cold. Similarly, the transient layer, affected by climate over decades to centuries, can play a protective or buffering role for underlying permafrost, though extremely hot summers may partially thaw the transient layer. Shur et al. (2005) discuss evidence from both the American and Russian Arctic concerning the existence of a transition zone that alternates between seasonally frozen ground and permafrost over subdecadal to centennial timescales. This zone is ice-enriched and functions as a buffer between the active layer and long-term permafrost by increasing the latent heat required for melting. The upper part of the transition zone is known as the transient layer.

The second stage, referred to as "climate-driven, ecosystem-protected permafrost," is where enhanced permafrost degradation is initiated after the complete thaw of ground ice in the transient layer. As discussed in Nitze et al. (2020) concerning thaw lake drainage, peat dampens the effects of atmospheric warming. Accelerated thaw results in increased near-surface temperatures followed by the permafrost table lowering. Cold winters with little snow accumulation may provide conditions for short-term permafrost recovery, but the absence of developed segregation ice typically found in the transient layer may only temporarily slow rather than reverse long-term degradation. Ice segregation takes place during the formation of discrete layers or lenses of isolated ice in freezing mineral or organic soils, as a result of the migration and subsequent freezing of pore water.

The third and final stages are when the progressive lowering of the permafrost table reduces the active layer to a seasonally frozen layer. This is associated with mean annual surface temperatures that exceed the permafrost threshold of 0°C. This third stage occurs with progressive warming or disturbance, such as surface fires.

Other Important Factors

While regional climate warming is the major driver of changes associated with permafrost degradation, other variables are also important. These include ecosystem change, particularly changes in vegetation and soil properties. Such differences determine where permafrost is more resilient or susceptible to warming. For instance, sites in degradation stages 1 and 2 may recover, given the right ecosystem-driven factors, such as type of vegetation cover and organic matter accumulation. Microclimatologic factors may also be important.

Eventual Loss of Regional Permafrost

The long-term effects of progressive climate warming will result in permanent permafrost degradation. This trend has been shown in the long-term permafrost monitoring records presented by Vasiliev et al. (2020) from the western Russian Arctic. These records demonstrate drastic permafrost system degradation from the mid-1970s to 2018 in response to rapid climate change. Warming air temperatures and added winter insulation from increased snow cover are driving observed permafrost degradation, including active layer thickening, permafrost table lowering, and increasing mean annual ground temperatures. Within the past few years, mean annual near-surface soil temperatures have increased above 0°C throughout the region, suggesting that permafrost melting is an ongoing process throughout the regional monitoring network.

DEGRADING PERMAFROST AND ARCTIC INFRASTRUCTURE

Gudmestad (2020) provided an excellent overview of the effects of permafrost on Arctic infrastructure. Melting permafrost exposes coastal soft sediment and sand fronts to increased erosion. This erosion depends on the action of the sea. Every coastal flooding event causes the waves to "take a bite" from the coast. In recent years, there have been an increasing number of storm events, causing more rapid and devastating coastal erosion. This erosion happens between May and September. As the coastal soils freeze in the autumn, the erosion process becomes negligible until the following spring. Generally speaking, a storm surge hits the shore at the end of each season (spring, summer, autumn), eroding any melted soil.

Some Arctic coasts have suffered enormous erosion in recent years. For instance, from 2005 to 2007, the average coastal erosion rate in Varandey, Nenets Autonomous Region of Russia was, 2.7 m per year. Some regions fared worse than this. Coastal erosion measured at Drew Point Alaska was 15 m per year from 2008 to 2011. When storm surges strip the thawed sediments from coastal embankments, the frozen soil is then exposed to heat and melting.

Submarine Permafrost Melting

Melting of underwater permafrost takes place as water temperatures rise during the summer. There is a certain probability of seafloor slides in the case of sloping seafloors. If these slides are sufficiently large, they create tsunamis that may devastate low-lying coastal settlements. For instance, on June 17, 2017, an Arctic tsunami destroyed the settlement Nuugaatsiaq, Greenland. All the village inhabitants were evacuated and moved to another location, due to the threats of further sea slides and tsunamis. The fjords of western Greenland are deep, and sea slides are fairly common.

Permafrost Melting Effects on Built Structures

Built structures in the Arctic are perched on permafrost, so unless the permafrost is insulated from the heat generated by the structures, they will settle into a pool of melted permafrost during the summer. Uninsulated buildings (now becoming quite rare in most Arctic settlements) will suffer serious structural damage. Concrete buildings in particular become uninhabitable. Even buildings made of wood become unsafe. Likewise, Arctic roads that are uninsulated from underlying permafrost will require extensive repairs every summer when the permafrost melts beneath the roadbed. There are similar problems with Arctic airports. For instance, the main airport on Greenland at Kangerlussuaq may become unusable past the year 2024. The runway pavements are already showing significant cracks. To make a long story short, Arctic infrastructure is at risk. The problems are complex; the fixes are extremely expensive and difficult.

Many Arctic settlements are situated on the coast, or at the mouth of a major river. This is because native peoples of the North rely heavily on both marine and freshwater food resources. The hunting of large marine mammals, especially whales, is one of the defining characteristics of Yupik, Inupiat, and Inuit peoples of North America and Greenland. Some of the Native groups of the Old World Arctic, including the Chukchi people, also rely on marine mammal hunting as an intrinsic part of their traditional cultures, while other Eurasian Native groups (the Koryak, Even, Dolgan, Nganasan, Enets, Nenets, and some of the Saami peoples who are not reindeer herders) rely heavily on a combination of freshwater and marine food resources. Thus, it should not be surprising that most of these Arctic natives are essentially coastal inhabitants. This worked well in prehistoric times, partly because the people were not tied to specific coastal locations but were free to travel along coastlines in search of the best fishing and hunting localities. Beginning in the 19th century, the governments of the United States, Canada, Denmark (exercising sovereignty over Greenland), the other Scandinavian countries, and Russia began to limit the movements of native Arctic peoples, forcing them to settle in permanent communities. Nowadays, village life has come to rely on the dominant (outside) society, importing structures and practices that developed in the temperate latitudes. These imports are rarely the "best fit" for Arctic environments. Ironically, it is this modern infrastructure that is under threat from unstable coasts.

Multiple sources are contributing to coastal erosion in the Arctic. One is the shrinking ice cover of the Arctic Seas. Before the sea-ice margins started to retreat northward, the ice cover remained along Arctic coastlands for much of the year, and it protected coastal bluffs from ocean storms by keeping large waves away from the coast. As we have seen in Chapter 1, Arctic coastal settlements no longer enjoy this protection during the stormy autumn months. The date of coastal sea-ice formation (known as the "ice-up" date) has been getting progressively later in the year. For instance, at Utqiaġvik (Barrow) Alaska, ice-up dates in the 1980s were in September (NOAA, 2017). Within the past decade, ice-up dates have moved 2 months later in the year, to early November (Fig. 14.9). In addition to the change in ice-up dates, the sea ice has changed in character through recent decades. In 1982, sea ice along the coast at Utqiaġvik was up to several years old and 9 m thick. In 2018, there was open water in November and the sea ice that eventually settled along the coast was all-new ice that formed that autumn—about 60 cm thick.

Management strategies to limit coastal erosion are being developed worldwide for different types of coastal deposits. The combination of rapid shoreline erosion from melting permafrost and the increasing number and intensity of storms are forcing some villages to depart their current locations and seek relocation on safer ground. Relocation of Arctic villages is an extremely expensive option since it requires the building of all the infrastructure. A cost of US$ 1M per person is quoted for the case of Kivalina northwestern Alaska. The option to repair and protect will cause high "operational costs"; however, investments will be lesser until relocation or transfer to a larger cluster will be necessary. National governments have been highly reluctant to spend such amounts of money to move only a few hundred people or less from an Arctic village location to another. Therefore, the governments of all the Arctic nations have attempted to solve these coastal erosion problems through a variety of expensive engineering measures, such as the use of sandbags, protection by stones or concrete. Unfortunately, most of these attempts to shore up coastal towns have failed, despite the millions of dollars that have been invested

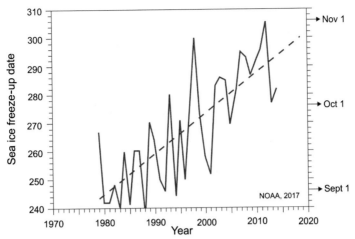

FIG. 14.9 Sea ice freeze-up date at Utqiaġvik (Barrow) Alaska (*purple line*), and long-term trend in the data (*dashed blue line*). (Data from NOAA., 2017. Barrow's Annual Snow Cycle; Ecological Responses to a Lengthening Snow-Free Season. https://www.esrl.noaa.gov/gmd/grad/snomelt.html.)

in their construction. If I may make an analogy, these engineering efforts to reinforce coastal defenses are akin to repeatedly placing bandages on a very deep, bleeding wound. The bandages will temporarily stanch the flow of blood, but they fail to address the real problem, which requires surgery.

Coastal Defense Case Studies, Alaska

Liew et al. (2020) reviewed the effectiveness of coastal erosion control methods in high-latitude locations. For instance, revetments are sloping structures used in coastal engineering. They are placed on banks or cliffs to absorb the energy of incoming water. Revetments are used as a low-cost solution for coastal erosion defense in areas where crashing waves may otherwise deplete the coastline. Liew et al. (2020) noted that revetments built with rocks have the least reported failures and are the most common measures applied along northern high-latitude coastlines including permafrost coasts, while riprap is the most common material used. Riprap is made of boulder-sized rock, blocks of concrete, or other hard material used by engineers to protect shoreline structures against scour and water, wave, or ice erosion. Revetments have been successfully implemented at sites with a wide range of mean annual erosion rates (0.3−2.4 m per year) and episodic storm-surge erosion (6.0−22.9 m) due to their low costs and easy construction, inspection, and decommissioning. Other types of coastal defenses against storm surges and ice damage include seawalls, bulkheads, and groin systems. However, failures of these structures are commonplace, associated with displacement, deflection, settlement, vandalism, and material ruptures. No successful case history has been reported for temporary expedient measures built in the face of an imminent emergency, such as the placement of sandbags to reinforce coastal bluffs and willows.

As is well known, it is difficult to hit a moving target. This applies to the cost of maintaining and developing new infrastructure in Arctic settlements. A recent estimate of climate-related damages to Alaskan infrastructure from 2015 to 2099 was estimated at $5.5 billion under the IPCC RCP8.5 (worst-case) scenario and $4.2 billion under the RCP4.5 (middle of the road) scenario (Liew et al., 2020).

In addition to coastal erosion, the freshwater supply of coastal Arctic villages sometimes becomes contaminated by seawater during storm surges. The situation may eventually force investment in desalinization equipment to ensure freshwater supplies. The release of organic-rich waters from local melting permafrost can also pollute freshwater supplies.

Village Relocation

Liew et al. (2020) also discussed the viability of Native village relocation in Alaska. Some villages have formed committees to consider relocation. The inadequacy of many attempts by the US Army Corps of Engineers to limit coastal erosion has brought state and national governments to agree that erosion, which had hitherto been considered a slow process, should now be included under the statutory definition of disaster. Relocations should be government-supported but community-led so that the displacement efforts are in agreement with the culture and traditional values of local communities. Even when relocation is the only viable option, such as the case in Alaska, which is affected by rapid barrier island migration and coastal erosion, the planning process has already lasted more than a decade. The cost to relocate Shishmaref (Fig. 14.10, below) to a new locality is now projected to exceed $180 million. With a population of just under 500 people, this works out to about $360,000 per person. In the case of the village of Kivalina located on the Chukchi Sea coast and somewhat north of Shishmaref, a cost of US$ 1 million per person has been projected (current population 683).

The US Army Corps of Engineers (2009) identified 31 coastal and riverside villages threatened by erosion. At least 12 of the 31 threatened villages have decided to relocate—in part or entirely—or to explore relocation options (Fig. 14.11). Of the 12 villages exploring relocation options, Newtok, a coastal village in southwestern Alaska at the mouth of the Ningaluk River, was forced to abandon its traditional village site because of repeated, heavy river flooding. The Newtok Planning Group has been a model of multipartner cooperation. Formed in 2006 by federal, state, regional, and village partners, the group has helped to accelerate the relocation process that the village initiated in 1994. About 130 people

FIG. 14.10 Above: Collapsed permafrost block of coastal tundra on Alaska's Arctic Coast; Below: A house in Shishmaref, Alaska, undermined by coastal erosion, finally falls over. Coastal erosion has increased dramatically in the 21st century, forcing Shishmaref residents to consider relocating the entire village. (Images courtesy of Courtesy the USGS Alaska Science Center, in the public domain. From U.S. Geological Survey, 2019. Collapsed Permafrost Block of Coastal Tundra on Alaska's Arctic Coast. https://www.usgs.gov/centers/pcmsc/science/climate-impacts-arctic-coasts?qt-science_center_objects=0#qt-science_center_objects.)

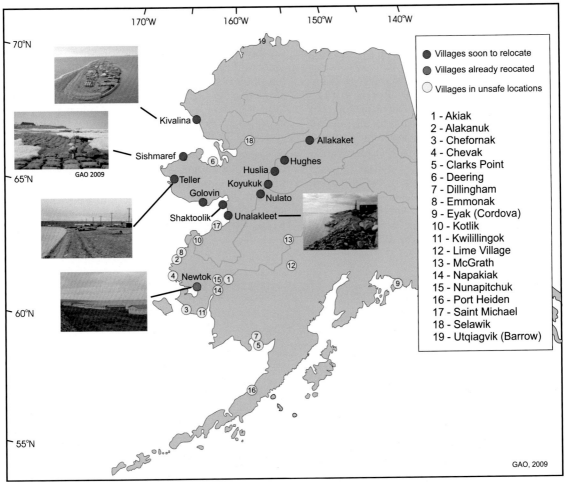

FIG. 14.11 Map of Alaska showing the location of Alaska Native villages pursuing relocation. (From US Government Accountability Office, 2009. Alaska Native Villages: Limited Progress Has Been Made on Relocating Villages Threatened by Flooding and Erosion. (GAO-09-55 1st ed.). US Government. https://www.poa.usace.army.mil/Portals/34/docs/operations/EFC/2017UnalakleetErosionOverview.pdf?ver=2018-12-31-113936-123.)

moved from Newtok to the new community of Mertarvik in October 2019 (Kim, 2020). As of July 2020, most of the village's residents were still in Newtok, waiting for additional houses to be built for them. Construction season in Mertarvik began late in 2020, due to the coronavirus pandemic. The houses are made of wood, but the lumber companies in Anchorage and Seattle were shut down because of the pandemic. During the fall of 2020, a few more families relocated to Mertarvik, but that still leaves about half the village residents in Newtok. Despite the frustratingly slow progress, many of the residents living at Mertarvik say that they are healthier and living a more traditional Yup'ik lifestyle.

In the fall of 2020, construction of a new airport began - a vital assett for this remote community.

The three other villages that will likely need to relocate all at once—Kivalina, Shaktoolik, and Shishmaref—have yet to identify sites that federal, state, and village officials agree are safe, sustainable, and desirable for the subsistence lifestyle of the villagers. Eight other villages have begun to gradually migrate to new locations over time or are evaluating options for doing so. Meanwhile, waves from a violent storm on November 9, 2020, tore away a road in Shishmaref, cutting back into the coastline by 7–15 m. The same storm flooded most of the townsite of Unalakleet, as the water rose 2.5 m above the high-tide line, and it delivered high waves and 80 kph winds to Kivalina (Schwing, 2020).

A short-term solution to the problem of relocating the village of Kivalina in northwestern Alaska has been the building of a road and bridge system that connects the village with the mainland (Fig. 14.12). The road-building took 2 years of construction, but now the Kivalina Access Road is useable for the community (Early, 2020). It is just the first step in the process to relocate the entire village from a rapidly eroding barrier island to the mainland. The current village is under constant threat from coastal erosion and rising sea levels. Kivalina sits on a barrier island, surrounded by the Chukchi Sea. The island is 4.23 km^2 and shrinking. For comparison, the nearby village of Kotzebue is more than 15 times the size of Kivalina. The island is prone to

heavy wind and rainstorms. The protection of sea ice has become less reliable, and sea levels have risen with climate change. That has let the ocean chew into the shore. "It was always eroding," said Tribal Administrator Millie Hawley. "We're on a small spit of land that has diminished in size over the last century." Storms also bring the risk of flooding. Hawley said it had been difficult to evacuate people from the village during emergencies. "According to the U.S. Coast Guard, it takes at least a day or two to come assist us. In a state of emergency, that's just not acceptable." The solution was to build an 8-mile gravel road from the northern part of the community near the airport over the Kivalina Lagoon, allowing evacuees to head to a higher-ground area known as K-Hill. After 3 years of planning and 2 years of construction, the lion's share of the road was completed at the end of October 2020.

Jonathan Hutchinson/Alaska Dept. of Transportation

FIG. 14.12 Above: Newly built road bridge and connecting Kivalina to the mainland. Jonathan Hutchinson/ Alaska Department of Transportation. In the public domain. Below: Aerial view of Kivalina, courtesy of Flickr. com under Creative Commons Attribution 2.0 Generic license.

Specific problems are encountered for these relatively small Arctic settlements as the entire settlements are located close to the sea. The erosion affects the entire settlements, including the access airstrips that may have to be relocated further inland. This will also cause the need to build new roads. An alternative would be to use helicopters more frequently with gathering airports located away from the exposed locations.

Water supply contamination

In addition to coastal erosion, the freshwater supply of coastal Arctic villages sometimes becomes contaminated by seawater during storm surges. The situation may eventually force investment in desalinization equipment to ensure freshwater supplies. For instance, the freshwater reservoirs of Utqiaġvik may soon be contaminated by floodwaters due to the erosion and the high seas. Relocation may be needed with subsequent needs for piping of freshwater. The costs must be compared with investments in a desalination plant. The release of organic-rich waters from local melting permafrost can also pollute freshwater supplies. The unsafe drinking water at Newtok was one of the reasons that the residents have started relocating to Mertarvik. The people who made this move thought their water quality problems were behind them, but even though a freshwater system was installed for the new village, the houses in the new village have yet to be hooked up to the system. There are considerable costs involved and making these connections, and the village construction project does run out of money for such things.

Human Infrastructure in Permafrost Regions

The development of Arctic endeavors and communities requires highly durable infrastructure such as supply roads, pipelines, fuel storages, airports, and other buildings constructed in the permafrost zone—highly sensitive frozen ground (Schneider von Deimling et al., 2020). Functional, safe infrastructure is important for the livelihood of Arctic populations and is directly dependent on the thermal stability of the underlying and surrounding permafrost.

Schneider von Deimling et al. (2020) recently presented an overview of these Arctic infrastructure problems. They noted that permafrost degradation is not limited to top-down soil warming and thawing but is often accompanied by thermokarst and a variety of erosional and mass wasting processes that operate on different spatial and temporal scales. Landslides, thermokarst features, and thermoerosion gullies are prominent landscape features that reveal the dynamics of destabilized permafrost landscapes. Infrastructure may also cause thermokarst and other processes resulting from thawing ground ice. In well-planned Arctic road construction projects, significant efforts are made to identify and quantify ground ice. Knowledge of ground ice distribution along the planned route of a road is key to predicting thermokarst processes that will affect the new infrastructure. A knowledge of *what lies below* is combined with knowledge of local climate, terrain, and geomorphic processes to generate a more complete understanding of how road infrastructure will respond to permafrost. These tasks sound as if they might take years to be accomplished, but road-building projects in the North work on short summer timetables, so these assessments must be both reliable and prompt.

Hjort et al. (2019) modeled future permafrost conditions in the Arctic under the IPCC RCPs 2.6, 4.5, and 8.5 for the years 2041−60 and 2061−80. The permafrost regions associated with the greatest hazards are in the unstable areas characterized by relatively high ground-ice content and thick deposits of frost-susceptible sediments, as well as other regions with an increased potential for permafrost thaw. By 2050, these high hazard environments will be home to human settlements containing nearly a million people and 25%−45% of the existing pan-Arctic infrastructure. The infrastructure in this unstable zone includes more than 36,000 buildings, 3000 km of roads, and 100 airports. Moreover, 45% of the globally important oil and natural gas production fields in the Russian Arctic are located in areas with high hazard potential because of adverse ground conditions and the thawing of near-surface permafrost.

Other than railways, most types of infrastructure in Arctic Eurasia are under greater threat from permafrost degradation than other Arctic regions. For example, the results from the study by Hjort et al. (2019) highlight critical areas such as the Pechora region, the northwestern parts of the Ural Mountains, northwest and central Siberia, much of the Yakutsk basin, and the central and western parts of Alaska. Local-scale infrastructure hazard assessments should be performed for all these regions as a matter of priority. The Yamal-Nenets region in northwestern Siberia is important because it is the primary natural gas extraction area in Russia, and accounts for more than one-third of the European Union's pipeline imports.

Between 19 and 34 large Siberian towns (population > 5000) are predicted to fall within the highest hazard zone in the coming decades. The potential harm to industrial facilities could be larger than previously estimated. For instance, predicted damage to Arctic pipelines has been estimated at 32% in the coming decades, but this value may be as great as 70%. The estimated damage to industrial areas is 25% but may be as high as 64%). This is important because damage to pipelines and industrial facilities may lead to significant regional ecosystem damage, such as large-scale oil spills. Recall from Chapter 13 that the major diesel oil spill in the summer of 2020, linked with thermokarst beneath an oil storage tank at the Norilsk site, is estimated to cost the company US $1.5 billion to clean up, and the entire remediation process is estimated to take up to 10 years.

To successfully manage climate change impacts in sensitive permafrost environments, a better understanding is needed about which elements of the infrastructure are likely to be affected by climate change, where they are located, and how to implement adaptive management most effectively, considering the changing environmental conditions. Such locally and regionally applied mitigation strategies for existing infrastructure and future development projects are paramount for sustainable development in the Arctic.

As reported by Hjort et al. (2019), more than two-thirds of Arctic residential, transportation, and industrial infrastructure is in areas with high potential for near-surface permafrost thaw by 2050. Consideration of ground properties in addition to permafrost thaw showed that 33% of the infrastructure is in areas where ground subsidence and loss of structural load-bearing capacity could severely damage the integrity of infrastructure. The threat to hydrocarbon extraction and transportation in the Russian Arctic is particularly troubling.

Road engineering problems, North Slope of Alaska

Recent studies indicate that many areas are already experiencing issues, including Alaska. The study by Schneider von Deimling et al. (2020) showed that 18% (34 out of 187) of rural communities evaluated should be considered as "high-risk" settlements, with over half located in the continuous permafrost zone. Industrial sites have fared no better. The oilfields of Prudhoe Bay on the North Slope of Alaska have suffered permafrost degradation for more than 62 years, and the rate of degradation has accelerated since 1990, tied with

rising summer air and permafrost temperatures. Projected future climate changes modeled under the RCP8.5 scenario indicate that infrastructure built on permafrost will be negatively affected by decreasing soil load-bearing capacities combined with increasing ground subsidence by 2050. These problems will result in steeply rising infrastructure replacement costs. The risks of disasters caused by damage to sensitive infrastructure such as pipelines, fuel storage facilities, and industrial plants have yet to be calculated.

It is important to note that infrastructure constructed on permafrost is not only being affected by climate change but is itself affecting thermal and hydrological conditions below ground, leading to infrastructure impacts on permafrost. For instance, road embankments strongly affect the thermal properties of the soil beneath and alter surface albedo. They result in increased winter insulation where additional snow accumulates. In contrast, snow plowing of road surfaces exposes the road and underlying soils to increased winter cooling by removing the insulating effects of the snow.

A key limitation of modeling infrastructure in current land-surface models is their coarse spatial resolution ($\sim$100 km). Up till now, no land-surface models have incorporated elements that would help assess the consequences of permafrost degradation on Arctic infrastructure. Here is yet another example of "What the modelers are getting wrong."

The results of the study by Schneider von Deimling et al. (2020) that consider 21st-century climate change scenarios reveal pronounced differences in the dynamics of projected thaw depths among the different structural units (road center, shoulder, toe, and adjacent tundra). Thawing rates can be split into two periods: a period of slow, gradual increase in maximum thaw depths, followed by a sharp increase in thawing rates once critical warming of the ground has been reached. Because of snow accumulation and subsequent ground warming, the toe structural unit is the first location where an increase in thawing rates is seen. Further warming ultimately leads to road destabilization if the ground below the embankment base is ice-rich. In contrast, the road center is much more stable; its thaw rate increase occurs many decades later. In the model runs, the outer road edge becomes destabilized by about 2060. The adjacent tundra is predicted to experience active layer deepening without showing accelerated thaw in the 21st century. The occurrence of individual cold winters can result in a strong winter refreeze in a particular year, causing large year-to-year changes in the maximum thaw depths.

Russian railway engineering on permafrost

In a report by Fazilova (2020), the importance of railroad infrastructure in the permafrost regions of Arctic Russia is spelled out. Nearly all petroleum extraction and accompanying towns exist within this territory. Gas and oil pipelines, railways, and highways pass through the permafrost regions. Train service between Arctic regional centers such as Murmansk, Arkhangelsk, Syktyvkar, Vorkuta, Salekhard, Novy Urengoy, Surgut, Nizhnevartovsk, and Tyumen remains poorly developed. However, the Russian government is committed to the development of the Siberian transport network, as demonstrated by the construction of the Northern Latitudinal line.

The original Tyumen-Tobolsk-Surgut railway was built in the 1960s, using sand washed out from lake bottoms and rivers for the sections of the route where there was a lack of suitable soil for the construction of the roadbed. This reduced the transportation distance of the sandy ground as well as reducing the cost of construction. A significant part of the Northern railway is characterized by troublesome conditions: permafrost, polygonal-veined ice, peaty-soil regions, rapid changes in temperature, and significant amounts of snow. The traditional approach to construction (i.e., the building of roadbeds made of sand) failed to take into account the effects of thermokarst on the rail line. Eventually, this led to nearly a quarter of the roadbed being inundated by water. Not surprisingly, because of the shortcomings in the design and construction, the roadbed began to collapse from thermokarst.

Today, the main objective of rail roadbed construction in Arctic Russia is to preserve the frozen state of the underlying ground. Once achieved, this greatly improves the reliability of the rail network. However, despite significant financial resources being committed to the repair and strengthening of the subgrade, its condition has changed very little. In fact, the amount of subgrade deformation has only increased. This deterioration comes at an inopportune time because rail traffic on the Northern Road is expected to increase by a factor of 2.5−3 times in response to the construction of the Northern latitudinal passage, and increased transportation of goods from the Yamal-Nenets District. The increased rail traffic brings higher axial loads that further exacerbate thermokarst deformation of the roadbed. In short, without the costly rebuilding of the roadbed, the railway will falter.

Rail beds built on alternating patches of drier, permafrost substrates and unfrozen wetlands require careful preparation. These substrates can change many times over a given stretch of roadbed. As of now, none of the roadbed is thermally insulated. Thus, heat flows in the summer, causing increases in soil foundation temperatures. Intersections of railbeds and the longitudinal troughs of ice-wedge polygonal terrain are highly susceptible to ground ice melting, leading to significant thermokarst deformation.

Culverts in catchment basins are currently only laid at points where there is significant water accumulation. However, during the operation of railways and roads in the Arctic region, surface water accumulates wherever the road passes through curved sections of polygonal relief, due to the lack of runoff. Water drains slowly through the embankment soils, leading either to soil deformation or to sudden slope failure. Despite the enormous costs involved, the only way to reduce the operating costs, in the long run, is to tear out sections of the rail line and rebuild the roadbed to insulate it from underlying permafrost. This alone would provide the necessary engineering remedy to prevent roadbed deformation. Ultimately, this will reduce the cost of roadbed maintenance as well as preventing cases of complete operational failure. Current methods of railbed thermal stabilization are being used after the fact.

Russian Arctic building construction

As reviewed by Tregubov et al. (2020), up to 80% of the buildings in towns and large industrial centers in Arctic Russia have cracked, sunk, or otherwise become deformed. As with Alaskan coastal site conservation efforts, recurrent problems with infrastructure deformation plague these Arctic towns and cities, despite repeated applications of various geotechnical solutions. Crumbling concrete buildings, especially multistory buildings, have not been built with the proper construction materials or methods to withstand the extreme conditions to which they are exposed. Furthermore, maintenance costs in Arctic towns are far higher than in European Russia. Of course, a major hazard is associated with thermokarst in permafrost soils. Despite the history of construction regulations for buildings in the permafrost zone that date back to the mid-1950s, current climatic conditions are far different from those of 60 years ago. For instance, rising temperatures and altered moisture regimes are driven by climate change and have not been considered by government officials responsible for updating Arctic building standards. It is true that changes in permafrost conditions are not easily observed from the surface, except perhaps for frost heaving (causing visible upward deformation) or thermokarst (causing the surface to collapse). However, the application of ground-penetrating radar (GPR) to permafrost studies has allowed researchers to recognize thawed and frozen layers, massive ground ice, and rocks in continuous and discontinuous permafrost.

Tregubov et al. (2020) have employed GPR to map frozen ground features beneath a native village in Chukotka. The village of Lorino (65.498°N, 171.725°W) is located on the Bering seacoast on a narrow coastal plain surrounded inland by mountains. The GPR survey revealed that about one-third of the town lies on either areas of active thermokarst and thermal erosion, or areas of unstable ice-rich permafrost and frost-susceptible sediments. The ice-rich sediments that lay close to the surface have been stabilized through the application of ground filling with gravel. The amelioration of permafrost problems begins with the control of human impacts. Improved maintenance of the buildings and infrastructure, the demolition or reconstruction of old buildings, and repair works of the communication infrastructure in Lorino in the 2000s have created an environment favorable for permafrost aggradation.

Permafrost soil engineering studies on the Tibetan Plateau

Permafrost on the Tibetan Plateau is currently losing stability under warming temperatures. Notably, permafrost has been rapidly warming along the Qinghai-Tibet Highway. Lin et al. (2020) examined the distribution and characteristics of near-surface ground ice as a means of assessing the negative impacts of permafrost degradation. They aimed to prevent road damage along this vital highway and to examine terrain responses to increased thermokarst processes.

The results of the research by Lin et al. (2020) indicate that surface vegetation cover and topography exert significant effects on moisture conditions in the active layer. These conditions, in turn, control ground ice content near the top of permafrost. The results indicate that the soils of the study region, the Beiluhe Basin, are characterized by ice-rich permafrost, with ground ice concentrated in the upper 2–3 m. The warm, ice-rich permafrost is very sensitive to thermal disturbance. Permafrost responses to warming include subsidence, thermokarst lake initiation and expansion, and thaw slumping. All these present hazards to the Qinghai-Tibet Highway and other regional infrastructure.

Strength of frozen soils during freeze-thaw cycles

Maslikova et al. (2020) described the criteria for the strength and stability of soils suitable for construction in permafrost regions. The calculation of loads during construction in the permafrost zone requires a special approach in each specific case since it is impossible to devise a single model that fits the variety of soils and the permafrost history. Typical soil classification schemes do reveal soil response to subfreezing temperatures since the transitions between water and ice in the soil change the behavior of the soil. Furthermore, the unpredictability of climatic changes that affect this process greatly complicates the task. The authors performed a series of experiments in a hydraulic laboratory, subjecting various types of soil to loads to determine the main parameters affecting soil stability during the thawing process. They aimed to develop recommendations to road engineers concerning the use of these soils as road bases.

The strength of the thawing soil is affected by many factors, such as ice content, connectivity, flowability, and porosity. As a rule, when soils freeze, the structure of coarse-grained particles changes smoothly as the water in the pores freezes. As the water phase changes, the soil volume increases by 9%. Because of the high hydraulic conductivity of coarse-grained soil, excess water flows away, leaving the soil structure unchanged. On the other hand, fine-grained soils are characterized by lower hydraulic conductivity, which leads to moisture accumulation. This causes the volume of fine-grained soils to increase during freezing. This increase in soil volume causes the soil to rise as it freezes. In contrast, the excess pore water that is generated during thawing leads to loss of soil strength, leading to significant deformations. Therefore, these soils are classified as heaving, i.e., sensitive to frost and thaw.

The experiments by Maslikova et al. (2020) led to the conclusion that a large ice content in thawing soils is closely linked with greater soil strength. In below-freezing temperatures, increased ice content conveys a strength like that found in hard rocks. The connectivity and strength of the soils are important aspects of this process. Porous, cohesive (coarse-grained) soils can retain large amounts of moisture. This leads to increased ice content and heaving of the bases, but such soils keep their strength properties during thawing due to the adhesion of soil particles, which the authors observed in their experiments.

Melting permafrost landslides, Denali highway, Alaska

Landslides pose a persistent hazard in high-latitude regions where permafrost is degrading rapidly. Although models of forecasted permafrost loss are highly variable, regional models estimate that 40%–60% of permafrost by area will be lost by the end of the century. Near Denali National Park (DNP), local monitoring suggests that permafrost temperatures are already near 0°C. The current rapid climate warming and subsequent permafrost thaw are creating landslide hazards that pose significant management challenges for DNP. Modeled permafrost response to climate change in DNP suggests that although 75% of the park was underlain by permafrost in the 1950s, only 1% of the park will be underlain by permafrost by the end of the 21st century. Changing precipitation patterns worldwide are affecting permafrost degradation, and the transition to freeze-thaw regimes will contribute to landslides by increasing landslide frequency and magnitude. Changes to landslide regimes in Alaska and other high-latitude regions make landslide hazard assessments more difficult. As such, the need for a mechanistic understanding of landslide initiation and up-to-date landslide inventory data is greater than ever. Patton et al. (2020) performed a study of permafrost and other controls on recent landslides in the Alaska Range. They noted that a disproportionate number of landslides (81%) are initiated in permafrost terrains. This demonstrates the influence of permafrost in landslide initiation, especially ice-rich permafrost. Furthermore, the slope angles where landslides occur are about 7° shallower in permafrost terrain. The authors concluded that ice-rich permafrost facilitates landslide initiation on shallow hillslopes due to perched groundwater and elevated pore pressure in the active layer and increases in pore pressure during ice melt. Furthermore, there is decreased soil cohesion and friction along permafrost boundaries. This can also reduce the shear strength of a slope and facilitate landslide initiation. Within the study area, multiple types of landslides mostly happen on high-elevation slopes, regardless of aspect.

Impacts of melting permafrost on roadbuilding in northern Siberia

Porfirieva et al. (2019) discussed three model scenarios concerning road infrastructure sustainability under permafrost thawing and degradation due to global climate change. They considered nine Russian Arctic regions in their project. An economic appraisal of the impacts of climate change in these regions was based on six model climate assessments of cryogenic conditions, reflecting the worst-case scenario (RCP8.5) of the IPCC global climate change forecasts, which, sadly, best fits the conditions of the Russian Arctic. The Russian government has developed a Transport Strategy that runs through 2035. This document was updated by the authors and serves as the basis for predicting future road infrastructure development. A conservative cost estimate of road infrastructure development from 2020 to 50 shows that capital costs to maintain road infrastructure sustainability and reduce damage risks under permafrost thawing and degradation will average at least 14 billion rubles ($US 188 million) a year and will exceed ₽21 billion ($US 281 million) and ₽28 billion ($US 375 million), respectively, under the moderate and full modernization scenarios. The high-end costs (full modernization) apply to the Republic of Sakha (Yakutia), Magadan Oblast, and the Chukotka Autonomous Okrug. The implementation of the modernization scenario will require revision of the existing standards, technologies, and the entire economy of road infrastructure and capital construction. However, full modernization will stimulate the development of innovative standards and construction technologies.

As discussed by Maslikova et al. (2020), the bearing capacity of permafrost soils primarily depends on temperature and the mechanical characteristics of the soil. Therefore, the planning of construction projects in permafrost areas must consider geological, climatic, and physical geographic factors. Unfortunately, little consideration was given to these conditions when the oilfields of northwest Siberia were being developed. As a result, the number of structures affected by permafrost degradation has increased in recent decades. According to some estimates, permafrost soil thawing and degradation at the oilfields of Western Siberia have caused an average of 7400 infrastructure emergencies a year, including about 1900 in the Khanty-Mansi Autonomous Okrug. Despite the severity of this financial drain on the system, no thorough economic assessments of permafrost degradation effects have been made. There are rather limited macroeconomic valuations of this problem, but one placed the average annual damage caused by permafrost degradation at about ₽150 billion ($US 2 billion), or 0.16% of the Russian GDP). More specifically, cost estimates for the reduction of damage to pipelines indicate that up to ₽55 billion ($US 737 million) are spent annually to maintain pipeline working capacity and eliminate mechanical deformations related to permafrost soil disturbance. While these costs seem staggering, it is simply a fact of life that drilling and pumping oil through pipelines in the Arctic will always be a very expensive proposition. For comparison, annual costs for the extraction and pipeline transport of oil from Prudhoe Bay to an ice-free port in Valdez, Alaska, are about $3 billion per year (King, 2017). Meanwhile, the cost estimates of road maintenance in the Russian permafrost zone from 2020 to 50 range from a minimum of ₽422.7 rubles ($US 5.7 billion) to ₽865 billion rubles ($US 11.6 billion).

Pipeline permafrost engineering

As discussed by O'Leary et al. (2018), the harsh Arctic environment presents unique challenges for the oil and gas and transportation sectors that are not found in more southern latitudes, including permafrost and permafrost degradation. It is well known that the extent of permafrost in northern environments remains incompletely known and mapped. New tools are being used to help determine the extent of permafrost and to identify areas that are more susceptible to permafrost degradation as climates continue to warm. One such tool is the use of softcopy mapping to help map terrain- and geological-modifying processes such as permafrost. Softcopy uses traditional stereo aerial photographs in a digital environment to allow scientists to view the landscape at scales of 1:1,000, converting images from traditional aerial photography that were captured at scales of 1:24,000 to 1:40,000. The high resolution of the digitized imagery allows researchers to zoom in on previously poorly defined land features. This ability facilitates better soil type determination (sand, silt, or clay), drainage conditions (rapid to very poor), and ongoing geological processes such as permafrost. Permafrost features visible in this high-definition system include frost boils and permafrost degradation as evidenced by the presence of thermokarst and thaw slides. Another method often utilized where stereo aerial photography is not available is the use of remote sensing data sets such as high-resolution digital elevation models and satellite imagery, which are becoming generally available in Arctic regions. These elevation models are used to create hill shade images of varying aspects and photorealistic 3D models to help map terrains.

Oil pipeline monitoring in permafrost regions of the Tibetan Plateau

In addition to remote sensing through aerial photographs or satellite imagery, field monitoring (aka, "ground-truthing") retains a critical role in studying various engineering and environmental problems during the design, construction, operation, and maintenance phases. Because of the harsh environment, this is especially the case for long-distance pipelines built in permafrost regions, as engineers must deal with more frost hazards, increased sensitivity of permafrost to temperature change, and the complex interaction between the pipeline and permafrost. Wang et al. (2019) describe a pipeline-permafrost interaction monitoring system installed along the China-Russia crude oil pipeline. Similar monitoring systems have been set up along other pipeline routes across permafrost regions in the world, including the Norman Wells pipeline in northwest Canada, the Trans-Alaska oil pipeline in the United States, and the Nadym-Pur-Taz natural gas pipeline in Russia. These monitoring systems were all installed to obtain field data for engineering research on the performance of permafrost soils when a pipeline goes through or above them.

A real-time monitoring system is especially important for the safe operation of pipeline engineering, which not only accumulates valuable field data revealing complicated pipeline–permafrost interactions but also monitors pipeline operational status. This monitoring may serve as an early warning system for various frost hazards on pipelines. The need for real-time pipeline monitoring was made clear when permafrost degradation damaged the Golmud–Lhasa pipeline on the Tibetan Plateau. To satisfy China's massive need for crude oil, four long-distance cross-border pipelines have been built for energy security in the past several years. The China-Russia Crude Oil Pipeline is 1030 km long, transporting crude oil from Russia to Northeast China. It is buried underground along a 518-km long stretch of permafrost soils, including 77 km in Russian territory. In the south, it runs through a 512-km long stretch of seasonally frozen ground in China. The real-time monitoring system records meteorological data, ground temperature and water content within the upper layer of permafrost, vertical settlement of the pipeline, ground surface deformation on the pipeline right-of-way, as well as measuring the thaw bulb around the pipeline at four sites with different engineering geological conditions.

The results of the study by Wang et al. (2019) show that heat dissipated from the pipeline is causing the rapid degradation of permafrost on the pipeline right-of-way, creating various sizes of thawed areas around the oil pipelines. The base of the thaw bulb (melted soils surrounding the pipeline) sank from 4.9 to 9.7 m deep, and the mean annual ground temperature increased by 0.3°C from 2014 to 2018. Following significant thaw settlement of permafrost on the pipeline right of way, the top of the buried pipe dropped by 1.4 m from 2011 to 2015 at a test site. Afterward, mitigative measures were installed in some permafrost sections where permafrost degradation was particularly common, including the addition of an insulation layer, a two-phase closed thermosiphon, and U-shaped air-ventilated pipes were found to be effective in minimizing permafrost degradation. These measures also ensured the thermal stability of permafrost soils along the pipeline.

Modeling permafrost interactions with oil and gas pipelines in Siberia

Operation of long-haul trunk pipelines in permafrost areas has shown that they are subject to such cold region processes as sweating, thermokarst, and ice formation. Permyakov et al. (2020) developed a mathematical model to simulate the heat-moisture regime of the soil base of the Eastern Oil Pipeline in Siberia during ice formation. They also developed a model of heat and moisture transfer that takes into account the process of freeze-thaw of the pore solution of soil. As a result of their numerical experiments, it was established that permafrost groundwater increases the average annual temperature of the soil around the pipeline and has a warming effect.

The Eastern Oil Pipeline (ESPO) is 4740 km long and was built in 2015. The construction of the Power of Siberia gas pipeline was completed at the end of 2019. Its length is about 3000 km. It is currently delivering 38 billion m^3 per year. Currently, the eastern route of the Power of Siberia gas trunkline supplies gas from the Chayandinskoye field—the basis for the Yakutia gas production center—to domestic consumers in Russia's Far East and to China. In late 2022, the pipeline will start to receive gas from an additional field—Kovyktinskoye, which serves as the basis for the Irkutsk gas production center (Gazprom, 2021). The gas pipeline route passes through the Irkutsk Region, Yakutia, and the Amur Region. All these regions experience extreme climatic conditions: minimum winter air temperatures along the pipeline route range from −62°C in Yakutia to −41°C in the Amur region. Most of this route lies in the permafrost zone. Where the pipeline crosses permafrost, the pipes are coated with external insulation made of innovative nanocomposite materials. This coating also provides corrosion resistance.

The design features and temperature conditions of pipelines depend on the nature of the product being pumped. Thus, oil pipelines can only be operated when the oil is heated to a minimum of 5−10°C. Below this temperature threshold, the oil congeals and becomes unsuitable for transportation. Gas pipelines can have both positive and negative temperatures. Pipelines laid underground have the greatest thermal effect on frozen soils because the main pipes are buried below the depth of seasonal thawing. Obviously, depending on the combination of regional climate and gas temperatures, the pattern of thermal interaction of the pipeline with frozen and thawed soils varies. Thus, during the transportation of gas that is above freezing, perennial thawing halos form during the entire period of operation in the areas of permafrost. To avoid ground instability, the pipeline routes were chosen to bypass the areas with active permafrost processes, such as frost-heaved mounds, areas of active thermokarst, icing, stone runs, and solifluction. Thermal insulation of the pipe is used to reduce the thermal interaction of hot and warm pipelines on the enclosing soils. Over many years of gas pipeline operation in which the temperature of the gas remains above freezing, the temperature of the underlying permafrost soils increases, but tests have shown that the thawing halo around the gas pipeline is small.

CONCLUSIONS

Results from various kinds of research indicate that reducing GHG emissions and stabilizing atmospheric concentrations, under a scenario consistent with the Paris Agreement, could stabilize risks to Arctic infrastructure after midcentury. In contrast, higher GHG levels would probably result in continued detrimental climate change impacts on the built environment and economic activity in the Arctic (Hjort et al., 2019). The people who live in the Arctic, 10 million of them, need the daily support of infrastructure, which requires the installation and maintenance of buildings, factories, roads, railroads, ports, and airports. Unfortunately, the legacy of failing Arctic infrastructure still plagues nearly all Arctic communities around the world. Specifically, much of the existing Arctic infrastructure was built with little or no regard for the melting of the ground beneath. One pervasive problem seems to be that regional and national governments are highly reluctant to replace crumbling (or drowning) infrastructure with fully modern materials and methods. Instead, governments typically resort to funding much smaller (but generally ineffective) temporary fixes. Sadly, this tactic simply forestalls the inevitable failure of the patched-up infrastructure.

Then there is the companion problem of human-induced thermokarst, which I have casually named "anthropokarst." This human-caused degradation of permanently frozen soils abounds in many Arctic countries, but especially in Russia. How have we arrived at this juncture? I think there are several contributing factors, as follows.

The Frontier Mentality

No matter the region, the interest of nonnative peoples in Arctic resources began to grow rapidly after the end of World War II. As discussed in Chapter 14, the postwar period launched new endeavors to extract minerals, gemstones, and petroleum products from remote Arctic localities. The frontier mentality is a world view that considers humans as superior above all other forms of life, rather than as an integral part of nature. It sees the world as an unlimited supply of resources for human use regardless of the impacts on indigenous peoples or plants and animals. Implicit in this view is the notion that bigger is better, increasing material wealth is an essential life goal, and nature must be tamed—put under subjection. Throughout history, people who scurried to remote regions to "get rich quick" by staking their claim to extract precious metals (gold rushes) have done so with precious little consideration for Native peoples (e.g., the Native inhabitants of the Arctic, American Indians, Australian Aboriginals, inhabitants of sub-Saharan Africa). This world view dictates that those seeking quick riches look on the lands holding the riches simply as a means to an end. Their intentions are to get to the goldfields (or diamond mines, petroleum reserves, deposits of rare earth elements, etc.), find a fortune, and go back to civilization.

Central Governments are not Located in Remote Regions

Even with the best will in the world, most politicians will never visit the Arctic. Back in 1983, the citizens of Anchorage were thrilled when Ronald Reagan passed through there and spent a few hours on the ground before flying off to Asia. The only other American presidents to visit this largest state in the Union were Warren Harding (1923), Franklin Roosevelt (1944), Dwight Eisenhower (1960), Lyndon Johnson (1966), Richard Nixon (1971), and Barack Obama (2015). Most of these were very short visits. But more to the point,

even the politicians who make the effort to visit the Arctic do not live there, so they cannot be fully aware of the living conditions, livelihoods, and problems of the residents of Arctic communities.

The "Otherness" of Arctic Native Cultures

Human beings tend to shun people and cultures with which they are not familiar. This has certainly been true for most of the history of interactions between Europeans and Native peoples of the Arctic, before the era of self-determination of Arctic Native peoples (much of it still a work in progress). One of the clearest indications of Western society's view of Arctic natives can be seen in the education of Native children. Take, for example, the case of Alaskan Native schooling. As reviewed by Barnhardt (2001), policymakers at local, state, and federal levels have initiated education reforms for Alaska and Alaska Native students solely based on short-term localized considerations, or research conclusions drawn from conditions outside of Alaska. This has been a theme throughout the history of reforms in the state, and it continues today as the state looks to the "Lower 48" for quick-fix solutions to long-standing schooling challenges.

In the late 1800s, the federal government established day schools in Alaskan villages and a limited number of state vocational boarding schools. The instruction was provided in the three "R's," in industrial skills, and patriotic citizenship. A strict "English-Only" policy prevailed, and students caught speaking their native language were punished. As a result of this, a whole generation of Alaskan Natives grew up not knowing their own language or cultural connections.

In the early 20th century, a dual system of education was initiated, with schools for Alaska Native students run by the Federal Bureau of Education, and schools for white children and a small number of "civilized" Native children operated by the Territory of Alaska and incorporated towns. In recent decades, groups of Alaska Native educators have attempted to find alternatives to federal and state educational reforms by building on the past and the wisdom of their elders.

Such was the novelty of the natives of Arctic North America, lumped together under the name "Eskimos," that individuals and small groups were put on display in exhibitions in Europe and North America from 1824 to 1909.

Mistrust of natives, such as the Inupiat people of northwest Alaska, can turn up in surprising and shocking ways. On a visit to a research station in this region, run by federal employees from the lower 48 states, I was warned that it was too dangerous to go into town on my own because the Natives were likely to cut my throat and rob me. As far as I could tell, this fear of Native Alaskans was not based on anything rational. Rather, it represented an extreme case of distrust of the Natives because of their "otherness."

The combination of the frontier mentality, the remoteness of federal governments, and the distrust of the "otherness" of Arctic Natives by people of European origin has wrought havoc with cultural relations between groups. It has also taken a heavy toll on the natural environments of the Arctic. Sadly, these attitudes continue to cast a long shadow over the North. They still shape the attitudes of infrastructure engineers. At first, these people tried to ignore the permafrost swells upon which they were attempting to build structures, roads, railbeds, oil and gas pipelines, and oil drilling sites. Not surprisingly, ignoring such major problems did not make them go away. The ground was going to be unstable and therefore from an engineering sense "uncooperative," unless the engineers gave it the respect it deserved. In fact, the only way to keep frozen soils stable is to make sure that they do not thaw. The only way to prevent their thawing is to keep them cold through various insulation materials and techniques. I find it startling that the engineers from one country have failed to pay attention to proven engineering techniques in other countries where permafrost has posed problems. For instance, consider the Chinese attempt to bury oil pipelines beneath the permafrost surface on the Tibetan plateau. The Golmud-Lhasa oil pipeline was completed in 1977—the same year as the Trans-Alaska Pipeline. The engineers building the latter pipeline in Alaska employed thermosiphon technology and placed the pipeline on stilts to further ensure the stability of the underlying permafrost. The pipeline itself was built on a zigzag pattern so that expansion and contraction of the pipe would not cause breakage. Keep in mind that thermosiphon technology was not new in the 1970s. American inventor William Bailey developed a working model in 1909. More specifically, thermosiphon systems have been used for stabilizing permafrost since 1960. Why, then, did the Chinese ignore this seemingly obvious solution to permafrost degradation, rather than burying their oil pipeline in the frozen soil and hoping for the best? Was it because the technology came from the United States? The building of the Chinese pipeline took place just after the 'Cultural Revolution' in China, so political considerations may have played the dominant role in this decision.

Self-Sustained Permafrost Melting

Randers and Goluke (2020) recently reported that in the ESCIMO climate model, the world is already past a point-of-no-return for global warming. The authors ran this climate model for the interval of 1850–2500. Using this model, they observed the self-sustained melting of the permafrost for the next few hundred years, even if global society stops all emissions of human-made GHGs immediately. The projected permafrost melting is the result of self-sustained increases in global temperatures. This warming stems from the combination of three physical processes: (1) declining surface albedo, driven by melting of the Arctic ice cover; (2) increasing amounts of water vapor in the atmosphere, driven by higher temperatures; and (3) changes in the concentrations of GHGs in the atmosphere, driven by the absorption of CO_2 in biomass and oceans, and the emission of carbon in methane and carbon dioxide from melting permafrost. This self-sustained melting process is a causally determined, physical process that will evolve over time. It started with anthropogenic warming of the atmosphere up to the 1950s. This warming led to a rise in water vapor levels in the atmosphere, which caused further heating of the atmosphere (remember that water vapor is an important contributor to GHG warming). This additional atmospheric warming is now causing increased release of carbon from melting permafrost, while it simultaneously contributes to a decline in the surface albedo by melting ice and snow cover in the Arctic. To stop the self-sustained warming in the model, enormous quantities of CO_2 must be removed from the atmosphere. The authors urged other Earth System modelers to test their results in bigger, more sophisticated models, and to report their findings.

I hate to end this chapter on such a discouraging note. In fact, I do not believe our situation to be hopeless. The natural world, including permafrost soil, is probably more adaptable than we know. As this chapter has shown, the permafrost world is certainly more complex than we have previously understood. Our increasing knowledge combined with changing societal attitudes and the development of new technologies may yet provide solutions to at least some of these problems. In short, I choose to be a well-educated optimist. As both Imam Ali (599–661) and Francis Bacon (1561–1626) said, "Knowledge is power."

REFERENCES

Barnhardt, C., 2001. A history of schooling for Alaska Native people. Journal of American Indian Education 40 (1), Fall, 2001.

Cyman, M., 2010. Wind Direction and Oriented Thaw Lakes. An Investigation Determining if Dominant Wind Direction is Solely Responsible for Thaw Lake Orientation. VDM Verlag Dr. Müller, Düsseldorf.

Deimling, S. von, Lee, H., Ingeman-Nielsen, T., Westermann, Romanovsky, V., et al., 2020. Consequences of permafrost degradation for Arctic infrastructure - bridging the model gap between regional and engineering scales. The Cryosphere Discussions. https://doi.org/10.5194/tc-2020-192.

Early, W., 2020. Kivalina Emergency Access Road Now Open for Use. Alaska's Energy Desk - Kotzebue. https://www.alaskapublic.org/2020/11/30/kivalina-emergency-access-road-now-open-for-use/.

Fazilova, Z., 2020. Construction and operation of linear constructions at the polygonal land relief, in the conditions of distribution of permafrost soil. IOP Conference Series: Materials Science and Engineering 919, 022029. https://doi.org/10.1088/1757-899x/919/2/022029.

Gazprom, 2021. 98 kilometers built within Power of Siberia gas pipeline's section between Kovyktinskoye and Chayandinskoye fields. Gazprom. January 26, 2021. https://www.gazprom.com/press/news/2021/january/article522770/.

Gudmestad, O.T., 2020. Technical and economic challenges for Arctic Coastal settlements due to melting of ice and permafrost in the Arctic. IOP Conference Series: Earth and Environmental Science 612, 012049. https://doi.org/10.1088/1755-1315/612/1/012049.

Harris, A., 2021. Permafrost and related landforms. In: Alderton, D., Elias, S. (Eds.), Encyclopedia of Geology, second ed. Elsevier, pp. 385−411. https://doi.org/10.1016/b978-0-08-102908-4.00190-9.

Hassol, J., 2004. Arctic Climate Impact Assessment: Impacts of a Warming Arctic. University of Cambridge Press.

Hjort, J., Karjalainen, O., Aalto, J., Westermann, S., Romanovsky, V.E., Nelson, F.E., Etzelmüller, B., Luoto, M., 2018. Degrading permafrost puts Arctic infrastructure at risk by mid-century. Nature Communications 9 (1), 5147. https://doi.org/10.1038/s41467-018-07557-4.

Holloway, J.E., Lewkowicz, A.G., Douglas, T.A., Li, X., Turetsky, M.R., Baltzer, J.L., Jin, H., 2020. Impact of wildfire on permafrost landscapes: a review of recent advances and future prospects. Permafrost and Periglacial Processes vol. 31 (Issue 3), 371−382. https://doi.org/10.1002/ppp.2048. John Wiley and Sons Ltd.

Kim, G., 2020. After Moving to New Village, Mertarvik Residents Say They are Living Healthier, More Traditional Lives. KYUK, Bethel Alaska. https://www.alaskapublic.org/2020/07/28/after-moving-to-new-village-mertarvik-residents-say-they-are-living-healthier-more-traditional-lives/.

King, E., 2017. How Profitable was Alaska's North Slope? https://kingeconomicsgroup.com/north-slope-profits/.

Lee, H., 2020. Permafrost Thaw Risks to Nature and Society. Norce. https://www.norceresearch.no/en/insight/permafrost-thaw-risks-to-nature-and-society.

Liew, M., Xiao, M., Jones, B.M., Farquharson, L.M., Romanovsky, V.E., 2020. Prevention and control measures for coastal erosion in northern high-latitude communities: a systematic review based on Alaskan case studies. Environmental Research Letters 15 (9), 093002. https://doi.org/10.1088/1748-9326/ab9387.

Lin, Z., Gao, Z., Fan, X., Niu, F., Luo, J., Yin, G., Liu, M., 2020. Factors controlling near surface ground-ice characteristics in a region of warm permafrost, Beiluhe Basin, Qinghai-Tibet Plateau. Geoderma 376, 114540. https://doi.org/10.1016/j.geoderma.2020.114540.

Martens, J., Wild, B., Muschitiello, F., O'Regan, M., Jakobsson, M., Semiletov, I., Dudarev, O.V., Gustafsson, Ö., 2020. Remobilization of dormant carbon from Siberian-Arctic permafrost during three past warming events. Science Advances 6 (42), eabb6546. https://doi.org/10.1126/sciadv.abb6546.

Maslikova, O.Y., Debolsky, V.K., Ionov, D.N., Gritsuk, I.I., Sinichenko, E.K., 2020. Strength of frozen soils during thawing. Journal of Physics: Conference Series 1687, 012039. https://doi.org/10.1088/1742-6596/1687/1/012039.

Murton, J.B., Goslar, T., Edwards, M.E., Bateman, M.D., Danilov, P.P., Savvinov, G.N., Gubin, S.V., Ghaleb, B., Haile, J., Kanevskiy, M., Lozhkin, A.V., Lupachev, A.V., Murton, D.K., Shur, Y., Tikhonov, A., Vasil'chuk, A.C., Vasil'chuk, Y.K., Wolfe, S.A., 2015. Palaeoenvironmental interpretation of Yedoma silt (ice complex) deposition as cold-climate loess, Duvanny yar, northeast Siberia. Permafrost and Periglacial Processes 26 (3), 208−288. https://doi.org/10.1002/ppp.1843.

NASA, 2020. NASA's Aqua satellite shows Siberian fires filling skies with smoke. NASA. https://www.nasa.gov/image-feature/goddard/2020/nasas-aqua-satellite-shows-siberian-fires-filling-skies-with-smoke/. July 2, 2020.

National Park Service (US), 2020. Nature Geoscience Editorial Board, 2020. Editorial: When permafrost thaws. https://www.nps.gov/subjects/arctic/wildfire.htm.

Nature Geoscience Editorial Board, 2020. Editorial: When permafrost thaws. Nature Geoscience 13, 765. https://doi.org/10.1038/s41561-020-00668-y.

Nichols, J., Peteet, D., Andreev, A., Stute, F., Ogus, T., 2019. Holocene ecohydrological variability on the east coast of kamchatka. Frontiers in Earth Science 7, 106. https://doi.org/10.3389/feart.2019.00106.

Nitzbon, J., Westermann, S., Langer, M., Martin, L.C.P., Strauss, J., Laboor, S., Boike, J., 2020. Fast response of cold ice-rich permafrost in northeast Siberia to a warming climate. Nature Communications 11 (1), 2201. https://doi.org/10.1038/s41467-020-15725-8.

Nitze, I., Cooley, S.W., Duguay, C.R., Jones, B.M., Grosse, G., 2020. The catastrophic thermokarst lake drainage events of 2018 in northwestern Alaska: fast-forward into the future. The Cryosphere 14 (12), 4279−4297. https://doi.org/10.5194/tc-14-4279-2020.

NOAA, 2017. Barrow's Annual Snow Cycle; Ecological Responses to a Lengthening Snow-free Season. https://www.esrl.noaa.gov/gmd/grad/snomelt.html.

O'Leary, D., Garrigus, A., Krzewinski, T., 2018. Importance of detailed terrain and geohazard information for pipeline and infrastructure developments in arctic environments. In: OTC Arctic Technology Conference 2018. Offshore Technology Conference. https://doi.org/10.4043/29146-ms.

Permyakov, P.P., Vinokurova, T.A., Popov, G.G., 2020. Effect of ice on the heat-moisture regime of soil foundation of gas pipeline. IOP Conference Series: Materials Science and Engineering 753, 052005. https://doi.org/10.1088/1757-899x/753/5/052005.

Patton, A.I., Rathburn, S.R., Capps, D., Brown, R.A., Singleton, J.S., 2020. Lithologic, geomorphic, and permafrost controls on recent landsliding in the Alaska range. Geosphere 16 (6), 1479–1494. https://doi.org/10.1130/GES02256.1.

Porfiriev, B.N., Eliseev, D.O., Streletskiy, D.A., 2019. Economic assessment of permafrost degradation effects on road infrastructure sustainability under climate change in the Russian Arctic. Herald of the Russian Academy of Sciences 89, 567–576. https://doi.org/10.1134/s1019331619060121.

Randers, J., Goluke, U., 2020. An earth system model shows self-sustained melting of permafrost even if all man-made GHG emissions stop in 2020. Nature Scientific Reports 10, 18456. https://doi.org/10.1038/s41598-020-75481-z.

Rocha, A.V., Loranty, M.M., Higuera, P.E., MacK, M.C., Hu, F.S., Jones, B.M., Breen, A.L., Rastetter, E.B., Goetz, S.J., Shaver, G.R., 2012. The footprint of Alaskan tundra fires during the past half-century: implications for surface properties and radiative forcing. Environmental Research Letters 7 (4), 044039. https://doi.org/10.1088/1748-9326/7/4/044039.

Schwing, E., Alaska Public Radio, 2020. Weekend Storms Pummel Arctic Coastal Villages. https://www.alaskapublic.org/2020/11/09/weekend-storms-pummel-arctic-coastal-villages/.

Shur, Y., Hinkel, K.M., Nelson, F.E., 2005. The transient layer: implications for geocryology and climate-change science. Permafrost and Periglacial Processes 16 (1), 5–17. https://doi.org/10.1002/ppp.518.

Tregubov, O., Kraev, G., Maslakov, A., 2020. Hazards of activation of cryogenic processes in the arctic community: a geopenetrating radar study in Lorino, Chukotka, Russia. Geosciences 10 (2), 7. https://doi.org/10.3390/geosciences10020057.

Turetsky, M.R., Abbott, B.W., Jones, M.C., Anthony, K.W., Olefeldt, D., Schuur, E.A.G., Grosse, G., Kuhry, P., Hugelius, G., Koven, C., Lawrence, D.M., Gibson, C., Sannel, A.B.K., McGuire, A.D., 2020. Carbon release through abrupt permafrost thaw. Nature Geoscience 13 (2), 138–143. https://doi.org/10.1038/s41561-019-0526-0.

U.S. Army Corps of Engineers, 2009. Alaska community erosion survey results summary Appendix A. Alaska Baseline Erosion Assessment (AK: USACE). USACE.

U.S. Geological Survey, 2019. Collapsed Permafrost Block of Coastal Tundra on Alaska's Arctic Coast. https://www.usgs.gov/centers/pcmsc/science/climate-impacts-arctic-coasts?qt-science_center_objects=0#qt-science_center_objects.

US Government Accountability Office, 2009. Alaska Native Villages: Limited Progress Has Been Made on Relocating Villages Threatened by Flooding and Erosion. (GAO-09-551). US Government. https://www.poa.usace.army.mil/Portals/34/docs/operations/EFC/2017UnalakleetErosionOverview.pdf?ver=2018-12-31-113936-123.

Vasiliev, A.A., Drozdov, D.S., Gravis, A.G., Malkova, G.V., Nyland, K.E., Streletskiy, D.A., 2020. Permafrost degradation in the western Russian arctic. Environmental Research Letters 15 (4), 045001. https://doi.org/10.1088/1748-9326/ab6f12.

Wang, F., Li, G., Ma, W., Wu, Q., Serban, M., Vera, S., Alexandr, F., Jiang, N., Wang, B., 2019. Pipeline–permafrost interaction monitoring system along the China–Russia crude oil pipeline. Engineering Geology 254, 113–125. https://doi.org/10.1016/j.enggeo.2019.03.013.

CHAPTER 15

Threats to Native Ways of Life

INTRODUCTION

A few years ago, the Smithsonian Institution staged an exhibit about environmental change in the Arctic. In their exhibit, they quoted Zacharias Aqqiaruq, an Inuit elder from the village of Igloolik in Arctic Canada, who described the Arctic weather in recent years as *uggianaqtuq*—an Inuit word suggesting strange, unexpected behavior, like a friend acting strangely. The climate of the Arctic dictates that the weather will be harsh and unforgiving for much of the year, with snow and ice the norm, accompanied by freezing temperatures (average January temperatures in Igloolik range from −28.4 to −34.8°C). Arctic natives learned to cope with their frozen lands and ice-covered seas long ago, developing technologies such as the dog sled and kayak that worked well in the snow and on sea ice, respectively. They relied on winter snowfall to cover the ground, making overland transport (by dog sled) far easier in winter than in summer. They relied on the fast ice to get them from the land out to the sea ice from about September through May. This easy access to the frozen-over sea allowed them to hunt for seals and walrus, beluga whales, and narwhals during the freezing season. In short, Arctic Natives, rather than cursing the cold, made friends with it and learned to work with it to make their living.

But, as described in detail in Chapters 1−3 of this book, in recent decades, Arctic Native peoples have had climate change thrust upon them from the outside world. The climate has become unpredictable, and the land and sea are unfamiliar. Their environment has become "a friend acting strangely." This chapter describes many of these unsettling changes and their effects on native peoples of the North.

Solastalgia

As discussed at length by Michelin (2020), for the Inuit of the Canadian Arctic, the term "Solastalgia" encapsulates the psychological impact of the current climate crisis. It represents a kind of homesickness for the way things used to be—but it is more than just a longing for the "good old days." Nowadays, Arctic environments are changing beyond recognition.

Take, for example, the goose hunting season on Baffin Island in the eastern Canadian Arctic. Across the island, Inuit try to get some geese before autumn turns to winter. The transition periods between seasons, known as the "shoulder" seasons, are becoming increasingly unpredictable and therefore more dangerous. The shore-fast ice and sea ice are too thin and unreliable to allow for the safe use of snowmobiles, but the early season ice may be too much for safe boating. The unpredictability of the autumn shoulder season is increasing as the climate warms. It can be difficult now to determine the safest route to the good hunting sites—and that has a worrying if predictable, psychological result. Anxiety over travel safety to and from hunting grounds is causing an increase in seasonal depression of the Baffin Island Inuit population. There are abrupt, large-scale shifts in precipitation and temperature that make hunting trip planning nearly impossible.

This fear and grief associated with a rapidly changing environment define the mental condition of solastalgia. Ashlee Cunsolo, the dean of Arctic and sub-Arctic studies at Memorial University of St John's, Newfoundland, has been working with Inuit to examine the mental, physical, spiritual, and emotional impacts of the climate crisis. "You don't have to move to mourn the loss of your home: sometimes the environment changes so quickly around us that that mourning already exists" (quoted in Michelin, 2020).

Native Arctic peoples are experiencing some of the worst effects of solastalgia. Climate fluctuations aggravate existing social problems, including food security, mental health, and addiction problems. The mental anguish is also exacerbated by a chronic housing shortage in the Canadian Arctic, with up to three or four generations all living under one roof. Finally, as discussed in the following, environmental change is disrupting Native access to traditional foods from both land and sea. Native peoples across the Arctic who used to rely almost exclusively on hunting or fishing are now forced to pay exorbitant prices for manufactured food products from "outside" that are only available at a grocery store.

Threats to the Arctic. https://doi.org/10.1016/B978-0-12-821555-5.00017-6

Knowledge transfer

For several reasons, particularly environmental change, there has been a decline in the number of people and time spent participating in hunting, fishing, and harvesting of natural foods in the Arctic. The knowledge of successful food procurement used to be passed on seamlessly from one generation to the next. Knowledge transference is how Inuit have survived the Arctic, and how they thrived there. But Arctic climate is severe and unforgiving. The younger generation does not have the luxury of time it takes to reinvent hunting methods and technologies. They would either starve or freeze to death first. Now Native elders in communities across the Arctic have come to realize that climate change is placing those long-honed skills in jeopardy. Will traditions vital to community culture and welfare, such as the annual caribou hunt on Baffin Island, ever return to the way they were when the old men were young? Here again, is that longing for the past, but this is not an exercise in nostalgia—it is a legitimate fear that the traditions essential to the wellbeing of the community will be lost. Once the older generation is gone, the chain of knowledge is broken, and there is no one left to restore it.

ARCTIC NATIVE WORLDVIEWS

As the cultures of Native Arctic peoples developed over millennia of living in lands of ice, snow, bitter temperatures, and strange solar conditions (i.e., continuous nights in winter followed by continuous days in summer), it is only natural that their worldviews became remarkably different from those of the midlatitudes, such as Europeans and North Americans of European descent. Agriculture is impossible in the Arctic, and the land is not very biologically productive. Other than berry picking in the fall and an annual caribou hunt, the ancestors of the modern Inuit and other Native groups looked mostly to the sea for their food. Even today in the Canadian High Arctic, 53 of the 54 Native communities are coastal.

As discussed in the last chapter, Western history books describe European explorers as the "discoverers" of the Arctic, including such features as the Northwest Passage. But the Inuit of the Canadian High Arctic not only lived where the European explorers ventured, but also they had full knowledge of this passage and made good use of it when sea ice conditions allowed. Furthermore, they were not only aware of the presence of these European explorers, but they also helped many of them to survive the hardships of the Arctic so that

they might return home when the ice pack broke up in the spring. They provided emergency food rations from their own supplies, and they acted as guides to help the Europeans find their way across the trackless ice. This "Native" aspect of Western Arctic exploration got little or no mention in the history books.

Incorporating Alaskan Native Worldviews in Global Change Initiatives

Robards et al. (2018) and the US Global Change Research Program (2018) both discussed the profound impacts of global change on the Arctic environment and inhabitants. These changes are affecting climate, connectivity, and commerce from outside the North. In recent decades, scientists and policymakers have come to recognize the value of incorporating different worldviews and perspectives when dealing with Arctic climate impacts and their consequences. When the outside world pays attention to indigenous needs, perspectives, and cultures, it promotes effective adaptation planning in response to climate change. Arctic Native peoples and cultures are nothing if not resilient, and they have often demonstrated great adaptability in response to changing conditions. This ability has been honed through millennia of knowledge acquired in the daily struggle for survival in a harsh environment.

Robards et al. (2018) presented a series of Alaskan case studies on the coproduction of locally relevant actionable knowledge. Three elements are consistent with earlier work, including (1) evolving communities of practice; (2) iterative processes for defining problems and solutions; and (3) the presence of boundary organizations, such as government agencies, universities, or comanagement councils. In addition, the authors found that it is critical to incorporate two additional factors: (1) the consistent provision of sufficient funds and labor that may transcend any one specific project goal or funding cycle, and (2) the setting of long time spans (sometimes decades) for the achievement of these cooperative ventures.

There are at least 400,000 indigenous residents living around the Arctic. Securing their food supply while helping preserve their cultural identity is incredibly challenging, given the current pace and intensity of social, commercial, and environmental changes. Also, Arctic Native resilience and food security rely on mixed economies. Local indigenous interests encompass more than traditional hunting and fishing. They also seek to maintain a wider market economy and culturally embedded social relationships, sustained by wild foods and other resources.

Case study 1: safe travel on sea ice near Utqiaġvik

As discussed by Robards et al. (2018), there are considerable dangers when Natives head out onto the sea ice, particularly during a period of rapid environmental change. Sea-ice researchers have joined forces with local Iñupiat experts to develop a hazard assessment that can guide both routine operations and emergency responses for all peoples living near and working on sea ice. The practical connection between academic research on sea-ice dynamics and morphology to the type of coastal sea-ice breakout events that pose a substantial hazard to local subsistence hunters has been shaped by a long and varied history of local Iñupiat experts working with scientists. The tools to travel safely over coastal sea ice were developed over millennia by indigenous communities, who have established a keen understanding of ice and ocean dynamics. Indigenous experts track a broad range of ice variables at specific locations, embedded in a broader worldview characterized by a long-term perspective and a system of cultural values, which can now guide and inform locally relevant sea-ice research toward a more complete and applied understanding of sea-ice dynamics.

Collaboration between local indigenous experts and sea-ice scientists began at Utqiaġvik in the late 1970s when geophysicists from the University of Alaska Fairbanks (UAF) worked with local experts in Utqiaġvik to conduct oral history interviews with hunters about their experiences with ice shoves—events in which wind or the ocean pushes ice onto land, presenting a hazard to infrastructure and human life. This research captured the attention of local hunters and of North Slope Borough (NSB) biologists, who, like hunters, spent long periods of time on shore-fast ice, exposing themselves to the risk of ice shove or breakout events. Collaboration between Utqiaġvik whaling captains and scientists from UAF continues today. The mapping and surveying of Utqiaġvik's ice trails have provided Iñupiat whale hunters and scientists a reoccurring focal activity around which to share knowledge. Ice safety emerged as a successful focus for knowledge sharing because of its vital and direct importance to the community, based on many years by a diverse and interdisciplinary group of scientists engaging with the community. Throughout the years, a "Community of Practice" had developed. With sea ice and hazard mitigation as the domain, the emerging Community of Practice can be an effective means of addressing how local and traditional knowledge can now inform both applied research and emergency response protocols on the North Slope and beyond.

Case study 2: protecting walrus haul-outs near Point Lay

Since 2007, Pacific walruses (*Odobenus rosmarus divergens*) have been hauling out in increasingly large numbers on the northwest coast of Alaska, particularly in the vicinity of Point Lay (Fig. 15.1). Prior to that year, walruses rarely hauled out on northwest Alaskan shores. But starting in 2007, walruses hauled out in their tens of thousands. This radical shift in walrus behavior was triggered by the melting of sea ice. Walruses prefer to haul out on the ice to rest, feed, and have their young. Adult females and young migrate to the Chukchi Sea for the summer and prefer to rest on ice floes until they begin to migrate back to the Bering Sea for the winter. Ice floes provide protection from predators, allow herds to haul out in smaller groups, and provide easy access to shallow water benthic feeding areas. During the southern migration, females and young use coastal haul-outs along the way as they move in advance of the developing ice, but groups are relatively small, and they only remain on land for only a few days (Robards et al., 2018).

Chukchi Sea ice typically reaches its annual minimum extent in mid-September. Before 2007, some ice always remained over the shallow and productive continental shelf of the Chukchi Sea, providing the preferred habitat for walruses. In 2007, sea ice completely melted out of the Chukchi Sea and the ice edge retreated far to the north over deep Arctic Basin waters. Walruses dive for mollusks and other benthic food resources, but only in water less than about 90 m deep. The waters of the Arctic Basin are far too deep (average of almost 1000 m) for this kind of foraging. The only alternative for Chukchi Sea walruses was to haul out on the shore. Starting in 2007, several groups of walruses came to shore to haul out in various places from Cape Lisburne to Wainwright along the Alaska coast as well as several spots along the Russian coast. According to the US Fish and Wildlife Service (2019), ice has completely melted out of the continental shelf waters of the Chukchi Sea in late summer in 10 of the last 12 years, forcing walruses to haul out to rest on the Alaska coast in large numbers. Many walruses seem to prefer to haul out on the barrier islands just north of the Native Village of Point Lay, Alaska. This site has been occupied by as many as 40,000 animals. In recent years, walruses have started to arrive at the Point Lay haul-out site in August. They either feed close by or make foraging trips to the Hanna Shoal area, a round trip of almost 600 km. These long-distance foraging trips are especially difficult for females with dependent young. By the middle of October, the walruses abandon

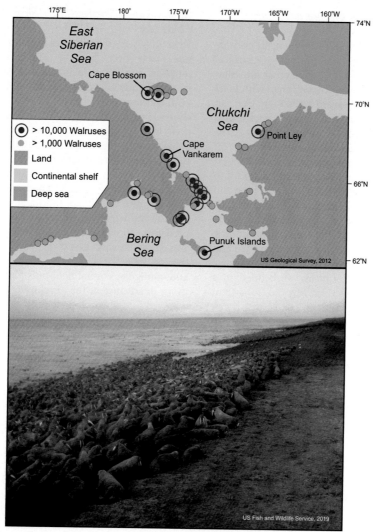

FIG. 15.1 Above: Pacific Walrus haul-out map from Oakley et al. 2012. Walruses hauled out on a Chukchi Sea beach near Point Lay, Alaska, 2015. (Photo courtesy of USFWS, in public domain)

the Point Lay haul-out site, and most of them head south toward Cape Lisburne and then turn west to reach the coast of Chukotka.

In response to the large populations of walruses hauling out near Point Lay, the Eskimo Walrus Commission passed a resolution encouraging hunters not to hunt or disturb walruses while hauled out on land, and the village of Point Lay took extra steps to protect walruses from all human disturbance. Disruptions from tourists encroaching from the land or in boats, as well as low altitude overflights, may cause the walruses to stampede on the overcrowded beaches, leading

to injuries and death, particularly calves and yearlings. Walrus hunting by Point Lay villagers plays a vital part in local food security, so keeping the herd as healthy as possible is regarded as a critical part of walrus management, both locally and by the US Fish and Wildlife Service (USFWS). To address this problem, Point Lay built on a tradition of stewardship in which village elders work closely with wildlife scientists (especially on beluga and bowhead whales). The entire community took the initiative to encourage respect toward the thousands of Pacific walruses coming ashore near the village and to provide stewardship for the walrus population.

The foundation for this initiative was built on existing relationships with the Eskimo Walrus Commission, NGOs (nongovernmental organizations), USFW staff, and scientists. In the spring of 2010, the World Wildlife Fund (WWF) and the USFWS worked with Point Lay leaders to invite a small group of Russian Chukchi hunters and a Russian walrus biologist to Alaska to collaborate with US marine mammal biologists involved in walrus conservation. The scientists developed strong relationships with the Iñupiat hunters and their local communities. These relationships were key to conducting effective field research. The international group traveled together to visit research sites, helping create comradery as they shared experiences and information from either side of the Bering Strait.

The group toured Point Hope, Point Lay, Wainwright, and Utqiaġvik. Russian delegates shared information with the Alaskan delegates concerning measures they had taken to protect walrus haul-outs on the Chukotka coast. For example, the Siberian village of Vankarem worked with the Russian aviation agency to create a no-fly zone to prevent noise disturbance above the haul-out sites. Vankarem hunters also issued rules for the community members on how to unobtrusively observe the walruses and prohibited dogs near the haul-out sites. These Russian examples of best practice helped prepare the people of Point Lay to handle the situation when thousands of walruses hauled out there in 2010. They repeated many of the actions they had learned from the Russians, such as staving off air traffic, and keeping people (tourists, journalists, and curious visitors) well away from the beach, thus preventing walrus stampedes.

The rules established by the Eskimo Walrus Commission and the USFWS were as follows:

1. Pilots of single-engine aircraft and UAS devices should not knowingly fly over or fly within 1/2 mile of walruses hauled out on land or ice to avoid causing a disturbance. If weather or aircraft safety require flight operations within 1/2 mile of walruses, small single-engine aircraft should maintain a 2000 ft minimum altitude.

2. Pilots of helicopters and multiengine aircraft should not knowingly fly over or fly within 1 mile of walruses hauled out on land or ice to avoid causing a disturbance. If aircraft safety requires flight operations within 1 mile of walruses, helicopters and multiengine aircraft should maintain a 3000 ft minimum altitude.

3. Please be aware that some locations have stricter requirements (such as Round Island which has a 5000 ft altitude and 3-mile restriction April 1–November 30) check with the applicable land manager.

4. Landings and take-offs should take place at least 1/2 mile distance from hauled-out walruses. Please be aware that some locations have prohibitions on landing and takeoffs, check with the applicable land manager. Walruses are particularly sensitive to changes in engine noise and are more likely to stampede off beaches when planes turn or fly low overhead. Aerial photography and/or circling of manned aircraft or unmanned aircraft systems (drones) within the vicinity of a walrus haul-out poses a high potential for disturbance and is specifically discouraged.

5. Marine vessels should maintain a buffer from walruses hauled out on land or ice to avoid disturbance: Marine vessels 50 ft in length or less should remain at least 0.5 miles away from hauled-out walruses. Larger marine vessels should remain at least 1 mile from hauled out walruses. Due to the unusually large number of walruses using the Point Lay haul-out, the community requests that marine vessels remain a minimum of 5 miles offshore when walrus are present.

The village's stewardship was recognized in 2011 when the USFWS presented Point Lay with an "Outstanding Partner" Award.

Inuit Perspective on a Sustainable Arctic Ocean

The Arctic Council (2020) conducted an Interview with Jim Stotts, president of the Alaskan chapter of the Inuit Circumpolar Council, concerning the Native view of Arctic Ocean sustainability. The following quotes are from this interview:

Question: Why is the marine environment so important for Inuit?

Stotts: "We live right by the coast, we basically are looking more toward the sea than looking inland. Fish, marine mammals, sea birds are a big part of the hunting activities that we are engaged in. We have a great knowledge about ice and ice conditions because we go out on the ice for big parts of the year to hunt seals, walrus, whales, and polar bears. So, we are very interested in maintaining the cleanliness and pristine nature of our part of the world. I recall, when I was a little boy, growing up at the coast, it was a much cleaner place. There was a lot more wildlife, in particular a lot more shorebirds. There were a couple of species that I haven't seen for years. In Barrow, Utqiaġvik, the main cultural and social event is our annual bowhead whale hunting activity. If successful, it results in a big community feast down at the beach with a dance, sharing of food and fellowship. These are the things that we grew up with and that we would like to keep. That is why, when we talk about the environment and biodiversity, we are talking about keeping things as they used to be. But, it has changed."

Question: What changes have you observed?

Stotts: "Environmentally, everybody knows, there's much less ice than there used to be. The ice, actually, acts as a barrier in the fall time and protects the coastline from erosion from storms. The ice will pile up on the beach and when the big waves come it keeps them from washing on shore. That's a big change. We have tremendous erosions along the Arctic coast these days. The hunting for marine mammals was also better in the past, there were more animals. Ice floes used to move North from the Bering Strait to my part of the Arctic. Today, there's not as much ice as there used to be. The ice used to bring up walruses that come from quite a bit South of Utqiaġvik. It's very hard to find walruses now as there's no ice for them to travel on. Oddly enough, bowhead whale hunting is better now than it was in the past. It seems that the bowhead whale is one of the few species that is actually enjoying climate change. It has expanded its range and there's more to eat. Bowhead whales are the only true Arctic whales, they do not go down South, they stay in the Arctic waters year round."

Question: How can the Inuit Circumpolar Council contribute to a sustainable Arctic Ocean through its work in the Arctic council?

Stotts: "We are not about to roll over and see our world turned upside down because of mistakes being made somewhere else. We intend to be involved. At the first thematic webinar session with a focus on the Arctic Marine Strategic Plan, I presented—as I called it—some suggestions from the coast. We are now half-way through our 10-year strategic plan, and we think we need to make some adjustments. We believe any plan should be flexible enough to accommodate change and there have been significant changes since the strategic plan was written in 2015. We wanted to highlight these changes and suggested five possible adjustments and solutions from the perspective of those of us living along the northern coast that could possibly make the plan better:

First, the Strategic Plan needs to acknowledge the reality of COVID-19. When the Council decided to address the impacts of the virus, ICC emphasized the many glaring infrastructure deficits that were exposed by the virus. These deficits were not only physical in nature, but also include systemic deficits in areas such as food chains, air transportation, communications, health and social services. We believe COVID impacts will be with us for the foreseeable future, especially economic impacts. We think the strategic plan should recommend the creation of a series of task forces to address the marine infrastructure deficits in Inuit and other Indigenous communities. Without national investment, Inuit can't be expected to meaningfully participate in the sustainable development of the Arctic.

Second, the strategic plan needs to more strongly consider the notion that Inuit will participate in the management of Arctic marine biodiversity. Inuit are working to develop a network of wildlife management authorities to carry this out at the scale that it should be, at the scale of large marine ecosystems. Large marine ecosystems with their biodiversity are not respectful of national boundaries. Biodiversity should be managed not only from the principles of preservation and conservation. It should also be managed from the principles of sustainable use. Food security is what Inuit are after. We will not get cut off from the ability to feed ourselves. There is a coming paradigm shift in the way people think about biodiversity. The strategic plan could reflect this change in thinking.

Third, the strategic plan needs to consider the notion that Inuit will participate more fully in the development and management of Marine Protected Areas. A key element of this approach would be the right for Inuit to continue using those resources that they have traditionally used within any marine protected area. Again, the principles of preservation and conservation must be balanced with the principles of sustainable use by Inuit.

Fourth, the strategic plan needs to renew its focus on pollution, marine litter, and microplastics. This summer, in the Bering Sea, there was a deluge of plastic litter from what appears to be Russian and Korean fishing vessels. Those involved should be called out. This has to stop. Stronger language in the strategic plan could help.

Fifth, the strategic plan needs to acknowledge the northern movement of killer whales into the Arctic Ocean as the result of declining sea ice. ICC brought this up some years ago and now it seems to have been recognized by the rest of the world. This invasion northward of killer whales has a potential for dire consequences for other marine species, particularly other marine mammals. The Strategic Plan could call for a new project to investigate this phenomena and report back any findings."

Canadian Inuit Worldviews

The Inuit of the Canadian Arctic likewise have much to say about the condition of the Arctic Ocean and how it affects their daily lives. The *Inuit Tapiriit Kanatami* (Inuit Knowledge Center) published a booklet entitled "Nilliajut 2: Inuit perspectives on the northwest passage shipping and marine issues," edited by Karen Kelley (2017). Several of the articles in this publication speak directly to the issues discussed in this section.

In the introduction to the book, editor Karen Kelley summed up the Canadian Inuit view of the development of the Northwest Passage for commercial use (page i):

> Inuit are a marine people. Our culture and way of life is inextricably linked to the ocean. The marine environment is central to our identity, the way that we perceive the world, and the way that we think of ourselves. The Northwest Passage is a part of Inuit Nunangat, and future activity has implications for our communities and way of life. Inuit considerations must be central to any conversation about how the Northwest Passage is utilized by Canada and other countries.

The choices Inuit take will decide our destiny

Alfred E.R. Jakobsen of Nuuk, Greenland, had this to say about the importance of Inuit decision making concerning their future (pp. 323–334):

> Over these many years, I have realized that Inuit are in the middle of all the many global issues mainly because of global warming, warmer waters, melting ice, and mineral resource development on land and off-shore. Compounding this, Inuit direct dependency on a clean and healthy environment and utilization of the natural wildlife resources force all Inuit societies with these difficult choices we must make. No matter what choices we take, they are full of conflicts. It is all up to us to decide whether to be in favor of or against offshore oil and gas development in our homelands etc. The choices Inuit take will decide our destiny.

Ecology, Cultural Identity—It Is All Connected

Okalik Eegeesiak of Iqaluit, Baffin Island emphasized the interconnectedness of the Arctic world and its people (pp. 52–54).

> There is a need to understand the Arctic through a food security lens, understand how everything is connected in order to understand cumulative impacts. We cannot consider the health of a walrus population, without understanding sea ice thickness, the state of the benthos, the ecology, Inuit cultural identity, language, and so on - it is all connected. And we all need knowledge to understand and to formulate policies that will enhance the future of the Arctic and its peoples. We must use the best knowledge for evidence based decision making.

Marine pollution: entire food chain devastation

Editor Karen Kelley summed up the written statements in the publication, along with the results of Workshop discussions and taped interviews, as follows (pp. 73–78):

> Melting ice, and longer open water seasons, means there will be an increase in traffic along the Northwest Passage. This will present Inuit communities with benefits and problems. In every workshop, Inuit spoke about how more shipping along the Northwest Passage increases the chances of an oil spill or other disaster. Such a disaster would have a devastating impact on the northern environment. Oil spills are not short-term problems: as one participant reminded others, "our vegetation and environment are so fragile that it could take years, decades, or even generations to recover." The combined effect of garbage and ballast waste being dumped by ships could be just as damaging as an oil spill. Inuit worry especially about how fuel spills and excess waste might affect their food supply in the future. Any toxins or waste in the ocean could show up in everything from small krill to larger mammals such as polar bears and whales. The entire food chain—and therefore the Inuit food supply—could be devastated for years.

> Just as the movement of animals is impacted by ships on the Passage, so too is Inuit travel. When a ship passes through and breaks the ice after the initial fall freeze, the ice remains thinner along that route and makes travel dangerous. Sometimes Inuit don't know when boats and ice breakers will be coming, so people who go out on the ice during the day get caught on the other side of a freshly opened waterway and are unable to return home in the evening. Ship traffic may cause problems for northern wildlife too. If there are too many ships in the Passage at once, animals such as whales, seals, polar bears, caribou, and muskox might be unable to travel along their usual routes.

Alaskan Natives and Climate Change Research

As reported by Stone (2020), efforts by the US National Science Foundation (NSF) to integrate Indigenous knowledge into Arctic climate research have not always run smoothly. Despite scientists' concerns about the impacts of climate change on Arctic Native lifeways, many scientists have overlooked the potential value of

Native inputs. A recently launched NSF initiative was meant to change that. The Navigating the New Arctic (NNA) initiative awarded grants totaling $37.5 million in October 2019, a doubling of the amounts NSF previously spent on Arctic research. The NNA initiative aimed to improve understanding of Arctic change by encouraging scientists and Indigenous communities in the "coproduction of knowledge" by involving them in planning and executing projects. Perhaps the NSF and the researchers who received this funding expected the full cooperation of Iñupiat and Yúpik peoples, but there was a backlash (Stone, 2020).

Despite the program's intentions to the contrary, some NNA project scientists either ignored potential input from Indigenous Alaskans or only included it as an afterthought. This feedback was communicated by a Native Alaskan coalition in a letter sent to the NSF in 2020. "We continue to lack meaningful access and voice in the vast landscape that is the 'research process,'" wrote Kawerak, Inc., a consortium of 20 tribes in the Bering Strait region and 3 other organizations representing dozens of Indigenous communities (as reported in Stone, 2020).

For its part, the NSF is urging outside scientists to take such concerns very seriously. To begin with, the climatologists need to come to grips with the concept of knowledge coproduction. "We made a mistake in assuming that scientists knew what that meant," says anthropologist Colleen Strawhacker, program officer for NSF's Arctic System Science Program. "We definitely have a lot of work to do to make sure that Arctic sciences is diversified and equitable." In an open letter on August 3, 2020, NSF's Arctic Sciences Section, which funds a separate research slate from NNA, called for proposals "that will enrich interactions and improve collaboration between Arctic residents," including indigenous-led projects (as cited by Stone, 2020).

As we have seen in this chapter and elsewhere in this book, the environmental challenges facing Arctic communities are many and growing. As these challenges have appeared in recent decades, Native Alaskans have sought to contribute their knowledge to finding solutions. "For many decades," the coalition wrote to NSF, "we have asked to be active partners with agencies and academics that wish to come onto our lands and waters to conduct research" (as cited by Stone, 2020).

That plea is often ignored, says Lauren Divine, director of the Ecosystem Conservation Office for the Aleut Community of St. Paul Island. The island could serve as a poster child for the kinds of environmental upheavals the region is enduring, with heavy coastal erosion and mass die-offs of puffins and other seabirds. Scientists studying these woes sometimes seem to view

Indigenous participation as an exercise in ticking a box, says Divine, who is a marine biologist by training. "We ended up just getting cold-called. Solicitations to hop onto a proposal without any thought for what funding would be directed to the tribe" (as cited by Stone, 2020).

In their letter to NSF, Indigenous leaders recommended NNA focus on projects that address the sustainability of Arctic communities—food security and infrastructure, in particular—and set aside 25% of NNA funds for Indigenous-led projects. The NSF Arctic System Science Program has stated that they would love to see more proposals submitted on those topics. On the other hand, the NSF has no plans to set aside funding for Indigenous-led projects.

EFFECTS OF CLIMATE CHANGE

Climate change is probably the single largest factor causing environmental degradation in the Arctic today. This has been borne out many times and in many ways in this chapter and elsewhere in this book. In the research notes that I compiled for this chapter alone, I find 95 references to climate change in the books and papers I reviewed. Of course, climate change is amplified in the Arctic. Due to the delicate balance between the cryosphere, atmosphere, land, and sea, this region is most susceptible to environmentally driven ecological catastrophes.

Arctic Climate Change Communications

How do we communicate the gravity and severity of Arctic climate change? As Breum (2018) so eloquently put it, "It is very difficult to have a conversation with an ice floe." The topic was deliberated by more than 2000 participants representing many Arctic communities at a 2018 Arctic Circle Conference in Iceland. It was likewise discussed in Finland a few weeks earlier among the 400 people attending a Conservation of Arctic Flora and Fauna Arctic Council (CAFF) working group. As reported by Breum (2018), several principles were established at these meetings—ground rules for discourse on Arctic climate change. Perhaps the most pertinent to this book is the concept that "science should not stand alone." In the concluding chapter of the book "Competing Arctic Futures," Sörlin (2018) warns of the many projections of the future of the Arctic based on simplistic or reductionist perspectives. Sörlin maintains that the world's current perspectives on the future of the Arctic are "science-informed projections of waning sea ice and irreversible warming." As climate change accelerates, natural science and its preoccupation with ice, snow, permafrost, and other natural phenomena quite sensibly come to dominate our thoughts on the future. But we must not forget that in the end it is us—people—who decide how to react.

Once you introduce societies and their complexities, contrasts, and creativities, there will always be a healthy balance between the necessity to simplify and predict on the one hand and the necessity to doubt and point to alternatives on the other. Above all, it will be necessary to allow for democratic and collective, indeed political agency which is another word for freedom.

SÖRLIN (2018).

Native Migration and Climate Change

As discussed in Chapter 15, many Native communities on or near the west coast of Alaska face immediate threats from coastal erosion and flooding associated with reduced sea-ice protection of shorelines, thawing coastal permafrost, and river floods. Hamilton et al. (2016) investigated the underlying causes for Native people, especially young people, leaving their home villages and moving elsewhere. Hamilton et al. (2016) used the annual 1990–2014 time series on 43 Arctic Alaska towns and villages to trace the history of these movements. An examination of Arctic Alaska high school student surveys done in the 1990s found that more than half the students expected to migrate permanently away from their home regions. Although many of these students ended up remaining in their hometowns or leaving for only a short period of time before returning, moving away from home remains a major life choice for young people. The authors' analysis suggests that the problem is increasing. Relocation or individual movements will certainly occur from the most exposed locations, and hopefully, before disasters occur.

Climate change is not the only driver of migration away from native villages, but it affects many other drivers. Migration decision-making can be based on many aspects of village life; the changing physical environment acts in concert with economic, political, social, and demographic drivers in this process (Table 15.1). For instance, climate change is altering the composition

TABLE 15.1
Total Population and Native Population Change by Region, 1990–2018.

| Region | TOTAL POPULATION | | | % POPULATION CHANGE | |
	1990	2018	Total	Natural Increase	Net Migration
ALASKA					
Iñupiat	17,105	21,094	23.3	N/A	N/A
CANADA					
Yukon	27,797	35,874	29.1	N/A	N/A
Northwest Territories	40,845	44,597	9.2	N/A	N/A
Nunavut	27,498	37,996	38.2	41.4	−6.1
Greenland	55,558	55,877	0.6	24.8	−23.1
Norway: Finnmark	74,148	76,167	2.7	12.2	−9.5
Finnish Lapland	1,99,973	1,79,223	−10.4	2.1	−12.5
RUSSIA					
Komi Republic	12,48,891	8,50,000	−31.9	−1.9	−30.1
Nenets Autonomous Okrug	51,993	44,000	−15.4	8.8	−24.2
Yamal-Nenets Autonomous Okrug	4,89,161	5,36,000	9.6	24.8	−15.2
Taymyr Autonomous Okrug	51,867	34,432	−33.6	7.8	−41.4
Evenki Autonomous Okrug	24,005	16,253	−32.3	6.5	−38.8
Sakha Republic (Yakutia)	11,11,480	9,63,000	−13.4	15.5	−28.8
Chukotka Autonomous Okrug	1,62,135	50,000	−69.2	4.3	−73.5
Kamchatka oblast	4,76,911	3,15,000	−33.9	0.5	−34.4
Koryak Autonomous Okrug	37,622	18,759	−50.1	−1.8	−48.3

Data from Heleniak et al. (2020) and Hamilton et al. (2018).

and functioning of regional ecosystems, which in turn are causing declines in subsistence food harvests. The lack of sufficient snow cover in winter is limiting travel between villages during that season. Melting permafrost is damaging village infrastructure including homes, schools, water, and power supplies.

Overall, Hamilton et al. (2016) concluded that migration away from Alaskan home villages was not linked with environmental change during the study period. This finding emphasizes that village social bonds are often sufficiently strong to counteract the urge to migrate to somewhere new. In other words, human society can work against migration as well as for it. Pressures to move are countered by reasons people might not want to leave their ancestral homes, or by a lack of appealing alternatives.

Climate Change Resilience in the Canadian Arctic

Canadian Inuit villages are home to more than 70,000 people. These communities are spread across the northern mainland coast and the islands of the Canadian Arctic Archipelago, in small, isolated communities. Canada's northern coastline extends more than 176,000 km, most of it north of the Arctic Circle. The entire region faces multiple challenges because of global change. As discussed by MacDonald and Birchall (2020), human-induced climate changes have caused increases in the frequency and severity of extreme weather events in the Arctic, the impacts of which are becoming increasingly difficult to manage under the current government policies. The continued resilience of Arctic communities in the face of climate change will depend largely on their ability to adapt. In the view of the authors, the development of any Arctic regional action plans on climate change must include full collaboration between government bodies and Native residents, to reflect the views and experience of those living in the Arctic. The authors worked with native communities across the Canadian Arctic to develop a set of key statements (messages) to guide this process. Here are the three key messages:

1. Climate change impacts affect core infrastructure and resources in the Arctic. The resilience of northern communities depends heavily on proper government leadership and the successful implementation of long-term adaptation policies.
2. Determining the adaptive capacity of Arctic communities and developing appropriate policies requires collaboration with local stakeholders who

can speak to vulnerability in the context of climate- and non–climate-related factors.
3. Climate change policies must be mainstreamed so that they are integrated into existing procedures and policy goals allow local governments to manage present-day issues while enhancing long-term community resilience to climate change.

Every Canadian Arctic Territory is approaching climate change adaptation differently, with support and oversight from the federal government in Ottawa. In 2016, the government partnered with the provincial and territorial governments to develop the Pan-Canadian Framework on Clean Growth and Climate Change (PCF). The PCF addresses the importance of climate change resilience through adaptation and lays out action items highlighting the need to develop climate-resilient codes and standards. Underlying this legislation are government commitments to invest in infrastructure projects that strengthen climate resilience. Many previous federal and territorial programs intended to target adaptation to climate change, but meaningful action has most often lacked long-term government commitments (MacDonald and Birchall, 2020). As climate change effects get worse, Arctic communities must implement adaptation plans urgently and efficiently. A key approach to facilitate immediate and significant change involves engaging local stakeholders, who are often most aware of vulnerabilities unique to their community.

Developing resilience to the effects of Arctic climate change means strengthening the capacity of communities to withstand stresses and shocks through sound management. However, local governments have limited resources, so they tend to focus on short-term issues with more immediate results. As would be expected when it comes to infrastructure, the local governmental approach is more reactive than proactive. Determining effective long-term adaptations is a lengthier and more expensive process, best handled by collaboration between territorial and federal governments and Native communities. It requires an ongoing assessment of community vulnerabilities and adaptive capacities. A helpful approach would be to mainstream federal climate change policies, integrating them with existing community procedures and planning. This would help the implementation of community adaptation plans, enhancing their resilience to climate change.

As we have seen, the Inuit are keen observers of all aspects of nature, including climate change. They know the risks they face and are certainly capable of developing realistic adaptation plans. In light of this, government and scientific collaboration with community members is far from just box-ticking exercises. Native peoples have much to offer in this process, and their inclusion in the planning process allows the most vulnerable individuals to have a voice. As Mac-Donald and Birchall (2020) concluded, policymakers should engage local stakeholders (i.e., Inuit villagers) throughout the planning and development process to ensure that their policies fully address vulnerabilities while taking Native cultural values into consideration.

Case study: Tuktoyaktuk

As discussed elsewhere in this and other chapters, as climate change progresses, the seasons in the Arctic are becoming highly variable, with frequently oscillating daily temperatures and shorter spans of extremely cold weather. Just as in western coastal Alaska, melting permafrost and increased exposure to storm waves are causing increased coastal erosion in Arctic Canada, made worse by sea level rise and increased storm surge frequency. The community of Tuktoyaktuk, located in the Inuvik Region of the Northwest Territories, has implemented measures to address vulnerabilities in local infrastructure. The community has been stockpiling gravel from winter roads to maintain and repair building foundations during the warmer months; boulders and concrete slabs have been used to help protect eroding shorelines. In areas where the shoreline is in the process of collapsing, nearby buildings have been removed or relocated. Prior to establishing protective measures, Tuktoyaktuk was experiencing coastal retreat of about 2 m per year.

Arctic Canada Health-Related Indicators of Climate Change

Healey Akearok et al. (2019) considered the effects of climate change on the health of Canadian Inuit populations. They were inspired by a report by the Lancet Commission on Climate Change and found that climate change is endangering the health of Native peoples in several ways, including changing patterns of disease and mortality, extreme weather events, food insecurity, water scarcity, heat waves, threats to houses and other structures, and threats to public infrastructure (Watts et al., 2017).

To set about their study, Healey Akearok et al. (2019) used the Piliriqatigiinniq Community Health Research Model as a guiding framework. This helped them develop a scoping review of health-related indicators of climate change. From this review, an initial list of 30 indicators was produced (see Table 15.2). Following

TABLE 15.2
Areas of Concern or of No Concern to Arctic Canadian Natives.

Areas of Concern by Villagers	Areas of Little or No Concern
ENVIRONMENTAL HEALTH INDICATORS	
Changing temperature and humidity	Greenhouse gas emissions and air quality
Harmful algae blooms and shellfish poisonings	Allergies to pollen
Permafrost (distribution, melt, shifting limits)	Wildfires
Sea ice (thickness, areal extent, location, duration)	Drought
Water security (e.g., river hydrology)	Slopes—coastal erosion, landslide frequency
The health of terrestrial and marine ecosystems	
MORBIDITY AND MORTALITY INDICATORS	
Extreme weather event injuries and mortality	Morbidity and mortality from extreme heat
Number of injuries or deaths from unstable sea ice	
Rates of environmental infectious diseases	
Respiratory disease related to air quality	
Stress, depression, or anxiety about climate change	
POPULATION VULNERABILITY INDICATORS	
Disability status (physical or mental)	Preexisting chronic disease
Elderly living alone	Heat vulnerability

Continued

TABLE 15.2 Areas of Concern or of No Concern to Arctic Canadian Natives.—cont'd	
Areas of Concern by Villagers	**Areas of Little or No Concern**
Concern for the welfare of children and infants	Flooding risk
Poverty status	Weather unpredictability and variability
Loss of access to fish, marine mammals, country foods	
CLIMATE MITIGATION INDICATORS	
Development of renewable energy	Municipal heat island mitigation plans
CLIMATE ADAPTATION INDICATORS	
Climate change public health workforce availability	Community access to cooling systems in summer
Community participation, climate change initiatives	Heat wave early warning systems
Nunavut organization participation, climate change initiatives	Surveillance systems collecting public health data
Community-based education initiatives about climate change	

Data from Healey Akearok et al. (2019)

on from this, individuals from several Inuit Arctic communities were invited to participate in a consensus-building process to identify their own health-related concerns related to regional climate change. A vital concept in this process is described by the Inuit concept of *Aajiqatigiinniq*: an agreement to set common goals.

Individual feedback and group discussions yielded a final set of 20 indicators chosen by workshop participants (Table 15.2). The identified indicators centered on four key themes: (1) environmental health; (2) morbidity and mortality; (3) population vulnerability; and (4) mitigation, adaptation, and policy.

Indigenous peoples of the North are already affected by climate change, and future climate changes are likely to exacerbate human health challenges. Direct impacts on health in the Arctic have come from rising temperatures and increases in the frequency and strength of storms, floods, droughts, and heatwaves. Indirect effects include increased vulnerability to coastal erosion, linked with declining sea-ice cover. The loss of sea ice is also causing major changes in the timing and success of marine mammal hunting. Winter sea-ice travel is critical to Inuit communities, as they use the ice pack to gain access to wildlife resources and traveling between coastal and island communities. Unpredictable weather patterns, increased storminess, and unreliable ice conditions are undermining Native travel safety and hunting or fishing activities. The increased safety risks are forcing Inuit to travel greater distances to reach harvesting sites.

Longer trips burn more snowmobile fuel, adding to living expenses. Many Native villagers are now avoiding traditional hunting trips because of travel danger. To paraphrase a saying about Alaskan Bush pilots, "There are old Inuit hunters and there are bold Inuit hunters, but there are no old-bold Inuit hunters." Finally, as we have seen elsewhere in this chapter, the unreliability of Arctic weather has made the planning of hunting and fishing trips almost impossible.

Taken in combination, these factors are contributing to a decline in the harvesting of country foods. This has led to decreasing access to those foods for some community members, as reported by the inhabitants of the Canadian Arctic villages of Arctic Bay, Kugaaruk, Naujaat, and others (Ford, 2009). Native health concerns associated with Arctic climate change aggravate existing health issues, including mental health and wellness, nutritional deficiencies, rates of respiratory illness, livelihood and economic stability, safety, and the spread of disease.

Indigenous Observations of Changes in Animals and Insects

The level of native awareness and keen observation of nature has been discussed earlier. The following information highlights the acuity of Native nature observations. As part of the assessment of the impacts of warming in the Arctic, Hassol (2004, p. 71) cited several Indigenous observations on recent changes in regional animals and insects, based on notes taken during interviews and texts submitted by informants. Here are my edited versions of these statements (i.e., not direct quotes):

Bering Strait region, Alaska

Seals: Spotted seal numbers declined from the late 1960s/early 1970s to the present. The spring breakup in 1996 and 1997 came early, resulting in more strandings of baby ringed seals on the beach. There are fewer seals in the Nome area these days, perhaps because of less ice for their dens.

Walrus: The physical condition of walrus was generally poor from 1996 to 1998. The animals were skinny, and their productivity was low. One cause was reduced sea ice, which forced the walruses to swim farther each day between feeding areas. Walruses returned to good condition in the spring of 1999 after a cold winter with good sea-ice formation in the Bering Sea.

Birds: Spring bird migrations took place early in the year. Geese and songbirds arrived in late April, earlier than in the past. In August of 1996 and 1997, there were large die-offs of kittiwakes and murres. Other birds are doing well. The lack of snow has been helpful to the ptarmigan. They also benefitted from a lower hare population that year, as hares compete with ptarmigan for food.

Fish: Chum salmon numbers crashed in Norton Sound in the early 1990s and have been down ever since.

Insects: New insects have been appearing in the region that have not been seen before. Mosquito numbers remain the same.

Sachs Harbor, Nunavut, Canada

Seals: There has been an increasing occurrence of skinny seal pups seen at spring breakup.

Caribou: Increased forage availability has helped the caribou. There have been changes in the timing of caribou migration between islands.

Bears: Fewer polar bears have been seen in the autumn due to delays in sea-ice formation.

Muskox: There has been increased forage availability for muskox.

Birds: It has been difficult to hunt geese in the spring because of rapid ice melt. Robins, and other previously unknown species of small birds, have been observed.

Fish: Different species of fish have been observed. More "least cisco" fish are being caught now. Two species of Pacific salmon have been caught near the community for the first time.

Kitikmeot region, Canada

Caribou: Caribou are changing migration routes due to early cracks in the sea ice. Changes in vegetation types and abundance are affecting caribou foraging strategies. Massive caribou drownings have been increasing due to thinner sea ice. For instance, a massive drowning event was observed in 1996. Lower water levels may mean that caribou can save energy by not having to swim as far. However, changing shorelines are affecting caribou forage behavior, though it is unclear how this is happening. Caribou deaths due to exhaustion from extreme heat have been observed, as well as exhaustion from attempts to escape the increased numbers of mosquitoes.

Seals: Seals have been surfacing through the unusually high number of cracks in the sea ice in early spring around Hope Bay. This attracts polar bears.

Bears: Grizzly bears were seen for the first time crossing from the mainland northward to Victoria Island in 1999. During spring 2000, unusually high numbers of grizzly bears and grizzly tracks were seen there.

Birds: New birds have recently been seen here for the first time, such as the robin and an unidentified yellow songbird.

Insects: The number of mosquitoes has been increasing with temperature, but this occurs only up to a certain threshold, after which their numbers collapse.

Baker Lake, Nunavut, Canada

Caribou: The caribou have not been as healthy in recent years. Their meat is tough. More liquid has been seen in their joints, and more white pustules have appeared on their meat. Caribou are thin and undernourished due to the heat and aridity of recent summers. Caribou skins are weak and tear easily during field dressing. There are more diseased caribou in the region, as shown by sores in mouths and on the tongue. However, the links between deteriorating caribou health and climate change are not always clear.

Bears: There have been more grizzly bear sightings and encounters in the Baker Lake area.

Muskox: There have been more muskox sightings around Baker Lake.

Birds: The birds seem smaller and not as lively and energetic (happy). Redpolls and white-throated sparrows have become more common, but Lapland longspurs are hardly seen any more. There are now more ravens in the area.

Fish: There have been changes in local fish, mainly char and trout. The trout are now darker in color, and little fat is observed between meat layers when they are boiled. Fewer fish are being caught in the traditional fishing spots. It appears that the fish are eating things they are not supposed to eat. Regional fish have become thinner, their meat has become mushy, and they smell different than they used to. This new smell has been described as "like earth."

Insects: Mosquitoes are decreasing in numbers in some areas as summer temperatures rise since there is less standing water. At least 10 new kinds of insects have been seen in the area. They are all winged insects, and some of them are recognized as having come from the treeline area. Strange kinds of flying insects are also being observed. Warmer temperatures may be responsible for the arrival of flying insects from the south and for insects remaining active longer in the year.

Northwestern Hudson Bay, Canada

Whales: There has been a decrease in whale numbers in the Arviat and Repulse Bay area.

Walrus: There has been a decrease in walrus numbers near Arviat and Whale Cove. However, there has been an increase in walrus numbers near Coral Harbor and Chesterfield Inlet.

Caribou: There has been an increase in caribou numbers. They do not appear to be intimidated by oil exploration activities in the region. In fact, they are even feeding close to the exploration camps. There has been a change in their diet.

Bears: Inuit in northwestern and eastern Hudson Bay report increasing numbers of polar bears. These bears appear to be leaner and more aggressive than usual.

Birds: There were more Canada geese in Repulse Bay during the summers of 1992 and 1993.

Fish: There has been a decrease in Arctic cod in near-shore areas. Arctic cod can no longer be found in near-shore areas off Cape Smith and Repulse Bay.

Saami observations from northern Finland

Birds: Many types of birds have declined in numbers including crows, buzzards, and some falcons. Arctic terns, long-tailed duck, and osprey have disappeared in some areas.

Fish: Fish populations have gone into decline in many lakes, partly to due overharvesting, but also due to unknown factors. For example, in the reindeer village of Kaldoaivi, perch have disappeared, but pike remain.

Insects: The number of insects has decreased, including mosquito populations.

Transportation Hazards

In addition to increasing dangers to native peoples attempting to traverse sea ice in winter, there is also the problem from how to get out onto the sea ice from the land. Traditionally, land-fast ice formed a bridge connecting the land to sea ice. However, just as the edge of the sea ice itself has been retreating, so also the land-fast ice has been forming later in the autumn and melting earlier in the spring. This section explores the comings and goings of land-fast ice in the Arctic.

Land-fast sea ice

Land-fast ice, also called "shore-fast ice," is an essential component of the coastal sea-ice system (Mahoney, 2018). Unfortunately, the extent and seasonal duration of land-fast ice on Arctic shores has been declining in recent years (Fig. 15.2), as shown in satellite image interpretations by Mahoney (2018).

Acting as a floating extension of the land, land-fast ice is immobile sea ice frozen to the shore that forms along the Arctic coastline during winter and spring. Land-fast ice is a stable, critically important transportation platform, connecting isolated communities and providing access to traditional hunting and fishing grounds for 3−9 months each year. The presence of land-fast ice also limits coastal erosion, which threatens many Arctic communities located along subsiding coastal margins. Land-fast ice is a critical habitat for marine mammals such as seals and polar bears, and the polynyas that form seaward of land-fast ice edges create hot spots of high ecological productivity. Land-fast ice, rather than drift ice, therefore, provides most of the "sea-ice services" utilized by Arctic communities.

Members of Arctic coastal communities travel across land-fast ice to hunt marine mammals and birds commonly found at its seaward edge. Ringed seals are an important prey species for both polar bears and Arctic subsistence hunters. These seals are uniquely adapted to maintain breathing holes in land-fast ice, which otherwise lacks the openings (e.g., cracks and leads) used by marine mammals in the pack ice. Land-fast ice also plays an important role in the sediment dynamics of coastal waters. It buffers coastal embankments against the erosive action of waves. Finally, land-fast ice allows river plumes to extend farther than they would under open water or pack ice by isolating the underlying ocean from wind mixing.

Land-fast sea ice Canada

Land-fast sea ice comprises only about 12% of global sea-ice cover, yet it has enormous importance for Arctic

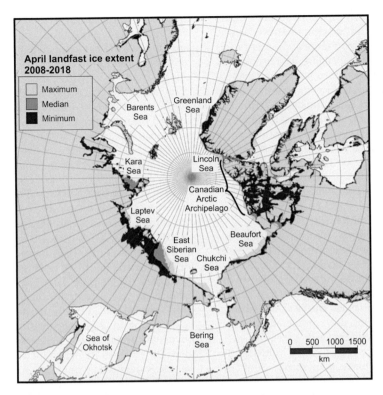

FIG. 15.2 Above: Changes in Arctic landfast ice cover from 2008–2018, after NOAA 2018. Below: Two community members returning by snowmobile from a flaw lead at the seaward edge of the landfast ice near Utqiaġvik, Alaska. (Photo courtesy of NOAA, 2018, in the public domain.)

societies and ecosystems. Relatively little is known, however, about the dominant drivers of its breakup or how it will respond to future climate warming. Since the early 2000s, residents of many Arctic communities have reported that land-fast ice has become thinner, freezes later, and breaks up earlier than in the 1990s. These changes increase travel risk, reduce hunting success, and threaten traditional activities, further exacerbating food insecurity in communities already experiencing socioeconomic and cultural stress. Despite its critical socioeconomic importance and reports of its decline, land-fast ice is challenging to observe and model and therefore has remained poorly understood. Long-term changes in land-fast ice and controls on its decline, therefore, remain largely unknown, especially at the community scale outside of Alaska (some in-depth studies have been performed around Point Barrow). Cooley et al. (2020) utilized 19 years of satellite imagery to document the timing of land-fast ice breakup in 28 communities in northern Canada and western Greenland. Their data came from daily cloud-filtered imagery acquired by NASA's Moderate Resolution Imaging Spectroradiometer (MODIS).

The authors combined past observations with numerical models of future Arctic warming. The results and projections presented by Cooley et al. (2020) provided helpful insights into the varying spatial patterns of Arctic climate change and the exposure of coastal communities to environmental change. For instance, based on their model results, they considered that other factors beyond air temperature (such as winds, ocean temperatures, and surface waves) likely influence the timing of land-fast ice breakup in spring. Similarly, breakup timing is just one of several land-fast ice metrics that affect community ice usage. For example, the two warmest communities in the study, Uummannaq (West Greenland) and Sanikiluaq (southern Hudson Bay), already experience much earlier breakup than other communities (May 29 and June 21 on average, respectively) and a shorter ice season overall. These communities are thus even more susceptible to even comparatively small changes in breakup timing, as they are already operating on a very limiting land-fast ice schedule. For instance, sea ice never formed during the winter of 2005–06 in Uummannaq. This meant that local people could not gain access to the sea ice, so they and their dog teams remained on land for nearly 2 years and marine mammal hunting ceased. In warming climate scenarios, such episodes may become more frequent and lead to the loss of traditional cultural activities such as dog sledding and seal hunting on the ice.

The variability in land-fast sea ice observed by Cooley et al. (2020) appears to be localized and is only weakly correlated with larger-scale Arctic ice records. This result corroborates previous reports by both Native residents and scientists. An unexpected result was the discovery that the coldest, northernmost communities of northern Canada and western Greenland appear to be the most vulnerable to future land-fast ice reductions. This emphasizes the importance of the local nature of climatic changes alongside community-level differences, a fact that might otherwise be overlooked when climate change policy decisions are being made in national capitals. There simply is no substitute for local knowledge, whether it is based on first-person Native accounts or decades of accumulated satellite data. Finally, breakup timing is strongly correlated with springtime air temperature, but the relationship is not straightforward.

Pacific Arctic Sea Surface Temperature and Sea-Ice Change Since 2017

According to the National Snow and Ice Data Center (NSIDC), in the spring of 2017, sea ice retreated from the Chukchi Sea earlier than any other year in the satellite data record (Fig. 15.3). Satellites have continuously monitored sea ice since 1979, so the events of 2017 break a 38-year record. By the third week in May, open water extended north to Utqiaġvik, Alaska. The rapid retreat in 2017 apparently marked a turning point—the beginning of a trend that has persisted since then. The early sea-ice departure from the Chukchi Sea was likely triggered by unusually high regional air temperatures through the previous winter.

Consistent with warm conditions, extensive open water in the Chukchi Sea region persisted into December 2016. Delayed ice growth that winter contributed to thinner ice than usual in the following spring. Another driver that year was the action of strong southerly winds at the end of March and early April. These winds broke up pack ice in the Chukchi Sea, pushing it through Bering Strait into the Bering Sea. Based on collaborative research by the NSIDC and the University of Washington, the pattern of spring sea-ice retreat also indicates the increasing role of oceanic heat flow from the Bering to the Chukchi Sea via Bering Strait.

The early sea-ice retreats that began in 2017 challenge previous notions that winter sea ice in the Bering Sea has been stable over the instrumental record, although long-term records remain limited. To work out the long-term history of winter sea-ice extent in the Bering Sea, Jones et al. (2020) analyzed a 5500-year-old oxygen isotope from St. Matthew Island. They found that sea ice in the Bering Sea decreased in response to increasing winter insolation and atmospheric CO_2 over the past 5500 years. This result suggests that the North Pacific is very sensitive to small

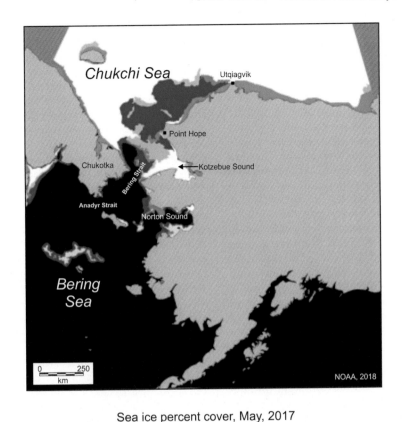

Sea ice percent cover, May, 2017

No ice   <10   10-30   40-60   70-80   90-100   fast ice   no data

FIG. 15.3 Bering and Chukchi Sea ice extent in May 2017. (From Mahoney (2018), in the public domain.)

changes in radiative forcing. Interestingly, Jones et al. (2020) found that the sea-ice conditions in 2018 were likely the lowest of the past 5500 years. They also concluded that sea-ice loss may lag behind changes in CO_2 concentrations by several decades. This, in turn, suggests that the large-scale decline in sea ice in the Bering Sea in the past few years may signal the beginning of a long-term trend, ultimately driven by the increased CO_2 concentrations of modern times.

Pacific Arctic Ecosystem Change Since 2017

As reported by Huntington et al. (2020), the sea-ice retreat from the Pacific Arctic region reported by the NSIDC (2017) formed part of a major upheaval in regional oceanography and marine ecology. The Pacific Arctic region comprises the Chukchi and northern Bering seas. To understand the importance of this ecological upheaval, it is necessary the grasp how the Arctic Pacific has functioned as an ecosystem until recently. This region encompasses one of the world's

most productive ocean ecosystems, characterized by high benthic biomass resulting from persistent, nutrient-rich flow through Bering Strait, which fuels high primary production. The region is home to millions of nesting and migratory seabirds in summer and autumn, with hot spots of foraging activity shared with marine mammals. Coastal Yúpik and Iñupiat communities rely extensively on these food resources. The structure of this ecosystem has been characterized by the oceanic transport of nutrients in combination with the extent and timing of the sea ice. The freeze-up of sea ice in autumn and winter eliminates large expanses of open water, causing whales, Pacific walruses (*Odobenus rosmarus divergens*), as well as seals, and seabirds to migrate southward into the Bering Sea and beyond.

The Bering Sea cold pool consists of near-bottom shelf waters cooler than 2°C, located south of Bering Strait. This has long served as a thermal barrier that prevents northward migration of sub-Arctic groundfish (e.g., Pacific halibut, pollock, cod, sablefish, turbot, flounder, sole, plaice, and rockfish). These fish supply the majority of stocks for the southeastern Bering Sea's US $2 billion fishery and, amazingly, account for about half of all the seafood landings in the United States (Huntington et al., 2020).

In addition to the northward retreat of pack ice beyond the Chukchi shelf in recent summers, warmer shelf waters have delayed sea-ice formation in autumn. Simultaneously, regional ocean hydrology has changed. Huntington et al. (2020) reported that the northward flow of water through the Bering Strait used to occur on a smaller scale, but the rate of this flow has increased, as has its temperature. Hence, the North Pacific is now delivering more heat, freshwater, nutrients, and biota northward into the Arctic. Near-bottom water temperatures now exceed $0°C$ for a larger portion of the year. These radically altered conditions have facilitated the movement of more salmon in the Chukchi and Beaufort seas. As discussed before, walruses are now hauling-out onshore instead of remaining on sea ice in northwestern Alaska in late summer. There has been an increase in the frequency and seasonal duration of killer whale (*Orcinus orca*) presence in the Chukchi Sea, as well as an increase in planktivorous seabirds feeding in the Chukchi Sea and a northward distributional shift of other seabird species. In the northern Bering Sea, the whale hunting season for Native peoples has shorted in spring but lengthened into early winter.

Since the sea-ice changes of 2017 have persisted in all subsequent years, this suggests that 2017 was not a passing oddity of brief consequence to marine-ecological systems, but a sign of what is to come. In early January 2017, the sea-ice edge had barely progressed south of Bering Strait, and for the entire winter, its extent remained at least $200{,}000 \text{ km}^2$ below the long-term average. In June, ship-based observations found that the near-bottom ocean temperatures in the Bering Strait were nearly $4°C$, more than $3°C$ above (and four standard deviations warmer than) the 1991–2016 mean (Huntington et al., 2020). These massive temperature changes were not restricted to the Bering Strait region. Waters in Norton Sound exceeded $10°C$ before the end of June 2018, and the size of the cold pool remained minimal through late summer. These thermal alterations have had pervasive ecological ramifications. For instance, the reduced ice cover almost certainly limited oceanic primary production by influencing temperatures, light levels, and stratification conditions. In the spring of 2018, the southern Bering Sea phytoplankton bloom was delayed due to a lack of freshwater input from melting sea ice, causing

chlorophyll concentrations to be an order of magnitude lower than usual. Meanwhile, in the northern Bering Sea, the ice-associated bloom was early and extensive. Furthermore, toxins found in a few stranded and harvested walruses from Bering Strait villages led to concern about harmful algal blooms (saxitoxin) and food safety for the Indigenous residents.

This ecosystem transformation will certainly include a cascade of sequential changes that take place over multiple years, although changes to individual ecosystem components may be sudden and dramatic. Huntington et al. (2020) argue that the ongoing changes may result in the transformation of an Arctic marine ecosystem into one that is characterized by sub-Arctic conditions, sub-Arctic species, and sub-Arctic interactions. In other words, the Chukchi Sea may come to resemble the Bering Sea shelf in structure and function, characterized by annual sea-ice cover, warmer bottom water temperatures, and ecosystem productivity derived from forage fishes and pelagic zooplankton rather than from benthic organisms and detritus. Changes in the historically strong benthic-pelagic coupling have already been observed in the southeastern Chukchi Sea. Here, overall epibenthic biomass (organisms that live on or just above the sea-floor) declined by an order of magnitude from 2004 to 2017, but despite this, there has been little species turnover in the region, suggesting overall changes in ecosystem productivity and functioning rather than specific habitat changes.

Where, then, is this marine ecosystem makeover headed? It is clearly more complex than a simple northward migration of an ecosystem. There are some physical characteristics of the Chukchi Sea that may cause it to retain its characteristics that distinguish it from the Bering Sea shelf, including its higher latitude and its downstream location relative to nutrients coming through Bering Strait. Many aspects of the ongoing transformation remain to be seen, but it seems likely that the sea-ice season will continue to grow shorter and that sea-ice coverage will also diminish. Waters will become warmer, and warm water will persist longer into autumn and winter. In summary, physical oceanic conditions that were once anomalous may become normal. The biological response to the new set of oceanic conditions may be delayed for several years until the species that thrive under the new regime can migrate and persist in their new territories.

And what of the Natives of the Arctic Pacific region? How will they cope with these changes? Whale hunters and fishers will have to adjust to their modus operandi to some degree. For instance, as we have seen, the changes in sea-ice and fast-ice conditions have already started to force changes in the timing of regional whale hunts. They have had to be curtailed in spring and lengthened in the fall. Fishers may have to switch targeted species as new fish replace old ones. This is quite serious business. The well-being of coastal communities and the management of human activities in the region, including subsistence as well as commercial fisheries, depend on what happens next.

In Alaskan waters, researchers are planning strategies to reduce commercial interference with Alaska Native subsistence harvests, and conscientious vessel operators are talking to Native communities about the need to adjust their plans to avoid areas where whale hunters are active. Concerns about food security will likely grow alongside uncertainties concerning the timing, scope, and paths of fish and marine mammal migrations and the continuing need to maintain optimal harvest conditions. Coastal communities are likely to face difficult choices between taking advantage of some increased economic opportunities and limiting commercial fishing interference with their subsistence activities (Huntington et al., 2020).

STRATEGIES FOR CLIMATE-INDUCED COMMUNITY RELOCATIONS IN ALASKA

An article by Bronen and Chapin (2013) presented a different perspective on governance and institutional strategies for climate-induced community relocations in Alaska. As discussed at length in Chapter 14, repeated extreme weather events coupled with climate change—induced coastal erosion have greatly damaged the habitability of several communities on the west coast of the state. While community residents and government agencies agree that relocation is the only viable strategy to protect lives and infrastructure, the carrying out of proposed relocations will be extremely costly, stretching the financial resources of local, regional, and state governments. The authors examined relocation constraints for three Alaskan Native communities: Kivalina, Newtok, and Shishmaref.

Outsiders might be forgiven for questioning the need of the people of these villages to relocate to new places rather than to simply merge with nearby native communities, thereby saving the money it would take to develop new infrastructure. The pressures to move because of environmental hazards are countered by reasons people are reluctant to leave their ancestral homes. La Ganga (2015) quotes a longtime Kivalina resident: When asked why her people don't move—somewhere,

anywhere to be safe—she is polite but firm. The land and the water make the Iñupiat [of Kivalina] who they are. If they moved to Kotzebue, they would be visitors. Moving to Anchorage or Fairbanks, she said, "would be like asking us not to be a people anymore."

Notwithstanding these sentiments that are certainly genuine and heartfelt, the need for action is urgent in many cases. As discussed in Chapter 14, several Alaska Native communities will not be able to remain where they are much longer, and in the case of Newtok, many residents have already fled the flooded townsite. In Alaska, the lack of an overarching institutional relocation framework has caused the relocation of Kivalina, Shishmaref, and Newtok to proceed in an ad hoc manner. Each community took a somewhat different approach to its relocation planning process. Newtok began their relocation planning with the Alaska Dept. of Commerce, Community, and Economic Development (DCCED). Kivalina and Shishmaref worked primarily with federal agencies, including the US Army Corps of Engineers (USACE). Kivalina attempted to use legal challenges to fund initial infrastructure, whereas Newtok engaged with a complex group of agencies, some of which were able to access funds not specifically designated for relocation. Communities also differed in local governance structure. Newtok has only one governing body (the Newtok Traditional Council); Shishmaref formed a working group comprised of elders and tribal and city government representatives; Kivalina worked through its two local governing bodies, the city government, which is a political subdivision of the State of Alaska, and the tribal council, which has a direct relationship with the federal government.

2020 GAO Report

In July 2020, the US Government Accountability Office (GAO) published a report discussing a pilot program for climate mitigation that would enhance the nation's resilience to natural disasters. Interestingly, this report used Newtok, Alaska, as a case study. The GAO report examined an underlying problem with federal support for climate mitigation projects.

Federal programs provide only limited support for climate migration efforts because they are designed to address other priorities, according to a GAO literature review and interviews with stakeholders and federal officials. Thus, the situation discussed by Bronen and Chapin (2013) remains unresolved. The GAO report concluded that federal programs are not designed to address the scale and complexity of community relocation. Federal support has generally gone to fund government purchase and subsequent demolition of properties at high risk of damage from disasters. However, even this program is reactive in nature, not proactive. It always comes in response to a specific event such as a hurricane or tornado. Unclear federal leadership is the key challenge to climate migration as a resilience strategy because no single federal agency has the authority to take charge of assistance for climate migration. This has led to a lack of long-term planning so that support for climate migration efforts has only been provided on an ad hoc basis. For example, it has taken over 30 years to begin relocating Newtok, in large part because no federal agency has the authority to coordinate assistance, according to village inhabitants.

According to the relocation project manager, the Newtok relocation effort received about $64 million in funds from federal agencies, the state of Alaska, and other organizations through December 2019. Infrastructure and housing construction continue at the new townsite of Mertarvik, and 135 people have moved there permanently. However, the June 2019 Master Implementation Plan for relocation and the Denali Commission (a federal organization operating in Alaska) estimated that the project would need around $115 million to complete the development of the new site, provide sufficient infrastructure, and perform cleanup of the Newtok site. Meanwhile, residents remaining in Newtok face increased disaster risks because the relocation to Mertarvik will not be complete before coastal erosion and flooding make Newtok uninhabitable.

According to the US Global Change Research Program (2018), indigenous populations, some of which have lived on their traditional lands for thousands of years, have place-based cultural, religious, economic, and traditional knowledge systems that are foundational to their identities and physical and mental health. For example, Native elders in Newtok said they view the loss of their land as an existential threat to their cultural identity and place-based traditions. Consequently, community migration efforts should be community driven with coordination across all the relevant levels of government.

Recent Developments in Relocation Policy

In June 2020, Federal Emergency Management Agency (FEMA) officials told GAO investigators that "managed retreat" (i.e., community relocation) is a politically unpopular option unless there is a particularly catastrophic event that changes peoples' willingness to stay in place and creates unified state and local support for relocation. Additionally, these officials said that

political will for pursuing climate migration would need to be sustained over time for relevant policies to be implemented.

The federal government's resistance to relocation policies finally started to break down in August 2020. Flavelle (2020) reported in the New York Times (August 26, 2020) that federal policies concerning relocation are finally changing. As we have seen, through July 2020, policymakers in Congress have refused to consider paying for the relocation of entire communities away from danger zones, regardless of the severity of hurricanes and tornadoes. However, in August 2020, the FEMA released plans for a new program designed to pay for large-scale community relocations nationwide. The funding for this new initiative will initially be $500 million, with additional billions to come. The Department of Housing and Urban Development followed suit with a similar $16 billion program. The impetus for these new policies was, in part, a decision by the Army Corps of Engineers to start telling local officials in flood-prone areas that they must agree to force people out of their homes or forfeit federal money for flood-protection projects.

The increasing acceptance of relocation (aka "managed retreat") represents a major shift in the national psyche. It has also changed the minds of lawmakers and federal emergency program administrators. The concept goes against the grain long-entrenched American ideals. Even the word "retreat" has connotations of defeat, which sits uncomfortably with the American values of self-reliance and rugged individualism. But the criticism that relocation represents defeat has been blunted by years of violent hurricanes, floods, and other disasters, as well as the scientific reality that rising waters ultimately will claim waterfront land. In the US Global Change Research Program (2018), 13 federal science agencies characterized the need to retreat from parts of the coast as "unavoidable" in "all but the very lowest sea-level rise projections."

According to that same assessment, increasing public awareness of the current and future impacts of climate change, coupled with the growing cost of recovery (federal spending on disaster recovery has totaled almost half a trillion dollars since 2005), has led Americans to realize that some coastal localities simply cannot be protected. The shift is even more remarkable since it took place during the presidency of Donald Trump, who has called climate change a hoax and rolled back programs to fight global warming (Flavelle, 2020).

As of June 2021, no new assistance has been given by the FEMA to support the relocation project Newtok.

Inuit Governance and Cooperation with the Canadian Government

Thus far in this chapter, we have focused almost exclusively on the political situation in the United States as it relates to Alaskan native groups. In the following sections, we will review the political situation in Canada, regarding Inuit legal rights and privileges.

Do Native peoples of the United States and Canada enjoy the same rights?

A review of this question was published by the American Bar Association (ABA) in 2016.

Federal law concerning Native Americans in the United States is both convergent with and divergent from the law that applies to First Nations in Canada. Although originating through similar colonial histories and eras of national policy, distinct foundational law in each nation has resulted in dissimilar approaches to tribal authority and jurisdiction, application of state and provincial laws, and economic development on reservations and reserves. In both the United States and Canada, federal law and policy affecting indigenous peoples has chiefly focused on land claims and government recognition and has followed mostly parallel legal and policy development since the 19th century.

In the United States, federal power over tribes is balanced against tribal sovereignty or the inherent governmental authority of tribes as pre- and extraconstitutional nations. The Supreme Court's modern conception of tribal sovereignty stems from cases published by Chief Justice John Marshall in the early 19th century. The guiding principle has been that tribes retain limited sovereignty, with the greatest authority over tribal members and tribal lands. The relationships are between the federal government and Native tribes—US states generally do not have legal authority over tribes under US law (ABA, 2016).

In Canada, the federal government is assigned power through the Constitution Act of 1867. As in the United States, federal authority in Canada is both exclusive and broad, as reflected in the federal Indian Act, the principal statute governing the status of Indians and First Nations and the management of reserve lands.

The Canadian Constitution Act of 1982 specifies concerning the Rights of the Aboriginal Peoples of Canada that the "existing aboriginal and treaty rights" of Indian, Inuit, and Métis peoples are "recognized

and affirmed." More recently, the Canadian Supreme Court examined the extent of aboriginal rights and concluded that they include a range of cultural, social, political, and economic rights, including land rights, hunting, and fishing rights, and the right to practice one's own culture. However, they do not include any aboriginal rights that had been nullified by the Canadian government before the 1982 Constitution Act. The Supreme Court must determine that the right is "integral to a distinctive culture," under agreed guidelines, including the following:

1. The court must take into account the perspective of aboriginal peoples themselves.
2. To be integral, a practice, custom, or tradition must be of central significance to the aboriginal society in question.
3. The practices, customs, and traditions that constitute aboriginal rights are those that have continuity with the practices, customs, and traditions that existed prior to contact.
4. The court must take into account both the relationship of aboriginal peoples to the land and the distinctive societies and cultures of aboriginal peoples.

Federally acknowledged American Indian tribes and Canadian First Nations share a similar status in relation to their respective federal governments. Indigenous legal and political rights ultimately are constrained and may be curtailed by federal authority. There are some interesting differences, however. Native American tribes derive their indigenous rights from their status as political sovereigns. The US federal government's acknowledgment of a tribe's sovereign status carries a full range of sovereign rights unless specifically limited. In Canada, First Nations rights follow from "traditions" and are a "defining feature of the culture." Therefore, in Canada, a recognized First Nation has a defined set of aboriginal rights that may be supplemented through First Nation petitions to the Supreme Court. In theory, constitutionally recognized sovereignty should empower American Indian tribes to a greater degree than the culturally defined practices that guide the relationship between First Nations and the Canadian government. The overriding principle is that the federal governments of both countries retain the power to abrogate indigenous sovereignty.

A key difference between Native rights in the United States and the First Nations of Canada is the way in which the federal government conceptualizes tribal rights. In the United States, tribal sovereignty is considered a "full box" of legal and political rights. In Canada, aboriginal rights begin with an "empty box" with the burden of proof placed on First Nations groups to demonstrate to the court that the group has a protected aboriginal right. Filling the empty box with an aboriginal right requires a First Nations group "to have that aboriginal right clearly identified in a formal judgment of the Supreme Court of Canada, which would require the First Nations group to undergo a very lengthy process taking between 10 and 15 years to complete. The legal costs of such prolonged deliberations are well beyond the means of most First Nations groups" (ABA, 2016).

Canadian Inuit Rights From Constitutional and Land Claim Process

Mary Simon comes from Kuujjuaq, Nunavik. She is the past president of the Inuit Tapiriit Kanatami (2006—12) and has recently completed a term as Chairperson of the National Committee on Inuit Education (2012—14). She has advanced critical social, economic, and human rights issues for Canadian Inuit regionally, nationally, and internationally. Here are her views on Inuit rights to participate in the governance of shipping through the Northwest Passage, as published in Kelley (2017, pp. 26—28):

> Certainly, as Inuit who live in fifty-three coastal communities except for one that is more inland, we see and deal with climate change season over season, year over year. We now realize that it is much more difficult to predict freeze up and break up and ice conditions. We also know very well the increasing number of ships that are coming through Arctic waters including large cruise ships that ply through our waters. We know also that the resource developers and the rest of the world are also looking northward to the Arctic's riches that warming temperatures and growing open water are making increasingly accessible. These are great challenges for all of us, but I am confident that we are up for the challenges confronting us because so much more than the climate has changed. In fact, I am optimistic because of the legal, political and intellectual foundation we have gained through the Constitutional and land claim process. Consider, collectively across Inuit Nunangat we have settled four comprehensive land treaties and an offshore agreement covering thousands of small regional islands. Through these five agreements, we have acquired constitutionally entrenched rights, instruments and organizations to protect our rights and share in the economic opportunities that a changing Arctic will offer.
>
> As Inuit, I believe we must push Canada to become a global leader in resolving the challenges of climate change through strategic investments in technologies to adapt to the changing conditions. We can become a centre of excellence for international scientific and Inuit traditional knowledge to understand the extreme changes that are occurring. Above all, we need to secure the basis of our relationship and use of the land, ice and ocean that produces our traditional food and defines us as a people.

Inuit Opposition to Oil Development

Peter F.K. Ittinuar lives in the village of Kangiqtugaapik (Clyde River) inland from the west coast of Hudson Bay. He currently works for the Negotiations and Reconciliation Division of the Ministry of Indigenous Relations and Reconciliation in the Ontario Government. He had some interesting things to say about the inclusion of Native people in the government decision-making process, especially concerning the governance of oil and gas development and shipping through the Northwest Passage, published in Kelley (2017, pp. 43—47):

> Witness the very public happening as I write this, the passing of the 1,600 passenger mega cruise ship Crystal Serenity through these waters, which ironically and perhaps inadvertently celebrates the opening up of the once-fabled Northwest Passage due to climate change and its effect on ice conditions in the waters of the Far North. This is not good news for the Indigenous population. We prefer it the way it is, and moreover, the way it was. The Crystal Serenity is not the only creature assaulting the waters of the Northwest Passage, there are many others in many forms: physical, in the form of cruise ships and private watercraft owners; in oil and gas interests from industry; and geopolitical territoriality from major powers interested in the newly accessible arctic waters.
>
> … since there appears to have been no attempt to settle [Inuit of Clyde River wanting to block exploratory activities by oil companies] based on equal and mutual respect, one must resort to the hard road of litigation; and there is the feeling that reconciliation was not achieved, and that the decisions in this context will be short-lived. This decision goes to the very core of the meaning of reconciliation for Inuit and Indigenous Peoples across Canada - reconciliation and consultation are inextricable.
>
> One thing is certain, Inuit feel that they have the first right of access to those waters, and that we must be fully consulted before activities take place that may irretrievably and irrevocably harm the current life of those waters. Life in the water sustains life onshore, both for its people and its animals. Inuit do not say no to development, but we want it done in a way that perhaps has never been done before, without harm to the environment, and hence, ourselves.

NATIVE RUSSIAN RIGHTS AND AUTONOMY

Sitdikov et al. (2019) discussed the history and current issues concerning Territorial (Native) and National autonomy in Russia. They pointed to the year 2000 as the turning point in the ongoing process of government centralization. The Kremlin renounced mutual agreements with the individual territories that remained part of Russia after the breakup of the Soviet Union. Then there was an enlargement of political regions and the creation of seven federal districts. Elections to decide regional leaders were largely replaced by the nomination of leaders chosen by the President of Russia, Vladimir Putin. Presidents of Republics became Heads of Regions. While preserving certain outer, formal signs of the former administrative-territorial system, post-2000 regional standardization led to the significant reduction in the political power and influence of regional bodies, and cultural communities ceased lobbying for independence (separation from Russia) and power-sharing schemes. To prevent protests, however, it is not necessary to destroy autonomy but to suggest an alternative— something different. The proffered alternative from the central government was the concept of "cultural autonomy." The Russian multinational state does not impede the cultural self-determination of ethnic minorities, so long as they do not attempt to claim political autonomy as a separate territory or state.

The narrow interpretation of culture in Russia, combined with the lack of meaningful governance below the national level, leads to the frequent exclusion of Siberian Indigenous groups from the processes that form the law and state policy. According to Western criteria, European leaders consider that ethnic minorities in Russian have lost their territorial autonomy.

It is not difficult to delineate this systematic power drainage. Any demand for autonomy by Native groups must pass through the filter of political and economic interests in the Kremlin. Although theoretically Territorial (Native) and National autonomy do not conflict, there are still no examples of the simultaneous prosperity of territorial and federal autonomy in Russia.

There has been an inversion of autonomy priorities since the dissolution of the Soviet Union (Sitdikov et al., 2019). Pleas for territorial autonomy sometimes succeeded during the Soviet era, but now such claims fall on deaf ears. Regime change has led to the implementation of the kind of national cultural autonomy that held sway in Tsarist Russia. In any case, Sitdikov et al. (2019) argue that the idea of "cultural autonomy" remains unpopular with the European Russian population. Its legal basis has been dissolved. It is convenient for the national government to maintain different administrative-territorial units in Russia, including republics, regions, cities of federal significance, and

autonomous regions. These divisions give the outward appearance of regional autonomy, and they serve practical purposes of regional administration, but they lack any real political power.

Siberian Reindeer Herding (Fig. 15.4)

Burtseva et al. (2020) studied the impacts of large-scale mining operations on the indigenous peoples of the Russian North and Far East. By the time European Russians entered the vast territory of Yakutia, the region had been inhabited by the Evenks for millennia. The northeastern part was inhabited by Lamuts (ancestors of modern Evens) and Yukaghirs, who led a nomadic and seminomadic lifestyle and were engaged in reindeer breeding, hunting, fishing, and gathering wild plants. Today this lifestyle remains the central aspect of the lives of these indigenous peoples. Indigenous peoples have long inhabited the harsh Arctic and sub-Arctic territories and have conducted their activities according to their own customs. Before becoming part of Russia, the Sakha people and indigenous minorities of the North—Evens, Evenks, Dolgans, Yukaghirs, and Chukchi—governed themselves and established agreements with neighboring groups (see Fig. 15.5). With the Russian annexation of Eastern Siberia and the North in the 1630s, the life of indigenous peoples radically changed. Their customary laws were superseded by the state laws of Russia.

Jumping forward several centuries, the management of indigenous peoples before 1917 hinged on an act called the "Charter on the Management of Indigenes" (1822), in which tribal and neighboring communities were recognized as self-governing bodies and the subject of land tenure. The charter protected Native groups from cultural assimilation, land seizures, violation of their customs and traditions, cultural degeneration, and extinction. After the establishment of Soviet power, the way of life of the peoples of the North changed completely. In the 1920s, self-government bodies were created (tribal councils, tribal meetings, executive committees). All the land became the property of the Soviet state.

In the 1930s, collective reindeer herding farms were organized, and national-territorial formations (national regions, autonomous districts) were created. Under the rules of a planned economy, the state carried out purposeful work to transfer the nomadic population to a settled way of life. That is, they forced previously nomadic groups to settle in permanent villages—not unlike what happened decades earlier in the United States and Canada. In the late 1950s to early 1960s, individual farms were forced into collective and state farm

FIG. 15.4 Top: Reindeer herd, northern Sweden. Middle: Dolgan man with reindeer, early 20th century. Bottom: Nomadic Sami people with reindeer skin tents and clothing, early 20th century. ((Top) Photo by Silje Bergum Kinsten, Nordic Co-operation website (norden.org). Licensed under CCA 2.5 Denmark. In the public domain. (Middle) Photographer unknown, in public domain. (Bottom) Photo by Granbergs Nya Aktiebolag, in public domain)

production. Until the 1970s, there was a decline in the number of farms engaged in reindeer husbandry. In fact, reindeer husbandry disappeared in Amginsky (1958), Verkhnevilyuisky (1957), Mirninsky (1976), Suntarsky (1956), Megino-Kangalassky (1967) city, Tattinsky (1963), Ust-Aldan (1956) regions of Yakutia.

The breakup of the Soviet Union in 1991 led to the degradation of the traditional economy of indigenous

FIG. 15.5 Map of the Indigenous peoples of the Arctic. Arctic Centre, University of Lapland.

peoples of the North. During this period, the socioeconomic sphere of the aboriginal peoples essentially fell back by a century. During the 21st century, there has been some improvement in the standard of living of indigenous peoples, but the state of traditional reindeer husbandry has continued to deteriorate. The number of reindeer in Yakutia in 1991 was 361,556; by 2005, the herds declined to 144,476 reindeer, less than half the number of the previous decade (Fig. 15.6).

As reindeer herds declined, there was also a sharp decline in the living standards of indigenous peoples, which led to the deterioration of human health. Currently, sociological studies of the interactions between Native groups and neighboring industrial mining companies are being used in the political struggle for the rights of Native peoples. This struggle reflects the conflict between traditional and industrial use of natural resources. As reviewed by Burtseva et al. (2020), the culture and traditions of indigenous peoples are at odds with the predatory use of natural resources and environmental pollution (Table 15.3) (see the conclusions section of Chapter 14: the Frontier Mentality). Arctic industrial development has led to the destruction of much of the surrounding natural environment. Among Native peoples, it has led to unemployment, alcoholism, and the loss of traditional values and native languages.

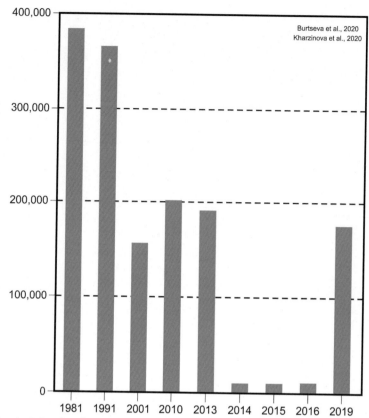

FIG. 15.6 Bar chart showing changes in domesticated reindeer populations of northern Siberia from 1981 to 2019. (Data from Burtseva, E., Sleptsov, A., Bysyina, A., Fedorova, A., Dyachkovskii, G., 2020. Mining and indigenous peoples of the North: assessment and development prospects. Resources 9, 95. https://doi.org/10. 3390/resources9080095 and Kharzinova, V., Dotsev, A., Solovieva, A., Sergeeva, O., Bryzgalov, G., et al., 2020. Insight into the current genetic diversity and population structure of domestic reindeer (Rangifer tarandus) in Russia. Animals, 10, 1309. https://doi.org/10.3390/ani10081309.)

Siberian Native health and the oil and gas industry

An ethnological examination was carried out in northern Russia for the first time in 2002, ordered by the Association of Indigenous Peoples of the Yamal-Nenets Autonomous District (Bogoyavlenskiy et al., 2002). The study was carried out by a team of researchers from the Research Institute of Anthropology of Moscow State University, the Institute of Economic Planning and Forecasting of the Russian Academy of Sciences, Tula State University, and the Moscow Regional Bar Association. This team of experts concluded that the implementation of Gazprom's exploration program in Obskaya and Tazovskaya bays is being carried out in violation of Russian legislation and poses a threat to the preservation of the indigenous peoples' homeland. The report also stipulated that the state license issued to Gazprom

for the development of the oil and gas project must provide socioeconomic guarantees to the indigenous population in the event of negative environmental impacts during project implementation.

The findings of the ethnology study included the following:

1. The Arctic and northern territories of Yakutia are characterized by a steady stream of outflowing migration of the population. There is also a high level of morbidity and a decrease in the quality of educational and medical services. The most acute problems are under development of social infrastructure, dilapidation, and accident rate of social facilities. An assessment of the ties between the state of health and the social environment was done through statistical medical and demographic indicators, focusing on tuberculosis.

TABLE 15.3
Vulnerability of Native Regions of Yakutia to Industrial Development.

Natural and Economic Zones of Yakutia	Native Settlements	Area (ha)	Vulnerability to Industrial Impacts
Arctic tundra hunting and reindeer breeding	23	45,084,925	Extremely high
Northwestern northern taiga hunting and reindeer herding zone	8	40,351,461	High
Northeastern lacustrine taiga reindeer and cattle breeding	6	22,224,024	Very high
Northeastern mountain taiga reindeer, horse, and cattle breeding	17	38,318,314	High
Southern mountain taiga land farming, hunting, and reindeer breeding	3	11,733,792	High
Total	57	157,712,516	High

Data from Burtseva, E., Sleptsov, A., Bysyina, A., Fedorova, A., Dyachkovskii, G., 2020. Mining and indigenous peoples of the North: assessment and development prospects. Resources 9, 95. https://doi.org/10.3390/resources9080095.

Tuberculosis is on the rise among the Sakha and Evenk peoples of Yakutia. A public health survey conducted in 2010 established that the people in four out of six native regions were suffering from medium to high levels of TB infection. Those in the "high" group had more than 20% mortality. Molecular genetic tests were performed on Evenks, Evens, and Yakuts who lived close to mining centers, as part of a larger assessment of their health problems. It was found that these people have relatively low genetic diversity in comparison with other East Eurasian populations. Small population size and geographic isolation have led to a reduction in genetic diversity—a problem inherent in many Arctic Native groups. The Siberian study concluded that the lack of genetic diversity made these small populations more susceptible to infection.

The study by Bogoyavlenskiy et al. (2002) also concluded that the main reasons for the decline in reindeer husbandry were as follows: (1) the republic was not interested in the development of reindeer husbandry in the Yakut regions, where the local population has long been engaged in cattle breeding and horse breeding; (2) the descendants of the indigenous peoples of the North who had previously been engaged in reindeer husbandry were gradually assimilated into the Russian culture. Accordingly, there were no Native peoples who retained the knowledge of reindeer husbandry in the region.

The findings of the ethnology report were summed up in four main points:

1. The number of indigenous minorities in Yakutia, according to the 2010 census, was 39,936 (4.2% of the total population of the republic). This is the combined number of Evenks and Evens people. The number of indigenous peoples in 2010 increased by 2.4. times compared with 1970,

2. The indigenous population of Yakutia (Evenks, Evens, and Yakuts) is characterized by low genetic diversity compared with other East Eurasian populations. This can lead to negative consequences concerning their health when exposed to industrial pollution.

3. The planned increasing industrial development of the territory of the republic may cause large-scale disturbances of the Earth's surface, depletion of biological resources, and pollution of the environment, which will ultimately lead to a deterioration in the quality of life of the indigenous population;

4. When implementing new mining projects in the regions of traditional nature management (i.e., indigenous homelands), it is necessary to conclude a tripartite agreement on cooperation and financing of specific programs between industrial companies, government bodies, and representatives of the indigenous minorities.

TABLE 15.4

Concentrations of Persistent Organic Pollutants (POP) and Heavy Metals in Yup'ik Maternal and Infant Blood From Alaska.

Year	2000	2005	2010 Adults	2010 Infants
POP Compound	Concentration (µg/kg plasma lipid)			
Oxychlordane	6.7	7.7	3.6	N/A
DDE	135	124	82.7	80.7
PCB138	35	14	9.1	8.3
HCB	14	22	15.9	19.6
HCH	6.7	7.7	3.6	N/A
PBDE47	N/A	26	19.8	16.1
Heavy Metal	Concentration (µg/kg whole blood)			
Lead	11	18	7.4	N/A
Mercury	1.1	5	2.2	N/A
Cadmium	0.6	0.4	0.2	N/A

Data from AMAP Assessment (2015).

COUNTRY FOOD PROCUREMENT AND CONSUMPTION

In the face of rapid environmental change, food insecurity is affecting Native communities throughout the Arctic. One solution for Native peoples is to return to a diet containing a greater proportion of locally harvested foods, also called country foods. However, as we shall see in the following, some country foods are not being eaten as often as they were before, and a return to the traditional Native diet is not straightforward. These days some country foods are contaminated with pollutants, such as heavy metals and persistent organic pollutants (POPs) (Table 15.4). Also, environmental changes threaten the health of Arctic fauna and thus its suitability for local consumption. Finally, changing sea-ice and fast-ice conditions are endangering Native travel to traditional hunting grounds for marine mammals.

Food Security of Inuit in the Western Canadian Arctic

Rosol et al. (2016) reported on widespread food insecurity and a shift away from nutrient-rich country foods in the diets of Inuit living in the Canadian Arctic. These dietary problems represent a growing public health challenge in Native communities. As discussed elsewhere in this chapter and Chapter 12 environmental factors, especially climate change, have limited the accessibility and availability of many country food species. Also, new species have entered Inuit homelands that have

not been seen there before, competing for resources with native species.

Compared with the rest of Canada, Inuit populations suffer disproportionately from food insecurity. The results from the Inuit Health Survey (Rosol et al., 2011) showed that adults living in Arctic Canada experienced levels of food insecurity at rates of 69% in Nunavut, 43% in the Inuvialuit Settlement Region (ISR), and 46% in Nunatsiavut (46%) compared with just 9.2% for the Canadian national average. Food-insecure adults showed reduced consumption of country food, leading to significantly lower intakes of energy, vitamin C, iron, zinc, and vitamin D. The Inuit level of food insecurity was far greater than that found among Alaska Natives, where 12% of households were food insecure in 2012. But like Canada, Indigenous Alaskans represented the largest proportion of food insecurity. Currently, many Inuit are unable to access safe, sufficient, nutritionally adequate, and socially acceptable food (Fig. 15.7). Many barriers impede Arctic Native ability to access healthy food. These include low income, the high cost of hunting equipment and fuel, insufficient food quantities, and compromised quality of market foods. Limited access to sufficient and affordable nutritious food makes healthy eating difficult to achieve and increases the risk of diet-sensitive chronic diseases.

Rosol et al. (2016) argue that a viable solution to these problems would include turning to alternate country food species as substitute sources for the

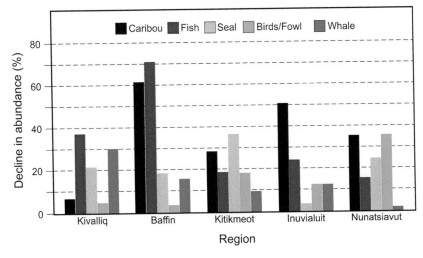

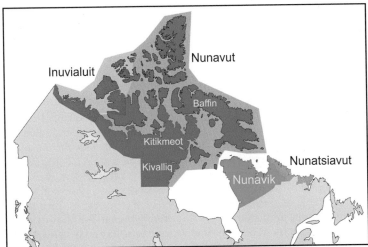

FIG. 15.7 Above: Native food declines by region. Below: Canadian Inuit regions map. ((Above) Data from Rosol, R., Powell-Hellyer, S., Chan, H.M., 2016. Impacts of decline harvest of country food on nutrient intake among Inuit in Arctic Canada: impact of climate change and possible adaptation plan. International Journal of Circumpolar Health 75, 31127. https://doi/org/10.3402/ijch.v75.31127.)

nutrients previously provided by traditional country foods. The authors estimated the impact on nutrient intake using hypothetical scenarios in which traditional country foods were reduced by 50% and replaced with alternate food species. In their study, they found that a 50% reduction in the consumption of food that was reported to be in decline, including fish, whale, ringed seals, and birds, resulted in a significant decrease in essential nutrient intake. Possible substitute foods were identified, but some nutrients such as zinc and especially vitamin D were most often found lacking in the alternative foods. They concluded that if the

alternative species are not available or feasible, more expensive, and less nutritionally dense, store-bought foods may have to be consumed. This alternative is regrettable, given the superior quality of country foods and their association with food security, Inuit cultural health, and personal identity. Further to this, the harvesting of country foods teaches young people the necessary skills for survival. It also promotes regional sharing networks, a central feature of Inuit culture since time immemorial.

Country foods are rich in nutrients and contribute to a higher-quality diet. Studies have shown that country food consumption facilitates higher intakes of protein and vitamins A and C while lowering the intake of refined carbohydrates, saturated fat, and sodium. A study of Native nutrition in a population from Western Alaska found that the participants maintained a steady, sufficient vitamin D concentration in their bloodstream throughout the year through the consumption of locally harvested foods, specifically fish. They found that those individuals (mostly youths) who consumed less local fish were more likely to have low levels of serum vitamin D. Wild-harvested country foods, including land animals, berries, birds, fish, and sea mammals, are rich in antioxidant compounds, omega-3 fatty acids, monounsaturated fatty acids, and protein. These foods are also high in vital micronutrients such as iron, riboflavin, zinc, copper, magnesium, potassium, selenium, thiamine, niacin, and vitamins A, D, E, B6, and B12.

In recent decades, there has been a shift in Arctic native diets away from country foods in favor of manufactured foods from the South. The shift is particularly noticeable among the young, while the diets of many older people continue to emphasize country foods. This trend was noticed in the Alaskan Native group study, and it is likewise a part of the changing youth cultures of Inuit of Canada and Greenland. They are also undergoing a dietary transition in which they continue to consume some country foods with their families but increasingly rely on highly processed and expensive market foods shipped from the south. Although it is widely accepted that the benefits of country foods far outweigh the benefits of available market foods, there are significant challenges that Inuit must overcome to obtain healthier market and country foods. In the view of Rosol et al. (2016), strong efforts should be made to preserve the nutritional integrity of Inuit diet through the consumption of country foods, as this diet is known to maintain the physical and cultural health of the community.

As Wesche and Chan (2010) observed, Arctic Native food security centers on access to food derived from the natural environment through Native cultural practices (Table 15.5). As such, it is essential to realize the importance of the various aspects of traditional food acquisition, from harvesting to preparing and consuming the foods. In a resource-limited environment where the harsh climate dictates many human activities, food sharing has always been an important cultural practice. It helps to limit the risk of food deficiency and acts to strengthen social bonds. The Inuit relationship to traditional food is more complex than just the consumption of healthy foods. As such, it must be recognized not only for its importance for physical health but also for its connection to emotional, spiritual, social, and cultural well-being.

These authors compiled data from five sets of climate change observations, impacts, and adaptations in 12 Inuit communities (2005–06), and a dietary report of food use from 18 Inuit communities (1997–2000). Changing access to, availability of, quality of, and ability to use traditional food resources has implications for the quality of diet. Nutritional implications of lower traditional food use include likely reductions in iron, zinc, protein, vitamin D, and omega-3 fatty acids, among others. The vulnerability of each community to changing food security is differentially influenced by a range of factors, including current harvesting trends, levels of reliance on individual species, opportunities for access to other traditional food species, and exposure to climate change hazards. Understanding linkages between climate change and traditional food security provides a basis for strengthening adaptive capacity and determining effective adaptation options to respond to future change.

Benefits and Risks of Country Food Consumption in Nunavik

Inuit diets have recently been affected by several environmental changes, including pollutant contamination, natural resource depletion, and climate change, as well as several social changes, such as Inuit population increase, challenges in Inuit Knowledge transmission, and local food transition toward a larger proportion of market foods. As discussed by Rosol et al. (2016), from a human health perspective, attempts to improve food security and dietary quality rely on the promotion of Inuit Knowledge about country foods, combined with efforts to diversify the diet with other locally harvested foods. Country foods such as caribou, beluga, Arctic char, and tundra berries continue to play a key part in the Inuit diet and culture in Nunavik, northern Quebec. On the other hand, some traditional country foods, such as walrus, are not eaten as often as before. A survey taken in Nunavik in 2004 revealed that 83% of the population had not consumed walrus during the previous year although it had previously been an important part of the regional diet.

Martinez-Levasseur et al. (2020) explored the benefits and risks of country food consumption using walrus meat consumption in Nunavik as a case study. They interviewed 34 hunters and Elders from Nunavik. They also performed laboratory studies of tissue samples from locally harvested walruses to investigate levels of mercury, omega-3 polyunsaturated fatty acids, and

TABLE 15.5
Environmental Changes Affecting Canadian Arctic Communities.

	COMMUNITY			
	Aklavik	Tuktoyaktuk	Paulatuk	Ulukhaktok
Population (2020)[a]	676	991	322	465
% Native	95.1	97.1	82.4	98.7
% hunters and fishers	49.3	56.9	49.5	76.1
% country foods in diet	35.5	49.5	51.9	45.8

Condition	Aklavik	Tuktoyaktuk	Paulatuk	Ulukhaktok	Consequences
Unpredictable weather	●	●	●	●	Travel danger, limits access to hunting areas
Thinner sea ice	●	●	●	●	Limits access to hunting areas; shorter, fewer hunting trips; decline in seal and polar bear health
Decreased snowpack	●	●	●	●	Reduced travel, especially in spring and fall
Earlier ice breakup	●	●	●	●	Reduced travel safety and hunting access; shorter fishing and waterfowl hunting season; decline in seal health
Later ice freeze-up	●	●	●	●	Reduced travel safety and access to hunting areas; delayed fishing and caribou hunting season; extended whale and seal hunting seasons in fall and early winter
Increased erosion on shore	●	●	●		Reduced access to hunting areas; reduced fish spawning due to sediment load
Lower freshwater levels	●	●		●	Reduced travel safety; reduced access to harvesting areas
New plant and wildlife species	●	●	●	●	Displacement of existing species; potential increase in harvesting opportunities
Declines in animal health	●	●	●	●	Reduced consumption due to contaminant concerns

[a] NWT Bureau of Statistics (2021).
Data from Wesche, S.D., Chan, H.M., 2010. Adapting to the impacts of climate change on food security among Inuit in the Western Canadian Arctic. EcoHealth 7, 361–373. https://doi.org/10.1007/s10393-010-0344-8.

selenium. They also tested 755 Atlantic walrus samples, collected between 1994 and 2013, for *Trichinella nativa*, a parasitic roundworm known to infect Atlantic walruses. They reviewed databases for suspected cases of botulism following consumption of walrus meat. Their laboratory analyses revealed that walruses had elevated levels of omega-3 fatty acids and selenium, but low levels of mercury compared with some other wildlife. Only 3% of the 755 walruses were infected with *T. nativa*. Most infected walruses were found in the South East Hudson Bay population. Inuit hunters decided to stop hunting walrus there in the mid-2000s. Finally, although the number of outbreaks of trichinellosis (round-worm infection) related to the consumption of walrus meat has significantly declined in Nunavik, botulism may continue to be a threat, especially when the meat from aged walruses is not properly prepared.

Greenland Inuit Intake of Refined Carbohydrates

DiNicolantonio (2016) reviewed the literature concerning Native health in Greenland, focusing particularly on coronary artery disease (CAD). Research done in the 1940s indicated that Greenland natives had remarkably few incidents of CAD. Specifically, a comparative study showed that the number of people with atherosclerosis was about four times lower in the Native community of Umanak than in the population of Korpo, Finland. Only 7.5% of the Umanak population had atherosclerosis (80 out of 1073 people examined), whereas 29% of adults in the Finnish test group suffered from this disease (300 out of 887 people examined). However, this situation changed in the following decades. Fodor et al. (2014) systematically reviewed the available data to determine the validity of claims made by researchers in the 1970s that Greenland Inuit had a low incidence of CAD. The study by Fodor et al. (2014) concluded that, on the contrary, the incidence of CAD among Greenland Inuit was at a rate as high or higher than that of the non-Native population of Greenland. It is important to note that almost all the evidence used in the study by Fodor et al. (2014) came from studies done since 1979. Based on this evidence, there can be little doubt that the Greenland Inuit were indeed at a similar or even greater risk of atherosclerosis or CAD than their European counterparts (Table 15.6).

Historic data show that the Greenland Inuit had virtually no refined sugar in their diet in earlier times, but by the 1950s, "sugar was taken five times a day, mostly in coffee or tea." Concurrent with this was a sharp decline in oral health. The Greenland Inuit had virtually no dental caries (cavities) in earlier times, but by the 1950s, public health officers noted that the Natives had "terrible teeth from their habit of sipping their coffee through a sugar cube."

DiNicolantonio (2016) concluded that the sharp increase in consumption of refined carbohydrates and sugar paralleled the rise in CAD in the Greenland Inuit population (Table 15.6). While the total carbohydrate intake of the Greenland Eskimos was just 2%–8% of total calories in 1855, this increased to around 40% of calories by 1955. The Greenland Inuit studied by Bang et al. (1976) no longer consumed a traditional healthy Eskimo diet. Indeed, the intake of refined sugar in Greenland Inuit increased almost 30-fold from 1855 (average of 1½ teaspoons of sugar) to the 1970s (40–44 teaspoons of sugar). Moreover, the intake of refined carbohydrates increased five- to sevenfold from 1855 (18 g/day from bread) to the 1970s (84–134 g/day from bread, biscuits, and rye flour).

In Greenland, as elsewhere in the world, the increased consumption of refined carbohydrates and sugar since the 1950s has been linked with the rise in atherosclerotic disease and other ailments. For instance, in the United States, the overconsumption of refined sugar has been closely linked with the remarkable rise in type 2 diabetes, hypertension, and coronary heart disease.

Changes in the Alaskan Inland Native Diet

As reviewed by DiNicolantonio and O'Keefe (2018), a study of the diet of the Alaskan Inland Inuit residents of Anaktuvuk Pass, Alaska, showed a ~50% increase in carbohydrate intake and a ~50% decrease in the intake of protein between 1955 and 1965. As was the case in Greenland, this change in diet may have led to the dramatic rise in dental caries and subsequent increase in atherosclerosis and CAD. The traditional Alaskan Iñupiat lived a seminomadic life hunting and fishing, taking marine mammals and birds. However, beginning around 1920, a group of Iñupiat developed a permanent settlement in the Brooks Range of northern Alaska, finally settling in Anaktuvuk Pass in 1950. Bang and Kristoffersen (1972) performed studies from two different time periods on the diet and dental health of the Alaskan Inland Iñupiat. They analyzed the Native diet in 1955–57 by determining the fat, protein, and carbohydrate content of the food being consumed, as well as the total calories consumed by each individual. In a follow-up study, a decade later, Bang and Kristoffersen (1972) noted the dramatic increase in carbohydrate intake and a decrease in the intake of protein.

TABLE 15.6
Summary of Changes Affecting Nunavut Inuit Since the 1950s.

Date	Change	Impact
1950s	Resettlement into centralized communities	Reduced harvesting, increased participation in wage economy, changing sociocultural norms, reduced mobility, increasing predominance of English
1960s	Mechanized transportation replaces dogs	Travel further and faster to hunt and fish, dependence on imported technology, need for cash income
	Increasing individualization in hunting	Reduction in group hunting
1970s	Increasing commercialization of harvesting	Increasing commercialization of hunting to support subsistence, participation in market economy, increasing importance of money
1980s	Economic development promoted	Big resource development affecting wildlife health and population, increasing integration into global markets
1990s	Transportation improvement	Development of regular scheduled flights improves access and importation of food
	Wildlife regulations implemented	Affects the flexibility of harvesting by controlling what harvested, when, and where for certain species, conflict among hunters regarding quotas
	Inuit land claims negotiations begin	Increased political power and decision-making capability for Inuit, increasing community involvement in decision-making
	Contaminants affecting traditional foods	Anxiety over traditional food consumption
2000s	Climate change impacts begin to be noted by scientists and communities	Change in access and availability of traditional foods, increasing danger, international political actors begin to demand decreased hunting activity
	Decline in intercommunity food sharing	Increased food insecurity; weakening of social ties between communities; changing social conditions

After Ford and Beaumier (2011).

The 1965 study showed that all village residents were suffering from dental caries, whereas there had been no evidence of carries in the previous study. Alaskan Natives eating a traditional low-carbohydrate, high-fat/protein diet had a much lower incidence of atherosclerosis, hypertension, and dental caries compared with more westernized populations. The health decline of the Alaskan Inland Iñupiat was likely connected to the rise in the intake of refined carbohydrates and sugar.

Bang and Kristoffersen (1972) also discussed how the Alaskan Inland Iñupiat lifestyle had changed from the 1950s to the 1960s. In the 1950s and before, the traditional Iñupiat lifestyle was being maintained. The authors observed that all able-bodied men in the village spent much of their time engaged in the procurement of country foods through hunting, trapping, and fishing. By 1965, only a few young men were still actively engaged in these activities. Before the 1950s, the Inland Iñupiat had lived a seminomadic life, stopping where there were good hunting and trapping opportunities. They migrated to the coast in summer to fish and hunt marine mammals and birds. Since the 1950s, however, they have stayed permanently in the mountains at Anaktuvuk Pass.

The residents of the village had abandoned their subsistence-based economy in favor of a cash-based economy centered on the manufacture of souvenirs. The newly acquired income financed the purchase of refined foods from the local store. Consequently, by 1965, only 20% of the food intake consisted of native foods, mainly caribou meat. Also, hunting was now mostly limited to the short periods when the caribou came close to the village.

Community-Based Subsistence Resources in Northwest Alaska

Gorokhovich et al. (2013) examined the subsistence resources available to the Indigenous communities in the Kotzebue Sound region of northwest Alaska (Table 15.7). Their study integrated physical, anthropological, and survey data to identify areas of concern for local and regional planning and environmental protection. The results of the study identify the villages of Kivalina and Deering as particularly vulnerable to coastal erosion and loss of subsistence resources. Shoreline degradation is unlikely to affect the availability of fish and caribou, the two most important subsistence resources in these communities. Many subsistence resources in northwestern Alaska are seasonal (e.g., caribou, berries, and seals). The seasonal phenology of these organisms is subject to rapid change. As shown in Fig. 15.8, the pace of Arctic environmental changes is predicted to increase to 2050 and beyond.

TABLE 15.7
Subsistence Resource Categories Ranked by Importance in Alaska Native Community Surveys.

Resource/Ranking (1–5)	VILLAGE			
	Kivalina	Kotzebue	Selawik	Deering
Beluga	X	5	X	3
Potential loss rated "very great"	56	41	65	73
Berries	2	1	X	2
Potential loss rated "very great"	71	80	78	65
Caribou	1	2	1	1
Potential loss rated "very great"	70	77	87	78
Birds, waterfowl	X	3	X	3
Potential loss rated "very great"	57	66	65	57
Birds' eggs	X	5	3	3
Potential loss rated "very great"	50	57	68	48
Fish	1	1	1	1
Potential loss rated "very great"	71	82	89	70
Furbearers	X	4	X	3
Potential loss rated "very great"	55	66	35	50
Moose	X	4	X	4
Potential loss rated "very great"	54	71	46	35
Seals	2	4	X	2
Potential loss rated "very great"	59	39	78	65
NONFOOD RESOURCES				
Camps, cultural sites	X	+	+	+
Trails	+	+	+	+
Wood	X	+	X	X

X, not included by villagers; +, included but not ranked.

Data from Gorokhovich, Y., Leiserowitz, A., Dugan, D., 2013. Integrating coastal vulnerability and community-based subsistence resource mapping in Northwest Alaska. Journal of Coastal Research 30, 158–169. https://doi.org/10.2112/JCOASTRES-D-13-00001.1.

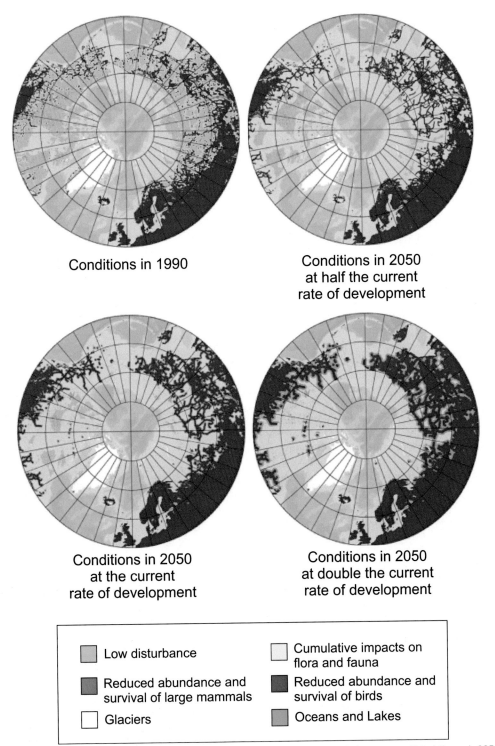

FIG. 15.8 Arctic projection maps showing modeled Arctic environmental changes predicted through 2050 from Grid Arendal.

According to Gorokhovich et al. (2013), an assessment of the potential impacts of sea level rise and coastal erosion on subsistence resources requires data on where these resources are geographically located and, crucially, harvested or gathered by local people, especially when obtained from community-based research. GIS modeling and analysis can be used to integrate diverse geographic data, including historic and projected physical changes in coastlines and the locations of various subsistence resources to identify potentially vulnerable areas, which can provide valuable information for regional coastal management and planning.

CONCLUSIONS—HOW TO COPE WITH A FRIEND ACTING STRANGELY

There are at least 400,000 indigenous residents living around the Arctic. They are now living in a rapidly changing world. As some Inuit have described it, their environment has become "a friend acting strangely." This chapter covers many of these unsettling changes and their effects on Native peoples. An issue that came up repeatedly in interviews with Yúpik, Iñupiat, and Inuit peoples is the unpredictability of the modern Arctic climate. There is widespread anxiety over travel safety to and from hunting grounds, and this is causing an increase in seasonal depression in Native villages. The most important aspects of this unpredictability concern sea ice. Sea ice has always been the Native's winter highway to reach seal hunting areas and to travel between villages. It was an aspect of life that native peoples could always rely on during winter, but now it has become unreliable or absent for much of the year. Land-fast ice is forming later in the year and melting earlier, further blocking Native access to marine food supplies.

Climate change is disrupting Native access to traditional foods from both land and sea. Native peoples across the Arctic who used to rely almost exclusively on hunting or fishing are now forced to eat manufactured food products loaded with sugars and other carbohydrates. These packaged foods are easier and therefore less expensive to transport to Arctic grocery stores because they are essentially nonperishable. The better-quality foods, those that are rich in vitamins, minerals, and protein, are fresh fruits and vegetables, eggs, fresh milk, and other dairy products. While consumption of these fresh food items would bolster the nutritional quality of native diets, they are also the most expensive items in the stores, because they are highly perishable and must be flown to the Arctic.

Key Points on Arctic Climate Change

The resilience of northern communities in the face of climate change impacts depends heavily on sound government leadership and the successful implementation of long-term adaptation policies. Developing the adaptive capacity of Arctic communities cannot rely on external policies. It requires collaboration with local people who can speak to vulnerability issues based on real-life experiences. What Inuit communities need most to enact adaptation plans is long-term financial support from the federal government.

This chapter briefly touched on the topic of "Solastalgia"—a longing for the past. While not explicitly mentioned throughout the chapter, the concept pervades almost every section. Adult human beings around the world tend to long for the "Good old days." But in the case of Arctic natives, this feeling represents a legitimate fear that the traditions essential to the wellbeing of the community will be lost. Once the older generation is gone, the chain of knowledge is broken, and there is no one left to restore it.

Arctic Native Worldview

Arctic Native cultures are almost exclusively maritime, and this lifestyle is central to their worldview. They live on the coast and are basically looking more toward the sea than toward the land. Fish, marine mammals, and sea birds are a big part of their hunting activities. Inuit worry especially about how fuel spills and the dumping of waste into the Arctic Ocean will affect their food supply. In addition to their holistic view of nature, there is a very practical reason for these concerns. Their food security depends on clean ocean waters and a healthy marine environment. As Okalik Eegeesiak said in her contribution to Kelley (2017), "We cannot consider the health of a walrus population, without understanding sea ice thickness, the state of the benthos, the ecology, Inuit cultural identity, language, and so on—it is all connected." The bottom line here is that marine pollution will cause devastation to the entire marine food chain, a legacy of poisoned food supply that will likely persist for many years after the fact.

Native Participation in Global Change Science

Inuit leaders have said that they refuse to "rollover" while their world is turned upside down because of mistakes being made somewhere else. They intend to be involved in climate change research and adaptation. For this to happen, there must be collaborative efforts between scientists and native groups. The history of such collaboration is spotty at best. There have been notable success stories, such as the collaboration between Iñupiat leaders and sea-ice scientists that began at Barrow (now Utqiaġvik) in the late 1970s. Another successful collaboration included inputs from local Iñupiat leaders, Chukchi walrus experts, and the US Fish and Wildlife Service. These three groups came together to brainstorm ways of protecting the huge walrus herds that began hauling out near Point Lay during the past few years. A very practical set of rules and regulations evolved from this collaborative effort. However, there have also been some setbacks in this process. The NSF "Navigating the New Arctic" (NNA) initiative was set up to facilitate cooperation between scientists and Indigenous communities in the "coproduction of knowledge" by involving them in planning and executing projects. But most of the scientists who received these grants simply carried on doing their research without seeking Native input. Despite assurances to the contrary, the NSF itself dropped the ball in its response to suggestions from Alaskan Natives. Indigenous leaders recommended that NSF set aside 25% of NNA funds for Indigenous-led projects in food security and infrastructure—two topics vital to Native concerns in Alaska. While the NSF Arctic System Science Program said that they would love to see more proposals submitted on those topics, the NNA program administrators have said that they have no plans to set aside funding for Indigenous-led projects.

Migration and Relocation

The motivation of young people to leave their home villages may be partly due to environmental change, but village social bonds are often strong enough to counteract the urge to migrate to somewhere new. Human society can work against migration as well as for it. Pressures to move are countered by reasons people might not want to leave their ancestral homes, or by a lack of appealing alternatives.

The GAO report of July 2020, using Newtok, Alaska, as a case study, concluded that federal programs are not designed to address the scale and complexity of community relocation. Federal programs have been designed to react to catastrophes, not to spend money on avoiding problems beforehand. This the main reason why the relocation of Newtok has taken over 30 years and remains half-finished. Until now, no federal agency has the authority to coordinate assistance, according to village inhabitants. Fortunately, federal policies concerning relocation are finally changing. In coordination with the Department of Housing and Urban Development, FEMA released plans for a new program that will pay for large-scale community relocations.

Native Governance Issues

A review of a legal structure of Native rights in the United States versus Canada revealed that the Native peoples of the two countries do not enjoy the same rights and privileges. The underlying difference between Native rights in the United States and the First Nations of Canada is the way in which the two federal governments conceive tribal rights. In the United States, tribal sovereignty is considered a "full box" of legal and political rights. If there is a clash between federal and tribal laws, it is up to the US government to prove that the Native group is in the wrong. In Canada, aboriginal rights begin with an "empty box" with the burden of proof placed on First Nations groups to demonstrate to the court that the group has a protected aboriginal right.

When it comes to the rights and autonomy of Indigenous groups in northern Russia, the top-down control of politics has subjugated regional and local governance to the will of the Kremlin. Since the dissolution of the Soviet Union, this change in Russian authority structure has essentially erased any vestige of Indigenous self-governance left over from the days of the USSR. The government may "talk the good talk" when passing legislation meant to protect Native rights in Siberia, but when vast fortunes can be made from oil, gas, and mining operations, native lands that previously were given protective status by Moscow have simply been confiscated and given over development. As we have seen in this chapter and elsewhere in this book, Arctic industrial development has led to the destruction of much of the surrounding natural environment. Teams of European Russian sociologists may complain in their publications that further industrial development in the North will lead to a deterioration in the quality of life of the indigenous population and that the implementation of new industrial mining projects in indigenous homelands requires the agreement of the regional indigenous minorities. So far, reports such as these failed to even slow the development of Arctic industrial projects, much less stop them.

Native Health Issues

Throughout the Arctic, many factors are contributing to a decline in the harvesting and consumption of country foods. In many regions, birds, fish, and mammals that once formed the core of the Native diet are becoming scarce, or the quality of their meat is deteriorating due to environmental change. The replacement of country foods by manufactured foods from the South is a cause for concern. Poor diet aggravates existing health issues, including mental health and wellness, nutritional deficiencies, rates of respiratory illness, livelihood and economic stability, safety, and the spread of disease.

One very troubling indicator of the decline in Native health in northern Siberia has been the rise of tuberculosis (TB) among the Sakha and Evenk peoples of Yakutia. The fact that the people in four out of six native regions were suffering from medium to high levels of TB infection indicates that these people are also suffering from several underlying conditions. As discussed by Figueroa-Munoz and Ramon-Pardo (2008), throughout the world, the conditions associated with high rates of TB infection include poverty, lack of education, overcrowding, inadequate living conditions, malnutrition, social exclusion, stigmatization, poor access to health care, low immunization rates, and poor treatment compliance.

Final Thoughts

Somewhat akin to the Arctic itself, the native peoples of this region display a combination of toughness and fragility. Their survival in perhaps the least hospitable region demonstrates their resilience in the face of lifelong adversity. They have adapted amazingly to an environment that most human beings would find incredibly challenging if not immediately fatal. But their intimate ties with nature are now making them vulnerable to changes in that nature, ranging from ecological upheaval to ocean pollution. Other than jeopardizing their health through total reliance on manufactured foods from the South, Native peoples of the Arctic have no "Plan B" to fall back on. When they first entered the Arctic thousands of years ago, they relied on their hunter-gatherer lifestyle to survive in a land where terrestrial biological productivity is minimal. The land itself had scant food resources to offer except caribou meat, lake fish, and berries. The sea held nearly all the nutritious foods they required, and they were reliant on the ocean's bounty. Now that same ocean is failing to deliver what they need, mostly because the physical and ecological nature of the ocean have changed. If such changes had taken place over many decades or centuries, native Arctic peoples would have found ways of adapting their culture. But the modern pace and amplitude of environmental change in the Arctic have been far too great for even these most adaptable people to accommodate.

REFERENCES

AMAP, 2015. AMAP Assessment 2015: Human Health in the Arctic. Arctic Monitoring and Assessment Programme (AMAP). vii +, Oslo, Norway, p. 165.

American Bar Association, September 20, 2016. American Indian tribes and Canadian First Nations: the Impact of gaming law and policy on the industry. Business Law Today. https://www.americanbar.org/groups/business_law/publications/blt/2016/09/04_rand/.

Arctic Council, 2020. Suggestions from the Coast: An Interview with Jim Stotts, President of the Alaskan Chapter of the Inuit Circumpolar Council (ICC). https://arctic-council.org/en/news/suggestions-from-the-coast-an-inuit-perspective-on-a-sustainable-arctic-ocean/.

Bang, H.O., Dyerberg, J., Hjoorne, N., 1976. The composition of food consumed by Greenland Eskimos. Acta Medica Scandinavica 200, 69−73.

Bang, G., Kristoffersen, T., 1972. Dental caries and diet in an Alaskan Eskimo population. Scandinavian Journal of Dental Research 80, 440−444.

Bogoyavlenskiy, D.D., Martynova, E.P., Murashko, O.A., Khmeleva, E.N., Yakel, Y.Y., Yakovleva, O.A., 2002. Experience in Conducting Ethnological Expertise. Assessment of the Potential Impact of the Gazprom Program of Exploration and Prospecting in the Waters of the Obskaya and Tazovskaya Bays on the Components of Sustainable Development of Ethnic Groups of the Indigenous Peoples of the North. Radunitsa Publishing, Moscow, 132 pp.

Burtseva, E., Sleptsov, A., Bysyina, A., Fedorova, A., Dyachkovskii, G., 2020. Mining and indigenous peoples of the North: assessment and development prospects. Resources 9, 95. https://doi.org/10.3390/resources9080095.

Breum, M., November 15, 2018. Why people matter when we talk about Arctic climate change. Arctic Today. https://www.arctictoday.com/people-matter-talk-climate-change/.

Bronen, R., Chapin, F.S., 2013. Adaptive governance and institutional strategies for climate-induced community relocations in Alaska. Proceedings of the National Academy of Sciences 110, 9320−9325. www.pnas.org/cgi/doi/10.1073/pnas.1210508110.

Cooley, S.W., Ryan, J.C., Smith, L.C., Horvat, C., Pearson, B., et al., 2020. Coldest Canadian Arctic communities face greatest reductions in shorefast sea ice. Nature Climate Change 10, 533−538. https://doi.org/10.1038/s41558-020-0757-5.

DiNicolantonio, J.J., 2016. Increase in the intake of refined carbohydrates and sugar may have led to the health decline of the Greenland Eskimos. Open Heart 3, e000444. https://doi.org/10.1136/openhrt-2016-000444.

DiNicolantonio, J.J., O'Keefe, J.H., 2018. The introduction of refined carbohydrates in the Alaskan Inland Inuit diet may have led to an increase in dental caries, hypertension and atherosclerosis. Open Heart 5, e000776. https://doi.org/10.1136/openhrt-2018-000776.

Figueroa-Munoz, J., Ramon-Pardo, P., 2008. Tuberculosis control in vulnerable groups. Bulletin of the World Health

Organization 86, 733–735. https://doi.org/10.2471/BLT.06.038737.

Flavelle, C., August 26, 2020. U.S. flood strategy shifts to 'Unavoidable' relocation of entire neighborhoods. New York Times. https://www.nytimes.com/2020/08/26/climate/flooding-relocation-managed-retreat.html.

Fodor, J.G., Helis, E., Yazdekhasti, N., Vohnout, N., 2014. "Fishing" for the origins of the "Eskimos and heart disease" story: facts or wishful thinking? Canadian Journal of Cardiology 30, 864–868. https://doi.org/10.1016/j.cjca.2014.04.007.

Ford, J.D., 2009. Vulnerability of Inuit food systems to food insecurity as a consequence of climate change: a case study from Igloolik, Nunavut. Regional Environmental Change 9, 83–100. https://doi.org/10.1007/s10113-008-0060-x.

Ford, J.D., Beaumier, M., 2011. Feeding the family during times of stress: experience and determinants of food insecurity in an Inuit community. The Geographical Journal 177, 44–61. https://doi.org/10.1111/J.1475-4959.2010.00374.x.

Gorokhovich, Y., Leiserowitz, A., Dugan, D., 2013. Integrating coastal vulnerability and community-based subsistence resource mapping in Northwest Alaska. Journal of Coastal Research 30, 158–169. https://doi.org/10.2112/JCOASTRES-D-13-00001.1.

Hamilton, L.C., Saito, K., Loring, P.A., Lammers, R.B., Huntington, H.P., 2016. Climigration? Population and climate change in Arctic Alaska. Population and Environment 38, 115–133. https://doi.org/10.1007/s11111-016-0259-6.

Hamilton, L.C., Wirsing, J., Saito, K., 2018. Demographic variation and change in the Inuit Arctic. Environmental Research Letters 13 (115007). https://doi.org/10.1088/1748-9326/aae7ef.

Hassol, S.J., 2004. Impacts of a Warming Arctic. Arctic Climate Impact Assessment. Cambridge University Press, 146 pp.

Healey Akearok, G., Holzman, S., Kunnuk, J., Kuppaq, N., Martos, Z., et al., 2019. Identifying and achieving consensus on health-related indicators of climate change in Nunavut. Arctic 72, 289–299. https://doi.org/10.14430/arctic68719.

Heleniak, T., Turunen, E., Wang, S., 2020. Demographic Changes in the Arctic. In: Coates, K., Holroyd, C. (Eds.), The Palgrave Handbook of Arctic Policy and Politics. Palgrave Macmillan, Cham. https://doi.org/10.1007/978-3-030-20557-7_4.

Huntington, H.P., Danielson, S.L., Wiese, F.K., Baker, M., Boveng, P., et al., 2020. Evidence suggests potential transformation of the Pacific Arctic ecosystem is underway. Nature Climate Change 10, 342–348. https://doi.org/10.1038/s41558-020-0695-2.

Jones, M.C., Berkelhammer, M., Keller, K.J., Yoshimura, K., Wooller, M.J., 2020. High sensitivity of Bering Sea winter sea ice to winter insolation and carbon dioxide over the last 5500 years. Science Advances 6, eaaz9588. https://doi.org/10.1126/sciadv.aaz9588.

Kelley, K. (Ed.), 2017. Nilliajut 2: Inuit Perspectives on the Northwest Passage and Marine Issues. Inuit Tapiriit Kanatami, Ottawa.

Kharzinova, V., Dotsev, A., Solovieva, A., Sergeeva, O., Bryzgalov, G., et al., 2020. Insight into the current genetic diversity and population structure of domestic reindeer (*Rangifer tarandus*) in Russia. Animals 10, 1309. https://doi.org/10.3390/ani10081309.

La Ganga, M.L., August 30, 2015. This is climate change: Alaskan villagers struggle as island is chewed up by the sea. LA Times. https://www.latimes.com/nation/la-na-arctic-obama-20150830-story.html.

MacDonald, S., Birchall, S.J., 2020. Climate change resilience in the Canadian Arctic: the need for collaboration in the face of a changing landscape. Canadian Geographer 64, 530–534. https://doi.org/10.1111/cag.12591.

Mahoney, A.R., 2018. Landfast Sea Ice in a Changing Arctic. NOAA Arctic Program, 2018 Arctic report card. https://arctic.noaa.gov/Report-Card/Report-Card-2018/ArtMID/7878/ArticleID/788/Landfast-Sea-Ice-in-a-Changing-Arctic.

Martinez-Levasseur, L.M., Simard, M., Furgal, C.M., Burness, G., Bertrand, P., et al., 2020. Towards a better understanding of the benefits and risks of country food consumption using the case of walruses in Nunavik (Northern Quebec, Canada). Science of the Total Environment 719, 137307. https://doi.org/10.1016/j.scitotenv.2020.137307.

Michelin, O., October 16, 2020. 'Solastalgia': Arctic Inhabitants Overwhelmed by New Form of Climate Grief. The Guardian. https://www.theguardian.com/us-news/2020/oct/15/arctic-solastalgia-climate-crisis-inuit-indigenous.

Northwest Territories Bureau of Statistics, 2021. Current Indicators. https://www.statsnwt.ca/

NSIDC, 2017. Arctic Sea Ice News and Analysis. http://nsidc.org/arcticseaicenews/2017/06/.

Oakley, K., Whalen, M., Douglas, D., Udevitz, M., Atwood, T., Jay, C., 2012. Changing Arctic ecosystems: polar bear and walrus response to the rapid decline in Arctic sea ice. US Geological Survey Fact Sheet 2012–3131, 4.

Robards, M.D., Huntington, H.P., Druckenmiller, M., Lefevre, J., Moses, S.K., et al., 2018. Understanding and adapting to observed changes in the Alaskan Arctic: actionable knowledge co-production with Alaska Native communities. Deep-Sea Research Part II 152, 203–213. https://doi.org/10.1016/j.dsr2.2018.02.008.

Rosol, R., Huet, C., Wood, M., Lennie, C., Osborne, G., Egeland, G.M., 2011. Prevalence of affirmative responses to questions of food insecurity: international polar year Inuit health survey, 2007-2008. International Journal of Circumpolar Health 70, 488–497. https://doi.org/10.3402/ijch.v70i5.17862.

Rosol, R., Powell-Hellyer, S., Chan, H.M., 2016. Impacts of decline harvest of country food on nutrient intake among Inuit in Arctic Canada: impact of climate change and possible adaptation plan. International Journal of Circumpolar Health 75, 31127. https://doi.org/10.3402/ijch.v75.31127.

Sitdikov, V.T., Abdullina, D.F., Shergeng, N.A., 2019. The legal regulation of cultural and national autonomy in Russia. Journal of Legal, Ethical and Regulatory Issues 22 (1), 1–7. https://www.abacademies.org/articles/the-legal-regulation-of-cultural-and-national-autonomy-in-russia-7952.html.

Sörlin, S., 2018. Conclusion: Anthropocene Arctic—reductionist imaginaries of a "new north.". In: Wormbs, N. (Ed.),

Competing Arctic Futures: Historical and Contemporary Perspectives. Palgravre, Cham, Switzerland, pp. 243–269. https://doi.org/10.1007/978-3-319-91617-0.

Stone, R., September 9, 2020. As the Arctic thaws, Indigenous Alaskans demand a voice in climate change research. Science. https://doi.org/10.1126/science.abe7149. https://www.sciencemag.org/news/2020/09/arctic-thaws-indigenous-alaskans-demand-voice-climate-change-research.

US Fish and Wildlife Service, 2019. Pacific Walrus Use of Coastal Haulouts along the Chukchi Sea Coast. https://www.fws.gov/alaska/sites/default/files/2019-08/coastal%20haulout%20walrus%20factsheet_0.pdf.

US Global Change Research Program, 2018. In: Reidmiller, D.R., Avery, C.W., Easterling, D.R., et al. (Eds.), Impacts, Risks, and Adaptation in the United States: Fourth National Climate Assessment, vol. II. USGCRP, Washington, DC, USA. https://doi.org/10.7930/NCA4.2018, 1515 pp.

Watts, N., Adger, W.N., Ayeb-Karlesson, S., Bai, Y., Byass, P., et al., 2017. The Lancet countdown: tracking progress on health and climate change. Lancet 389, 1151–1164. https://doi.org/10.1016/s0140-6736(16)32124-9.

Wesche, S.D., Chan, H.M., 2010. Adapting to the impacts of climate change on food security among Inuit in the Western Canadian Arctic. EcoHealth 7, 361–373. https://doi.org/10.1007/s10393-010-0344-8.

FURTHER READING

Dallmann, W.K., 2017. Indigenous Peoples of the Arctic Countries. Norwegian Polar Institute. https://ansipra.npolar.no/image/Arctic04E.jpg.

Fischbach, A., Laustsen, P., 2016. More than 160 Years of Walrus Haulout Observations Reported by Russians and Americans Published as Database. US Geological Survey. https://www.usgs.gov/news/more-160-years-walrus-haulout-observations-reported-russians-and-americans-published-database.

US Government Accountability Office, July 2020. GAO-20-488, a Report to Congressional Requesters. A Climate Migration Pilot Program Could Enhance the Nation's Resilience and Reduce Federal Fiscal Exposure.

CHAPTER 16

Changing Political Landscape of the Arctic

INTRODUCTION

As Steinberg (2015) pointed out, the Arctic is not normal. While Antarctica is a continent centered around the South Pole, the Arctic is not a single piece of land, but rather the northernmost parts of two continents and the island of Greenland. Furthermore, the northern part of the Arctic is an ocean, not land. The world paid little attention to this peculiar state of affairs until recently. The Arctic used to be considered the back of beyond: frozen, dark, and inhospitable. It caught the attention of the two superpowers during the Cold War, as Russia and the United States faced off at the point where the two countries practically meet. Billions of dollars were spent on the installation of long-range radar sites, missile batteries, submarines capable of launching nuclear missiles, and other elements of armed force. Military expenditures by the United States during the Cold War years have been estimated at $8 trillion. The Soviet Union kept its military spending a tightly held secret, but by the mid-1980s, the USSR spent about 16% of its annual gross national product on military spending, according to US government estimates. When the Cold War was over, most of that militarization was scaled down to a minimum and the Arctic went back to being a relatively insignificant part of the world's imagination.

In the 1990s, there was a drop in the Arctic's geopolitical and geostrategic relevance. This enabled various regional cooperative schemes to be established in the Arctic. However, the strategic importance of the North has risen again (Østhagen, 2019). When the Arctic sea ice started to melt in the 21st century, it changed everything. Suddenly there was a new, pressing interest in the Arctic by many different countries, corporations, and strategic planners. Suddenly, the Arctic matters a great deal, but things are complicated. The situation would be far more simple if the entire Arctic fell within the bounds of a single country, but it does not. The Arctic has never been a single political entity, but rather contains the northernmost bits and pieces of five major countries. Unlike Antarctica that has been set aside as a demilitarized, apolitical region where no resource extraction is allowed, the Arctic is encumbered by a hodge-podge of international treaties, national governance, resource extraction by private- and government-run enterprises, and endless disputes over the sovereignty of land and sea.

The Ilulissat Declaration

As various governments have attempted to come to grips with the political situation in the Far North, they have found themselves grappling with the very concept of the Arctic. What is it? Where does it fit in the world political system? How should it be treated? Is it the domain of the five countries (Russia, Norway, Denmark-governed Greenland, Canada, and the United States that "own" parts of it), or is it rather a unique region where the normal political and economic rules do not apply? These topics were discussed by the "Big Five" Arctic coastal states, meeting during an Arctic Ocean Conference in Ilulissat, Greenland in 2008. The meeting resulted in the release of the Ilulissat Declaration, which has helped shape the Arctic policies of many nations in the following years. As evidence of the poorly thought-out politics of the North, the Arctic Council (AC), the only Arctic international forum, was not invited to participate. Likewise, Arctic indigenous peoples had no seat at the table in the Ilulissat negotiations.

As discussed by Steinberg (2015), the Ilulissat meeting was convened largely to develop the Big Five's responses to two very different scenarios. One of these scenarios envisions the Arctic emerging as a site where war could break out, as countries' efforts to map their outer continental shelf rights would result in an all-out "land-grab" and "sea-grab." The perceived threat of a "race for the Arctic" was not actually happening, but just the idea that such a race was occurring could lead to an actual race, ending in general instability and conflict. Under the other scenario, which apparently was at least as frightening to representatives of the "Big Five," even a tentative "race for the Arctic" could lend weight to calls for an Arctic treaty that would change the political status of the Arctic Ocean. The

Threats to the Arctic. https://doi.org/10.1016/B978-0-12-821555-5.00018-8

Ocean has thus far fallen within the jurisdiction of the United Nations Convention for the Law of the Sea (UNCLOS), and the participants in the Ilulissat meeting were very concerned that the rapidly evolving situation in the Arctic might lead to the Arctic Ocean being ruled by something other than a UNCLOS-governed global commons. Such a development would have the potential to undermine the near-universal acceptance of UNCLOS, possibly leading some to question core assumptions of the modern state system.

The Arctic of the Imagination

It may seem that conceptions of the Arctic play only a small role in what happens there, but this is not the case. As Steinberg et al. (2015) have pointed out in their book, "Contesting the Arctic: politics and imaginaries in the circumpolar North," these conceptions—termed imaginaries in the book—drive the political and economic actions that shape Arctic policies. If some remain unconvinced about the abnormality of the Arctic, Steinberg (2015) refers the reader to the wording of the Ilulissat Declaration, affirming the Arctic's normality. The felt need to state this in an international declaration suggests that the Arctic is not truly normal, or that there are sufficient perceptions of it being exceptional that its normality is questioned, and therefore must be defended. In short, conceptions of the nature of the Arctic, and what it ought to be, matter profoundly, and increasingly so as climate change transforms the Arctic's underlying character. Ultimately, the actors (national and regional governments, private enterprises, etc.) involved in making, interpreting, implementing, or responding to Arctic policies are the ones who decide whether the Arctic is a "normal" space that follows the standards of the rest of the world, or is somehow "exceptional."

FOUR DOMINANT ARCTIC CONCEPTS

As described by Steinberg et al. (2015), there are several schools of thought competing for prominence in ways of thinking of the Arctic. It may seem a purely academic exercise to formally frame these concepts, as most people do not adhere to any single school of thought. However, a discussion of the background and reasoning behind these four Arctic concepts helps to clarify how the conceptions shape our views of the Arctic, from individuals to States to the United Nations.

Arctic Conceptions and Climate Change

Depending on one's perspective, the Arctic may be considered in several ways. It may represent an integral part of the existing nation-states. It could also be considered as indigenous groups' homeland. As things change, some may consider the Arctic as a lost part of the national soul. Many would prefer to return to the days when the Arctic was off the international radar screen and was generally left alone by the outside world. Indeed, this view prevails among many Native groups throughout the North, with good reason. Some governments and most commercial enterprises consider the Arctic as only a potential resource colony, essentially empty of humans. This view holds that the Arctic exists to be exploited, whether through mining, petroleum extraction, or Arctic cruise ship tourism. For Native groups, the Arctic must be preserved as their homeland—the space of their everyday activities.

As the Ilulissat conference demonstrates, the Arctic may either be considered as "normal" or "exceptional." It can be considered "exceptional" on the basis of its physical characteristics—a dynamic region of ice, land, and water—or it may be exceptional because of its social characteristics, distant from southern capitals and populated by historically nomadic peoples who do not necessarily acknowledge national borders. Although national governments forced Arctic Native groups to settle in permanent communities during the 20th century, Native groups have always viewed the Arctic as a seamless entity. For them, the sea-ice cover that forms in winter serves as a convenient bridge linking various pieces of land. For instance, except during the Cold War era, it has been common practice for the Inupiat of northwestern Alaska to visit Chukchi friends and family across the Bering Strait in Chukotka and vice versa. Greenland and Canadian Inuit likewise cross the frozen seas of Baffin Bay to the shores of Greenland, for various reasons. Finally, more southerly based resource firms take little interest in the region's long-term preservation.

In fact, most people who spend time thinking about the Arctic are not influenced by just one of the mind-sets described earlier. Rather, most all are influenced by a combination of conceptions about the Arctic's physical and political properties and potentialities. In other words, there are constellations of ideas concerning the identity of the Arctic, and what it might become. As I said before, it is complicated.

Imaginaries are not stable. For centuries, Europeans from outside the Arctic saw the planet's northern periphery as a surface to cross, not a destination. The first European explorers to the North, like those throughout the Americas, sought neither territorial acquisition nor the establishment of permanent settlements. Rather, 18th- and 19th-century European explorers such as

Bering, Franklin, and Nansen were primarily driven by a desire to chart a sea route through the Arctic to the Orient. As Arctic landscapes and seascapes change, residents, states, corporations, and others active in the North respond and adapt, but rarely without contestation and disagreement. In the process, all parties make recourse to imaginaries—ideas about what the Arctic is and about what it can, or should, be. Arctic imaginaries, like the Arctic itself, are never settled.

The Status Quo as Imaginary

To explore the dynamism and ambiguity of imaginaries in the Arctic, it is useful to return to the Ilulissat Declaration. The declaration was intended to reaffirm the status quo, reproducing an imaginary inherited from the modern system of sovereign states. And yet, according to many of its critics, the declaration, especially the process by which it was achieved, went well beyond the status quo, setting a new course for the Arctic, one that could ultimately lead to the region being placed within a new imaginary that would sanction a higher level of appropriation and enclosure, whether by individual countries with Arctic ambitions or by the "Big Five" Arctic states acting as a collective entity that would exclude outsiders.

Imaginaries Exemplified by the Ilulissat Declaration

In summary, these six imaginaries can be placed into two groups of three. The imaginaries in the first group extend the basic system of bounded, sovereign, territorially defined states of the Arctic, albeit with specific modifications that account for the region's unique cultural and geophysical character. In the first of these imaginaries, the Arctic is conceived as a *terra nullis*, an unclaimed but potentially claimable space beyond the usual regulations of international law, where individual states are free to exercise their expansionist tendency, whether claiming land, water, ice, or seabed. Although few, if any, states or nonstate actors in the region directly advocate this ideal, this perspective on the Arctic—or the fear that others hold this perspective—lies just beneath the surface, shaping not just media coverage but also the statements and actions of those responsible for making, interpreting, and implementing Arctic policy.

The second imaginary is in a sense a modified version of the first. In this imaginary, the underlying matter of the Arctic is not irrelevant (as it is in the first imaginary), but it is different. Therefore, different norms and legal regimes are needed to govern this region where ice cover reworks the division between land and water that elsewhere orders the world. States thus are presented not with a *terra nullis* where there are limited opportunities for territorial expansion, but with a unique space of special and different opportunities.

The third imaginary is perhaps the most conventional. Here, although the Arctic may not present opportunities for new levels of territorial formation, it does present opportunities for replicating existing forms. This is seen in the various calls for Indigenous, especially Inuit and Greenlandic, statehood.

Some imaginaries conceive of the Arctic as suitable for bounding by sovereign territorial states, contrasting with ones that highlight processes and alliances that transcend boundaries. The Arctic can be imagined as an evolving frontier, a trove of opportunities for states, corporations, and individuals whose roots are elsewhere and who seek not the incorporation of territory but the extraction of riches. In this scenario, corporations, states, and a host of partners work together to facilitate economic activity to realize their financial goals.

Most of these imaginations allow only limited space for Indigenous peoples. Also, the pace and scale of current climate change are squeezing this room even further. As if the field of Arctic players were not sufficiently crowded, Paglia (2018) observed that the current understanding of the Arctic is increasingly incorporating countries and actors who are not in any usual sense of the word Arctic at all. The entrance of China and other Asian and European countries from the midlatitudes onto the stage of Arctic affairs is discussed in the following.

WHO OWNS THE ARCTIC?

As discussed by Bryce (2019), the question of Arctic sovereignty has increasingly appeared in the news of recent years. For instance, in August 2019, US President Donald Trump made international headlines when he expressed an interest in buying Greenland. The Danish government responded politely but firmly that Greenland is not for sale. Trump was widely ridiculed for this diplomatic blunder. What was behind this unprecedented offer? Was it something to do with the United States' growing interest in expanding its slice of the Arctic?

All the nations surrounding the Arctic such as the United States, Canada, Denmark, Finland, Iceland, Norway, Russia, and Sweden are currently jostling for ownership of the region's frozen seas. Several of the countries have already submitted formal papers to a

United Nations body, claiming portions of the vast Arctic seabed. As we have seen, climate change is opening the formerly ice-locked waters of the Arctic, making the region more accessible than ever before. Current predictions of an ice-free Arctic consider dates between 2040 and 2050.

This surge of interest in the Arctic has been called the "scramble for the Arctic," or more sensationally, "the new Cold War," because Russia and the United States are facing off once again. But despite the opportunities the region presents, can the Arctic Ocean truly be owned by anybody? And why do so many countries want a stake in this seascape of drifting icebergs and polar bears? Perhaps it is not so much the waters of the Arctic Ocean that interest the world, but what lies beneath. The bed of the Arctic Ocean holds massive oil and gas reserves—an estimated 90 billion barrels of oil, about 13% of the world's undiscovered oil reserves. Moreover, about 30% of the planet's untapped natural gas is sitting beneath Arctic lands and seabed. Recent technological advances in transportation and petroleum extraction are making remote regions of the Arctic Ocean increasingly accessible. This has forced international lawmakers to quickly develop and expand definitions of where countries can legally explore (starting with the Ilulissat Declaration, discussed earlier). Currently, UNCLOS-signatory countries are allowed to exploit resources from the seabed within 370 km (200 miles) of their shorelines. However, more than a decade ago, the UN encouraged Arctic countries to provide evidence that certain features on the seabed beyond the 200-mile limit represent geologic extensions of their nation's continental landmass. If a good case can be made, then a country's jurisdiction can be expanded toward the North Pole. Russia, Canada, Denmark (on behalf of Greenland), and the United States have compiled the data and submitted their claims to the Commission on the Limits of the Continental Shelf, a UN-appointed body, who will adjudicate the claims.

Much Ado About the Seafloor

As discussed by Henriques (2020), the sovereignty of the Arctic Ocean seafloor has become a three-way geopolitical tug-of-war to decide which country owns a ridge of undersea mountains, the Lomonosov Ridge. The winner will bring a permanent change to Arctic political and economic geography. For such a seafloor feature to count in a country's favor, there must be conclusive evidence that it is a piece of submerged land—not an oceanic ridge that has always been underwater and has little to do with the country's land mass.

A study by scientists from the Geological Survey of Denmark and Greenland (GEUS) shows that the Lomonosov Ridge is indeed submerged land, and not formed by seafloor spreading, like the mid-Atlantic ridge. The same finding is backed up by other studies, including seismic research into the structure of the crust, led by the Geological Survey of Canada, and other crucial evidence, such as extensive efforts to map the regional seafloor. According to GEUS geologist Christian Knudsen, the "Lomonosov Ridge" … is definitely continent, and it is a continent that is similar to what we find in eastern Greenland—it is a continuation of Greenland, which is their main point. This claim is very likely to be true, but rocks of a very similar nature have also been found by the Russians on Franz Josef Land, north of Novaya Zemlya (Russian territory). Furthermore, Canada also has evidence that the Lomonosov Ridge is a prolongation of land extending from Ellesmere Island. This is not surprising, given that Ellesmere is only 20 km from the northwest coast of Greenland, across the narrow Davis Strait. All things considered, it is possible that the Lomonosov Ridge is Russian, Canadian, and Greenlandic all at once.

Four countries have submitted evidence to extend their sovereignty much closer to the pole. Meanwhile, the United States claims that the Lomonosov Ridge is an oceanic ridge—not an extension of any country's continental shelf. The US government, therefore, refutes any territorial claims based on connections to the ridge. The US claim falls on the opposite side of the Arctic Ocean (Fig. 16.1).

So far, the UN commission has not ruled on any of these submissions, though Russia, which was first in line to make its submission in 2001, appeared to hear positive early signals toward the end of 2018. Klaus Dodds, professor of geopolitics at Royal Holloway University of London, said "The commission had clearly released an indication to the Russians that they were sympathetic to their submission. That is exceptionally exciting to the Russians. After a mere 20-year wait, I think they're going to get what they want—which is confirmation that the continental shelf and those submarine ridges belong to one another." That puts Russia in a strong negotiating position, according to Dodds. "You can imagine what's going to happen. President Putin will stand somewhere very grand—you'll have an enormous map of the Arctic beside him—and he's going to say: 'The Arctic is ours.'" It is important to keep in mind that the UN commission has no legal powers. It will only rule on the scientific credibility of the evidence. Furthermore, the commission will not recommend where to draw the lines on the map. That must be done through diplomacy.

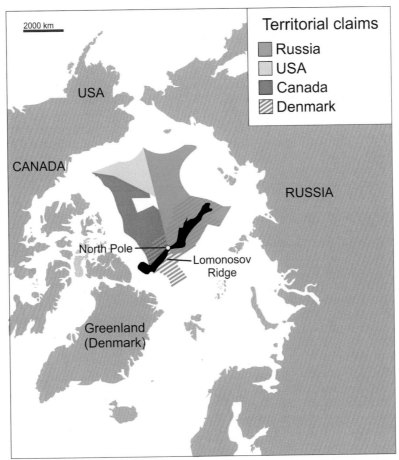

FIG. 16.1 Map of the Arctic Ocean region, showing the territorial claims submitted to the United Nations by Russia, the United States, Canada, and Denmark, by the author.

The long game

All this research and political posturing point toward a future in which various nations will indeed own pieces of the Arctic Ocean floor. Political power will accompany each piece of the Arctic pie, commensurate with the size of the piece. Russia and Canada, for instance, as claimants of the two largest pieces, will inevitably gain more regional influence. The process of winning claims to the Arctic seabed will be long and costly. Even if countries get UN approval, they will have to shoulder the huge expense of getting their ships to the Arctic, building deep-sea infrastructure, and extracting oil, gas, and minerals from many km beneath the surface.

The Arctic Ocean remains an isolated environment, far from the mainland. The sea ice may be retreating, but there are still dangerous sea conditions and icebergs with which to contend. It will be very difficult to get such projects insured. As cited by Bryce (2019), Amy Lauren Lovecraft of the University of Alaska, Fairbanks, has commented that "At this stage… countries' claims to the Arctic are mostly anticipatory. A lot of what's being divvied up doesn't have anything to do with immediate need. It's about 'let's get what we can under UNCLOS so that we have access to all of that space in the future." However, Lovecraft remains cautiously optimistic. "If I put on my absolute environmentalist's hat, it's true, the Arctic will be used more. However, I don't think it's a race to the bottom." In other words, the Arctic will be owned and explored—but that does not necessarily mean it will be destroyed.

THE PURPOSE OF THE ARCTIC COUNCIL

The AC is an intergovernmental forum established in 1996 to promote cooperation, coordination, and interaction among the Arctic states, Indigenous communities, and other groups on issues of common importance. The rising geopolitical importance of the Arctic and the onset of climate change has resulted in the Council becoming a focus of increasing interest from both inside and beyond the Arctic. This has resulted in new demands placed on the Council, attracting an increasing number of participants and instigating a period of transformation as Arctic states work to find a way to balance conflicting demands for improving the effectiveness of the Council and taking care of national interests. Barry et al. (2020) discussed whether the Council is having an impact today on the issues it was formed to address in the 1990s, such as environmental protection and sustainable development. He examined drivers and barriers to the Council's effectiveness, particularly its problem-solving abilities.

For the most part, non—Arctic states did not object to claims on the part of the Arctic states to dominance in the Arctic arising from their self-proclaimed "sovereignty, sovereign rights, and jurisdiction" in the region. This *laissez-faire* attitude worked well as long as Arctic concerns did not spill over into global arenas. In other words, the rest of the world saw no harm in leaving Arctic affairs to the Arctic states (Young, 2019).

The AC initially focused its attention on environmental protection and sustainable development issues. It has evolved in the 21st century and now also addresses social, cultural, and economic issues. Its founding documents explicitly exclude any focus on military/ security issues, although it currently includes a working group on Emergency Prevention, Preparedness, and Response (EPPR). The broadening of the Council's agenda has been in response to the increasing geopolitical importance of the Arctic, as well as the onset of rapid regional climate change. At the start, the AC was composed solely of representatives from the Arctic countries. The increasing geopolitical and economic importance of the Arctic has resulted in the attraction of growing numbers of representatives from well outside the Arctic, most notably China. The AC is currently working to find a way to balance conflicting demands for improving its effectiveness while and maintaining its care of diverse national interests.

It is clear that the AC has changed significantly since its formation. Its structure and operation are evolving in response to the increasing global attention on Arctic issues. This has resulted in expanded focus by the Arctic states in pursuing the geopolitical agendas in the region. Some factors are facilitating this change, such as the willingness of member states—both Permanent Participants and Observers—to commit resources to support the AC. The AC remains the primary facilitator of meetings between Arctic stakeholders aimed at reaching a consensus. That being said, challenges to ensuring effective outcomes from the Council's activities remain. The AC is more responsive than proactive. This approach leads to an ad hoc designation of new components in its agenda and the creation of new working groups. These groups have sometimes formed without clear direction or mandates, which can lead to wasted resources and a lack of clarity concerning the division of responsibilities.

In recent years, then, the Arctic has moved towards the center of attention with regard to matters of global concern. The increasing accessibility of the Arctic has generated a surge of interest in extracting the Arctic's natural resources, especially its world-class reserves of oil and natural gas. For example, the extraction of natural gas on Russia's Yamal Peninsula and its shipment to Asian and European markets using specially designed liquefied natural gas (LNG) tankers has caught the attention of financial markets throughout the world. Some see the Arctic as an arena that will attract investments on the order of $1 US trillion within the next few years.

The institutional problems within the AC reflect the uncertainty among the Arctic states concerning the ability of AC institutional structure to fulfill their needs. There is also a desire by some representative countries to exert more direct control on specific issues rather than working through the established AC rules and procedures in an effort to redirect Council priorities. According to Barry et al. (2020), the AC lacks an overall strategy to address broader issues such as climate change and sustainable development. The AC is not obligated to report its findings to any other authority (e.g., the United Nations). This lack of transparency especially affects the ways in which States act on the outcomes from Council decisions. So, while the Council has often been effective at the global scale, it is difficult to assess its influence at the national or local level.

The Arctic Council's Role in Geopolitics

Conditions in the Arctic today differ from those prevailing during the 1990s when the AC was formed. The current Arctic transformation is having far-reaching implications for the architecture of Arctic governance, especially the AC. As discussed by Young (2019), what was once a peaceful, peripheral region has turned into ground zero for climate change and the scene of geopolitical jostling in which Russia is flexing its muscles as a resurgent great power. China is launching economic initiatives, and the United States is reacting defensively as it attempts to reassert its regional power. Specifically, Russia has taken steps to reclaim its status as a great power, articulating renewed claims to a leading role in the affairs of the Far North and strengthening military assets in the Russian Arctic. China, the emerging superpower, has taken steps to include the Arctic within the scope of its Belt and Road Initiative, leading, among other things, to a series of bilateral moves relating to economic investments or proposed investments in Russia, Finland, Iceland, Greenland, and Canada. The United States has become sensitive about its superpower status and has begun to treat the Arctic activities of other countries as hostile initiatives threatening America's interests. All these developments are playing out within a shifting global context characterized by the decline of the postwar world order, but lacking clarity regarding the future global order. The relative tranquility associated with the Arctic's peripheral status has vanished like the pack ice.

The AC is struggling to cope with these problems. For one thing, it is hamstrung by the lack of any governing authority over Arctic countries or Arctic affairs. It can only advise and hope to influence countries that go against its policies. In this sense, the AC is like the United Nations, or the old League of Nations that formed after World War I. When countries such as Germany and Italy were held to account by the League of Nations, they simply walked out of the organization and did what they pleased. Nowadays, this seems a dramatic response to a rebuke, so countries called on the carpet for misbehavior in the Arctic often choose to disregard the scolding and carry on.

Don't rock the boat

Østhagen (2019) observed that the Arctic states have shown a preference for a "Don't rock the boat" stance in which they maintain dominance over the region. This is supported by the importance they collectively give to the Law of the Sea and issue-specific agreements made by the AC. These arrangements preferentially benefit the Arctic states, and they also ensure that Arctic issues are generally dealt with by the Arctic states themselves. Although the Law of the Sea regime is meant to guarantee the primacy of the Arctic states concerning their offshore resources, the Arctic states must protect their maritime rights through maintaining high-profile visibility in the region. This is being accomplished by the perpetual presence of naval and air forces in and around the Arctic Ocean. However, Østhagen (2019) cautions that the status quo of Arctic geopolitics can

no longer be maintained in the face of broadening international interest in the region and that the Arctic states must come to grips with this fact while staying focused on their own national policies. Those policies include Arctic security concerns and economic investments, both of which require the cooperation of their allies. The rise of the Arctic on the world's agenda is no passing trend: it is here to stay.

The Arctic Council and climate change

The Arctic states have agreed that climate change is the most serious threat to the Arctic's biodiversity and that decisive action would help sustain Arctic ecosystems and ecosystem services, as expressed in a statement by the AC's working group on the Conservation of Arctic Flora and Fauna (CAFF). Through CAFF, the AC has developed a comprehensive plan of actions needed to address the loss of biodiversity, but the lack of AC reporting to any partners or umbrella organizations means that the extent and effectiveness of any resulting actions remain unclear. The development in 2017 of a legally binding instrument under UNCLOS concerning the conservation and sustainable use of marine biological diversity of areas beyond national jurisdiction may have significant impacts on how the AC deals with Arctic biodiversity. These impacts are already being reflected in the increased emphasis placed on the need for improved coordination on ocean governance, e.g., the AC's Task Force on Arctic Marine Cooperation, or TFAMC. What role might the Council play in implementing this agreement? Without clear strategies, many of the Council's efforts can appear ad hoc, reactive rather than responsive, and without due recourse to forward planning. However, when clear and detailed plans are in place to guide the work of the Council as in their 2015 Action plan for Arctic biodiversity and the Arctic Migratory Birds Initiative (AMBI), then glimpses can be seen of the potential of the Council to act as an agent of change.

The architecture of Arctic governance, such as it is, has focused on the role of the AC, treated as a "high-level forum" designed to promote "cooperation, coordination, and interaction among the Arctic states." The original design of the AC reflects the prevailing conditions of the 1990s, a period marked by the end of the Cold War, the collapse of the Soviet Union, and the reduction in tension in the High North that followed these events. Most analysts agree that the AC has performed well during its 25-year history. The accomplishments of the council have exceeded the expectations of most Arctic scholars and politicians (Young, 2019).

National Arctic Strategies

As discussed earlier, the rapidly disappearing sea-ice cover on the Arctic Ocean has facilitated an increase in the unclaimed ocean and land territory beyond any nation's control. Several countries are attempting to gain jurisdiction over the pieces of the Arctic seafloor for purposes such as resource extraction and the development of trade routes. Control over Arctic territory has given nations the ability to settle it, extract its resources, and establish military outposts, making legal rights and claims over such territory incredibly valuable (Gross, 2020). Beyond economic motivations, nations such as the United States, Russia, and China are competing in the region to project military supremacy and gain power. The Arctic also sits at a critical position between North America and Eurasia, making it a powerful strategic region from which to project military strength. As the liberal international order is increasingly under threat by great power conflict (a scenario in which the already-contentious relationship among the United States, Russia, and China could devolve into a war), the Arctic Circle faces a crossroads: Will the underlying boiling tensions explode to the point of larger global conflict or can diplomacy ensure peace in the region? In the face of these geopolitical pressures, Russia, China, and the United States have been developing new national Arctic strategies.

Andreas Østhagen is a Senior Research Fellow at the Fridtjof Nansen Institute of Norway and the Arctic Institute of Canada. His recent paper (Østhagen, 2019) discussed what he considers an important difference between overarching national security strategies and those that specifically concern the Arctic region. First, security dynamics in the Arctic have remained anchored to the subregional level: the Barents area, the Bering Sea/Strait area, and the Baltic Sea. Thus, it is futile to generalize about security interests and challenges across the whole northern circumpolar region. It makes more sense to discuss security in the different parts of the Arctic, not in the Arctic as a whole. Of these different parts, the European Arctic is undoubtedly the most active and the most challenging.

National Arctic Strategy of the United States

International politics are not exempt from the consequences of climate change, and yet it is commonly acknowledged that the United States, the superpower that emerged from the Cold War, has until recently devoted scarce attention to the Arctic geopolitical scenario. This is evident in the gap in both media coverage and academic literature addressing the US position on the Arctic. In January 2009, the United States issued its first Arctic strategy of the 21st century as a presidential directive, one of its most traditional security documents. In the same year, Canada and Norway published multifaceted documents addressing numerous aspects of Arctic politics ranging from national security to indigenous communities. It was not until 2013 that the US government under President Barack Obama issued its first full-fledged Arctic strategy. Since then, US strategy has evolved significantly, and the number of US defense documents addressing the Arctic has risen sharply.

Root causes of policy change

The change in the US Arctic posture can be explained by several factors. First, the Arctic can be a fertile ground for unanticipated teleconnections. For instance, it has been shown that international crises taking place in a totally different theater, such as Russia's military actions in Ukraine, can have complex repercussions on Arctic security (through sanctions). Second, climate change in the Arctic is affecting the economic and political interests of Arctic and non–Arctic states alike. Increased access to Arctic waters is providing economic benefits beyond the oil and gas industries, namely in the extraction of rare earth minerals, in the development of maritime shipping routes, and in the expansion of fisheries and nature tourism. The national competition over Arctic resources may do more than exacerbate existing issues such as land rights claims of Indigenous peoples and maritime sovereignty disputes. The rivalry between nations has the propensity to escalate into major international disputes.

US naval strategies

As discussed in DefenseNews.com in March 2020, the US Navy released details of a new strategy for operations in the Arctic as competition for resources among nations increases in the region. The Navy statement titled "A Blue Arctic" provides an outline of planned operational changes for the military's sea services in and around Alaska, including the Navy, Coast Guard, and Marine Corps. The services will operate a full range of missions, with operations adapted to Arctic conditions, and work with local and Indigenous communities to build regional security. The Navy highlighted the regional strategy as receding sea ice makes mineral and biological resources more accessible to the nations that can exploit them. Another national security issue concerns Arctic Ocean trade routes that are expected to open, allowing access to shipping lanes previously closed by sea ice.

The United States and China have invested in expanded icebreaker fleets to support northern operations, while Russia has expended vast amounts of resources to modernize its northern fleet. "The coming decades will witness significant changes to the Arctic Region," the strategy statement said. "Encompassing about six percent of the global surface, a Blue Arctic will have a disproportionate impact on the global economy given its abundance of natural resources and strategic location."

While acknowledging the need to evolve US forces in the region, Senator Dan Sullivan from Alaska considers that the strategy "lacks some of the urgency needed to drive the development of critical capabilities that are required to effectively compete with our rivals in the Arctic," Sullivan said. Sullivan advocates further changes—including making Alaska the home port for one or more Coast Guard icebreakers.

Climate change impacts on US Arctic military strategy

As reported by Lavorio (2020), while most of the US military establishment did not integrate climate change assessments into its actual strategy, the US Navy was the exception. The Navy procedures for monitoring and charting sea-ice conditions began in 1948 through reconnaissance flights, and 3 years later, the Navy launched a sea-ice observation and forecasting program. To gather more data on the warming of the Arctic, the Navy established a partnership with Canada, starting an intensive collaboration with the Canadian Defense Research Board (DRB) in expanding the network of Arctic observing stations to report on sea-ice conditions. Average Arctic sea ice thickness decreased from 365 cm in 1893–96 to 218 cm in 1937, and in the 1970s, US Navy oceanographers predicted that ice-free Arctic summers might occur 20–30 years hence. This prediction was shared with both Central Intelligence Agency analysts and DRB officials, but, unfortunately, it failed to be acted upon by successive presidential administrations.

After the end of the Cold War, the first steps toward the inclusion of climate change in national security planning were characterized by vagueness rather than clarity. Environmental security began to be formally recognized in the 1991 National Security Strategy (NSS) issued under President Bush. Economic considerations were given more attention than national security by the Bush administration, however.

Major changes in US policy occurred in 2007, the year in which sea ice retreated to the lowest extent ever recorded in the satellite era. In the same year, the Russian *Arktika* expedition planted the Russian flag in a titanium tube on the ocean floor beneath the North Pole. Envisaging dramatic geopolitical scenarios in a seasonally ice-free Arctic in the 2040–50s, public opinion and governments began to focus on a new edition of the Gold Rush (the Arctic Rush), which replaced the well-established cooperative climate of Arctic relationships that had emerged among international and national actors since the past decade of the Cold War. The following year, 2008, Chief of Naval Operations Admiral Gary Roughead convened an Executive Board to discuss the changing environment. The initiative culminated in the foundation of the Navy's Task Force on Climate Change (2009) meant to detect and consequently address the implications of climate change, especially in the Arctic.

A 2011 US Department of Defense (DoD) report to Congress concluded that the Arctic is currently considered of peripheral interest by much of the existing national security community, a situation not likely to change significantly in the next decade or more, notwithstanding some external forcing event, such as a major environmental or human disaster or activity in the Arctic viewed as threatening US interests in the region. Østhagen (2019) contends that the Arctic does not play the same seminal role in national security considerations in the United States as it does in Russia or Northern Europe. Although the military and political rhetoric might suggest otherwise, for the United States, the Arctic has served primarily as the location for missile defense capabilities, surveillance infrastructure, and a limited number of strategic forces. It is also of importance to the US Navy and Coast Guard, although the DoD has yet to invest significantly in Arctic capabilities and infrastructure.

The Collapse of the "Arctic Ice Shield"

The Arctic Ocean north of Alaska has been easily protected and of limited strategic importance due to the ice that has shielded it, impeding both access and use. Now the ice is melting, creating new opportunities and potential threats to US national interests. In 2013, the Obama assessed the situation in the Arctic as follows:

> On the one hand, opportunities are constituted by the abrupt effects of climate change, exposing natural resources such as oil, gas, rare earth elements, iron ore, and nickel, all suitable for some original development of infrastructure and commercial initiatives in the region. On the other hand, challenges include the impact on indigenous populations, fish, and wildlife due to the melting of sea ice and the stability of ice sheets, as well as pollution caused by the thawing of permafrost.

2013 US Department of Defense Assessment

The evidence of more favorable climate conditions, attributable to climate change and leading to a (moderate) growing human presence in one of the most adverse regions in the world, called for the inclusion of the Arctic in the perimeter of homeland defense, a remarkable change in the US Arctic posture. Increased access to the Arctic Ocean and adjacent lands caused by the decreasing seasonal sea-ice cover has driven increases in the Arctic's suitability as a jumping-off point from which to launch attacks on the United States. Given this change in Arctic conditions, the DoD plans to increase US preparedness to detect, deter, prevent, and defeat such threats. An additional section of the 2013 document declares that the DoD will continue to support the exercise of US sovereignty.

In the long term, the US Navy envisages a scenario where the Arctic Ocean will be completely ice-free, a fact that inevitably calls for an urgent strategic recalibration. As the reduction of sea ice is expected to continue, waterways such as the Transpolar Route and the Northwest Passage (NWP) may no longer be limited to 30–45 day periods but may be navigable for 130 days per year.

The Trump Era of Climate Change Denial

Concerning climate security, it is worth noting that after the release of the peer-reviewed 2018 National Climate Assessment, President Trump appointed the climate denialist physicist William Happer to the National Security Council. Trump's controversial and ambiguous position, close to climate change denial, alarmed many members of the security community. In 2018, national security leaders, retired military officials, and lawmakers expressed the urgency of comprehensive identification of climate change effects in reference to national security and action, addressing the President and the Secretary of Defense. In 2019, a letter was sent to the President, signed by independent think tanks (the American Security Project, and the Center for Climate and Security). The letter urged the President to reconsider his policy on climate change. Denying climate change (and its anthropogenic cause), it argued, may have harmful effects on national security regardless of political beliefs. During the same month, and in spite of these multiple warnings from expert groups, the White House saw to it that the Navy's Task Force on Climate Change was dismantled without public announcement or the release of a final report.

In June 2019, the US DoD released an Arctic strategy updating the 2016 version. Much of the document is devoted to achieving "competitive military advantages" in various theaters to protect national interests in the Arctic. A military advantage is thus considered functional to the maintenance of "a credible deterrent for the Arctic region." It clearly states that the main global (and the Arctic) competitors are Russia and China. The first issue concerns sea routes: "Russia and Canada claim the right to regulate Arctic waters in excess of the authority permitted under international law," respectively, in the Northern Sea Route (NSR) and the NWP. The strategy observes that the NWP is "subject to the complete sovereignty of Canada." Then, conflictual relations are envisaged in the increasing military activity of Russia, concerning territorial defense on the coastline (in particular, the establishment of the Northern Fleet Joint Strategic Command) and China, whose regional presence is still limited but characterized by an intensifying trend "potentially including the deployment of submarines to the region." The involvement of competitors in the Arctic is not limited to the military level. There are also issues at the economic level, as proven by China's Polar Silk Road initiative announced in January 2018 and the establishment of research stations in partner states, such as Svalbard. In light of these events, the strategy document declared that "despite China's claim of being a 'Near Arctic State,' the United States does not recognize any such status."

National Arctic Strategies—China

China has transformed itself from a ramshackle, quasi-feudal empire at the beginning of the 20th century into one of the Great Powers of the 21st century. The driving force behind this push has been China's historical experiences, most notably those of the 19th and early 20th centuries, known to the Chinese as the "Century of Humiliation" (百年国耻), during which China lost both its territory and its prestige to the imperial powers of the day. These experiences continue to influence China's relationships with the wider world. They also have served as a unifying force within China, the legacy of which persists today (Harper, 2019). China's past experiences have guided its modern international strategies, including the desire to avoid falling behind the Western powers as they did during the Century of Humiliation (see the late 19th-century political cartoon, Fig. 16.2).

A David-and-Goliath story has recently been developing between the nation of Greenland (2020 population of 56,081) and China (2020 population of 1,439,323,776). The Government of Greenland has been promoting the development of a large mining operation in the southwest of the country, called the Kvanefjeld Rare Earth Project (Sevunts, 2020). The

FIG. 16.2 French political cartoon by Henri Meyer, 1898, entitled "Foreign powers carving up China." The image is in the public domain.

proposed mine is situated near the southern tip of Greenland, adjacent to a fjord (Fig. 16.3). The project has been billed as being "ideally placed to meet growing rare earth demand and has the potential to become Greenland's first world-class mining operation." Proponents of the project say it will help break China's stranglehold on strategic rare earth minerals. Experts estimate that China controls more than 80% of the global supply chain of rare earth elements, a group of 17 metals that are critical for the development of new clean energy technologies, as well as cutting-edge space and defense materials.

As discussed in Sharma (2020), apart from the numerous economic opportunities that the rapidly changing Arctic is bringing, the importance of its strategic geographic position must not be underestimated. In case of any direct military confrontation between the United States and China, the Arctic will offer the shortest missile trajectory routes for Chinese and the US ICBM's to reach each other's mainland. If launched from the North Pole, China's Dongfeng A (long-range missile with an estimated range of 11,200 km) can easily destroy a series

of large US cities on the East Coast and New England, such as Ann Arbor, Philadelphia, New York, Boston, Portland, Baltimore, and Norfolk. The rapid modernization of the Chinese People's Liberation Army-Navy and its ambitious plans of becoming a global maritime power with blue water capabilities also have an Arctic dimension. This notion has been justified by President Xi Jinping when in November 2014 he stated that, "Polar affairs have a unique role in our maritime development strategy, and the process of becoming a polar power is an important component of China's process to become a maritime great power." Current Arctic ice conditions offer an ideal opportunity for the Chinese navy to develop a hidden base for submarines operating in the region. For now, at least, the thick ice overlying the North Pole and adjacent regions makes the detection of submarines operating beneath the ice difficult, enabling the potential operational success of their missions. In any war-like situation with either the United States or Russia, if China successfully deploys any submarine in the Arctic waters with either nuclear or even conventional warheads, it can pose a direct threat to Europe, Russia, and United States. China now has two icebreakers (one domestically built) for its commercial and strategic operational needs. The United States currently possesses only one operational ice breaker in service (the *Polar Star*), which is quite old and suffering from operational limitations and budget cuts. Recently, Beijing also revealed its future plans for the construction of its first nuclear-powered icebreaker, at its own shipyard. It is estimated that this proposed new icebreaker will be 152 m long, 30 m wide, and 18 m deep with a displacement of 30,000 tons. The new icebreaker is expected to be powered by two nuclear reactors with the capability of crushing Arctic ice at a maximum speed of 11.5 knots.

China has more than just military ambitions in the Arctic. Gross (2020) observed that the Chinese Government has dubbed their involvement in the Arctic region as the "Polar Silk Road," which is an allusion to the trade routes that they intend on pursuing after the acquisition of Arctic natural resources. China's legal claim to the region differs from many other nations due to their lack of a land border with the Arctic, complicating their efforts. Chinese companies such as Shenghe Resources, China Nuclear Hua Sheng Mining, China National Petroleum Corp., and China National Offshore Oil Corporation have attempted to expand their mining and petroleum development projects in the region but have found bureaucratic barriers and political opposition coming from the Danish, who control Greenland. Furthermore, China is outnumbered by its primary competitors, the United States and Russia, who have been operating in

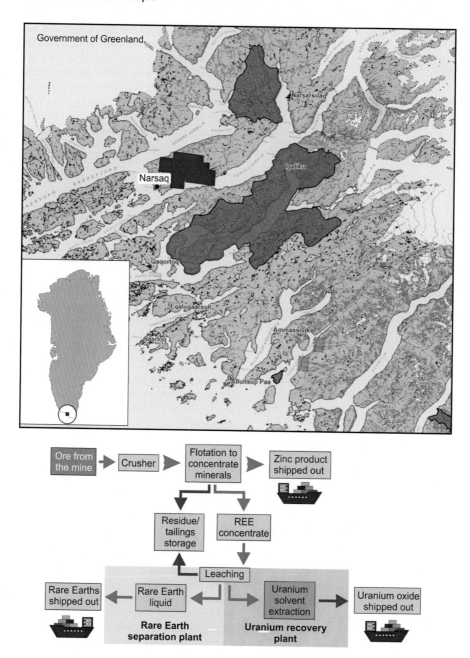

FIG. 16.3 Above: A map showing the site of the proposed Kuannersuit mine (*red*) and its location in proximity to Narsaq and other towns and hamlets that could be affected by its operations (*dark gray*). Below: Flowchart illustrating the processes proposed by Greenland Minerals, Inc., to extract uranium and rare earth minerals from the mineral seenstrupin. (Above Image courtesy of Naalakkersuisut (Government of Greenland), in the public domain. Below: After Greenland Minerals, Inc.)

the region far longer and have more experience in resource extraction. China's first icebreaker, known as the *"Xuelong"* or *"The Snow Dragon,"* pales in comparison with Russia's 51 icebreakers. But the recent construction of *"Xuelong 2"* (Fig. 16.4) has motivated the United States to ramp up the production of new icebreakers, including a plan to launch three more by 2029. The icebreakers are representative of the larger cat-and-mouse game that is emerging among the three nations, in which they attempt to preempt the other two countries and respond in kind to an opposing nation's expansion in the region.

China has few domestic energy resources. Most of its fossil fuel supply comes from the Middle East, the United States, and Russia, which drives its ambition to tap Arctic natural gas to secure energy independence. China's Arctic ambitions lay beyond mere energy, however, as they have recently expanded their plans to take advantage of other nation's resources while there remains a state of international legal confusion over Arctic sovereignty.

China and the Arctic Council

In 2013, the AC accepted as Observers 12 non–Arctic states, all key players in the current international arena. These included France, Germany, the Netherlands, Poland, Spain, the United Kingdom, Italy, China, Singapore, Korea, Japan, and India. China's acceptance in the AC has been seen by them as an acknowledgment of its international participation as an equal power by

FIG. 16.4 Above: Chinese icebreaker Xue Long. Photograph by Bahnfrend, Creative Commons Attribution-Share Alike 4.0 International license. Below: Russian Icebreaker Arktika. Photo courtesy of Rosatomflot Communications Department. (Creative Commons Attribution-Share Alike 4.0 International license.)

those actors, in particular, Russia and the United States, who have historically stressed their superiority since the time of China's "Hundred Years of Humiliation" and during the Cold War.

As discussed by Mazurier et al., 2020, Chinese involvement with the AC began well before its acceptance as an Observer state. Before its acceptance, the Chinese government promoted a 5-year polar plan to increase China's "status and influence" in polar affairs and to protect its interests as a "near-Arctic state." The Chinese government urged that "an end to the Arctic states' monopoly of Arctic affairs is now imperative." As soon as China gained a seat at the participants' table in 2013, it signed an agreement with Iceland, representing its first free-trade agreement with a European country. At the same time, Beijing also announced the investment of $2 billion in Greenland's mining sector, allowing that autonomous island to become a large exporter of uranium and generating a massive political debate in Denmark. In fact, the export of uranium to China, agreed by the Native government of Greenland (the *Inatsisartut*), represents a careful balancing act between its competence as the authority responsible for the extraction of Greenland's minerals, and the Danish government's security concerns about putting more uranium in the Chinese nuclear arsenal. China's entrance into Greenlandic affairs was so intrusive that the New York Times published an article in 2013 titled: "No, Greenland Does Not Belong to China."

While climate change is the principal reason put forth to justify the Asian states' interest in the Arctic, the discovery of new natural resources and sea routes is also one of the reasons for the curiosity of those states. Indeed, since the late 2000s, China, Japan, South Korea, India, and Singapore have shown an interest in all these issues. However, those states do not possess any Arctic territory and, hence, must play along with the Arctic states to be integrated into Arctic debates, although the Chinese government and Chinese authors have articulated the concept of "near-Arctic state" in an effort to justify China's interest. For the record, the northernmost point in China lies at 53° 29′N, about the same latitude as the central Netherlands and Philadelphia, Pennsylvania. Furthermore, China is not accepted as an Arctic state by other countries and has largely been excluded from regional politics (Østhagen, 2019).

Western views of China's Arctic ambitions

The US government has rejected China's self-characterization as a "near-Arctic" state. Distrust of China's Arctic intentions goes beyond Washington, DC.

As discussed by Babin and Lasserre (2019), several newspaper articles published in American and European newspapers after China's acceptance as an AC Observer depicted China as a greedy country ready to challenge the sovereignty of the true Arctic states to gain access to Arctic resources. Chinese officials have declared that "countries closer to the Arctic, such as Iceland, Russia, Canada, and a few other European countries may tend to wish the Arctic, were private or that they had priority to develop it, but China insists that the Arctic belongs to everyone, just like the Moon." Statements such as these have only increased the existing concerns among the Arctic states about China's intentions.

Two views emerge from Western academic articles on "Polar Orientalism" as defined by Dodds and Nuttall (2015) as "a way of representing, imagining, seeing, exaggerating, distorting and fearing 'the East' and its involvement in Arctic Affairs," which is only amplified by the application and admission of Asian states to the AC. On view conceives of Asian countries, led by China, wanting to invade the AC and reshape it. The second view considers that Asian countries wish only to participate in an interstate forum for collaboration while respecting the rule of Indigenous peoples and the Arctic states. Since 2011, many popular and scholarly articles have expressed concern about the "Arctic appetite" of Asian countries with sensationalist titles such as "What is China's Arctic Game plan?" "Cold, Hard Facts: Why the Arctic is the World's Hottest Frontier?" "China's Interests in the Arctic: Threat or Opportunity?" or "Enter Asia: The Arctic Heats Up." The potential value of Arctic natural resources could represent an extremely attractive source of supply for the Asian States. Other articles mentioned the "race" of the Asian States to acquire or even seize these resources.

National Arctic Strategies—Russia

As discussed by Østhagen (2019), the Arctic's strategic importance has evolved primarily because Russia is intent on reestablishing its military power, and the Arctic is one region where it can do so without facing disapprobation (and obstruction) from Western governments. This has little to do with the Arctic itself but stems from Russia's ambition to build and maintain its military strength in the North. Russia's Northern Fleet is based on the Kola Peninsula, including strategic submarines essential to the county's status as a nuclear power on the world stage. So, it is not the melting of the sea ice that has spurred Russia's military emphasis on the Arctic, it is the importance of the Arctic for Moscow's more general strategic plans and ambitions.

In opposition to Østhagen's (2019) view that Russia's military ambitions in the Arctic are not driven by climate change, Burke (2021) asserted that climate change *is* stirring up the underlying frictions that are geopolitical in scope and involve some of the most advanced and well-armed societies in the world. While World War III is unlikely to start at the North Pole, there is a risk of miscalculation, which could lead to military escalation. Moreover, the friction in the North between the United States, China, and Russia is part of a larger pattern. What happens in the Arctic reflects what is happening in the South China Sea, the Crimea, and Ukraine, and vice versa. Climate change has the potential to light a fuse in a formerly frozen place, with impacts across the geopolitical landscape. Alternatively, this formerly frozen place may be where fuses lit elsewhere cause a northern explosion. Admittedly, the latter is unlikely for now, given the highly challenging operating conditions.

Burke (2021) took particular aim at the Trump administration's dealings with Russia. "When it comes to Russia, though, the Trump administration did a lousy job of allocating resources to the risk. A good polar strategy should reflect the overall approach to the U.S. relationship with Russia. Deterrence in the north was never going to be especially productive when it was paired with impunity everywhere else." Burke (2021) also commented on the effects of Arctic warming on Russian policy decisions. The "Great Melt" has generally been seen by the Russian government as a very favorable development. Their formerly inaccessible Arctic resources and sea lines of communication are becoming increasingly available. On the other hand, Russia ratified the Paris climate agreement in 2016, though Russia has so far failed to meet its voluntary commitments. One of Russia's principal concerns about climate change is the melting of permafrost. Frozen ground lies under roughly two-thirds of the country, including around 80% of the wells of Russia's gas industry. One study estimates that permafrost melt will affect 20% of all Russia's infrastructure assets and more than 50% of all residential structures in the permafrost zone in the next few decades. Attempts to remediate these problems will cost an estimated $100 billion or more. This massive outlay of money will certainly limit Russia's ability to fund its military ambitions in the Far North. Disturbing evidence of things to come was discovered in July 2020, when a scientific expedition to the Yamal Peninsula discovered a 50 m crater that had suddenly opened. In fact, this was the 17th crater found in the region since 2014, and another large Siberian crater opened up in February 2021. Geologists hypothesize that these massive sinkholes are the result of subterranean methane gas that likely accumulated over tens or hundreds of thousands of years. The current thawing of the permafrost allowed the methane to come near the surface, where it exploded.

Even though the relationship between the United States and Russia has deteriorated, cooperation between the two countries has continued. Cooperative endeavors include efforts to reduce the risk of collision between military aircraft in international airspace by the coordination of their movements (sometimes, but not always abided by Russian pilots). Also, Russian and American scientists continue to work together on joint projects—something unheard of during the Cold War.

As discussed by Gross (2020), Russia's need to preserve its geopolitical and economic interests has motivated the government to boost its Arctic military presence. Russia's military objective, called "Red Arctic," has seen a large increase in its Arctic military presence. In recent years, the government has built numerous airbases and developed a large network of border outposts. Russia flaunted its military might in its August 2019 operation "Ocean Shield." This training exercise demonstrated Russia's ability to carry out military operations in the Arctic and was apparently designed to deter other nations from interfering. Russia's motivation to expand its Arctic military presence stems from its perceived need to project strength and defend its northern border. Its protection of economic interests in the region is also a vital factor in the equation.

The dynamic nature of US-Russian confrontations above the Arctic Circle is reminiscent of the Cold War, and while the armed forces of the two countries have not come into direct conflict, their constant parade of military exercises and increasing military investment in the region set the collision course for a conflict that might erupt (Gross, 2020). Evidence of increasing tension between Russia and the NATO alliance is provided by the numerous incursions of the Russian Air Force into the NATO countries' air space. NATO (2020) has reported that their air forces across Europe scrambled more than 400 times in 2020 to intercept unknown aircraft approaching Alliance airspace. Almost 90% of these missions—around 350—were in response to flights by Russian military aircraft. This represents a modest increase from 2019. Russian military aircraft often do not transmit a transponder code indicating their position and altitude; they do not file flight plans, nor do they communicate with NATO countries' air traffic controllers, posing a potential risk to civilian airliners. Russian incursions into US airspace focus on Alaska—the closest frontier between the two countries. Snow (2020), reporting in the Military Times, reported that two Russian IL-38 maritime reconnaissance aircraft were recently intercepted near Alaska by F-22s supported by KC-135 and E-3 airborne early warning aircraft, according to North American Aerospace Defense Command (NORAD). This particular Russian aircraft is used to hunt submarines. A string of Russian patrol aircraft intercepts were repulsed by US fighter planes near Alaska in March 2020. Their apparent objective was to spy on ICEX (Fig. 16.8), the US military's 3-week biennial exercise that tests the readiness of US submarines to their ability to operate in the Arctic.

These Russian incursions into US and NATO allies' airspace are not accidental. The Russian aircraft are not lost. On the contrary, they know exactly where they are going. The United States and Russia have a long history of aerial staring matches over Alaskan coastal waters, with the Russians frequently encroaching on Alaska's air defense identification zone (ADIZ) without identifying themselves through transponder signals or radio (voice) contact. Within the past few years, Russian bomber intercepts are becoming more common for the United States and its allies.

Stepping up the confrontations

The Kremlin has clearly issued directives to increase these incursions, underlining their commitment to this strategy by sending some of their most capable (and expensive) aircraft on these missions. What is Russia trying to accomplish? Like many military operations, these flights are motivated by multiple internal and external factors. Russia routinely conducts long-distance bomber practice missions all over the world, sometimes prompting an intercept response from nations that feel threatened by their bomber presence. According to the BBC, Royal Air Force intercept fighters have ushered away Russian bombers and other aircraft encroaching on their airspace no fewer than 10 times since the beginning of 2019. The main reason for these Russian incursions into NATO airspace, particularly of heavy payload bombers, is simply for training. To be able to execute long-range bombing missions in the event of real war, Russian pilots conduct training flights that closely resemble actual combat operations. Long-range missions also require a great deal of logistical planning. The bombers are very often accompanied by fighters that have much more limited fuel range than the massive bombers they escort. This means that escort fighter flights must be coordinated from multiple airbases. It also requires the coordination of support from airborne refuelers and flights of Advanced Warning and Control (AWAC) planes. Executing such complex operations takes practice, no matter which nation is conducting them.

Endless posturing

Since the beginning of the Cold War, Russia has always used such posturing in the face of opponents as part of its military and political strategy. An important part of Russia's foreign policy is maintaining the threat they represent to countries they consider their opponents, especially the United States and its NATO allies. Deterrence is the main goal of many military operations and Russia's constant demonstrations of the capability to launch long-range strikes against national opponents. Incursions into "enemy" airspace are meant to reinforce that doctrine.

MILITARY GEOPOLITICS

As discussed by Saxena (2020), most scholars writing in the post—Cold War era exulted over the geopolitical stability and constructive cooperation by the Arctic nations. This Northern bonhomie was labeled "Arctic exceptionalism," characterized by the absence of competition between regional powers. However, this peaceful coexistence did not last long. After the collapse of the Soviet Union in December 1991, new players with Arctic aspirations entered the scene. Chief among these was the rise of China and its unprecedented claims in the Arctic, as well as its self-declared status of being the near-Arctic state. China's economy had been growing extremely rapidly for several years before the global financial crisis of 2007—09, but the crisis marked an important change in its growth (Arora and Vamvakidis, 2010). During the 6 years leading up to 2007, China's gross domestic product (GDP) grew at an impressive rate of 11% per year. Since then, the Chinese economy, like that of the rest of the world, has been either holding steady or growing far more slowly. However, recent increases in Chinese investments have reached more than 50% of their GDP. The Chinese prosperity has allowed the government to earmark money for the development of their Arctic plans. In the past few years, a great deal of money has been spent on military hardware, such as fighter planes, surface ships, and submarines. Chinese military ambitions in the Arctic were clearly on display when they attempted to buy a defunct naval base in Greenland, presumably to establish their own naval base in the Arctic. The Chinese also expressed eagerness to build a new airport in Nuuk, the capital city of Greenland. This might be seen as a nice humanitarian gesture toward an underdeveloped country. Alternatively, it might be viewed as an opportunity to improve the only international airport runways in the country to facilitate takeoffs and landings of Chinese military aircraft.

Military Geopolitics—US Navy

The US military is taking the increased threats from Russia and China seriously. As an example of their response, consider the Associated Press article (2021) concerning the US Navy's release of a new strategy for operations in the Arctic as competition for resources among nations increases in the region. The Navy statement titled "A Blue Arctic" provides an outline of planned operational changes for the military's sea services in and around Alaska, including the Navy, Coast Guard, and Marine Corps. The services will operate a full range of missions, adapt to Arctic operations, and work with local and Indigenous communities to build regional security, the Navy said.

The Navy highlighted the regional strategy as receding sea ice makes mineral and biological resources more accessible to nations that can exploit them. Trade routes are expected to open, allowing access to shipping lanes previously closed by sea ice. "The coming decades will witness significant changes to the Arctic Region," the strategy statement said. "Encompassing about six percent of the global surface, a Blue Arctic will have a disproportionate impact on the global economy given its abundance of natural resources and strategic location."

US Rep. Don Young of Alaska said in a statement that the United States must ensure its waters are navigable and the military has the necessary equipment and training "to keep the peace in a rapidly evolving climate." "We must remember that the United States is not the only country working to pursue new opportunities in the Arctic—our adversaries are as well," Young said. Republican US Sen. Dan Sullivan of Alaska said in a statement that the Navy recognizes the critical importance his state has in the region.

Military Geopolitics—NATO Exercises

According to Sevunts (2020), a joint NATO exercise involving the United States, the United Kingdom, Norwegian, and Danish armed forces that took place during the summer of 2020 off the Russian coast in the Barents Sea is the latest example of rapidly growing military tensions in the Arctic. It appears that NATO and Russia are revisiting their Cold War rivalry and saber-rattling. Shortly after the NATO exercise, the Royal Navy announced in September 2020 that it led a multi-national task group of warships and aircraft into the Arctic for the first time in more than 20 years. *HMS Sutherland*, supported by *RFA Tidespring*, commanded a task group comprising the US destroyer *USS Ross* and the Norwegian frigate *Thor Heyerdahl* on a deployment to the Barents Sea. The exercise was held in the waters of Russia's exclusive economic zone (EEZ) in the Barents Sea, off the coast of the strategically important Kola Peninsula, home to Russia's Northern Fleet and a large part of the Russian nuclear arsenal, as reported by the Barents Observer. More than 1200 military personnel from the United States, the United Kingdom, Norway, and Denmark took part, supported by US P-8 Poseidon and Danish Challenger Maritime Patrol Aircraft along with Royal Air Force (RAF) Typhoon fighter jets and the refueling tanker *RAF Voyager*, the Royal Navy said in a press release. The exercise marked the first time the United Kingdom has operated Typhoons in the High North, the statement said. "The UK is the closest neighbor to the Arctic states. In addition to preserving UK interests we have a responsibility to support our Arctic allies such as Norway to preserve the security and stability of the region," UK Defense Secretary Ben Wallace said in a statement. "It is vital to preserve the freedom of navigation when melting ice caps are creating new shipping lanes and increasing the risk of states looking to militarize and monopolize international borders."

The joint NATO exercise came on the heels of a huge Russian exercise in the Bering Sea, off Alaska's coast in late August, involving more than 50 warships and about 40 aircraft. At almost the same time, the United States flew its B-52 strategic bombers close to Russian airspace in a show of force that included six nuclear-capable aircraft making symbolic overflights of all 30 NATO allies, including Canada. Canadian defense expert Rob Huebert said the decision to carry out an exercise in the Russian EEZ, which adjoins a country's territorial waters but is considered international waters, was meant to send a signal to Moscow. "From a political perspective, remember the NATO allies haven't sailed that close within the Russian EEZ since the end of the

Cold War," Huebert said. "I can't find any example of a NATO-based group doing that. There is a political symbolism in fact of NATO going into these waters." NATO's increasingly assertive moves in the Arctic are a response to Russia's growing military presence in the Arctic and Northern Atlantic, Huebert said. "This is something that the Russians have been doing for a fairly long time, pushing into the West," Huebert said. The joint US, UK, Norwegian, and Danish operation in the Barents Sea is the latest example of the West pushing back against Russia at sea, Huebert said. "To a certain degree what we are seeing is a return to great power politics and the games that are then associated with that," Huebert said. "They're deadly games."

Sevunts (2020) interviewed Rebecca Pincus, assistant professor at the US Naval War College, who said that while the Royal Navy claimed the exercise was intended to assert freedom of navigation, the multilateral exercise was a classic maritime security operation. "A freedom of navigation operation would be an operational assertion that counters excessive maritime claims and there are no excessive maritime claims in the Barents that are being challenged, none of the countries challenge Russia's EEZ in the Barents," Pincus said. It is important to underline the difference between maritime security operations and freedom of navigation operations because the latter is targeted at coastal nations, she said. "A maritime security operation is not specifically targeted at anyone, a maritime security operation—like what is going in the Barents Sea—is aimed at showcasing the level of coordination among allies and their technical skill at conducting multilateral operations in a challenging maritime environment," Pincus said. "A freedom of navigation operation is a strong assertion against a specific coastal state that is intended to boldly contradict that coastal state's claims." For example, the United States conducts highly visible freedom of navigation operations in the South China Sea, she said. "Those are bold, they are highly visible, they are very targeted against China and they are part of a much broader campaign to uphold freedom of navigation in the South China Sea that's targeted squarely at China," Pincus said.

Growing risk of accidents and unintended consequences

Nevertheless, the tit-for-tat war games NATO allies and Russia have recently been playing in the Arctic carry huge risks of accidents or other unintended consequences, Pincus said. "Last week Russia conducted a joint military exercise in the Bering Sea, off the coast of Alaska, and they drove naval warships right through the Alaskan fishing fleet," Pincus said. "It was incredibly dangerous - if something had gone wrong, who knows what would have happened." These exercises create a situation of a "classic security escalation," she said. "The thing is that these exercises confirm the worst fears of each side," Pincus said. "And that's why each side feels compelled to respond. Russia is increasing military activity and its exercises, its overflights, patrols—all send a signal that it has the intention and capability of having an active military presence, which makes its neighbors uncomfortable." But when NATO responds with their own exercises, that validates Russian fears that NATO is trying to encircle them, she said. "It's this mirroring where neither side feels comfortable, so it acts in a way that increases the anxiety on the opposite side," Pincus said. Within the last year, there has been a dramatic increase in these military exercises within the Arctic context, Huebert said. "One could make the case that what we're seeing is the next level of militarization that's occurring," Huebert said. All this points to the necessity of a dialogue between the Arctic states and other countries that have been active in the region lately, Pincus said. "Having more channels for communication and dialogue would help provide some reassurance, communication, and transparency that would bring down some of these tensions," Pincus said. Providing advance notice of military exercises would also help alleviate some of the tensions, she added.

Military Geopolitics—Media Representations

Padrtova (2019) assessed international media coverage of recent events in the Arctic. The article noted that there is a general agreement concerning numerous challenging issues in the Arctic, including the consequences of climate change, oil and gas extraction, mining, and fisheries. However, there is limited public awareness of regional security issues.

In a parallel study, Babin and Lasserre (2019) used a case study focused on the United States, identifying media articles focusing on Arctic security issues and ranking them numerically by topic (Fig. 16.5). Generally speaking, an event becomes a security issue when it is presented by national authorities (politicians, government bureaucrats, and business leaders) as an existential threat. Headlines in newspapers or online news sites often employ words that trigger public attention, including "dangerous," "threatening," and "alarming."

In one sense, the news media become participants in the promulgation of security issue concerns. In some cases, the media play only an intermediary role as a "channel" or "vehicle" for transferring information

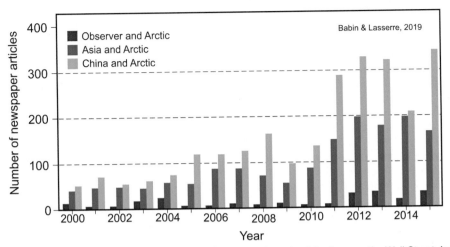

FIG. 16.5 Chart illustrating the results of a survey of the number of publications on the Wall Street Journal, the New York Times, the Telegraph, and the Washington Post between 2000 and 2016 for the following terms: Asia and Arctic, China and Arctic, Japan and Arctic, Italy and Arctic, Observer and Arctic. (Modified from Babin, J., Lasserre, F., 2019. Asian states at the Arctic Council: perceptions in Western States. Polar Geography 42(3), 145–159. https://doi.org/10.1080/1088937X.2019.1578290 150.)

from the political leadership or other influential groups to a general audience. However, most media outlets add some elements to the information that inevitably helps shape public opinion. Depending on the nature of the "added element," the media can further accelerate the securitization or initiate the whole securitization process. Padrtova (2019) argued that all five analyzed US media outlets (the Washington Post, New York Times, Wall Street Journal, Los Angeles Times, and USA Today) frame different issues related to the Arctic region as security issues.

The raising of national security issues by governments or the media are simply considered to be part of routine duties. The issuance of such security warnings is rarely discussed in-house; they are considered a part of the continuation of normal procedures. Therefore, security-conscious actors may take advantage of the fact that the national security topic had already been published. These actors are tasked with managing existing persistent threats and identifying new ones.

By investigating the media, Padrtova (2019) identified four key security issue narratives:

(1) The Arctic as a resource base (economic security)
(2) The Arctic as a nature reserve (environmental security)
(3) The Arctic as an area for the protection of national interests (political and military security)
(4) The Arctic as a region of traditional livelihood (societal security)

Despite the fact that the Arctic is rarely the center of US media attention, Padtrova's research shows that Arctic issues were repeatedly covered by the most influential media outlets from 2008 to 2016. Also, the empirical study showed that Arctic issues covered by national newspapers displayed a clear tendency to present these issues as threats to national security.

POST–COLD WAR GEOPOLITICS

The US-Russia dynamic within the Arctic Circle region is reminiscent of the Cold War, and while the two militaries have not come into direct conflict, their constant parade of military exercises and increasing military investment in the region sets the collision course for a conflict that might erupt, especially as the territory is rapidly decreasing in land mass (Gross, 2020). According to Saxena (2020), the current international security situation in the Arctic combines of three intersecting factors:

(1) The shifting balance of power toward China at both global and regional (Arctic) levels
(2) Increasing Cold War rivalries between the United States and Russia
(3) Growing strategic convergence between Russia and China

The security and prosperity of Arctic nations beyond these "Big Three" powers are intertwined with global security developments, for the simple reason best summed up in what Thucydides wrote in "A History of the Peloponnesian Wars" in the 5th century BC, "The strong do what they can and the weak suffer what they must." It is highly doubtful that the less powerful Arctic nations will cooperate around issues such as environmental degradation, climate change, and sustainable development when the "Big Three" are setting up camp in the region. Great power rivalries are governed not by cooperation, but by perceived security threats. In the absence of international Arctic governance, no one can trust anyone. Relative military and economic gains matter, not absolute ones. Increasingly, realists have started to declare the demise of Arctic exceptionalism discussed earlier.

At the meeting of the AC held in Rovaniemi, Finland in May 2019, US Secretary of State Michael Pompeo described the Arctic as "an arena of global power and competition." There is evidence to back this assessment. In 2018, China released its first white paper on Arctic strategy, making the now-famous declaration of being a "near-Arctic state." China also unveiled the ambitious "Polar Silk Road" component of its grandiose Belt and Road Initiative. Meanwhile, as discussed earlier, Russia is continuing to strengthen its presence in the Arctic by developing new military and naval bases, refurbishing old ones, and expanding its already extensive fleet of Nuclear-powered icebreakers and submarines. The race to conquer resources in the world's "final frontier" has already begun.

Shifting Balance of Power

The global balance of power is shifting toward China while the American share of global trade and GDP is declining. During the Trump administration, America receded from its global commitments and turned inward. China is increasingly challenging the US hegemony. China is in the process of obtaining a stronghold on major international organizations. Beijing is fast bridging the military gap with Washington, and the globe is moving toward a New World Order that is nonpolar in nature. With the shift of the strategic balance of power, China is simultaneously becoming both emboldened and assertive. Although it is highly unlikely that China will get into a shooting war with the United States, its salami-slicing techniques and debt-trap strategy are a cause of concern for the Arctic stability in large and the United States interests in particular (Saxena, 2020). Greenland and Iceland are the centers of a diplomatic power struggle between the United States and China. As is evident from US policy papers and diplomatic behavior, the American government is highly concerned about Chinese aspirations in the Arctic, as evidenced by their investments in the mineral resources of Greenland and a failed attempt to buy a defunct naval base there, the joint development of geothermal energy in Iceland, and a joint project with Finland to develop a "data silk road."

According to Thomas Ayres (2020), general counsel to the US Air Force, the United States must work with long-time Arctic partners to increase the vigilance of its security in the Arctic, an increasingly vital region. Stepping up national security is necessary in light of China's Arctic strategy that makes clear the government's desire to control infrastructure along Arctic routes.

As reported by Woody (2021), the US military's renewed focus on the Arctic has prompted new operations in the North. The latest force headed to the Arctic in 2021 is an expeditionary B-1B bomber squadron of more than 200 flight crew and support team members. The squadron will go to Orland Air Base in central Norway to conduct Bomber Task Force missions around Europe. US European Command did not disclose the squadron's arrival date or how long they will stay. The European Command did say the squadron's training regime will range from operating in the high Arctic to improving interoperability with allies and partners in Europe. The term "interoperability" is military-speak for "the ability of military equipment or groups to operate in conjunction with each other." This is the first time US bombers have deployed to Norway, but the US Air Force has been conducting joint exercises with the Norwegian Air Force in recent months, flying B-1Bs, B-2s, and B-52s over northern Europe (Figs. 16.6 and 16.7). "The Arctic remains a key area for us to continue to best understand how we will operate up there, and key to me for that is how we operate with our partners," General Jeff Harrigan, commander of US Air Forces in Europe, told Business Insider during a Defense Writers Group event in June 2020. Harrigan cited Norway, Sweden, and Finland as partners from whom the Air Force is learning "how we can best leverage what they've been doing for years to support operations up there." "It is crystal clear to us that our partners have the best understanding of how to do that, so our reliance on them and the interaction … is really going to be key to our success," Harrigan added. The Air Force has the military's largest Arctic presence, and they underscored its importance last summer in its first Arctic Strategy document, which labeled it "an increasingly vital region for US national security interests."

FIG. 16.6 Above: Joint NATO naval task force in the Barents Sea. Middle: US Airforce fighter intercepts a Russian IL-38 aircraft over US airspace. Below: A US Airforce F-15 fighter intercepts a Russian Tu-95 Bear Bomber. (Above: Photo courtesy of the Royal Navy, in the public domain. Middle: Photo courtesy of Wikimedia Commons, in the public domain. Below: Photo courtesy of the U.S. Department of Defense, in the public domain.)

A near-Arctic state?

As stated earlier, no part of China is in the Arctic. China's northernmost city is close to the latitude of Philadelphia. However, China remains undeterred by the reality of geography. In their doublespeak fashion, the Chinese Communist Party has declared China a "near-Arctic State"—an assertion not made by any other country and not factually supportable. In response to the return to strategic competition and the publication of the National Defense Strategy, the US Air Force recently published its own Arctic Strategy. The strategy seeks to further peace and security in the Arctic as it

FIG. 16.7 Above: A B-1B bomber over the Arctic Circle, September 25, 2020. Below: Three Norwegian F-16s and a US B-52H bomber during a Bomber Task Force mission over the Barents Sea, November 6, 2019. US Air Force/Airman 1st Class Duncan C. Bevan. (Both photos courtesy of the US Air Force, Department of Defense, in the public domain.)

recognizes substantial political changes that affect land, sea, and sky.

The Chinese are committed to taking advantage of recent technological advancements and Arctic warming that provide increased access to the Arctic's sea routes, rare earth materials, hydrocarbons, and fisheries. China's interest in the Arctic is linked to its One Belt, One Road initiative—President Xi's plan for a Chinese centered and dominated global trade, cyber, and space network.

Security of Arctic airspace (level 2)

As reviewed by Ayres (2020), Arctic air routes are essential to aircraft operating in the Northern Hemisphere.

Accordingly, the Arctic has long been key to the US Air Force's objectives of securing air routes and ensuring nuclear deterrence. The US Air Force and Army began monitoring "over-the-pole" air traffic in the 1950s, with the installation of Distant Early Warning network (the DEW line) of radar sites across Arctic Alaska, Canada, Greenland, Iceland, and the Faroe Islands. It was set up to detect incoming Soviet bombers during the Cold War and provide early warning of any sea-and-land invasion. The remaining DEW line radar sites, such as the facility at Thule Base in northwest Greenland (Fig. 16.8), are enduring symbols of America's vigilance during the Cold War and the importance of the Air Force's air defense mission.

FIG. 16.8 Above: Two Coast Guardsmen and a scientist walk on the frozen Arctic Sea from the Coast Guard Cutter Healy to conduct an ice survey on October 2, 2018, about 1150 km north of Barrow, Alaska. Photo by NyxoLyno Cangemi, US Coast Guard, courtesy of the US Department of Defense. Below: Radar installation at Thule US Air Force Base, Greenland. Photo by JoAnne Castagna, US Army Corps of Engineers, courtesy of the US Department of Defense. (Modified from Cangemi, U.S. Coast Guard, N., 2018. The Coast Guard Cutter Healy. https://www.navytimes.com/news/your-navy/2020/12/15/congress-oks-new-arctic-icebreakers-for-coast-guard/.)

The US Air Force Arctic Strategy addresses new realities, recognizing the increasing importance of the region. The Strategy seeks the cooperation of America's closest allies while maintaining an eye toward strategic competitors' predatory and bullying behaviors. James Woolsey, former Director of the CIA, noted that for nations such as China and Russia, the law is a malleable instrument of power. They engage in legal, cyber, and publicity campaigns, known as "lawfare," which seek to subdue, conquer, or gain access to territories by bullying. For them, lawfare produces the same effects previously attained only by conventional warfare. For instance, there is no basis in international law for Russia to formally annex Crimea or for China to declare the high seas between the Philippines and Vietnam as Chinese territorial waters. China has not yet completely revealed any legal arguments to bolster its claim to Arctic resources. A Chinese state-owned firm, Shandong Gold Mining, is seeking to buy out a Canadian mining company with operations almost 200 km north of the

Arctic Circle. If successful, China would further consolidate its control of gold and rare earth resources while also establishing a physical toehold in the Arctic that could be used for satellite communications.

ICEBREAKER FLEETS

Drewniak et al. (2018) described the history and current state of the icebreaker fleets of the Arctic nations. Even with the decline in Arctic sea-ice cover, there will be ice-packed regions of Arctic waters that will require icebreakers to facilitate the passage of other ships. The so-called "Arctic Passages" will not become completely free of ice in summer until the second half of this century. Given that sea ice will remain a concern in the future, the United States, Russia, Canada, and Norway are maintaining or increasing their icebreaker fleets.

Russia has already built and operates a fleet of icebreakers fully capable of handling its current navigational demands, as well as the expected future increase. On the other hand, both Canada and the United States are struggling to meet current icebreaker demands, much less any future increases in polar navigation. Their lack of icebreaking capabilities is somewhat disconcerting given the expected demand in the Arctic region in coming years, especially considering the lengthy process of building new icebreakers. This discussion starts with a consideration of China's icebreaker fleet, as the Chinese have made clear that they intend to become a major player in Arctic navigation.

Icebreaker Fleets—China

Bennett et al. (2020) described the 2017 voyage of China's Ukrainian-built icebreaker, *MV Xue Long* across the Transpolar Sea Route (TSR)—a trip lasting 83 days. This icebreaker was built in 1990 for Soviet logistics and resupply along the NSR. It was purchased by China in 1994 and rebuilt by them in 2007 to support polar research and logistics. Chinese state media heralded this journey as the country's first crossing of the Central Arctic Ocean (CAO).

Kulik et al. (2020) reported on the state of China's icebreaker fleet. After the completion of the refitting of the *MV Xue Long* in 2007, China launched another icebreaker in 2014, the *Xuenlong-2*, built with the participation of a Finnish company. Recently the Chinese authorities have declared they are going to design an Arctic fleet for research and exploration. This has been the cause for concern for several Arctic countries including the United States. The stated intention to design an Arctic fleet proves China's determination to gain better

access to the NSR and thereby explore Siberian hydrocarbons. Since it is developing its own fleet, China is very likely to use the NSR without any help from Russian icebreakers, meaning that China will be independent and will not pay Russia for assistance, thereby lowering cargo transportation costs. Topping things off, Beijing also recently revealed its plans for the construction of its first nuclear-powered icebreaker, at its own shipyard. This proposed new icebreaker will likely be 152 m long, 30 m wide, and 18 m deep with a displacement of 30,000 tons. The new icebreaker is expected to be powered by two onboard nuclear reactors with the capability of crushing Arctic ice at a maximum speed of 11.5 knots (Sharma, 2020).

Icebreaker Fleets—Russia

Russia has 53 icebreakers. With this large fleet, Russia is strongly positioned to offer its expertise and services along the TSR (Sharma, 2020). According to Drewniak et al. (2018), Russia is far better prepared than any other country to capitalize on the many untapped resources in the Arctic region. As the competition for control and exploitation of Arctic resources increases, the Russians are showing their national power and presence through their icebreaker fleet. While the United States seems to be conflicted about how to bolster its lackluster icebreaker fleet, Russia has taken aggressive steps in the past few decades, investing large amounts of money to build the world's largest icebreaker fleet. In addition to their own shipbuilding facilities, they have utilized shipyards and facilities in Germany and Finland to add to their fleet. Russia now has 12 nuclear-powered vessels. These icebreakers do not need refueling, so they may stay at sea far longer than conventionally powered ships. The *Arktika*, commissioned in 2019, is the largest nuclear icebreaker in history, with engines developing 80,000 horsepower. This ship is capable of breaking ice up to 3 m thick.

Russian multipurpose icebreakers

Drewniak et al. (2018) reported that Russia is building a series of icebreaking multipurpose vessels (MPSVs) that are being constructed in three classes. The largest class includes two vessels each weighing 1370 tons (dwt). Deadweight tonnage (dwt) is a measure of the weight a ship can carry. This includes the weights of cargo, fuel, freshwater, ballast water, provisions, passengers, and crew. These large MPSVs are furnished with two large cranes for salvage capabilities, as well as a forward-mounted landing platform for helicopters. The slightly smaller MPSV07 class includes four vessels launched by 2015. These medium-sized vessels can carry approximately 1171 tons dwt. They can stay at sea for roughly 20 days and can support underwater diving operations down to depths of 300 m. The third class of icebreakers, designated MPSV12, is yet to be completed. These vessels have a projected capacity to carry 1820 tons dwt. They are designed to operate at a speed of 26 km per hour (14 knots) and have a maximum draft of 4.50 m. This makes them ideal shallow-draft vessels that can use Russia's major river estuary ports in the Arctic.

Icebreaker Fleets—The United States

Drewniak et al. (2018) observed that during the Cold War era, the US fleet possessed eight icebreakers, but that number has dropped to just two icebreakers during decades of neglect and lack of attention. Currently, the country has only two polar icebreakers: one heavy (the *Polar Star*) and one medium (the *Healy*) (Fig. 16.9). The *Polar Star* is more than 40 years old and acquires spare parts cannibalized from her grounded sister ship, the *Polar Sea*, which experienced an engine failure in 2010. This poses the question: Do two functioning icebreakers constitute a fleet? Before February 2017, the US Coast Guard had hoped to refurbish the *Polar Sea* and put the heavy icebreaker back in service. The *Healy* cannot penetrate deep ice and is not capable of handling future needs in the Arctic. To make matters worse, the only heavy icebreaker, the *Polar Star*, cannot take on Arctic missions in the near future, according to US Coast Guard Commandant, Admiral Paul Zukunft, because it is obligated to support Antarctica's McMurdo Sound for the US National Science Foundation for at least four more years. The US icebreaker fleet is operated by the Coast Guard, but the exorbitant cost of funding the new generation of icebreakers falls well outside their budget. New icebreakers now cost at least $1 billion each. Even if funding were not a problem, US shipbuilders long ago ceased to build icebreakers. Having to start from scratch, it is estimated that a new American icebreaker would take several years to build. Given the bleak outlook for the US fleet of polar icebreakers, the respective American capabilities for the time-being remain extremely limited. This is a significant national disadvantage for the United States, concerning operations in the NWP. In December 2020, the US Congress passed a bill authorizing the construction of a new icebreaker. Ongoing construction work on a new icebreaker is not expected to be finished until 2024.

Icebreaker Fleets—Canada

According to Drewniak et al. (2018), Canada is experiencing a crisis similar to that of the United States in relation to its current and projected future icebreaking capabilities. The Canadian icebreaker fleet has five functional vessels. However, the existing fleet consists of aging icebreakers that are well past their shelf life and are very costly to maintain. In 2013, the Canadian Coast Guard's most capable icebreaker, the *Louis S. St-Laurent* was credited with escorting the first bulk carrier to transit the NWP. This transit highlighted the importance of escort ships to take advantage of Arctic opportunities for maritime transport. However, the *Louis S. St-Laurent* is a 45-year-old vessel and one of only two Canadian icebreakers capable of making the NWT transit. Transport Canada and the Canadian Coast Guard plan to construct a new icebreaker, the *John G. Diefenbaker*, to replace the aging *Louis S. St-Laurent*, which was originally slated to be decommissioned in 2010. Until the *John G. Diefenbaker* is completed, the Canadian Coast Guard believes it may need as many as five extra icebreakers at various times over the next few years, as the current fleet goes through repairs and upgrades and a new polar icebreaker is built. As of February 2021, the Canadian government had not awarded the contract to build the *John G. Diefenbaker*. Beyond the initial plan to replace one icebreaker, Canada does not appear to have a long-term plan to replace its remaining icebreakers, most of which are 30 to nearing 40 years of age.

Meanwhile, joint US-Canadian exercises have been carried out by the US Coast Guard, the US Navy, and the Royal Canadian Navy (Fig. 16.10). The three organizations signed a joint 5-year strategic plan in May 2020. The new plan lays the foundation for future coordination and joint operations (Hakirevic, 2020). Vice Admiral Linda Fagan, commander of the Coast Guard Pacific Area, commented, "This strategic plan improves interoperability [there's that word again] and coordination across our sea services, allowing us to more effectively secure our countries' shared maritime safety, security, and economic interests."

GEOPOLITICS OF CLIMATE CHANGE

According to Sharon Burke (2021), former US assistant secretary of defense, climate change must be considered a national security matter in its own right, because it threatens the safety and well-being of the American people. According to the US National Climate Assessment document of 2018, communities throughout the United States are already feeling the effects of climate change, which will damage "human health and safety, quality of life, and the rate of economic growth." There are a variety of effects, including an increase in natural disasters. The United States hit a record 22 natural

FIG. 16.9 North American Icebreakers. Above: Canadian Coast Guard Icebreaker Louis S. St-Lauren. Middle: US Coast Guard Icebreaker, Healy. Below: US Coast Guard Icebreaker Polar Star. (All photos courtesy of the US Department of Defense, in the public domain.)

disasters that cost $1 billion or more in 2020, the sixth year in a row with more than 10 billion-dollar disasters.

The Arctic, as the National Climate Assessment puts it, is on the "front line" of climate change. According to the NSIDC, the Arctic has lost more than 2.6 million km^2 of ice at its seasonally lowest point since satellite tracking began in 1979. To put this in a geographic perspective, that area represents more than twice the

FIG. 16.10 Above: Coast Guard Cutter Maple crew member keeps a lookout from the bridge as the Maple follows Canadian Icebreaker Terry Fox into Bellot Strait in Nunavut, Canada, August 12, 2017. Below: The US Coast Guard buoy tender Maple follows the Canadian Coast Guard icebreaker, Terry Fox, through Franklin Strait, Nunavut, August 12, 2017. (Both photos by Petty Officer 2nd Class Nate Littlejohn, US Coast Guard District 17, courtesy of the US Department of Defense, in the public domain.)

size of Alaska. This is not disputed science. These are conditions that can be seen and measured through satellite imagery and physical observations from sailors, automated buoys, and other devices.

Climate change is a national security issue for all nations, not just the United States. Unfortunately, it catalyzes instability. Much of the international research on instability risks has focused on states with a history of conflict, weak governments, and poor economic development. In such circumstances, climate change can exacerbate bad governing decisions or complicated social and economic relationships, affecting access to resources. It generally worsens natural disasters. All these climate change impacts can foment social unrest, forced migration, and tension between political factions or countries, and even military conflict. Even if climate change does not cause war directly, it can stir up the underlying tensions that lead to armed conflict.

Arctic warming is also agitating underlying frictions between some of the most advanced and well-armed societies in the world. As such, it becomes geopolitical in scope. World War III is unlikely to start at the North Pole, but there is a risk of miscalculation, which could lead to escalation. For example, consider the cat-and-mouse games played by the Russian and NATO air forces in recent years. As Burke (2021) sees it, the friction between the superpowers over Arctic issues is part of a larger pattern. What happens in the Arctic reflects what is happening with China's ambitions in the South China Sea and the Russian takeover of Crimea, and vice versa. The effects of Arctic climate change are reverberating far to the south, causing shock waves across the geopolitical landscape. All things considered, the Arctic is the one place where armed conflict is least likely to take place, because of the logistic difficulties of operating military equipment in very cold temperatures. For instance, diesel fuel starts to gel at temperatures just below 0°C, potentially clogging fuel lines. Jet fuel stored in tanks on the ground gels at temperatures below −40°C. The human body is likewise not built to withstand Arctic winter temperatures without highly insulated clothing. Exposure of an unclothed human body to temperatures less than −35°C causes the victim to enter hypothermia in less than 10 min; any exposed skin becomes frostbitten within the same time frame.

These highly challenging operating conditions put a damper on even the most bellicose nations. Burke (2021) went on to criticize the Russian policy of the Trump administration, saying that it did a lousy job of allocating resources to the risk. A good polar strategy should reflect America's overall approach to its relationship with Russia. Deterrence in the north cannot be especially effective when it was paired with a *laissez-faire* attitude toward Russian activities elsewhere in the world.

As discussed in Chapter 16, another pressing issue for Indigenous peoples and commercial interests across the Arctic is the thawing of permafrost and flooding along coasts and rivers. The United Nations estimates that by 2050, 70% of Arctic infrastructure will be at risk from permafrost thaw and subsidence. One study of the economic impacts of climate change on Alaska predicted $5.5 billion in costs by the end of the century to repair damaged public infrastructure. High maintenance costs are already hitting military bases in Alaska, including Eielson Air Force Base, home to two F-35 squadrons. In planning for additional military bases in Alaska, the departments of Defense and Homeland Security will have to include construction costs to pay for the adaptation of infrastructure to prevent permafrost thaw.

Russian Climate Policy

As we have seen, permafrost degradation is a huge concern for Russia, despite the Kremlin's view that generally considers Russia to be a winner in the climate change sweepstakes. They are winning in the sense of gaining access to previously inaccessible Arctic resources and sea routes. However, this rosy view is starting to change. Russia adopted a Climate Adaptation Plan

and ratified the Paris climate agreement in 2020, although it has yet to meet its voluntary carbon-reduction commitments. The government is facing the reality of the potential impacts of permafrost melting, looking at a price tag of over $100 billion to maintain housing and infrastructure in the North. This will not be a one-time fix, but rather a long-term commitment that will keep draining the Russian economy for decades to come. In light of this, Russia's military plans for the Far North may have to be scaled down.

As discussed in Burke (2021), all the Arctic nations maintain pragmatic cooperative relationships with Russia and China, despite the aggressive posturing of both countries in recent years. Friction between the two authoritarian states and the other Arctic nations has given the latter group cause for concern on many occasions. As a result, there has been an increase in cooperation between the Arctic nations facing off with Russia and China. The Trump administration essentially wasted that opportunity, with its "America First" message from the White House, but the Biden administration gets the benefit of a fresh start. All of America's NATO and Nordic partners place great importance on dealing with climate change, but President Trump's lack of willingness to engage more broadly with Arctic allies on clean energy, climate change adaptation, and climate negotiations put considerable strain on those vital alliances.

European Climate Policy (Level 2)

Mazurier et al. (2020) discussed the response of European countries to the climate change situation. In their view, European climate change policies are an example of the "Tragedy of the Commons." The concept of the Tragedy of the Commons was coined by British economist William Forster Lloyd in 1833. He observed that the lack of governmental regulation over the grazing of livestock on common lands in Britain was causing these patches of land to be overused, while parcels of land in private hands were in relatively good condition. Each cattle herder, guided by individual incentives, prefers to increase the exploitation of commonly shared lands to feed his cattle. In so doing, all members of the community erode common lands by putting themselves first. The international competition for the vast resources of the Arctic is an obstacle to achieving a broad consensus on global warming. On the World Stage, the Tragedy of the Commons is thus linked to the self-interested behavior of individual states, at the expense of endangering common resources. As we have seen, the Arctic has a multitude of resources to be developed or overexploited.

In the past decade, there has been a change in nation-states' perceptions of the resource opportunities emerging in the Arctic. In 2013, the AC accepted as Observers 12 non–Arctic states, all key players in the international arena. These Observer states include France, Germany, the Netherlands, Poland, Spain, the United Kingdom, Italy, China, Singapore, Korea, Japan, and India. The European countries validated their participation in the AC because of their strong polar research traditions, stretching back to the 19th century.

China's Arctic Gambit

The acceptance of Asian countries as Observer nations in the AC was based on recognition of their status as emerging global powers. The acceptance of China as an Observer is seen by the Chinese as an acknowledgment of its international status as a power equal to Russia and the United States. The Chinese harkened back to their "Hundred Years of Humiliation" when foreign powers held sway over their country. Before its acceptance as an Observer, the Chinese government published a 5-year Polar Plan to increase China's "status and influence" in Polar Affairs, and to protect its interests as a "near-Arctic state." The Chinese government urged that "an end to the Arctic states' monopoly of Arctic affairs is now imperative." Acceptance by the AC was just one chess piece in the Chinese game to increase their influence in the Arctic. Beijing's $2 billion bid for Greenland's uranium triggered a wave of anxiety and heated political debate in Denmark. The export of uranium to China, as agreed by Greenland, forms a nexus between its autonomous sovereignty in making this agreement and the Danish central administration's security concerns related to uranium as an explosive and radioactive raw material. The autonomous Greenland government argued that China had the technology and funds to responsibly extract its minerals.

GEOPOLITICS OF RESOURCE EXTRACTION

As discussed by Gross (2020), the Arctic's treacherous ice caps are melting away, and many nations are considering opportunities to engage in a modern "gold rush" over the region's unclaimed territory and natural resources. Some countries are attempting to gain jurisdiction over newly emerging lands or newly ice-free seascapes for resource extraction and the development of trade routes. Sharma (2020) discussed how climate-driven changes in the Arctic are facilitating access to an abundance of natural resources. In addition to the Arctic's phenomenal reserves of oil and gas (estimates of quantities use numbers in the billions and the

trillions), the Arctic holds immense quantities of iron, gold, nickel, lead, platinum, diamonds, uranium, platinum, and various rare earth elements such as titanium, cobalt, copper, and manganese that are vital to modern technology. These rare earth elements find their applications in the manufacture of products ranging from smartphones to wind turbines and advanced weapons. China's hunger and assertiveness for all these resources are increasing, and Beijing wants to dominate all these strategically important elements of the Polar North.

Advancement in technology and the rapid industrial growth since 1990 has made China the world's largest importer of raw materials. To meet its industrial raw material needs, China seeks to maintain a strong foothold in resource-rich areas, including the Arctic. China is pursuing these interests in the Arctic through its State-Owned Enterprises, by investing billions of dollars on various resource extraction projects in Nordic countries that hold sovereignty over such Arctic resources.

Greenland Rare Earth Mining

As reported by Sevunts (2020), in December of 2020, the government of Greenland, Naalakkersuisut, began the process of establishing rare earth mining at Kuannersuit in the southwest of the island. The process began with public consultation of the Environmental Impact Assessment and Social Impact Assessment of the proposed project. A series of five public consultations were announced in five communities surrounding the proposed project: Igaliku, Narsaq, Narsarsuaq, Qaqortoq, and Qassiarsuk. The start of the consultation process had to be postponed early in 2021 because of concerns about COVID-19 infection (McGwin, 2021). The Inuit Ataqatigiit political party, which opposes a mining project in southern Greenland, secured a plurality of votes in an April, 2021 national election. Its leader said on Wednesday that the Kvanefjeld mine, home to major deposits of rare minerals, would not go ahead. Many locals had raised concerns about the potential for radioactive pollution and toxic waste in the farmland surrounding the proposed mine (Jonassen, 2021). The principal ore mineral at Kvanefjeld is called steenstrupine. Complex metallurgy makes extraction of rare earth elements from this mineral compound extremely difficult on an industrial scale.

The Greenland rare earth mining project is not the only enterprise faced with the problem of how to deal with radioactive waste products, such as thorium and low-grade uranium. Some environmentalist groups point to what has happened to rare earth mining and processing in Malaysia. Because of domestic pressure over environmental concerns related to the radioactive pollution caused by rare earth processing, last year Malaysia demanded that the mining company, Lynas, find another location for processing ores containing radioactive materials such as thorium. As part of the license granted by the Malaysian government, Lynas is committed to building a cracking and leaching facility outside Malaysia before July 2023, after which the company will not be allowed to import raw materials containing naturally occurring radioactive material.

Russian offshore petroleum development

Shapovalova et al. (2020) discussed the geopolitics of Russian Arctic offshore petroleum. Russia accounts for more than half of the coastline of the Arctic Ocean and holds sovereign rights over most of the Arctic's offshore petroleum resources, especially natural gas. The NSR along the Siberian coast is projected to become the primary shipping lane for the transport of hydrocarbons to other countries. While Russia is an active participant in the AC and has ties with neighboring Norway, its conflict with Ukraine led to the adoption of economic sanctions by the European Union and the United States. These sanctions prohibit Western companies from selling, supplying, transferring, or exporting technology and finance to Russian companies for offshore oil developments in the Arctic. Furthermore, they impose restrictions on certain companies and individuals. While Russian energy policies delineate ambitious plans for Arctic resource development, the limitations on foreign capital and technology since the enforcement of Western sanctions have slowed the pace and limited the scale of these developments. The sanctions have also forced Russia to improve its own technological capabilities and infrastructure, incurring increased government spending to support Russian oil and gas companies. Isolation from western technologies and investment has also increased Russia's cooperation with partners in the East.

Oil and gas prospecting is essentially over in the Russian Arctic. The Russians already know most of the locations where oil and gas extraction will yield profitable enterprises. This is because most of the oil and gas field discoveries were made by Russian petroleum geologists during the Soviet era. So, discovery is not an obstacle, but the development of resources is now called into question. For instance, Shapovalova et al. (2020) concluded that Western sanctions have forced changes in Russian resource governance, namely the establishment of import substitution priorities through the development of domestic technologies, cooperation with Asian countries, and liberalization of continental

shelf access. Russia's ties with China form the key element of this cooperation. China, however, is not a petroleum-producing country, so it lacks the technical expertise to aid the Russian petrochemical industry. China can only provide money to fund projects, rather than large-scale technology and equipment. Furthermore, the building of cooperative agreements with Eastern countries will take time, especially considering the numerous channels that feed into different aspects of the Russian energy sector.

Conventional sources of oil and gas outside the Arctic are starting to be depleted, leaving only the highly challenging development of Arctic petroleum resources. Russia depends heavily on petroleum revenues and feels the need to develop its Arctic assets to sustain high production rates. That being said, even if the Russian expansion of offshore oil and gas production succeeds fully, this will happen against a backdrop of a global economy that is moving away from reliance on fossil fuels toward the development of clean, renewable energy. This transition is already well under way. International interest in and finance of petroleum drilling is waning throughout the world, due to falling oil prices and increasing environmental concerns over Arctic oil development and shipping. Thus, Arctic petroleum development projects are losing their economic attractiveness. Witness the recent failed attempt by the Trump administration to sell oil-drilling leases in the Arctic National Wildlife Refuge in northern Alaska.

Setback for oil drilling in Arctic Alaska

One of the Trump administration's biggest efforts to support the fossil fuels industry suffered a major setback in January 2021, as a decades-long push to drill for oil in Alaska's Arctic National Wildlife Refuge ended with a lease sale that attracted just three bidders, one of which was the state of Alaska itself. Alaska's state-owned economic development corporation was the only bidder on nine of the parcels offered for lease in the northernmost swath of the Refuge, known as the coastal plain. The legality of this lease purchase by a state government is being disputed in the courts. Two small companies each picked up a single parcel. Half of the offered leases drew no bids at all (NPR, 2021). All major oil companies share a lack of interest in the further development of Arctic oil enterprises, citing falling oil and increasing expenses of oil drilling in remote regions.

In another sea change in the world of petroleum financing, Cushing (2020) reported that as of December 2020, all six major US banks have committed not to finance Arctic oil drilling projects. This includes the Bank of America, Goldman Sachs, JP Morgan Chase, Wells Fargo, Citibank, and Morgan Stanley. The banks were under intense pressure from shareholders, leaders of the Gwitchin and Inupiat peoples of northern Alaska, and many wildlife conservation societies. Together, these groups made the case that exploiting the Arctic Refuge for oil would violate the rights of the Gwich'in people, who rely on it for subsistence and the continuation of their spiritual and cultural lifeways. It would also cause irreparable harm to the hundreds of species that use the refuge as a home and spawning ground, including caribou, moose, polar bears, and migratory birds. It would exacerbate the climate crisis, which is already wreaking havoc and melting the very places that oil companies now want to invest in more drilling. Finally, the coalition of antidrilling groups addressed the banks' bottom lines. They reminded the banks' boards of directors that drilling in the Arctic would be bad for business, as this would trigger a major public relations backlash from the majority of Americans who are opposed to drilling in Arctic Alaska.

Native geopolitics—Canada

Kleist (2017) discussed the need for Canadian Inuit to respond promptly to the threats looming from the current rush for Arctic resources and control of the North Pole—the process of dividing the northernmost marine areas between the so-called Arctic states, under the auspices of the 1982 United Nations Convention on the Law of the Sea. Kleist also expressed concern about the potential for the expansion of marine transportation, resource extraction, tourism, and high sea fisheries. It takes little imagination to picture the Arctic Ocean as the new battleground between Inuit, the Arctic states, industry, and environmental organizations. In Kleist's view, it is essential for Inuit to act now, to take the lead in shaping a new regulatory regime to manage the seas sustainably. Otherwise, the outside world is more than ready to take over and once again set the agenda for Inuit lands, seas, and lives.

A Canadian Arctic Highway

As reviewed by Bennett (2018), by 2050, the world's combined road and rail network will grow an estimated 60%. National governments are building many of these roads, which are often perceived as disenfranchising Indigenous communities. However, a joint venture between two Indigenous-owned construction and transportation companies has recently broken this mold. They aimed to build a paved road in northwestern Canada: the Inuvik-Tuktoyaktuk Highway. They leveraged opportunities afforded by land claims, treaties, and

shifting geopolitics to find a way to build this highway in the Canadian Arctic's Mackenzie Delta region. It was completed in 2017 and was the first public highway in North America to reach the Arctic Ocean. This road-building project challenges the idea that roads are invariably top-down initiatives that negatively affect Indigenous peoples and their lands. Inuvialuit community leaders lobbied for this road project and succeeded in winning CAD $299 million in government funding to construct the highway. Arctic warming has turned this region into a frontier of renewed national and global interest. The joint venture group used this fact to win financial support.

Strategically, the consortium sought to promote economic development and improve local mobility between two communities, thereby creating a highway of national importance. Inuvialuit community leaders, having gained political power from the land claims process, successfully drew the Canadian Federal government's attention to their proposed highway project through strategic lobbying. They appealed to government imperatives to gain access to offshore oil and gas resources, and to the government's perceived need to strengthen Canada's sovereignty along its Arctic frontier. Although the consortium's arguments focused on national priorities, their intent throughout was to accrue financial benefits locally rather than nationally. Thus, Inuvialuit-owned joint ventures won the contracts for the highway's design and construction, and hundreds of new jobs were created to construct the highway. The highway project was not the typical state incursion onto Indigenous lands. Rather, it exemplifies a case of Indigenous actors marshaling national arguments to achieve their immediate local agenda of road-building and economic development. Cases such as this that illustrate Indigenous peoples' abilities to initiate development projects and bring them to completion with or without state or private assistance show how Indigenous peoples can prosper within the existing power structures. This is particularly important in the current era in which national governments are generally scaling back their financial support for Indigenous public services as they seek to encourage economic self-sufficiency.

Native geopolitics—Greenland

Jacobsen (2020) notes that Greenland is a state-in-the-making. Almost 90% of its population is Inuit, and there have been calls for Greenland's independence from Denmark in recent decades. Greenland is currently stuck between two opposing views of sovereignty. On the one hand, their self-government is part of the transnational Inuit community. On the other hand, Greenland aspires to become a fully sovereign state in the traditional sense.

To understand how Greenland arrived at its current geopolitical conundrum, it is necessary to take a brief look at the history of Greenland's governance, which is somewhat complicated. During the early 18th century, Norway and Denmark were politically unified, and the joint government considered Greenland part of its territory. As a result of the Napoleonic Wars, the Danish-Norwegian unification ended in 1814 when Norway was ceded from Denmark. The Treaty of Kiel in that year gave Denmark full control of Greenland. From 1814 to 1953, Greenland was considered a colony of Denmark—not independent and not part of Denmark. Greenland's affairs were under the direct control of the Danish government.

During World War II, Denmark was occupied by Nazi Germany. As a result, the United States persuaded the Danish government in exile, represented by the Danish ambassador to the United States, to relinquish defense and control of Greenland to the United States for the duration of the war. The first American troops arrived in Greenland in July 1941, where they built two military airports with full-length runways. Coincidentally, these two facilities remain the only international-class airports in Greenland. Greenland acted independently during the war years, allowing the United States to build bases on its territory, despite Danish prewar neutrality. After the war, Danish control of Greenland was restored. The American bases remained there, and Denmark, with Greenland as a part of the Danish Kingdom, joined NATO.

In 1953, a new Danish Constitution incorporated Greenland into Denmark. The island thereby gained representation in the Danish Parliament for the first time in its history and was recognized as a Danish province, called the County of Greenland, even though Greenland is 2,166,000 km^2, more than 800 times larger than the average size of the domestic Danish counties (2687 km^2). In 1979, the Danish government granted Greenland home rule, although the Danes retained control of foreign relations, defense, currency matters, and the legal system.

In 2008, Greenland's citizens approved a Greenlandic self-government referendum with a 75% vote in favor of a higher degree of autonomy. Greenland took control of law enforcement, the coast guard, and the legal system. The official language changed from Danish to Greenlandic in 2009. The act cedes control of foreign relations from Denmark to Greenland. Greenland then began to send national representatives

to other national capitals, including Copenhagen, Brussels, Reykjavik, and Washington, DC. The self-rule law of 2009 gives Greenland the freedom to declare full independence if they wish to pursue it through a public referendum. A 2016 poll identified a clear majority (64%) in favor of full independence among the Greenlandic people, but a poll in the following year reflected majority opposition (78%) if independence would mean a fall in living standards.

Greenland's development of increasing sovereignty over foreign policy is enhancing their international status and aiding their ability to attract external investments. The governmental arrangement with Denmark allows the Government of Greenland some foreign policy competence on areas of exclusive concerns, which entirely relate to fields of responsibility that Greenland has taken over. As this definition is open to interpretation, it gives rise to the question: Do Greenland's government representatives expand their country's room to maneuver in foreign policy by challenging Danish sovereignty?

Greenland's "Arctic Advantage"

As discussed by Jacobsen (2020), it is useful to consider Greenland's sovereignty issues in light of the new global geopolitical interest in the Arctic. This increased interest has been welcomed by successive semiautonomous Greenlandic administrations as an opportunity for establishing new international relations—moves that contribute to their goal of lessening dependence on Denmark. Meanwhile, the Arctic has also become one of Denmark's top-five foreign policy priorities, causing a somewhat reverse dependency between the two countries. Greenland's geographic location and membership in the Danish Realm is the only thing legitimizing Denmark's Arctic state status. If Greenland were to break completely from Denmark, then the most Denmark could hope for would be a demotion from full membership in the AC to Observer status. This gives Greenland substantial leverage, termed an "Arctic advantage" in negotiations with Denmark. This leverage allows the Greenlandic government a good deal of latitude in choosing strategic arenas in which to wield its sovereignty, all the while seeking to expand the boundaries of its international relations without Danish involvement.

In 2014, Denmark submitted data to the United Nations' Commission on the Limits of the Continental Shelf (CLCS), claiming 895,000 km^2 of Arctic Ocean seabed. The claimed area is nearly half the size of Greenland and is about 19 times the area of Denmark. The Greenlandic claim overlaps significantly with

Russia's and Canada's claims. The Danish government funded the seafloor mapping and geological research needed to support Greenland's claim to the UN commission. Reportedly, part of the Danish motivation for financing the venture was to dampen the Greenlandic independence movement by showing how continued membership in the Danish Realm has tangible benefits beyond standard economic support.

Jacobsen (2020) has described Greenland's current governance as being "situated on the mezzanine between their past as a Danish colony and an envisioned future as an independent nation-state." As the most autonomous self-governing Arctic territory, Greenland enjoys a special place in Arctic governance. No other Native groups enjoy the level of political autonomy found in Greenland. By proclaiming that Greenland is a state-in-the-making, the self-government subscribes to the traditional either/or definition of sovereignty shared by the Arctic states, well-exemplified by the wording of the Ilulissat Declaration. One of the chief goals written into the declaration was blockage of any "new comprehensive international legal regime to govern the Arctic Ocean."

Native geopolitics—the native voice

As reviewed in several places in this chapter, the concerns of Native Arctic peoples were rarely considered by European or North American Arctic countries until quite recently. As discussed by Dodds (2019), in the 2018 Agreement to Prevent Unregulated High Seas Fisheries in the CAO, the "Big Five" Arctic nations were eager to protect their interests, sovereign rights, and special role as coastal state parties. Indigenous representation in some of the delegations was in part a reaction to the 2009 Circumpolar Inuit Declaration on Sovereignty in the Arctic, which did not challenge the sovereign rights of the Arctic states but did demand respectful consultation with Indigenous peoples. In 2011, the Circumpolar Inuit Declaration on Resource Development Principles in Inuit homelands (*Nunaat*), which called for indigenous peoples to be active participants in state-led projects and collective resource management. Native groups reminded audiences representing federal government offices that "resource development in Inuit *Nunaat* must proceed only with the free, prior, and informed consent of the Inuit of that region" (Inuit Declaration, 2011). In 2014, the Inuit Circumpolar Council adopted the Kitigaaryuit Declaration, which called for Inuit to be included in any commissions, committees, and councils addressing Arctic fisheries. Underpinning those demands lies the UN Declaration on the Rights of Indigenous Peoples (UN DESA, 2007), which notes that Indigenous communities have rights to lands, territories, and resources that they have traditionally owned, occupied, and/or used.

Native geopolitics—native rights

In an overview of Arctic Native rights, Coates and Broderstad (2019) remarked that the situation represents one of the most remarkable and peaceable political transitions in recent decades. Indigenous peoples across much of the Arctic have been gradually reasserting their rights within their traditional homelands. Fifty years ago, most Indigenous peoples had been pushed to the political, economic, and social margins of the Arctic nations, dominated by the resource economy, and stifled by the southern-based welfare states. By the beginning of the 21st century, Indigenous peoples established a substantial international presence, captured considerable media attention, and, except in Russia, secured a significant degree of self-government. Arctic Native groups have also begun to exert influence over northern policy in national and international arenas. Throughout much of the Arctic, Indigenous peoples have gained political power and authority that belies their small numbers. Despite these political advances, Native peoples still lack the financial and political resources to resume the control over traditional lands that they lost centuries ago.

As described by Coates and Broderstad (2019), recent Native advances in homeland autonomy were a long time coming. After several centuries of marginalization by national governments, Indigenous peoples started to push back in the 1960s and early 1970s, capitalizing on the greater acceptance in Western democracies of social and cultural protests and the growing frustration with the effects of development and non-Indigenous protests. Approaches varied widely. Greenlanders criticized Denmark's administration of their island. It was in many ways a typical European colonial administration, in that it was stable, financially significant, but culturally insensitive. The Saami, led by Saami-environmental alliance protests against the Alta Dam in northern Norway, carried their frustrations to the streets of Oslo and the grounds of the Norwegian parliament. In 1960s Alaska, the Inupiat recoiled at plans to use hydrogen bombs to create a port near the Bering Strait. In the 1970s, they joined forces with inland Native American groups to protest the development of North Slope oil and the proposed pipeline from Point Barrow to Valdez. Starting in the 1970s, Canadian Native groups, as part of a nation-wide effort to combine political protests with legal challenges, took the Government of Canada to court on numerous occasions, particularly on matters related to resource development. In the 1990s, they also launched demands for the negotiation of treaties across the entire northern reaches of the country, resulting in the Nunavut Land Claims Agreement of 1993. In sharp contrast, Indigenous peoples in Siberia, collectively described as the 'Small Peoples of the North,' struggled to get government attention and to cope with the rapid industrialization and resource development of their homelands, marked by some of the most severe environmental pollution ever seen in the Arctic—millions of gallons of oil and other toxic chemicals spilled on land and in freshwater (see Chapter 14).

Shifting attitudes. The efforts by national authorities were sincere, albeit often caught up in residual colonial attitudes about Indigenous peoples and cultures. Various government initiatives sought to improve the lives of Arctic peoples (Coates and Broderstad, 2019). Although such efforts were often well-funded and broad in aspiration, they rarely included any aspects that would empower Indigenous peoples to make their own decisions and operate their own programs. Despite government intentions to address social, economic, and cultural issues facing the Indigenous communities, Arctic Native peoples of several countries came to see these government programs as heavy-handed, interfering in their ways of life and culture. Before the late 20th century, Arctic Indigenous communities had been subjected to benign neglect from national governments. They had largely been left alone, particularly the Inuit of Arctic Canada. Funding for Indigenous governments in Canada was typically subject to federal oversight. Toward the end of the 20th century, government programs and strategies proliferated. These initiatives placed additional administrative and governance burdens on Indigenous communities, causing even greater Inuit dependency on Ottawa. However, more recently, one of the most important political transitions in recent history has taken place. Indigenous peoples and communities have become increasingly involved in the codevelopment of public policy and the general management of government programs. As history has shown, colonialism dies hard, and profound policy-making challenges remain. But largely due to the tireless efforts of Indigenous peoples, the Far North has produced an outstanding example of the reestablishment of Native rights, described by Coates and Broderstad (2019) as "a living laboratory for the understanding of Indigenous reempowerment."

Although the process remains far from complete and quite unevenly distributed across the Arctic, there are promising signs that Indigenous engagement in the coproduction of Arctic policy in all levels of government can produce substantial improvements in Indigenous conditions.

Clearly, in the 21st century, Arctic Indigenous peoples must respond to a series of upcoming issues and challenges, particularly concerning resource development, economic marginalization, and cultural loss. They will have to cope with the environmental disorder caused by climate change. They face a double-edge sword of maintaining their cultural integrity while at the same time adapting to major shifts in work, commerce, and society brought on by the introduction of new technologies. Because Arctic climate change is so dramatic, Arctic Indigenous peoples are now global symbols of 21st-century environmental vulnerability (Breum, 2018). Arctic Indigenous political leaders, arguably some of the most accomplished Indigenous leaders in the world, have long fought to secure a greater role in decision-making, governance, land rights, and economic development. The reempowerment of Indigenous communities is at the top of the list of urgent Arctic issues but remains a work in progress.

Case Study: Political Action on Baffin Island

According to a news story published by the Canadian Press in early February 2021, a group of Inuit hunters traveled up to 150 km by snowmobile over 2 days to set up a blockade near an iron mine on Baffin Island. Naymen Inuarak and other hunters from Pond Inlet and Arctic Bay blocked the road and an airstrip used to fly workers in and out of the Mary River mine site (Fig. 16.11). The mine, owned by Baffinland Iron Mines Corp., wants to double its annual output from 6 million to 12 million metric tons of iron ore. They also want to build a 110-km railway to transport the ore to the coast. The railway would be the only one in Nunavut and the northernmost in Canada. To prevent thermokarst beneath the rail line, the proposed line would sit on a platform of gravel, several meters tall. Inuarak, his group, and other community members are against the proposed expansion based on concerns about its effects on wildlife and the environment. Inuarak said he also believes community voices are not being heard by the mining company and their regional Inuit organization. "We've been left out and ignored. We're taking our own steps now," Inuarak said in an interview with the Canadian Press (2021). Seven hunters have set up tents on the road leading to the mine and are also taking shelter in a hunter's nearby cabin, and others will be joining

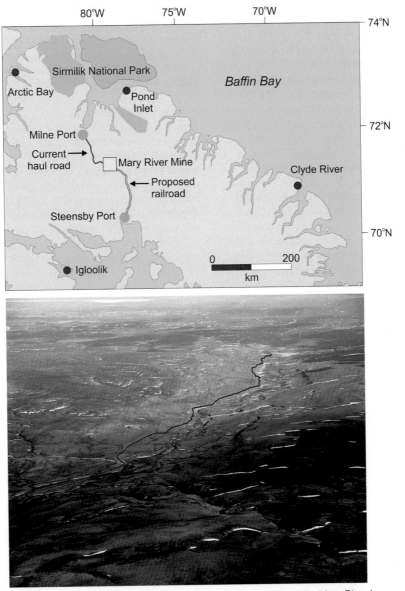

FIG. 16.11 Above: Map of northern Baffin Island, showing the location of the Mary River iron ore mine, the existing haul road to Milne Port, and the proposed rail line to Steensby port. Source: the author. Below: Photo of the Road (highlighted in red) connecting the Mary River Mine to Milne Port, northern Baffin Island. Photo by Timkal, Creative Commons Attribution-Share Alike 3.0 Unported license, in the public domain. (Bell, J., 2021. Mary River mine blockade highlights Nunavut Agreement's fatal flaw. Nunatsiaq News. https://nunatsiaq.com/stories/article/mary-river-mine-blockade-highlights-nunavut-agreements-fatal-flaw/.)

them, he said. They do not have a timeline for how long they plan to block the road and the airstrip. "This is just the beginning for us."

The most recent chapter of this story was written on February 11, 2021, when Justice Susan Charlesworth signed an interim injunction in the Nunavut Court of Justice that authorized the Royal Canadian Mounted Police (RCMP) to physically remove and detain anyone who continues to obstruct access to the Mary River mine. The stand-off situation was relieved when the

Mayor Pond Inlet, Joshua Arreak, proposed a face-saving option that let the protesters abandon their blockade without having to admit defeat. All parties agreed to a future meeting with officials from QIA and Nunavut Tunngavik Inc. (NTI). The protesters at the mine agreed to that offer and started heading home.

According to Bell (2021), the root cause of this situation likely lies in the Nunavut Land Claims Agreement of 1993 (NLCA), which NTI now prefers to call the "Nunavut Agreement." The Nunavut Agreement contains very little for Inuit communities. Rather, it is more of a land management system designed to empower Inuit organizations. The Agreement was specified to work through "designated Inuit organizations." In other words, it will work with any organization that board of NTI authorizes to act on its behalf. It is through this system that the Qikiqtani Inuit Association (QIA) became the manager of the Nunavut's Inuit-owned lands. Since the land being mined for iron on Baffin Island is Inuit-owed, it fell within the jurisdiction of the QIA, who became the landlord of Baffinland Iron Mines Corp. Financially, the QIA has done quite well out of Baffinland, Inc. The Association is the sole recipient of the mining company's commercial lease payments, royalties, and fees. In 2018, these Baffinland payments to the QIA totaled $33.4 million. By the end of the 2019–20 fiscal year, the market value of QIA's legacy fund was $61.7 million.

Pond Inlet and the other regional communities see very little of this money, only what the QIA decides to give them. That includes some benefits contained in the Inuit Certainty Agreement, which QIA struck with Baffinland during negotiations over the proposed mine expansion. However, except for municipal lands held by the local Pond Inlet government, the community owns nothing. This lack of ownership means a lack of bargaining power. So, the conflict between the protesters and the mining company stems from a belief of the local people that they remain the rightful owners of the Nuluujaat (Mary River region) lands and are therefore entitled to the royalty payments from mine. Unfortunately, because of terms of the NLCA, the citizens of Pond Inlet and other regional communities they are not legally entitled to anything. That fatal flaw in the NLCA was made clear in 2014, when the mining company's expansion proposal was entered into Nunavut's regulatory system. According to Bell (2021), the QIA enjoys nearly unlimited resources to participate in the regulatory process. On the other hand, Inuit communities are forced to make do with whatever legal funds they can get Canada's federal government. By using these limited resources, they have been able to introduce questions about the potential impacts of iron mine expansion on marine mammals and caribou into the negotiations.

Bell (2021) reminded readers that this regulatory system was not devised in Ottawa. It came directly out of the NLCA. It was created by Inuit negotiators, and endorsed by Inuit voters in 1992. The Nunavut Impact Review Board, tasked with the assessment of the Mary River expansion, resides at the center of this Inuit-designed system. The blockade of the mine access by Native protesters was an expression of their frustration with the system. Hopefully, the promised meeting between the protesters, the NTI, the QIA, and representatives from Pond Inlet and other regional communities will yield results acceptable to all stakeholders.

On the day after the start of the blockade, Baffinland released a statement saying that the company has been in communication with the hunters and is aware of the blockades. "Baffinland respects the right to peacefully protest and continues to work with the hunting party to maintain the safety of everyone at Mary River," it said. Public hearings before the Nunavut Impact Review Board concerning the mine's expansion plans have begun in two Baffin communities: Pond Inlet and Iqaluit. Inuit protesters insist that existing mine operations have already affected regional wildlife, lowering the numbers of animals they hunt and rely upon for food. Among the communities' concerns are that caribou will not be able to cross the proposed railway and that increased shipping traffic will drive away marine mammals.

The mine's shipping port at Milne Inlet opens onto narwhal habitat and lies within a national marine conservation area (Tallurutiup Imanga). Much of the surrounding landscape is part of Sirmilik National Park. The proposed expansion would see 176 ships travel in and out of Milne Inlet each year. "There's hardly any marine mammals now in the Pond Inlet area. There used to be so many," Inuarak said. "If they start doing 12 million tonnes a year, our marine mammals will be completely extinguished in our area." Baffinland has said their expansion plans will include mitigation measures to protect wildlife, such as caribou crossings on the railway and reduced ship speeds to minimize disturbances to marine life. The company has also signed a benefits agreement with the Qikitani Inuit Association worth $1 billion over the life of the mine. Qikitani, the regional Inuit organization that represents the affected communities. The agreement will only take effect if the expansion is approved.

Geopolitics of the Northwest Passage

In September 2007, the NWP was ice-free for the first time in recorded history. Since then, voyages using the NWP have increased steadily, and the passage has become quite busy in recent years. A snapshot of the level of shipping activity is shown in Fig. 16.12. Ulmke et al. (2017) used Automatic Identification System (AIS) data to track shipping traffic at the eastern end of the NWP for August 2013. The AIS system also records vessel type. The traffic included fishing boats, passenger cruisers, cargo ships, an icebreaker, oil and chemical tankers, and Canadian warships.

Byers and Lalonde (2009) review the consequences of the opening of the passage, in terms of the security and environmental risks that would result from international shipping. Their paper presents analyses of the positions of the Canadian and US governments regarding the legal status of the waterway. As with much else in today's Arctic, the end of the Cold War and the pace of climate change have put North American geopolitics in a state of flux and uncertainty. The Canadian position is that the NWP falls within Canadian internal waters and is therefore subject to Canadian domestic law. A joint US-Canadian panel meeting on the topic of sovereignty and jurisdiction over the NWP was held in February 2008 in Ottawa. The participants were nongovernment experts in the fields of international security and maritime law. They concluded that the Canadian position coincides with US interests, as well as the interests of other responsible countries and shipping companies. The political situation, instead of constituting a major international conflict between two friendly nations, should be considered a unique opportunity to resolve a longstanding dispute as well as a chance for the two countries to cooperate in the protection of the broader regional security and environment of the continent.

The panel recognized that the United States would not easily be persuaded that Canadian control over the NWP serves its interests. As a result, they made nine recommendations for intermediate steps that the two countries could take to address their common concerns about shipping via the NWP. The aim of these recommendations was to build American confidence in Canada's commitment to developing the NWP as a safe and efficient waterway for everyone's benefit.

Recommendations of the joint US-Canadian panel meeting:

(1) The United States and Canada should collaborate in developing parallel rules, standards, and cooperative enforcement mechanisms for notification and interdiction zones in the northern waters of both Alaska and Canada. This recommendation would see the United States adopt a mandatory Arctic shipping registration scheme that would protect, among other things, the western approaches of the NWP, thus keeping suspect vessels at bay and alerting Canada about foreign ships headed its way.
Actions taken: In 2010, the Canadian government enacted mandatory vessel registration for use of the NWP. The registration system falls within NORDREG—the Canadian Coast Guard's Arctic Canada Traffic System.

(2) The United States and Canada should share maritime surveillance in northern waters and cooperatively develop further surveillance capabilities. This recommendation is consistent with the May 2006 expansion of the functions of the North American Aerospace Defense Command to include surveillance over maritime approaches and "internal waterways."
Actions taken: Vessel surveillance in the Canadian Arctic was established by 2011 through the satellite-based AIS. This system provides important though not comprehensive information about ships in the area. The modular architecture of the maritime surveillance and risk management system continues to be adapted to Arctic conditions and focused on the NWP.

(3) The two countries should build on Canada's already strict Arctic marine environmental protection laws by developing even more advanced navigation, safety and ship construction, and operation standards. This recommendation accepts the legitimacy of the current application of Canada's Arctic Waters Pollution Prevention Act to the NWP and seeks to improve upon it, including by promoting the adoption of equally strict rules in the waters north of Alaska.
Actions taken: In 2014, the Canadian Parliament revised the Canada Shipping Act regulations to meet its obligation for sustainable development of the resources of Canada's Arctic waters, to ensure that the waters are navigated in a way that accommodates the "welfare of the Inuit and other inhabitants. Canada holds that the new rules are applied to shipping in a way that preserves the 'peculiar ecological balance' between the open water, ice, and land."

(4) The United States and Canada should cooperate on the establishment of shipping lanes, traffic management schemes, and oil spill response plans for the northern waters of both Alaska and Canada.

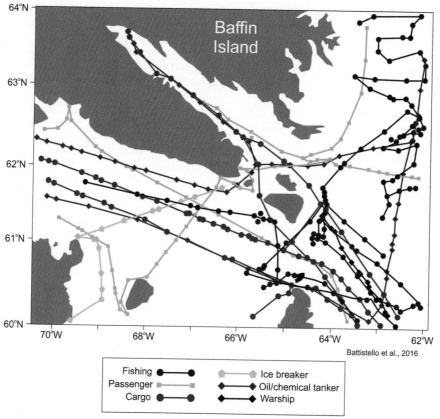

Battistello et al., 2016

FIG. 16.12 AIS data of shipping traffic at the eastern end of the Northwest Passage, August 2013, with vessel type. Groups of crosses represent the data bursts, segments between the bursts are gaps in the data. *AIS*, Automatic Identification System. (Modified from Ulmke, M., Battistello, G., Biermann, J., Mohrdieck, C., Pelot, R., Koch, W., 2017. Multi-sensor maritime monitoring for the Canadian Arctic: case studies. 19th International Conference on Information Fusion. DOI: 978-0-9964527-4-8©2016 ISIF 2.)

Actions taken: In the Canada Shipping Act (2014), the Canadian government clarified their definition of the country's "Arctic waters" as the internal waters of Canada and the waters of the territorial sea of Canada and the exclusive economic zone of Canada, within the area enclosed by the 60th parallel of north latitude, the 141th meridian of west longitude and the outer limit of the exclusive economic zone (Kraska, 2016). The United States has repeatedly challenged Canada on its regulations and interpretation of Canadian maritime authority over foreign-flagged ships in its EEZ, stating that both go too far. For instance, during the Canadian government's public comment period on NORDREG in 2010, the US Embassy in Ottawa sent a letter to the Canadian Department of Transport to express concerns about the draft regulations. The letter complimented Canada on its "efforts to provide for the safety of navigation and protection of the marine environment in the Arctic area" while suggesting that the regulations being implemented were inconsistent with international law.

(5) The two countries should cooperatively address the immigration and search-and-rescue concerns arising from the increasing number of cruise ships in northern waters. One obvious step would be for each country to require the submission of the full crew and passenger lists as part of a mandatory ship registration scheme-consistent with our first recommendation.

Actions taken: Search and rescue response is given by Canada to vessels that comply with NORDREG regulations by registering with Canadian authorities before beginning their NWP journey. NORDREG registration also qualifies vessels for access to current data on ice conditions, routing suggestions, and icebreaker assistance. These benefits come with a price, however. Once NORDREG registration and ship reporting became mandatory by the Canadian government, mariners were subject to criminal penalties under the Canada Shipping Act for noncompliance (Kraska, 2016).

(6) Both Canada and the United States need to acquire new icebreakers to replace their aging coast guard vessels.
Actions taken: As discussed elsewhere in this chapter, both countries are in the process of building new icebreakers, though both Canada and the United States lag far behind the Russians in this.

(7) The two countries need to work together to develop safety infrastructure, including navigation aids and perhaps even new port facilities in support of northern shipping. This recommendation, again, is aimed at promoting US action in the waters north of Alaska that mirrors and supports Canadian action in the NWP.
Actions taken: Little has changed, and there has been no cooperation between the United States and Canada. However, the Chinese Maritime Safety Administration published a document entitled "Guidances [sic] on Arctic Navigation in the Northwest Route" in which it criticizes Canada for the "numerous risks [that] remain with regard to the safety of the route amid lack of infrastructure and inclement weather" (Safety4Sea, 2016).

(8) Canada and the United States should make maximum use of the considerable legal powers they already possess over vessels, either sailing to or from Canadian or US ports or registered in one or the other country.
Actions taken: The 2014 Canadian Parliament Canada Shipping Act regulations strengthened control of many aspects of shipping in the NWP. There are no existing policies agreed upon between the two nations that determine who can and cannot pass through the NWP.

(9) A US-Canada Arctic Navigation Commission should be created to promote dialogue, conduct studies, and make policy proposals on matters of navigation, environmental protection, security, safety, and sustainable economic development. Like the International Joint Commission, this would be a purely recommendatory body, though

it could be granted an ad hoc arbitration role, if and when the two governments desired.
Actions taken: On October 6, 2010, NOAA led a US delegation that formally established a new Arctic Regional Hydrographic Commission (ARHC) with four other Arctic coastal nations. The commission includes Canada, Denmark, Norway, and the Russian Federation and seeks to promote cooperation in hydrographic surveying and nautical charting. The Commission provides a forum for better collaboration to ensure the safety of life at sea, protect the increasingly fragile Arctic ecosystem, and support the maritime economy (NOAA, 2020).

GEOPOLITICS OF THE ARCTIC OCEAN

To begin this section in a positive vein (i.e., not all Arctic geopolitics are confrontational), consider the arguments of Roest and Haworth (2019) that the Arctic Ocean is an area where geoscience and geopolitics are closely intertwined, and the relations between the Arctic states are mainly harmonious. The Arctic Ocean is not the only oceanic region where sovereignty claims are creating tensions between countries, but of all the disputed regions, the Arctic Ocean presents the most severe, challenging maritime environment. The harsh conditions of the Arctic Ocean have required immense efforts by its coastal States to delineate the outer limits of their continental shelves. In fact, these logistic difficulties have forced the Arctic nations to collaborate and share information. During the 21st century, significant advances in geoscience research have been achieved through the process of mapping and imaging the Arctic seafloor and sediments, and many of these advances have come through collaborative efforts between Arctic states. This includes the development of technology to facilitate deepwater operations in remote regions and under harsh conditions. Much of this was achieved through the use of remotely operated equipment and autonomous underwater vehicles. As a result of these efforts to document the extent and nature of Arctic continental shelf regions, geologists and oceanographers have enjoyed tremendous increases in collective knowledge and understanding of the Arctic.

Geopolitics of the Transpolar Sea Route

Bennett et al. (2020) reviewed the geopolitics of the TSR. The route remained closed by pack ice until the past few years, long after the deployment of Russian and American nuclear submarines for occasional Cold

War military maneuvers. By the middle of the 21st century, however, the TSR is likely to become much more accessible to various kinds of marine traffic ranging from cargo ships to cruise ships. The opening up of the TSR is predicted to increased use of this route by more than 50%, which represents a far bigger change to Arctic navigation than the predicted expansions in the use of the NWP (30% more accessible) and NSR (16% more accessible). Currently, the NSR reduces sailing times between Europe and East Asia from about 30 days via the Suez Canal (blockages by huge container ships notwithstanding) to about 18 days. The TSR offers even greater time savings, cutting an additional 1−5 days off the journey.

One of the TSR's main advantages is that, in the absence of ice, it offers an alternative to the NWP and NSR routes by being simpler to navigate and less politically complicated (the TSR does not pass through any country's territorial waters). In spite of these advantages, sovereignty issues and other geopolitical aspects of the TSR remain complex. The route was previously traversed only by submarines and icebreakers involved in the Great Polar Stand-Off between Russia and the United States. From the 1950s until the end of the Cold War, submarines from both navies used to play hide-and-seek in and around the CAO, using the pack ice to conceal their whereabouts. This Cold War legacy will likely continue to affect relations between governments both within and outside the region, especially maritime states. In their review, Bennett et al. (2020) address three geopolitical concerns regarding the TSR: international governance, Russia, and China.

International governance of the Transpolar Sea Route

Arctic shipping is regulated by a mix of international and national regulations under the auspices of UNCLOS (the United Nations Convention for the Law of the Sea), the IMO (International Maritime Organization), and the Polar Code of Trans-Arctic Shipping. However, unlike the NWP, over which Canada claims sovereignty and jurisdiction, and the NSR, lying completely within Russia's EEZ and controlled by the Russian government, the TSR crosses the high seas, where international regulations apply. Chief among them is the UNCLOS, established in 1982, and the IMO's Polar Code, recognized in 2017. UNCLOS governs the use of the world's oceans, including the 4.7 million km^2 of the CAO. UNCLOS Article 87 allows all states the use of the high seas for freedom of navigation, overflight, laying submarine cables and pipelines, constructing artificial islands and other installations

permitted under international law, fishing (subject to future negotiations), and scientific research. Assuming the TSR will be sufficiently ice-free to open for shipping in the 2030s or 2040s, the newly available route will undoubtedly stimulate the development of many of the maritime activities listed earlier.

Because of its geography, the Arctic Ocean is unique among the world's oceans because it has just one high seas point of access: the Fram Strait between the Greenland and Norwegian Seas. On the Pacific side of the Arctic, shipping regulations are more complex due to political tensions created by the proximity of the Russian and US territories, separated only by about 80 km at the Bering Strait. Nevertheless, as discussed in Chapter 8, the Bering Strait is a very busy route for international navigation. In the less geopolitically constrained regions of the seas, a strait is defined as connecting one part of the high seas or a state's EEZ with another part of the high seas or state's EEZ. The EEZ of each country with coastal waters extends outward by 370 km from a country's coastline. Vessels consequently enjoy the right of transit passage under UNCLOS Article 37. However, the EEZs of the Pacific coasts of Russia and the United States overlap considerably. This means that the Bering Strait's two main navigational channels pass through the territorial seas of both Russia and the United States. UNCLOS Article 42 allows nations bordering international straits to adopt regulations on maritime traffic and pollution prevention so long as they do not impede the right to transit. Because of this, vessels using the Bering Strait route to navigate between the North Pacific and the Arctic Ocean are bound to cross both American and Russian waters and may be subject to differing laws at various points during their journeys. Vessels not in transit generally hew to one side or the other of the maritime boundary. The United States and Russia, motivated by their observations of decreasing sea ice and increasing economic activity in the region, have established a two-way shipping system through the narrow Bering Strait to improve navigation safety and protect the environment. In 2018, the IMO approved the two countries' joint proposal to implement six two-way routes, six precautionary areas, and three areas to be avoided in the Bering Sea and Bering Strait, which took effect later that year (Bennett et al., 2020) (Fig. 16.13).

Parts of the TSR fall within the EEZs of Canada, Greenland, Norway, and Iceland. The boundaries between four of the five EEZs of the Arctic Ocean coastal states have thus far been established through multinational agreements. In the meantime, competing

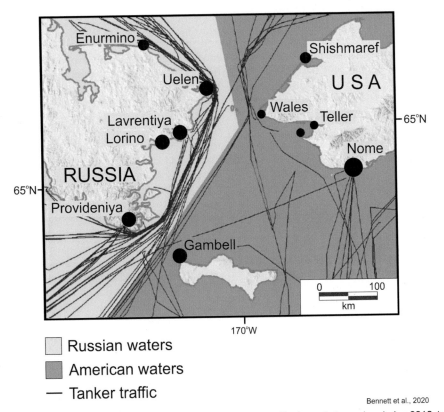

Russian waters

American waters

— Tanker traffic

Bennett et al., 2020

FIG. 16.13 Map of the Bering Strait region, showing the tanker traffic through the region during 2012, tracked by the Automatic Identification System (AIS). (Modified from Bennett, M., Stephenson, S., Yang, K., Bravo, M., De Jonghe, B., 2020. The opening of the Transpolar Sea Route: logistical, geopolitical, environmental, and socioeconomic impacts. Marine Policy 121, 104179. https://doi.org/10.1016/j.marpol.2020.104178 8.)

territorial claims to parts of the CAO seafloor have been submitted to the UN Commission on the Limits of the Continental Shelf by Canada, Russia, the United States, and Denmark, but because the waters that cover extended continental shelf regions are considered part of the high seas, they remain free to navigate regardless of how the claims are resolved. Nevertheless, in the case of an emergency or shifting ice conditions, a vessel may have to enter the waters of an Arctic coastal state, potentially falling under national regulations (Bennett et al., 2020).

The IMO's International Code for Ships Operating in Polar Waters (International Maritime Organization, 2017) mandates certain safety precautions, such as the possession of a Polar Ship Certificate and careful voyage planning to ensure safety at sea and pollution prevention. This Code covers the TSR, and no additional requirements apply to shipping within the CAO.

The key attraction of the TSR for the shipping industry is that the route passes mostly through international waters. However, this hinders the enforcement of environmental regulations, as no particular country is "in charge" of the CAO, and the United Nations is neither willing nor able to patrol this ocean. Indeed, a 2010 AC report asserted that "The greatest risk to the Arctic comes not from traffic originating or ending in the Arctic region, but from shipments that are simply passing through Arctic waters."

Russia's role in the Transpolar Route

Bennett et al. (2020) noted that warming climates are starting to melt the year-round sea-ice cover on the CAO, soon to be replaced by far thinner seasonal ice cover. When this transition takes place (by midcentury, if not sooner), the TSR will open up to commercial traffic. It is important to keep in mind that the complete disappearance of Arctic sea ice for just one summer would eliminate the meters-thick multiyear sea ice once and for all. Once the CAO is covered only by thin seasonal sea ice, ships will cease to need the support of icebreakers to make the TPR passage safely. This has significant implications for the design, construction, and operational standards of future Arctic marine shipping. Icebreakers may eventually become obsolete. Within the next few decades, summer voyages across the TPR are likely to become a common occurrence.

The predicted change in Arctic shipping will be a game-changer. It will very likely damage Russia's dominance of northern shipping, currently maintained through the world's largest fleet of icebreakers, as well as control over the only Arctic Asiatic passage, the NSR. As of now, shipping companies are saving a great deal of time and money by using the NSR, which only recently became a viable route. It was not until the summer of 2017 that the first ship traversed the NSR without the help of icebreakers. For journeys between Europe and Asia, the NSR passage from Europe to the Far East shaves 2—3 weeks off the time it takes European ships to sail south through the Mediterranean and the Suez Canal and cuts several thousand km off the southern route, saving large amounts of money on fuel. However, the use of the TSR will shorten such a trip even further. The combination of this cost-saving and elimination of the need to pay Russia for the privilege of using the NSR might make sailing the NSR "obsolete," as suggested by Ostreng et al. (2013).

Geopolitics of the Northern Sea Route

Sergunin and Hoogensen (2020) examined the various geopolitical factors that have shaped Moscow's policy on the NSR. The authors began by discussing how Russia's security perceptions of Arctic shipping evolved in the post—Cold War era, including the influence of Western sanctions in response to the Ukrainian crisis and other ongoing tensions between Russia and the West. Russia's security and geopolitical policies include considerations of the role of the NSR in ensuring the country's economic security, its cohesiveness, and connectivity between its territories. Perceived external threats to the NSR and Arctic Russia include NATO military exercises in the regions adjacent to the western terminus of the NSR. Russia refutes the view of the United States and its allies that the Arctic sea lanes, including the NSR, are part of the "global commons" where there should be freedom of navigation. Russian naval resources are being stretched by the need to control vast maritime spaces and coastline to prevent potentially

illegal activities ranging from poaching and smuggling to illegal migration and attacks against critical industrial and military objects. Likewise, Russia is struggling to fund the development of search and rescue (SAR) capabilities in Arctic waters, in addition to achieving preparedness to prevent and/or fight oil spills.

Some practical measures are being taken by the Russian government to ensure the security of Arctic shipping. These include the reopening of old Soviet military bases along the NSR and the construction of new ones. Most of these bases serve multiple roles. Many have SAR capabilities. Some are undergoing development as border guard stations, filling gaps in the regional security network. In concert with military installations, Russia is in the process of modernizing its Coast Guard fleet. There has also been an initiative to improve communication and navigation systems along the Arctic Ocean coastline, thereby improving maritime safety (Sergunin and Hoogensen, 2020).

Sergunin and Hoogensen (2020) drew several conclusions about the geopolitics of the NSR.

(1) Despite the overriding economic significance of the NSR in shaping Moscow's Arctic shipping policies, the role of security and geopolitical factors should not be completely ignored.

(2) Among Russia's hard security concerns are the NATO countries' military modernization programs, especially modernization of naval and air forces deployed in the Arctic.

(3) There has been an increasing number of NATO military exercises near the Russian borders and the NSR.

(4) The Kremlin is also concerned about the deployment of US submarines with ballistic missile defense systems in Arctic waters.

(5) Russia's soft security threat perceptions include illegal activities in NSR waters, including poaching, smuggling, illegal migration, marine pollution, oil spills, ballast and gray water discharges, violation of Polar Code requirements (Fig. 16.13), and potential attacks on critical industrial objects (oil and gas rigs, pipelines, cables, floating nuclear power plants, etc.).

Moscow aims to solve NSR-related problems via cooperation with other regional players and international organizations such as the AC, Barents-Euro-Arctic Council, Nordic institutions, the IMO, and other specialized UN bodies. Russia still has much work to do to solve the NSR's problems and make the route safe and desirable to foreign shipping companies. Among the issues that need to be settled, Russian authorities must establish clear and transparent navigation rules for the NSR. Operations need to be streamlined through an improved division of labor among the various governmental agencies concerned. Port infrastructure, SAR, and communication along the NSR need substantial improvement. Icebreaker and pilot escort services should be made affordable for all potential customers. Information systems on ice conditions and weather forecasts should be improved as well. In other words, if the Russian government can bring the NSR up to modern standards of convenience and safety, then the route can become a strong asset to the country, contributing to Moscow's goal to strengthen its control of the geopolitical aspects of Arctic shipping (Sergunin and Hoogensen, 2020).

CONCLUSIONS

As we have seen throughout this chapter, the Arctic is truly not "normal." It is the northernmost part of two continents and the island of Greenland. Furthermore, the northernmost part of the Arctic is ocean, not land. Until quite recently the world did not turn its attention to the North, except for the military standoffs of the Cold War. But many things changed as global warming caused Arctic temperatures to rise to record levels, melting the sea ice for the first time in millennia. As the southern edge of Arctic sea ice retreated year by year and navigation in Arctic waters became safer and easier, many nations gazed northward and started planning ways to exploit this change. Those plans featured the exploitation of a wide variety of resources, including petroleum, diamonds, iron ore, nickel, and rare earth minerals. By the turn of the 21st century, something akin to a gold rush was taking place in the Arctic. Access to Arctic resources became so important to many nations that some of them worked hard to gain an Observer seat on the AC, believing that this would give them increased visibility and a say in what was happening in the North. The Chinese strategy was to label itself as a "near-Arctic" country. Most nations scoffed at this assertion, but the Chinese considered it one of several steps to achieve their goal of gaining a strong foothold in the Arctic.

The Arctic is perhaps unique in the level of connectivity between various issues faced by its people. The World Economic Forum sponsored a paper by Stroeve et al. (2019) that discussed the major Arctic issues, including climate change, transport and shipping, environmental risk and degradation, natural resources, infrastructure gaps, local communities, and governance (Fig. 16.14). Each issue has five or six aspects to be considered within the major issues. For instance, the

Governance section contains subdivisions including geopolitics, regional autonomy, self-governance, human rights, and international security.

Military presence in the Arctic has increased dramatically in recent years, with Russia squaring off against the NATO alliance. Both sides have launched numerous exercises that either encroach or come very close to the opponent's borders. These provocations feature Russians incursions into the airspace of NATO countries, including US territory in Alaska. So far, these incursions have mostly been considered an annoyance to the NATO air forces. But the risk of fully equipped military airplanes coming within a few meters of each other cannot be overlooked. Someday there may be a midair collision between opponents, or a trigger-happy pilot might shoot down an opponent's plane. So, there is the danger that these incursions might trigger a shooting war in the Arctic.

The AC (Fig. 16.15) sponsors six working groups within their organization, three of which are directly concerned with environmental problems. These are the Arctic Contaminants Action Program (ACAP), the Conservation of Arctic Flora and Fauna (CAFF), and the Protection of the Arctic Marine Environment (PAME). But because the AC has no authority to enforce any of its rulings, they have not stemmed the tide of the "Arctic Rush." Meanwhile, many states expressing an interest in the North (particularly Russia and China) seem to have overlooked the very real environmental problems of the region. As we have seen throughout this book, the Arctic is in trouble, with environmental problems affecting the land, sea, and ice. In a perfect world, the nations would first seek to limit the environmental damage that humans have inflicted on the Arctic by educating the public and modifying harmful behaviors. Chief among these would be a global commitment to greatly curtail the emission of greenhouse gases. This has not happened.

On the contrary, the world seems intent on exploiting the newly available opportunities created by global warming. It must be admitted that this kind of resource exploitation has been the pattern of human behavior since time immemorial. We have always sought to grab what we can when an opportunity arises. If we get there before the rest of the world discovers these hidden treasures, so much the better. The concept is embodied in the phrase "Looking out for number one." This concept is so pervasive in human society that it governs much of international geopolitics. But in the end, it is extremely shortsighted and self-defeating. Moving from the global to a local scale, imagine that a house has caught fire, and the next-door neighbors, rather than trying to help

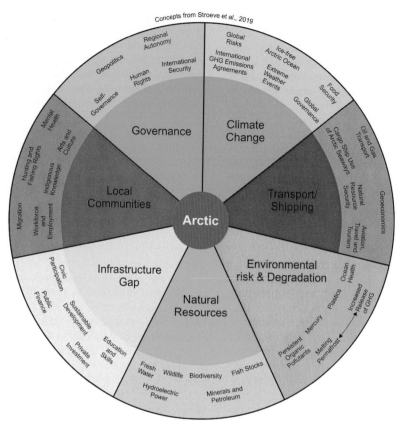

FIG. 16.14 World Economic Forum diagram showing the diverse elements within Arctic issues, including climate change, transport and shipping, environmental risk and degradation, natural resources, infrastructure gaps, local communities, and governance. (Concepts from Stroeve et al. (2019).)

FIG. 16.15 Arctic Council Ministers meeting in Rovaniemi, Finland, on May 7, 2019. (Photo courtesy of the US State Department, in the public domain.)

quench the fire, think only of ways to loot the property before the fire gets totally out of control. We need to refocus the concept of "looking out for number one." Our first concern should be the health of the planet. This was not a big concern when the human population was less than a billion people (i.e., before 1804). In those days the Earth seemed to be able to take care of itself. But the ability of the planet to heal its wounds has been lost in the Anthropocene. There are now almost 8 billion of us, and we are collectively using up the Earth's natural resources at unprecedented rates. The rate of human population growth has reached exponential levels and is constantly increasing. This fact has been borne out to me because the world's population has tripled in my lifetime. As of 2021, humans are using up the natural resources of 1.6 Earths. In other words, we are overusing the planet's resources by 60%. We can only carry on using the planet's resources if we decrease our use of them considerably, while also working hard to improve the environmental health of our planet.

FINAL THOUGHTS ON THREATS TO THE ARCTIC

The aim of this book has been to present and explain the threats that face the Arctic. These threats are numerous, often convoluted, and highly interactive. The one common denominator is that all these threats come directly from human actions. The Arctic is not ruining itself—we are ruining it. Our greenhouse gas emissions have triggered unprecedented Arctic warming that is endangering the cryosphere, including sea-ice cover, ice sheets, glaciers, and permafrost soils. The glaciers are disappearing, the ice sheets are shrinking, and the permafrost is melting. The loss of sea-ice cover, combined with other elements of climatic warming, has produced a one-two punch that is causing Arctic ecosystems to stagger. The warming of Arctic seas is wreaking havoc with marine ecosystems, from microscopic organisms to the largest of marine mammals. The plants and animals on Arctic landscapes are responding to changes in temperature and precipitation in manifold ways, some of which were expected by ecologists, while others were not.

The Human Element

Of course, the Arctic is not devoid of human life, despite the misconceptions of early European explorers and some military "experts" and government bureaucrats operating as recently as the 1960s. The Indigenous peoples of the Arctic are now forced to cope with all the changes discussed in this book, whether they like it or not. Before the sea ice started to thin and then melt away, it served Arctic natives for millennia as their winter transportation highway and their platform for harvesting birds, fish, and marine mammals. This essential aspect of their way of life is vanishing, never to return. The loss of sea-ice cover has also left coastal villages vulnerable to massive erosion from storm waves that hit the shoreline unabated. Coastal villages are slumping into the sea as frozen permafrost soils melt and turn to mud.

Pollution and Contamination

Turning from climate change to pollution, it is sad to say that people well South of the Arctic Circle have unintentionally sent wave after wave of poisonous chemical and heavy metal pollutants to the northernmost reaches of the Arctic, by ocean and wind currents. Many of these poisons accumulate in the tissues of Arctic organisms (bioaccumulation), starting with phytoplankton and ending in polar bears and whales. Unfortunately, their toxic excursion through Arctic ecosystems is accompanied by biomagnification, in which the levels of toxic substances increase up the food chain, bringing the strongest poisoning effects to people and top-level predators. The once-pristine natural foods that met all the nutritional needs of Arctic peoples have now been contaminated by persistent organic pollutants and heavy metals such as mercury.

The most recent invasion of Arctic waters has been microplastic (MP) pellets and fibers that are finding their way into Arctic ecosystems through ingestion by all ranks of animals. As discussed by Ross et al. (2021), MPs have been detected in Arctic pack ice, seawater, and seafloor sediments, penetrating all the way to the North Pole (Fig. 16.16). Ingestion of MPs by numerous species have been documented across the various Arctic marine habitats, from benthic to pelagic, and across all levels of the marine food web. Their ingestion is causing potentially significant harm to these organisms, although their effects are yet to be fully understood. Preliminary data on MP contamination in seafood have raised concerns about potential human health problems tied to their ingestion. For Indigenous peoples such as the Inuit who rely heavily on foods from the ocean, these concerns represent an increasing threat.

THE RUSH FOR RESOURCE EXPLOITATION

Sandwiched in the middle of all these troubles is the recent shift in the world's geopolitical focus toward the Arctic. It would be wonderful if the world's attention

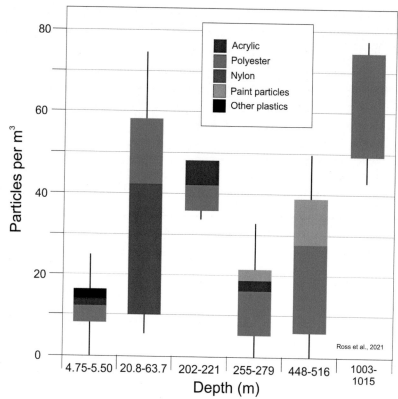

FIG. 16.16 Microplastic compounds found in test samples from the water column in the Beaufort Sea. Figure by the author. (Data from Ross, P.S., Chastain, S., Vassilenko, E., Etemadifar, A., Zimmermann, S., Quesnel, S.A., Eert, J., Solomon, E., Patankar, S., Posacka, A.M., Williams, B., 2021. Pervasive distribution of polyester fibres in the Arctic Ocean is driven by Atlantic inputs. Nature Communications 12(1). https://doi.org/10.1038/s41467-020-20347-1.)

were focused on solving the environmental problems we have caused in the North. Sadly, this is not the case. Many countries are interested in the Arctic, but only because the retreating sea ice is giving potential access to natural resources that were hitherto inaccessible. Again, whether Native peoples of the Arctic like it or not, the North is currently experiencing something like a gold rush as countries vie with each other to gain access to petroleum products, metals, rare earth minerals, and even diamonds. Thus, the Arctic has been transformed from the unknown, frozen North that few people cared about to a new Land (and seabed) of Opportunity.

I do not apologize for all the bad news I have presented in this book. It is what it is. The old saying "Ignorance is bliss" does not apply to the modern Arctic. Ignoring these problems will not only lead to the destruction of the Arctic environments we have known. When the frozen Arctic world melts, the whole planet will suffer. Even though the situation appears to be an instance of "the tail wagging the dog (in this case, the Arctic tail wagging the global dog)," the Arctic plays an oversized role in steering the course of global environments. That is why its preservation is so important.

REFERENCES

Arora, V., Vamvakidis, A., 2010. China's economic growth: international spillovers. In: International Monetary Fund Working Paper 10/65. International Monetary Fund, pp. 1–25.

Associated Press, 2021. US Navy Rolls Out New Strategy for Expanding Arctic Operations. https://www.defensenews.com/digital-show-dailies/surface-navy-association/2021/01/11/us-navy-rolls-out-new-strategy-for-expanding-arctic-operations/ (Original work published 2021).

Ayers, T., 2020. China's Arctic Gambit a Concern for U.S. Air and Space Forces. Space News. https://spacenews.com/op-ed-chinas-arctic-gambit-a-concern-for-u-s-air-and-space-forces/.

Babin, J., Lasserre, F., 2019. Asian states at the arctic council: perceptions in western states. Polar Geography 42 (3), 145−159. https://doi.org/10.1080/1088937X.2019.1578290.

Barry, T., Davíðsdóttir, B., Einarsson, N., Young, O.R., 2020. The Arctic Council: an agent of change? Global Environmental Change 63, 102099. https://doi.org/10.1016/j.gloenvcha.2020.102099.

Bell, J., 2021. Mary River mine blockade highlights Nunavut Agreement's fatal flaw. Nunatsiaq News. https://nunatsiaq.com/stories/article/mary-river-mine-blockade-highlights-nunavut-agreements-fatal-flaw/.

Bennett, M.M., 2018. From state-initiated to Indigenous-driven infrastructure: the Inuvialuit and Canada's first highway to the Arctic Ocean. World Development 109, 134−148. https://doi.org/10.1016/j.worlddev.2018.04.003.

Bennett, M.M., Stephenson, S.R., Yang, K., Bravo, M.T., De Jonghe, B., 2020. The opening of the Transpolar Sea Route: logistical, geopolitical, environmental, and socioeconomic impacts. Marine Policy 121, 104179. https://doi.org/10.1016/j.marpol.2020.104178.

Breum, M., 2018. Why people matter when we talk about Arctic climate change. Arctic Today. https://www.arctictoday.com/people-matter-talk-climate-change/.

Bryce, E., 2019. Who Owns the Arctic? Live Science. October 13, 2019. https://www.livescience.com/who-owns-the-arctic.html.

Burke, S., 2021. The Arctic threat that must not be named. War on the Rocks, January 28, 2001. https://warontherocks.com/2021/01/the-arctic-threat-that-must-not-be-named/

Byers, M., Lalonde, S., 2009. Who controls the northwest passage? Vanderbilt Journal of Transnational Law 42 (4), 1133−1210.

Canadian Press, 2021. Hunters Block Mine Road. Nunavut Hunters Block Mine Road, Airstrip; Say Their Voices Aren't Being Heard. https://www.castanet.net/news/Canada/324216/Nunavut-hunters-block-mine-road-airstrip-say-their-voices-aren-t-being-heard (Original work published 2021).

Cangemi, U.S. Coast Guard, N, 2018. The Coast Guard Cutter Healy. https://www.navytimes.com/news/your-navy/2020/12/15/congress-oks-new-arctic-icebreakers-for-coast-guard/.

Coates, K.S., Broderstad, E.G., 2019. Indigenous peoples of the arctic: re-taking control of the far north. In: The Palgrave Handbook of Arctic Policy and Politics. Palgrave Macmillan, pp. 9−25. https://doi.org/10.1007/978-3-030-20557-7_2.

Cushing, B., 2020. Not a Single Major US Bank Is Now Willing to Finance Arctic Drilling. Sierra Club. https://www.sierraclub.org/articles/2020/12/not-single-major-us-bank-now-willing-finance-arctic-drilling.

Dodds, K., 2019. 'Real interest'? Understanding the 2018 agreement to prevent unregulated high seas fisheries in the central Arctic Ocean. Global Policy 10 (4), 542−553. https://doi.org/10.1111/1758-5899.12701.

Dodds, K., Nuttall, M., 2015. The Scramble for the Poles: The Geopolitics of the Arctic and Antarctic. Wiley. ISBN: 978-0-745-65245-0.

Drewniak, M., Dalaklis, D., Kitada, M., Ölçer, A., Ballini, F., 2018. Geopolitics of Arctic shipping: the state of icebreakers and future needs. Polar Geography 41 (2), 107−125. https://doi.org/10.1080/1088937X.2018.1455756.

Gross, M., 2020. Geopolitical competition in the arctic Circle. Harvard International Review. https://hir.harvard.edu/the-arctic-circle/.

Hakirevic, N., 2020. US, Canadian Navies and USCG Strengthen Partnerships through New Strategic Plan. Navaltoday.Com. https://www.navaltoday.com/2020/05/29/us-canadian-navies-and-uscg-strengthen-partnerships-through-new-strategic-plan/.

Harper, T., 2019. How the Century of Humiliation Influences China's Ambitions Today. Imperial & Global Forum, Centre for Imperial and Global History. https://imperialglobalexeter.com/2019/07/11/how-the-century-of-humiliation-influences-chinas-ambitions-today/.

Henriques, M., 2020. The Rush to Claim an Undersea Mountain Range. BBC News. https://www.bbc.com/future/article/20200722-the-rush-to-claim-an-undersea-mountain-range.

International Maritime Organization, 2017. Resolution MEPC 68/21/Add. 1, Annex 10. International Code for Ships Operating in Polar Waters (Polar Code), p. 55.

Jacobsen, M., 2020. Greenland's Arctic advantage: articulations, acts and appearances of sovereignty games. Cooperation and Conflict 55 (2), 170−192. https://doi.org/10.1177/0010836719882476.

Kleist, K.V., 2017. The Arctic seas - the history & destiny of inuit. In: Nilliajut 2: Inuit Perspectives on the Northwest Passage and Marine Issues. Inuit Tapiriit Kanatami, Ottawa, pp. 48−51.

Kraska, J., 2016. Canadian arctic shipping regulations and the law of the sea. In: Berry, D., Bowles, N., Jones, H. (Eds.), Governing the North American Arctic: Sovereignty, Security, and Institutions. Palgrave Macmillan, pp. 51−73.

Kulik, A., Kulik, S.V., Lagutin, O.V., Chistalyova, T., 2020. Comparative analysis of China's and Singapore's policies in the Arctic. IOP Conference Series: Earth and Environmental Science 012040. https://doi.org/10.1088/1755-1315/539/1/012040.

Lavorio, A., 2020. Geography, climate change, national security: the case of the evolving US arctic strategy. The International Spectator. https://doi.org/10.1080/03932729.2020.1823695.

Mazurier, P.A., Delgado-Morán, J.J., Payá-Santos, C.A., 2020. The meta-tragedy of the commons. Climate change and the securitization of the arctic region. In: Ramírez, J.M., Biziewski, J. (Eds.), Advanced Sciences and Technologies for Security Applications. Springer, pp. 63−74. https://doi.org/10.1007/978-3-030-12293-5_5.

McGwin, K., 2021. Greenland officials extend hearing period, add meetings ahead of Kuannersuit mine decision. Arctic Today. https://www.arctictoday.com/greenland-officials-extend-hearing-period-add-meetings-ahead-of-kuannersuit-mine-decision/.

National Public Radio, 2021. Major Oil Companies Take a Pass on Controversial Lease Sale in Arctic Refuge. https://www.npr.org/2021/01/06/953718234/major-oil-companies-take-a-pass-on-controversial-lease-sale-in-arctic-refuge.

NATO, 2020. NATO intercepted hundreds of Russian fighters in 2020. NATO, 28 December 2020. https://www.nato.int/cps/en/natohq/news_180551.html.

NOAA, 2020. Arctic Navigation. https://oceanservice.noaa.gov/economy/arctic/.

Nunaat, 2011. A Circumpolar Inuit Declaration on Resource Development Principles in Inuit Nunaat. http://inuitcircumpolar.com/files/uploads/iccfiles/Declaration_on_Resource_Development_A3_FINAL.pdf.

Østhagen, 2019. The different levels of geopolitics of the Arctic. Georgetown Journal of International Affairs. https://gjia.georgetown.edu/2019/12/05/different-levels-of-arctic-geopolitics/.

Ostreng, W., Eger, K.M., Fløistad, B., Jørgensen-Dahl, A., Lothe, L., et al., 2013. Shipping in Arctic Wters: A Comparison of the Northeast. Northwest and Trans Polar Passages. Springer Science & Business Media. ISBN, 13, 978−3642167898.

Padrtova, B., 2019. Frozen narratives: how media present security in the Arctic. Polar Science 21, 37−46. https://doi.org/10.1016/j.polar.2019.05.006.

Paglia, E., 2018. The telecoupled Arctic: assessing stakeholder narratives of non-arctic states. In: Wormbs, N. (Ed.), Competing Arctic Futures: Historical and Contemporary Perspectives Perspectives. Palgrave - Springer International, pp. 189−212.

Roest, W.R., Haworth, R.T., 2019. The Arctic Ocean: advances in geopolitics and geoscience. In: Goel, P., Ravindra, R., Chattopadhyay, S. (Eds.), Climate Change and the White World. Springer International Publishing, pp. 19−29. https://doi.org/10.1007/978-3-030-21679-5_3.

Ross, P.S., Chastain, S., Vassilenko, E., Etemadifar, A., Zimmermann, S., Quesnel, S.A., Eert, J., Solomon, E., Patankar, S., Posacka, A.M., Williams, B., 2021. Pervasive distribution of polyester fibres in the Arctic Ocean is driven by Atlantic inputs. Nature Communications 12 (1), 106. https://doi.org/10.1038/s41467-020-20347-1.

Safety4Sea, 2016. China Issues 'Guidances on Arctic Navigation in the Northwest Route.'. https://safety4sea.com/china-issues-guidances-on-arctic-navigation-in-the-northwest-route/.

Saxena, A., 2020. The Return of Great Power Competition to the Arctic. Arctic Institute. https://www.thearcticinstitute.org/return-great-power-competition-arctic/.

Sergunin, A., Hoogensen, G., 2020. The Politics of Russian Arctic shipping: evolving security and geopolitical factors. The Polar Journal 10 (2), 251−272. https://doi.org/10.1080/2154896x.2020.1799613.

Sevunts, L., 2020. Experts warn of potentially 'deadly' great power games in the Arctic. In: Eye on the Arctic; Barents Observer. https://thebarentsobserver.com/en/security/2020/09/experts-warn-potentially-deadly-great-power-games-arctic.

Shapovalova, D., Galimullin, E., Grushevenko, E., 2020. Russian Arctic offshore petroleum governance: the effects of western sanctions and outlook for northern development. Energy Policy 146, 11753. https://doi.org/10.1016/j.enpol.2020.111753.

Sharma, B., 2020. China's emerging Arctic engagements: should India reconsider its approach towards the polar north? Maritime Affairs: Journal of the National Maritime Foundation of India 16, 46−67. https://doi.org/10.1080/09733159.2020.1772532.

Snow, S., 2020. US Navy Surface Ships Enter the Barents Sea for the First Time since Mid-1980s. Navy Times. https://www.navytimes.com/flashpoints/2020/05/04/us-navy-surface-ships-enter-the-barents-sea-for-the-first-time-since-mid-1980s/.

Steinberg, P.E., 2015. Chapter 1: imagining the arctic. In: Steinberg, P.E., Tasch, J., Gerhardt, H., Keul, A., Nyman, E.A. (Eds.), Contesting the Arctic: Politics and Imaginaries in the Circumpolar North. I.B. Tauris, London and New York, pp. 1−17.

Steinberg, P.E., Tasch, J., Gerhardt, H., 2015. Contesting the Arctic: Politics and Imaginaries in the Circumpolar North. I. B. Tauris, London and New York. ISBN 978-1780761480.

Stroeve, J., Whiteman, G., Wilkinson, J., 2019. The shrinking Arctic ice protects us all. It's time to act. In: World Economic Forum Annual Meeting; World Economic Forum. https://www.weforum.org/agenda/2019/01/the-shrinking-arctic-ice-protects-all-of-us-its-time-to-save-it/.

Ulmke, M., Battistello, G., Biermann, J., Mohrdieck, C., Pelot, R., Koch, W., 2017. Persistent maritime traffic monitoring for the Canadian Arctic, Proceedings of SPIE 10190, Ground/Air Multisensor Interoperability, Integration, and Networking for Persistent ISR VIII, 101900Q. https://doi.org/10.1117/12.2267323.

UN DESA, 2007. United Nations Declaration on the Rights of Indigenous Peoples. https://www.un.org/development/desa/indigenouspeoples/declaration-on-the-rights-of-indigenous-peoples.html.

Woody, C., 2021. The US Is Sending Bomber Crews and Destroyer Captains North to Learn How to Fight in the Arctic. Business Insider. https://www.businessinsider.com/us-bombers-navy-ships-learning-how-to-fight-in-arctic-2021-2.

Young, O.A., 2019. Is it time for a reset in arctic governance? Sustainability 11, 4497. https://doi.org/10.3390/su11164497.

FURTHER READING

Global Footprint Network, 2021. 2021 is here. We do not need a pandemic to #MoveTheDate. Global Footprint Network, p. 2021. https://www.footprintnetwork.org/.

Hamilton, L.C., Wirsing, J., Saito, K., 2018. Demographic variation and change in the Inuit arctic. Environmental Research Letters 13 (11), 115007. https://doi.org/10.1088/1748-9326/aae7ef.

Heleniak, T., Turunen, E., Wang, S., 2019. Demographic changes in the arctic. In: Coates, K.S., Holroys, C. (Eds.), The Palgrave Handbook of Arctic Policy and Politics. Palgrave Macmillan, pp. 41−59. https://doi.org/10.1007/978-3-030-20557-7_4.

International Marine Organization information sheet, 2020. How the Polar Code Protects the Environment. https://www.imo.org/en/MediaCentre/HotTopics/polar/.

Wormbs, N., 2018. Competing Arctic futures: historical and contemporary perspectives. Palgrave Studies in the History of Science and Technology. Cham, Switzerland: Springer International Publishing. ISBN 978-3-319-91616-3.

Index

Note: Page numbers followed by "f" indicate figures and "t" indicate tables.